ART OF SINGLE-WELL PUSH-PULL TEST

Unconventional Subsurface Flow and Transport

Series Editor: Hongbin Zhan (*Texas A&M University*)

Study of groundwater flow and solute transport has been advanced into new territories that is beyond the conventional theories such as Darcy's law and Fick's law in recent decades. The studied media change from permeable porous and fractured ones to much less permeable ones such as clay and shale. The studied pore sizes also change from millimetres to micro-meters or even nano-meters. The objective of this series is to report recent advances in subsurface flow and solute transport that pushes the knowledge boundary into new territories which include, but not limited to topics in forthcoming titles.

Published

Vol. 2 *Art of Single-Well Push-Pull Test*
by Quanrong Wang, Hongbin Zhan and Wenguang Shi

Vol. 1 *Complexity of Groundwater Systems*
by Teng Ma and Liuzhu Chen

Unconventional Subsurface Flow and Transport – Volume 2

ART OF SINGLE-WELL PUSH-PULL TEST

Quanrong Wang
China University of Geosciences, Wuhan, China
Hongbin Zhan
Texas A&M University, USA
Wenguang Shi
China University of Geosciences, Wuhan, China

World Scientific

NEW JERSEY • LONDON • SINGAPORE • BEIJING • SHANGHAI • TAIPEI • CHENNAI

Published by

World Scientific Publishing Co. Pte. Ltd.
5 Toh Tuck Link, Singapore 596224
USA office: 27 Warren Street, Suite 401-402, Hackensack, NJ 07601
UK office: 57 Shelton Street, Covent Garden, London WC2H 9HE

Library of Congress Control Number: 2025005555

British Library Cataloguing-in-Publication Data
A catalogue record for this book is available from the British Library.

Unconventional Subsurface Flow and Transport — Vol. 2
ART OF SINGLE-WELL PUSH-PULL TEST

ISBN 978-981-12-9558-4 (hardcover)
ISBN 978-981-12-9559-1 (ebook for institutions)
ISBN 978-981-12-9560-7 (ebook for individuals)

For any available supplementary material, please visit
https://www.worldscientific.com/worldscibooks/10.1142/13912#t=suppl

Desk Editors: Selas Hamilton/Amanda Yun

Typeset by Stallion Press
Email: enquiries@stallionpress.com

Preface

Understanding the processes of reactive transport and heat transport in the subsurface around injection and extraction wells is of critical importance in numerous hydrogeological and hydro-geothermal investigations, as well as in a range of practical applications. An accurate description of reactive transport in porous or fractured media is essential for the effective assessment of aquifer contamination risks and the development of remediation strategies. Similarly, accurate modeling of heat transport behavior near the test well is essential for geothermal resource evaluation, development, and thermal energy storage, as well as stimulating and managing the utilization of geothermal energy. Single-Well Push-Pull (SWPP) Test has been widely used to characterize *in situ* aquifers related to the reactive transport and the geothermal transport in the subsurface. The scientist and engineers prefer the SWPP test due to its low cost and high efficiency. However, the interpretation of the observed SWPP data limited the application of this effective method since existing mathematical models often rely on oversimplified assumptions that do not adequately represent many real-world conditions, restricting their applicability to a broad range of experimental data.

This book focuses on the mathematical modeling of groundwater flow, solute transport, and geothermal heat transport around wells used in SWPP, a method that has been widely used but is often limited by the interpretation of its data. By addressing this limitation, the book offers a significant contribution to the field. The models discussed in this book account for critical factors such as wellbore mixing effects, skin effects around the well, intro-well hydraulics, the influence of aquitards, and non-Darcian flow behavior in the aquifer. The new models are validated against both experimental and numerical data, offering a more accurate framework for interpreting results from radial groundwater flow, reactive transport, and heat transport experiments.

Chapter 1 provides an overview of the purpose and scope of the book. It introduces the basic properties of groundwater, aquifers, and boreholes and defines the hydraulic parameters of aquifers.

Chapter 2 presents general models of radial groundwater flow and hydraulic parameter values for representative aquifers. It also introduces advanced models of pumping and slug tests conducted in both Darcian and non-Darcian flow regimes.

Chapter 3 first introduces general reactive transport models and typical parameter values related to reactive transport and then discusses analytical and numerical models for radial tracers. These models cover reactive transport models for injection wells, two-stage SWPP analytical models, multi-stage SWPP analytical models, and groundwater circulation well systems. The numerical solutions presented focus on multi-species reactive transport problems under non-Darcian flow and regional groundwater flow conditions.

Chapter 4 begins with an introduction to the general heat transport model and the classical parameter values related to thermal properties. It then proceeds to discuss three analytical models for thermal SWPP testing in fractured and porous aquifers, providing a detailed exploration of the methodologies used to simulate heat transfer in such subsurface media.

This book's primary goal is to guide readers in developing new models for the groundwater flow and SWPP test, thereby improving the accuracy and reliability of data interpretation. It provides a comprehensive toolbox for the design, execution, and analysis of field tests, making it an invaluable resource for researchers, engineers, and practitioners working in the fields of aquifer testing, solute tracer testing, heat tracer testing, and the prediction of groundwater flow, solute migration, geothermal energy transfer, and gas transport around wellbores.

About the Authors

Quanrong Wang is a Professor in Hydrogeology at China University of Geosciences, Wuhan, China. He was awarded the titles of the Excellent Young Scholar by the National Science Fund of China and the Outstanding Youth Scholar by Hubei Province, China. His research mainly focused on multiphase flow modeling in subsurface, including groundwater flow, reactive transport, geothermal energy, CCUS, and so on. He hosted or is hosting five projects of the National Natural Science Foundation of China (NSFC). He has published more than 50 high-quality papers in international peer-reviewed journals, such as *Water Resource Research*, *Hydrology and Earth System Sciences*, *Journal of Hydrology*, *Advances in Water Resources*, and *Groundwater*. He is the co-editor of *Principles and Dynamics of the Critical Zone*, which was published by Elsevier. He is an Associate Editor for the *Journal of Hydrology* and *Groundwater*, as well as a member of the editorial boards for *Geoscience and Technology Bulletin*, *Earth Science*, and *Journal of Earth Sciences*.

Hongbin Zhan is a Professor of Geology and Geophysics and a Professor of Water Management and Hydrological Science at Texas A&M University (TAMU), College Station, USA. He is the Holder of the Endowed Dudley J. Hughes '51 Chair in Geology and Geophysics at TAMU. His teaching and research interests are in fundamental processes of groundwater hydrology, flow and transport in geological formations, and their applications in water resources management and geological, environmental, and energy engineering. He is an elected Fellow of the Geological Society of America and a Fellow of the American Association for the Advancement of Science (AAAS). He was the recipient of the Distinguished Achievement Award in Faculty Teaching and the Distinguished Achievement Award in Faculty Research from TAMU. He has published more than 370 peer-reviewed articles with a Google Scholar citation of 9506 and an H-index of 54. He has graduated over 60 graduate students and has supervised over 25 visiting scholars and postdoctoral scientists. He has served in many national and international committees and panels, such as the US DOE Subsurface Biogeochemical Research program; AAAS Research Competitiveness Program; US DOD/EPA/DOE Strategic Environmental Research and Development Program; American Society of Mechanical Engineers, and the Institute for Regulatory Science peer review board.

Wenguang Shi is a Postdoctoral Fellow in Hydrogeology, interested in groundwater flow and its impact on solute and heat transport in porous and fractured media at the China University of Geosciences (CUG), Wuhan, China. His research focuses primarily on the fundamental processes of groundwater flow, solute and thermal transport in porous and fracture aquifers, with applications extending to water resources management, geological, environmental, and geothermal energy engineering. His research has been funded by the scholarship from China Scholarship Council, the outstanding Doctoral Innovation Fund Program of China University of

Geosciences (Wuhan), the National Scholarship of China, the First Prize for a Bachelor Thesis by CUG, and PhD fellowship of the Chinese government. Wenguang has published over 20 papers in international peer-reviewed journals, with him serving as the first author for 9 of these papers.

Acknowledgments

This book was partially supported by Programs of the National Natural Science Foundation of China (Grant No. 42222704 and Grant No. U24A20596). The authors thank World Scientific Publishing for making the publication of this manual possible. We also express sincere gratitude to all persons who contributed directly or indirectly to the completion of this book, such as Changcheng Ke, Yingfan Zhao, Qihui Yang, Aohan Jin, Tao Zhang, and Yuechun Zhao, who are the MS or PhD students of Dr. Quanrong Wang.

Contents

List of Figures

List of Tables

Chapter 1

Introduction

1.1 Purpose and Scope

Groundwater is defined as the water that soaks into the soil and is stored between pore spaces, fractures and joints found within rocks and other geological formations. It is one of the most valuable resources, as it is not only the largest source of freshwater for mankind but also the carrier of chemicals (e.g., Nutrients for contaminants) and energy (e.g., geothermal). However, different from surface water, groundwater is stored in the subsurface, and it is invisible. It is not easy to determine how much groundwater is available due to the heterogeneity and anisotropy of hydraulic conductivity. Due to the influence of human activities, the groundwater environment has become worse and worse since industrialization, and therefore, aquifer remediation is a hot spot in Environmental science. In addition to resource attributes of groundwater, it also has energy attributes, like geothermal energy.

To understand what happens in aquifers, many techniques have been developed, such as isotope techniques, geophysical methods, drilling means, and so on, where the information obtained in the wellbore is closer to the reality, like the hydraulic parameters estimated by the pumping test, and parameters related to the contaminant transport and heat transport estimated by the tracer test in the wellbore.

This book mainly covers mathematical modeling of groundwater flow, solute transport, and geothermal transport around wells and their applications. It is intended to serve as a basic introductory textbook and to provide instructions and guidance to scholars and engineers engaged in aquifer tests, solute tracer tests, heat tracer tests, and predictions of groundwater flow, solute, geothermal, and gas transport around the wellbore.

1.2 Aquifer and Boreholes

Aquifer, aquitards and aquicludes are terms used to characterize hydrogeologic systems. An aquifer was defined by the US Geological Survey (Lohman *et al.*, 1972) as a formation, group of formations, or part of a formation that contains sufficient saturated permeable material to yield significant quantities of water to wells and/or springs, such as sandstones, conglomerates, fractured limestone and unconsolidated sand, gravels and fractured volcanic rocks (Maliva, 2016). Aquifers are generally classified into two main categories, namely confined aquifers and unconfined aquifers. Confined aquifers are referred to as "Artesian aquifers", which are those bodies of water in a permeable rock and are overlain and underlain by aquicludes; meanwhile, the groundwater level never drops below the aquifer top. Unconfined aquifer is generally found located near the land surface and have no layers of clay (or other impermeable geologic material) above the water table although they are found lying relatively above impermeable clay rock layers.

An aquitard is any geological formation of a rather semipervious nature that transmits water at slower rates than an aquifer. Freeze and Cherry (1979) describe an aquitard as the less permeable bed in a stratigraphic sequence. The aquitard may be permeable enough to transmit water in quantities that are significant in the study of regional groundwater flow, but their permeability is not sufficient to allow the completion of production wells. Aquicludes are layers or beds that generally have extremely low if any, porosity and permeability (de Marsily *et al.*, 1986).

A borehole is a narrow shaft and is the only way to gain access to natural resources, like oil, gas and water in the subsurface. It is not only used for resource extraction (especially wellbore) but also used extensively for monitoring water levels, groundwater quality and water temperature. A wellbore is usually a straight vertical shaft that "bores" into the ground to allow the recovery of natural resources. It also includes the open hole and uncased portions of the well. The wellbore could be fully or partially penetrating the aquifer. For the fully penetrating wellbore, its screen length is equal to the thickness of the aquifer.

1.3 Petro-Physical Properties of Aquifers

Aquifers are characterized by petrophysical properties such as hydraulic conductivity (alternatively called permeability), transmissivity (the product of hydraulic conductivity and aquifer thickness) and storage coefficient.

1.3.1 *Hydraulic conductivity*

The hydraulic conductivity proportionality constant, K, can be conceptualized as the relative ease of fluid passage through a porous material. It has direction and magnitude and is represented as a vector; however, the first part of this discussion presents it as a scalar value. Rearranging Darcy's law could be used to solve for hydraulic conductivity:

$$K = -\frac{Q}{A}\frac{\Delta L}{\Delta h} \tag{1.1}$$

where Q is flow rate, A is area, h is the hydraulic head, and L is length. Hydraulic Conductivity is quite different from intrinsic permeability:

$$K = k\frac{\rho g}{\mu} \tag{1.2}$$

where k is the intrinsic permeability, ρ is the density of fluid, μ is the dynamic viscosity of the fluid.

Generally, for unconsolidated porous media, hydraulic conductivity varies with particle size, as clayey materials exhibit low values of hydraulic conductivity compared to sands and gravels, which exhibit high values.

1.3.2 *Porosity*

Porosity of a geological formation or rock or soil, θ, could be described as the measure of the contained voids or interstices expressed as a ratio of the volume of voids to the total volume:

$$\theta = \frac{V_p}{V_t} \tag{1.3}$$

where V_p is the pores of rock or soil sample and V_t is the total volume of pores and solid material.

1.3.3 *Transmissivity*

Transmissivity (T) could be mathematically defined as

$$T = Kb \tag{1.4}$$

where K is the hydraulic conductivity and b is the saturated thickness of the aquifer.

1.3.4 *Storage coefficient*

In equations describing the flow of fluids in a porous medium (e.g., groundwater in an aquifer), the storage coefficient is the quantity that relates the change in fluid potential (i.e., hydraulic head) in the aquifer to the change in the amount of water stored in the aquifer at a given point. In practical terms, the storage coefficient tells us the volume of water that can be removed from an aquifer for a given drop in hydraulic head. There are three storage coefficients used in groundwater flow equations (the units of each are given in [brackets] following each definition):

Specific storage (S_s) is the volume of water removed from a unit volume of a confined aquifer for a unit drop in hydraulic head.

Storativity (S) is the volume of water removed from a unit area of an aquifer for a unit drop in the hydraulic head; in confined aquifers, it is equal to the specific storage times the thickness of the aquifer, in unconfined aquifers, it is equal to the specific storage times the thickness of the aquifer plus the specific yield.

Specific yield (S_y) is the volume of water removed from a unit volume of an unconfined aquifer for a unit drop in water level; this quantity is also defined as the volume of water that will drain by gravity from a saturated unit volume of an aquifer and is equivalent to the effective porosity.

References

Freeze, R.A., Cherry, J.A., 1979. *Groundwater.* Englewood Cliffs: Prentice-Hall.

Maliva, R.G., 2016. Aquifer characterization and properties. In *Aquifer Characterization Techniques. Springer Hydrogeology.* Springer: Berlin.

de Marsily *et al.*, 2005. Dealing with spatial heterogeneity. *Hydrogeology Journal*, 13: 161–183.

Lohman, S.W. *et al.*, 1972. Definition of selected ground-water terms — Revisions and conceptual refinements. *U.S. Geological Survey Water-Supply Paper 1988.*

Chapter 2

Radial Groundwater Flow

2.1 Background

The aquifer parameters, such as hydraulic conductivity and specific storage, are critical in understanding the water resource distributions and the flow mechanism. As these parameters cannot be measured directly, they are estimated and calculated using mathematical models of pumping or slug tests through best fitting the hydraulic head/drawdown observed in boreholes during the test. Therefore, the accuracy of parameter estimation is not only dependent on the experimental operations but also on the choice of mathematical models. Pumping and slug tests are widely conducted in subsurface hydrology and hydraulic engineering to determine the hydraulic parameters of the target aquifers or reservoirs.

To date, abundant analytical models of pumping tests have been developed and implemented for aquifer parameters estimation. The most typical model is Theis well model (Theis, 1935), which was used to describe the groundwater flow around a well that fully penetrates a confined aquifer. Most of the previous analytical models contained several oversimplified assumptions. For instance, the well radius was assumed to be infinitely small, the skin effect of the wellbore was ignored, the aquifer was isolated from the overlain and underlain aquitards, the well was assumed to fully penetrate the confined aquifer, the pumping rate was assumed to be constant during the entire test, and the groundwater flow was assumed to follow Darcy's law. These assumptions cannot be satisfied easily in the real aquifer test; therefore, many scholars revisited these assumptions to develop the model for various scenarios, as summarized in Table 2.1.

Among these analytical models, the well radius was assumed to be finite in the analytical models of Cassiani *et al.* (1999), Yang and Yeh (2005), Yang *et al.* (2006), Tsai and Yeh (2012) and Lin *et al.* (2016), Papadopulos and Cooper (1967), Lai and Su (1974), Novakowski *et al.* (1989), Mathias and Todman (2010), Wen *et al.* (2011), Chen *et al.* (2020) and Lin and Yeh (2021). The skin effect was found to be an important factor affecting the groundwater flow in the wellbore and was described by a two-zone model (Chen *et al.*, 2020; Lin *et al.*, 2016; Novakowski *et al.*, 1989; Tsai and Yeh, 2012; Yang and Yeh, 2005) or represented by a skin factor (Agarwal *et al.*, 1970; Cassiani *et al.*, 1999; Lin and Yeh, 2017, 2021) in the analytical models.

Lai and Su (1974), Mathias and Todman (2010), and Chen *et al.* (2020) developed the analytical models of pumping tests with time-dependent pumping rates. When the pumping rate is very large, so that the Reynolds number is greater than 10 around the wellbore, the flow may not follow Darcy's law. Wen *et al.* (2008a, 2011), Mathias and Todman (2010) and Wang *et al.* (2021) derived approximating analytical models based on non-Darcian equations. Hantush (1964), Cassiani *et al.* (1999), Yang and Yeh (2005), and Yang *et al.* (2006) extended the Theis well model by considering the partially penetrating well. Many of these models have been packaged into the software of Aquifer Test (Istok and Dawson, 2014; Sindalovskiy and Leonid, 2017; Walton, 2006) for parameter estimations.

When modeling flow into a well, an important issue is how to handle the vertical screened well face. Presently, there are two ways to deal with it: a uniform-flux boundary condition (UFBC) or a uniform-head boundary condition (UHBC). UFBC assumes that the flux distribution along the screened well face is uniform and has the advantage of relatively simple implementation in the analysis. The results of UFBC show that the head distribution varies along the screened well face and the lowest head is at the midpoint of the screen, which implies that water is removed from the midpoint of the screen. UHBC assumes that the hydraulic head is uniformly distributed along the screened well face, which implies that the wellbore hydraulic conductivity is infinite. This seems more reasonable than UFBC since the wellbore hydraulic conductivity is much larger

Table 2.1. The analytical models of published pumping test.

References	Well radius		Skin effect		Leakage effect	Darcy's law	Pumping type	Well type
	$r_w = 0$	Wellbore storage	Two zone	Skin factor				
Theis (1935)	Y	N	N	N	N	Y	Constant rate	Fully
Hantush and Jacob (1955); Hantush and Mahdi (1960); Hantush (1967)	Y	N	N	N	Y	Y	Constant rate	Fully
Hantush (1964)	Y	N	N	N	Y	Y	Constant rate	Partially
Papadopulos and Cooper (1967)	N	Y	N	N	N	Y	Constant rate	Fully
Agarwal *et al.* (1970)	N	Y	N	Y	N	Y	Constant pressure	Fully
Neuman and Witherspoon (1969)	Y	N	N	N	Y	Y	Constant rate	Fully
Lai and Su (1974); Feng and Zhan (2015)	N	Y	N	N	Y	Y	Time-dependent rate	Fully
Moench (1984); Feng and Zhan (2016)	N	Y	Y	N	Y	Y	Constant rate	Fully
Moench (1985)	N	Y	N	N	Y	Y	Constant rate	Fully
Javandel and Witherspoon (1983)	Y	N	N	N	N	Y	Constant rate	Partially

(*Continued*)

Table 2.1. (*Continued*)

References	Well radius		Skin effect					
	$r_w = 0$	Wellbore storage	Two zone	Skin factor	Leakage effect	Darcy's law	Pumping type	Well type
Novakowski *et al.* (1989)	N	Y	Y	N	N	Y	Constant rate	Fully
Cassiani *et al.* (1999)	N	N	N	Y	N	Y	Constant head	Partially
Yang and Yeh (2005)	N	N	Y	N	N	Y	Constant head	Partially
Yang *et al.* (2006)	N	N	N	N	N	Y	Constant rate	Partially
Wen *et al.* (2008a); Wang *et al.* (2021)	Y	N	N	N	N	N	Constant rate	Fully
Mathias and Todman (2010)	N	Y	N	N	N	N	Time-dependent rate	Fully
Wen *et al.* (2011)	N	Y	N	N	Y	N	Constant rate	Fully
Tsai and Yeh (2012); Lin *et al.* (2016)	N	N	Y	N	N	Y	Constant head	Fully
Lin and Yeh (2017)	N	Y	N	Y	Y	Y	Constant head	Fully
Chen *et al.* (2020)	N	Y	Y	N	Y	Y	Time-dependent rate	Fully

Note: Denote "Y" and "N" represent whether the term is considered or not.

than the aquifer vertical hydraulic conductivity. UHBC is also named non-uniform flux boundary condition (Perina and Lee, 2006) or uniform drawdown boundary condition (Hemker, 1999). It has been widely used to describe the flow into a partially penetrating well, e.g., Hemker (1999), Cassiani *et al.* (1999), Perina and Lee (2006), and Chang and Chen (2003).

However, neither UFBC nor UHBC specifically considers the physical processes of head losses in the wellbore, which will be considered in this section. We name the approach considering the intra-wellbore losses along the screened well face as the mixed-type boundary condition (MTBC). It is different from the definition of MTBC in some previous studies (Chang and Chen, 2003; Mathias and Wen, 2015), in which UHBC is sometimes used to refer to MTBC.

The influence of the intra-wellbore head losses on the aquifer test has been observed in many previous studies. It creates complex flow patterns around the well. For instance, with respect to the single-well pumping test in which the pumping well is also used as a monitoring well to measure the water level, Chen and Jiao (1999) developed a model to describe the intra-wellbore head losses from the frictional effect along the well screen using the Darcy-Weisbach formula (Munson *et al.*, 1995) for pipe flow. Hu *et al.* (2011) investigated the intra-wellbore head losses in the well using a similar approach.

Due to the nature of radial inflow into a pumping vertical well through a screen, the mechanics of flow along the well screen are somewhat different from that of the conventional pipe flow. In addition to the frictional effect, head losses along the wellbore might be further altered by the kinematic effect. This includes both the change from horizontal to vertical flow and the acceleration due to the increasing flow velocity towards the point of intake inside the vertical well. There is evidence that both the frictional effect and the kinematic effect are important factors to consider for a vertical well (Dikken, 1990; Tarshish, 1992, 1993). Tarshish (1992, 1993) proposed a model for steady-state flow into a well, by considering the intra-wellbore head losses from both the frictional effect and the acceleration effect. In addition, numerous studies on the intra-wellbore head losses have been conducted in the petroleum engineering, such as Dikken (1990), Ozkan *et al.* (1995), Sarica *et al.* (1994) and Ozkan *et al.* (1999), who used MTBC to evaluate the well production performance.

As a particular type of aquifer test, slug test has been commonly applied to determine the near-well aquifer properties because of its logistical and economic advantages over other aquifer test methods (Butler, 1998; Yeh and Chang, 2013). In a slug test, only a single well is needed, in which the water level is instantaneously increased or decreased. Then, the water level recovery (WLR) or pressure recovery in the wellbore is monitored with time (Ferris *et al.*, 1962; Hvorslev, 1951).

Since the 1950s, research on groundwater flow associated with slug tests has attracted great attention among hydrogeologists (Ferris and Knowles, 1954a; Hvorslev, 1951), and the first solution of slug test might be derived by Hvorslev (1951), which was named the Hvorslev solution. Because of its efficiency and simplicity, this solution has been widely used to estimate the in-situ aquifer parameters. However, the Hvorslev solution was obtained based on some assumptions, including that the aquifer compressibility, the finite effective radius and the skin effect of the well were negligible, the inertia effects of flow along the wellbore could be ignored, and the groundwater flow was Darcian. Therefore, the Hvorslev solution does not work well for some actual problems with complex flow processes (such as non-Darcian flow). Subsequent studies following Hvorslev (1951) have attempted to remove one or more of the above simplifying assumptions. For instance, Bouwer and Rice (1976) extended the model of Hvorslev (1951) by considering a partially penetrating well in unconfined aquifers. Cooper *et al.* (1967) found that the aquifer compressibility may not be ignored and then developed a transient solution of a slug test with the wellbore storage for confined aquifers. Moench and Hsieh (1985) presented a solution with a finite radius well skin by extending the solution of Cooper *et al.* (1967). All investigations mentioned above treated the well-aquifer flow process through the mass balance equation and ignored the inertia effect of groundwater flow, so WLR of slug test would exhibit monotonic (non-oscillatory) responses. However, some observations showed that the water level in the well may have oscillated, which was mainly ascribed to the inertia effect of moving water inside the well (Cooper *et al.*, 1965; van der Kamp, 1976). van der Kamp (1976) presented an approximate solution of slug test for confined aquifers considering the under-damped well response. Wylie and Magnuson (1995) developed a spreadsheet modeling technique by simplifying the solution of van der Kamp

(1976). Springer and Gelhar (1991) employed a simplified equation of pipe hydraulics to describe the inertial effect of moving water in the wellbore and presented a solution to determine the hydraulic conductivity of an unconfined aquifer, where the aquifer groundwater flow was modeled by the solution of Bouwer and Rice (1976). McElwee and Zenner (1998) derived a new solution for the slug test by using the Navier-Stokes equation for flow in the wellbore and the Hvorslev model for flow in the aquifer, of which the robustness of solution was subsequently tested by field data (McElwee, 2001). Malama *et al.* (2011) modeled slug tests in unconfined aquifers considering water table kinematics, wellbore skin and inertial effects. In addition to the above-mentioned studies, other scientists also devoted significant efforts to slug test research by incorporating the effects of partial penetration, anisotropy, short well screens, aquifer compressibility and finite radius well skins (Brown *et al.*, 1995; Hyder and Butler, 1995; Hyder *et al.*, 1994; Mills, 2010; Singh, 2007; Widdowson *et al.*, 1990). The book of Butler (1998) provided a good recent summary on this subject.

All the above studies contained an assumption that flow in the aquifer was Darcian, regardless of oscillatory or non-oscillatory WLR inside the well. An instantaneous change in water level inside a slug test well can create a significant hydraulic gradient near the well screen that is rarely seen in naturally occurring groundwater flow systems. Consequently, aquifer flow near the well screen may deviate considerably from naturally occurring groundwater flow that usually follows Darcy's law, particularly during the early stage of the slug test.

Recently, quite a few studies demonstrated that non-Darcian flow might occur in unconsolidated deposits during slug tests (McElwee, 2001; McElwee and Zenner, 1998; Quinn *et al.*, 2012; Zenner, 2009). The research of Zenner (2009) indicated that groundwater flow in the formation was non-Darcian. Quinn *et al.* (2012) concluded that much of the non-Darcian flow identified in aquifer tests was caused by flow in the formation rather than by the effect of the equipment plumbing. Quinn *et al.* (2013) further validated the conclusions of Quinn *et al.* (2012) by a series of rising and falling head slug tests in open fractured dolostone and sandstone boreholes. Ji and Koh (2015) investigated the nonlinear flow regime and linear regime during the slug tests using a representative Reynolds number. To resolve the problems related to non-Darcian flow, McElwee and Zenner (1998), McElwee (2001) and Ji and Koh (2015) used the Forchheimer equation (Forchheimer, 1901) to describe the non-Darcian flow in the aquifer. Furthermore, Ji and Koh (2015) pointed out that the Forchheimer equation can be used to estimate the hydraulic conductivity in a linear flow regime using data collected from the slug tests in a nonlinear flow regime. By carefully reviewing the literature above, we found that all these models assumed aquifer compressibility was negligible.

In addition to the non-Darcian effect on parameter estimation of slug test, groundwater seepage from adjacent aquitards might also be one important factor influencing groundwater flow during the slug test in an aquifer-aquitard system, particularly for a system where WLR occurs over a long period of time (like days). The consideration of aquitard on other types of aquifer tests was quite common. For instance, Hantush and Jacob (1955) derived a generalization of the Theis solution in a leaky confined aquifer for pumping tests. Hantush (1964) presented a solution for a constant-head test in leaky confined aquifers. Furthermore, Rehbinder (1995) pointed out that the solution of a slug test in a non-leaky aquifer might be unphysical since it lacked a stationary part. Therefore, it may be significant to test the influence of aquitard leakage on slug tests. Wang *et al.* (2015b) developed a transient nonlinear model of slug test in a leaky confined aquifer considering non-Darcian near well flow and presented a finite-difference solution for such a model. By analyzing the sensitivities of dimensionless hydraulic parameters on slug tests, the influences of aquitard leakance and non-Darcian behavior will be quantified.

The purpose of this chapter is to present some advanced models of pumping or slug tests in Darcian or non-Darcian flow fields. More specifically, we introduced the general models of groundwater flow in Section 2.2, followed by discussions on radial pumping test in Darcian flow fields and pumping test in a vertical well with intra-wellbore head losses in Sections 2.3 and 2.4, respectively. Subsequently, we further present the model of radial pumping test and the model of radial slug tests in the non-Darcian flow fields in Sections 2.5 and 2.6, respectively. These models can be applied as rapid-screening tools to estimate hydraulic parameters by analyzing on-site hydraulic head measurements.

2.2 Mathematical Models of Groundwater Flow

2.2.1 *General models of groundwater flow*

The three-dimensional movement of groundwater of constant density through porous earth material may be described by the partial-differential equation:

$$\frac{\partial}{\partial x}\left(K_{xx}\frac{\partial h}{\partial x}\right)+\frac{\partial}{\partial y}\left(K_{yy}\frac{\partial h}{\partial y}\right)+\frac{\partial}{\partial z}\left(K_{zz}\frac{\partial h}{\partial z}\right)+W=S_S\frac{\partial h}{\partial t} \quad (2.1)$$

where K_{xx}, K_{yy} and K_{zz} are values of hydraulic conductivity [m/s] along the x, y, and z coordinate axes, respectively, which are assumed to be parallel to the major axes of hydraulic conductivity [m/s]; h is the potentiometric head [m]; W is a volumetric flux per unit volume representing sources and/or sinks of water, with $W < 0.0$ for flow out of the ground-water system, and $W > 0.0$ for flow into the system [1/s]; S_S is the specific storage of the porous material [1/m]; and t is time [s].

In general, S_S, K_{xx}, K_{yy} and K_{zz} may be functions of space ($S_S = S_S(x, y, z)$, $K_{xx} = K_{xx}(x, y, z)$, and so forth) and Wmaybe a function of space and time ($W = W(x, y, z, t)$). Table 2.2 lists some respective values of hydraulic conductivity or transmissivity.

Equation (2.1) describes ground-water flow under nonequilibrium conditions in a heterogeneous and anisotropic medium, provided the principal axes of hydraulic conductivity are aligned with the coordinate directions. Equation (2.1), together with the specification of flow and/or head conditions at the boundaries of an aquifer system and specification of initial-head conditions, constitutes a mathematical representation of a ground-water flow system.

2.2.2 *General model of radial flow*

Using a cylindrical coordinate system is more practical for axisymmetric flow problems (well flow in a single well typically falls into this category), where $H = f(r, z)$ and H is connected to θ unrelated. This allows the Eq. (2.1) to be rewritten as

$$\frac{\partial}{\partial r}\left(K_{rr}r\frac{\partial h}{\partial r}\right)+K_{zz}\frac{\partial^2 h}{\partial z^2}=S_S\frac{\partial h}{\partial t} \quad (2.2)$$

Table 2.2. Summary of hydraulic parameters from the literature.

References	Aquifer material	Hydraulic conductivity [m/s] or transmissivity [m^2/s]	Effective porosity [%]
Adams and Gelhar (1992); Columbus, Mississippi	Very heterogeneous sand and gravel	10^{-5} to 10^{-3} m/s	35
Ahlstrom *et al.* (1977); Hanford, Washington	Glaciofluvial sands and gravels	5.7×10^{-4} to 3.0×10^{-2} m/s	/
Bentley and Walter (1983); WIPP	Fractured dolomite	/	18
Bierschenk (1959), Cole (1972); Hanford, Washington	Glaciofluvial sands and gravels	1.7×10^{-1} m^2/s	10
Bredehoeft and Pinder (1973); Brunswick, Georgia	Limestone	6.5×10^{-7} to 8.6×10^{-7} m^2/s	35
Claassen and Cordes (1975); Amargosa, Nevada	Fractured dolomite and limestone	5×10^{-2} to 11×10^{-2} m^2/s	6~60
Daniels (1981); Nevada Test Site	Alluvium derived from tuff	1.7×10^{-5} m/s	/
Dieulin (1981); Le Cellier (Lozere, France)	Fractured granite	3×10^{-4} to 9×10^{-4} m/s	2~8
Dieulin (1981); Torchy, France	Alluvial deposits	3×10^{-4} m/s	/
Egboka *et al.* (1983); Borden	Glaciofluvial sand	10^{-5} to 10^{-7} m/s	38
Fenske (1973); Tatum Salt Dome, Mississippi	Limestone	4.7×10^{-6} m/s	23

Freyberg (1986); Borden	Glaciofluvial sand	7.2×10^{-5} m/s	33 (total)
Fried (1975); Rhine aquifer (salt mines) southern Alsace, France	Alluvial; mixture of sand, gravel, and pebbles with clay lenses	10^{-3} m/s	/
Garabedian (1987); Cape Cod, Massachusetts	Medium to coarse sand with some gravel overlying silty sand and till	1.3×10^{-3} m/s	39
Goblet (1982); Site B, France	Fractured granite	10^{-5} to 10^{-7} m/s	
Grove (1977); NRTS, Idaho	Basaltic lava and sediments	1.4×10^{-1} to 1.4×10^{1} m^2/s	10
Grove and Beetem (1971); Eddy County (near Carlsbad), New Mexico	Fractured dolomite	/	12
Halevy and Nir (1962), Lenda and Zuber (1970); Nahal Oren, Israel	Dolomite	/	3.4
Hoehn (1983); lower Glatt Valley, Switzerland	Layered gravel and silty sand	9.2×10^{-4} to 6.6×10^{-3} m/s	/
Hoehn and Santschi (1987); Lower Glatt Valley, Switzerland	Layered gravel and silty sand	8.1×10^{-5} to 6.6×10^{-3} m/s	/

(*Continued*)

Table 2.2. (*Continued*)

References	Aquifer material	Hydraulic conductivity [m/s] or transmissivity [m^2/s]	Effective porosity [%]
Huyakorn *et al.* (1986); Mobile, Alabama	Layered medium sand	/	0.35
Iris *et al.* (1985); Campuget (Gard), France	Alluvial deposit	3.6×10^{-3} m^2/s	/
Ivanovich and Smith (1978); Dorset, England	Fractured chalk	2.2×10^{-3} m/s (fast pulse)	0.5
	Chalk	3.6×10^{-4} m/s (slow pulse)	2.3
Kies (1982); New Mexico State University, Las Cruces	Fluvial sands	9.55×10^{-5} m/s	42 (total)
Konikow (1976); Rocky Mountain Arsenal	Alluvium	/	30
Konikow and Bredehoeft (1974); Arkansas River Valley (at La Junta, Colorado)	Alluvium, Inhomogeneous clay, silt, sand and gravel	2.4×10^{-4} to 4.2×10^{-3} m/s	20
Kreft *et al.* (1974); Poland	Sand	3.1×10^{-5} to 1.5×10^{-4} m/s; 1.2×10^{-4} m^2/s	24
Kreft *et al.* (1974); Zn-Pb deposits, Poland	Fractured dolomite	2.5×10^{-4} to 4.7×10^{-4} m/s	2.4
Kreft *et al.* (1974); sulfur deposits, Poland	Limestone	1.1×10^{-4} m/s	12.3
Lau *et al.* (1957); University of California, Berkeley	Sand and gravel with clay lenses	9×10^{-4} m/s	30

Lee *et al.* (1980); Perch Lake, Ontario, (lakebed)	Sand	3.2×10^{-5} m/s	/
Leland and Hillel (1981); Amherst, Massachusetts	Fine sand and glacial till	2.4 to 3×10^{-5} m/s	40
Mercado (1966); Yavne region, Israel	Sand and sandstone with some silt and clay	2.1×10^{-8} to 2.4×10^{-8} m^2/s	23.3
Molinari and Peaudecerf (1977), Sauty (1977); Bonnaud, France	Sand	8.3×10^{-4} to 1.1×10^{-3} m^2/s	/
Moltyaner and Killey (1988); Twin Lake aquifer (Chalk River)	Fluvial sand	/	40.8 (total)
Naymik and Barcelona (1981); Illinois (Morgan County)	Unconsolidated sand and gravel	2.2×10^{-2} to 4.3×10^{-2} m^2/s	/
New Zealand Ministry of Works and Development (1977) Heretaunga aquifer, New Zealand: Roys Hill site	Gravel with cobbles	0.29 m^2/s	22
Flaxmere site 2	Alluvium (gravels)	0.37 m^2/s	22
Hastings City rubbish dump	Alluvium (gravels)	0.14, 0.35 m^2/s	
Oakes and Edworthy (1977), Clip stone, United Kingdom	Sandstone	2.4×10^{-6} to 1.4×10^{-4} m/s	32~48

Table 2.2. (*Continued*)

References	Aquifer material	Hydraulic conductivity [m/s] or transmissivity [m^2/s]	Effective porosity [%]
Papadopulos and Larson (1978); Mobile, Alabama	Medium to fine sand interspersed with clay and silt	5×10^{-4} m/s (horizontal) and 5.1×10^{-5} m/s (vertical)	25
Pickens and Grisak (1981); Chalk River	Sand	2×10^{-5} to 2×10^{-4} m/s	38
Pinder (1973); Long Island	Glacial outwash	7.5×10^{-4} m/s	35
Rabinowitz and Gross (1972); Roswell Basin, New Mexico	Fractured limestone	1.1×10^{-2} to 2.9×10^{-1} m^2/s	1
Rajaram and Gelhar (1991); Borden	Glaciofluvial sand	7.2×10^{-5} m/s	33 (total)
Roberts (1985); Palo Alto Bay lands	Sand, gravel, and silt	1.25×10^{-3} m^2/s (lower aquifer); 5.0×10^{-4} m^2/s (upper aquifer)	25
Robertson (1974), Robertson and Barraclough (1973); NRTS, Idaho	Basaltic lava and sediments	1.4×10^{-1} to 1.4×10^{1} m^2/s	10
Robson (1974); Barstow, California	Alluvial sediments	2.1×10^{-4} to 1×10^{-2} m^2/s	40
Robson (1978); Barstow, California	Alluvial sediments	5×10^{-4} m/s	40
Rousselot *et al.* (1977); Byles-Saint Vulbas near Lyon, France	Clay, sand, and gravel	6.5×10^{-3} to 1.5×10^{-2} m/s	14 2.1~18 1.8~5.9 11~24

Sauty *et al.* (1978); Bonnaud, France	Sand	8.3×10^{-4} to 1.1×10^{-3} m^2/s	/
Segol and Pinder (1976); Cutler area, Biscayne Bay aquifer, Florida	Fractured limestone and calcareous sandstone	0.45×10^{-2} m/s (horizontal) and 0.09×10^{-4} m/s (vertical)	25
Sudicky *et al.* (1983); Borden	Glaciofluvial sand	4.8×10^{-5} to 7.6×10^{-5} m/s	38
Sykes *et al.* (1982, 1983); Borden	Sand	5.8 to 7.2×10^{-5} m/s	35
Sykes *et al.* (1983); Mobile, Alabama	Sand, silt, and clay	5×10^{-4} m/s (horizontal) and 2.5×10^{-5} m/s (vertical)	25
Vaccaro and Bolke (1983); Spokane aquifer, Valocchi *et al.* (1981); Palo Alto Bay lands	Glaciofluvial sand and gravel	9×10^{-5} m^2/s to 6.5 m^2/s	7~40
Valocchi *et al.* (1981); Palo Alto Bay lands	Sand, gravel, and silt	1.25×10^{-3} m^2/s (lower aquifer); 5.0×10^{-4} m^2/s (upperaquifer)	25
Walter (1983); WIPP	Fractured dolomite	8.0×10^{-5} m^2/s	0.7 and 11 (along separate paths)

(*Continued*)

Table 2.2. (*Continued*)

References	Aquifer material	Hydraulic conductivity [m/s] or transmissivity [m²/s]	Effective porosity [%]
Webster *et al.* (1970); Savannah River Plant, South Carolina	Crystalline, fractured schist and gneiss	3.6×10^{-7} m/s	/
Werner (1983); Hydrothermal Test Site, Aefligen, Switzerland	Gravel	6×10^{-3} m/s	17
Wiebenga *et al.* (1967); Lenda and Zuber (1970); Burdekin Delta, Australia	Sand and gravel	5.5×10^{-3} m/s	32
Wilson (1971), Robson (1974); Tucson, Arizona	Unconsolidated gravel, sand, and silt	5.75×10^{-3} m²/s	38
Wood (1981); Aquia Formation, southern Maryland	Sand	2.9×10^{-4} to 8.7×10^{-4} m²/s	35
Wood and Ehrlich (1978); Bassett *et al.* (1981); Lubbock, Texas	Sand and gravel	3.2×10^{-3} to 4.4×10^{-3} m²/s	/

Note: Adapted from Gelhar *et al.* (1992).

where r is radial distance [m]; K_{rr} is the value of horizontal hydraulic conductivity [m/s].

The general model of groundwater flow is a three-dimensional model that can simulate complex flow fields and variations in groundwater levels in groundwater systems. On the other hand, the two-dimensional radial flow model better describes how surface water bodies or well casings interact with groundwater movement.

The general models of groundwater flow are generally more computationally complex, requiring more computational resources and time. On the other hand, the two-dimensional radial flow model is relatively simple to compute and is more suitable for dealing with local groundwater problems such as well pumping, groundwater recharge, etc.

2.2.3 *General models of non-Darcian flow*

Darcy's law indicates a linear relationship between the fluid velocity and the hydraulic gradient (Bear, 1972), which is a basic assumption used to handle a great deal of problems related to flow in porous and fractured media. However, many pieces of evidence from the laboratory and field experiments show that this linear law may be invalid in some situations, especially when the groundwater flow velocity is sufficiently high or sufficiently low, where non-Darcian flow prevails (Basak, 1977; Bordier and Zimmer, 2000; Engelund, 1953; Forchheimer, 1901; Izbash, 1931; Liu *et al.*, 2012; Soni *et al.*, 1978).

Darcy's law considers kinematic forces but excludes inertial forces of flow. However, the inertia forces become significant with respect to the kinematic forces when the velocity is great, leading to non-Darcian flow (Engelund, 1953; Forchheimer, 1901; Irmay, 1959; Izbash, 1931). Forchheimer (1901) proposed a heuristic Forchheimer law describing the non-Darcian flow, which is an extension of Darcy's law by adding a second-order velocity term, representing the inertial effect. To verify the applicability of the Forchheimer law, many approaches were introduced, such as the dimensional analysis (Ward, 1964), the capillary model (Dullien and Azzam, 1973), the hybrid mixture theory (Hassanizadeh and Gray, 1987), and the volume averaging method (Whitaker, 1996). Giorgi (1997) and Chen *et al.* (2001) analytically derived the Forchheimer law from the Navier–Stokes equation. Another widely used model describing the non-Darcian flow is the Izbash equation (Izbash, 1931). This equation is a fully empirical power-law function obtained through analyzing experimental data. The Izbash equation is preferred for modeling purposes since the power index in the Izbash equation can be parameterized depending on flow conditions (Basak, 1977). Some researchers demonstrated that the Forchheimer and Izbash equations were identical for some cases (George and Hansen, 1992).

In general, a flow that cannot be characterized by Darcy's law is often considered a non-Darcian flow. There is great evidence showing that flow in porous media and/or fractured rocks can be non-Darcian under certain circumstances, not only for flows with high velocities but also for flows with low velocities (Hansbo, 1960; Lutz and Kemper, 1959; Miller and Low, 1963; Mitchell and Younger, 1967; Soni *et al.*, 1978; von Engelhardt and Tunn, 1955). For instance, Soni *et al.* (1978) conducted a series of experiments and found that fluid flow can be divided into three distinct regimes, i.e., pre-linear, linear, and post-linear, as the hydraulic gradient increases. Figure 2.1 shows three flow regimes in which both pre-linear flow and post-linear flow can be regarded as non-Darcian flow. In addition to the velocity, some characteristics of the flow and medium can also become the induction factors for non-Darcian flow, such as gas phase flow (Gao and Li, 2016; Gensterblum *et al.*, 2015), multiphase flow (Vincent and Caltagirone, 1999; Wu, 2002a, 2002b), media containing fractures (Berkowitz, 2002; Novakowski *et al.*, 1995; Qian *et al.*, 2005) or deforming with large strain consolidation (Hansbo, 1960, 1997, 2001).

2.2.3.1 *High-velocity non-Darcian flow*

High-velocity non-Darcian flow is common in anthropogenic flow systems such as pumping wells, dam foundations, tunnels, etc. For example, massive groundwater pumping (such as for municipal water supply or dewatering of mining sites) and large hydraulic gradients generated near a dam site often invalidate Darcy's law due to the locally high Reynolds number. In this section, we highlight the significance of high-velocity non-Darcian flow as a judging criterion and present its mathematical expressions and specific applications.

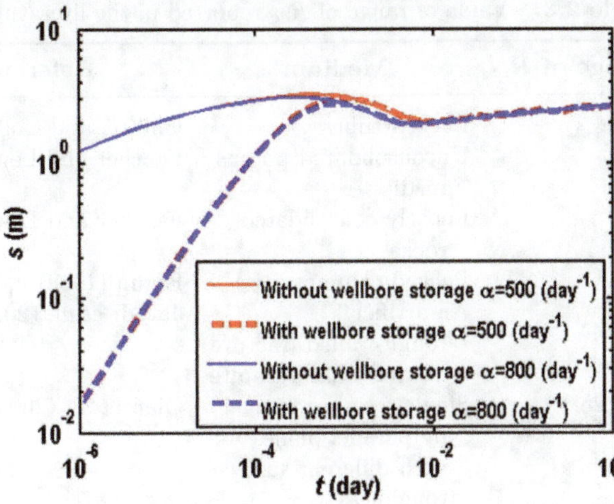

Figure 2.1. The-drawdown-time behavior with wellbore storage-for-the-different α values.

(a) Criteria for high-velocity non-Darcian flow

The deviation from Darcy's law was originally attributed to the occurrence of turbulence, until Green and Duwez (1951) addressed that the onset of high-velocity non-Darcian flow could not be explained accurately by turbulence in porous media. Since then, extensive heuristic studies investigated how microstructure and fluid properties trigger the mechanism that drives high-velocity non-Darcian flow (Hayes *et al.*, 1995; Lasseux *et al.*, 2011; Sivanesapillai *et al.*, 2014; Veyskarami *et al.*, 2016). There are two prevailing criteria that distinguish high-velocity non-Darcian flow from Darcy flow at the macroscopic scale. One is Reynolds number (R_e) while the other is the Forchheimer number (F_0). The Reynolds number is a dimensionless number indicating the ratio of inertial to viscous forces that act on the fluid, defined as $Re = \rho dv/\mu$, where ρ is the fluid density and d is the characteristic length of flow which may be defined as the diameter of the medium particle. The critical Reynolds number (R_{ec}), which is defined as the range of R_e that triggers the transition from Darcy flow to high-velocity non-Darcian flow has been reported in the literature listed in Table 2.3.

Nevertheless, an indefinable characteristic length makes R_{ec} only applicable for columns composed of packed particles, unconsolidated or loosely consolidated sands. This establishes the need for

Table 2.3. Value of range of R_{ec} reported in the literature.

Value or range of R_{ec}	Medium	References
40–80	Packed tubes	Chilton and Colburn (1931)
10–1000	Unconsolidated porous media	Fancher and Lewis (1933)
0.4–3	Loosely consolidated rocks	Fancher and Lewis (1933)
3–10	Packed column	Ergun (1952)
0.001–25	An artificial rough-walled fracture of intact hard granite	Javadi *et al.* (2014)
Less than 336.26	Single fracture formed by parallel planes with different surface roughness	Qian *et al.* (2005)
Approximately 0.32	Natural outcrop core of high production reservoir	Yao *et al.* (2015)

Forchheimer number F_0 which has wider applicability to indicate the occurrence of high-velocity non-Darcian flow. The Forchheimer number F_0 indicates the ratio of the pressure drop induced by liquid-solid interactions to that prompt viscous resistance. It is defined as $F_0 = \frac{k\beta\rho v}{\mu}$, where β is the non-Darcian coefficient or inertial resistance, which indicates the resistance to the fluid acceleration or local changes in the flow direction. Ma and Ruth (1993) gave F_0 a range of 0.005–0.02, corresponding to $R_e = 1$–10 with F_0 being more favored because it addresses macroscopic parameters. However, few studies have experimentally measured the values of critical Forchheimer number (F_{0c}), because γ is often not readily available. Zeng and Grigg (2006) described F_{0c} mathematically as a function of the critical value of the non-Darcian effect, which yields $F_{0c} = 0.11$ at 10% non-Darcian effect. This non-Darcian expression proposed by Zeng and Grigg (2006) enables users to define the value of F_{0c} through experiments (Macini *et al.*, 2011).

(b) High-velocity non-Darcian flow laws

As Darcy's law can be invalid under a high Reynolds number, it is necessary to find new laws that describe the relationship between the specific discharge q and the hydraulic gradient i. Many scientists have attempted to explain the laws governing post-linear non-Darcian flow either theoretically or empirically. The following sections briefly review some representative non-Darcian flow laws. Forchheimer (1901) proposed a second-order polynomial function to describe the non-Darcian flow, which can be expressed as:

$$-\frac{dp}{dX} = aq + bq^2 \tag{2.3}$$

where $a = \mu/k_0$ and $b = \beta\rho$. If β is equal to zero, then Eq. (2.3) becomes Darcy's law. The second term on the right-hand side of Eq. (2.3) can be regarded as a perturbation or correction term of the first-order linear term. Forchheimer (1901) also presented another extended equation to describe non-Darcian flow, that is:

$$-\frac{dp}{dX} = aq + bq^2 + cq^3 \tag{2.4}$$

where $c = \varphi\rho^2$ with φ to be an empirical constant. As R_e increases, a weak inertia regime appears first at the onset of nonlinearity with an inertial correction scaled by the term cq^3. This weak inertial regime behaves as a transition to a strong inertial regime (Fourar *et al.*, 2004; Lasseux *et al.*, 2011) and is usually overlooked in practice due to its limited macroscopic influences (Lo Jacono *et al.*, 2005) and short range of validity (Agnaou *et al.*, 2017). When the R_e increases to 1–10, one can observe a strong inertial regime, which is also known as the Forchheimer regime and scaled by the term bq^2 (Agnaou *et al.*, 2017). The strong inertial regime transforms to a turbulence regime when *Re* exceeds the upper limit of the Forchheimer regime (Lage *et al.*, 1997; Lasseux *et al.*, 2011). In addition to the polynomial function, there are also many power-law functions proposed to describe the non-Darcian flow, and the most commonly used one might be the Izbash equation (Izbash, 1931), expressed as

$$q = Mi^n \tag{2.5}$$

where M and n are empirical parameters. Other formulas proposed to describe non-Darcian flow have been summarized by Basak (1978) and are listed in Table 2.4. The equations listed in Table 2.4 are often used to describe post-linear flow under relatively high Reynolds numbers.

(c) Applications of high-velocity non-Darcian flow laws

Non-Darcian flow to a pumping well has caught scientists' attention since the 1970s. In recent decades, the studies of non-Darcian flow to a pumping well have solved practical problems with respect to water resources estimation (Wen *et al.*, 2008a, 2008b, 2009), environmental protection (Wang *et al.*, 2017; Zhang *et al.*, 2014), foundation pit dewatering (Wang *et al.*, 2017), hydraulic fracturing associated with gas production (Wang *et al.*, 2016), and reduction of fluid production (Mathias and Todman, 2010; Mathias and Wen, 2015), among others. Generally speaking, the applied studies in well hydraulics prefer Forchheimer equation (Mathias and Wen, 2015; Wang *et al.*, 2017; Wen *et al.*, 2011) and its refined versions (Barree and Conway, 2009; Zhang *et al.*, 2014) compared to the Izbash equation, which is only popular in water resources estimation (Wen *et al.*, 2008a, 2008b, 2009). These high-velocity non-Darcian flow laws as applied in well hydraulics aim to contribute mainly in two aspects, one, to uncover the problems by assuming the validation of Darcy's law when pumping wells are used, such as the misestimation of aquifer parameters (Wen *et al.*, 2008a, 2008b, 2009), the estimation of nonlinear energy losses and well performance in fluid production (Mathias and Wen, 2015); the other aspect is to quantitatively determine the influencing mechanism of well parameters and aquifer properties on the non-Darcian effects, particularly the non-Darcian coefficient γ, for the purpose of optimizing the design of a pumping or injection engineering work such as CO_2 sequestration (Zhang *et al.*, 2014), foundation pit dewatering (Wang *et al.*, 2017), hydraulic fracturing through pumping or injection well (Wang *et al.*, 2016), etc.

Due to the coarse size of particles in the foundation of various rockfill dams (traditional dams, embankments dams, tailing dams, etc.), the high-velocity seepage flow through these dams often shows turbulent behavior that is inconsistent with Darcy's law.

Table 2.4. Published non-Darcian flow laws at high R_e (revised from Basak, 1978).

Type of equation	Equation	Original proposer	Explanation of terms	Comments
Polynomial equation	$i = aq + bq^2$	Forchheimer (1901)	a and b are constants	Empirical, but theoretical basis was found later
	$i = aq + bq^2 + cq^3$	Forchheimer (1901)	a, b and c are constants	Empirical
	$i = aq + bq^{1.5} + cq^2$	Rose (1951)	a, b and c are constants	Empirical
	$i = aq + bq^2 + c\dfrac{\partial v}{\partial t}$	Polubarinova-Kochina (2015)	a, b and c are constants	Empirical
	$i = aq + bq^m$	Muskat (1938), Harr (1962)	a, b and m are constants	Empirical
Power equation	$q = Mi^n, \cdot n < 1$	Izbash (1931)	M and n are constants	The empirical value of n lies between 1 and 0.5
	$q = (Bi)^{1/2}$	Escande (1953)	B is constant	empirical, B varies between 80 and 290 (cm/s) 2 for particles of diameter 2.54 cm.
	$a = 32.9m^{1/2}t^{0.54}$	Wilkins (1955)	M is hydraulic radius	Semiempirical based on test results with particles of 1.905–7.62 cm
	$q = \alpha(\mu/\sigma)^{f-1}(ki)^f$	Slepicka (1961)	α, f and k are constants; μ is Viscosity; σ is surface tension	Semiempirical, derived from dimensional analysis
	$q = n(igd/K_{st})^{1/2}$	Stephenson (1979)	d is particle diameter; n is porosity; g is gravitational constant; K_{st} is Stephenson's friction **factor**	Semiempirical and better used in field-scale

Investigating the non-Darcian flow inside the dam has great engineering benefits, like improving the design and selection of aggregate material (Samani *et al.*, 2013), predicting the hydraulic head to assess dam stability (Bari and Hansen, 2002; Sarkhosh *et al.*, 2018) and computing leakage from reservoir storage (Chen *et al.*, 2019). Overall, there are more equations to choose from to characterize non-Darcian behavior in subsurface flow in dam foundations or coarse granular material compared to the vicinity surrounding a pumping well. These include the Forchheimer equation used in Kells (1996), Izbash equation used in Bordier and Zimmer (2000) and Samani *et al.* (2013), Wilkins and Stephenson's equation used in Bari and Hansen (2002). In contrast to the widely distributed nonlinear flow near a pumping well, flow inside a dam often shows gradual transitions from laminar to turbulent stages. George and Hansen (1992) investigated the interchangeability between Darcian and non-Darcian flows. Afterwards, Hansen *et al.* (1995) presented six empirical non-Darcian correlations incorporating particle geometry and porosity, which were categorized into three different degrees of turbulence to cover the whole process of flow regime transition.

2.2.3.2 *Low-permeability non-Darcian flow*

Low-permeability media are mostly associated with tight configurations composed of fine-grained sediments such as clays and silts Neuzil (1986). The practical problems of low-permeability non-Darcian flow stem from various earth science disciplines, such as the production of tight oil/gas reservoirs, aquifer parameter estimation, environmental protection and plant physiology. Unlike the high-velocity non-Darcian flow, which is often induced by anthropogenic factors, the low-permeability non-Darcian flow is strongly related to the intrinsic properties of the medium and fluid.

(a) Low-permeability non-Darcian flow equations

Note that the effects of pore structure and solid–liquid interaction are microscopic in nature and are often inadequately described by a mathematical expression of low-velocity non-Darcian flow law. Instead, permeability and TPG are two measurable macroscopic parameters commonly used in practice which are based on experimental data. For example, oil or gas flow in shale and clay reservoirs does not apparently move when the existing hydraulic gradient is smaller than the TPG. Such a relationship between the specific discharge and the hydraulic gradient can be expressed as (Bear, 1979):

$$q = \begin{cases} -K(i - I), & i > I \\ 0, & i < I \end{cases} \tag{2.6}$$

where q is the specific discharge; K is the hydraulic conductivity; i is the pressure gradient and I is the TPG. In some cases, ultralow velocity flow indicates nonlinear behavior rather than stagnant state, which requires more accurate pre-linear non-Darcian expressions. The relationships between the specific discharge (q) and the hydraulic gradient (i) for low-permeability non-Darcian equations have been thoroughly summarized by Liu *et al.* (2012) in Table 2.5. The motivations for designing these formulas are many, such as better fitting a specific experimental observation (Hansbo, 1960, 2001), keeping the empirical or uncertain parameters as less as possible (Bear, 1979), using continuous instead of piecewise functions (Swartzendruber, 1962), covering controlling factors as much as possible (Zou, 1996), applying for both saturated and unsaturated media (Liu *et al.*, 2012), to name a few. All these experiment-based formulas attempt to generate a transition from the nonlinear relationship of $q-i$ to the linear one as the excess of i over the TPG increases, and the greatest difference between them is the characterization of the $q-i$ relationship before i exceeds the TPG significantly. Each proposed formula has its advantages and drawbacks as well.

2.3 Radial Pumping Test in Darcian Flow Fields

Actual field pumping tests often involve variable pumping rates which cannot be handled by the classical constant-rate or constant-head test models, and often require a convolution process to interpret the test data. A striking new feature of the pumping test with an exponentially decayed rate is that the drawdowns will decrease over a certain period during the intermediate pumping stage, which has never been seen before in constant-rate or constant-head pumping tests.

Table 2.5. Published non-Darcian flow laws in low-permeability medium.

Equation	Original proposer	Explanation of terms	Comments
$q = \kappa i^n$ for $i \leq i_1$ $q = \kappa n i_1^{n-1}(i - l)$ for $i \geq i_1$	Hansbo (1960); Hansbo (2001)	κ is coefficient of permeability $i_1 = In/(n-1)$ I is TPG and n is constant	Mathematical expressions are separated and the unknown parameters
$q = K[i - I(1 - e^{-i/l})]$	Swartzendruber (1962)	K is hydraulic conductivity, I is TPG	Less unknown parameters and covering all domain of definition of i
$q = 0$ for $i \leq I$ $q = K(i - I)$ for $i \geq I$	Bear (1979)	K is hydraulic conductivity, I is TPG	Less unknown parameters; Flaws: excluding the nonlinear behavior
$q = \frac{Ki}{1 + b\exp(-ci)}$	(Zou, 1996)	B and c are empirical parameters	Unable to account for both saturated and unsaturated conditions
$q = K\left[i - \frac{I}{\gamma\left(\frac{1}{\alpha}\right)}\gamma\left(\frac{1}{\alpha}, \left(\frac{i}{I^*}\right)^{\alpha}\right)\right]$	Liu *et al.* (2012)	$I = \frac{I^*}{\alpha}\gamma\left(\frac{1}{\alpha}\right)$, $\gamma(a)$ $= \int_0^{\infty} t^{a-1}e^{-t}dt$ $\gamma(ax) = \int_0^{x} t^{a-1}e^{-t}dt$	Able to account for both saturated and unsaturated conditions

In this section, the analytical model of pumping tests developed by Wen *et al.* (2017) considering an exponentially decreasing pumping rate started at a certain (higher) rate and eventually stabilized at a certain (lower) rate for cases with or without wellbore storage is introduced.

2.3.1 *Mathematical model*

The assumptions used in the study of Wen *et al.* (2017) are the same as those of the Theis model, except that the pumping rate is assumed to vary with time and can be described by a certain form of function, such as an exponential function. After establishing the mathematical model for such an infinitesimal pumping well, Wen *et al.* (2017) further expanded the model to a finite-diameter pumping well with wellbore storage. The mathematical model for the pumping test with an exponentially decayed rate function and an infinitesimal well radius can be expressed as follows.

2.3.1.1 *Governing equation*

The mathematical model for pumping test with an exponentially decayed rate function and an infinitesimal well radius can be expressed as

$$\frac{\partial^2 s}{\partial r^2} + \frac{1}{r}\frac{\partial s}{\partial r} = \frac{S}{T}\frac{\partial s}{\partial t} \tag{2.7}$$

where s is the drawdown [m] at distance r and time t, S is the storage coefficient [dimensionless]; T is the transmissivity [m^2/s].

2.3.1.2 *Initial condition*

For the flow model, the initial condition is

$$s(r, 0) = 0 \tag{2.8}$$

2.3.1.3 *Boundary conditions without wellbore storage*

The assumptions used in this section are the same as those of the Theis model, except that the pumping rate is assumed to vary with time and can be described by a certain form of function, such as an exponential function. Then, the boundary conditions can be expressed as

$$s(\infty, t) = 0 \tag{2.9}$$

$$\lim_{r \to 0} -2\pi r T \frac{\partial s}{\partial r} = Q_2 + (Q_1 - Q_2)e^{-\alpha t} \tag{2.10}$$

where Q is the pumping rate [m^3/s], Q_1 is the initial pumping rate [m^3/s], and Q_2 is the final pumping rate [m^3/s], and α is a decay constant which can be estimated from the actual pumping data $[1/s]$.

2.3.1.4 *Boundary conditions with wellbore storage*

When the wellbore storage is considered (for a finite-diameter pumping well), the boundary condition Eq. (2.9) should be rewritten as

$$\lim_{r \to r_w} \left[-2\pi r T \frac{\partial s}{\partial r} + \pi r_w^2 \frac{ds_w(t)}{dt} \right] = Q_2 + (Q_1 - Q_2)e^{-\alpha t} \tag{2.11}$$

where r_w is the radius [m] of the pumping well; $s_w(t)$ is the drawdown [m] inside the pumping well.

2.3.2 *Analytical solution without wellbore storage*

Applying Laplace transform, Eqs. (2.7)–(2.10) can be rewritten in the Laplace domain as

$$\frac{d^2\bar{s}}{dr^2} + \frac{1}{r}\frac{d\bar{s}}{dr} = \frac{S}{T} p\bar{s} \tag{2.12}$$

$$\bar{s}(\infty, p) = 0 \tag{2.13}$$

$$\lim_{r \to 0} -2\pi r T \frac{d\bar{s}}{dr} = \frac{Q_2}{p} + \frac{(Q_1 - Q_2)}{p + \alpha} \tag{2.14}$$

where p is the Laplace variable, and the over bar means the variables in the Laplace domain. The initial condition Eq. (2.8) has been used in the Laplace transformation of Eq. (2.12). Equation (2.13) is a

Bessel equation whose general solution can be expressed as

$$\bar{s}(r,p) = C_1 I_0\left(r\sqrt{\frac{pS}{T}}\right) + C_2 K_0\left(r\sqrt{\frac{pS}{T}}\right) \tag{2.15}$$

where $I_0(x)$ and $K_0(x)$ are the first and second kinds of the modified Bessel functions with order 0, respectively. C_1 and C_2 are constants that can be determined by the boundary conditions. Considering Eq. (2.13) and the properties of the modified Bessel function, one has $C_1 = 0$. Then Eq. (2.14) can be simplified as

$$\bar{s}(r,p) = C_2 K_0\left(r\sqrt{\frac{pS}{T}}\right) \tag{2.16}$$

Applying Eq. (2.14) in Eq. (2.16), one has:

$$\lim_{r\to 0} -2\pi r T C_2 K_0'\left(r\sqrt{\frac{pS}{T}}\right)\sqrt{\frac{pS}{T}} = \frac{Q_2}{p} + \frac{(Q_1 - Q_2)}{p+\alpha} \tag{2.17}$$

With the properties of the modified Bessel functions: $dK_0(x)/dx = -K_1(x)$ and $K_v(x) \approx \frac{\Gamma(v)}{2}\left(\frac{x}{2}\right)^{-v}$, $v > 0$ when x goes to zero, one can obtain:

$$C_2 = \frac{Q_2}{2\pi T}\frac{1}{p} + \frac{(Q_1 - Q_2)}{2\pi T}\frac{1}{(p+\alpha)} \tag{2.18}$$

Therefore, the solution of the drawdown in the Laplace domain can be determined accordingly:

$$\bar{s}(r,p) = \frac{Q_2}{2\pi T}\frac{1}{p}K_0\left(r\sqrt{\frac{pS}{T}}\right) + \frac{(Q_1 - Q_2)}{2\pi T}\frac{1}{(p+\alpha)}K_0\left(r\sqrt{\frac{pS}{T}}\right) \tag{2.19}$$

Applying the convolution theorem of the Laplace transform, one can obtain the solution of the drawdown in the real-time domain as

$$s(r,t) = \frac{Q_2}{2\pi T}\int_0^t \frac{1}{2t'}e^{-\frac{r^2 S}{4Tt'}}dt' + \frac{(Q_1 - Q_2)}{2\pi T}\int_0^t \frac{1}{2t'}e^{-\frac{r^2 S}{4Tt'}}e^{-\alpha(t-t')}dt' \tag{2.20}$$

where t' is a dummy variable of integration [T]. The first term of Eq. (2.20) is the same as the Theis model after using a variable substitution of $u = \frac{r^2 S}{4Tt'}$, indicating that our solution can be simplified to the Theis solution under the special condition of $Q_1 = Q_2$.

2.3.3 Analytical solution with wellbore storage

For the case with wellbore storage, the boundary condition is replaced by Eq. (2.11). Applying Laplace transform, Eq. (2.11) can be transformed to

$$\lim_{r\to r_w} -2\pi rT\frac{d\bar{s}}{dr} + \pi r_w^2 p\bar{s}_w = \frac{Q_2}{p} + \frac{(Q_1 - Q_2)}{p+\alpha} \tag{2.21}$$

Substituting Eq. (2.21) to Eq. (2.16), one can obtain:

$$C_2 = \frac{\frac{Q_2}{p} + \frac{Q_1-Q_2}{p+\alpha}}{\pi r_w^2 pK_0\left(r_w\sqrt{\frac{PS}{T}}\right) + 2\pi r_w TK_1\left(r_w\sqrt{\frac{PS}{T}}\right)\sqrt{\frac{PS}{T}}} \tag{2.22}$$

Therefore, the drawdowns considering wellbore storage in the Laplace domain can be expressed as

$$\bar{s}(r,p) = \frac{\left(\frac{Q_2}{p} + \frac{Q_1-Q_2}{p+\alpha}\right)K_0\left(r_w\sqrt{\frac{pS}{T}}\right)}{\pi r_w^2 K_0\left(r_w\sqrt{\frac{pS}{T}}\right) + 2\pi r_w TK_1\left(r_w\sqrt{\frac{pS}{T}}\right)\sqrt{\frac{pS}{T}}} \tag{2.23}$$

where overbar represents the term in the Laplace domain, p is the Laplace variable [1/s], $K_0(x)$ and $K_1(x)$ are the second kind of the modified Bessel functions with order 0 and 1, respectively.

It seems that Eq. (2.23) cannot be inverted analytically. Thus, the Stehfest method is used to invert Eq. (2.23) numerically (Stehfest, 1970). Eventually, drawdowns in the real-time domain can be obtained. The Stehfest method uses 18 terms.

2.3.4 Comparisons of models with and without wellbore storage

It is notable that the mathematical model without wellbore storage in the model of Wen *et al.* (2017) becomes the conventional Theis model if α becomes infinity. Similarly, the case considering wellbore storage changes to the model of Papadopulos and Cooper Jr (1967) when a becomes infinity. In addition, considering a decayed rate as long as the pumping starts ($t = 0$), the solutions of Wen *et al.* (2017) can be easily extended to a case with the pumping rate starts to decay at a pre-defined time using a piecewise approach. The details can be found in several previous studies (Abu-Zied and Scott, 1963; Abu-Zied *et al.*, 1964; Mishra *et al.*, 2013).

In many field applications, one must rely on drawdown data collected from the pumping well itself, i.e., the so-called single-well test, when observation wells do not exist or are not available. This is particularly true when the aquifer is deep (more than several hundred meters below the ground surface) and drilling additional observational wells is not feasible. The wellbore storage effect is almost always unavoidable when interpreting the single-well test data. Figure 2.1 depicts the drawdowns inside the pumping well while considering the wellbore storage effect. The parameters are given as $Q_1 = 500\,\mathrm{m}^3/\mathrm{day}$, $Q_2 = 200\,\mathrm{m}^3/\mathrm{day}$, $T = 100\,\mathrm{m}^2/\mathrm{day}$, $S = 0.001$, $r = rw = 0.1\,\mathrm{m}$, $a = 500\ \mathrm{day}^{-1}$ and $800\ \mathrm{day}^{-1}$. The case without wellbore storage has also been depicted in this figure as a reference.

It is found that the drawdowns inside the well increase rapidly with time, and drawdown-time curves exhibit straight lines in log-log scale at early times. This is a widely recognized wellbore storage effect in constant-rate pumping tests before Papadopulos and Cooper (1967). When the pumping time is large enough, the wellbore storage effect disappears, and the drawdowns approach the same values without wellbore storage. Once again, a decreasing feature of the drawdowns is also found at intermediate times when the wellbore storage effect is included. It also should be pointed out that there are great discrepancies at early times between the cases with and without the wellbore storage.

Figure 2.2 shows a comparison between the solution without wellbore storage (Eq. (2.20)) and the results of Sen and Altunkaynak (2004) (referred to as the SA solution hereafter and in subsequent figures), who studied the same problem without rigorous mathematical derivation. The parameters are given as $Q_1 = 500\,\mathrm{m}^3/\mathrm{day}$, $Q_2 = 200\,\mathrm{m}^3/\mathrm{day}$, $= 100\,\mathrm{m}^2/\mathrm{days}$, $S = 0.001$, $\alpha = 500\ \mathrm{day}^{-1}$, $r = 0.1\,\mathrm{m}$, $1.0\,\mathrm{m}$, $2.0\,\mathrm{m}$ and $5.0\,\mathrm{m}$, respectively. Obviously, there are great differences between our solution and the SA solution, especially at late times.

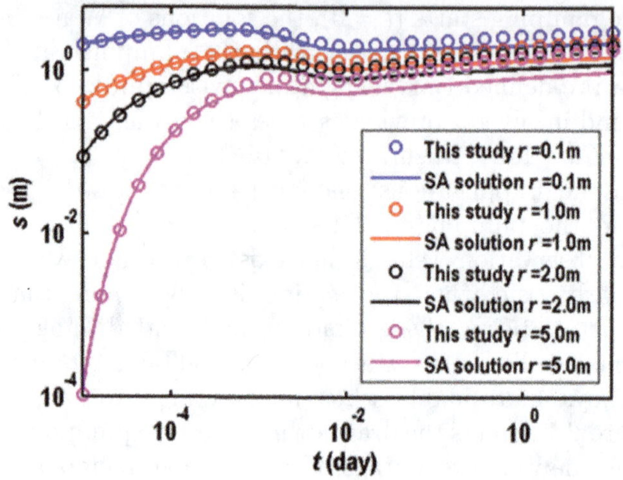

(a)

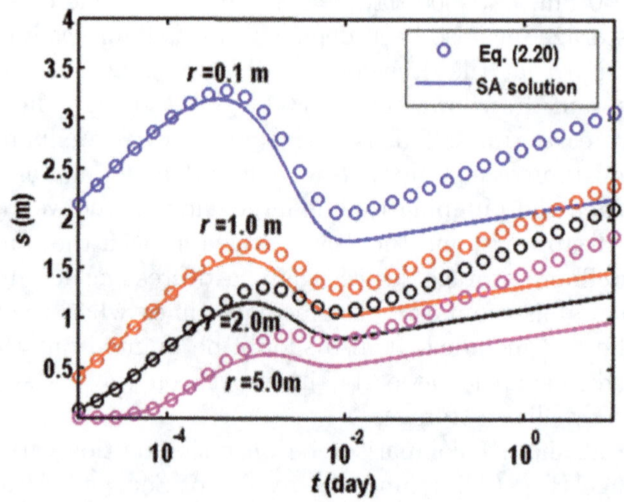

(b)

Figure 2.2. Comparison between the solutions of Eq. (2.20) and Sen and Altunkaynak (2004). (a) Log–log scale and (b) semi-log scale.

2.3.5 *Results and discussions*

The impact of the parameter α on the drawdown with $a = 100, 200$, 500 and 800 day^{-1}, respectively, is analyzed in this section. The other parameters are given as: $Q_1 = 500\,\text{m}^3/\text{day}$, $Q_2 = 200\,\text{m}^3/\text{day}$,

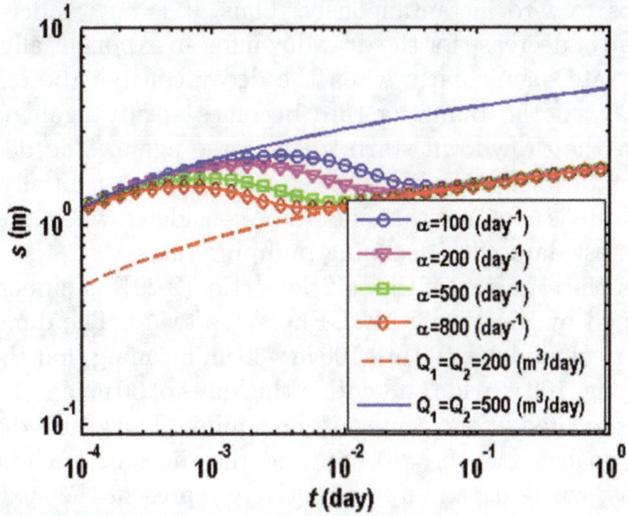

Figure 2.3. The drawdown-time behavior without wellbore storage for the different α values.

$T = 100\,\text{m}^2/\text{day}$, $S = 0.001$, $r = 1\,\text{m}$. The cases of constant-rate pumping are also depicted in this figure as a reference with $Q_1 = Q_2 = 500\,\text{m}^3/\text{day}$ and $Q_1 = Q_2 = 200\,\text{m}^3/\text{day}$, as shown in Figure 2.3. It is interesting to see that all the curves approach the same asymptotic values of the constant-rate pumping case with $Q_1 = Q_2 = 500\,\text{m}^3/\text{day}$ at early times, and decrease to approach the same asymptotic values of the constant-rate pumping case with $Q_1 = Q_2 = 200\,\text{m}^3/\text{day}$ at late times. However, the curves deviate from the cases of constant-rate pumping at intermediate times. It is also found that the drawdown will decrease at intermediate times for a short period and a larger value of α results in an earlier decrease of the drawdown.

To understand this behavior of the drawdown-time curves, one may recall the recovery test in which the pumping rate is reduced to zero after shutting down the pumping well. It is well known that the drawdown will decrease (or the water level will start to recover) during the recovery period. A decreasing pumping rate is somewhat like the recovery stage of a constant-rate pumping test, particularly when the decay constant α is very large. This is because a very large α value implies that the pumping rate drops very rapidly over time, which is somewhat like the shutdown case in which the pumping rate drops to zero instantaneously. Thus, it is no wonder to see a short stage of decrease for the drawdown for an exponentially decayed pumping rate, particularly when the decay constant becomes large enough. When the pumping rate becomes steady again (equal to Q_2), then the drawdown starts to increase again. The decrease of drawdown during the intermediate stage for an exponentially decayed pumping rate is a feature that should be considered when interpreting pumping test data with declining pumping rates.

To test the accuracy of the solutions (Eq. (2.20)), a numeric model is developed by the Visual MODFLOW software. The dimension of the numerical model is $100\,\mathrm{m} \times 100\,\mathrm{m} \times 30\,\mathrm{m}$, meaning that the aquifer is 100 m long, 100 m wide and with a thickness of 30 m, and the pumping well is located at the center of the aquifer. The grid is discretized to a grid space $dx = dy = 0.1\,\mathrm{m}$, and the timestep is chosen to be 0.0001 day, where dx and dy respectively represent the spatial steps in the x- and y-axis in Visual MODFLOW. In order to keep the consistency of the conceptual models of the numerical solution and the analytical solution, especially for the boundary condition, we choose a small pumping rate (such as $Q_1 = 5\,\mathrm{m}^3/\mathrm{day}$, $Q_2 = 2\,\mathrm{m}^3/\mathrm{day}$, and $\alpha = 500\ \mathrm{day}^{-1}$), a short observation duration (less than 1 day), and the other parameters are the same as those used in Figure 2.2. Such parameters could result in an extremely small drawdown (smaller than $10^{-4}\,\mathrm{m}$) at four peripheral boundaries of the aquifer, so that the influence of the boundary conditions can thus be neglected. In other words, one can set either a constant-head boundary or no-flux boundary for such numerical models. The no-flux boundary is used in this section. The comparison between the proposed solutions with the numerical solution is shown in Figure 2.4.

It is found that the solution of Eq. (2.20) agrees perfectly with the numerical solution. This indicates that the solution of Eq. (2.20) is reliable, and the SA solution has unacceptable errors and should not be used to estimate the aquifer parameters.

2.3.6 *Parameter estimation*

Using the aquifer tests to estimate the hydraulic parameters of the aquifer, a common approach is to best fit the observed time-drawdowns at the well using the mathematic model, in which the observed drawdown data are not collected until the groundwater flow

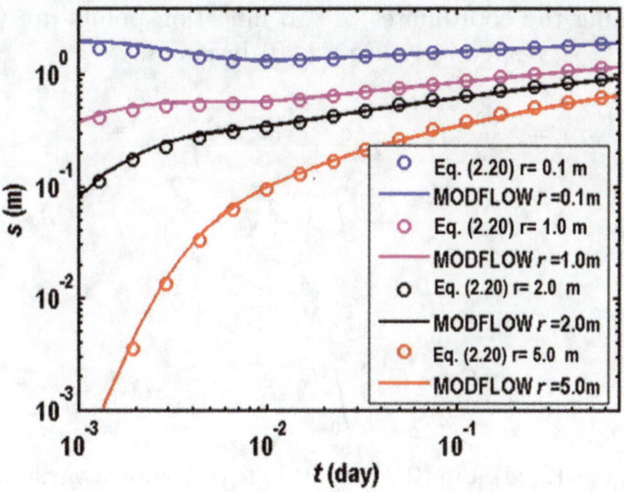

Figure 2.4. Comparison between the solution of Eq. (2.20) and the numerical simulation of Visual MODFLOW.

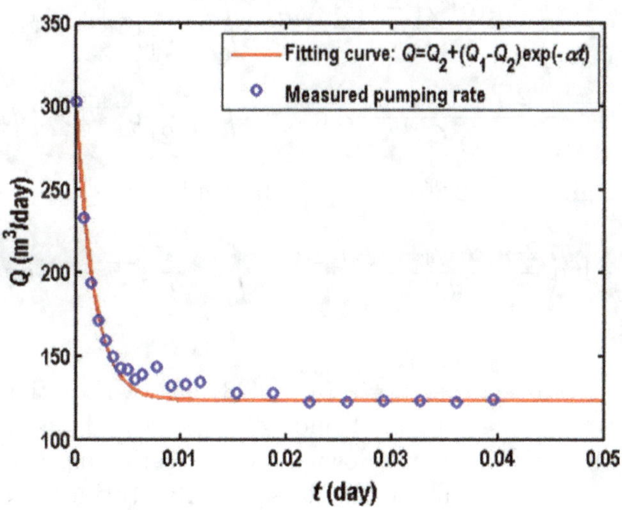

Figure 2.5. The fitting curve of the pumping rate from a borehole in the Jianghan Plain of China.

is up to the steady state. As for the decreasing pumping rate aquifer test, the time-drawdown curves have two inflection points where the slopes are zero there, as shown in Figures 2.3 and 2.5. Such two points might be used for parameter estimation.

Assuming the coordinates of two inflection points are (t_{P1}, s_{P1}) and (t_{P2}, s_{P2}), respectively, Eq. (2.20) becomes:

$$s_{P1} = \frac{Q_2}{2\pi T}\int_0^{t_{P1}} \frac{1}{2x} e^{-\frac{r^2 S}{4Tx}} dx + \frac{(Q_1 - Q_2)}{2\pi T}\int_0^{t_{P1}} \frac{1}{2x} e^{-\frac{r^2 S}{4Tx}} e^{-\alpha(t_{P1}-x)} dx \tag{2.24}$$

$$s_{P2} = \frac{Q_2}{2\pi T}\int_0^{t_{P2}} \frac{1}{2x} e^{-\frac{r^2 S}{4Tx}} dx + \frac{(Q_1 - Q_2)}{2\pi T}\int_0^{t_{P2}} \frac{1}{2x} e^{-\frac{r^2 S}{4Tx}} e^{-\alpha(t_{P2}-x)} dx \tag{2.25}$$

Equations (2.24) and (2.25) contain two unknown variables (T, S), which can be calculated using the iteration method. Derivative Eqs. (2.24) and (2.25), one has:

$$0 = e^{-\frac{r^2 S}{4Tt_{P1}}} - \frac{2t_{P1}\alpha(Q_1 - Q_2)}{Q_1} e^{-\alpha t_{P1}} \int_0^{t_{P1}} \frac{1}{2x} e^{-\frac{r^2 S}{4Tx} + \alpha x} dx \tag{2.26}$$

$$0 = e^{-\frac{r^2 S}{4Tt_{P2}}} - \frac{2t_{P2}\alpha(Q_1 - Q_2)}{Q_1} e^{-\alpha t_{P2}} \int_0^{t_{P2}} \frac{1}{2x} e^{-\frac{r^2 S}{4Tx} + \alpha x} dx \tag{2.27}$$

Rearranging Eq. (2.26) or Eq. (2.27) leads to:

$$\chi + \frac{4t_{P1}}{r^2} \ln\left(\frac{2t_{P1}\alpha(Q_1 - Q_2)}{Q_1} e^{-\alpha t_{P1}} \int_0^{t_{P1}} \frac{1}{2x} e^{-\frac{r^2}{4x}\chi + \alpha x} dx\right) = 0 \tag{2.28}$$

where $\chi = \frac{S}{T}$. In another word, Eq. (2.28) has two solutions of time if χ is given. If the (t_{P1}, s_{P1}) and (t_{P2}, s_{P2}) could be determined from the observed time-drawdown curves, substituting t_{P1} or t_{P2} to calculate χ. Then T could be calculated by substituting the computed χ into Eq. (2.24) or (2.25), and finally $S = \chi T$.

Due to the complexity of the analytical solutions, it seems impossible to directly get mathematic expressions of χ. In this section, a genetic algorithm (GA method) will be employed to estimate the aquifer parameters, in which the adaptive Simpson quadrature method will be called in for numerical integration in Eq. (2.28). The

GA method is chosen because it is more powerful than root-finding algorithms such as bisection or secant methods, especially for nonlinear problems. Meanwhile, it is also efficient for the problem investigated in this section.

The GA method is an adaptive heuristic search algorithm based on the evolutionary concept of natural selection and genetics, which has been widely used for parameter estimation in hydrological sciences and many other disciplines. The basic technique of the GA method is designed to simulate processes in natural systems for evolution, following the principle of "survival of the fittest", which was first proposed by Charles Darwin. The chromosome was used to represent the solution of the problems, and the population was randomly generated from the range of the parameters to be estimated. For the problem proposed in this section, the range of χ is set at (10^{-10} day/m^2, 10^{-1} day/m^2). After an initial population is randomly generated, the algorithm evolves three processes: selection, crossover, and mutation. The selection process represents the survival of the fittest, the crossover process represents mating between two stochastic individual chromosomes, and the mutation process introduces random modifications of the chromosomes. The crossover probability and mutation probability are chosen to be 0.3 and 0.25, respectively. The range of those values is from 0 to 1, such values used in this section are proven to be appropriate. Each chromosome of the GA is encoded by using a binary string in this section. The population size of the GA depends on the number of observation points (N).

The objective function of the genetic algorithm is

$$\varepsilon = \min \left| \chi + \frac{4t_{P1}}{r^2} \ln \left(\frac{2t_{P2}\alpha(Q_1 - Q_2)}{Q_1} e^{-\alpha t_{P1}} \int_0^{t_{P1}} \frac{1}{2x} e^{-\frac{r^2}{4x}\chi + \alpha x} dx \right) \right| \tag{2.29}$$

The ranges of χ is set at (10^{-10} day/m^2, 10^{-1} day/m^2) in this section.

To test the accuracy of the parameter estimation approach proposed in this section, we use three examples, listed in Tables 2.6 and 2.7.

Assuming S and T are known (shown in the first column of Table 2.6), one can easily obtain the two inflection points (t_{P1}, s_{P1})

Table 2.6. Comparison of the computed and given χ for t_{P1}.

t_{P1} [day]	Given $\chi = S/T$ [day/m^2]	Computed χ [day/m^2] χ (t_{P1})	χ ($t_{P1} \times 1.003$)	χ ($t_{P1} \times 0.997$)
0.8052	$\chi = 002/1 = 20 \times 10^{-3}$	2×10^{-3}	21×10^{-3}	2×10^{-3}
0.5831	$\chi = 002/10 = 20 \times 10^{-4}$	20×10^{-4}	20×10^{-4}	2×10^{-4}
0.4546	$\chi = 002/100 = 20 \times 10^{-5}$	21×10^{-5}	19×10^{-5}	21×10^{-5}

Table 2.7. Comparison of the computed and given χ for t_{P2}.

t_{P2} [day]	Given $\chi = S/T$ [day/m^2]	Computed χ [day/m^2] χ (t_{P2})	χ ($t_{P2} \times 1.003$)	χ ($t_{P2} \times 0.997$)
11.9406	$\chi = 002/1 = 20 \times 10^{-3}$	21×10^{-3}	1.9×10^{-3}	24×10^{-3}
12.5835	$\chi = 002/10 = 20 \times 10^{-4}$	21×10^{-4}	20×10^{-4}	2×10^{-4}
13.1559	$\chi = 002/100 = 20 \times 10^{-5}$	1.9×10^{-5}	18×10^{-5}	2×10^{-5}

and (t_{P2}, s_{P2}). Then one can calculate (T, S) using the method proposed in this section. The parameters used in this case are given as $Q_1 = 500\,\text{m}^3/\text{day}$, $Q_2 = 200\,\text{m}^3/\text{day}$, $r = r_w = 0.1\,\text{m}$, $\alpha = 500\,\text{day}^{-1}$. However, measurement errors of the observed drawdowns are generally inevitable for actual problems, partially due to the accuracy of the equipment. Therefore, four cases of $t_{P1} \times (1 \pm 0.3\%)$ and $t_{P2} \times (1 \pm 0.3\%)$ are designed to examine the robustness of the method in analyzing data with measurement errors, in which t_{P1} and t_{P2} are calculated using the parameters mentioned above. The results are shown in Tables 2.6 and 2.7. The calculated results are very accurate, meaning that the proposed method is reliable for estimating the aquifer parameters.

2.4 Pumping Test in a Vertical Well with Intra-Wellbore Head Losses

Among numerous studies on aquifer tests, a great deal of effort has been devoted to studying a partially penetrating vertical well because of its complex three-dimensional flow nature near the well, and its common application in the field, especially when aquifer thickness is large (Driscoll, 1986; Kruseman *et al.*, 1970; Wang *et al.*, 2015a; Yeh and Chang, 2013). Butler and Zhan (2004) presented a semi-analytical solution of groundwater flow in response to a slug or pumping test in a highly permeable aquifer. Perina and Lee (2006) used Laplace and finite Fourier-cosine transforms to develop a general well function for pumping tests in an unconfined or leaky aquifer, considering the effects of a finite-thickness skin and well partial penetration. This general well function could be used to describe the transient hydraulic head response for constant-head tests, constant-rate tests, and slug tests. Shaw-Yang and Yeh (2012) presented similar solutions for the case of a confined aquifer.

Wang and Zhan (2017) advances the present knowledge of pumping test models by developing new MTBC solutions based on the UFBC solutions. This solution is not only capable of interpreting the data of the transient hydraulic head response for constant-head tests and constant-rate tests but can also be applied to well design, pump selection and pump placement in a confined aquifer. In this section, the model of Wang and Zhan (2017) will be introduced.

2.4.1 *Flow into a partially penetrating vertical well*

2.4.1.1 *Well boundary models*

With respect to the treatment of the well screen of a partially penetrating vertical well in aquifer tests, three types of boundary conditions are relevant: UFBC, UHBC and MTBC. UFBC and UHBC can be respectively described by:

$$\left.\frac{\partial H}{\partial r}\right\|_{r=r_w} = F_1(t) \quad \text{when } z_{\text{bot}} \leq z \leq z_{\text{top}} \tag{2.30}$$

$$H\|_{r=r_\omega} = F_2(t), \quad \text{when } z_{\text{bot}} \leq z \leq z_{\text{top}} \tag{2.31}$$

where H is the hydraulic head inside the wellbore [m]; $F_1(t)$ and $F_2(t)$ functions of time t[s]; r is radial distance [m]; r_w is the radius of the well screen [m]; z_{top} and z_{bot} are the vertical coordinates [m] of the top and bottom of the well screen, respectively.

Applying a finite-difference scheme in which the well screen is divided into N equal segments with segment 1 at the top of the screen and segment N at the bottom as shown in Figure 2.6, Eqs. (2.30) and (2.31) become:

$$q_{i+1} = q_i, \quad \text{for UFBC, } i = 1, 2, \ldots, N-1 \tag{2.32}$$

$$H_{i+1} = H_i, \quad \text{for UHBC, } i = 1, 2, \ldots, N-1 \tag{2.33}$$

where q_i and H_i respectively represent the mean flow rate per unit length from the aquifer to the well [m/s] and the mean hydraulic head [m] at the well screen over the well segment i. If k_{well} is the hydraulic conductivity of the wellbore [m/s], one can conclude from

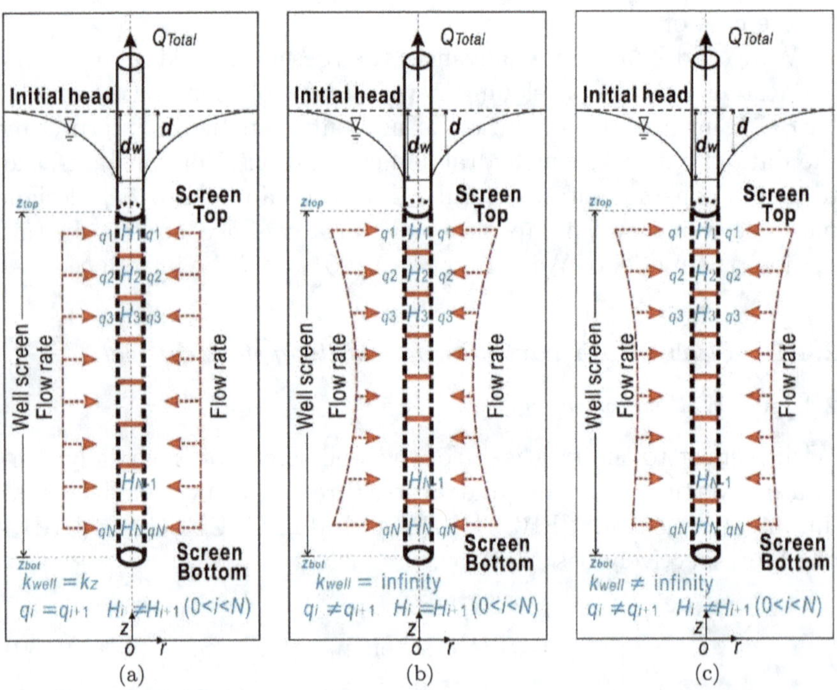

Figure 2.6. Schematic diagram of the problem setting. (a) UFBC model; (b) UHBC model and (c) MTBC model.

Eqs. (2.30) and (2.31) that approaches infinity for UHBC, while for UFBC k_{well} is equal to the vertical hydraulic conductivity of the aquifer, k_z [m/s]. Note that UFBC and UHBC are identical for a fully penetrating well because there is no vertical flow in the aquifer, and the flow within the well is not considered. The well hydraulic conductivity of the MTBC model is higher than that of UFBC and lower than the infinity of UHBC, and MTBC also depends on the pump location inside the well. Thus, MTBC is never identical to UFBC or UHBC.

Due to the complexity of the UHBC and MTBC models, analytical solutions might not be available. Alternatively, the UHBC and MTBC solutions can be obtained by a hybrid analytical-numerical method, in which the analytical solution is based on the UFBC solutions. In the following, we first introduce the UFBC solution.

2.4.1.2 *The UFBC solutions*

A general well function of UFBC in a confined aquifer was derived by Shaw-Yang and Yeh (2012) via the Laplace and finite Fourier transforms, considering the effect of well partial penetration. Assumptions include an aquifer of infinite radial extent with uniform thickness. This UFBC solution can solve the problems of different types of aquifer tests (like constant-head tests, constant-rate tests, or slug tests) for confined aquifers. A cylindrical coordinate system has been adopted with the r-axis horizontal, and z-axis vertically upward. The origin of the coordinate system is at the intercept point of the center of the well and the aquifer bottom. The UFBC solution of Shaw-Yang and Yeh (2012) in the Laplace domain is

$$\bar{d}(r,z,p) = \frac{(\bar{Q} + \pi r_c^2 \overline{d_w})(z_{\text{top}} - z_{\text{bot}}) K_0\left(r\sqrt{\frac{pS_s}{k_r}}\right)}{\pi r_c^2 b p K_0\left(r_w\sqrt{\frac{pS_s}{k_r}}\right) + 2\pi r_w k_r (z_{\text{top}} - z_{\text{bot}})\sqrt{\frac{pS_s}{k_r}} K_1\left(r_w\sqrt{\frac{pS_s}{k_r}}\right)}$$
$$+ \sum_{n=1}^{\infty} \frac{2(\bar{Q} + \pi r_c^2 d_w)\frac{sin(z_{\text{top}}) - sin\left(\frac{z_{\text{bot}} n\pi}{b}\right)}{n\pi}}{\pi r_c^2 p K_0(\xi_n r_w) + 2\pi r_w k_r (z_{\text{top}} - z_{\text{bot}}) \xi_n K_1(\xi_n r_w)} \cos\left(\frac{zn\pi}{b}\right) \quad (2.34)$$

where the overbar implies the variables in the Laplace domain; p is the Laplace transform variable; b is the saturated thickness [m]; d and dw are the drawdowns [m] in the aquifer and the well, respectively;

Q is the well pumping rate [m^3/s]; k_r is the horizontal hydraulic conductivity [m/s]; S_s is the specific storage [1/m]; r_c is the radius of the well casing [m]; K_0 is the modified Bessel functions of the first kind and zeroth order; K_1 is the modified Bessel functions of the first kind and first order; z is the vertical coordinate (upward positive); $\xi_n = \sqrt{\frac{k_z}{k_r}\left(\frac{n\pi}{b}\right)^2 + \frac{S_s p}{k_r}}$.

Equation (2.34) for constant-head tests becomes:

$$\bar{d}(r,z,p) = \bar{Q}\left[\frac{1}{2\pi r_w k_r b}\frac{K_0(r\sqrt{pS_s/k_r})}{\sqrt{pS_s/k_r}K_1(r_w\sqrt{pS_s/k_r})} + \sum_{n=1}^{\infty}\frac{\sin(z_{\text{top}}n\pi/b) - \sin(z_{\text{bot}}n\pi/b)}{n\pi} \times \frac{K_0(\xi_n r)}{\xi_n K_1(\xi_n r_w)}\frac{\cos(zn\pi/b)}{\pi r_w k_r(z_{\text{top}} - z_{\text{bot}})}\right] \quad (2.35)$$

As for constant-rate tests, Eq. (2.34) becomes:

$$\bar{d}(r,z,p) = \frac{Q}{p}\left[\frac{1}{2\pi r_w k_r b}\frac{K_0(r\sqrt{pS_s/k_r})}{\sqrt{pS_s/k_r}K_1(r_w\sqrt{pS_s/k_r})} + \sum_{n=1}^{\infty}\frac{\sin(z_{\text{top}}n\pi/b) - \sin(z_{\text{bot}}n\pi/b)}{n\pi} \times \frac{K_0(\xi_n r)}{\xi_n K_1(\xi_n r_w)}\frac{\cos(zn\pi/b)}{\pi r_w k_r(z_{\text{top}} - z_{\text{bot}})}\right] \quad (2.36)$$

Equations (2.34), (2.35), and (2.36) are the same as Eqs. (17), (18), and (21) of Shaw-Yang and Yeh (2012), respectively. Note that Eq. (2.36) is identical to Eq. (2.35) except that Q/p replaces $\bar{Q}$ because Q is constant with time.

2.4.1.3 *The UHBC solutions*

Different from UFBC, UHBC implies that the well permeability approaches infinity, or intra-wellbore head losses are negligible, as shown in Figure 2.6. Many studies show that this treatment is closer to physical reality than UFBC, but it is difficult to handle analytically (Kawecki, 2000; Steward and Jin, 2001) since the flux of individual segments might be different. Perina and Lee (2006) proposed an approach for computing the well flux for a UHBC.

As introduced previously, the UFBC assumes that the flux distribution along the screened well face is uniform, and the analytical solutions of Eqs. (2.35) and (2.36) could be taken as the line source/sink solution, which could be obtained by integrating the point source/sink solution (Penmatcha and Aziz, 1999):

$$\bar{d}(r,z,p) = \int_{z_{\text{bot}}}^{z_{\text{top}}} \bar{Q}\Psi(z_0,r,z,p)dz_0 \tag{2.37}$$

where $\bar{Q}\Psi(z_0,r,z,p)$ is the general point source/sink solution and z_0 is the point sink/source vertical location [m]. Using the finite-difference method, (2.37) can be approximated as

$$\bar{d}(r,z,p) = \sum_{j=1}^{N} \int_{z_{j-}}^{z_{j+}} \bar{q}_j \Delta z \Psi(z_0,r,z,p)dz_0 = \sum_{j=1}^{N} \bar{q}_j \Delta z \Psi_j(r,z,p) \tag{2.38}$$

where $\Psi_j(r,z,p) = \int_{z_{j-}}^{z_{j+}} \Psi(z_0,r,z,p)dz_0$; z_{j+} and z_{j-} represent the coordinates of the ending and starting points of $\Delta z = \mathrm{L/N}$; L is the length of the well screen, and the wellbore is discretized into N equal segments. $\int_{z_{j-}}^{z_{j+}} \bar{q}_j \Delta z \Psi(z_0,r,z,p)dz_0$ represents the line source/sink solution for the screen from z_{j-} to z_{j+}. A well test can vary both the drawdown and/or the discharge rate. Two special cases are a specified constant-head test (with varying discharge) and a specified constant-rate test (with varying drawdown). Constant-head tests are often carried out in free-flowing wells without a pump, while constant-rate tests are usually with a pump.

(a) UHBC for constant-head tests

Substituting (2.35) into Eq. (2.38), one could obtain the values of Ψ_j for constant-head tests:

$$\Psi_j(r,z,p) = \frac{1}{2\pi r_w k_r b} \frac{K_0(r\sqrt{pS_s/k_r})}{\sqrt{pS_s/k_r} K_1(r_w\sqrt{pS_s/k_r})}$$

$$+\sum_{n=1}^{\infty}\frac{\sin(z_{j+}n\pi/b)-\sin(z_{j-}n\pi/b)}{n\pi}$$
$$\times\frac{K_0(\xi_n r)}{\xi_n K_1(\xi_n r_w)}\frac{\cos(zn\pi/b)}{\pi r_w k_r(z_{j+}-z_{j-})} \tag{2.39}$$

For a constant-head aquifer test where the water level of the well (d_w) is given (Perina and Lee, 2006), one has:

$$\bar{d}_{w,i}=\bar{d}_{w,i+1}=\bar{d}_w,\quad i=1,2,\ldots,N-1 \tag{2.40}$$

where $d_{w,i}$ represents the mean drawdown [m] at the well screen over the well segment i, and it can be calculated using Eqs. (2.38) and (2.39), one has:

$$\bar{d}_{w,i}=\bar{d}(r_w,z_i,p)=\sum_{j=1}^{N}\bar{q}_j\Delta z\Psi_j(r_w,z_i,p) \tag{2.41}$$

in which r_w is the radius of the well screen; z_i is the mid-point coordinate of well segment i and $z_i=(z_{i+}+z_{i-})/2$. Denote $\Psi_j(r_w,z_i,p)$ by $\Psi_{i,j}$, one has:

$$\begin{bmatrix}\Psi_{1,1} & \Psi_{1,2} & \cdots & \cdots & \Psi_{1,N}\\ \Psi_{2,1} & \Psi_{2,2} & \vdots & \vdots & \Psi_{2,N}\\ \vdots & \vdots & \vdots & \vdots & \vdots\\ \Psi_{N,1} & \Psi_{N,2} & \cdots & \cdots & \Psi_{N,N}\end{bmatrix}\begin{bmatrix}\Delta z\bar{q}_1\\ \Delta z\bar{q}_2\\ \vdots\\ \Delta z\bar{q}_N\end{bmatrix}=\begin{bmatrix}\bar{d}_w\\ \bar{d}_w\\ \vdots\\ \bar{d}_w\end{bmatrix} \tag{2.42}$$

where the term of $\Psi_{i,j}$ [s/m] calculates the drawdown in the well segment i created by flow into the well segment j from the aquifer. Substituting Ψ_j from Eq. (2.39) and $\bar{q}_j$ ($j=1,2,\ldots,N$) By converting Eq. (2.42) into Eq. (2.38), the UHBC solutions of the constant-head aquifer test can be obtained.

(b) UHBC for constant-rate tests

For constant-rate tests, Eqs. (2.37) and (2.38) reduce into:

$$\bar{d}(r,z,p)=\int_{z_{\text{bot}}}^{z_{\text{top}}}Q\Psi(z_0,r,z,p)dz_0 \tag{2.43}$$

$$\bar{d}(r,z,p)=\sum_{j=1}^{N}\int_{z_{j-}}^{z_{j+}} q_j\Delta z\Psi(z_0,r,z,p)dz_0=\sum_{j=1}^{N} q_j\Delta z\Psi_j(r,z,p) \tag{2.44}$$

In a similar way to constant-head tests, one could obtain the values of Ψ_j for the constant-flux tests by substituting Eq. (2.36) into Eq. (2.44), one has:

$$\Psi_j(r,z,p)=\frac{1}{p}\left[\frac{1}{2\pi r_w k_r b}\frac{K_0(r\sqrt{pS_s/k_r})}{\sqrt{pS_s/k_r K_1(r_w\sqrt{pS_s/k_r})}}\right.$$
$$+\sum_{j=1}^{N}\frac{\sin(z_{j+}n\pi/b)-\sin(z_{j-}n\pi/b)}{n\pi}$$
$$\left.+\frac{K_0(\xi_n r)}{\xi_n K_1(\xi_n r_w)}\frac{\cos(zn\pi/b)}{\pi r_w k_r(z_{j+}-z_{j-})}\right] \tag{2.45}$$

For an aquifer test with a constant pumping rate, the pumping rate (Q_{Total}) is given while $\bar{d}_w$ is unknown, and

$$Q=Q_{\mathrm{Total}}=\sum_{i=1}^{N}\Delta z q_i \tag{2.46}$$

Therefore, one could yield (Ouyang *et al.*, 1998; Penmatcha and Aziz, 1999):

$$\begin{bmatrix} 1 & 1 & \cdots & \cdots & 1 \\ \Psi_{2,1}-\Psi_{1,1} & \Psi_{2,2}-\Psi_{1,2} & \vdots & \vdots & \Psi_{2,N}-\Psi_{1,N} \\ \vdots & \vdots & \vdots & \vdots & \vdots \\ \Psi_{N,1}-\Psi_{N-1,1} & \Psi_{N,2}-\Psi_{N-1,2} & \cdots & \cdots & \Psi_{N,N}-\Psi_{N-1,N} \end{bmatrix}\begin{bmatrix}\Delta z q_1 \\ \Delta z q_2 \\ \vdots \\ \Delta z q_N\end{bmatrix}$$
$$=\begin{bmatrix} Q_{\mathrm{Total}} \\ 0 \\ \vdots \\ 0 \end{bmatrix} \tag{2.47}$$

where $\Psi_{i,j}$ is $\Psi_j(r_w z_i p)$ as given by Eq. (2.41).

The first row of the coefficient matrix on the left side of Eq. (2.47) is obtained from Eq. (2.46), and the other rows are obtained from Eqs. (2.40) and (2.44). Substituting q $(j = 1, 2, \ldots, N)$ (determined from Eq. (2.47)) into Eq. (2.44), one can obtain the UHBC solutions of the constant-rate aquifer test.

2.4.2 *The MTBC solutions*

2.4.2.1 *Wellbore flow model*

Using both UFBC and UHBC to treat a vertical well may not reflect the physical processes of flow inside the well accurately since intra-wellbore head losses might be significant due to the great flow rate inside the wellbore (Hu *et al.*, 2011; Perina and Lee, 2006). Therefore, the MTBC model is proposed. In this section, we will develop the solution of flow into a partially penetrating vertical well for the MTBC model based on the UFBC and UHBC solutions.

The partially penetrating well's wellbore can be divided into two parts: a screened wellbore section and a cased (non-screened) wellbore section. For the solutions presented here, the pump (if any) is assumed to be set above the screened section of the well, so there is no pump or pump pipe within the screen that would affect the intra-wellbore head loss calculation.

Many previous studies show that the frictional, acceleration and fluid-inflow effects might cause head losses along the screened wellface (Dikken, 1990; Penmatcha and Aziz, 1999; Tarshish, 1992). The fluid-inflow effect refers to the energy losses when the flow changes direction from horizontal (before entering the wellbore) to vertical (after entering the wellbore). Assuming the flow direction is from the $(i+1)$th well segment to the ith well segment inside the wellbore, the difference of drawdown between the adjacent well segments can be expressed as (Penmatcha and Aziz, 1999):

$$\begin{aligned}\Delta d_{w,i} &= d_{w,i+1} - d_{w,i} \\ &= \Delta d_{w,f,i} + \Delta d_{w,c,i} + \Delta d_{w,a,i}, \quad i = 1, 2, \ldots, N-1 \qquad (2.48)\end{aligned}$$

where $\Delta d_{w,i}$ is the total difference of the well drawdown [m] between two adjacent well segments; $\Delta d_{w,f,i}$, $\Delta d_{w,c,i}$ and $\Delta d_{w,a,i}$ represent the differences of the drawdown [m] caused by the frictional, fluid-inflow, and acceleration effects, respectively:

$$\Delta d_{w,c,i} + \Delta d_{w,f,i} = -f\frac{\Delta z}{2r_w}\frac{u_i^2}{2g} \tag{2.49}$$

$$\Delta d_{w,a,i} = -\frac{2Q_{w,i-1}\Delta z q_i + (\Delta z q_i)^2}{gA^2} \tag{2.50}$$

in which $A(=\pi r_w^2)$ is the cross-section of the vertical well [m^2]; f is the coefficient of friction (dimensionless) (Munson *et al.*, 1995; Ouyang *et al.*, 1998); g is the gravity acceleration [m/s^2]; u_i is the mean flow velocity [m/s] inside the well at the well segment i:

$$u_i = \frac{Q_{w,i+1} + Q_{w,i}}{2\pi r_w^2} \tag{2.51}$$

where $Q_{w,i}$ is the flow rate [m^3/s] into the well segment $(i+1)$ from well segment i, and it is:

$$Q_{w,i} = \sum_{j=N}^{i} \Delta z q_j \tag{2.52}$$

in which $Q_{w,i} = 0$ when $i = N$.

Equation (2.45) is the Darcy-Weisbach equation (Munson *et al.*, 1995), which incorporates $\Delta d_{w,c,i}$ by modifying f to Ouyang *et al.* (1998):

$$f = \begin{cases} \dfrac{64}{N_{Re}}(1 + 0.04304 N_{Re,w}^{0.6142}), & \text{for laminar flow} \\ f_0(1 - 0.0153 N_{Re,w}^{0.3978}), & \text{for turbulent flow} \end{cases} \tag{2.53}$$

where f_0 is the friction factor calculated when there is no fluid-inflow effect through the well-face (Munson *et al.*, 1995; Ouyang *et al.*, 1998); N_{Re} is the Reynolds number, and $N_{Re} = 2r_w u_i \rho/\mu$, in which ρ is the density of water [Kg/m^3] and μ is the dynamic viscosity of water [Kg/(ms)]; $N_{Re,w}$ is wall Reynolds number and $N_{Re,w} = q_i\rho/(\pi\mu)$. The flow may be divided into three regimes: laminar, transitional and turbulent. To simplify the problem, only two regimes are considered in this section. It is a laminar flow when $N_{Re} \leq 2300$, and otherwise turbulent flow. Equation (2.50) is derived from the momentum conservation principle, and it describes the effect of flow acceleration in the wellbore as more water enters the well along the flow direction (Penmatcha and Aziz, 1999).

As for the cased wellbore section, the flow inside the well is pipe flow from the top of the screen to the pump or, in a free-flowing well, the fluid-inflow effect does not exist. Also, with no inflow, the flow rate is the same at all points, so there is no acceleration effect. The head loss $\Delta d_{w,f}$ over a distance D of the casing become:

$$\Delta d_{w,f} = -f\frac{D}{2r_c}\frac{u_c^2}{2g} \tag{2.54}$$

where r_c is the inside radius of the casing; $u_c = Q_c/(\pi r_c^2)$ is the flow velocity inside the casing resulting from the total inflow, Q_c, into the well from the aquifer. Modification of the coefficient of friction f as in Eq. (2.49) is not needed, i.e., $f = 64/N_{Re}$ for laminar flow and $f = f_0$ for turbulent flow.

Equations (2.50), (2.51), and (2.53) can be modified for annular flow allowing consideration of the pump location within the screen but that is beyond the scope of this analysis. Moreover, it is rare in practice to set the pump within the well screen in a confined aquifer.

Following the same approach as for UHBC, we shall present the MTBC for the two special cases of constant-head and constant-rate tests by considering the intra-wellbore head losses.

2.4.2.2 *MTBC for constant-head tests*

For a constant-head aquifer test in a confined aquifer, Eq. (2.42) becomes:

$$\begin{bmatrix} \Psi_{1,1} & \Psi_{1,2} & \cdots & \cdots & \Psi_{1,N} \\ \Psi_{2,1} & \Psi_{2,2} & \vdots & \vdots & \Psi_{2,N} \\ \vdots & \vdots & \vdots & \vdots & \vdots \\ \Psi_{N,1} & \Psi_{N,2} & \cdots & \cdots & \Psi_{N,N} \end{bmatrix} \begin{bmatrix} \Delta z\bar{q}_1 \\ \Delta z\bar{q}_2 \\ \vdots \\ \Delta z\bar{q}_N \end{bmatrix} = \begin{bmatrix} \bar{d}_{w,1} \\ \bar{d}_{w,2} \\ \vdots \\ \bar{d}_{w,N} \end{bmatrix} \tag{2.55}$$

As can be seen, there are $2N$ variables ($\bar{q}_i$ and $\bar{d}_{w,i}$) but only N equations. The drawdown at the pump is defined as

$$\bar{d}_p = \bar{d}_{w,1} + \Delta\bar{d}_{w,f} \tag{2.56}$$

where $\Delta\bar{d}_{w,f}$ is given by Eq. (2.54), in which D is the distance from the pump to the top of the screen (or from surface to the screen for a free flowing well) and $\bar{Q}_c = \sum_{i=1}^{N} \Delta z \bar{q}_i$. Combining Eq. (2.48) (producing $N-1$ equations), Eq. (2.55) (N equations) and Eq. (2.56) (1 equation) compose a set of $2N$ equations, which could be used to determine $\bar{q}_i (i = 1, 2, \ldots, N)$. Another point to note is that $\bar{d}_{w,i}$ on the right side of Eq. (2.55) contains variables $\bar{q}_i$ (as shown in Eqs. (2.48)–(2.53), thus it cannot be solved directly. In this section, we employ the Gauss-Seidel method to iteratively calculate $\bar{q}_i$. Substituting $\bar{q}_i$ (which is determined from Eq. (2.55)) into Eq. (2.38), one can obtain the MTBC solutions of the constant-head tests.

2.4.2.3 *MTBC for constant-rate tests*

For a constant-rate test in a confined aquifer, Eq. (2.47) becomes:

$$\begin{bmatrix} 1 & 1 & \cdots & \cdots & 1 \\ \Psi_{2,1} - \Psi_{1,1} & \Psi_{2,2} - \Psi_{1,2} & \vdots & \vdots & \Psi_{2,N} - \Psi_{1,N} \\ \vdots & \vdots & \vdots & \vdots & \vdots \\ \Psi_{N,1} - \Psi_{N-1,1} & \Psi_{N,2} - \Psi_{N-1,2} & \cdots & \cdots & \Psi_{N,N} - \Psi_{N-1,N} \end{bmatrix} \begin{bmatrix} \Delta z q_1 \\ \Delta z q_2 \\ \vdots \\ \Delta z q_N \end{bmatrix}$$
$$= \begin{bmatrix} Q_{\text{Total}} \\ \Delta\bar{d}_{w,1} \\ \vdots \\ \Delta\bar{d}_{w,NM} \end{bmatrix} \tag{2.57}$$

Like constant-head tests, the values of q_i for the case of constant-rate tests can be determined using Eqs. (2.48) and (2.57). The drawdown at the pump is $d_p = d_{w,1} + \Delta d_{w,f}$, f where $\Delta d_{w,f}$ is given by Eq. (2.50), D is the distance from the pump to the top of the screen and $Q_{\text{C}} = Q_{\text{Total}}$. Because Q_{Total} is constant with time, it is easiest to perform the calculation of d_p after inverting the Laplace transform to determine $d_{w,1}$.

2.4.3 *Comparison with the numerical solution of Chen et al. (2003)*

It is desirable to compare the semi-analytical solutions with numerical solutions derived here of more complicated conditions since the Theis solution is only a special case. The boundary condition of the wellbore used in Chen *et al.* (2003) was also a form of MTBC, as they coupled a laminar flow model describing the aquifer flow process with a pipe-flow model describing the intra-wellbore flow using the Darcy–Weisbach equation, which is applicable to both turbulent and laminar flows. Although both fluid-inflow and acceleration effects have been ignored in Chen *et al.* (2003), their solution is closer to the MTBC solution than the Theis solution used in the previous section, so it is used for comparison.

Figure 2.7 represents the comparison between the semi-analytical solutions and the finite-difference solution of Chen *et al.* (2003) at the bottom of the screened section of the well ($r = r_w$ and $z = z_{\text{bot}}$). The parameters used in this case are as follows: $b = 20\,\text{m}$, $k_r = 10\,\text{m/day}$, $k_z = 1\,\text{m/day}$, $r_w = r_c = 0.1\,\text{m}$, $S_s = 0.001\,\text{m}^{-1}$, $z_{\text{top}} = 15\,\text{m}$, $z_{\text{bot}} = 5\,\text{m}$, $Q_{\text{Total}} = 300\,\text{m}^3/\text{day}$. The location of the pump intake

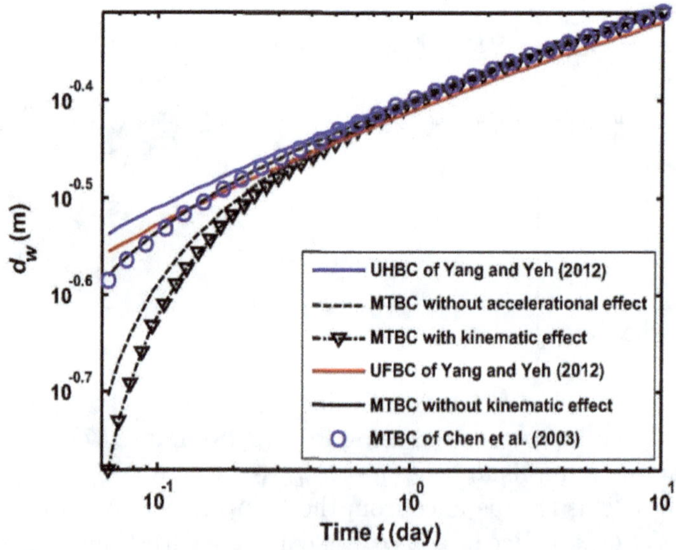

Figure 2.7. The time-drawdown curves of the UFBC, UHBC, and MTBC solutions at the bottom of the screened section of the well.

point is at $z = 17.5\,\text{m}$, which is 2.5 m above the top of the screen. Obviously, the agreement between the solution of MTBC without kinematic effect and the solution of Chen *et al.* (2003), which also excludes the kinematic effect, is very good.

2.4.4 *Comparison of the UFBC, UHBC and MTBC solutions*

Figures 2.8(a) and 2.8(b) compare the contour map of the drawdown near the well at an early time for UFBC, UHBC, and MTBC. The parameters are the same as in the previous example.

Figure 2.8(a) shows the comparison between the UFBC and UHBC solutions. The difference between them is obvious in the aquifer near the well. It becomes smaller with distance away from the well and is negligible when $r > 15\,\text{m}$ (i.e., $r/b > 0.25$), which is smaller than the typical influence radius of $1.5b\sqrt{k_r/k_z}$ as proposed by Hantush (1964). The main difference is the shape of the equal-potential contours. The UHBC equal-potential contour is nearly parallel to the well screen near the well because UHBC assumes that the drawdown is uniformly distributed along the wellbore. However, the UFBC equal-potential contour is symmetrical to the middle plane of the screen at $z = 10\,\text{m}$, and the greatest difference between the UHBC and UFBC models is at $z = 10\,\text{m}$.

Figure 2.8(b) represents the comparison between the UFBC and MTBC solutions. The difference between the UFBC and MTBC solutions near the well is obvious, and the MTBC equal-potential contours are asymmetric about the middle plane of the screen. This is because the flow direction in the well screen is upward to the pump at the top of the screen, as shown in Figure 2.8(b). The UFBC solution underestimates the drawdown in the aquifer above the elevation of the screen midpoint (i.e., $z > 10\,\text{m}$), and overestimates the drawdown in the aquifer below such a horizontal section (i.e., $z < 10\,\text{m}$).

2.4.5 *Results and discussions*

The MTBC solutions developed in this section can interpret the data of transient hydraulic head responses for constant-head tests and constant-rate tests. In this section, we shall investigate here the case

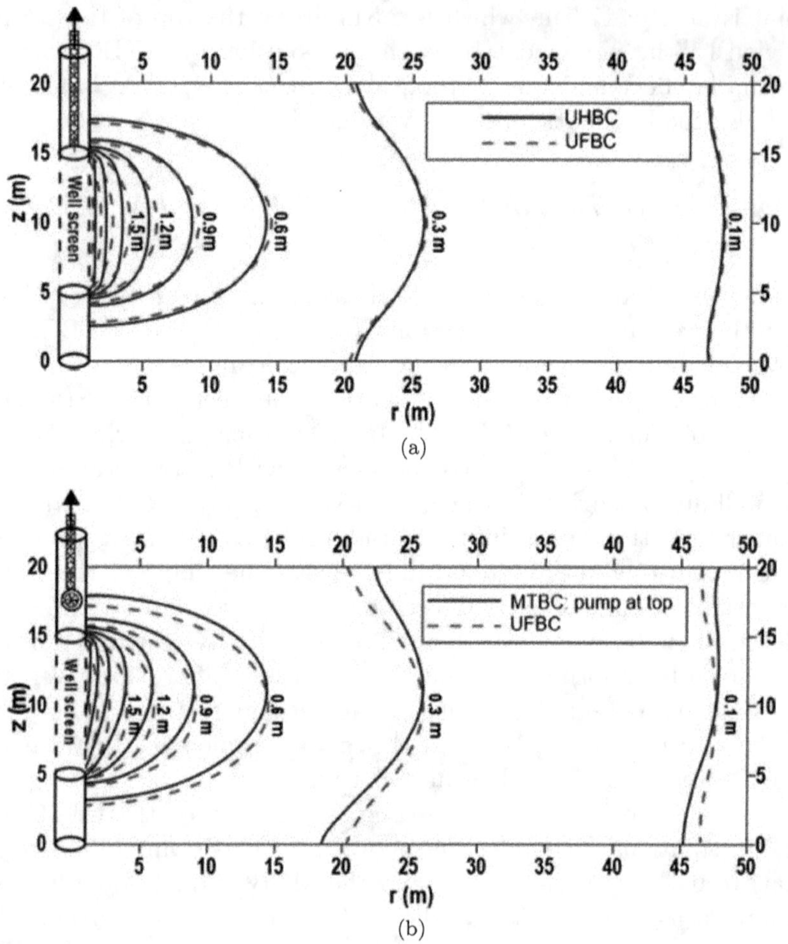

Figure 2.8. Contour map of the drawdown for different pump placements when $t = 0.1$ day. (a) Comparison of UFBC and UHBC solutions. (b) Comparison of UFBC and MTBC solutions when the pump is at the top of the well screen.

of a constant-rate test, since this type of test is most common in practice. At first it is necessary to verify the solutions.

2.4.5.1 *Verification of the Stehfest method for numerical Laplace inversion*

It seems difficult, if not impossible, to conduct analytical inverse Laplace transforms of Eqs. (2.34), (2.35), and (2.36) to obtain the drawdown in the real-time domain, so a numerical inverse Laplace method is called in. There are several available numerical inverse Laplace methods with each method best fitted for a particular type of problem (Hassanzadeh and Pooladi-Darvish, 2007; Wang and Zhan, 2015). Many studies demonstrate that the Stehfest method (Stehfest, 1970) works well for diffusion problems such as groundwater flow (Hassanzadeh and Pooladi-Darvish, 2007). Therefore, the Stehfest method is employed to invert the Laplace solutions into the real-time solutions in this section, due to its easy implementation and accuracy.

The Stehfest method contains a parameter, N_{Stehfest}, which is an even integer representing the optimal number of summation terms. Previous studies have determined $8 \leq N_{\text{Stehfest}} \leq 24$ (Hassanzadeh and Pooladi-Darvish, 2007; Wang and Zhan, 2015), but the optimal number for the problem in this section is unknown.

To determine the optimal value of N_{Stehfest} for the problem of this section, it is necessary to compare the numerically inverted Laplace solution with a closed-form analytical solution, which is unfortunately not available. However, one can simplify the problem to the classical Theis case by considering a fully penetrating well in a confined aquifer. Under such a special circumstance, Eq. (2.32) reduces to

$$\bar{d}_a(r,p) = \frac{Q_{\text{Total}}}{2\pi b k_r p} K_0\left(r\sqrt{\frac{S_{st}p}{k_r}}\right) \tag{2.58}$$

where Eq. (2.58) is the Theis solution in the Laplace domain.

Figure 2.9 compares time-drawdown curves for the standard Theis curve with the numerically inverted Laplace solution of Eq. (2.58) for $N_{\text{Stehfest}} = 6, 12$ and 24. The parameters used in this case are as follows: $b = 8\,\text{m}$, $k_r = k_{rz} = 1\,\text{m/day}$, $S_s = 0.0001\,\text{m}^{-1}$, and $Q_{\text{Total}} = 10\,\text{m}^3/\text{day}$. The results confirm the findings of previous studies that the accuracy of the Stehfest method becomes less satisfactory for $N_{\text{Stehfest}} > 20$ (Hassanzadeh and Pooladi-Darvish, 2007; Wang and Zhan, 2015). In this section, we choose $N_{\text{Stehfest}} = 12$.

2.4.5.2 *Significance of the kinematic effect*

From Figure 2.7, one can see that the solution of Chen *et al.* (2003) is closer to the MTBC solutions than to the UHBC and UFBC

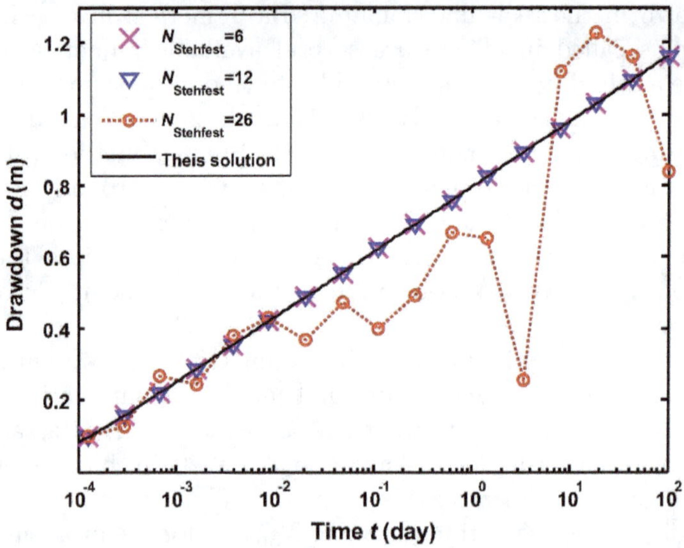

Figure 2.9. The time-drawdown curves of the classical Theis solution and the solutions by the Stehfest method at $r = 1\,\text{m}$.

solutions. The difference among the solutions at the early pumping stage is obvious, and the MTBC solution with the kinematic effect yields the smallest drawdown at the bottom of the screened wellbore section. This is because the kinematic effect also causes the greatest head losses inside the wellbore. The pressure computed by the UHBC solution, the MTBC solution without the acceleration effect, or the MTBC solution without the full kinematic effect could overestimate the drawdown at the early pumping stage at the bottom of the screened wellbore section. The differences between the UHBC solution and the MTBC solutions with and without the kinematic effect will decrease with time and finally become negligible after 2 days.

2.4.5.3 *Effect of aquifer anisotropy on the wellbore flux*

Besides the analysis of drawdown distribution and intra-wellbore head losses, wellbore flux distribution has also received great attention in previous investigations of horizontal well hydraulics (e.g., Penmatcha and Aziz, 1999). From the mathematical models shown in Sections 2.4.1 and 2.4.2, one can see that the intra-wellbore head losses are closely related to the wellbore flux. In this section, the influence of aquifer anisotropy (κ) on the wellbore flux will be investigated, as shown in Figures 2.10(a) and 2.10(b). κ is defined as a ratio of the vertical hydraulic conductivity over the horizontal conductivity ($\kappa = k_z/k_r$). The parameters used in this case are as follows: $b = 20\,\text{m}$, $k_r = 10\,\text{m/day}$, $r_w = r_c = 0.2\,\text{m}$, $S_s = 0.001\,\text{m}^{-1}$, $z_{\text{top}} = 17.5\,\text{m}$, $z_{\text{bot}} = 2.5\,\text{m}$, $Q_{\text{Total}} = 100\,\text{m}^3/\text{day}$. The location of the pump intake point is at $z = 17.5\,\text{m}$. Three values of κ are examined: $\kappa = 1, 0.1$, and 0.01.

Two observations can be made from Figures 2.10(a) and 2.10(b), which represent the comparison among the UFBC, UHBC and MTBC wellbore fluxes for time $t = 0.05$ day and 5 days, respectively. First, a smaller κ makes the wellbore flux curves for both UHBC and MTBC solutions flatter at the early pumping stage and closer to the UFBC wellbore flux curve, which is a horizontal line, as shown in Figures 2.10(a) and 2.10(b). This is because the flow is mostly horizontal for a partially penetrating vertical well with low vertical hydraulic conductivity. Second, the UHBC wellbore flux is symmetric with respect to the screen midpoint. However, the MTBC wellbore flux is asymmetric, and the maximum value is located at the top of the screen, next to the pump intake. At the late time of $t = 5$ day, the MTBC wellbore flux becomes like that of UHBC.

2.4.5.4 *Characteristics of the MTBC drawdowns*

Figure 2.11 illustrates the time-drawdown curves in the pumping well for different vertical locations. The parameters used in this case are: $b = 20\,\text{m}$, $k_r = 10\,\text{m/day}$, $k_z = 1\,\text{m/day}$, $r_w = r_c = 0.1\,\text{m}$, $S_s = 0.0001\,\text{m}^{-1}$, $z_{\text{top}} = 15\,\text{m}$, $z_{\text{bot}} = 5\,\text{m}$, and $Q_{\text{Total}} = 100\,\text{m}^3/\text{day}$. The pump is at $z = 17.5\,\text{m}$ for all the curves presented in Figure 2.11. The MTBC drawdown increases towards the top of the well screen because of the upward flow to the pump. A notable point is that the drawdown at the top of the screen initially decreases with time and then starts to increase again with time in Figure 2.11. This phenomenon will be observed regardless of the pump location if the pump is within the cased section of the well.

This can be understood as follows. After the pumping starts, water is immediately removed from the point of intake, and this process will lead to an almost instantaneous increase of drawdown (like a step-function) at such a point. The rapid increase of drawdown will

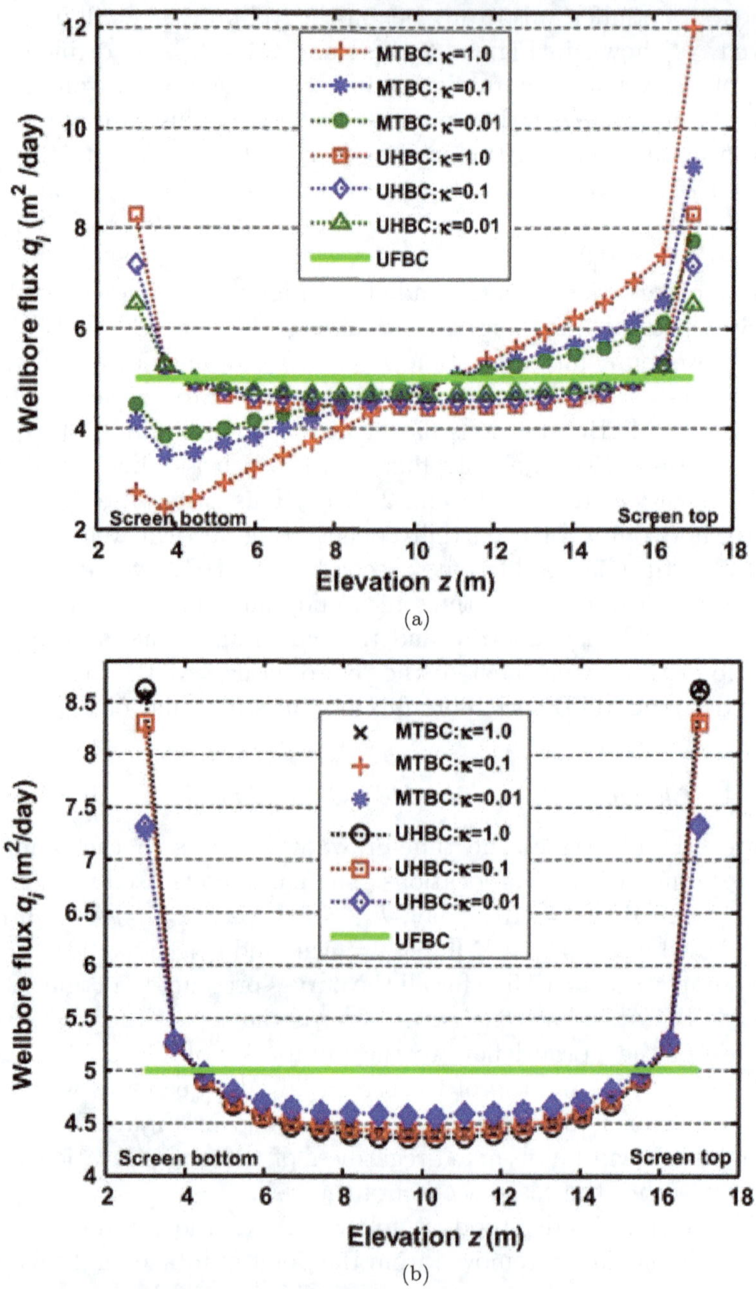

Figure 2.10. Wellbore flux distribution for different κ. (a) $t = 0.05$ day and (b) time $t = 0.05$ day.

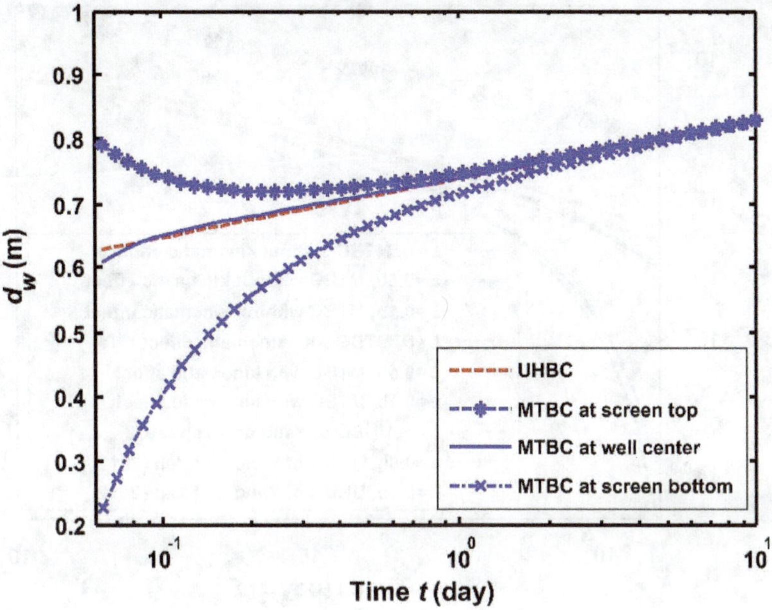

Figure 2.11. Comparison of the time-drawdowns in the pumping well between the MTBC and UHBC solutions.

generate a significant hydraulic gradient, which will initiate intra-wellbore flow. Mobilization of water inside the wellbore will subsequently initiate aquifer flow towards the wellbore. Such intra-wellbore flux will partially release the stress imposed by the pumping, thus reducing the drawdown at the top of the screen. However, such a reduction of drawdown does not last forever. Intra-wellbore flow will eventually decrease with time, and the drawdown at the point of intake also starts to increase with time, just like the drawdown at the bottom of the wellbore.

The difference between the MTBC solution and the UHBC solution becomes smaller with time, and finally, the MTBC solution approaches the UHBC solution at the late pumping stage after 5 days. Generally, the pumping tests can last from hours to days or even weeks in duration, depending on the purpose of the test, but usually last from 24 to 72 hours. Therefore, when using the time-drawdowns in the pumping well for aquifer parameter estimation, the kinematic and frictional effects along the wellbore may not be negligible.

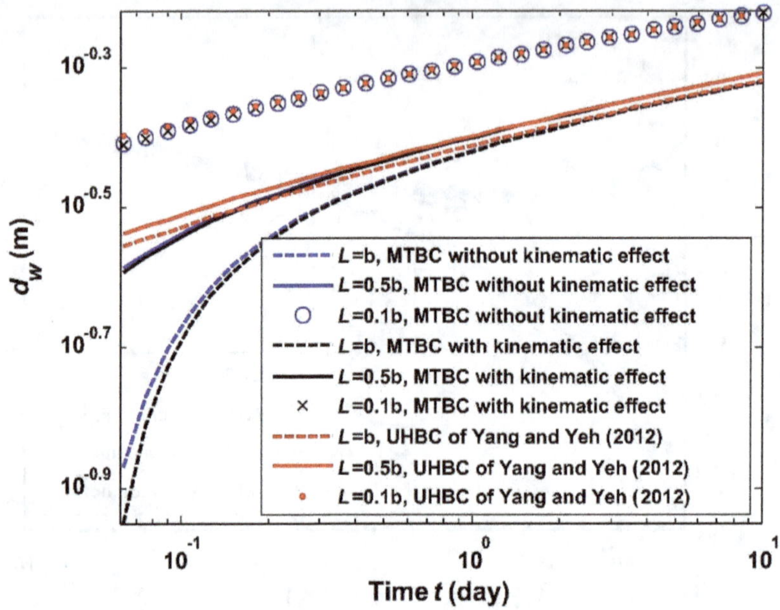

Figure 2.12. MTBC time-drawdowns for different well screen lengths.

2.4.5.5 *The effect of screen length on MTBC drawdowns*

It is interesting to analyze the impact of the well screen length on the MTBC drawdown for a partially penetrating vertical well. Figure 2.12 shows the MTBC drawdown at the pump's intake for $L0.1b, 0.5b$, and b. The case of $L = b$ represents a fully penetrating well. The parameters used in this figure are as follows: $b = 20\,\text{m}$, $k_r = 10\,\text{m/day}$, $k_z = 1\,\text{m/day}$, $r_w = r_c = 0.1\,\text{m}$, $S_s = 0.001\,\text{m}^{-1}$, and $Q_{\text{Total}} = 300\,\text{m}^3/\text{day}$. The well screen is in the center of the aquifer along the z-axis and the location of the pump intake is above the screen. From this figure, one can obviously see that the effect of the partial penetration on the MTBC drawdowns is significant. The difference between the MTBC and UHBC drawdowns decreases with the decreasing L, and becomes negligible when $L < 0.1b$. That is because the intra-wellbore loss depends on the wellbore length.

In this section, we developed the MTBC solutions of transient flow into a partially penetrating vertical well in a confined aquifer using a hybrid analytical-numerical method. The MTBC model of this section includes both kinematic and frictional effects along the wellbore, where the kinematic effect refers to the accelerational and fluid inflow effects. The new solution can be used not only to interpret the aquifer-test data but also applied to well design and pump selection in a confined aquifer.

2.5 Radial Pumping Test in Non-Darcian Flow Fields

Although many efforts have been devoted to studying the non-Darcian flow around the well, the exact solutions have not been obtained due to the nonlinearity of the problem (Mathias *et al.*, 2008; Yeh and Chang, 2013). For example, Sen (1990, 2000) employed the Boltzmann transform to analytically solve the problems related to the non-Darcian flow. This method was shown to be problematic since both initial and boundary conditions cannot be simultaneously transformed into a form only containing the Boltzmann variable (Camacho and Vasquez, 1992; Wen *et al.*, 2008b). Wen *et al.* (2008a, 2008b) obtained the semi-analytical solutions of the non-Darcian flow model by combining the linearization procedure and the Laplace transform method (LL method). This LL method assumed that transient flow in the non-Darcian flow region can be taken as a quasi-steady state flow. Wen *et al.* (2008a, 2008b) pointed out that solutions by the Boltzmann transform and the LL method coincided at late time. To test the accuracy of the semi-analytical solutions (Sen, 2000; Wen *et al.*, 2008b), Mathias *et al.* (2008) and Wen *et al.* (2009) employed the finite-difference method to study the non-Darcian flow problems, and their results showed that the semi-analytical solution only agreed very well with the numerical solution at the late pumping stage.

The above-mentioned investigations assume that the non-Darcian flow occurs over the entire domain, which is called a fully non-Darcian (F-ND) flow model hereinafter. In fact, the regime of the flow to the pumping well can be divided into two regions: non-Darcian flow occurs within a narrow region around the well due to the relatively high velocity of flow there, and Darcian flow prevails over the rest domain. One may think that such a two-region flow could be described by the Forchheimer law, which would automatically reduce to Darcy's law at the location far from the well (because the second-order velocity term in the Forchheimer law will be negligible if velocity approaches zero). However, Mackie (1983) demonstrated that the two-region model could fit the experimental data in the laboratory better than the F-ND model. Huyakorn and Dudgeon (1976) employed a two-region model to study flow near a pumping well. Basak (1978) presented analytical solutions of the two-region model for steady-state flow to a fully penetrating well. Sen (1988) and Wen *et al.* (2008a) derived the analytical solutions of the two-region model for transient flow to a pumping well.

Previous studies on the non-Darcian flow into a pumping well assumed that critical radius (R_{CD}) was a constant or infinity, where R_{CD} represents the location of the interface between the non-Darcian flow region and the Darcian flow region. Wang *et al.* (2014) established a two-region model considering time-dependent R_{CD}, where the non-Darcian flow was described by the Forchheimer equation. A new iteration method based on the finite-difference method was also proposed to estimate the moving critical radius.

Using the model of Wang *et al.* (2014) in this section, we will investigate non-Darcian flow into a fully penetrating pumping well considering a time-dependent critical radius using the finite-difference method. When the critical radius is infinite, this model reduces to the F-ND model, and it becomes the fully Darcian flow model when it is 0.

2.5.1 *Mathematical model*

Previous research showed that the porous media flow may be divided into four regimes, such as (a) non-Darcy pre-linear laminar flow, (b) Darcy flow, (c) non-Darcy post-linear laminar flow, and (d) non-Darcy post-linear turbulent flow (Basak, 1977; Bear, 1972). For radial flow to a pumping well, the velocity in the aquifer decreases with the distance from the well. Therefore, the radial flow might experience all four-flow regimes. To simplify the problem, we use a two-region model that considers a non-Darcian flow region near the well and a Darcian flow region away from the well. A unique feature of the two-region model used in this book is that the critical radius is allowed to vary with time, whereas it was assumed to be constant in previous studies (Dudgeon *et al.*, 1972; Huyakorn and Dudgeon, 1976, 1978; Mackie, 1983; Sen, 1988; Wen *et al.*, 2008a).

Generally, the start of the non-Darcian flow can be determined by the critical Reynolds number (R_{ec}), where the Reynolds number is defined as

$$R_e(r,t) = \frac{D_p q(r,t)}{\nu} \tag{2.59}$$

where ν is the kinematic viscosity of the fluid [m^2/s]; D_p is the characteristic grain diameter [m]; $q(r,t)$ is specific discharge [m/s] at distance r [m] and time t [s]; R_e is Reynolds number which depends on time and space (dimensionless). The critical Reynolds number (R_{ec}) refers to R_e at the start of non-Darcian flow. Up to the present, there is still considerable debate on R_{ec} for the initiation of non-Darcian flow in porous media. Bear (1972) pointed out that R_{ec} varied between 3 to 10; Scheidegger (1974) gave R_{ec} to be 0.1 to 75; Zeng and Grigg (2006) suggested the range of R_{ec} as 1 to 100. R_{ec} will be set as 10 in this section. According to Eq. (2.55), one can see that the specific discharge has a linear relationship with R_e. Therefore, the critical specific discharge (q_c) can also be used to determine the start of the non-Darcian flow, since one can calculate q_c for a given R_{ec}. When the specific discharge is less than or equal to q_C (or $R_e \leq R_{ec}$, the flow is considered as Darcian. When the specific discharge is greater than q_c (or $R_e > R_{ec}$), the flow is taken as non-Darcian. Denoting $R_C(t)$ as the critical radius at which $q = q_c$ (or $R_e = R_{ec}$), then it is non-Darcian flow when $r \leq R_C(t)$ and Darcian flow when $r > R_C(t)$, as shown in Figure 2.13.

For the quasi-steady state flow, one has

$$R_C = Q/(2\pi B q_C) \tag{2.60}$$

where B is the thickness of the aquifer [m]; Q is the well discharge [m^3/s]. For the case with a constant pumping rate, R_C is also a constant for a specific R_{ec}. This constant R_C was used in previous two-region models of transient non-Darcian flow (Sen, 1988; Wen *et al.*, 2008a). Actually, R_C is not a constant for transient flow, and it cannot be determined directly since the velocity distribution changes with time. In this section, a new iteration method will be proposed to determine R_C as described below.

Figure 2.13 shows the physical model investigated in this section, where a pumping well fully penetrates a confined aquifer. The origin

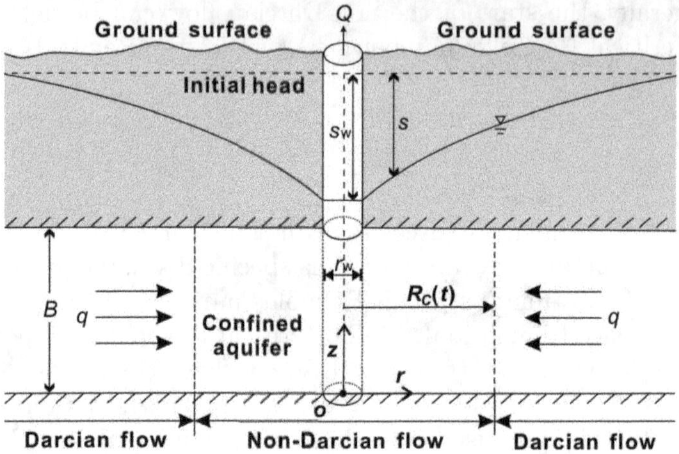

Figure 2.13. The schematic diagram of the non-Darcian flow into a fully penetrating pumping well considering the time-dependent critical radius.

of the cylindrical coordinate system is at the center of the well. The r-axis is horizontal and outward from the well, and the z-axis is upward vertical. Two assumptions are made in this section. First, the non-Darcian and Darcian flow may coexist, and the critical radius is time-dependent, and the Forchheimer law governs the non-Darcian flow. Second, the system is hydrostatic before the pumping starts, so $R_C(t = 0) = 0$. These assumptions, although quite idealized, are standard in the well hydraulic studies (Papadopulos and Cooper, 1967; Sen, 1988; Wen *et al.*, 2008a).

2.5.1.1 *Governing equation*

Based on the above assumptions, the governing equations of the mathematical model of the two-region flow model can be described as follows:

$$\frac{\partial q_N(r,t)}{\partial r} + \frac{q_N(r,t)}{r} = \frac{S}{B}\frac{\partial s_N(r,t)}{\partial t}, \quad \text{if } r \le R_C(t) \tag{2.61}$$

$$\frac{\partial q_Y(r,t)}{\partial r} + \frac{q_Y(r,t)}{r} = \frac{S}{B}\frac{\partial s_Y(r,t)}{\partial t}, \quad \text{if } r > R_C(t) \tag{2.62}$$

where $q_Y(r,t)$ and $q_N(r,t)$ are specific discharge [m/s] at distance r and time t in Darcian flow and non-Darcian flow regions, respectively;

$s_Y(r,t)$ and $s_N(r,t)$ are drawdowns [m] at distance r and time t in Darcian flow and non-Darcian flow regions, respectively.

In the non-Darcian flow region, the Forchheimer law is used to describe the flow (Forchheimer, 1901):

$$q_N + \beta q_N|q_N| = K_\beta \frac{\partial s_N}{\partial r}, \quad r \leq R_C \tag{2.63}$$

For the Darcian flow region, one has:

$$q_Y(r,t) = K\frac{\partial s_Y(r,t)}{\partial r}, \quad r > R_C \tag{2.64}$$

where β is called the inertial force coefficient [m/s], which is an empirical constant depending on the properties of the medium (Sidiropoulou *et al.*, 2007); K_β is called the apparent hydraulic conductivity [m/s] and it reduces to the hydraulic conductivity when $\beta = 0$ (Chen *et al.*, 2001; Sidiropoulou *et al.*, 2007), which is likewise an empirical constant depending on the properties of the medium (Chen *et al.*, 2001; Sidiropoulou *et al.*, 2007).

2.5.1.2 *Initial condition*

For the two-region flow model, the initial condition is

$$s_Y(r,0) = s_N(r,0) = 0 \tag{2.65}$$

2.5.1.3 *Boundary conditions*

Assuming that the pumping rate is large enough to induce non-Darcian flow near the well, the boundary condition at the wellbore, considering the wellbore storage with a finite diameter well, can be written as

$$2\pi r B q_N(r,t)\|_{r\to r_w} - \pi r_w^2 \frac{ds_w(t)}{dt} = -Q \tag{2.66}$$

where Q is positive for the pumping rate [m^3/s]; r_w is the radius of the well [m]; s_w is the drawdown inside the well [m]. Note that well loss is not considered so the drawdown is continuous across the well screen:

$$s_w(t) = s_N(r_w,t) \tag{2.67}$$

The drawdown and the discharge from the Darcian flow region into the non-Darcian flow region are continuous at the critical radius:

$$s_N[R_C(t), t] = s_Y[R_C(t), t] \tag{2.68}$$

$$q_N[R_C(t), t] = q_Y[R_C(t), t] \tag{2.69}$$

The outer boundary condition is:

$$s_Y(\infty, T) = 0 \tag{2.70}$$

Equations (2.61)–(2.70) compose of the mathematical model of the two-region model with a time-dependent critical radius $R_C(t)$. This model is an extension of the previous model by Şen (1988). When $R_C(t) \to \infty$, this model becomes the F-ND model. When $R_C(t) = 0$, it reduces to the fully Darcian flow model.

2.5.1.4 *Dimensionless transformation*

For the purposes of analysis, dimensionless variables are defined as follows:

$$r_{\mathrm{D}} = \frac{r}{B}, \quad r_{\mathrm{wD}} = \frac{r_{\mathrm{w}}}{B}, \quad R_{\mathrm{CD}} = \frac{R_{\mathrm{C}}}{B}, \quad t_{\mathrm{D}} = \frac{K_\beta t}{SB}, \quad \beta_{\mathrm{d}} = \frac{Q\beta}{2\pi B^2},$$
$$s_{\mathrm{wD}} = \frac{2\pi K_\beta B}{Q} s_{\mathrm{w}}, \quad s_{\mathrm{YD}} = \frac{2\pi K_\beta B}{Q} s_{\mathrm{Y}}(r,t), \quad s_{\mathrm{ND}} = \frac{2\pi K_\beta B}{Q} s_{\mathrm{N}}(r,t),$$
$$q_{\mathrm{ND}} = -\frac{2\pi B^2}{Q} q_{\mathrm{N}}(r,t), \quad q_{\mathrm{YD}} = -\frac{2\pi B^2}{Q} q_{\mathrm{Y}}(r,t), \quad q_{\mathrm{CD}} = -\frac{2\pi B^2}{Q} q_{\mathrm{C}} \tag{2.71}$$

Then, Eqs. (2.61)–(2.70) can be rewritten as

$$\frac{\partial q_{ND}}{\partial r_D} + \frac{q_{ND}}{r_D} = -\frac{\partial s_{ND}}{\partial t_D}, \quad r_D \le R_{CD} \tag{2.72}$$

$$\frac{\partial q_{YD}}{\partial r_D} + \frac{q_{YD}}{r_D} = -\frac{\partial s_{YD}}{\partial t_D} r_D > R_{CD} \tag{2.73}$$

$$s_{ND}(r_D, 0) = s_{YD}(r_D, 0) = 0 \tag{2.74}$$

$$s_{YD}(\infty, t_D) = 0 \tag{2.75}$$

$$s_{ND}[R_{CD}(t_D), t_D] = s_{YD}[R_{CD}(t_D), t_D] \tag{2.76}$$

$$q_{ND}[R_{CD}(t_D), t_D] = q_{YD}[R_{CD}(t_D), t_D] \tag{2.77}$$

The boundary condition with the wellbore storage in the dimensionless form is

$$(r_D q_{ND})|_{r_D \to r_{wD}} + \frac{r_{wD}^2}{2S}\frac{ds_{wD}(t_D)}{dt_D} = 1 \tag{2.78}$$

The dimensionless Forchheimer law becomes

$$q_{ND} + \beta_D q_{ND}|q_{ND}| = -\frac{\partial s_{ND}}{\partial r_D}, \quad r_D \leq R_{CD} \tag{2.79}$$

where β_D is the dimensionless inertial force coefficient. When $r_D > R_{CD}$, groundwater flow follows the Darcy's law in the dimensionless format as

$$q_{YD}(r, t) = -\lambda \frac{\partial s_{YD}}{\partial r_D}, \quad r_D > R_{CD} \tag{2.80}$$

where λ is the ratio of the hydraulic conductivity and apparent (Sidiropoulou *et al.*, 2007).

2.5.2 *Numerical solutions*

Because of the nonlinearity of the problem, it is not easy to obtain the analytical solution of drawdown, even if $R_{CD}(t_D)$ is constant. Therefore, the finite-difference method is employed to investigate the problem considering a time-dependent $R_{CD}(t_D)$. Due to the axisymmetric nature of the problem, the numerical simulation was conducted with a non-uniform grid system, where the spatial steps are smaller near the well and become progressively greater away from the well. Similar to previous studies (Mathias *et al.*, 2008; Wen *et al.*, 2009), the dimensionless space r_D was discretized logarithmically. The dimensionless space domain $[r_{wD}, r_{eD}]$ is discretized into N nodes excluding the two boundary nodes r_{wD} and r_{eD}, where r_{eD} is a relatively large dimensionless distance used to approximate the infinite boundary (Mathias *et al.*, 2008; Wen *et al.*, 2009). For any node of r_i, $r_{wD} < r_i < r_{eD} i = 1, 2, N$, one has:

$$r_i = (r_{i-1/2} + r_{i+1/2})/2, \quad i = 1, 2, \ldots, N \tag{2.81}$$

where $r_{i+1/2}$ is calculated as follows:

$$\log_{10}\left(r_{i+\frac{1}{2}}\right) = \log_{10}(r_{wD}) + i\left[\frac{\log_{10}(r_{eD}) - \log_{10}(r_{wD})}{N}\right], \quad i = 0, 1, \ldots, N \tag{2.82}$$

After spatial discretization, Eqs. (2.72) and (2.73) become:

$$\frac{ds_{YD,i}}{dt_D} \approx \frac{r_{i-\frac{1}{2}}q_{YD,i-\frac{1}{2}} - r_{i+\frac{1}{2}}q_{YD,i+\frac{1}{2}}}{r_i\left(r_{i+\frac{1}{2}} - r_{i-\frac{1}{2}}\right)}, \quad i = 2, 3, \ldots, N_s - 1, \quad r_D \leq R_{CD} \tag{2.83}$$

$$\frac{ds_{ND,i}}{dt_D} \approx \frac{r_{i-\frac{1}{2}}q_{ND,i-\frac{1}{2}} - r_{i+\frac{1}{2}}q_{ND,i+\frac{1}{2}}}{r_i\left(r_{i+\frac{1}{2}} - r_{i-\frac{1}{2}}\right)}, \quad i = N_s, N_s + 1, \ldots, N - 1, \; r_D > R_{CD} \tag{2.84}$$

where $q_{YD,i}$ and $s_{YD,i}$ are the dimensionless specific discharge q_{YD} and dimensionless drawdown q_{YD} at node i for the Darcian flow, respectively, $q_{ND,i}$ and $s_{ND,i}$ are the dimensionless specific discharge q_{ND} and dimensionless draw down s_{ND} at node i for the non-Darcian flow, respectively. In terms of the Forchheimer equation of Eq. (2.79), one can obtain:

$$q_{ND,i-\frac{1}{2}} \approx \frac{1}{2\beta_D}, \quad i = 2, 3, \ldots, N_s - 1 \tag{2.85}$$

$$q_{ND,i+\frac{1}{2}} \approx \frac{1}{2\beta_D}, \quad i = 2, 3, \ldots, -1 \tag{2.86}$$

where node N_s means the location of $R_{CD}(t_D)$. At the well aquifer boundary, one has:

$$q_{ND,1-\frac{1}{2}} \approx \frac{1}{2\beta_D}\left\{-1 + \left[1 + 4\beta_D\left(\frac{s_{WD} - s_{ND,1}}{r_1 - r_{wD}}\right)\right]^{\frac{1}{2}}\right\} \tag{2.87}$$

where s_{WD} is the dimensionless drawdown inside the well. Considering Eq. (2.74), s_{WD} can be approximated as follows:

$$\frac{ds_{wD}}{dt_D} \approx \frac{2S}{r_{wD}^2}\left(1 - r_{wD}q_{ND,1-\frac{1}{2}}\right) \tag{2.88}$$

When $r_D > R_{CD}$ the finite-difference scheme of the specific discharge can be obtained from Eq. (2.80):

$$q_{YD,i-\frac{1}{2}} \approx \lambda \frac{s_{YD,i-1} - s_{YD,i}}{r_i - r_{i-1}}, \quad i = N_s, N_s + 1, \ldots, N - 1 \tag{2.89}$$

$$q_{YD,i+\frac{1}{2}} \approx \lambda \frac{s_{YD,i} - s\Upsilon D, i+1}{r_{i+1} - r_i}, \quad i = N_s N_s + 1, \ldots, N - 1 \tag{2.90}$$

As for the boundary at the infinity, the finite-difference scheme is

$$q_{YD,N+1/2} \approx \lambda \frac{s_{YD,N}}{r_{eD} - r_N} \tag{2.91}$$

Now one obtains a set of ordinary differential equations. It is notable that R_{CD} or N_S which is related to the index i in Eqs. (2.85)–(2.86) and Eqs. (2.89)–(2.90) is time-dependent. In the following section, a new iteration method will be proposed to determine the values of R_{CD} or N_S.

2.5.3 *Iteration method to determine critical radius*

Before introducing the new iteration method, the relationship between R_{CD} and the velocity distribution will be investigated first, based on the two-region model with a constant R_{CD}. The values of the constant R_{CD} are set to 0, 0.02, 0.04, 0.08 and 0.50. The other parameters are $r_{wD} = 1 \times 10^{-4}$, $\beta_D = 20$, $\lambda = 1$. The mathematic model with a constant R_{CD} will be solved by the finite-difference method.

Figure 2.14 shows the specific discharge distributions with different R_{CD} of 0, 0.02, 0.04, 0.08 and 0.50. The curve of $R_{CD} = 0$ represents the fully Darcian flow model. One can find that the specific discharge decreases with increasing R_{CD} at a given r_D, starting from its maximum at $R_{CD} = 0$ (Darcian flow). This observation is understandable. The increasing R_{CD} implies a stronger contribution of the inertial effect, which also means a larger resistance to flow. Thus, it leads to a smaller specific discharge. After trying many different sets of aquifer parameters, numerical simulation indicates that this observation is universally valid. This observation will serve as the basis for the new iteration method to seek the location of $R_{CD}(t_D)$.

Like the use of R_{ec} to determine the start of the non-Darcian flow, one can use q_{CD} for the initiation of the non-Darcian flow, where q_{CD}

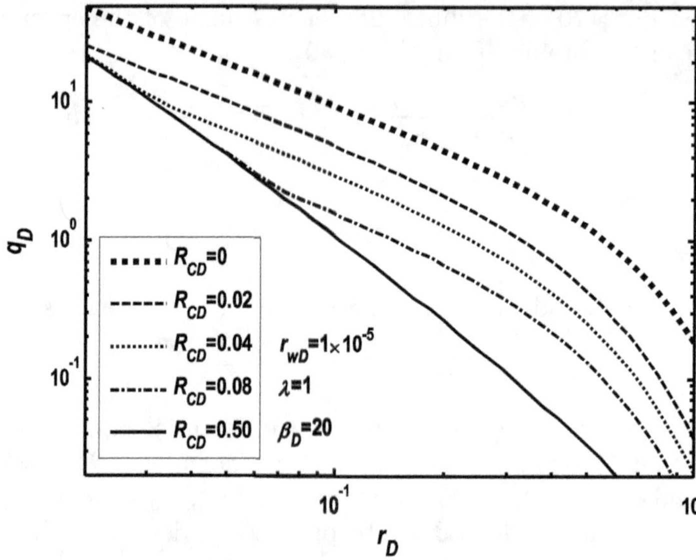

Figure 2.14. Specific discharge distributions with different critical radius R_{CD}.

is the dimensionless critical specific discharge defined in Eq. (2.71). We denote r_{jCD} as the newly computed critical radius at the jth step of the new iteration method, where $j = 1, 2, 3, \ldots$. Since the aquifer system is initially hydrostatic, the initial critical radius r_{0CD} is set to 0. For a given dimensionless time t_{1D}, the detailed procedures of the iteration method for searching $R_{CD}(t_{1D})$ will be introduced as follows. First, the specific discharge distribution in the aquifer can be calculated using Eqs. (2.83)–(2.91) with $R_{CD}(t_{1D}) = r_{0CD}$, as shown in Figure 2.15.

Based on the computed specific discharge distribution, one can find the new critical radius r_{1CD} according to a given constant q_{CD}. Second, the new specific discharge distribution can be similarly calculated using Eqs. (2.83)–(2.91) with $R_{CD}(t_{1D}) = r_{1CD}$, and the new critical radius r_{2CD} can be obtained according to q_{CD}. It is notable that r_{1CD} and r_{2CD} serve as the upper and lower limits for searching $R_{CD}(t_{1D})$, as illustrated in Figure 2.15. Similarly, one can estimate the new critical radius r_{3CD} using r_{2CD}, where r_{3CD} is located somewhere between r_{1CD} and r_{2CD}. Following the same procedures, a new critical radius r_{4CD} can be calculated based on r_{3CD}, and r_{4CD} is between r_{2CD} and r_{3CD}. One can repeat the above computation

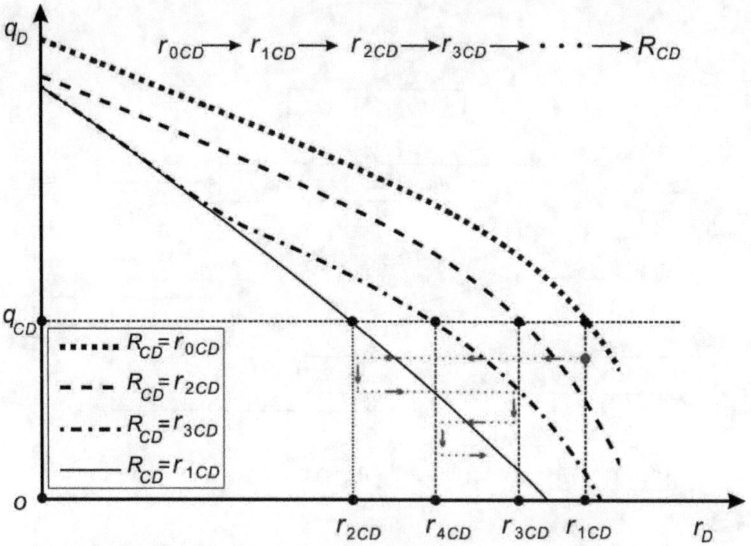

Figure 2.15. The schematic diagram showing the iterative process of seeking R_{CD}.

until the new critical radius finally converges. For the actual problems, we define a convergence criterion $|R_{CD}^{\text{old}} - R_{CD}^{\text{new}}| \leq \xi$, where R_{CD}^{old} and R_{CD}^{new} are the critical radius for the previous step and present step, respectively, ξ is a small positive value, such as 0.001. If this criterion is satisfied, the new critical radius r_{jCD} is thought as the estimation of $R_{CD}(t_{1D})$. A MATLAB program named as two-region model with moving critical radius (MTRM) to facilitate the computation is developed. This iteration method is convergent. Figure 2.16 represents the flow chart of the MTRM algorithm, where t_k is the time at the time step k, $k_{\max}$ is the total number of time steps; d_{tD} the dimensionless time step; $s_{D,i}$ and $q_{D,i}$ are the dimensionless specific drawdown and dimensionless discharge at the node i in the aquifer, respectively.

2.5.4 *Comparison with the previous solutions*

To test the new solution, the fully Darcian flow solution of Papadopulos and Cooper (1967), the F-ND flow solution of Mathias *et al.* (2008) and the two-region model of Şen (1988) will be introduced. Using the notations of this section, the original solutions of Şen

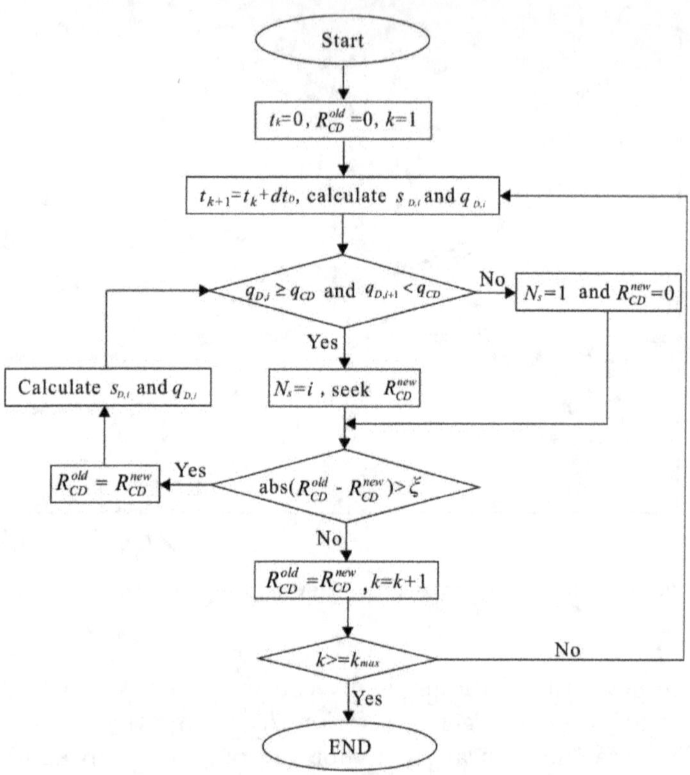

Figure 2.16. Flow chart of the MTRM algorithm.

(1988) in the Darcian flow region ($r_D \geq R_{CD}$) become:

$$s_{YD}(r_D, t_D) = \frac{1}{\lambda}F(r_{cD}, t_D)\exp\left[\frac{r_{CD}^2(1-\lambda)}{4\lambda t_D}\right]\int_{r_D}^{\infty}\frac{\exp\left(-\frac{\zeta^2}{4t_D\lambda}\right)}{\zeta}d\zeta \tag{2.92}$$

$$q_{YD}(r_D, t_D) = F(r_{cD}t_D)\text{exp}\left[\frac{r_{CD}^2(1-\lambda)}{4\lambda t_D}\right]\frac{\exp\left(-\frac{r_D^2}{4t_D\lambda}\right)}{r_D} \tag{2.93}$$

where $F(r_D, t_D) = \frac{1}{1+\frac{\sqrt{\pi}\beta_D}{2}\left(\frac{1}{t_D}\right)^{1/2}\text{erf}\left[\left(\frac{r_D^2}{4t_D}\right)^{1/2}\right]}$, and erf$(\cdot)$ is the error function sign.

In the non-Darcian flow region ($r_D < R_{CD}$), one has:

$$s_{ND}(r_D, t_D) = s_{YD}(r_{cD}, t_D) + \int_{r_D}^{r_{cD}} F(\zeta, t_D) \frac{\exp\left(-\frac{\zeta^2}{4t_D}\right)}{\zeta} d\zeta$$

$$+ \beta_D \int_{r_D}^{r_{cD}} F^2(\zeta, t_D) \frac{\exp\left(-\frac{\zeta^2}{2t_D}\right)}{\zeta^2} d\zeta \qquad (2.94)$$

Figures 2.17(a) and 2.17(b) show the distance-drawdown curves of the four mentioned above models in the early and late pumping stages, respectively. In these two figures, Papadopulos and Cooper (1967) represent the analytical solution of the fully Darcian flow model, Şen (1988) is the analytical solution of the two-region model by the Boltzmann transform method, and Mathias *et al.* (2008) represents the numerical solution of the F-ND flow model. The critical radius is located at the curve's deflection point.

In the early stage, the differences between the three previous solutions and the new solution of this study are obvious, as shown in Figure 2.17(a). First, the solution of Papadopulos and Cooper (1967) is smaller than the others near the well. This is because the inertial forces of the non-Darcian flow increase the resistance for flow, thus resulting in a drawdown greater than those for the Darcian flow near the well. The second is that the F-ND solution agrees with the new solution near the well. The third is that the solution of Şen (1988) does not agree with the new solution near the well at the early time. This is probably because of the Boltzmann transform method used by Şen (1988) to deal with the non-Darcian flow at the early time, which has been discussed in several previous studies (Camacho-V and Vásquez-C, 1992; Wen *et al.*, 2008a).

In the late pumping stage, the transient flow approaches the quasi-steady state, and the specific discharge distribution is invariant with time, according to Eqs. (2.90) and (2.91) or Eqs. (2.72) and (2.73), regardless of the Darcian flow or non-Darcian flow. Under the quasi-steady state flow condition, the critical radius obtained by this new solution becomes a constant, which is the same as the one used by previous two-region models such as Sen (1988) and Wen *et al.* (2009). Therefore, it is not surprising to see that the new solution agrees with that of Sen (1988) very well at late time (see Figure 2.17(b)). Another observation is that the new solution agrees with the solution

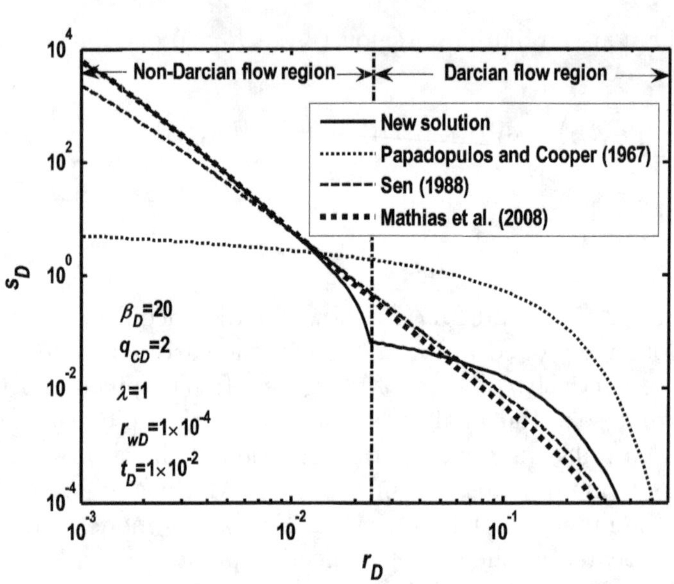

(a)

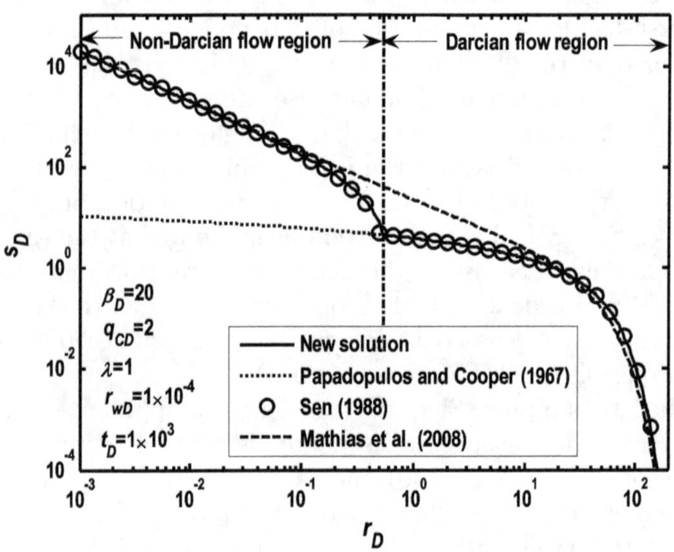

(b)

Figure 2.17. (a) Comparison of the distance-drawdown curves by the fully Darcian flow model, the F-ND flow model, the two-region flow model, and the new model in the early pumping stage. (b) Comparison of the distance-drawdown curves by the fully Darcian flow model, the F-ND flow model, the two-region flow model, and the new model in the early pumping stage.

of Papadopulos and Cooper (1967) in the Darcian flow region and with the solution of the F-ND model in the non-Darcian flow region near the well.

2.5.5 *Results and discussions*

2.5.5.1 *Effect of the inertial force coefficient to the critical radius*

For the non-Darcian radial flow near the pumping well investigated here, the dimensionless inertial force coefficient (β_D) is of primary concern. To test the effect of β_D to the groundwater flow, the representative parameters are chosen as follows: the value of ν is $1.004 \times 10^{-6}\,\text{m}^2\text{s}^{-1}$ at 20°C, $R_{ec} = 10$, $D_p = 0.0012\,\text{m}$, $Q = 0.628\,\text{m}^3\text{s}^{-1}$, and $B = 5\,\text{m}$. According to Eqs. (2.88) and (2.89) and the definition of the dimensionless critical specific discharge, one has $q_{CD} = 2$ and $R_C = 2.4\,\text{m}$. The values of β_D of this section are the same as the ones used by Wen *et al.* (2011), and they are 1, 10, and 50. The dimensionless radius of the well is $r_{wD} = 1 \times 10^{-4}$.

Figure 2.18 shows the critical radius changes with time for different dimensionless inertial force coefficients. Several observations can be seen. First, R_{CD} increases with time until the flow approaches the quasi-steady state condition. In the early pumping stage, the specific discharge is very large near the well and decreases quickly with the distance from the well, so R_{CD} is very small. With ongoing time, the cone of depression will expand along the radial direction, and the slope of the cone of depression becomes flatter, so R_{CD} becomes greater. Second, a larger β_D would reduce the rate of change R_{CD} versus time, thus resulting in a longer time to approach its asymptotic value, and consequently leading to a smaller R_{CD} at a specific time in the transient state (see Figure 2.18). This is because a larger β_D implies a stronger inertial force, which increases the resistance of flow. The third interesting observation is that the asymptotic value of R_{CD} is the same for different β_D. This can be explained using Eq. (2.89). Based on the definition of the dimensionless parameters defined in Eq. (2.83), Eq. (2.89) becomes:

$$q_{CD} = 1/R_{CD} \tag{2.95}$$

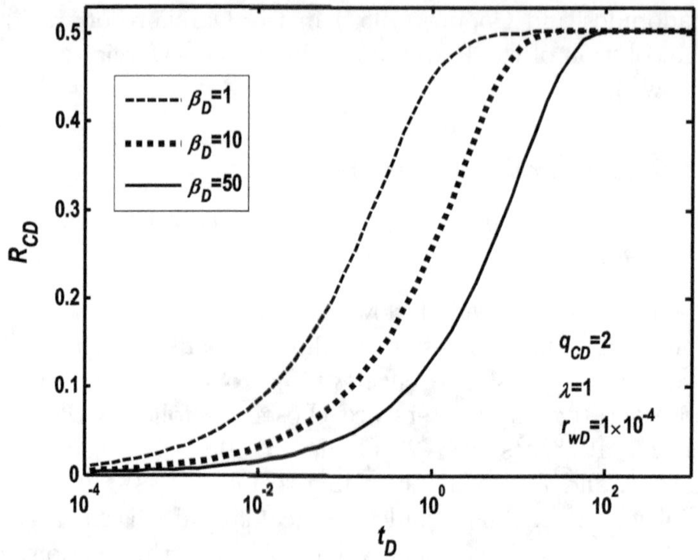

Figure 2.18. Critical radius changes with time for different inertial force coefficients.

Therefore, the value of R_{CD} has no relationship with β_D under the quasi-state state flow condition, while it only reciprocally depends on the critical specific discharge.

2.5.5.2 *Effect of the critical specific discharge to the critical radius*

The criterion to judge the initiation of the non-Darcian flow is an important factor of concern. Up to now, there is still considerable debate on what value of R_{ec} to use for the start of non-Darcian low. The recommended values of R_{ec} range from 0.1 to 100 for porous media flow (Bear, 1972; Scheidegger, 1974; Zeng and Grigg, 2006). To check the influence of R_{ec} on R_{CD} during the transient flow, the values of q_{CD} are chosen as 1, 2, and 5, considering the direct relationship of q_{CD} and R_{CD} in Eq. (2.89). The other parameters are $\beta_D = 1$, and $r_{wD} = 1 \times 10^{-4}$.

Figure 2.19 represents the effect of q_{CD} on R_{CD}. It is obvious that the asymptotic value of R_{CD} is equal to $1/q_{CD}$, as reflected in Eq. (2.95). Another interesting observation is that R_{CD} decreases with increasing q_{CD}, and it takes a shorter time for R_{CD} to

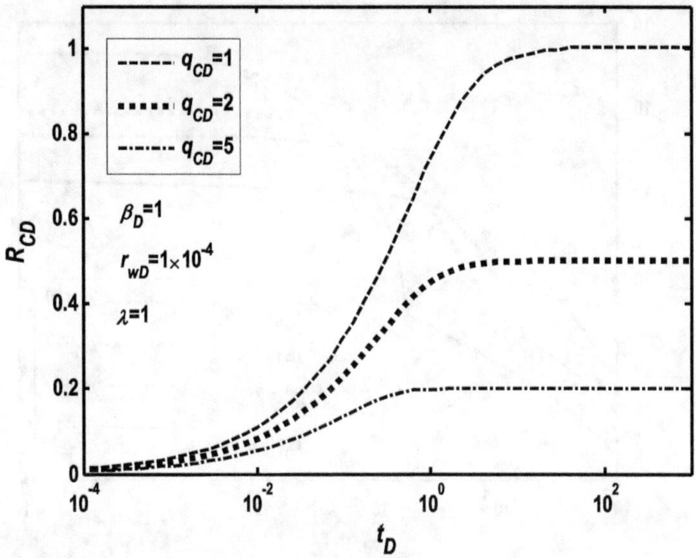

Figure 2.19. Critical radius changes with time for different critical specific discharges.

approach its asymptotic value. The reason can be also explained using Eq. (2.95).

2.5.5.3 *Type curves in the non-Darcian flow region and Darcian flow region*

Type curves are a series of curves that reveal the functional relationship between the well functions (or drawdown) and the dimensionless time factors (Sen, 1988; Wen *et al.*, 2011). A type curve is one of the common approaches to identifying the aquifer parameters or predicting the drawdown (Sen, 1988; Wen *et al.*, 2011). Sen (1988) presented different type curves in the Darcian flow region and non-Darcian flow region based on a two-region model. In the model of Sen (1988), R_{CD} was a fixed value that only depends on the rate of pumping but is independent of time. In this section, R_{CD} changes with time, and the type curves might be different from the ones generated by Sen (1988). To investigate the behaviors of the type curves of the new solution, the two observation locations will be chosen, $r_D = 0.1$ and 1.0. According to Eq. (2.95), the maximum of R_{CD} is 0.5 at the

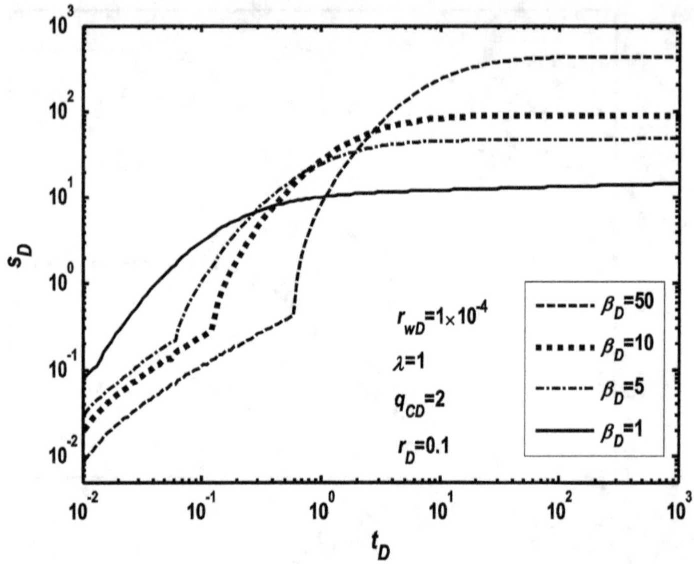

Figure 2.20. Time-drawdown at $r_D = 0.1$ for different inertial force coefficients in a log-log scale.

quasi-steady state, so the flow at $r_D = 0.1$ will experience both Darcian flow (at the early time) and non-Darcian flow (at a late time), while the flow at $r_D = 1.0$ is always Darcian.

Figure 2.20 shows the time-drawdown curves at $r_D = 0.1$ for different dimensionless inertial force coefficients in the log-log scale. Two interesting observations can be seen from this figure. The first observation is that there is a deflection point in the curve, and the time of this deflection point becomes longer with increasing β_D. This is because a larger β_D implies a stronger inertial effect, which leads to a larger drawdown and a longer time to approach the quasi-steady state condition. This observation is not found in the F-ND model (Wen *et al.*, 2011) and in the two-region model (Sen, 1988). The second observation is that the drawdown in the quasi-steady state increases with increasing β_D, and the reason for this has been explained in previous studies (Wen *et al.*, 2011).

Figure 2.21 represents the time-drawdown curves at $r_D = 1$ in the semi-log scale. One notable point is that flow at $r_D = 1$ is always Darcian, so there is no deflection point in the type curves. The differences among the curves with different β_D are very small at the beginning,

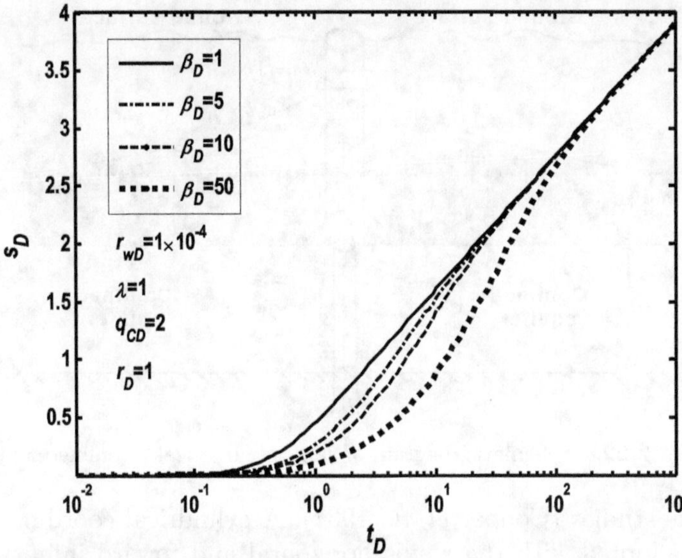

Figure 2.21. Time-drawdown at $r_D = 1$ for different inertial force coefficients in a semi-log scale.

then they become larger with time and finally approach the same value at the quasi-steady state. This is because the specific discharge far from the well is very small at the beginning, so the drawdown approaches 0. With time, the inertial force affects the groundwater flow at $r_D = 0.1$. At the quasi-steady state, the drawdown in the Darcian flow region does not change with β_D.

2.6 Radial Slug Test in Non-Darcian Flow Fields

As for the radial slug test in non-Darcian flow fields, Wang *et al.* (2015b) developed a transient nonlinear model of slug test in a leaky confined aquifer considering non-Darcian near well flow and presented a finite-difference solution for such a model. In this section, the model of Wang *et al.* (2015b) will be introduced.

2.6.1 *Mathematical model*

A schematic diagram of slug test in a leaky confined aquifer is depicted in Figure 2.22, and it is similar to what has been used in

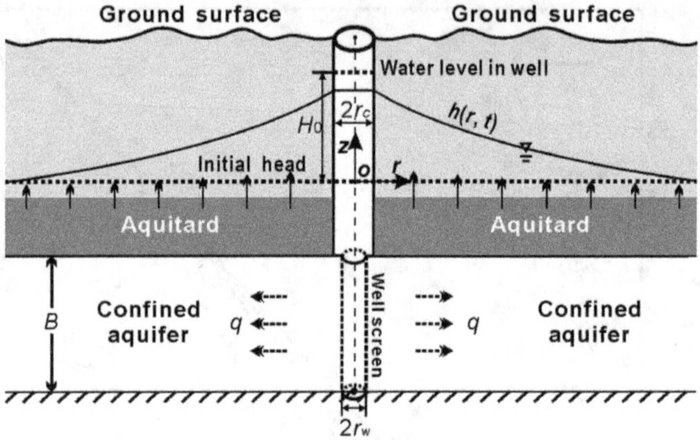

Figure 2.22. Schematic diagram of slug test in a leaky confined aquifer.

previous studies (Cooper *et al.*, 1967). A cylindrical coordinate system is adopted with the r-axis horizontal and toward infinity from the center of the well, and the z-axis vertical upward. The origin is at the center of the well, located at the plane designating the initial head, which is assumed to be horizontal. The main aquifer is bounded by an impermeable layer at the bottom and an aquitard at the top. An unconfined aquifer sits above the aquitard with a horizontal water table that is not affected by the slug test. The slug test well fully penetrates the main confined aquifer, which is assumed to be homogenous and horizontally isotropic and with a constant thickness.

2.6.1.1 *Governing equation*

Based on these assumptions, the governing equation of groundwater flow in the main aquifer can be described as

$$-\frac{\partial[q(r,t)]}{\partial r}-\frac{q(r,t)}{r}+\frac{\alpha}{B}(h_{unc}-h)=\frac{S}{B}\frac{\partial h(r,t)}{\partial t}, \cdot r_w \leq r < \infty \quad (2.96)$$

in which r is radial distance [m] from the well screen; t is time [s]; r_w is radius of well screen [m]; h and are hydraulic heads [m] of confined and unconfined aquifers, respectively; S is the storage coefficient (dimensionless); B is aquifer thicknesses [m]; $q(r,t)$ is horizontal specific discharge [m/s]; $\alpha = K'/B'$ is aquitard leakance [s/m] (Hantush, 1967; Sun and Zhan, 2006), where K' and B' are hydraulic conductivity [m/s] and thickness [m] of aquitard, respectively. Like previous studies, we use $\alpha = K'/B'$ to calculate the aquitard leakance by ignoring the aquitard storage effect, which can be seen from Eq. (2.96).

Volker and Zhang (2001) pointed out that this assumption was too simplified and might cause unacceptable errors when K'/B' from the aquitard was greater than 1 day^{-1}. However, Jiao and Tang (2001) pointed out that the errors, by this way of treating the aquitard leakance in this section, were negligible, since K'/B' was much smaller than 1 1/day] for most actual applications.

In this section, the non-Darcian flow was assumed to occur in the aquifer near the wellbore. By reviewing the literature on non-Darcian flow, we found that there were two commonly used formulas to describe non-Darcian flow. The first was the Forchheimer equation which showed that the hydraulic gradient was a second-order polynomial function of the specific discharge (Forchheimer, 1901). The second was the Izbash equation (Izbash, 1931), in which the hydraulic gradient was a power-law function of the specific discharge. Many experimental data indicated that both functions could describe non-Darcian flow very well (George and Hansen, 1992). In this section, the Izbash equation for non-Darcian flow is used:

$$q(r,t)|q(r,t)|^{n-1} = -k\frac{\partial h(r,t)}{\partial r} \tag{2.97}$$

where $k\,[\text{m/s}]^n$ and n (dimensionless) are constants. Since k could reflect how easily the aquifer can transmit water; it is named quasi-hydraulic conductivity. n is a non-Darcian index, which is generally between one and two. When n is less than 1, flow is pre-linear. When n is between 1 and 2, flow is post-linear, which occurs when the hydraulic gradient is very large, as is the case at the early stage of a slug test. When n is 2, flow is fully developed turbulent, as sometimes found in fractured media (Qian *et al.*, 2005). When n is 1, the flow becomes Darcian, and k becomes the hydraulic conductivity (Qian *et al.*, 2005).

2.6.1.2 *Initial condition*

The slug test can be conducted in two different ways: the water level is instantaneously lowered in the well by removing a certain volume of water or instantaneously raised by injecting a certain volume of water into the well. The mathematic models for both types of slug tests are similar. Therefore, we will focus on the latter in this section. The initial conditions of slug test for Eq. (2.96) can be written as

$$h(r, t=0) = \begin{cases} H_0, & \text{if } r \le r_w \\ h_0, & \text{if } r > r_w \end{cases} \tag{2.98}$$

where h_0 is the constant value [m] representing the hydrostatic condition prior to the start of the slug test, and it is 0 due to the coordinate system as shown in Figure 2.22; H_0 is the instantaneous water level increase (positive) in the wellbore due to the slug and can be calculated as

$$H_0 = \frac{V}{\pi r_C^2} \tag{2.99}$$

where V is the volume of initially added water in the wellbore [m^3]; r_c is the radius of well casing [m].

2.6.1.3 *Boundary conditions*

For the boundary condition at the well screen, Ferris and Knowles (1954b) employed the line source/sink model, which assumed that the well radius was infinitesimal by ignoring the storage of the well. Cooper *et al.* (1967) found that this model cannot be used to interpret the field data from tests on a well of Dawsonville, Georgia, and then developed a model considering the wellbore storage by

$$[2\pi r_w B q(r,t)]_{r \to r_w} = -\pi r_c^2 \frac{dh(r_c, t)}{dt} \tag{2.100}$$

When r_c and r_w approach zero, Eq. (2.100) reduces to the line source/sink model.

The outer boundary condition at an infinite distance can be expressed as

$$h(r \to \infty, t) = 0 \tag{2.101}$$

The mathematical model described by Eqs. (2.96)–(2.101) reduces to the model of Cooper *et al.* (1967) if n is 1 and α is zero. One point to note is that the sign (either negative or positive) of the specific discharge $q(r,t)$ in Eq. (2.97) always depends on $-\partial h(r,t)/\partial r$, and therefore one could get the value of $q(r,t)$ from Eq. (2.97) as follows:

$$q(r,t) = \text{sign}\left(-\frac{\partial h(r,t)}{\partial r}\right)\left(k\left|\frac{\partial h(r,t)}{\partial r}\right|\right)^{1/n} \quad (2.102)$$

where sign represents the sign of $-\partial h(r,t)/\partial r$. It is negative when $-\partial h(r,t)/\partial r$ is negative, while positive when $-\partial h(r,t)/\partial r$ is positive.

2.6.2 *Numerical solution*

Because of the nonlinearity of the problem, as shown in Eq. (2.97), it is not easy to obtain an analytical solution to the problem since the two commonly used analytical approaches for non-Darcian flow (Boltzmann transform by Sen (1990) and linearization procedure by Wen *et al.* (2008b)) do not work for the problem investigated in this section. Therefore, a finite-difference method will be developed here. Considering the flow velocity distribution along the r-axis, which is largest at the well screen and decreases with distance from the well, the numerical simulation will be conducted with a non-uniform grid system to reduce the numerical errors to a minimum and simultaneously save computational time. The spatial steps are smaller near the well screen and become progressively greater away from the well. Like previous studies (Mathias *et al.*, 2008; Wang *et al.*, 2014; Wen *et al.*, 2009), we discretize the spatial coordinate r logarithmically. The spatial domain $[r_w, r_e]$ is discretized into N nodes excluding the two boundary nodes r_w and r_e, where r_e is a relatively large radial distance [m] used to approximate the infinite boundary (Mathias *et al.*, 2008; Wen *et al.*, 2009). For any node of r_i, $r_w < r_i < r_e$, $i = 1, 2, \ldots, N$, one has:

$$r_i = \frac{r_{i-\frac{1}{2}} + r_{i+\frac{1}{2}}}{2}, \quad i = 1, 2, \ldots, N \quad (2.103)$$

where $r_{i+1/2}$ is calculated as follows:

$$\log_{10}(r_{i+1/2}) = \log_{10}(r_w) + i\left[\frac{\log_{10}(r_e) - \log_{10}(r_w)}{N}\right], \quad i = 0, 1, \ldots, N \quad (2.104)$$

Applying the finite-difference method to the right side of Eq. (2.96), one has:

$$\frac{dh_i}{dt} \approx \frac{B}{S}\frac{r_{i-1/2}q_{i-1/2} - r_{i+1/2}q_{i+1/2}}{r_i(r_{i+1/2} - r_{i-1/2})} + \alpha\frac{B}{S}(h_i - h_{\text{inc}}), \quad i = 1, 2, \ldots, N \tag{2.105}$$

where q_i and h_i are the specific discharge q and hydraulic head h at node-i, respectively. The values of q_i can be calculated by $q_i = (q_{i-1/2} + q_{i+1/2})/2$. In terms of the Izbash equation of Eq. (2.102), one can obtain:

$$q_{i-1/2} \approx \text{sign}\left(-\frac{h_i - h_{i-1}}{r_i - r_{i-1}}\right) k^{1/n} \left|\frac{h_i - h_{i-1}}{r_i - r_{i-1}}\right|^{1/n}, \quad i = 2, 3, \ldots, N_{i,a} \tag{2.106}$$

$$q_{i+1/2} \approx \text{sign}\left(-\frac{h_{i+1} - h_i}{r_{i+1} - r_i}\right) k^{1/n} \left|\frac{h_{i+1} - h_i}{r_{i+1} - r_i}\right|^{1/n}, \quad i = 1, 2, \ldots, N-1 \tag{2.107}$$

At the boundaries, the finite difference scheme of the specific discharge can be written as

$$q_{1-1/2} \approx \text{sign}\left(-\frac{h_1 - h_w}{r_1 - r_w}\right) k^{1/n} \left|\frac{h_1 - h_w}{r_1 - r_w}\right|^{1/n} \tag{2.108}$$

$$q_{N+1/2} \approx \text{sign}\left(\frac{h_N}{r_e - r_N}\right) k^{1/n} \left|\frac{h_N}{r_e - r_N}\right|^{1/n} \tag{2.109}$$

in which h_w is the hydraulic head inside the well, $h(r_c, \text{t})$, and can be approximated by the following equation based on Eq. (2.100):

$$\frac{dh_w}{dt} = -\frac{2r_w B}{r_c^2} q_{1-1/2} \tag{2.110}$$

With above preparation, one obtains a set of ordinary differential equations, which can be solved by the stiff integrator ODE15s (Shampine and Reichelt, 1997; Shampine *et al.*, 1999). ODE15s is available in any version of MATLAB. It employs an adaptive time grid automatically for temporal discretization when solving the ordinary differential equations, so the numerical error within each step could be controlled below a specified tolerance. In addition, this stiff integrator has been widely utilized to solve the non-Darcian flow around a well (Mathias *et al.*, 2008; Wang *et al.*, 2014; Wen *et al.*, 2009). In this section, a MATLAB program named STNE (Slug Test considering the Non-Darcian Effect) is developed to facilitate computation.

2.6.3 *Comparison with the Cooper et al. (1967) model*

To test the accuracy of numerical solutions in this section, we first considered a simple case in which an analytical solution is available. Cooper *et al.* (1967) employed the Laplace transform method to derive an analytical solution of slug test in which the groundwater flow was described by Darcy's law:

$$h(r,t) = \frac{2H_0}{\pi}\int_0^\infty e^{-\beta u^2/\varsigma}(J_0(ur/r_w)[uY_0(u) - 2\varsigma Y_1(u)] \\ -Y_0(ur/r_w)[uJ_0(u) - 2\varsigma J_1(u)])\frac{du}{\Delta u} \tag{2.111}$$

where $\Delta u = [uJ_0(u) - 2\varsigma J_1(u)]^2 + [uY_0(u) - 2\varsigma Y_1(u)]^2$, $\zeta = SBr_w^2/r_c^2$, $\beta = SBr_w^2/r_c^2$, and the other parameters are similarly defined in the model of this section. The head inside the well can be calculated by substituting $r = r_w$ into Eq. (2.111), one has:

$$h(r_w,t) = \frac{8\mathrm{GH}_0}{\pi^2}\int_0^\infty e^{-\beta u^2/\zeta}\frac{du}{u\Delta u} \tag{2.112}$$

It seems difficult to analytically conduct the integration of Eqs. (2.111) and (2.112), so a numerical approach is called in. In this section, we will use computer software AQTESOLV (version 1.0) to calculate the head inside the well of Eq. (2.112), which was coded by Geraghty and Miller (1990). AQTESOLV might have been the world's leading software package since 1989 for the design and analysis of aquifer tests, including pump tests, slug tests, and single well step tests. The accuracy of the solution incorporated in this software package has been tested in many different cases.

In comparison with the solution of Cooper *et al.* (1967), the parameters are set as $n = 1$, $S = 0.001$, $h_{unc} = 0$, $r_c = r_w = 0.1\,\mathrm{m}$,

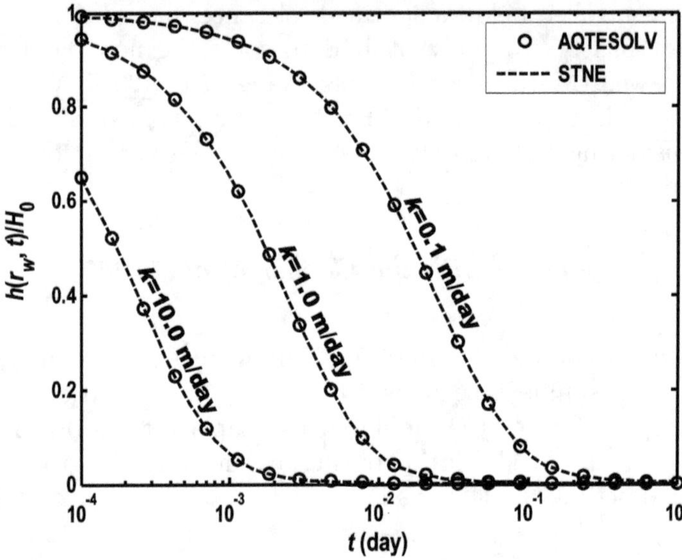

Figure 2.23. Comparison of the proposed solution with Cooper *et al.* (1967) under Darcian flow conditions.

$B = 10\,\text{m}$ and $\alpha = 0$. The reason for choosing $n = 1$ and $\alpha = 0$ are to remove the non-Darcian effect and the aquitard leakage effect on WLR, which are not considered in the model of Cooper *et al.* (1967). Figure 2.23 represents the comparison between the solutions of Cooper *et al.* (1967) and of this section for different k values. The analytical solutions agree very well with the numerical solutions, which indicates that this study's numerical solution and MATLAB program are valid for such a special case.

2.6.4 *Results and discussions*

From the mathematical model of this section, one can see that there are several parameters affecting the WLR of the slug test: S, α, B, r_w, $r_c\,k$, and n, as shown in Eqs. (2.96)–(2.102). Since these parameters may not independently affect the WLR of the slug test, it is not easy to directly figure out the relationship between each parameter variation and the WLR of the slug test. The dimensionless variable analysis offers a method for reducing multiple-parameter complex physical problems to the simplest form by transforming dimensional parameters into dimensionless ones. In this section, we will employ the dimensionless variable analysis to investigate the effect of parameters on WLR. The dimensionless groups are defined as

$$r_D = \frac{r}{r_w}, \quad t_D = \frac{tk^{1/n}B}{Sr_w^2}, \quad h_D = \frac{h(r,t)}{H_0}, \quad h_{uncD} = \frac{h_{unc}}{H_0},$$
$$H_{0D} = \frac{H_0}{r_w}, \quad q_D = \frac{q(r,t)r_w}{H_0 k^{1/n}}, \quad \alpha_D = \frac{r_w^2}{k^{1/n}B}, \quad \chi = \frac{r_w}{r_c} \tag{2.113}$$

in which $h_D = h(r_w, t)/H_0$ is also named the normalized water level in the wellbore in previous studies.

Substituting the dimensionless variables into Eqs. (2.96)–(2.102), one can obtain the dimensionless mathematical model of slug test as follows:

$$-\frac{\partial q_D}{\partial r_D} - \frac{q_D}{r_D} + \alpha_D(h_{uncD} - h_D) = \frac{\partial h_D}{\partial t_D}, \quad r_{wD} \le r_D < \infty \tag{2.114}$$

$$h_D(r_D, t_D = 0) = \begin{cases} 1, & \text{if } r_D \le r_{wD} \\ 0, & \text{if } r_D > r_{wD} \end{cases} \tag{2.115}$$

$$h_D(r_D \to \infty, t_D) = 0 \tag{2.116}$$

$$[2S\chi^2 q_D(r_D, t_D)]_{t_D \to r_w} = -\frac{dh_D(r_{cD}, t_D)}{dt_D} \tag{2.117}$$

$$q_D(r_D, t_D)|q_D(r_D, t_D)|^{n-1} = -H_{0D}^{1-n}\frac{\partial h_D(r_D, t_D)}{\partial r_D} \tag{2.118}$$

In the following section, we will investigate the effect of non-Darcian flow and aquitard leakance on the WLR of the slug test using the dimensionless model.

2.6.4.1 *Flow of non-Darcian effect on slug test*

Compared with Darcy's law, there are two primary parameters involved in the Izbash equation of non-Darcian flow: the quasi-hydraulic conductivity k and the non-Darcian index n as shown in Eq. (2.97). From Eqs. (2.113) and (2.118), one can see that k is incorporated into other dimensionless parameters in the dimensionless model, and only n is left to stand alone. Therefore, we will investigate the non-Darcian effect on the WLR of the slug test using five sets of n, including $1.0, 1.2, 1.5, 1.8$ and 2.0. The other parameters

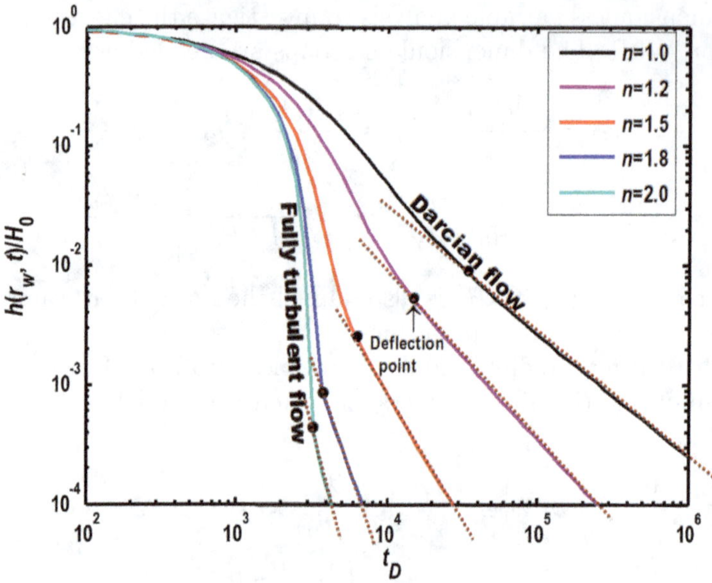

Figure 2.24. Normalized time-WLR for different non-Darcian indexes in log-log scales.

are set as $S = 0.001$, $h_{\text{unc}} = 0$, $r_c = r_w = 0.1\,\text{m}$, $B = 10\,\text{m}$, $\alpha = 0$, and $k = 1.0\ (\text{m/day})^n$. The definitions of the dimensionless variables in Eq. (2.113) make it easy to obtain their corresponding values.

Figure 2.24 represents the normalized water level in the wellbore for different non-Darcian index values in log-log scales. Several observations can be seen. First, the values of the water level increase with the decrease of non-Darcian index n. The WLR curve of Darcian flow is above the post-linear flow curves of $n = 1.2, 1.5$ and 1.8, and the WLR curve of the fully developed turbulent flow is below the Darcian flow and the post-linear flow curves. In another words, the WLR for non-Darcian cases is faster than the one for Darcian cases, particularly for a large n value. This observation can be explained as follows.

To illustrate, we use the slug test with an instantaneous increase of water level inside the well as an example. It is worthwhile to recall that the hydraulic head is merely the hydraulic energy per unit weight of water, so an instantaneous increase in water level inside the well implies that a certain amount of "extra" hydraulic energy is added there. Such extra hydraulic energy will be subsequently dissipated into the aquifer surrounding the well, resulting in the gradual decline of water level (or hydraulic energy) inside the well. Therefore, the WLR curve or the decline of water level as a function of time is simply a reflection of how fast the hydraulic energy inside the well will be dissipated into the aquifer. In this regard, turbulent flow ($n = 2$) is the most efficient way of dissipating hydraulic energy in comparison with laminar, Darcian flow ($n = 1$), as commonly known in fluid mechanics. Consequently, one will observe the quickest decline in water level for the turbulent flow case ($n = 2$), and the slowest decline of water level for the Darcian flow case ($n = 1$). For a general case ($1 < n < 2$) between these two extremes of $n = 2$ and $n = 1$, one will see the decline of water level is faster for a greater n value, as expected.

Second, the difference between curves of different n values increases with time. Thirdly, there is a deflection point for each curve, and the slope of the curve approaches a constant after this point in log-log scales.

2.6.4.2 *Effect of the aquitard leakance on slug test*

The dimensionless aquitard leakance (α_D) maybe also of primary concern for a slug test and the values of α_D are chosen as $0.4, 0.04, 0.004$ and 0. The case of $\alpha_D = 0.04$ corresponds to the following parameters: $\alpha = 0.01\,1/(\mathrm{m} \cdot \mathrm{day})$, $S = 0.001$, $h_{\mathrm{unc}} = 0$, $r_c = r_w = 0.1\,\mathrm{m}$, $B = 10\,\mathrm{m}$, $n = 1.3$ and $k = 0.01\,(\mathrm{m/day})^n$. The case of $\alpha_D = 0.4$ is for the purpose of showing the effect of the aquitard leakance on slug test, as such a value may be too large for actual problems (Jiao and Tang, 2001).

Figure 2.25 represents the normalized WLR for different dimensionless aquitard leakages in log-log scales, and the curves of Darcian flow are included as a reference. From this figure, one can obtain several interesting observations. First, the effect of the dimensionless aquitard leakance is very small at the early stage by comparing the curves with and without leakage effect, regardless of non-Darcian flow or Darcian flow. One can see that the normalized water level $h(r_w, t)/H_0$, corresponding to the time that the difference becomes obvious when $\alpha_D > 0.04$, is about 10^{-3} for non-Darcian flow and about 10^{-2} for Darcian flow. Therefore, the neglect of the aquitard

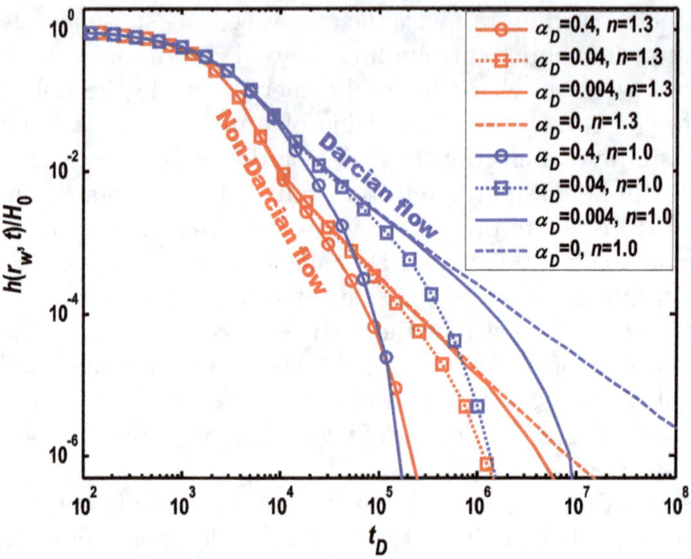

Figure 2.25. Normalized time-WLR for different aquitard leakages in log-log scales.

leakance on slug test may be reasonable at an early stage when $\alpha_D \leq 0.04$. Second, the WLR curve without the aquitard leakance is a linear function at a late stage in log-log scales, which can be seen in Figure 2.24, while it is a convex function when considering the aquitard leakance. The reason for the convex shape of the curves with aquitard leakance can be explained as follows. The model with the aquitard leakance used in Eq. (2.96) assumes that the hydraulic head in the unconfined aquifer is constant and the initial conditions of the hydraulic head in both unconfined and confined aquifer are the same. When instantaneously injecting a volume of water into the slug test well, the water in the well will recharge the aquifer to increase the hydraulic head in the formation, which produces a vertical hydraulic gradient over the aquitard thickness. Obviously, the directions of aquitard leakance are upwards since the head in the confined aquifer is greater than that in the upper unconfined aquifer. Aquitard leakance becomes a discharge sink for the underlying confined aquifer, resulting in the lower hydraulic head for greater α_D. Third, the difference between the curves of different leakance values is smaller for non-Darcian flow by comparing with the curves of Darcian flow. Fourthly, the effects of aquitard leakance on WLR for non-Darcian cases appear later than the Darcian case, i.e., non-Darcian flow reduces the impacts of aquitard on WLR.

2.6.4.3 *Sensitivity analysis*

Generally, input parameters of a groundwater flow model may be subject to various sorts of uncertainty caused by errors of measurements, absence of information, poor understanding of the flow mechanisms, and so on (Jiao and Rushton, 1995; Liou and Yeh, 1997). Therefore, to create a more accurate and reliable model, it is always helpful to know how the uncertainty in the output could be apportioned to its inputs, where sensitivity analysis is one of the most powerful techniques to meet this need (Jiao and Rushton, 1995; McElwee and Yukler, 1978). The sensitivity of a dependent variable can be defined as (Jiao and Rushton, 1995; Liou and Yeh, 1997):

$$SC_{i,j} = \frac{\partial O_i}{\partial I_j} \tag{2.119}$$

in which $SC_{i,j}$ is the sensitivity coefficient of the jth parameter I_j at the ith time; O_i is the output function of the model at the ith time, which is the water level in the wellbore in this section. However, Eq. (2.119) may not work well when the unit and/or the order of magnitude of the parameters are different, and then the normalized sensitivity analysis was proposed as follows (Kabala, 2001; Yang and Yeh, 2009):

$$SC'_{i,j} = I_j \frac{\partial O_i}{\partial I_j} \tag{2.120}$$

where $SC'_{i,j}$ is the *Normalized Sensitivity Coefficient* (NSC) of the jth parameter I_j at the ith time. In this section, a finite-difference formula will be employed to approximate the differentiation of Eq. (2.120), and one has:

$$SC'_{i,j} = I_j \frac{O_i(I_j + \Delta I_j) - O_i(I_j)}{\Delta I_j} \tag{2.121}$$

where ΔI_j is a small increment.

In this section, the sensitivity of three primary dimensionless parameters on WLR inside the wellbore will be investigated, including the dimensionless aquitard leakance (α_D), the non-Darcian

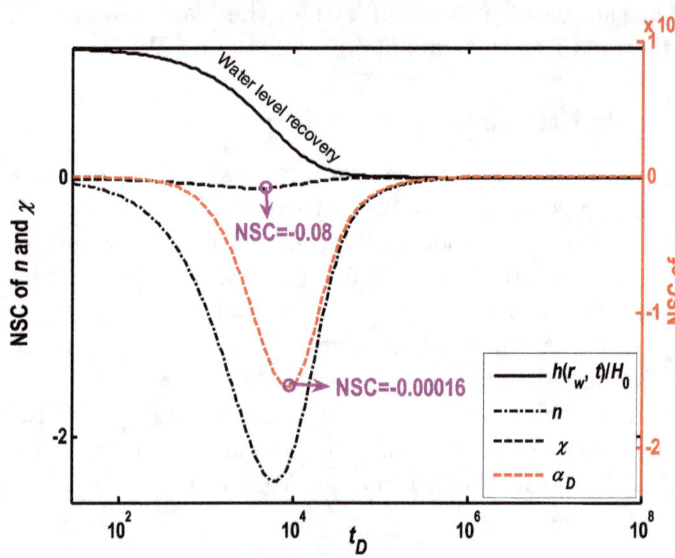

Figure 2.26. Dimensionless time-WLR curve and NSC of the parameters n, α_D, and χ in semi-log scales when $\alpha_D = 0.04$.

index (n), and the ratio of well screen radius versus well casing radius (χ). We have found that the aquitard leakance effect on the slug test is very small at early stage in earlier section, and hereby it is very interesting to see the sensitivity of α_D on slug test for different leakance values. Figure 2.26 shows NSC of the parameters n, α_D, and χ for $\alpha_D = 0.04$, where the normalized WLR is included as a reference. The other dimensionless parameters are $n = 1.3$, $S = 0.001$, and $\chi = 1$. For the demonstration purpose, Figure 2.26 is plotted in double vertical scales. The curves of normalized WLR, NSC of the parameters n and χ are subjected to the left vertical axis, and the NSC curves of α_D relates to the right vertical axis.

From Figure 2.26, several interesting observations can be seen. First, NSC of the water level with respect to n, α_D, and χ has a negative effect, which represents increasing n, α_D, and χ will decrease the values of recovery water level in the well. Second, the curves of n, α_D, and χ appear as convex shapes, in which NSC decreases with time firstly, and then increases with time until it becomes zero when the water level is very small. Since the initial conditions of the unconfined and underlying confined aquifer are the same, aquitard leakance is zero at first. After instantaneously raising the water level in the well, the hydraulic head in the confined aquifer becomes larger, initiating upward aquitard leakance. Subject to the infinite boundary condition of the aquifer, the hydraulic head will return to the initial hydraulic head eventually, which is the same as the hydraulic head in the unconfined aquifer, and finally, the aquitard leakance will become zero again. This is the reason for seeing convex functions in Figure 2.26. Third, one can find that WLR is most sensitive to n while least sensitive to α_D. It seems that the curve of α_D is below the curve of χ, but the curve of χ is below the curve of α_D when putting these two curves into the same vertical scale. The magnitude of NSC of α_D is as small as 10^{-5}, while it is 1 for χ in this plot using two different vertical scales. The minimum value of NSC of α_D is -1.6×10^{-4}, while is minimum NSC of χ is -0.08. Generally the effect of aquitard leakance is very limited for actual problems, reported by many previous studies (Jiao and Tang, 2001), therefore, it is may be reasonable for ignoring such an effect at early stage.

Figure 2.27 represents NSC for different n in semi-log scales. One can see that NSC has negative effect on recovery water level in the well, and peak values of the NSC curves increase with decreasing n.

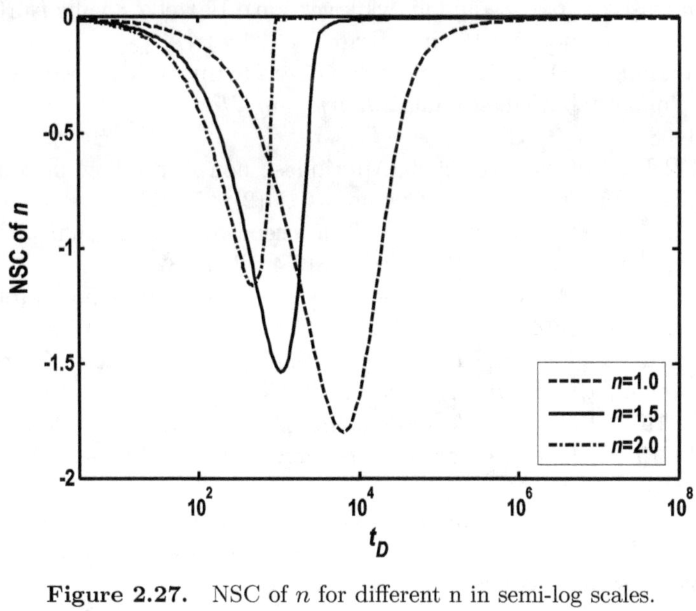

Figure 2.27. NSC of n for different n in semi-log scales.

The peak value of NSC is smallest when $n = 1$. Additionally, the dimensionless times corresponding to the peak values also increase with decreasing n.

References

Abu-Zied, M.A., Scott, V.H., 1963. Nonsteady flow for wells with decreasing discharge. *Journal of the Hydraulics Division*, 89(3): 119–132.

Abu-Zied, M.A., Scott, V.H., Aron, G., 1964. Modified solutions for decreasing discharge wells. *Journal of the Hydraulics Division*, 90(6): 145–160.

Adams, E.E., Gelhar, L.W., 1992. Field study of dispersion in a heterogeneous aquifer: 2. Spatial moments analysis. *Water Resource Research*, 28(12): 3293–3307.

Agarwal, R.G., Al-Hussainy, R., Ramey, H.J., 1970. An investigation of wellbore storage and skin effect in unsteady liquid flow: I. Analytical treatment. *SPE Journal*, 10(3): 279–290.

Agnaou, M., Lasseux, D., Ahmadi, A., 2017. Origin of the inertial deviation from Darcy's law: An investigation from a microscopic flow analysis on two-dimensional model structures. *Physical Review E*, 96(4): 043105.

Ahlstrom, S., Foote, H., Arnett, R., Cole, C., Serne, R., 1977. *Multicomponent Mass Transport Model: Theory and Numerical Implementation* (discrete-parcel-random-walk version). Battelle Pacific Northwest Labs., Richland, WA (United States).

Bari, R., Hansen, D., 2002. Application of gradually-varied flow algorithms to simulate buried streams. *Journal of Hydraulic Research*, 40(6): 673–683.

Barree, R.D., Conway, M., 2009. Multiphase non-Darcy flow in proppant packs. *SPE Production & Operations*, 24(2): 257–268.

Basak, P., 1977. Non-penetrating well in a semi-infinite medium with nonlinear flow. *Journal of Hydrology*, 33(3.4): 375–382.

Basak, P., 1978. Analytical solutions for two-regime well flow problems. *Journal of Hydrology*, 38: 147–159.

Bassett, R. *et al.*, 1981. *Preliminary Data from a Series of Artificial Recharge Experiments At Stanton.* US Geological Survey, Texas.

Bear, J., 1972. *Dynamics of Fluids in Porous Media.* Elsevier, New York.

Bear, J., 1979. *Hydraulics of Groundwater.* McGraw-Hill Intl. Book Co., New York.

Bentley, H., Walter, G., 1983. *Two-Well Recirculating Tracer Tests at H-2: Waste Isolation Pilot Plant (WIPP).* Southwest New Mexico, draft paper. Hydro Geochem., Inc., Tucson, Ariz.

Berkowitz, B., 2002. Characterizing flow and transport in fractured geological media: A review. *Advances in Water Resources*, 25(8–12): 861–884.

Bierschenk, W.H., 1959. *Aquifer characteristics and Ground-Water Movement at Hanford*, General Electric Co. Hanford Atomic Products Operation, Richland, Wash.

Bordier, C., Zimmer, D., 2000. Drainage equations and non-Darcian modelling in coarse porous media or geosynthetic materials. *Journal of Hydrology*, 228(3.4): 174–187.

Bouwer, H., Rice, R.C., 1976. A slug test method for determining hydraulic conductivity of unconfined aquifers with completely or partially penetrating wells. *Water Resources Research*, 12(3): 423–428.

Bredehoeft, J.D., Pinder, G.F., 1973. Mass transport in flowing groundwater. *Water Resources Research*, 9(1): 194–210.

Brown, D.L., Narasimhan, T.N., Demir, Z., 1995. An evaluation of the Bouwer and Rice method of slug test analysis. *Water Resources Research*, 31(5): 1239–1246.

Butler, J.J., Jr., 1998. *The Design, Performance, and Analysis of Slug Tests*. Lewis Publishers, New York.

Butler, J.J., Jr., Zhan, X., 2004. Hydraulic tests in highly permeable aquifers. *Water Resources Research*, 40(12).

Camacho, R.G., Vásquez-C, M., 1992. Comment on "analytical solution incorporating nonlinear radial flow in confined aquifers" by Zekâi Şen. *Water Resources Research*, 28(12): 3337–3338.

Camacho, R.G., Vasquez, M., 1992. Analytical solution incorporating nonlinear radial flow in confined aquifers — comment. *Water Resources Research*, 28(12): 3337–3338.

Cassiani, G., Kabala, Z., Medina, M. Jr, 1999. Flowing partially penetrating well: solution to a mixed-type boundary value problem. *Advances in Water Resources*, 23(1): 59–68.

Chang, C.-C., Chen, C.-S., 2003. A flowing partially penetrating well in a finite-thickness aquifer: A mixed-type initial boundary value problem. *Journal of Hydrology*, 271(1–4): 101–118.

Chen, C., Jiao, J.J., 1999. Numerical simulation of pumping tests in multilayer wells with non-Darcian flow in the wellbore. *Groundwater*, 37(3): 465–474.

Chen, C., Wan, J., Zhan, H., 2003. Theoretical and experimental studies of coupled seepage-pipe flow to a horizontal well. *Journal of Hydrology*, 281(1–2): 159–171.

Chen, C., Zhang, W., Hong, Z., Jakada, H., 2020. New Semi-analytical model for an exponentially decaying pumping rate with a finite-thickness skin in a leaky aquifer. *Journal of Hydrologic Engineering*, 25(8).

Chen, Y.-F., Li, B.-Y., Liu, M.-M., Hu, R., Yang, Z., 2019. A Forchheimer's law-based analytical model for constant-rate tests with linear flow pattern. *Advances in Water Resources*, 128: 1–12.

Chen, Z.X., Lyons, S.L., Qin, G., 2001. Derivation of the Forchheimer law via homogenization. *Transport in Porous Media*, 44(2): 325–335.

Chilton, T.H., Colburn, A.P., 1931. II — Pressure drop in packed tubes1. *Industrial & Engineering Chemistry*, 23(8): 913–919.

Claassen, H.C., Cordes, E.H., 1975. Two-well recirculating tracer test in fractured carbonate rock, *Nevada. Hydrology Science Bulletin*, 20(3).

Cole, J. A., 1972. Some interpretations of dispersion measurements in aquifers, *Groundwater Pollution in Europe*, 86–95, Water Research Association, Reading, England.

Cooper, H.H., Bredehoeft, J.D., Papadopulos, I.S., 1967. Response of a finite-diameter well to an instantaneous charge of water. *Water Resources Research*, 3(1): 263–269.

Cooper, H.H., Bredehoeft, J.D., Papadopulos, I.S., Bennett, R.R., 1965. The response of well-aquifer systems to seismic waves. *Journal of Geophysical Research*, 70(16): 3915–3926.

Daniels, W., 1981. Laboratory and field studies related to the radionuclide Migration Project, October 1, 1979 — September 30, 1980. Los Alamos Nat onal Laboratory report LA-8670-PR, February.

Dieulin, A., 1981. Lixiviation in situ d'un gisement d'uranium en mileu granitiqu. Draft Rep. LHM/RD/81/63. Ecole Natl. Super. des Mines de Paris...

Dikken, B.J., 1990. Pressure drop in horizontal wells and its effect on production performance. *Journal of Petroleum Technology*, 42(11): 1426–1433.

Driscoll, F.G., 1986. *Groundwater and Wells*. Johnson Filtration Systems Inc, St. Paul, Minnesota.

Dudgeon, C.R., Huyakorn, P.S., Swan, W.H.C., 1972. Hydraulics of flow near wells in unconsolidated sediments. *Theoretical and Experimental Studies*. The University of New South Wales, Australia.

Dullien, F.A.L., Azzam, M.I.S., 1973. Flow rate pressure gradient measurements in periodically nonuniform capillary tubes. *Aiche Journal*, 19(2): 222–229.

Egboka, B., Cherry, J., Farvolden, R., Frind, E., 1983. Migration of contaminants in groundwater at a landfill: A case study: 3. Tritium as an indicator of dispersion and recharge. *Journal of Hydrology*, 63(1–2): 51–80.

Engelund, F., 1953. On the laminar and turbulent flows of ground water through homogeneous sand, Tech. Univ. Denmark, Copenhagen, Denmark.

Ergun, S., 1952. Fluid flow through packed columns. *Chemical Engineering Progress*, 48(2): 89.

Escande, L., 1953. Experiments concerning the infiltration of water through a rock mass. In *Proceedings: Minnesota International Hydraulic Convention. ASCE*, pp. 547–553.

Fancher, G.H., Lewis, J.A., 1933. Flow of simple fluids through porous materials. *Industrial & Engineering Chemistry*, 25(10): 1139–1147.

Feng, Q., Zhan, H., 2015. On the aquitar-aquifer interface flow and the drawdown sensitivity with a partially penetrating pumping well in an anisotropic leaky confined aquifer. *Journal of Hydrology*, 521: 74–83.

Feng, Q., Zhan, H., 2016. Integrated aquitard-aquifer flow with a mixed-type well-face boundary and skin effect. *Advances in Water Resources*, 89: 42–52.

Fenske, P.R., 1973. Hydrology and radionuclide transport monitoring well HT-2m, Tatum Dome, Mississippi, Nevada Univ., Reno (USA). Desert Research Inst.

Ferris, J.G., Knowles, D.B., 1954a. The slub test for estimating transmissibility, U.S. Geological Survey. Ground Water Note 26.

Ferris, J.G., Knowles, D.B., 1954b. The slug test for estimating transmissibility. Open-File Report.

Ferris, J.G., Knowles, D.B., Brown, R.H., Stallman, R.W., 1962. Theory of aquifer tests, U.S. Geological Survey Water-Supply Paper 1536 E.

Forchheimer, P.H., 1901. Wasserbewegun durch Boden. *Zeitsch-rift des Vereines Deutscher Ingenieure*, 49: 1736–1749.

Fourar, M., Radilla, G., Lenormand, R., Moyne, C., 2004. On the non-linear behavior of a laminar single-phase flow through two and three-dimensional porous media. *Advances in Water Resources*, 27(6): 669–677.

Freyberg, D.L., 1986. A natural gradient experiment on solute transport in a sand aquifer: 2. Spatial moments and the advection and dispersion of nonreactive tracers. *Water Resources Research*, 22(13): 2031–2046.

Fried, J.J., 1975. *Groundwater Pollution*. Elsevier.

Gao, H., Li, H.A., 2016. Pore structure characterization, permeability evaluation and enhanced gas recovery techniques of tight gas sandstones. *Journal of Natural Gas Science and Engineering*, 28: 536–547.

Garabedian, S.P., 1987. Large-scale dispersive transport in aquifers: Field experiments and reactive transport theory, Massachusetts Institute of Technology.

Gelhar, L.W., Welty, C., Rehfeldt, K.R., 1992. A critical review of data on field-scale dispersion in aquifers. *Water Resources Research*, 28(7): 1955–1974.

Gensterblum, Y. *et al.*, 2015. Gas transport and storage capacity in shale gas reservoirs — A review. Part A: Transport processes. *Journal of Unconventional Oil and Gas Resources*, 12: 87–122.

George, G., Hansen, D., 1992. Conversion between quadratic and power law for non-Darcy flow. *Journal of Hydraulic Engineering*, 118(5): 792–797.

Geraghty & Miller, Inc., 1990. AQTESOLV-aqufier test design and analysis computer software (Version 1.0). Modeling Group. Enviromental Services, Geraghty & Miller, Inc.

Giorgi, T., 1997. Derivation of the Forchheimer law via matched asymptotic expansions. *Transport in Porous Media*, 29(2): 191–206.

Goblet, P., 1982. *Interpretation d'Experiences de Tracage en Milieu Granitique* (Site B), Report LHM/RD/82/11, Centre d'Information Geologique, Ecole Nationale.

Green Jr, L., Duwez, P., 1951. Fluid flow through porous metals. *Journal of Applied Mechanics*, 18(1): 39–45.

Grove, D., Beetem, W., 1971. Porosity and dispersion constant calculations for a fractured carbonate aquifer using the two well tracer method. *Water Resources Research*, 7(1): 128–134.

Grove, D.B., 1977. The use of Galerkin finite-element methods to solve mass-transport equations, US Dept. of the Interior, Geological Survey.

Halevy, E., Nir, A., 1962. The determination of aquifer parameters with the aid of radioactive tracers. *Journal of Geophysical Research*, 67(6): 2403–2409.

Hansbo, S., 1960. Consolidation of clay with special reference to influence of vertical sand drains. *Swedish Geotechnical Institute Proceeding*, 18: 45–50.

Hansbo, S., 1997. Aspects of vertical drain design: Darcian or non-Darcian flow. *Geotechnique*, 47(5): 983–992.

Hansbo, S., 2001. Consolidation equation valid for both Darcian and non-Darcian flow. *Geotechnique*, 51(1): 51–54.

Hansen, D., Garga, V.K., Townsend, D.R., 1995. Selection and application of a one-dimensional non-Darcy flow equation for two-dimensional flow through rockfill embankments. *Canadian Geotechnical Journal*, 32(2): 223–232.

Hantush, Jacob, C.E., 1955. Non-steady radial flow in an infinite leaky aquifer. Eos. *Transactions American Geophysical Union*, 36(1): 95–100.

Hantush, Mahdi, S., 1960. Modification of the theory of leaky aquifers. *Journal of Geophysical Research*, 65(11): 3713–3725.

Hantush, Mahdi, S., 1964. Hydraulics of wells. *Advances in Hydroscience*, 1: 281–432.

Hantush, Mahdi, S., 1967. Flow of groundwater in relatively thick leaky aquifers. *Water Resources Research*, 3(2): 583–590.

Harr, M., 1962. *Groundwater and Seepage*, McGraw-Hill Book Co. Inc. New York, p. 249.

Hassanizadeh, S.M., Gray, W.G., 1987. High-velocity flow in porous-media. *Transport in Porous Media*, 2(6): 521–531.

Hassanzadeh, H., Pooladi-Darvish, M., 2007. Comparison of different numerical Laplace inversion methods for engineering applications. *Applied Mathematics and Computation*, 189(2): 1966–1981.

Hayes, R., Afacan, A., Boulanger, B., 1995. An equation of motion for an incompressible Newtonian fluid in a packed bed. *Transport in Porous Media*, 18: 185–198.

Hemker, C., 1999. Transient well flow in vertically heterogeneous aquifers. *Journal of Hydrology*, 225(1–2): 1–18.

Hoehn, E., 1983. Advection-dispersion interpretation of tracer observations in a river/groundwater infiltration system. In *International Conference on Modern Approach to Groundwater Resources Management*, pp. 361–371.

Hoehn, E., Santschi, P.H., 1987. Interpretation of tracer displacement during infiltration of river water to groundwater. *Water Resources Research*, 23(4): 633–640.

Hu, L., Chen, C., Chen, X., 2011. Simulation of groundwater flow within observation boreholes for confined aquifers. *Journal of Hydrology*, 398(1–2): 101–108.

Huyakorn, P., Dudgeon, C.R., 1976. Investigation of 2-regime well flow. *Journal of the Hydraulics Division-ASCE*, 102(9): 1149–1165.

Huyakorn, P., Dudgeon, C.R., 1978. Investigation of 2-regime well flow. *Journal of the Hydraulics Division-ASCE*, 104(2): 302–304.

Huyakorn, P.S., Andersen, P.F., Molz, F.J., Güven, O., Melville, J.G., 1986. Simulations of two-well tracer tests in stratified aquifers at the Chalk River and the Mobile sites. *Water Resources Research*, 22(7): 1016–1030.

Hvorslev, M.J., 1951. Time lag and soil permeability in ground-water observations, Vicksburg, Mississippi: Waterways Experiment Station, United States Army Corps of Engineers.

Hyder, Z., Butler, J.J., 1995. Slug tests in unconfined formations: An assessment of the Bouwer and Rice technique. *Ground Water*, 33(1): 16–22.

Hyder, Z., Butler, J.J., McElwee, C.D., Liu, W., 1994. Slug tests in partially penetrating wells. *Water Resources Research*, 30(11): 2945–2957.

Iris, P., Adnot, J., Bonfils, R., Marchio, D., 1985. The heliogeothermal doublet. *International Journal of Solar Energy*, 3(2): 73–88.

Irmay, S., 1959. On the theoretical derivations of Darcy and Forchheimer formulas — Discussion — Reply. *Journal of Geophysical Research*, 64(4): 486–487.

Istok, J.D., Dawson, K.J., 2014. *Aquifer Testing: Design and Analysis of Pumping and Slug Tests.* CRC Press.

Ivanovich, M., Smith, D., 1978. Determination of aquifer parameters by a two-well pulsed method using radioactive tracers. *Journal of Hydrology*, 36(1–2): 35–45.

Izbash, S.V., 1931. *O filtracii V Kropnozernstom Materiale.* Leningrad, USSR (in Russian).

Javadi, M., Sharifzadeh, M., Shahriar, K., Mitani, Y., 2014. Critical Reynolds number for nonlinear flow through rough-walled fractures: The role of shear processes. *Water Resources Research*, 50(2): 1789–1804.

Javandel, I., Witherspoon, P.A., 1983. Analytical solution of a partially penetrating well in a two-layer aquifer. *Water Resources Research*, 19: 567–578.

Ji, S.H., Koh, Y.K., 2015. Nonlinear groundwater flow during a slug test in fractured rock. *Journal of Hydrology*, 520: 30–36.

Jiao, J.J., Rushton, K.R., 1995. Sensitivity of drawdown to parameters and its influence on parameter estimation for pumping tests in large-diameter wells. *Ground Water*, 33(5): 794–800.

Jiao, J.J., Tang, Z., 2001. Reply to "Comment on 'An analytical solution of groundwater response to tidal fluctuation in a leaky confined aquifer' by Jiu Jimmy Jiao and Zhonghua Tang". *Water Resources Research*, 37(1): 187–188.

Kabala, Z.J., 2001. Sensitivity analysis of a pumping test on a well with wellbore storage and skin. *Advances in Water Resources*, 24(5): 483–504.

Kawecki, M., 2000. Transient flow to a horizontal water well. *Groundwater*, 38(6): 842–850.

Kells, J.A., 1996. The analysis of flow through and over a gabion dam. Ph.D. thesis, Dept. of Civil Engineering, Univ. of Saskatchewan, Saskatoon, Canada.

Kies, B., 1982. Solute transport in unsaturated field soil and in groundwater. New Mexico State University.

Konokow, L.F., 1976. Modelling solute transport in groundwater. In: *International Conference on Environmental Sensing and Assessment Proceedings*, 20–30, Las Vegas, NV.

Konikow, L.F., Bredehoeft, J.D., 1974. Modeling flow and chemical quality changes in an irrigated stream-aquifer system. *Water Resources Research*, 10(3): 546–562.

Kreft, A., Lenda, A., Turek, B., Zuber, A., Czauderna, K., 1974. Determination of effective porosities by the two-well pulse method. In *Isotope Techniques in Groundwater Hydrology 1974, Vol. II. Proceedings of a Symposium.*

Kruseman, G.P., De Ridder, N.A., Verweij, J.M., 1970. Analysis and evaluation of pumping test data, 11. International Institute for Land Reclamation and Improvement Wageningen.

Lage, J., Antohe, B., Nield, D., 1997. Two types of nonlinear pressure-drop versus flow-rate relation observed for saturated porous media. *Journal of Fluids Engineering*, 119(3): 700–706.

Lai, R., Su, C.W., 1974. Nonsteady flow to a large well in a leaky aquifer. *Journal of Hydrology*, 22(3.4): 333–345.

Lasseux, D., Abbasian Arani, A.A., Ahmadi, A., 2011. On the stationary macroscopic inertial effects for one phase flow in ordered and disordered porous media. *Physics of Fluids*, 23(7).

Lau, L.-K.S., Kaufman, W.J., Todd, D.K., 1957. Studies of dispersion in a radial flow system. Sanitary Engineering Research Laboratory, Department of Engineering.

Lee, D.R., Cherry, J.A., Pickens, J.F., 1980. Groundwater transport of a salt tracer through a sandy lakebed 1. *Limnology and Oceanography*, 25(1): 45–61.

Leland, D., Hillel, D., 1981. Scale effects on measurement of dispersivity in a shallow unconfined aquifer. In *Chapman Conference on Spatial Variability in Hydrologic Modeling*. AGU. Fort Collins. CO., July, pp. 23.

Lenda, A., Zuber, A., 1970. Tracer dispersion in groundwater experiments. *Isotope Hydrology*, 1970: 619–641.

Lin, Y.C., Yang, S.Y., Fen, C.S., Yeh, H.D., 2016. A general analytical model for pumping tests in radial finite two-zone confined aquifers with Robin-type outer boundary. *Journal of Hydrology*, 540: 1162–1175.

Lin, Y.C., Yeh, H.D., 2017. A lagging model for describing drawdown induced by a constant-rate pumping in a leaky confined aquifer. *Water Resources Research*, 53(10): 8500–8511.

Lin, Y.C., Yeh, H.D., 2021. An analytical model with a generalized nonlinear water transfer term for the flow in dual porosity media induced by constant rate pumping in a leaky fractured aquifer. *Water Resources Research*, 57(8):e2020WR029186.

Liou, T.S., Yeh, H.D., 1997. Conditional expectation for evaluation of risk groundwater flow and solute transport: One-dimensional analysis. *Journal of Hydrology*, 199(3.4): 378–402.

Liu, H.H., Li, L.C., Birkholzer, J., 2012. Unsaturated properties for non-Darcian water flow in clay. *Journal of Hydrology*, 430: 173–178.

Lo Jacono, D., Plouraboué, F., Bergeon, A., 2005. Weak-inertial flow between two rough surfaces. *Physics of Fluids*, 17(6).

Lutz, J., Kemper, W., 1959. Intrinsic permeability of clay as affected by clay-water interaction. *Soil Science*, 88(2): 83–90.

Ma, H., Ruth, D., 1993. The microscopic analysis of high Forchheimer number flow in porous media. *Transport in Porous Media*, 13: 139–160.

Macini, P., Mesini, E., Viola, R., 2011. Laboratory measurements of non-Darcy flow coefficients in natural and artificial unconsolidated porous media. *Journal of Petroleum Science and Engineering*, 77(3.4): 365–374.

Mackie, C.D., 1983. Determination of Nonlinear formation losses in pumping wells. In *International Conference on Groundwater and Man.* Sydney, Australia, pp. 199–209.

Malama, B., Kuhlman, K.L., Barrash, W., Cardiff, M., Thoma, M., 2011. Modeling slug tests in unconfined aquifers taking into account water table kinematics, wellbore skin and inertial effects. *Journal of Hydrology*, 408(1–2): 113–126.

Mathias, S.A., Todman, L.C., 2010. Step-drawdown tests and the Forchheimer equation. *Water Resources Research*, 46(7).

Mathias, S.A., Wen, Z., 2015. Numerical simulation of Forchheimer flow to a partially penetrating well with a mixed-type boundary condition. *Journal of Hydrology*, 524: 53–61.

Mathias, S.A., Butler, A.P., Zhan, H., 2008. Approximate solutions for Forchheimer flow to a well. *Journal of Hydraulic Engineering*, 134(9): 1318–1325.

McElwee, C.D., 2001. Application of a nonlinear slug test model. *Ground Water*, 39(5): 737–744.

McElwee, C.D., Yukler, M.A., 1978. Sensitivity of groundwater models with respect to variations in transmissivity and storage. *Water Resources Research*, 14(3): 451–459.

McElwee, C.D., Zenner, M.A., 1998. A nonlinear model for analysis of slug-test data. *Water Resources Research*, 34(1): 55–66.

Mercado, A., 1966. Recharge and mixing tests at Yavne 20 well field. *Underground Water Storage Study Tech. Rep.* 12, Publ. 611. Tahal Water Plann. for Israel, Tel Aviv.

Miller, R.J., Low, P.F., 1963. Threshold gradient for water flow in clay systems. *Soil Science Society of America Journal*, 27(6): 605–609.

Mills, A.C., 2010. Applying parameter-estimation methods to recovery-test and slug-test analyses. *Ground Water*, 48(4): 604–608.

Mishra, P.K., Vessilinov, V., Gupta, H., 2013. On simulation and analysis of variable-rate pumping tests. *Groundwater*, 51(3): 469–473.

Mitchell, J., Younger, J., 1967. Abnormalities in hydraulic flow through fine-grained soils, Permeability and capillarity of soils. *ASTM International.*

Moench, A., Hsieh, P., 1985. Analysis of slug test data in a well with finite thickness skin. In *Memoirs of the 17th International Congress on the Hydrogeology of Rocks of Low Permeability.* USA Members of the International Association of Hydrologists Tucson, AZ, p. 1729.

Moench, A.F., 1984. Double-porosity models for a fissured groundwater reservoir with fracture skin. *Water Resources Research*, 20: 831–846.

Moench, A.F., 1985. Transient flow to a large... diameter well in an aquifer with storative semiconfining layers. *Water Resources Research*, 21: 1121–1131.

Molinari, J., Peaudecerf, P., 1977. Essais conjoints en laboratoire et sur le terrain en vue d'une approach simplefiée de la prévision des propagations de substances miscibles dans les aquifères réels. *Symposium on Hydrodynamic Diffusion and Dispersion in Porous Media.* Int. Assoc. for Hydraul. Res., Pavis, Italy.

Moltyaner, G., Killey, R., 1988. Twin Lake tracer tests: Transverse dispersion. *Water Resources Research*, 24(10): 1628–1637.

Munson, B.R., Young, D.F., Okiishi, T.H., 1995. Fundamentals of fluid mechanics. *Oceanographic Literature Review*, 10(42): 831.

Muskat, M., 1938. The flow of homogeneous fluids through porous media. *Soil Science*, 46(2): 169.

Naymik, T.G., Barcelona, M.J., 1981. Characterization of a contaminant plume in ground water, Meredosia, Illinois. *Groundwater*, 19(5): 517–526.

Neuman, S.P., Witherspoon, P.A., 1969. Theory of flow in a confined two aquifer system. *Water Resources Research*, 5(4): 803–816.

Neuzil, C.E., 1986. Groundwater flow in low-permeability environments. *Water Resources Research*, 22(8): 1163–1195.

Novakowski, K., Lapcevic, P., Voralek, J., Bickerton, G., 1995. Preliminary interpretation of tracer experiments conducted in a discrete rock fracture under conditions of natural flow. *Geophysical Research Letters*, 22(11): 1417–1420.

Novakowski, K.S., Brusseau, M.L., Jessup, R.E., Wallach, R., Israeli, M., 1989. A composite analytical model for analysis of pumping tests affected by well bore storage and finite thickness skin. *Water Resources Research*, 25(9): 1937–1946.

Ouyang, L.-B., Arbabi, S., Aziz, K., 1998. A single-phase wellbore-flow model for horizontal, vertical, and slanted wells. *SPE Journal*, 3(2): 124–133.

Ozkan, E., Sarica, C., Haciislamoglu, M., 1999. Influence of pressure drop along the wellbore on horizontal-well productivity. *SPE Journal*, 4(3): 288–301.

Ozkan, E., Sarica, C., Haciislamoglu, M., Raghavan, R., 1995. Effect of conductivity on horizontal-well pressure-behavior. *SPE Advanced Technology Series*, 3(1): 85–94.

Papadopulos, I.S., Cooper Jr, H.H., 1967. Drawdown in a well of large diameter. *Water Resources Research*, 3(1): 241–244.

Papadopulos, I.S., Larson, S., 1978. Aquifer storage of heated water: Part II — Numerical simulation of field results. *Groundwater*, 16(4): 242–248.

Penmatcha, V., Aziz, K., 1999. Comprehensive reservoir/wellbore model for horizontal wells. *SPE Journal*, 4(3): 224–234.

Perina, T., Lee, T.-C., 2006. General well function for pumping from a confined, leaky, or unconfined aquifer. *Journal of Hydrology*, 317(3.4): 239–260.

Pickens, J.F., Grisak, G.E., 1981. Scale-dependent dispersion in a stratified granular aquifer. *Water Resources Research*, 17(4): 1191–1211.

Pinder, G.F., 1973. A Galerkin-finite element simulation of groundwater contamination on Long Island, New York. *Water Resources Research*, 9(6): 1657–1669.

Polubarinova-Kochina, P.Y., 2015. *Theory of Ground Water Movement.* Princeton University Press.

Qian, J., Zhan, H., Zhao, W., Sun, F., 2005. Experimental study of turbulent unconfined groundwater flow in a single fracture. *Journal of Hydrology*, 311(1–4): 134–142.

Quinn, P., Cherry, J.A., Parker, B.L., 2012. Hydraulic testing using a versatile straddle packer system for improved transmissivity estimation in fractured-rock boreholes. *Hydrogeology Journal*, 20(8): 1529–1547.

Quinn, P.M., Parker, B.L., Cherry, J.A., 2013. Validation of non-Darcian flow effects in slug tests conducted in fractured rock boreholes. *Journal of Hydrology*, 486: 505–518.

Rabinowitz, D.D., Gross, G.W., 1972. Environmental tritium as a hydrometeorologic tool in the Roswell Basin, New Mexico.

Rajaram, H., Gelhar, L.W., 1991. Three-dimensional spatial moments analysis of the Borden tracer test. *Water Resources Research*, 27(6): 1239–1251.

Rehbinder, G., 1995. Theory of single well tests in a leaky aquifer. Analytical solution for non steady flow. *Applied Scientific Research*, 55(3): 211–225.

Roberts, P.V., 1985. Advection-dispersion sorption models for simulating the transport of organic pollutants. *Ground Water Quality*, 425–445.

Robertson, J., Barraclough, J., 1973. Radioactive and chemical waste transport in groundwater of National Reactor Testing Station: 20-year case history and digital model. In *Underground Waste Manage, Artificial Recharge Prepr. Paper. International Symposium. 2nd*, pp. 291–322.

Robertson, J.B., 1974. Digital modeling of radioactive and chemical waste transport in the aquifer underlying the Snake River Plain at the National Reactor Testing Station, Idaho. US Geological Survey, pp. 2331–1258.

Robson, S.G., 1974. Feasibility of digital water-quality modeling illustrated by application at Barstow, California, 46. US Geological Survey, WRD, California District.

Robson, S.G., 1978. Application of digital profile modeling techniques to ground-water solute transport at Barstow, California, 2050. US Government Printing Office.

Rose, H., 1951. Fluid flow through beds of granular material. *Some Aspects of Fluid Flow*: Papers Presented at a Conf. of the Institute of Physics. Edward Arnold & Co., pp. 136–163.

Rousselot, D., Sauty, J.P., Gaillard, B., 1977. Etude hydrogéologique de la zone industrielle de Blyes-Saint-Vulbas, rapport préliminaire no. 5: Caracteristiques hydrodynamiques du système aquifère. *Rep. Jal 77/33*. Bur. de Rech. Geol. et Min., Orleans, France.

Samani, J.M., Moknatian, M., Heidari, M., 2013. Two-dimensional multi-rockfill detention dam flow model in reservoir routing. *Proceedings of the Institution of Civil Engineers — Engineering and Computational Mechanics*, 166(2): 59–67.

Sarica, C., Ozkan, E., Haciislamoglu, M., Raghavan, R., Brill, J., 1994. Influence of wellbore hydraulics on pressure behavior and productivity of horizontal gas wells. In *SPE Annual Technical Conference and Exhibition*? SPE, pp. SPE-28486-MS.

Sarkhosh, P., Samani, J.M.V., Mazaheri, M., 2018. A one-dimensional flood routing model for rockfill dams considering exit height. In *Proceedings of the Institution of Civil Engineers — Water Management*. Thomas Telford Ltd, pp. 42–51.

Sauty, J.P., 1977. Contribution à l'identification des paramèters de dispersion dans les aquifères par interprètation des expèriences de tracage, dissertation. Univ. Sci. et Med. et Inst. Natl. Polytech. de Grenoble, Grenoble, France.

Sauty, J., Gringarten, A., Landel, P., 1978. The effects of thermal dispersion on injection of hot water in aquifers. In *Proceedings of Second Invitational Well Testing Symposium*. Lawrence Berkeley Laboratory, Berkeley, California, pp. 122–131.

Scheidegger, A.E., 1974. *The Physics of Flow through Porous Media*. University of Toronto Press, pp. 152–170.

Segol, G., Pinder, G.F., 1976. Transient simulation of saltwater intrusion in southeastern Florida. *Water Resources Research*, 12(1): 65–70.

Sen, Z., 1988. Type curves for two-region well flow. *Journal of Hydraulic Engineering*, 114(12): 1461–1484.

Şen, Z., 1988. Analytical solution incorporating nonlinear radial flow in confined aquifers. *Water Resources Research*, 24(4): 601–606.

Sen, Z., 1990. Nonlinear radial flow in confined aquifers toward large-diameter wells. *Water Resources Research*, 26(5): 1103–1109.

Sen, Z., 2000. Non-Darcian groundwater flow in leaky aquifers. *Hydrological Sciences Journal-Journal Des Sciences Hydrologiques*, 45(4): 595–606.

Sen, Z., Altunkaynak, A., 2004. Variable discharge type curve solutions for confined aquifers 1. *Journal of the American Water Resources Association (JAWRA)*, 40(5): 1189–1196.

Shampine, L.F., Reichelt, M.W., 1997. The matlab ode suite. *SIAM Journal on Scientific Computing*, 18(1): 1–22.

Shampine, L.F., Reichelt, M.W., Kierzenka, J.A., 1999. Solving index-1 DAEs in MATLAB and Simulink. *SIAM Review*, 41(3): 538–552.

Shaw-Yang, Y., Yeh, H.-D., 2012. A general semi-analytical solution for three types of well tests in confined aquifers with a partially penetrating well. *TAO: Terrestrial, Atmospheric and Oceanic Sciences*, 23(5): 577.

Sidiropoulou, M.G., Moutsopoulos, K.N., Tsihrintzis, V.A., 2007. Determination of Forchheimer equation coefficients a and b. *Hydrological Processes*, 21(4): 534–554.

Sindalovskiy, Leonid, N., 2017. *Aquifer-Test Analytical Methods.* Springer International Publishing, pp. 227–259.

Singh, S.K., 2007. New methods for aquifer parameters from slug test data. *Journal of Irrigation and Drainage Engineering-ASCE*, 133(3): 272–275.

Sivanesapillai, R., Steeb, H., Hartmaier, A., 2014. Transition of effective hydraulic properties from low to high Reynolds number flow in porous media. *Geophysical Research Letters*, 41(14): 4920–4928.

Slepicka, F., 1961. Hydraulic function of cylindrical well in an artesian aquifer with regard to new research on flow through porous media. In *Proceedings of the 9th World Congress of the International Association of Hydraulic Research*, Dubrovnik, Czechoslovakia.

Soni, J.P., Islam, N., Basak, P., 1978. Experimental evaluation of non-Darcian flow in porous-media. *Journal of Hydrology*, 38(3.4): 231–241.

Springer, R.K., Gelhar, L.W., 1991. Characterization of large scale aquifer heterogeneity in glacial outwash by analysis of slug tests with oscillatory response, Cape Cod, Massachusetts., U.S. Geological Survey, Water Res. Invest. Rep. 91-4034, Reston, Virginia.

Stehfest, H., 1970. Algorithm 368: Numerical inversion of Laplace transforms. *Communications of the ACM*, 13(1): 47–49.

Stephenson, D., 1979. *Rockfill in Hydraulic Engineering.* Elsevier.

Steward, D.R., Jin, W., 2001. Gaining and losing sections of horizontal wells. *Water Resources Research*, 37(11): 2677–2685.

Sudicky, E., Cherry, J., Frind, E., 1983. Migration of contaminants in groundwater at a landfill: A case study: 4. A natural-gradient dispersion test. *Journal of Hydrology*, 63(1–2): 81–108.

Sun, D., Zhan, H., 2006. Flow to a horizontal well in an aquitard-aquifer system. *Journal of Hydrology*, 321(1): 364–376.

Swartzendruber, D., 1962. Modification of Darcy's law for the flow of water in soils. *Soil Science*, 93(1): 22–29.

Sykes, J., Pahwa, S., Lantz, R., Ward, D., 1982. Numerical simulation of flow and contaminant migration at an extensively monitored landfill. *Water Resources Research*, 18(6): 1687–1704.

Sykes, J., Pahwa, S., Ward, D., Lantz, D., 1983. In R. Stapleman (ed.), *The Validation of SWENT, a Geosphere Transport Model; W: Scientific Computing, IMAES*. North Holland, Amsterdam, The Netherlands.

Tarshish, M., 1992. Combined mathematical model of flow in an aquifer-horizontal well system. *Groundwater*, 30(6): 931–935.

Tarshish, M., 1993. Combined mathematical model of flow in an aquifer-vertical well system. *Groundwater*, 31(4): 545–550.

Theis, C.V., 1935. The relation between the lowering of the Piezometric surface and the rate and duration of discharge of a well using groundwater storage. Eos. *Transactions American Geophysical Union*, 16(2).

Tsai, C., Yeh, H., 2012. Wellbore flow-rate solution for a constant-head test in two-zone finite confined aquifers. *Hydrological Processes*, 26(21).

Vaccaro, J., Bolke, E., 1983. Evaluation of water-quality characteristics of part of the Spokane aquifer, Washington and Idaho, using a solute-transport digital model. US Geological Survey, pp. 2331–1258.

Valocchi, A.J., Roberts, P.V., Parks, G.A., Street, R.L., 1981. Simulation of the transport of ion-exchanging solutes using laboratory-determined chemical parameter values. *Groundwater*, 19(6): 600–607.

van der Kamp, G., 1976. Determining aquifer transmissivity by means of well response tests: The underdamped case. *Water Resources Research*, 12(1): 71–77.

Veyskarami, M., Hassani, A.H., Ghazanfari, M.H., 2016. Modeling of non-Darcy flow through anisotropic porous media: Role of pore space profiles. *Chemical Engineering Science*, 151: 93–104.

Vincent, Caltagirone, 1999. Numerical simulation of single-phase and multiphase non-Darcy flow in porous and fractured reservoirs. *Transport in Porous Media*, 49: 209–240.

Volker, R.E., Zhang, Q., 2001. Comment on "An analytical solution of groundwater response to tidal fluctuation in a leaky confined aquifer" by Jiu Jimmy Jiao and Zhonghua Tang. *Water Resources Research*, 37(1): 185–186.

von Engelhardt, W., Tunn, W., 1955. The flow of fluids through sandstones. *Heidelberger Beitrage zur Mineralogie und Petrographie* (1954) 2: 12. Reprinted in English translation as Circular 194, Illinois State Geological Survey, Urbana, IL (1955).

Walter, G., 1983. Convergent flow tracer test at H-6: Waste isolation pilot plant (WIPP), southeast New Mexico (draft). Hydro Geochem, Inc., Tucson, Ariz.

Walton, W.C., 2006. *Aquifer Test Modeling.* CRC Press.

Wang, B., Jin, M., Liang, X., 2015a. Using EARTH model to estimate groundwater recharge at five representative zones in the Hebei Plain, China. *Journal of Earth Science*, 26: 425–434.

Wang, J., Xu, J., Wang, Y., Li, H., 2016. A new production predicting model of fractured horizontal wells in dual-porosity gas reservoirs on non-Darcy flow conditions. *Journal of Petroleum Science and Engineering*, 146: 670–676.

Wang, Q., Zhan, H., 2015. On different numerical inverse Laplace methods for solute transport problems. *Advances in Water Resources*, 75: 80–92.

Wang, Q., Zhan, H., 2017. The effect of intra-wellbore head losses in a vertical well. *Journal of Hydrology*, 548: 333–341.

Wang, Q., Zhan, H., Tang, Z., 2014. Forchheimer flow to a well-considering time-dependent critical radius. *Hydrology and Earth System Sciences*, 18(6): 2437–2448.

Wang, Q., Zhan, H., Wang, Y., 2015b. Non-Darcian effect on slug test in a leaky confined aquifer. *Journal of Hydrology*, 527: 747–753.

Wang, Q., Zhan, H., Wang, Y., 2017. Single-well push-pull test in transient Forchheimer flow field. *Journal of hydrology*, 549: 125–132.

Wang, Y., Zhan, H., Huang, K., Linqing, H., Wan, J., 2021. Identification of non-Darcian flow effect in double-porosity fractured aquifer based on multi-well pumping test. *Journal of Hydrology*, 600: 126541.

Ward, J.C., 1964. Turbulent flow in porous media. *Journal of Hydraulics Division, American Society of Civil Engineers*, 90: 1–12.

Webster, D., Proctor, J., Marine, I.W., 1970. Two-well tracer test in fractured crystalline rock, USGPO.

Wen, Z., Huang, G., Zhan, H., 2009. A numerical solution for non-Darcian flow to a well in a confined aquifer using the power law function. *Journal of Hydrology*, 364(1–2): 99–106.

Wen, Z., Huang, G., Zhan, H., 2011. Non-Darcian flow to a well in a leaky aquifer using the Forchheimer equation. *Hydrogeology Journal*, 19(3): 563–572.

Wen, Z., Huang, G., Zhan, H., Jian, L., 2008a. Two-region non-Darcian flow toward a well in a confined aquifer. *Advances in Water Resources*, 31(5): 818–827.

Wen, Z., Huang, G.H., Zhan, H.B., 2008b. An analytical solution for non-Darcian flow in a confined aquifer using the power law function. *Advances in Water Resources*, 31(1): 44–55.

Wen, Z., Zhan, H., Wang, Q., Liang X., Chen C., 2017. Well hydraulics in pumping tests with exponentially decayed rates of abstraction in confined aquifers. *Journal of Hydrology*, 548: 40–45.

Werner, A., 1983. Nutzung von Grundwasser fur Warmepumpen, Versickerrungstest Aefligen, Versuch 2, 1982/83, Water and Energy Manage. Agency of the State of Bern, Switzerland.

Whitaker, S., 1996. The Forchheimer equation: A theoretical development. *Transport in Porous Media*, 25(1): 27–61.

Widdowson, M.A., Molz, F.J., Melville, J.G., 1990. An analysis technique for multilevel and partially penetrating slug test data. *Ground Water*, 28(6): 937–945.

Wiebenga, W., Ellis, W., Seatonberry, B., Andrew, J., 1967. Radioisotopes as groundwater tracers. *Journal of Geophysical Research*, 72(16): 4081–4091.

Wilkins, J., 1955. Flow of water through rock fill and its application to the design of dams. *New Zealand Engineering*, 10(11): 382–387.

Wilson, L., 1971. Investigations on the subsurface disposal of waste effluents at inland sites. US Department of the Interior.

Wood, W.W., 1981. A geochemical method of determining dispersivity in regional groundwater systems. *Journal of Hydrology*, 54(1–3): 209–224.

Wood, W.W., Ehrlich, G.G., 1978. Use of baker's yeast to trace microbial movement in ground water. *Groundwater*, 16(6): 398–403.

Wu, Y.S., 2002a. Numerical simulation of single-phase and multiphase non-Darcy flow in porous and fractured reservoirs. *Transport in Porous Media*, 49: 209–240.

Wu, Y.S., 2002b. An approximate analytical solution for non-Darcy flow toward a well in fractured media. *Water Resources Research*, 38(3): 5-1–5-7.

Wylie, A., Magnuson, S., 1995. Spreadsheet modeling of slug tests using the van der Kamp method. *Ground Water*, 33(2): 326–329.

Yang, S.Y., Yeh, H.D., 2005. Laplace-domain solutions for radial two-zone flow equations under the conditions of constant-head and partially penetrating well. *Journal of Hydraulic Engineering*, 131(3): 209–216.

Yang, S.Y., Yeh, H.D., 2009. Radial groundwater flow to a finite diameter well in a leaky confined aquifer with a finite-thickness skin. *Hydrological Processes*, 23(23): 3382–3390.

Yang, S.-Y., Yeh, H.-D., Chiu, P.-Y., 2006. A closed form solution for constant flux pumping in a well under partial penetration condition. *Water Resources Research*, 42(5).

Yao, Y., Li, G., Qin, P., 2015. Seepage features of high-velocity non-Darcy flow in highly productive reservoirs. *Journal of Natural Gas Science and Engineering*, 27: 1732–1738.

Yeh, H.-D., Chang, Y.-C., 2013. Recent advances in modeling of well hydraulics. *Advances in Water Resources*, 51: 27–51.

Zeng, Z., Grigg, R., 2006. A criterion for non-Darcy flow in porous media. *Transport in Porous Media*, 63: 57–69.

Zenner, M.A., 2009. Near-well nonlinear flow identified by various displacement well response testing. *Ground Water*, 47(4): 526–535.

Zhang, R., Wu, Y.-S., Fakcharoenphol, P., 2014. Non-Darcy displacement in linear composite and radial aquifer during CO2 sequestration. *International Journal of Oil, Gas and Coal Technology*, 7(3): 244–262.

Zou, Y., 1996. A non-linear permeability relation depending on the activation energy of pore liquid. *Géotechnique*, 46(4): 769–774.

Chapter 3

Radial Reactive Transport

3.1 Introduction

Understanding reactive transport in the subsurface around an injection/extraction well is very important for the risk assessment of aquifer pollution and remediation since the well is a common source/sink through which contaminates could be introduced into (or withdrawn from) the aquifer (Anderson *et al.*, 2015; Istok, 2012). Radial dispersion refers to a process of reactive transport under the radial flow condition. One unique feature of radial dispersion (as compared to unilateral dispersion where the flow velocity is unilateral) is that the dispersive transport becomes progressively weaker when the radial distance from the injection/pumping well becomes larger (or the radial flow velocity becomes smaller). Thus, the relative importance of molecular diffusion (which is assumed to be constant) versus the dispersion becomes progressively more robust with a more significant radial distance. The radial dispersion problem is both theoretically interesting and practically important in many fields, like chemical engineering (Davis and Davis, 2002), environmental science (Chen *et al.*, 2016; Reinhard *et al.*, 1997), and hydrogeology (Webster *et al.*, 1970).

An accurate description of reactive transport in porous or fractured media is indispensable for groundwater quality monitoring and contamination remediation. For a simplified case, the transport equation includes the processes of advective and dispersive transports. The mathematical statement of the solute transport equation, also called the advective-dispersion equation (ADE), is the foundation upon which numerous mathematical analyses of solute transport in porous media have been based. The usual approach to reactive transport in porous media is based on the use of concentrations representing average amounts of solutes present in the pore fluid within a finite representative elementary volume (REV) of the porous medium (Bear, 1961). It is significant to note that most of the previous analytical studies on radial solute transport are based on an important assumption for the ADE, that is, Fickian dispersion behavior. The ADE was found to work very well in relatively homogeneous aquifers (Bear, 2013; Chen, 1985; Hsieh and Yeh, 2014; Huang *et al.*, 2014; Wang *et al.*, 2020). With regard to reactive transport in heterogeneous media, the estimated transport parameters obtained by best fitting the observed breakthrough curves (BTCs) with the ADE model frequently exhibit temporally and spatially dependent behavior. Meanwhile, the observed BTCs usually exhibit the non-Fickian phenomenon, such as early arrival and heavy tailing that are not captured by the ADE model, which is based on the Fickian transport presumption (Di Dato *et al.*, 2017; Gao *et al.*, 2010; Neuman and Tartakovsky, 2009). Alternatively, many non-Fickian transport models have been presented, such as the mobile-immobile model (MIM) (Gao *et al.*, 2010), fractional-derivative ADE (FADE) models (Yin *et al.*, 2020), continuous-time random-walk (CTRW) models (Le Vot *et al.*, 2017), multi-rate mass transfer (MRMT) model (Babey *et al.*, 2015), and so on (Zheng *et al.*, 2019). Besides the MIM, FADE, CTRW and MRMT models, another approach to represent the heterogeneity is to use a scale-dependent dispersivity (or dispersion) in the ADE or MIM models (Gao *et al.*, 2010; Gelhar *et al.*, 1992; Wang *et al.*, 2020; You and Zhan, 2013).

At present, significant efforts have been put forward over many decades on developing advanced analytical models of radial dispersion. Some examples include the works of Hoopes and Harleman (1967), Gelhar and Collins (1971), Tang and Babu (1979), Moench and Ogata (1981), Chen (1985, 1986, 1991), Hsieh (1986), Tang and Peaceman (1987), Yates (1988), Falade and Brigham (1989), Novakowski (1992a), Philip (1994), Veling (2001, 2012), Huang and Goltz (2006), Chen *et al.* (2007, 2011a, 2012, 2017), Gao *et al.* (2009a), Cihan and Tyner (2011), Wang and Zhan (2013), Hsieh and Yeh (2014), Zhou *et al.* (2017), Wang *et al.* (2018, 2020),

Huang *et al.* (2019), Li *et al.* (2020), Shi *et al.* (2023), etc. The differences among the currently available analytical solutions for radial dispersion have been reviewed and summarized in Table 3.1. As one can see from this table, the mixing effect in the wellbore was ignored in all of the models except for Novakowski (1992a), Wang *et al.* (2018, 2020), and Shi *et al.* (2020). Only Chen *et al.* (2012), Hsieh and Yeh (2014), Huang *et al.* (2019), and Li *et al.* (2020) took the skin effect into account. The differences among the solutions of Tang and Babu (1979), Moench and Ogata (1981), Hsieh (1986), Tang and Peaceman (1987), Yates (1988), Cihan and Tyner (2011), and Chen *et al.* (2012) mainly consist of boundary conditions, source-injection types (instantaneous or continuous), and initial conditions.

Such analytical models can fulfill a host of tasks such as (1) prioritizing the importance of different controlling parameters through a sensitivity analysis; (2) benchmarking the numerical solutions to elucidate the possible numerical errors such as numerical dispersion and artificial oscillation which are notorious for advection-dominated transport problems; and (3) providing a quick screening tool before implementing a full-scale comprehensive study.

The analytical solutions of the above-mentioned models can be obtained by employing various mathematical techniques, such as the Fourier series method (Bharati *et al.*, 2019; Güven *et al.*, 1984), the Laplace transform method (Li *et al.*, 2020; Wang *et al.*, 2020; You and Zhan, 2013), the Green's function method (Leij *et al.*, 2000; Rivera *et al.*, 2016) and the finite integral transforms (Cleary and Adrian, 1973). Based on these analytical solutions, many parameter estimation methods have also been developed such as CXTFIT (Parker and Genuchten, 1984; Toride, 1995), Extended Kalman Filter (Welch and Bishop, 1995), UCODE (Poeter and Hill, 1999), CXTANNEAL (Li *et al.*, 1999), PEST (Doherty, 2002), WLSM (Wang *et al.*, 2013), and so on. Although the analytical solutions are usually computationally more efficient than the numerical solutions, and can be easily coupled into optimization algorithms for problems related to parameter estimation (Neuman and Mishra, 2012), the numerical method is more powerful than the analytical method for problems with complex initial and boundary conditions and heterogeneous aquifers of interest. Numerical errors could not be avoided easily for the MIM models of concern here, such as numerical dispersion and

Table 3.1. Summary of the current models for the radial dispersion around the recharge well.

Authors	Conceptual models	GE	ME	SCE	SKE	Method
Hoopes and Harleman (1967)	Confined aquifer	ADE	N	N	N	Approximated solution and finite-difference solution
Gelhar and Collins (1971)	Confined aquifer	ADE	N	N	N	A boundary layer approximation
Tang and Babu (1979), Moench and Ogata (1981), Hsieh (1986), Tang and Peaceman (1987), Yates (1988), Cihan and Tyner (2011)	Confined aquifer	ADE	N	N	N	Laplace transform
Chen (1985, 1991)	Leaky-confined aquifer	ADE	N	N	N	Laplace transform
Chen (1986)	Fracture aquifer	ADE	N	N	N	Laplace transform
Falade and Brigham (1989)	Confined aquifer	MIM	N	N	N	Laplace transform
Novakowski (1992a)	Leaky-confined aquifer	ADE	Y	N	N	Laplace transform
Philip (1994)	Confined aquifer	ADE	N	N	N	Finite-difference solution
Veling (2001), Chen *et al.* (2011b)	Confined aquifer	ADE	N	N	N	Generalized Hankel transform
Chen *et al.* (2007), Gao *et al.* (2009a)	Confined aquifer	ADE	N	Y	N	Laplace transform
Chen *et al.* (2012), Hsieh and Yeh (2014)	Confined aquifer	ADE	N	N	Y	Laplace transform

Wang and Zhan (2013)	Leaky-confined aquifer	ADE	N	N	N	Laplace transform
Zhou *et al.* (2017)	Fracture aquifer	MIM	N	N	N	Laplace transform
Chen *et al.* (2017)	Confined aquifer	MIM	N	N	N	Laplace transform
Wang *et al.* (2018)	Confined aquifer	ADE	Y	N	N	Laplace transform and Green's function method
Huang *et al.* (2019)	Confined aquifer	ADE	N	N	Y	Laplace transform
Li *et al.* (2020)	Confined aquifer	MIM	N	N	Y	Laplace transform
Shi *et al.* (2020)	Confined aquifer	ADE	Y	N	N	Approximation
Wang *et al.* (2020)	Confined aquifer	MIM	Y	N	N	Laplace transform and Green's function method
Shi *et al.* (2023)	Confined aquifer	MIM	Y	Y	Y	Laplace transform

Note: "GE", "ME", "SCE", and "SKE" represent governing equation, mixing effect, scale effect, and skin effect, respectively; "Y" and "N" represent whether the effect is considered or not.

numerical oscillation issues (Wang and Zhan, 2013; Zheng and Wang, 1999). Actually, some simplified assumptions used in the analytical and semi-analytical solutions can be relaxed in the numerical models, such as MODFLOW/MT3DMS (Harbaugh *et al.*, 2000; Zheng and Wang, 1999), FEFLOW (Diersch, 2014), SUTRA (Voss, 1984), STOMP (Nichols *et al.*, 1997), COMSOL Multiphysics (COMSOL Inc., Burlington, MA, USA), etc.

The purpose of this chapter is to present some advanced reactive transport models. More specifically, in Section 3.1, we introduced the general reactive transport models, followed by discussions on analytical models of radial tracer in Section 3.2. In Section 3.3, we introduce classical reactive transport models, which include the analytical models of reactive transport around injection wells (including the models of Hsieh and Yeh (2014), Li *et al.* (2020), Wang and Zhan (2013)), the two-stage single-well push-pull (SWPP) analytical models (involving the models of Huang *et al.* (2010), Wang *et al.* (2018), Shi *et al.* (2020, 2023)), as well as the multi-stage SWPP analytical models (including the models of Chen *et al.* (2017) and Wang *et al.* (2020)). Subsequently, we introduce the numerical models of radial tracer test, which involve the numerical models of reactive transport around injection wells (developed by Gan and Wang (2024)), the numerical models of multi-stage SWPP tests without regional groundwater flow (proposed by Wang *et al.* (2017) and Wang and Zhan (2019)), as well as the numerical models of SWPP tests with regional groundwater flow (presented by Jin *et al.* (2024)). Finally, we introduce the model of reactive transport in a groundwater circulation well system, which was proposed by Zhu *et al.* (2020). These models can be used as rapid screening techniques to estimate aquifer parameters linked with reactive transport.

3.2 Mathematical Models of Reactive Transport

3.2.1 *General ADE models for describing reactive transport*

3.2.1.1 *Governing equations*

The governing equation describing the fate and transport of contaminants of species k in 3D, transient groundwater flow systems can be written as follows (Zheng and Wang, 1999):

$$\frac{\partial(\theta C^k)}{\partial t} = \frac{\partial}{\partial x_i}\left(\theta D_{ij}\frac{\partial C^k}{\partial x_j}\right) - \frac{\partial}{\partial x_i}(q_i C^k) + q_s C_s^k + \sum R_n, \quad t > 0 \tag{3.1}$$

where C^k is the dissolved concentration of species k [Kg/m^3]; θ is the porosity of the subsurface medium [-]; q_i is the specific discharge [m/s]; D_{ij} is the hydrodynamic dispersion coefficient tensor [m^2/s]; q_s is the volumetric flow rate per unit volume of the aquifer representing fluid sources or sinks [1/s]; C_s^k is the concentration of the source or sink flux for species k [Kg/m^3]; $\sum R_n$ is the chemical reaction term [Kg/(m$^3 \cdot$ s)].

Equation (3.1) is the mathematical expression to simulate reactive transport on the macroscopic scale, which is essentially a mass balance statement. Namely, the term on the left-hand side of Eq. (3.1) represents the change in the mass at any given time, which is equal to the difference in the mass inflow and outflow due to hydrodynamic dispersion, advection, sink/source, and chemical reactions. The first term on the right-hand side of Eq. (3.1) represents hydrodynamic dispersion, the second term on the right-hand side of Eq. (3.1) represents advection, which describes the transport of miscible contaminants at the same velocity as the groundwater. The fluid sink/source term of the governing equation, $q_s C_s^k$, represents solute mass entering the model domain through sources or leaving the model domain through sinks.

The left-hand side of Eq. (3.1) could be expended into two terms:

$$\frac{\partial(\theta C^k)}{\partial t} = \theta\frac{\partial C^k}{\partial t} + C^k\frac{\partial\theta}{\partial t} = \theta\frac{\partial C^k}{\partial t} + q_s' C^k \tag{3.2}$$

where q_s' is the rate of change in transient groundwater storage.

The chemical reaction term in Eq. (3.1) can be employed to describe the effect of general biochemical and geochemical reactions on solute fate and transport. If considering only two basic types of chemical reactions, namely the reaction between aqueous and solid surfaces (sorption) and the first-order reaction, the chemical reaction term can be expressed as follows:

$$\sum R_n = -\rho_b\frac{\partial S^k}{\partial t} - \lambda_1\theta C^k - \lambda_2\rho_b S^k \tag{3.3}$$

where ρ_b is the bulk density of the subsurface medium; S^k is the concentration of species k sorbed on the subsurface solids; λ_1 is the first-order reaction rate for the dissolved phase; λ_2 is the first-order reaction rate for the sorbed (solid) phase.

Substituting Eqs. (3.2) and (3.3) into Eq. (3.1) and dropping the species index for simplicity of presentation, Eq. (3.1) can be rearranged and rewritten as

$$\frac{\partial C^k}{\partial t} + \rho_b \frac{\partial S^k}{\partial t} = \frac{\partial}{\partial x_i}\left(\theta D_{ij} \frac{\partial C^k}{\partial x_j}\right) - \frac{\partial}{\partial x_i}(q_i C^k) + q_s C_s^k - q_s' C^k$$
$$-\lambda_1 \theta C^k - \lambda_2 \rho_b S^k \tag{3.4}$$

Local equilibrium is often assumed for the various sorption processes (i.e., sorption is sufficiently fast compared to the transport time scale). Referring to the local equilibrium assumption, it is common to express Eq. (3.4) in the following form:

$$R\theta \frac{\partial C^k}{\partial t} = \frac{\partial}{\partial x_i}\left(\theta D_{ij} \frac{\partial C^k}{\partial x_j}\right) - \frac{\partial}{\partial x_i}(q_i C^k) + q_s C_s^k - q_s' C^k$$
$$-\lambda_1 \theta C^k - \lambda_2 \rho_b S^k \tag{3.5}$$

where R is referred to as the retardation factor [–], which is defined as

$$R = 1 + \frac{\rho_b}{\theta} \frac{\partial S^k}{\partial C^k} \tag{3.6}$$

When the assumption of local equilibrium is inappropriate, the sorption processes are typically represented by a first-order kinetic mass transfer equation.

Note that in the transport-governing equations described above, only a single porosity has been assumed. This porosity has been commonly referred to as "effective" porosity, which is generally smaller than the total porosity of the porous medium, reflecting the fact that some pore spaces may contain immobile water with zero groundwater seepage velocity.

3.2.1.2 *Dispersion*

Dispersion in porous media refers to the spreading of contaminants over a greater region than would be predicted solely from the average groundwater velocity vectors (Zheng and Wang, 1999). Dispersion is caused by mechanical dispersion, a result of deviations of actual velocity on a microscale from the average groundwater velocity, and by molecular diffusion driven by concentration gradients. Molecular diffusion is generally secondary and negligible, compared with the effects of mechanical dispersion, and only becomes important when groundwater velocity is very low. The sum of mechanical dispersion and molecular diffusion is termed hydrodynamic dispersion, or simply dispersion. The following mechanisms contribute to pore-scale dispersion:

(a) Within the single pores, the velocity profile is $\pm$ parabolic. Averaging over the cross-section of a single pore, the net mass flux consists of the advective flux using the mean velocity and concentration, the diffusive flux in the longitudinal direction, and a dispersive flux caused by correlation of concentration and velocity fluctuations (so-called Taylor-Aris dispersion). However, for typical pore diameters, the molecular diffusion coefficient is larger than the coefficient according to the third process.
(b) The diffusive exchange of solute between pores with flow and stagnant regions, such as dead-end pores, leads to additional spreading.
(c) The pores differ in their diameter, leading to differing velocities. The velocity along a particular path line change. That is, a particle released into a low-velocity region may be transferred into a high-velocity region by advection alone. In general, this will be a random process since the exact geometry of the pore network is random. The random advective sampling of the velocity distribution by the solute particles results in a dispersion coefficient that is proportional to the velocity.
(d) The presence of the grains also requires transverse deviations from the mean flow direction. Stochastic fluctuations of the transverse velocity component led to a dispersive mass flux in the direction orthogonal to the mean flow direction.

Although the dispersion mechanism is generally understood, the representation of dispersion phenomena in a transport model is the subject of intense continuing research. The treatment of mechanical dispersion as a Fickian process (in effect, addictive to diffusion) represents a pragmatic approach through which realistic transport calculations can be made without fully describing the heterogeneous velocity field, which, of course, is impossible to do in practice (Chen *et al.*, 2007; Gelhar and Collins, 1971; Le Borgne and Gouze, 2008; Wang *et al.*, 2017).

The hydrodynamic dispersion tensor in Eq. (3.1) D_{ij}, for an isotropic porous medium, is defined, according to (Bear, 2012, 2013), in the following component forms:

$$D_{xx} = \alpha_L \frac{\nu_x^2}{|v|} + \alpha_T \frac{\nu_y^2}{|v|} + \alpha_T \frac{\nu_z^2}{|v|} + D^* \tag{3.7a}$$

$$D_{yy} = \alpha_L \frac{\nu_y^2}{|v|} + \alpha_T \frac{\nu_x^2}{|v|} + \alpha_T \frac{\nu_z^2}{|v|} + D^* \tag{3.7b}$$

$$D_{zz} = \alpha_L \frac{\nu_z^2}{|v|} + \alpha_T \frac{\nu_x^2}{|v|} + \alpha_T \frac{\nu_y^2}{|v|} + D^* \tag{3.7c}$$

$$D_{xy} = D_{yx} = (\alpha_L - \alpha_T)\frac{\nu_x \nu_y}{|v|} \tag{3.7d}$$

$$D_{xz} = D_{zx} = (\alpha_L - \alpha_T)\frac{\nu_x \nu_z}{|v|} \tag{3.7e}$$

$$D_{yz} = D_{zy} = (\alpha_L - \alpha_T)\frac{\nu_y \nu_z}{|v|} \tag{3.7f}$$

where D_{xx}, D_{yy}, D_{zz} are the principal components of the dispersion tensor [m^2/s]; $D_{xy}, D_{xz}, D_{yx}, D_{yz}, D_{zx}, D_{zy}$ are the cross terms of the dispersion tensor [m^2/s]; α_L is the longitudinal dispersivity [m]; α_T is the transverse dispersivity [m]; D^* is the effective molecular diffusion coefficient [m^2/s]; v_x, v_y, v_z are the components of the velocity vector [m/s] along the x, y, and z axes, respectively; $|v| = \sqrt{v_x^2 + v_y^2 + v_z^2}$ is the magnitude of the velocity vector [m/s].

When the velocity vector is aligned with one of the coordinate axes, all the cross terms become zero.

Strictly speaking, the dispersion tensor defined by two independent dispersivities for isotropic media as in Eq. (3.7) is not valid for anisotropic porous media, which require five independent dispersivities (Bear, 2012). However, it is generally not feasible to obtain all five dispersivities in the field. As a practical alternative, the usual practice in transport modeling is to assume that the isotropic dispersion coefficient is also applicable to anisotropic porous media. In addition to the isotropic dispersion described above, the MT3DMS transport model supports an alternative form that allows the use of two transverse dispersivities, a horizontal transverse dispersivity, α_{TH}, and a vertical transverse dispersivity, α_{TV}, as proposed by (Burnett and Frind, 1987):

$$D_{xx} = \alpha_L \frac{\nu_x^2}{|v|} + \alpha_{TH} \frac{\nu_y^2}{|v|} + \alpha_{TV} \frac{\nu_z^2}{|v|} + D^* \tag{3.8a}$$

$$D_{yy} = \alpha_L \frac{\nu_y^2}{|v|} + \alpha_{TH} \frac{\nu_x^2}{|v|} + \alpha_{TV} \frac{\nu_z^2}{|v|} + D^* \tag{3.8b}$$

$$D_{zz} = \alpha_L \frac{\nu_z^2}{|v|} + \alpha_{TV} \frac{\nu_x^2}{|v|} + \alpha_{TH} \frac{\nu_y^2}{|v|} + D^* \tag{3.8c}$$

$$D_{xy} = D_{yx} = (\alpha_L - \alpha_{TH}) \frac{\nu_x \nu_y}{|v|} \tag{3.8d}$$

$$D_{xz} = D_{zx} = (\alpha_L - \alpha_{TV}) \frac{\nu_x \nu_z}{|v|} \tag{3.8e}$$

$$D_{yz} = D_{zy} = (\alpha_L - \alpha_{TV}) \frac{\nu_y \nu_z}{|v|} \tag{3.8f}$$

Equation (3.8) becomes equivalent to Eq. (3.7) when the two transverse dispersivities are set equal. Table 3.2 lists some respective values of longitudinal dispersivity.

3.2.1.3 *Advection*

The advection term of the transport equation, $\frac{\partial}{\partial x_i}(q_i C^k)$, describes the transport of miscible contaminants at the same velocity as the groundwater. For many field-scale contaminant transport problems, the advection term dominates over other terms. To measure the degree of advection domination, a dimensionless Peclet number is usually used, which is defined as

$$P_e = \frac{|v| L}{D} \tag{3.9}$$

where $|v|$ is the magnitude of the seepage velocity vector [m/s]; L is a characteristic length [m], commonly taken as the grid cell width; D is the dispersion coefficient [m^2/s].

Table 3.2. Summary of dispersivities from literature.

References	Aquifer material	Tracer and input	Velocity [m/d]	Scale of test [m]	Dispersivity (α_L), (α_{TH}), (α_{TV}) [m]
Adams and Gelhar (1992); Columbus, Mississippi	Very heterogeneous sand and gravel	Br^- Pulse	0.03∼0.5	200	(7.5), (···), (···)
Ahlstrom *et al.* (1977); Hanford, Washington	Glaciofluvial sands and gravels	^3H Contamination	/	20000	(30.5), (18.3), (···)
Bentley and Walter (1983); WIPP	Fractured dolomite	PFB, SCN Step	0.3	23	(5.2), (···), (···)
Bierschenk (1959), Cole (1972); Hanford, Washington	Glaciofluvial sands and gravels	Fluorescein Pulse	26 31	3500 4000	(6), (···), (···) (460), (···), (···)
Bredehoeft and Pinder (1973); Brunswick, Georgia	Limestone	Cl^- Contamination	/	2,000	(170), (52), (···)
Claassen and Cordes (1975); Amargosa, Nevada	Fractured dolomite and limestone	^3H Pulse	0.14∼3.4	122	(15), (···), (···)
Daniels (1981); Nevada Test Site	Alluvium derived from tuff	^3H Contamination	0.04	91	(10∼30), (···), (···)
Dieulin (1981); Le Cellier (Lozere, France)	Fractured granite	Cl^-, I^- Pulse	3	5	(0.5), (···), (···)
Dieulin (1981); Torchy, France	Alluvial deposits	Cl^- Pulse	0.5	15	(3), (···), (···)
Egboka *et al.* (1983); Borden	Glaciofluvial sand	^3H Environmental	0.01∼0.04	600	(30∼60), (···), (···)

Fenske (1973); Tatum Salt Dome, Mississippi	Limestone	^{3}H Pulse	1.2	91	(11.6)), (···), (···)
Freyberg (1986); Borden	Glaciofluvial sand	Br^{-}, Cl^{-} Pulse	0.09	90	(0.43), (0.039), (···)
Fried and Ungemach (1971); Rhine aquifer	Sand, gravel, and cobbles	Cl^{-} Pulse	/	6	(11), (···), (···)
Fried (1975); Rhine aquifer (salt mines) southern Alsace, France	Alluvial; mixture of sand, gravel, and pebbles with clay lenses	Cl^{-} Contamination	/	800	(15), (1), (···)
(Fried, 1975); Lyons, France (sanitary landfill)	Alluvial, with sand and gravel and slightly stratified clay lenses	EC Contamination	5.0	600∼1000	(12), (4), (···)
Garabedian (1987); Cape Cod, Massachusetts	Medium to coarse sand with some gravel overlying silty sand and till	Br^{-} Pulse	0.43	250	(0.96), (0.018), (0.0015)
Gelhar and Leonhart (1982); Hanford, Washington	Brecciated Basalt interflow zone	^{131}I Pulse	/	17.1	(0.60), (···), (···)
Goblet (1982); Site B, France	Fractured granite	RhWt, SrCl Pulse	84	17	(2), (···), (···)
Grove (1977); NRTS, Idaho	Basaltic lava and sediments	Cl^{-} Contamination		20000	(91), (91), (···)
Grove and Beetem (1971); Eddy County (near Carlsbad), New Mexico	Fractured dolomite	^{3}H Step	3.5	55	(38.1), (···), (···)

(Continued)

Table 3.2. (*Continued*)

References	Aquifer material	Tracer and input	Velocity [m/d]	Scale of test [m]	Dispersivity (α_L), (α_{TH}), (α_{TV}) [m]
Gupta and Tanji (1976); Sutter Basin, California	Sandstone, shale, sand, and alluvial sediments	Cl^- Environmental	/	50000	(80∼100), (8∼20), (···)
Halevy and Nir (1962), Lenda and Zuber (1970); Nahal Oren, Israel	Dolomite	^{60}Co Pulse	4.0	250	(6), (···), (···)
Harpaz (1965); Southern coastal plain, Israel	Sandstone with silt and clay layers	Cl^- (step)	14	28	(0.1∼1.0), (···), (···)
Helweg; Helweg and Labadie (1977); Bonsall subbasin, California		TDS Contamination		14000	(30.5), (9.1), (···)
Hoehn (1983); lower Glatt Valley, Switzerland	Layered gravel and silty sand	Uranine Pulse	3.4	4.4	(0.1), (···), (···)
			1.8	4.4	(0.01), (···), (···)
			1.2	4.4	(0.2), (···), (···)
			8.6	10.4	(0.3), (···), (···)
			4.1	10.4	(0.04), (···), (···)
			1.7	10.4	(0.7), (···), (···)

Hoehn and Santschi (1987); Lower Glatt Valley, Switzerland	Layered gravel and silty sand	Uranine Pulse	1.5	4.4	(1.1), (···), (···)
			3.2	10.4	(1.2), (···), (···)
		^3H Environmental	5.6	100	(6.7), (···), (···)
			3.9	110	(10.0), (···), (···)
			3.2	500	(58.0), (···), (···)
Huyakorn *et al.* (1986); Mobile, Alabama	Layered medium sand	Br^- Pulse	/	38.3	(4.0), (···), (···)
Iris *et al.* (1985); Campuget (Gard), France	Alluvial deposit	Heat Pulse	/	40	(3), (1.5)), (···)
Ivanovich and Smith (1978); Dorset, England	Fractured chalk	^{82}Br Pulse	57.6	8	(3.1), (···), (···)
	Chalk	^{82}Br Pulse	9.6	8	(1.0), (···), (···)
Kies (1982); New Mexico State University, Las Cruces	Fluvial sands	NO_3^- Pulse	/	25	(1.6), (0.76), (···)
Klotz *et al.* (1980); Dormach, Germany	Fluvioglacial gravels	^{82}Br, Uranine Pulse	20	10	(5), (1.9), (···)
Konikow (1976); Rocky Mountain Arsenal	Alluvium	Cl^- Contamination	/	13000	(30.5), (···), (···)
Konikow and Bredehoeft (1974); Arkansas River valley (at La Junta, Colorado)	Alluvium, inhomogeneous clay, silt, sand and gravel	Dissolved solids Contamination	/	18000	(30.5), (9.1), (···)

(*Continued*)

Table 3.2. (*Continued*)

References	Aquifer material	Tracer and input	Velocity [m/d]	Scale of test [m]	Dispersivity (α_L), (α_{TH}), (α_{TV}) [m]
Kreft *et al.* (1974); Poland	Sand	^{131}I Pulse	29	5∼6	(0.18), (···), (···)
Kreft *et al.* (1974); Zn-Pb deposits, Poland	Fractured dolomite	^{131}I Pulse	7.5	22	(44∼110), (···), (···)
	Fractured dolomite	^{131}I Pulse	100 60.1 22.7	21.3	(2.1), (···), (···)
Kreft *et al.* (1974); Sulfur deposits, Poland	Limestone	^{58}Co Pulse	10	27	(2.7∼27), (···), (···)
	Limestone	^{58}Co Pulse	10.8 8.6	41.5	(20.8), (···), (···)
Lau *et al.* (1957); University of California, Berkeley	Sand and gravel with clay lenses	Cl^- Step	7	19	(2∼3), (···), (···)
Lee *et al.* (1980); Perch Lake, Ontario, (lakebed)	Sand	Cl^- Pulse	0.14	≤ 6	(0.012), (···), (···)
Leland and Hillel (1981); Amherst, Massachusetts	Fine sand and glacial till	Cl^- Pulse	0.3∼0.6	4	(0.05∼0.07), (···), (···)
Mercado (1966); Yavne region, Israel	Sand and sandstone with some silt and clay	^{60}Co, Cl^- Step	0.84∼3.4	≤ 115 (observation wells)	(0.5∼1.5), (···), (···) (injection phase)

Meyer *et al.* (1981); Koeberg Nuclear Power Station, South Africa	Sand	^{131}I Pulse	0.12	2~8	(0.01), (0.03), (0.01) for layers; (0.42), (···), (···) for depth average
Molinari and Peaudecerf (1977), Sauty (1977); Bonnaud, France	Sand	I^-	2.7	13	(0.79), (···), (···)
		^{3}H	1.0	13	(1.27), (···), (···)
		^{131}I	2.4	13	(0.72), (···), (···)
		^{131}I	1.0	26	(2.23), (···), (···)
		^{131}I	2.0	33.2	(1.94), (0.11), (···)
		^{131}I Pulse	2.0	32.5	(2.73), (0.11), (···)
Moltyaner and Killey (1988); Twin Lake aquifer (Chalk River)	Fluvial sand	^{131}I Pulse	1.2	40	(0.06~0.16), (···), (0.0006~0.002)
Naymik and Barcelona (1981); Illinois (Morgan County)	Unconsolidated sand and gravel	NH_3 Contamination	/	16.4	(2.13~3.35), (0.61~ 0.915), (···)
New Zealand Ministry of Works and Development [1977] Heretaunga aquifer, New Zealand: **Roys Hill site**	Gravel with cobbles	^{131}I, RhWt, ^{82}Br, Cl^-, *E.Coli* Pulse	150~200	54~59	(1.4~ 11.5), (0.1~3.3), (0.04~0.10)

(Continued)

Table 3.2. (*Continued*)

References	Aquifer material	Tracer and input	Velocity [m/d]	Scale of test [m]	Dispersivity (α_L), (α_{TH}), (α_{TV}) [m]
Flaxmere site 2	Alluvium (gravels)	RhWt, ^{82}Br Pulse	20∼25	25	(0.3∼1.5), (⋯), (0.06)
Hastings City rubbish dump	Alluvium (gravels)	Cl^- Contamination	20	290	(41), (10), (0.07)
Oakes and Edworthy (1977); Clipstone, United Kingdom	Sandstone	^{82}Br Pulse	5.6, 4.0	6	(0.16), (0.38), (⋯)
			9.6	3	(0.31), (⋯), (⋯)
		Cl^-, I^- Pulse	2.4, 3.6	6	(0.6), (⋯), (⋯)
				3	(0.6), (⋯), (⋯)
Papadopulos and Larson (1978); Mobile, Alabama	Medium to fine sand interspersed with clay and silt	Heat Step	0.05	57.3	(1.5), (⋯), (⋯)
Pickens and Grisak (1981b); Chalk River	Sand	^{51}Cr Step	0.15	8	(0.5), (⋯), (⋯)
	Sand	^{131}I Step	0.15	3	(0.03), (⋯), (⋯)
Pinder (1973); Long Island	Glacial outwash	Cr^{+6} Contamination	0.43	1000	(21.3), (4.2), (⋯)
Rabinowitz and Gross (1972); Roswell Basin, New Mexico	Fractured limestone	^3H Environmental	11∼21	32000	(20∼23), (⋯), (⋯)
Rajaram and Gelhar (1991); Borden	Glaciofluvial sand	Br^-, Cl^- Pulse	0.09	90	(0.50), (0.05), (0.0022)

Roberts (1985); Palo Alto Bay lands	Sand, gravel, and silt	Cl^- Step	15.5	11	(5), (···), (···)
			12.0	20	(2), (···), (···)
			3.5	40	(8), (···), (···)
			25.6	16	(4), (···), (···)
			7.9	43	(11), (···), (···)
Robertson (1974); Robertson and Barraclough (1973); NRTS, Idaho	Basaltic lava and sediments	Cl^- Contamination	1.5~8	20000	(910), (1370), (···)
Robson (1974); Barstow, California	Alluvial sediments	Cl^- Step	3	6.4	(15.2), (···), (···)
		TDS Contamination		10000	(61), (18), (···)
Robson (1978); Barstow, California	Alluvial sediments	TDS Contamination	/	3200	(61), (···), (0.2)
Rousselot *et al.* (1977); Byles-Saint Vulbas near Lyon, France	Clay, sand, and gravel	I^- Pulse	18	9.3	(6.9), (···), (···)
			11.5, 3.8	5.3	(0.3), (0.7), (···)
			46.7, 16	10.7	(0.46), (1.1), (···)
			24	7.1	(0.37), (···), (···)
Sauty (1977); Corbas, France	Sand and gravel	I^- Pulse	125, 100	25	(11), (1.25), (···)
			15.5, 78	50	(25), (6.25), (···)
			6.9	150	(12.5), (···), (···)

(*Continued*)

Table 3.2. (*Continued*)

References	Aquifer material	Tracer and input	Velocity [m/d]	Scale of test [m]	Dispersivity (α_L), (α_{TH}), (α_{TV}) [m]
Sauty *et al.* (1978); Bonnaud, France	Sand	Heat Step	/	13	(1.0), (···), (···)
Segol and Pinder (1976); Cutler area, Biscayne Bay aquifer, Florida	Fractured limestone and calcareous sandstone	Cl^- Environmental	20	490	(6.7), (···), (0.67)
Sudicky *et al.* (1983); Borden	Glaciofluvial sand	Cl^- Pulse	0.07∼0.25	11	(0.08), (0.03), (···)
				0.75	(0.01), (0.005), (···)
Sykes *et al.* (1982, 1983); Borden	Sand	Cl^- Pulse	/	700	(7.6), (···), (0.31)
Sykes *et al.* (1983); Mobile, Alabama	Sand, silt, and clay	Heat Pulse	0.05	57.3	(0.76), (···), (0.15)
Vaccaro and Bolke (1983); Spokane aquifer, Valocchi *et al.* (1981); Palo Alto Bay lands	Glaciofluvial Sand and gravel	Cl^- Contamination	0.003∼2.8	43400	(91.4), (27.4), (···)
Valocchi *et al.* (1981); Palo Alto Bay lands	Sand, gravel, and silt	Cl^- Step	27	16	(1.0), (0.1), (···)
Walter (1983); WIPP	Fractured dolomite	MTFMB, PFBMFB, para- FB Pulse	4.7, 2.4	30	(10∼15), (···), (···)

Webster *et al.* (1970); Savannah River Plant, South Carolina	Crystalline, fractured schist and gneiss	^{85}Sr, ^{85}Br Pulse	1.3 21.4	538	(134), (···), (···)
Werner (1983); Hydrothermal Test Site, Aefligen, Switzerland	Gravel	Heat Step	9.1	700	(130∼234), (···), (···)
				37	(131), (···), (···)
				105	(208), (···), (···)
				200	(234), (···), (···)
Wiebenga *et al.* (1967); Lenda and Zuber (1970); Burdekin Delta, Australia	Sand and gravel	^{131}I, ^{3}H Pulse	29	18.3	(0.26), (···), (···)
Wilson (1971); Robson (1974); Tucson, Arizona	Unconsolidated gravel, sand, and silt	Cl^- Step	/	79.2	(15.2), (···), (···)
		Cl^- Step		4.6	(0.55), (···), (···)
Wood (1981); Aquia Formation, southern Maryland	Sand	Na^+ Environmental	0.0003 ∼0.0007	100000	(5600–40000), (···), (···)
Wood and Ehrlich (1978); Bassett *et al.* (1981); Lubbock, Texas	Sand and gravel	I^- Pulse	78	1.52	(0.015), (···), (···)

Source: From Gelhar *et al.* (1992).

In advection-dominated problems, also referred to as sharp front problems, the Peclet number has a large value. For pure advection problems, the Peclet number approaches infinity.

3.2.1.4 *Sinks and sources*

The fluid sink/source term of the governing equation, q_sC_s, represents solute mass entering the model domain through sources or leaving the model domain through sinks. The term $q_s'C$ may be viewed as the "internal" sink/source term, which represents the change in solute mass storage caused by the change in transient groundwater storage. It does not cause the mass to leave or enter the model domain.

Sinks or sources may be classified as areally distributed sinks or sources or as point sinks or sources. The areally distributed sinks or sources include recharge and evapotranspiration. The point sinks or sources include wells, drains, and rivers. Constant-head and general head-dependent boundaries in the flow model are also treated as point sinks or sources because they function in the same fashion as wells, drains, or rivers in the transport model.

For sources, it is necessary to specify the concentration of source water. For sinks, the concentration of sink water is generally equal to the concentration of groundwater in the aquifer at the sink location and cannot be specified. However, there is one exception where the concentration of sinks may differ from that of groundwater. The exception is evapotranspiration, which may be assumed to take only pure water away from the aquifer so that the concentration of the evapotranspiration flux is zero.

3.2.1.5 *Chemical reactions*

The reaction transport equation is capable of describing equilibrium-controlled linear or nonlinear sorption, nonequilibrium (rate-limited) sorption, and first-order reaction that can represent radioactive decay or provide an approximate representation of biodegradation. The general formulation designed to model rate-limited sorption can also be used to model kinetic mass transfer between the mobile and immobile domains in a dual-domain advection-diffusion model. More sophisticated chemical reactions can be modeled through add-on reaction packages.

(a) Equilibrium-controlled linear or nonlinear sorption

Sorption refers to the mass transfer process between the contaminants dissolved in groundwater (aqueous phase) and the contaminants sorbed on the porous medium (solid phase). It is generally assumed that equilibrium conditions exist between the aqueous-phase and solid-phase concentrations and that the sorption reaction is fast enough, relative to groundwater velocity, to be treated as instantaneous. The functional relationship between the dissolved and sorbed concentrations under a constant temperature is referred to as the sorption isotherm. Equilibrium-controlled sorption isotherms are generally incorporated into the transport model through the use of the retardation factor as defined in Eq. (3.6). Three types of equilibrium-controlled sorption isotherms (linear, Freundlich, and Langmuir) are considered in the MT3DMS transport model.

(a) The linear sorption isotherm assumes that the sorbed concentration (S) is directly proportional to the dissolved concentration (C):

$$S = K_d C \tag{3.10}$$

where K_d is the distribution coefficient [m^3/Kg]. The retardation factor is thus given by

$$R = 1 + \frac{\rho_b}{\theta}\frac{\partial S}{\partial C} = 1 + \frac{\rho_b}{\theta}K_d \tag{3.11}$$

(b) The Freundlich isotherm is a nonlinear isotherm which can be expressed in the following form:

$$S = K_f C^a \tag{3.12}$$

where K_f is the Freundlich constant [(m^3/Kg)^a] and a is the Freundlich exponent [–].

Both K_f and a are empirical coefficients. When the exponent a is equal to unity, the Freundlich isotherm is equivalent to the linear isotherm. The retardation factor for the Freundlich isotherm is defined accordingly as

$$R = 1 + \frac{\rho_b}{\theta}\frac{\partial S}{\partial C} = 1 + \frac{\rho_b}{\theta}aK_f C^{a-1} \tag{3.13}$$

(c) Another nonlinear sorption isotherm is the Langmuir isotherm, which is described by the equation:

$$S = \frac{K_l S_t C}{1 + K_l C} \tag{3.14}$$

where K_l is the Langmuir constant [m^3/Kg]; S_t is the total concentration of sorption sites available.

The retardation factor defined for the Langmuir isotherm is

$$R = 1 + \frac{\rho_b}{\theta}\frac{\partial S}{\partial C} = 1 + \frac{\rho_b}{\theta}\left[\frac{K_l S_t}{(1 + K_l C)^2}\right] \tag{3.15}$$

The retardation factor is an important parameter for reactive transport, and it refers to the ratio of the velocity of a conservative tracer over the velocity of a partitioning tracer. Schroth *et al.* (2001) proposed a simplified method of the SWPP test to estimate the values of this parameter, and the computational efficiency was much better than the traditional method, by which the parameter was determined by directly best fitting the observed BTCs. Cassiani *et al.* (2005) pointed out that the simplified method of Schroth *et al.* (2001) was based on the approximate solutions of Gelhar and Collins (1971), and it was limited by the condition that the travel distance of chemicals was much larger than the longitudinal dispersivity. Then, they used a numerical solution instead of the analytical solution in the simplified model of Schroth *et al.* (2001). Actually, both models of Schroth *et al.* (2001) and Cassiani *et al.* (2005) ignored the mixing effect in the wellbore, which may be problematic.

(b) Nonequilibrium sorption

When the local equilibrium assumption is not valid, it is assumed that the sorption process can be represented through a first-order reversible kinetic reaction as follows:

$$\rho_b \frac{\partial S}{\partial t} = \beta\left(C - \frac{S}{K_d}\right) \tag{3.16}$$

where β is the first-order mass transfer rate between the dissolved and sorbed phases [1/s] and K_d is the distribution coefficient for the sorbed phase as defined previously for the linear sorption.

Equation (3.16) needs to be solved simultaneously with the transport governing equation to obtain solutions of solute transport affected by nonequilibrium sorption. As the mass transfer rate β increases (i.e., the sorption process becomes increasingly faster), and the nonequilibrium sorption approaches the equilibrium-controlled linear sorption as defined in Eq. (3.10). For very small values of β, the exchange between the aqueous and solid phases is so slow that sorption becomes essentially negligible.

3.2.1.6 *Radioactive decay or biodegradation*

The first-order irreversible rate reaction term included in the governing equation, $-(\lambda_1\theta C + \lambda_2\rho_b S)$, represents the mass loss of both the dissolved phase (C) and the sorbed phase (S). The rate constant is usually given in terms of the half-life:

$$\lambda = \ln(2)/t_{1/2} \tag{3.17}$$

where $t_{1/2}$ is the half-life of radioactive or biodegradable materials (i.e., the time required for the concentration to decrease to one-half of the original value).

For radioactive decay, the reaction generally occurs at the same rate in both phases. For biodegradation, however, it has been observed that certain reactions occur only in the dissolved phase. That is why two different rate constants may be needed. It should be noted that various biodegradation processes in the subsurface are usually more complex than those described by the first-order irreversible rate reaction. Other reaction packages for MT3DMS are either currently available or under development for modeling more complex types of biochemical and geochemical reactions (Zheng and Wang, 1999).

3.2.2 *General MIM models for describing reactive transport*

Solute transport in fractured media or extremely heterogeneous porous media may be conceptualized as a dual-domain system; transport is primarily by advection through the fractures or zones of high hydraulic conductivity filled with mobile water (mobile domain), whereas transport is primarily by diffusion through the nonfractured matrix, or zones of low hydraulic conductivity filled with immobile or relatively stagnant water (immobile domain). The governing equations for the dual-domain system can be expressed, without explicit consideration of sorption, as follows:

$$\theta_m \frac{\partial C_m}{\partial t} + \theta_{im} \frac{\partial C_{im}}{\partial t} = \frac{\partial}{\partial x_i}\left(\theta_m D_{ij} \frac{\partial C_m}{\partial x_j}\right) - \frac{\partial}{\partial x_i}(\theta_m \nu_i C_m) - \omega(C_m - C_{im}) + q_s C_s - q_s' C_m - \lambda_{1,m}\theta_m C_m - \lambda_{1,im}\theta_{im} C_{im} \tag{3.18a}$$

$$\theta_{im} \frac{\partial C_{im}}{\partial t} = \omega(C_m - C_{im}) - \lambda_{1,im}\theta_{im} C_{im} \tag{3.18b}$$

where C_m is the dissolved concentration in the mobile domain (referred to as the mobile liquid phase) [Kg/m^3]; C_{im} is the dissolved concentration in the immobile domain (referred to as the immobile liquid phase) [Kg/m^3]; θ_m is the porosity of the mobile domain [–]; θ_{im} is the porosity of the immobile domain [–], or the difference between the total and mobile porosities ($\theta_{im} = \theta - \theta_m$); $\lambda_{1,m}$ is the first-order reaction rate for the mobile liquid phase [1/s]; $\lambda_{1,im}$ is the first-order reaction rate for the immobile liquid phase [1/s], ω is the first-order mass transfer rate between the mobile and immobile domains [1/s].

Equation (3.18a) is a statement of mass conservation for the total mass, while Eq. (3.18b) is for the mass in the immobile domain. If sorption must be considered for solute transport in a dual-domain system, Eqs. (3.18a) and (3.18b) can be rewritten as

$$\theta_m \frac{\partial C_m}{\partial t} + f\rho_b \frac{\partial S_m}{\partial t} + \theta_{im} \frac{\partial C_{im}}{\partial t} + (1-f)\rho_b \frac{\partial S_{im}}{\partial t} = \frac{\partial}{\partial x_i}\left(\theta_m D_{ij} \frac{\partial C_m}{\partial x_j}\right) - \frac{\partial}{\partial x_i}(\theta_m \nu_i C_m) - \omega(C_m - C_{im}) + q_s C_s - q_s' C_m - \lambda_{1,m}\theta_m C_m - \lambda_{1,im}\theta_{im} C_{im} - \lambda_{2,m} f \rho_b S_m - \lambda_{2,im}(1-f)\rho_b S_{im} \tag{3.18c}$$

$$\theta_{im} \frac{\partial C_{im}}{\partial t} + (1-f)\rho_b \frac{\partial S_{im}}{\partial t} = \omega(C_m - C_{im}) - \lambda_{1,im}\theta_{im} C_{im} - \lambda_{2,im}(1-f)\rho_b S_{im} \tag{3.18d}$$

where S_m is the sorbed concentration in the mobile domain (referred to as the mobile sorbed phase) [Kg/m^3], S_{im} is the sorbed concentration in the immobile domain (referred to as the immobile sorbed phase), f is the fraction of sorption sites that are in contact with the mobile liquid phase dimensionless (f may be approximately set equal to θ_m/θ), $\lambda_{2,m}$ is the first-order reaction rate for the mobile sorbed phase [1/s]; $\lambda_{2,im}$ is the first-order reaction rate for the immobile sorbed phase [1/s].

If a linear equilibrium-controlled sorption isotherm can be assumed, Eqs. (3.18c) and (3.18d) are simplified as follows, assuming that the sorption coefficient is the same in the mobile and immobile domains:

$$\begin{aligned}
&\theta_m R_m \frac{\partial C_m}{\partial t} + \theta_{im} R_{im} \frac{\partial C_{im}}{\partial t} \\
&\quad = \frac{\partial}{\partial x_i}\left(\theta_m D_{ij} \frac{\partial C_m}{\partial x_j}\right) - \frac{\partial}{\partial x_i}(\theta_m \nu_i C_m) - \omega(C_m - C_{im}) \\
&\qquad + q_s C_s - q_s' C_m - \lambda_{1,m}\theta_m C_m - \lambda_{1,im}\theta_{im} C_{im} - \lambda_{2,m} f \rho_b K_d C_m \\
&\qquad - \lambda_{2,im}(1-f)\rho_b K_d C_{im}
\end{aligned} \tag{3.18e}$$

$$\begin{aligned}
&\theta_{im} R_{im} \frac{\partial C_{im}}{\partial t} \\
&\quad = \omega(C_m - C_{im}) - \lambda_{1,im}\theta_{im} C_{im} - \lambda_{2,im}(1-f)\rho_b K_d C_{im}
\end{aligned} \tag{3.18f}$$

where $R_m = 1 + f\rho_b K_d/\theta_m$ is retardation factor [-] for the mobile domain and $R_{im} = 1 + (1-f)\rho_b K_d/\theta_{im}$ is retardation factor [-] for the immobile domain.

With the dual-domain conceptual model, the dispersion coefficient D_{ij} in Eqs. (3.18a), (3.18c), and (3.18e) may be set equal or close to the molecular diffusion coefficient since mechanical dispersion is introduced through mass transfer between the mobile and immobile domains. As the mass transfer rate ω increases, i.e., the exchange between the mobile and immobile domains becomes increasingly fast; the dual-domain model functions more and more like a single-domain model whose porosity approaches the total porosity of the porous medium. On the other end of the spectrum, as the mass transfer rate ω approaches zero, the dual-domain model also becomes equivalent to a single-domain model but with the porosity of the single-domain model approaching the porosity of the mobile domain.

3.2.3 *The relationship between ADE and MIM models*

Since the effective porosity cannot be readily measured in the field due to the complexity of the pore structure, rather, it generally must be interpreted as the lumped parameter, which, in model calibration, gives the closest representation of both plume movement and observed solute accumulation effects. In some cases, such as transport in fractured media or extremely heterogeneous porous media, it may be more appropriate to use a dual-porosity approach, defining a primary porosity for those pore spaces filled with mobile water where advection is the predominant means of transport and a secondary porosity for those pore spaces filled with immobile water where transport is primarily by molecular diffusion. The exchange between the mobile and immobile domains can be defined through a kinetic mass transfer equation similar to that used to describe nonequilibrium sorption (Zheng and Wang, 1999).

In other words, the MIM model is an extension of the ADE model by considering both flowing and stagnant regions in porous media and mass transfer between them (van Genuchten and Wierenga, 1976). Namely, the MIM model becomes equivalent to the ADE model when the θ_{im} and ω are set to zero.

3.3 Analytical Models of Radial Tracer Test

3.3.1 *Reactive transport around the injection well*

The mechanism of radial dispersion is essential for understanding reactive transport in the subsurface and for estimating aquifer parameters required in the optimization design of remediation strategies. A well-aquifer system with radial dispersion comprises a wellbore, skin zone, and aquifer formation zone, as shown in Figure 3.1. The skin zone refers to the disturbed region around the well caused by drilling and construction practices or well completion (Chen *et al.*, 2012; Huang *et al.*, 2019; Li *et al.*, 2019, 2020; Yeh and Chang, 2013). It is spatially between the well screen and the aquifer formation zone. Correspondingly, the injected solute may experience three processes from the wellbore to the aquifer formation zone.

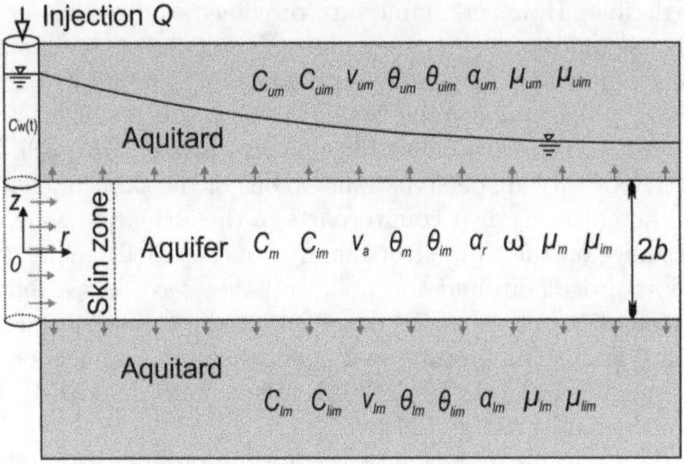

Figure 3.1. The schematic diagram of the radial dispersion in the aquifer-aquitard system.

First, the injected solute goes through a mixing process with native (or pre-injection) water in the wellbore at the early injection stage, which is called the mixing effect. Probably due to the small radius of the well, the mixing effect has been overlooked by almost all the analytical solutions mentioned above except Novakowski (1992a), Wang *et al.* (2018, 2020), and Shi *et al.* (2020, 2023), e.g., either by assuming that the well radius was infinitesimal or assuming that the solute concentration in the wellbore was the same as the concentration of the injected solution (Hoopes and Harleman, 1967; Veling, 2011; Zhou *et al.*, 2017). Consequently, the solutions developed without considering the wellbore mixing effect may overestimate concentration values in both the wellbore and the aquifer (Novakowski, 1992a; Wang *et al.*, 2018, 2020). The reason is that the solute concentration in the wellbore is initially zero (when the aquifer is free of solute before the injection) and then increases steadily until it reaches the maximum, which is equal to the concentration of the injected solution.

Second, the solute enters the skin zone after leaving the wellbore. Compared with the aquifer formation zone of interest, the dimension of the skin zone is much smaller, e.g., ranging from 0.1 m to several meters, and it is ignored or included in the wellbore. In other words, the effect of the skin zone on radial dispersion (named as skin effect) was negligible. However, numerous previous studies demonstrated that the existence of a skin zone might significantly alter the mechanism of groundwater flow and solute transport around a well (Chen *et al.*, 2012; Hsieh and Yeh, 2014; Li *et al.*, 2019, 2020; Yeh and Chang, 2013). This is because the physical properties (such as permeability, porosity, dispersivity, and so on) of the skin zone are often vastly different from their counterparts in the formation zone. Previously, studies on the skin effect mainly concentrated on the groundwater flow process around the well, and they gave less attention to solute transport processes. To date, few studies have considered the skin effect among the above-mentioned analytical models on radial dispersion, such as Chen *et al.* (2012), Hsieh and Yeh (2014), Huang *et al.* (2019), and Li *et al.* (2020).

Third, the solute moves into the formation zone from the skin zone by advective, diffusive, and dispersive processes.

Recognizing the importance of aquitards in subsurface transport, hydrogeologists have made considerable effort to investigate transport in aquifer-aquitard systems in the past several decades (Hendry *et al.*, 2004; Liu *et al.*, 2007; Sudicky *et al.*, 1985; Zhan *et al.*, 2009a, 2009b). The early investigation of transport in an aquifer-aquitard system adopts approaches commonly employed for studying solute transport in a fracture-matrix system, with two important assumptions. Firstly, mass transport in the rock matrix is diffusive only (without advection). This assumption is reasonable because fluid flow inside the rock matrix is often negligible. Secondly, solute mixing across the fracture aperture is instantaneous; i.e., any variation of solute concentration along the aperture thickness is ignored. As a result of these two assumptions, the diffusive mass flux across the fracture-matrix interface is described as a sink/source term averaged over the fracture aperture in the governing equation of transport in the fracture (Tang *et al.*, 1981; Tang and Aral, 1992). This kind of approach leads to the so-called AA solution in some recent studies (Zhan *et al.*, 2009a, 2009b), where "AA" is the abbreviation of the "averaged approximation".

However, some fundamental differences between an aquifer-aquitard system and a fracture-matrix system could make the above two assumptions invalid for solute transport in an aquifer-aquitard system. First, groundwater leakage across an aquitard is likely to occur if a considerable hydraulic gradient exists across the aquitard, so the advective transport in the aquitard cannot be neglected. Second, the thickness of an aquifer is much greater than the fracture aperture and solute across the aquifer thickness is rarely quickly mixed. Therefore, the variation of solute concentration along the aquifer thickness should be considered specifically. To overcome the problems of applying the models developed for a fracture-matrix system to an aquifer-aquitard system, several endeavors have been undertaken. Sudicky *et al.* (1985) and Starr *et al.* (1985) proposed a method that did not treat the solute transport across the aquifer-aquitard interface as a thickness-averaged source/sink term in the governing equation. Instead, they dealt with two coupled governing equations: one was the diffusive equation in the aquitard, and the other was the ADE in the aquifer, supplemented by the continuity of concentration and mass flux across the aquifer-aquitard interface. There are two weaknesses in this approach. One is that the advection in the aquitard is not considered. The second is that the longitudinal dispersion in the aquifer is not included. These two neglected processes could be important; in particular, the longitudinal dispersion in the aquifer could be the primary transport process. To improve the models of Sudicky *et al.* (1985) and Starr *et al.* (1985), Zhan *et al.* (2009a, 2009b) established a new approach that specifically took into account the advective transport in the aquitard and the longitudinal dispersion in the aquifer, in addition to other factors already considered by Sudicky *et al.* (1985) and Starr *et al.* (1985). All the studies mentioned above only dealt with unilateral flow in the aquifer.

As pointed out at the beginning of the introduction, radial rather than unilateral flow is of concern for many problems. A unique feature of radial dispersion in a homogeneous aquifer is that flow velocity is inversely proportional to the radial distance. Consequently, the mechanical dispersion coefficient becomes inversely proportional to the radial distance if the radial dispersivity is constant. Hoopes and Harleman (1967), Tang and Peaceman (1987), and Hsieh (1986) investigated radial dispersion in an aquifer and developed useful analytical and numerical solutions, but the aquitard bounding the aquifer was assumed to be non-penetrable for the solute, not even with the diffusion. This is certainly not true because molecular diffusion will always exist and will transport some solute mass between the aquifer and aquitard, even without any flow involved. Such a process was proven to be important for assessing the long-term groundwater remediation plans (Hendry *et al.*, 2004; Mazurek *et al.*, 2011). To consider the aquitard in radial dispersion, Chen (1985) developed an analytical solution for solute transport in an aquifer-aquitard system by treating the mass flux across the aquifer-aquitard interface as a thickness averaged source/sink term in the governing equation of transport in the aquifer (the AA model), similar to what has been done in the fracture-matrix system described above. Chen (1985) did not consider advection in the aquitard. Recently, Barbour *et al.* (2012) conducted an in-situ experiment to investigate the mechanism of advective-diffusive controls on solute transport in a clay-rich aquitard. Their results showed that the solute transport changed from advection dominating to diffusion dominating near a borehole.

In this section, four analytical models for simulating reactive transport around injection well are introduced. The first analytical model was developed by Hsieh and Yeh (2014). They extended the model of Chen *et al.* (2012b), who proposed an analytical solution of solute transport with skin effect to investigate the influences of dispersivity on radial dispersion by taking into account a third-type (Robin) condition. The second analytical model is an analytical model presented by Li *et al.* (2020) for radial reactive transport with skin effect to investigate the impacts of dispersivity, effective porosity, and mass transfer coefficient in the skin zone on radial dispersion. In the third analytical model, we further extend the model of Li *et al.* (2020) by considering the wellbore mixing effect. The last analytical model was developed by Wang and Zhan (2013), who developed semi-analytical solutions of concentrations and mass storage in the aquifer and aquitard via the Laplace transform and finite Fourier transform techniques. In addition, Wang and Zhan (2013) expanded the AA solution of Chen (1985) to include aquitard advection.

Since the first three models are investigated in the confined aquifer and have a progressive relationship, the fourth analytical solution model focuses on solute transport processes in aquifers and weakly permeable layers. This model has not taken the skin effect and mixing effect into account like the previous three models. Therefore, we will do the analysis separately for the fourth model. Moreover, we will continue to introduce the expansion of the fourth model in Section 3.3.2.

3.3.1.1 *Analytical solutions of Hsieh and Yeh (2014)*

Some assumptions were included in the mathematical model of Hsieh and Yeh (2014) for describing radial transport of the injected contaminant in a two-zone confined aquifer system: (a) the aquifer is homogeneous, isotropic and of uniform thickness; (b) the injection well has a finite radius and a finite thickness skin, and therefore the aquifer can be considered as a two-zone system; (c) the well fully penetrates the aquifer; and (d) the effect of molecular diffusion is negligible.

(a) Mathematical model

The governing equation describing the concentration distributions in the wellbore skin zone and formation zone are expressed, respectively, as

$$\frac{\partial C_1}{\partial t} = D_1 \frac{\partial^2 C_1}{\partial r^2} - v\frac{\partial C_1}{\partial r}, \quad r_w < r \leq r_1 \tag{3.19}$$

$$\frac{\partial C_2}{\partial t} = D_2 \frac{\partial^2 C_2}{\partial r^2} - v\frac{\partial C_2}{\partial r}, \quad r_1 < r < \infty \tag{3.20}$$

where C_1 and C_2 are the contaminant concentrations [Kg/m^3] in the skin zone (or called first zone) and formation zone (or called second zone, respectively; D_1 and D_2 are the dispersion coefficients [m^2/s] in the first zone and second zone defined by $D_1 = \alpha_1 v$ and $D_2 = \alpha_2 v$, respectively, α_1 and α_2 are the radial dispersivities [m] in the first zone and second zone, respectively; t is the time [s] since injection; r is a radial distance [m] from the center of the wellbore; r_w is the well radius [m]; r_1 is the radial distance [m] from the center of the well to the outer radius of the skin zone. v is groundwater velocity in a steady-state radial flow system, which can be written as

$$v = \frac{Q}{2\pi r B\theta} \tag{3.21}$$

where Q is a constant injection rate [m^3/s]; B is the aquifer thickness [m]; and θ is the aquifer porosity [–].

Initially, the aquifer system is considered to be free from contamination; i.e., the contaminant concentrations in both the skin and formation zones are equal to zero and expressed as

$$C_1(r, 0) = C_2(r, 0) = 0, \quad r \geq r_w \tag{3.22}$$

The Robin condition specified at the wellbore boundary is expressed as

$$C_1(r_w,t) - \alpha_1 \frac{\partial C_1(r_w,t)}{\partial r} = C_0, \quad t > 0 \tag{3.23}$$

where C_0 is the initial tracer concentration [kg/m^3].

The condition at the remote boundary is considered to be free from contamination and thus described as

$$C_2(r \to \infty, t) = 0 \tag{3.24}$$

The continuity requirements for the contaminant concentration and flux at the interface of the skin zone and formation zone are, respectively:

$$C_1(r_1,t) = C_2(r_1,t), \quad t > 0 \tag{3.25}$$

$$\kappa \frac{\partial C_1(r_1,t)}{\partial r} = \frac{\partial C_2(r_1,t)}{\partial r}, \quad t > 0 \tag{3.26}$$

where $\kappa = \alpha_1/\alpha_2$.

Equations (3.19)–(3.26) compose the mathematical model for contaminant transport due to well injection in a radial two-zone confined aquifer system, which is composed of a wellbore skin zone and a formation zone. The model contains two transient equations describing the contaminant concentration distributions; one is for contaminant transport in the skin zone, while the other is for transport in the formation zone. The relevant analytical solution is shown in the following section.

(b) Dimensionless transform

To simplify the analysis, the dimensionless parameters are defined as follows:

$$C_{1D} = C_1/C_0, \quad C_{2D} = C_2/C_0, \quad r_D = r/\alpha_2, \quad r_{wD} = r_w/\alpha_2,$$
$$r_{1D} = r_w/\alpha_2, \quad t_D = \frac{Qt}{2\pi B\theta\alpha_2^2} \tag{3.27}$$

Using the dimensionless variables, the dimensionless forms for Eqs. (3.19) and (3.20) can be formulated, respectively, as

$$\kappa \frac{\partial^2 C_{1D}}{\partial r_D^2} - \frac{\partial C_{1D}}{\partial r_D} = r_D \frac{\partial C_{1D}}{\partial t_D}, \quad r_{wD} < r_D \le r_{1D} \tag{3.28}$$

$$\kappa\frac{\partial^2 C_{2D}}{\partial r_D^2} - \frac{\partial C_{2D}}{\partial r_D} = r_D\frac{\partial C_{2D}}{\partial t_D}, \quad r_{1D} < r_D < \infty \tag{3.29}$$

The initial and boundary conditions in the dimensionless form are

$$C_{1D}(r_D, 0) = C_{2D}(r_D, 0) = 0, \quad r_D \geq r_{wD} \tag{3.30}$$

$$C_{1D}(r_{wD}, t_D) - \kappa\frac{\partial C_{1D}(r_{wD}, t_D)}{\partial r_D} = 1, \quad t_D > 0 \tag{3.31}$$

$$C_{2D}(r_D \to \infty, t_D) = 0 \tag{3.32}$$

$$C_{1D}(r_{1D}, t_D) = C_{2D}(r_{1D}, t_D), \quad t_D > 0 \tag{3.33}$$

$$\kappa\frac{\partial C_{1D}(r_{1D}, t_D)}{\partial r_D} = \frac{\partial C_{2D}(r_{1D}, t_D)}{\partial r_D}, \quad t_D > 0 \tag{3.34}$$

(c) Analytical model

Then, applying the Laplace transform to Eqs. (3.28)–(3.34) results in

$$\kappa\frac{d^2\overline{C}_{1D}}{dr_D^2} - \frac{d\overline{C}_{1D}}{dr_D} = r_D s\overline{C}_{1D} \tag{3.35}$$

$$\frac{d^2\overline{C}_{2D}}{dr_D^2} - \frac{d\overline{C}_{2D}}{dr_D} = r_D s\overline{C}_{2D} \tag{3.36}$$

$$\overline{C}_{1D}(r_{wD}, s) - \kappa\frac{d\overline{C}_{1D}(r_{wD}, s)}{dr_D} = \frac{1}{s} \tag{3.37}$$

$$\overline{G_2}(r_D \to \infty, s) = 0 \tag{3.38}$$

$$\overline{C}_{1D}(r_{1D}, s) = \overline{C}_{2D}(r_{1D}, s) \tag{3.39}$$

$$\kappa\frac{d\overline{C}_{1D}(r_{1D}, s)}{dr_D} = \frac{d\overline{C}_{2D}(r_{1D}, s)}{dr_D} \tag{3.40}$$

where $\overline{C}$ is the dimensionless Laplace-domain concentration and s is the transform parameter.

Letting

$$\overline{C}_{1D} = U_1 \exp(m r_D) \tag{3.41}$$

where $m = \frac{1}{2\kappa}$.

Substituting Eq. (3.41) into Eq. (3.35) results in

$$\frac{d^2U_1}{dr_D^2} - \left(\frac{1}{4\kappa^2} + \frac{r_D s}{\kappa}\right) U_1 = 0 \tag{3.42}$$

Defining

$$Z_1(r_D, s) = (s/\kappa)^{1/3}(r_D + 1/4\kappa s) \tag{3.43}$$

Then, Eq. (3.42) can be transformed to the Airy equation expressed as

$$\frac{d^2U_1}{dZ_1^2} - Z_1U_1 = 0, \quad \text{for } r_{wD} < r_D \leq r_{1D} \tag{3.44}$$

Similarly, letting

$$\overline{C}_{2D} = U_2 \exp(nr_D) \tag{3.45}$$

where $n = \frac{1}{2}$.

With Eq. (3.45), Eq. (3.36) leads to

$$\frac{d^2U_2}{dr_D^2} - \left(\frac{1}{4} + r_D s\right) U_2 = 0 \tag{3.46}$$

Setting $Z_2(r_D, s) = s^{1/3}(r_D + 1/4s)$, Eq. (3.46) becomes:

$$\frac{d^2U_2}{dZ_2^2} - Z_2U_2 = 0, \quad r_{1D} < r_D < \infty \tag{3.47}$$

The general solutions of Eqs. (3.44) and (3.47) are

$$U_1(r_D, s) = N_1A_i(Z_1) + N_2B_i(Z_1) \tag{3.48}$$

and

$$U_2(r_D, s) = N_3A_i(Z_2) + N_4B_i(Z_2) \tag{3.49}$$

where $A_i(\cdot)$ and $B_i(\cdot)$ are the Airy functions of the first and second kinds, respectively.

Based on the boundary conditions (Eqs. (3.37)–(3.40)), the coefficients N_1 and N_2 in Eq. (3.48) as well as N_3 and N_4 in Eq. (3.49) can be simultaneously determined as

$$N_1 = \frac{1}{s}\exp\left(-\frac{r_{wD}}{2\kappa}\right)\left[2A_i'\left(Z_{2,r_{1D}}\right)B_i\left(Z_{1,r_{1D}}\right)\right.$$
$$\left. - 2\kappa^{\frac{2}{3}}A_i\left(Z_{2,r_{1D}}\right)B_i'\left(Z_{1,r_{1D}}\right)\right]\frac{1}{\Psi} \tag{3.50}$$

$$N_2 = \frac{1}{s}\exp\left(-\frac{r_{wD}}{2\kappa}\right)\left[2\kappa^{\frac{2}{3}}A_i'\left(Z_{1,r_{1D}}\right)A_i\left(Z_{2,r_{1D}}\right)\right.$$
$$\left. - 2A_i\left(Z_{1,r_{1D}}\right)A_i'\left(Z_{2,r_{1D}}\right)\right]\frac{1}{\Psi} \tag{3.51}$$

$$N_3 = \frac{1}{s}\exp\left(\frac{r_{1D}-r_{wD}}{2\kappa} - \frac{r_{1D}}{2}\right)\kappa^{\frac{2}{3}}\left[2A_i'\left(Z_{1,r_{1D}}\right)B_i\left(Z_{1,r_{1D}}\right)\right.$$
$$\left. - 2A_i\left(Z_{1,r_{1D}}\right)B_i'\left(Z_{1,r_{1D}}\right)\right]\frac{1}{\Psi} \tag{3.52}$$

$$N_4 = 0 \tag{3.53}$$

where

$$\Psi = f(r_{wD}, r_{1D}) - \kappa^{2/3}g(r_{wD}, r_{1D}) + 2s^{1/3}\kappa^{2/3}h\left(r_{1D}, r_{wD}\right)$$
$$- 2s^{1/3}\kappa^{4/3}i\left(r_{1D}, r_{wD}\right) \tag{3.54}$$

The $f(x,y), g(x,y), h(x,y), i(x,y)$ and $j(x,y)$ are functions composed of the Airy functions and expressed as

$$f(x,y) = A_i'\left(Z_{2,r_{1D}}\right)\left[A_i(Z_1,x)B_i(Z_{1,y}) - A_i(Z_{1,y})B_i(Z_{1,x})\right] \tag{3.55}$$
$$g(x,y) = A_i\left(Z_{2,r_{1D}}\right)\left[A_i(Z_{1,x})B_i'(Z_{1,y}) - A_i'(Z_{1,y})B_i(Z_{1,x})\right] \tag{3.56}$$
$$h(x,y) = A_i'\left(Z_{2,r_{1D}}\right)\left[A_i(Z_{1,\chi})B_i'(Z_{1,y}) - A_i'(Z_{1,y})B_i(Z_{1,\chi})\right] \tag{3.57}$$
$$i(x,y) = A_i\left(Z_{2,r_{1D}}\right)\left[A_i'(Z_{1,x})B_i'(Z_{1,y}) - A_i'(Z_{1,y})B_i'(Z_{1,x})\right] \tag{3.58}$$
$$j(x,y) = A_i\left(Z_{2,r_D}\right)\left[A_i'(Z_{1,x})B_i(Z_{1,y}) - A_i(Z_{1,y})B_i'(Z_{1,x})\right] \tag{3.59}$$

where the arguments $Z_{1,r_{1D}}, Z_{1,r_{wD}}$ and $Z_{2,r_{1D}}$ represent $Z_1(r_{1D}, s)$, $Z_1(r_{wD}, s)$, and $Z_2(r_{1D}, s)$, respectively.

Hence, Eqs. (3.35)–(3.40) can be solved as

$$\overline{C}_{1D}(r_D, s) = \frac{1}{s}\exp\left(\frac{r_D - r_{wD}}{2\kappa}\right) \times \frac{2f(r_D, r_{1D}) - 2\kappa^{2/3}g(r_D, r_{1D})}{\begin{array}{l} f(r_{wD}, r_{1D}) - \kappa^{2/3}g(r_{wD}, r_{1D}) \\ \quad +2s^{1/3}\kappa^{2/3}h(r_{1D}, r_{wD}) - 2s^{1/3}\kappa^{4/3}i(r_{1D}, r_{wD}) \end{array}} \tag{3.60}$$

$$\overline{C}_{2D}(r_D, s) = \frac{1}{s}\exp\left(\frac{r_{1D} - r_{wD}}{2\kappa} + \frac{r_D - r_{1D}}{2}\right) \times \frac{2\kappa^{2/3}j(r_{1D}, r_{1D})}{\begin{array}{l} f(r_{wD}, r_{1D}) - \kappa^{2/3}g(r_{wD}, r_{1D}) \\ \quad +2s^{1/3}\kappa^{2/3}h(r_{1D}, r_{wD}) - 2s^{1/3}\kappa^{4/3}i(r_{1D}, r_{wD}) \end{array}} \tag{3.61}$$

where $f(x, y), g(x, y), h(x, y), i(x, y)$ and $j(x, y)$ are functions composed of the Airy functions $A_i(z), B_i(z)$, and the derivatives of the Airy functions $A_i'(z)$ and $B_i'(z)$.

For the absence of the wellbore skin (i.e., $r_{1D} = r_{wD}, \alpha_1 = \alpha_2$, and $\kappa = 1$), both Eqs. (3.60) and (3.61) reduce to the same result, expressed as

$$\overline{C}_{1D}(r_D, s) = \overline{C}_{2D}(r_D, s) = \frac{1}{s}\exp\left(\frac{r_D - r_{wD}}{2}\right) \times \frac{2A_i\left(z_1, r_D\right)}{A_i(z_1, r_{wD}) - 2s^{1/3}A_i'(z_1, r_{wD})} \tag{3.62}$$

which is identical to the solution of (Chen, 1987) for a single-zone aquifer. Obviously, the solution of (Chen, 1987) can be considered as a special case of the present solution.

Equations (3.60) and (3.61) are in the Laplace domain and expressed in terms of the Airy functions. The inversion of those two equations to the time domain may not be tractable due to the complexity of the Airy functions. Fortunately, many numerical inverse Laplace transform methods are available, and the De Hoog method and the Stehfest method have been found to be accurate and efficient in solving problems related to solute transport (Wang and Zhan, 2015). Therefore, the De Hoog method and the Stehfest method can be adopted to obtain the time-domain solution.

3.3.1.2 *Analytical solutions of Li et al. (2020)*

Similarly, a confined aquifer is assumed to be unbounded laterally with a uniform thickness B. Figure 3.2 is the conceptual model of Li *et al.* (2020). A cylindrical coordinate system is established with the origin at the center of the injection well, which has a finite radius r_w and a screen that fully penetrates the aquifer. In addition, the skin zone and outer aquifer are assumed to be homogeneous and isotropic. Thus, two zones are defined in the radial direction. First, the skin zone, which is located around the injection well and has a thickness of $r_1 - r_w$, where r_1 is the radius of the patchy zone and r_w is the well radius. The second zone is known as the aquifer formation zone.

In the model of Li *et al.* (2020), the following assumptions are applied in both the skin and aquifer formation zones: the liquid phase of porous medium includes mobile (flowing) and immobile (stagnant) regions, the advection-dispersion transport is restricted to the mobile liquid region, and the solute exchange between these two liquid regions can be described as a first-order mass transfer process. The solute adsorption by the solid phase is described with

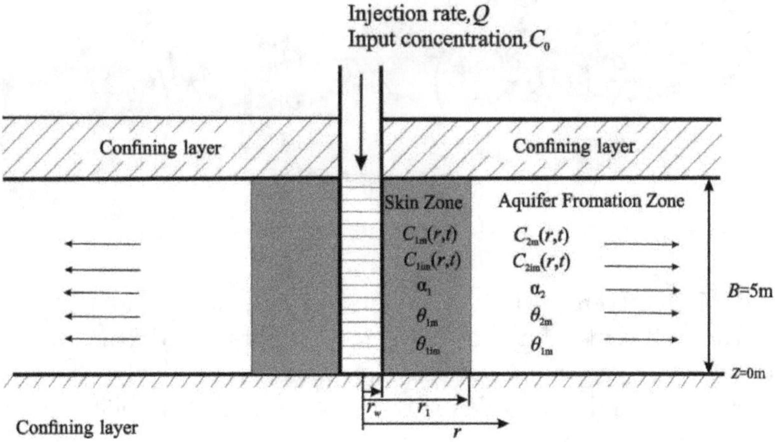

Figure 3.2. Conceptual model of a 2D in a two-zone skin confine aquifer system; the first zone is the skin zone; the second zone is the aquifer formation zone.

a linear isotherm, and a first-order reaction is considered as well. The mobile and immobile domains could have different retardation factors (Parker and Valocchi, 1986; Zhou *et al.*, 2017) and different reaction rates (Gao *et al.*, 2010).

(a) Mathematical model

Based on the above conceptual description, the MIM model of a reactive solute transport under radial steady-state flow in the skin zone can be described as

$$\theta_{1m}R_{1m}\frac{\partial C_{1m}}{\partial t}+\theta_{1im}R_{1im}\frac{\partial C_{1im}}{\partial t} = \theta_{1m}\frac{1}{r}\frac{\partial}{\partial r}\left(rD_1\frac{\partial C_{1m}}{\partial r}\right)-\theta_{1m}v_1\frac{\partial C_{1m}}{\partial r}-\theta_{1m}R_{1m}\lambda_{1m}C_{1m}, \quad r_w<r\le r_1 \tag{3.63}$$

$$\theta_{1im}R_{1im}\frac{\partial C_{1im}}{\partial t} = \omega_1(C_{1m}-C_{1im})-\theta_{1im}R_{1im}\lambda_{1im}C_{1im} \tag{3.64}$$

and in the aquifer formation zone as

$$\theta_{2m}R_{2m}\frac{\partial C_{2m}}{\partial t}+R_{2im}\theta_{2im}\frac{\partial C_{2im}}{\partial t} = \theta_{2m}\frac{1}{r}\frac{\partial}{\partial r}\left(rD_2\frac{\partial C_{2m}}{\partial r}\right)-\theta_{2m}v_2\frac{\partial C_{2m}}{\partial r}-\theta_{2m}R_{2m}\lambda_{2m}C_{2m}, \quad r_1<r<\infty \tag{3.65}$$

$$\theta_{2im}R_{2im}\frac{\partial C_{2im}}{\partial t} = \omega_2(C_{2m}-C_{2im})-\theta_{2im}R_{2im}\lambda_{2im}C_{2im} \tag{3.66}$$

where t is the transport time [s] and r indicates the radial distance from the well center [m]; the former subscripts 1 and 2 indicate the skin zone and the aquifer formation zone, respectively; the latter subscripts m and im indicate the mobile and immobile domains, respectively; C is the concentration; θ_m and θ_{im} are the mobile and immobile porosities, where $\theta_m + \theta_{im} = \theta$ with θ to be the total porosity [-]; λ is the first-order reaction rate [1/s]; ω is the first-order mass transfer coefficient [1/s]; R is the retardation factor including four species in this model, which are respectively the mobile and immobile ones in the skin zone calculated as $R_{1m} = 1 + \rho_1 K_m/\theta_m$ and $R_{1im} = 1 + \rho_1 K_{im}/\theta_{im}$ and in the aquifer formation zone, calculated as $R_{2m} = 1 + \rho_2 K_m/\theta_m$ and $R_{2im} = 1 + \rho_2 K_{im}/\theta_{im}$, where ρ is the bulk density [Kg/m^3], K is the equilibrium distribution coefficient [m^3/Kg]; D is the radial dispersion coefficient [m^2/s], which is velocity-dependent, depicted in the skin and aquifer formation zones respectively as

$$D_1 = \alpha_1 v_1 + D_0 \tag{3.67}$$

$$D_2 = \alpha_2 v_2 + D_0 \tag{3.68}$$

where α_1 and α_2 are the radial dispersivity in the skin and aquifer formation zones, respectively [m]; v_1 and v_2 are the pore velocity in the radial direction for the skin and aquifer formation zones, respectively [m/s]; D_0 is the effective molecular diffusion coefficient [m^2/s], which is usually much smaller than αv and can be ignored. Therefore, the radial dispersion coefficients in the skin and aquifer formation zones are simplified as $D_1 = \alpha_1 v_1$ and $D_2 = \alpha_2 v_2$, respectively. The flow of the injection well is assumed to be steady state. Thus, the groundwater flow velocities in the skin and aquifer formation zones can be expressed as

$$v_1 = \frac{Q}{2\pi\theta_{1m} B r}, \quad r_w < r \le r_1, \quad t > 0 \tag{3.69}$$

$$v_2 = \frac{Q}{2\pi\theta_{2m} B r}, \quad r_1 < r < \infty, \quad t > 0 \tag{3.70}$$

where B is aquifer thickness [m]; Q is injection rate [m^3/s]; r is the radial distance [m].

The initial concentration in the skin and aquifer formation zones is expressed as

$$C_{1m}(r,0) = C_{1im}(r,0) = C_{2m} = (r,0) = C_{2im}(r,0) = 0, \quad r_w < r \tag{3.71}$$

For the two-zone confined aquifer system, the concentration and the flux at the interface between the patchy and aquifer formation zones are continuous and can be expressed as (Hsieh and Yeh, 2014):

$$C_{1m}(r_1,t) = C_{2m}(r_1,t), \quad t > 0 \tag{3.72}$$

$$\theta_{1m}\left(v_1\alpha_1\frac{\partial C_{1m}}{\partial r} - v_1C_{1m}\right)\bigg|_{r=r_1} = \theta_{2m}\left(v_2\alpha_2\frac{\partial C_{2m}}{\partial r} - v_2C_{2m}\right)\bigg|_{r=r_1} \tag{3.73}$$

The condition at the infinite boundary can be described as

$$C_{2m}(\mathrm{r}\to\infty,t) = C_{2im}(\infty,t) = 0, \quad t > 0 \tag{3.74}$$

Three different types of boundary conditions are considered in this model.

The first one is the resident concentration continuity (RCC) model, which is expressed as

$$C_{1m}(r_w,t) = C_0, \quad t > 0 \tag{3.75}$$

where C_0 is constant source concentration.

The second boundary condition is the pulse-type boundary condition:

$$C_{1m}(r_w,t) = \begin{cases} C_0, & t \le t_0 \\ 0, & t > t_0 \end{cases} \tag{3.76}$$

where t_0 is the duration of injection [s].

The third boundary condition is the instantaneous solute input condition:

$$C_{1m}(r_w,t) = \frac{M\delta(t)}{v(r=r_w)}, \quad t > 0 \tag{3.77}$$

where M is the total amount of mass added to a unit area [Kg/m^2]; $\delta(t)$ is the Dirac function with respect to t [1/s].

(b) Dimensionless transform

The dimensionless variables are similar to those used by Hsieh and Yeh (2014) are defined as follows:

$$t_D = \frac{Qt}{2\pi B\theta_{2m}R_{2m}\alpha_2^2}, \quad t_{0D} = \frac{Qt_0}{2\pi B\theta_{2m}R_{2m}\alpha_2^2}, \quad k_1 = \frac{\alpha_1}{\alpha_2},$$

$$k_2 = \frac{\theta_{2m}}{\theta_{1m}}, \quad \beta_1 = \frac{\theta_{1im}}{\theta_{1m}}, \quad \beta_2 = \frac{\theta_{2im}}{\theta_{2m}}, \quad \omega_{1imD} = \frac{2\pi B\theta_{2m}\alpha_2^2}{Q}\frac{\omega_1}{\theta_{1im}},$$

$$\omega_{2imD} = \frac{2\pi B\theta_{2m}\alpha_2^2}{Q}\frac{\omega_2}{\theta_{2im}}, \quad C_{1mD} = \frac{C_{1m}}{C_0}, \quad C_{2mD} = \frac{C_{2m}}{C_0},$$

$$\varepsilon_1 = \frac{R_{1m}}{R_{2m}}, \quad \varepsilon_2 = \frac{R_{1im}}{R_{2m}}, \quad \varepsilon_3 = \frac{R_{2im}}{R_{2m}}, \quad \lambda_{1mD} = \frac{\lambda_{1m}2\pi B\theta_{2m}\alpha_2^2}{Q},$$

$$\lambda_{2mD} = \frac{\lambda_{2m}2\pi B\theta_{2m}\alpha_2^2}{Q} \tag{3.78}$$

The dimensionless variables of r_D, r_{wD}, and r_{1D} used in this section are the same as those defined in Eq. (3.27).

Substituting Eq. (3.78) into Eqs. (3.63)–(3.73), the dimensionless form can be written as

$$r_D\varepsilon_1\frac{\partial C_{1mD}}{\partial t_D} + r_D\beta_1\varepsilon_2\frac{\partial C_{1imD}}{\partial t_D} = k_1k_2\frac{\partial^2 C_{1mD}}{\partial r_D^2} - k_2\frac{\partial C_{1mD}}{\partial r_D} - R_{1m}\lambda_{1mD}r_DC_{1mD},$$
$$r_{wD} < r_D \leq r_{1D} \tag{3.79}$$

$$\varepsilon_2\frac{\partial C_{1imD}}{\partial t_D} = \omega_{1imD}(C_{1mD} - C_{1imD}) - R_{1im}\lambda_{1imD}C_{1imD},$$
$$r_{wD} < r_D \leq r_{1D} \tag{3.80}$$

$$r_D\frac{\partial C_{2mD}}{\partial t_D} + r_D\beta_2\varepsilon_3\frac{\partial C_{2imD}}{\partial t_D} = \frac{\partial^2 C_{2mD}}{\partial r_D^2} - \frac{\partial C_{2mD}}{\partial r_D} - R_{2m}\lambda_{2mD}r_DC_{2mD},$$
$$r_{1D} < r_D < \infty \tag{3.81}$$

$$\varepsilon_3\frac{\partial C_{2imD}}{\partial t_D} = \omega_{2imD}(C_{2mD} - C_{2imD}) - R_{2im}\lambda_{2imD}C_{2imD},$$
$$r_{1D} < r_D < \infty \tag{3.82}$$

$$C_{1mD}(r_D, 0) = C_{2mD}(r_D, 0) = 0, \quad r_D > r_{wD} \tag{3.83}$$

$$C_{2mD}(r_D \to \infty, \tau) = 0, \quad t_D > 0 \tag{3.84}$$

$$C_{1mD}(r_{1D}, \tau) = C_{2mD}(r_{1D}, \tau), \quad t_D > 0 \tag{3.85}$$

$$\left(k_1\frac{\partial C_{1mD}}{\partial r_D}\right)\bigg|_{r_D=r_{1D}} = \left(\frac{\partial C_{2mD}}{\partial r_D}\right)\bigg|_{r_D=r_{1D}}, \quad t_D > 0 \tag{3.86}$$

Three different types of boundary conditions in the dimensionless form can be re-written as

$$C_{1mD}(r_{wD}, t_D) = 1, \quad t_D > 0 \tag{3.87}$$

$$C_{1mD}(r_{wD}, t_D) = \begin{cases} 1, & t_D \le t_{0D} \\ 0, & t_D > t_{0D} \end{cases} \tag{3.88}$$

$$C_{1mD}(r_{wD}, t_D) = \frac{M\delta(\tau)}{C_0 v(r = r_w)}, \quad t_D > 0 \tag{3.89}$$

(c) Analytical model

Taking the Laplace transform of Eqs. (3.79)–(3.86) with respect to the dimensionless time t_D, one obtains:

$$k_1 k_2 \frac{\partial^2 \overline{C}_{1mD}}{\partial r_D^2} - k_2 \frac{\partial \overline{C}_{1mD}}{\partial r_D} - r_D s \varepsilon_1 \overline{C}_{1mD} - r_D s \varepsilon_2 \beta_1 \overline{C}_{1imD}$$
$$- R_{1m} \lambda_{1mD} r_D \overline{C}_{1mD} = 0, \quad r_{wD} < r_D \le r_{1D} \tag{3.90}$$

$$s\varepsilon_2 \overline{C}_{1imD} = \omega_{1imD}(\overline{C}_{1mD} - \overline{C}_{1imD}) - R_{1im}\lambda_{1imD}\overline{C}_{1imD},$$
$$r_{wD} < r_D \le r_{1D} \tag{3.91}$$

$$\frac{\partial^2 \overline{C}_{2mD}}{\partial r_D^2} - \frac{\partial \overline{C}_{2mD}}{\partial r_D} - r_D s \overline{C}_{2mD} - r_D s \varepsilon_3 \beta_2 \overline{C}_{2imD}$$
$$- R_{2m} \lambda_{2mD} r_D \overline{C}_{2mD} = 0, \quad r_{1D} < r_D < \infty \tag{3.92}$$

$$s\varepsilon_3 \overline{C}_{2imD} = \omega_{2imD}(\overline{C}_{2mD} - \overline{C}_{2imD}) - R_{2im}\lambda_{2imD}\overline{C}_{2imD},$$
$$r_{1D} < r_D < \infty \tag{3.93}$$

$$\overline{C}_{2mD}(r_D \to \infty, s) = 0 \tag{3.94}$$

$$\overline{C}_{1mD}(r_{1D}, s) = \overline{C}_{2mD}(r_{1D}, s) \tag{3.95}$$

$$\left(k_1 \frac{\partial \overline{C}_{1mD}}{\partial r_D} \right)\bigg|_{r_D = r_{1D}} = \left(\frac{\partial \overline{C}_{2mD}}{\partial r_D} \right)\bigg|_{r_D = r_{1D}} \tag{3.96}$$

Besides, three different types of boundary conditions in the Laplace form can be transformed as

$$\overline{C}_{1mD}(r_{wD}, s) = \frac{1}{s} \tag{3.97}$$

$$\overline{C}_{1mD}(r_{wD}, s) = \begin{cases} \dfrac{1}{s}, & t_D \le t_{0D} \\ \dfrac{1-\exp(-st_{0D})}{s}, & t_D > t_{0D} \end{cases} \tag{3.98}$$

$$\overline{C}_{1mD}(r_{wD}, s) = \frac{M}{C_0 v(r = r_w)} \tag{3.99}$$

where $\overline{C}_{1mD}, \overline{C}_{1imD}, \overline{C}_{2mD}$ and $\overline{C}_{2imD}$ represent the Laplace transforms of $C_{1mD}, C_{1imD}, C_{2mD}$ and C_{2imD}.

According to Eqs. (3.91) and (3.93), the correlations between $\overline{C}_{1imD}, \overline{C}_{1mD}, \overline{C}_{2imD}$ and $\overline{C}_{2mD}$ can be expressed as

$$\overline{C}_{1imD} = \frac{\omega_{1imD}}{s\varepsilon_2 + \omega_{1imD} + R_{1im}\lambda_{1imD}}\overline{C}_{1mD}, \quad r_{wD} < r_D \le r_{1D} \tag{3.100}$$

$$\overline{C}_{2imD} = \frac{\omega_{2imD}}{s\varepsilon_3 + \omega_{2imD} + R_{2im}\lambda_{2imD}}\overline{C}_{2mD}, \quad r_{1D} < r_D < \infty \tag{3.101}$$

Substituting Eqs. (3.100) and (3.101) into Eqs. (3.90) and (3.92) yields the following equation:

$$k_1\frac{\partial^2 \overline{C}_{1mD}}{\partial r_D^2} - \frac{\partial \overline{C}_{1mD}}{\partial r_D} - \frac{r_D}{k_2}\left(s\varepsilon_1 + R_{1m}\lambda_{1mD} + \frac{s\varepsilon_2\beta_1\omega_{1imD}}{s\varepsilon_2 + \omega_{1imD} + R_{1im}\lambda_{1imD}}\right)\overline{C}_{1mD} = 0,$$
$$r_{wD} < r_D \le r_{1D} \tag{3.102}$$

$$\frac{\partial^2 \overline{C}_{2mD}}{\partial r_D^2} - \frac{\partial \overline{C}_{2mD}}{\partial r_D} - r_D\left(s + R_{2m}\lambda_{2mD} + \frac{s\varepsilon_3\beta_2\omega_{2imD}}{s\varepsilon_3 + \omega_{2imD} + R_{2im}\lambda_{2imD}}\right)\overline{C}_{2mD} = 0,$$
$$r_{1D} < r_D < \infty \tag{3.103}$$

Similar to the derivations of Hsieh and Yeh (2014) (Eqs. (3.35)–(3.67)), the solutions of $\overline{C}_{1mD}$ and $\overline{C}_{2mD}$ considering constant concentration boundary condition in the Laplace domain can be expressed as

$$\overline{C}_{1mD}(r_D,s)=\frac{1}{s}\exp\left(\frac{r_D+r_{wD}}{2k_1}\right)\left[\frac{\psi_1 A_i\left(Z_{1,r_D}\right)}{B_i\left(Z_{1,r_{wD}}\right)\psi_2+A_i\left(Z_{1,r_{wD}}\right)\psi_1}\right.$$

$$\left.+\frac{\psi_2 B_i\left(Z_{1,r_D}\right)}{B_i\left(Z_{1,r_{wD}}\right)+A_i\left(Z_{1,r_{wD}}\right)\psi_1}\right] \tag{3.104}$$

$$\overline{C}_{2mD}(r_D,s)=\frac{1}{s}\exp\left(\frac{r_D-r_{1D}}{2}+\frac{r_{1D}-r_{wD}}{2k_1}\right)\frac{A_i\left(Z_{2,r_D}\right)}{A_i\left(Z_{2,r_{1D}}\right)}$$

$$\times\left[\frac{\psi_1 A_i\left(Z_{1,r_D}\right)}{B_i\left(Z_{1,r_{wD}}\right)\psi_2+A_i\left(Z_{1,r_{wD}}\right)\psi_1}\right.$$

$$\left.+\frac{\psi_2 B_i\left(Z_{1,r_D}\right)}{B_i\left(Z_{1,r_{wD}}\right)+A_i\left(Z_{1,r_{wD}}\right)\psi_1}\right] \tag{3.105}$$

where

$$Z_1(r_D,s)=\left(\frac{s\varepsilon_1+R_{1m}\lambda_{1mD}+\frac{s\varepsilon_2\beta_1\omega_{1imD}}{s\varepsilon_2+\omega_{1imD}+R_{1im}\lambda_{1imD}}}{k_1k_2}\right)^{1/3}$$

$$\times\left(r_D+\frac{k_2}{4k_1\left(s\varepsilon_1+R_{1m}\lambda_{1mD}+\frac{s\varepsilon_2\beta_1\omega_{1imD}}{s\varepsilon_2+\omega_{1imD}+R_{1im}\lambda_{1imD}}\right)}\right) \tag{3.106}$$

$$Z_2(r_D,s)=\frac{4r_D\left(s+R_{2m}\lambda_{2mD}+\frac{s\varepsilon_3\beta_2\omega_{2imD}}{s\varepsilon_3+\omega_{2imD}+R_{1im}\lambda_{1imD}}\right)+1}{4\left(s+R_{2m}\lambda_{2mD}+\frac{s\varepsilon_3\beta_2\omega_{2imD}}{s\varepsilon_3+\omega_{2imD}+R_{1im}\lambda_{1imD}}\right)^{2/3}} \tag{3.107}$$

$$\psi_1=B_i\left(Z_{1,r_{1D}}\right)A_i'\left(Z_{2,r_{1D}}\right)-(k_1)^{\frac{2}{3}}A_i\left(Z_{2,r_{1D}}\right)B_i'\left(Z_{1,r_{1D}}\right)$$

$$\times\left(\frac{s\varepsilon_1+R_{1m}\lambda_{1mD}+\frac{s\varepsilon_2\beta_1\omega_{1imD}}{s\varepsilon_2+\omega_{1imD}+R_{1im}\lambda_{1imD}}}{k_2\left(s+R_{2m}\lambda_{2mD}+\frac{s\varepsilon_3\beta_2\omega_{2imD}}{s\varepsilon_3+\omega_{2imD}+R_{2im}\lambda_{2imD}}\right)}\right)^{\frac{1}{3}} \tag{3.108}$$

$$\psi_2 = (k_1)^{\frac{2}{3}} A_i\left(Z_{2,r_{1D}}\right) A_i'\left(Z_{1,r_{1D}}\right)$$
$$\times \left(\frac{s\varepsilon_1 + R_{1m}\lambda_{1mD} + \frac{s\varepsilon_2\beta_1\omega_{1imD}}{s\varepsilon_2+\omega_{1imD}+R_{1im}\lambda_{1imD}}}{k_2\left(s + R_{2m}\lambda_{2mD} + \frac{s\varepsilon_3\beta_2\omega_{2imD}}{s\varepsilon_3+\omega_{2imD}+R_{2im}\lambda_{2imD}}\right)} \right)^{\frac{1}{3}}$$
$$- A_i\left(Z_{1,r_{1D}}\right) A_i'\left(Z_{2,r_{1D}}\right) \tag{3.109}$$

Using the same methods, the solutions of $\overline{C}_{1mD}$ and $\overline{C}_{2mD}$ considering pulse-type boundary condition in the Laplace domain can be expressed as

$$\overline{C}_{1mD} = \begin{cases} \text{Eq. (3.104)}, & t \le t_0 \\ [1 - \exp(-st_{0D})]\text{Eq. (3.104)}, & t > t_0 \end{cases} \tag{3.110}$$

$$\overline{C}_{2mD} = \begin{cases} \text{Eq. (3.105)}, & t \le t_0 \\ [1 - \exp(-st_{0D})]\text{Eq. (3.105)}, & t > t_0 \end{cases} \tag{3.111}$$

The solutions of $\overline{C}_{1mD}$ and $\overline{C}_{2mD}$ factoring instantaneous release of solutes in the Laplace domain can be expressed as

$$\overline{C}_{1mD} = \frac{Ms}{v(r = r_w)} \text{Eq. (3.104)} \tag{3.112}$$

$$\overline{C}_{2mD} = \frac{Ms}{v(r = r_w)} \text{Eq. (3.104)} \tag{3.113}$$

Table 3.3 lists the solutions of the constant source, the pulse source and the instantaneous source in the Laplace domain in the two-zone confined aquifer system.

It is notable that the solutions listed above are obtained in Laplace domain, which can be inversed into analytical solutions in the real time domain by applying analytical or numerical inversion techniques. Similarly, the numerical inverse Laplace method will be introduced to obtain semi-analytical solutions in the real time domain.

3.3.1.3 *The extension of Li et al. (2020)*

In this section, we further extend the model of Li *et al.* (2020) by considering the wellbore mixing effect. When taking into account the

Table 3.3. Solutions for different types of sources.

Source type		Solutions
Constant source $C_{1m} = C_0$	In skin zone	Eq. (3.104)
	In aquifer	Eq. (3.105)
Pulse source $C_{1m} = \begin{cases} C_0, & t \le t_0 \\ 0, & t > t_0 \end{cases}$	In skin zone	$\begin{cases} \text{Eq. (3.104)}, & t \le t_0 \\ [1-\exp(-st_{0D})]\,\text{Eq. (3.104)}, & t > t_0 \end{cases}$
	In aquifer	$\begin{cases} \text{Eq. (3.105)}, & t \le t_0 \\ [1-\exp(-st_{0D})]\,\text{Eq. (3.105)}, & t > t_0 \end{cases}$
Instantaneous source $C_{1m} = \dfrac{M\delta(t)}{v(r=r_w)}$	In skin zone	$\dfrac{Ms}{v(r=r_w)}$ Eq. (3.104)
	In aquifer	$\dfrac{Ms}{v(r=r_w)}$ Eq. (3.104)

mixing effect in the wellbore, the inner boundary condition of Li *et al.* (2020) (Eq. (3.75)) should be revised as

$$[C_{m1}(r,t)]|_{r=r_w} = [C_{\text{inj}}(t)]|_{r=r_w} \tag{3.114}$$

where $C_{\text{inj}}(t)$ represents the concentration variations in the wellbore, which can be determined by

$$V_{w,\text{inj}}\frac{dC_{\text{inj}}}{dt} = -2\pi r_w \theta_{m1} B v_w [C_{\text{inj}}(t) - C_0] \tag{3.115}$$

$$v_w = \frac{Q}{2\pi\theta_{1m} B r_w} \tag{3.116}$$

where $V_{w,\text{inj}}$ is the volume [m^3] of water in the wellbore, and $V_{w,\text{inj}} = \pi r_w^2 h_{w,\text{inj}}$; $h_{w,\text{inj}}$ is the water level [m] in the wellbore. The mass balance for the well in Eqs. (3.115)–3.116) is only relevant when velocity exceeds zero because it does not contain terms for possible diffusive losses.

According to Eq. (3.78), the dimensionless forms of Eqs. (3.114) and (3.115) are

$$[C_{m1D}(r_D,t_D)]|_{r_D=r_{wD}} = [C_{\text{inj},D}(t_D)]|_{r_D=r_{wD}} \tag{3.117}$$

$$\beta_w \frac{dC_{\text{inj},D}(t_D)}{dt_D} = 1 - C_{\text{inj},D}(t_D) \tag{3.118}$$

where

$$\beta_w = \frac{V_{w,\text{inj}}}{2\pi\theta_{2m}R_{2m}\alpha_2^2}.$$

Then, Eqs. (3.117) and (3.118) in the Laplace domain can be described as

$$\overline{C}_{m1D}(r_{wD}, s) = \frac{1}{s(s\beta_w + 1)} \tag{3.119}$$

Finally, the solutions of $\overline{C}_{1mD}$ and $\overline{C}_{2mD}$ considering wellbore mixing effect in the Laplace domain can be expressed as

$$\begin{aligned}\overline{C}_{1mD}(r_D, s) = {} & \frac{1}{s(s\beta_w + 1)}\exp\left(\frac{r_D + r_{wD}}{2k_1}\right)\\ & \times\left[\frac{\psi_1 A_i\left(Z_{1,r_D}\right)}{B_i\left(Z_{1,r_{wD}}\right)\psi_2 + A_i\left(Z_{1,r_{wD}}\right)\psi_1}\right.\\ & \left. + \frac{\psi_2 B_i\left(Z_{1,r_D}\right)}{B_i\left(Z_{1,r_{wD}}\right) + A_i\left(Z_{1,r_{wD}}\right)\psi_1}\right]\end{aligned} \tag{3.120}$$

$$\begin{aligned}\overline{C}_{2mD}(r_D, s) = {} & \frac{1}{s(s\beta_w + 1)}\exp\left(\frac{r_D - r_{1D}}{2} + \frac{r_{1D} - r_{wD}}{2k_1}\right)\frac{A_i\left(Z_{2,r_D}\right)}{A_i\left(Z_{2,r_{1D}}\right)}\\ & \times\left[\frac{\psi_1 A_i\left(Z_{1,r_D}\right)}{B_i\left(Z_{1,r_{wD}}\right)\psi_2 + A_i\left(Z_{1,r_{wD}}\right)\psi_1}\right.\\ & \left. + \frac{\psi_2 B_i\left(Z_{1,r_D}\right)}{B_i\left(Z_{1,r_{wD}}\right) + A_i\left(Z_{1,r_{wD}}\right)\psi_1}\right]\end{aligned} \tag{3.121}$$

The variables used in these two equations are the same as in Eqs. (3.104) and (3.105). The analytical solutions of Eqs. (3.120) and (3.121) in the real-time domain can be obtained using the numerical inverse Laplace method like the Stehfest method.

3.3.1.4 *Comparison among the three models*

As shown in Section 3.3.1.1, Hsieh and Yeh (2014) developed an ADE-based analytical solution considering skin effect and a third-type (Robin) condition. By adopting the MIM model, Li *et al.* (2020b) has extended the model of Hsieh and Yeh (2014), as shown in Section 3.3.1.2. However, both models mentioned ignore proficient mixing effects. We thus further extend these two models by considering the wellbore mixing effect. The comparison of these three analytical models is listed in Table 3.4.

Figure 3.3 shows the comparison of concentration distribution curves at $t_{\text{inj}} = 6\,\text{h}$. The parameters used in this figure are listed in Table 3.5. According to this figure, we can find that the wellbore mixing effect has a significant impact on the solute transport in porous media. When the wellbore mixing effect is ignored, the extended model of this book (Eqs. (3.120) and (3.121)) can reduce

Table 3.4. Comparison of reactive transport models around injection well with constant source.

Models	Governing equation	Boundary condition	Solutions
Hsieh and Yeh (2014)	ADE	Robin condition without mixing effect	Eqs. (3.60) and (3.61)
Li *et al.* (2020)	MIM	RCC without mixing effect	Eqs. (3.104) and (3.105)
Extended model of this book	MIM	RCC with mixing effect	Eqs. (3.120) and (3.121)

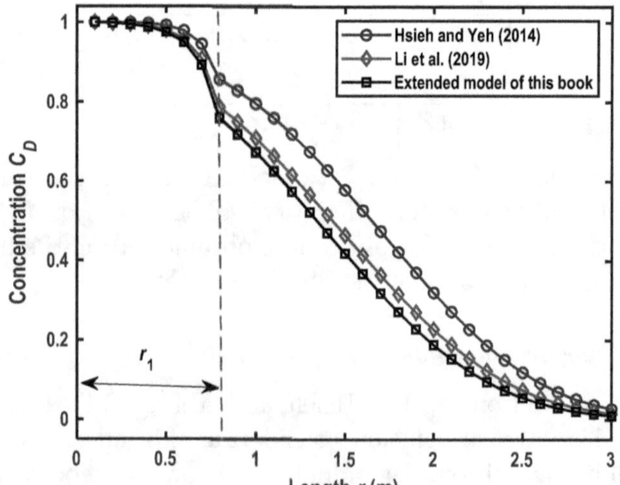

Figure 3.3. Comparison among three models listed in Table 3.4.

Table 3.5. The parameter values used in this section.

Parameter name	Symbols	Values
Aquifer thickness [m]	B	5
Radius of well screen [m]	r_w	0.1
Water level in the wellbore [m]	$h_{w,\text{inj}}$	50
Radius of skin zone [m]	r_1	0.8
Mobile porosity of aquifer [–]	θ_{2m}	0.3
Mobile porosity of skin [–]	θ_{1m}	0.3
Immobile porosity of aquifer [–]	θ_{2im}	0.1
Immobile porosity of skin [–]	θ_{1im}	0.1
Radial dispersivity of aquifer [m]	α_2	0.5
Radial dispersivity of skin [m]	α_1	0.1
Injection rate [m^3/h]	Q	2.5
Solute injection time [h]	t_0	0.5
Injection time [h]	t_{inj}	6
Mass transfer coefficient of aquifer [1/h]	ω_2	0.05
Mass transfer coefficient of skin [1/h]	ω_1	0.05

to the model of Li *et al.* (2020), and also, when ignoring the mass transfer between mobile and immobile zones, the model of Li *et al.* (2020) can reduce to the model of Hsieh and Yeh (2014).

3.3.1.5 *Parameter sensitivity analysis*

In this section, ADEs and MIMs solutions are used to illustrate the impacts of the skin zone on the radial solute transport process where the case of pulse-type boundary condition is considered as an example in the following analysis. For the ADEs solution, we assume the mobile porosity (i.e., effective porosity) is equal to the total porosity when all the pore spaces are well-connected with negligible immobile porosity; hence, the parameters are set as $\theta_1 = \theta_{1m}$, and $\theta_2 = \theta_{2m}$, respectively. In addition, for the sake of simplicity, the sorption and reaction of solute were not considered. Under this simplification, the values of $\lambda_{1m}, \lambda_{1im}, \lambda_{2m}$ and λ_{2im} are equal to 0.

(a) Effects of dispersivities and porosities of skin on the radial solute transport

The impacts of dispersivities and effective porosities in the skin zone on the radial solute transport using the ADE solution are analyzed first. The parameters used in models are $r_1 = 0.8\,\text{m}, \theta_1 = \theta_{1m} = 0.3, \alpha_2 = 0.5\,\text{m}, \alpha_1 = 0.1\,\text{m}, 0.5\,\text{m}$, and $2.5\,\text{m}$, respectively. Hence, the dimensionless parameters calculated in Eq. (3.78) are $k_2 = 1, k_1 = 0.2, 1$, and 5, respectively; $k_1 < 1$ represents $\alpha_1 < \alpha_2$ while $k_1 > 1$ represents $\alpha_2 < \alpha_1$. Note that in the case of $k_1 = 1$ and $k_2 = 1$ represents the case without the skin zone. The other parameters are listed in Table 3.5.

Figures 3.4(a) and 3.4(b) show the concentration distribution curves at $t_{\text{inj}} = 3\,\text{h}$ and the BTCs observed at $r = 2\,\text{m}$ with different dispersivities in the skin zone. As shown in Figure 3.4(a), at the same distance, a smaller dispersivity in the skin zone results in a shorter transport distance of concentration peak value at a specific time $t_{\text{inj}} = 3\,\text{h}$, and it can also be observed that a larger dispersivity in the skin zone leads to a lower concentration near the well and a higher concentration in the area far away from the well. In addition, the slopes of the concentration curves are significantly different at the interface between the skin and aquifer formation zones, and it is obvious that a smaller dispersivity in the skin zone leads to steeper concentration curves in the skin zone.

Figure 3.4(b) shows the BTCs observed at $r = 2\,\text{m}$ with different dispersivities in the skin zone, indicating that a larger dispersivity in the skin zone results in a higher concentration at the early injection stage but a lower concentration at the late stage. Furthermore, a larger dispersivity in the skin zone results in a higher peak value of BTCs, while a smaller dispersivity in the skin zone results in a longer tail of solute transport. This is because a larger dispersivity in the skin zone contributes to faster tracer transport through hydrodynamic dispersion, which leads to a lower concentration distribution near the well and higher values of BTCs at the early stage after the pulse-type injection. Therefore, the variation in dispersivity around the test well can result in significant impacts on the concentration distribution and the BTCs in a tracer injection test.

Figures 3.5(a) and 3.5(b) show the concentration distribution curves at $t_{\text{inj}} = 3\,\text{h}$ and the BTCs observed at $r = 2\,\text{m}$ with different effective porosities. The parameters are given as $r_1 = 0.8\,\text{m}, \alpha_1 = \alpha_2 = 0.5\,\text{m}, \theta_2 = 0.3, \theta_1 = 0.1, 0.3$ and 0.5. Hence, the dimensionless parameters calculated in Eq. (3.78) are $k_1 = 1, k_2 = 3/5, 1$ and 3, respectively; $k_2 < 1$ represents $\theta_1 > \theta_2$ while $k_2 > 1$ represents $\theta_1 < \theta_2$. As illustrated in Figure 3.5(a), at the same distance, it is

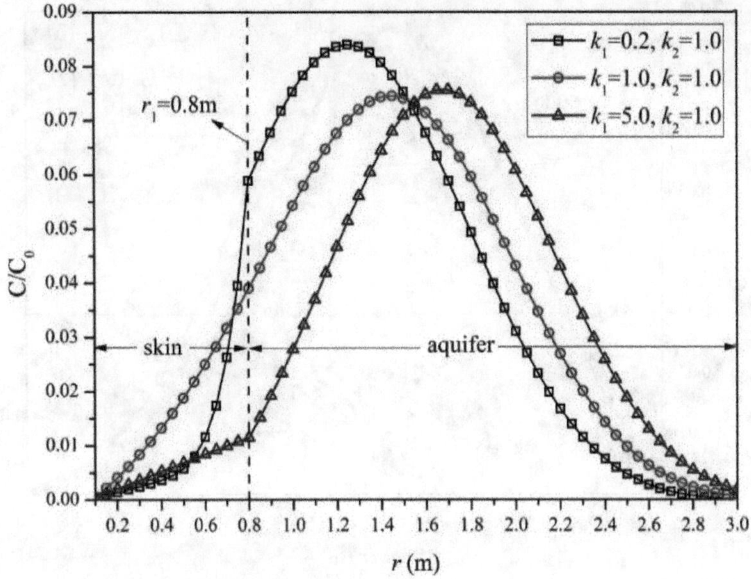

(a)

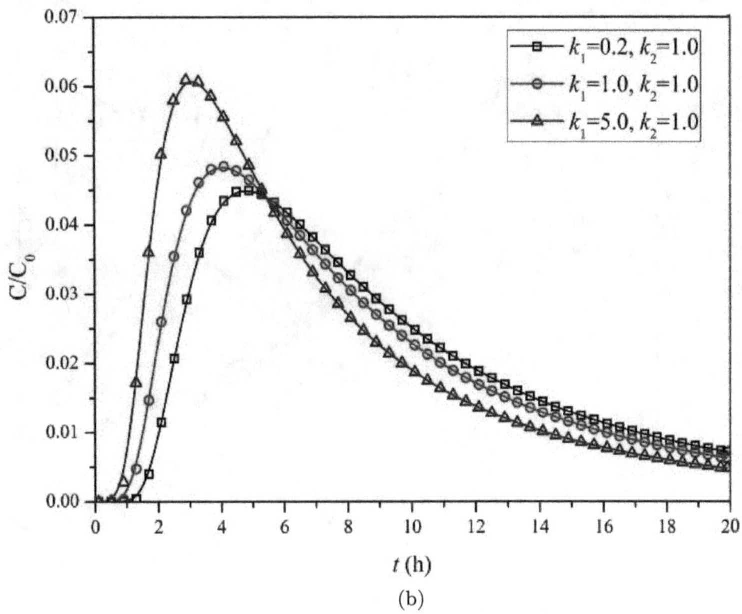

(b)

Figure 3.4. (a) Concentration distribution curves for different dispersivities in the skin zone at a specific time $t_{\text{inj}} = 3\,\text{h}$ with $r_1 = 0.8\,\text{m}, \theta_1 = \theta_2 = 0.3$, $\alpha_2 = 0.5\,\text{m}, \alpha_1 = 0.1\,\text{m}, 0.5$ m, and $2.5\,\text{m}$, respectively. (b) BTCs for different dispersivities in the skin zone at $r = 2\,\text{m}$ with $r_1 = 0.8\,\text{m}, \theta_1 = \theta_2 = 0.3$, $\alpha_2 = 0.5\,\text{m}, \alpha_1 = 0.1\,\text{m}, 0.5\,\text{m}$, and $2.5\,\text{m}$, respectively.

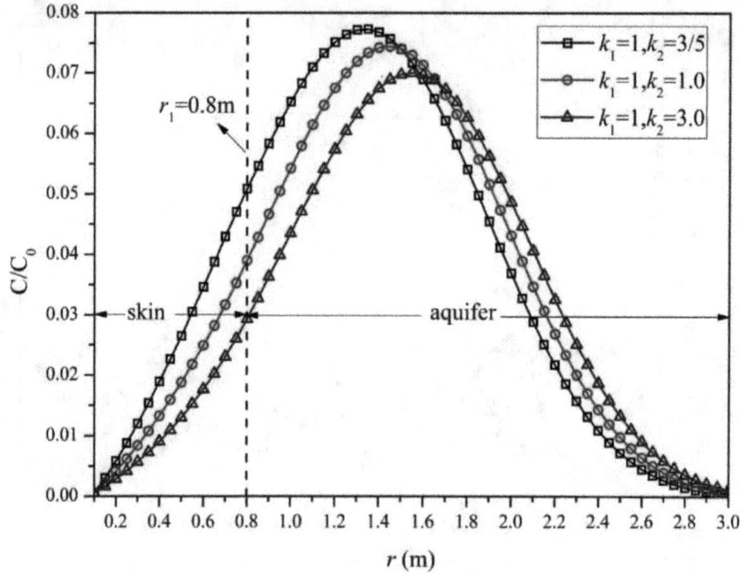

(a)

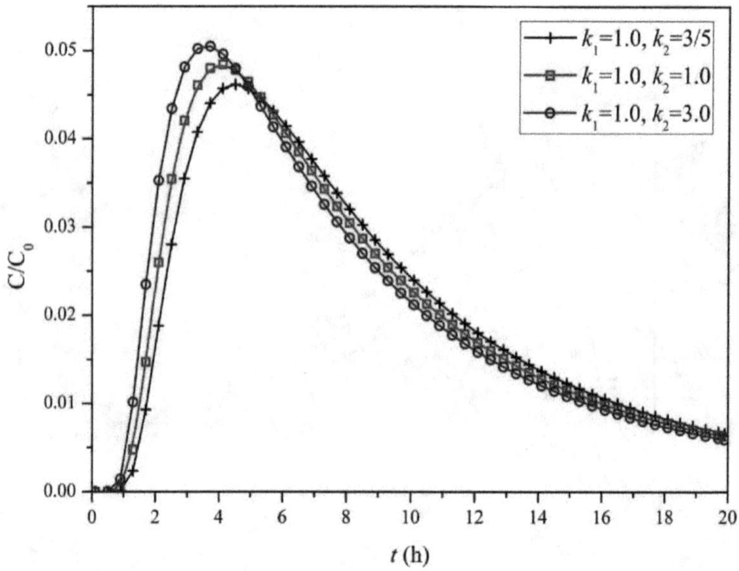

(b)

Figure 3.5. (a) Concentration distribution curves for different effective porosities at a specific time $t_{\text{inj}} = 3\,\text{h}$ in the skin zone with $r_1 = 0.8\,\text{m}, \alpha_1 = \alpha_2 = 0.5\,\text{m}, \theta_2 = 0.3, \theta_1 = 0.1, 0.3$, and 0.5, respectively. (b) BTCs for different effective porosities in the skin zone at $r = 2\,\text{m}$ with $r_1 = 0.8\,\text{m}, \alpha_1 = \alpha_2 = 0.5\,\text{m}, \theta_2 = 0.3, \theta_1 = 0.1, 0.3$, and 0.5, respectively.

notable that a larger effective porosity in the skin zone results in a shorter transport distance of concentration peak value at a specific time, $t_{\text{inj}} = 3\,\text{h}$, and it can also be observed that a larger effective porosity in the skin zone leads to a higher concentration near the well but a lower concentration away from the well, and conversely the case a smaller effective porosity. Similarly, we can also note the variation in effective porosity in the skin zone does not change the slopes of the concentration curve at the interface between the skin and aquifer formation zones, which is different from the impacts of dispersivity indicated by Figure 3.4(a).

Figure 3.5(b) shows the BTCs with different effective porosities in the skin zone at $r = 2\,\text{m}$, indicating that a larger effective porosity of the skin zone results in a lower concentration at the early injection stage and a higher concentration at the late stage. It is also noteworthy that a smaller effective porosity in the skin zone results in a larger peak value of the BTCs, while a larger one results in a longer tail of solute transport. This is because a larger k_2 means a smaller effective porosity in the skin zone. Comparably, from Eq. (3.69), variation of θ_1 means a different pore flow velocity in the skin zone. Therefore, a smaller effective porosity results in a faster solute transport process and vice versa in the skin zone, which generally results in a lower concentration distribution near the well and higher values of BTCs at the early stage after the pulse-type injection.

(b) Effects of skin thickness on radial solute transport

The impacts of skin thickness on the radial solute transport using the ADE solution are analyzed in this section. Figures 3.6(a) and 3.6(b) show the concentration distribution curves at $t_{\text{inj}} = 3\ h$ and the BTCs observed at $r = 2\,\text{m}$ with different skin thicknesses. The parameters are given as $r_1 = 0.4\,\text{m}, 0.8\,\text{m}, \theta_1 = 0.1, 0.3, 0.5, \theta_2 = 0.3, \alpha_2 = 0.5\,\text{m}, \alpha_1 = 0.25\,\text{m}, 0.5\,\text{m}$ and $1\,\text{m}$, respectively. Hence, the dimensionless parameters are $k_1 = 0.5, 1, 2, k_2 = 3, 1$ and $3/5$, respectively. Clearly, the discrepancy between the curve of "$k_1 = 2, k_2 = 3, r_1 = 0.8$" and that of "$k_1 = 0.5, k_2 = 3/5, r_1 = 0.8$" is larger than the discrepancy between the curve of "$k_1 = 2, k_2 = 3, r_1 = 0.4$" and that of "$k_1 = 0.5, k_2 = 3/5, r_1 = 0.4$", which means that a larger skin thickness strengthens the influence of dispersivity and porosity on the radial solute transport. Consequently, a larger skin thickness results in a larger transport distance of concentration peak value.

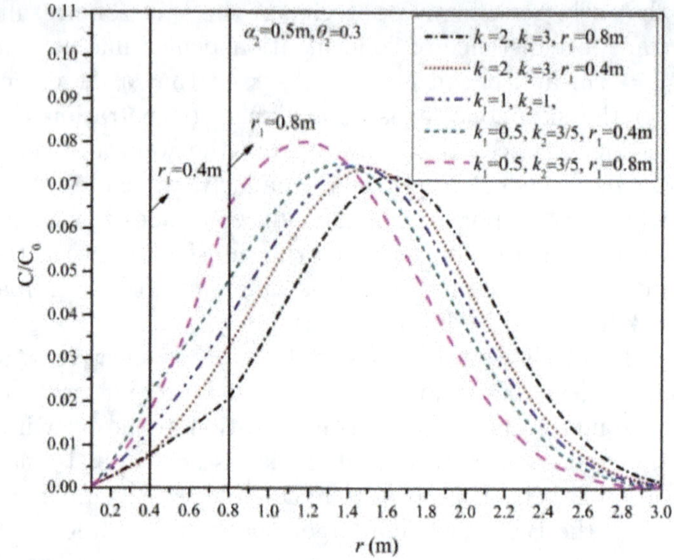

(a)

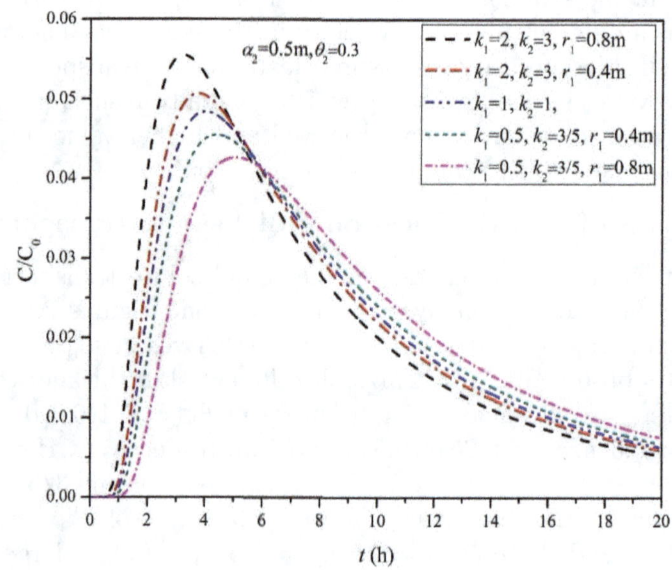

(b)

Figure 3.6. (a) Concentration distribution curves for different skin thicknesses at a specific time $t_{\text{inj}} = 3\,\text{h}$ with $r_1 = 0.4\,\text{m}, 0.8\,\text{m}, \theta_1 = 0.1, 0.3, 0.5, \theta_2 = 0.3, \alpha_2 = 0.5\,\text{m}, \alpha_1 = 0.25$ m, $0.5\,\text{m}$, and $1\,\text{m}$, respectively. (b) BTCs for different skin thicknesses at $r = 2\,\text{m}$ with $r_1 = 0.4\,\text{m}, 0.8\,\text{m}, \theta_1 = 0.1, 0.3, 0.5$, $\theta_2 = 0.3, \alpha_2 = 0.5\,\text{m}, \alpha_1 = 0.25\,\text{m}, 0.5\,\text{m}$, and $1\,\text{m}$, respectively.

Figure 3.6(b) shows the BTCs with different skin thicknesses at $r = 2\,\text{m}$, indicating that a larger skin thickness results in a higher concentration at the early injection stage but a lower concentration at the late stage in the cases of $k_1 = 2$ and $k_2 = 3$. Correspondingly, larger skin thickness leads to a larger peak value of BTCs with a shorter tail of solute transport. However, it is conversely true for the case where $k_1 = 0.5$ and $k_2 = 3/5$, where a larger skin thickness results in a lower concentration at the early injection stage but a higher concentration at the late stage and lower peak values with longer tails of BTCs. This is because a larger r_1 strengthens the influence of the skin on radial solute transport, resulting in a greater change in BTC shapes.

(c) Comparisons between MIM and ADE with skin effect

As mentioned above, the wellbore skin has a significant effect on concentration distribution curves and BTCs using the ADE. However, the use of ADE has been questioned by many hydrogeologists in recent years because this law cannot adequately account for anomalous transport in heterogeneous porous media, as stated earlier. Therefore, the characteristics of radial solute transport using the MIMs solution are investigated in this section.

Figure 3.7 shows the concentration distribution curves in the mobile and immobile domains at $t_{\text{inj}} = 3\,\text{h}$ with different dispersivities and mobile porosities of the skin zone. The parameters of the MIMs solution are given as $\theta_{1m} = 0.1, 0.3, 0.5, \theta_{1im} = 0.1, \theta_{2m} = 0.3, \theta_{2im} = 0.1, \alpha_2 = 0.5\,\text{m}, \alpha_1 = 0.25\,\text{m}, 0.5\,\text{m}$ and $1\,\text{m}$, respectively; hence, the dimensionless parameters are $k_1 = 0.5, 1, 2, k_2 = 3, 1$ and $3/5$, and the immobile/mobile ratios are $\beta_2 = 1/3, \beta_1 = 1/5, 1/3$ and 1, respectively. The results indicate that dispersivities and mobile porosities in the skin zone for MIMs have significant effects on the concentration distribution curves in the mobile domain. The explanation is similar to that for Figures 3.4 and 3.5. In addition, it can be observed that the solute transfer into the immobile region from the mobile region at a large distance, and the solute transfer between the mobile and immobile regions reach equilibrium in the middle, resulting in the concentration being equal in the two regions (Figure 3.7). Conversely, the solute transfer into the mobile region from the immobile region near the injection well results in a long-tail phenomenon of solute transport.

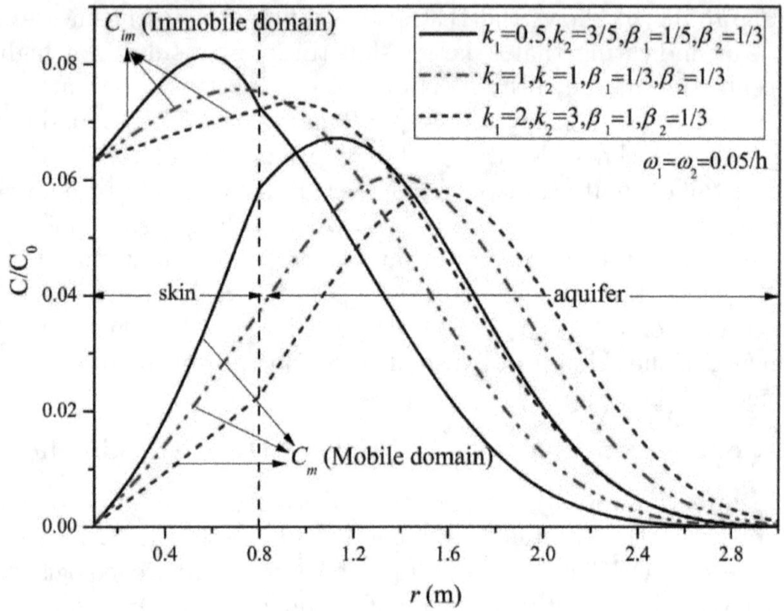

Figure 3.7. Concentration distribution curves for different dispersivities and mobile porosities in the mobile and immobile skin domains at $t_{\text{inj}} = 3\,\text{h}$ with $\theta_{1m} = 0.1, 0.3, 0.5, \theta_{1im} = 0.1, \theta_{2m} = 0.3, \theta_{2im} = 0.1, \alpha_2 = 0.5\,\text{m}, \alpha_1 = 0.25\,\text{m}, 0.5\,\text{m}$ and $1\,\text{m}$, respectively.

In further highlight the differences between the different transport models, Figures 3.8(a) and 3.8(b) show the concentration distribution curves at $t_{\text{inj}} = 3\,\text{h}$ and the BTCs observed at $r = 2\,\text{m}$ with different radial transport models. The parameters of the MIMs solution are given as $r_1 = 0.8\,\text{m}, \theta_{1m} = 0.5, \theta_{1im} = 0.1, \theta_{2m} = 0.3, \theta_{2im} = 0.1, \alpha_1 = 0.25\,\text{m}, \alpha_2 = 0.5\,\text{m}$, and $\omega_1 = \omega_2 = 0.05/\text{h}$, respectively; hence, the dimensionless parameters are $k_1 = 0.5, k_2 = 3/5, \beta_1 = 1/5$ and $\beta_2 = 1/3$, respectively. For the cases of ADE (no skin), and ADEs, the mobile porosity (or effective porosity) is assumed to equal to the total porosity, i.e., $\theta_1 = \theta_{1m} = 0.5, \theta_2 = \theta_{2m} = 0.3$.

Three interesting observations can be found in this figure. First, Figure 3.8(a) shows that the peak value of MIM is lower than that of the ADE because the MIM separates the liquid regime into mobile and immobile regions with the solute transfer between the two regions. Second, the comparison between the MIM and MIMs results shows evidently (Figure 3.8(a)) that the skin leads to a higher

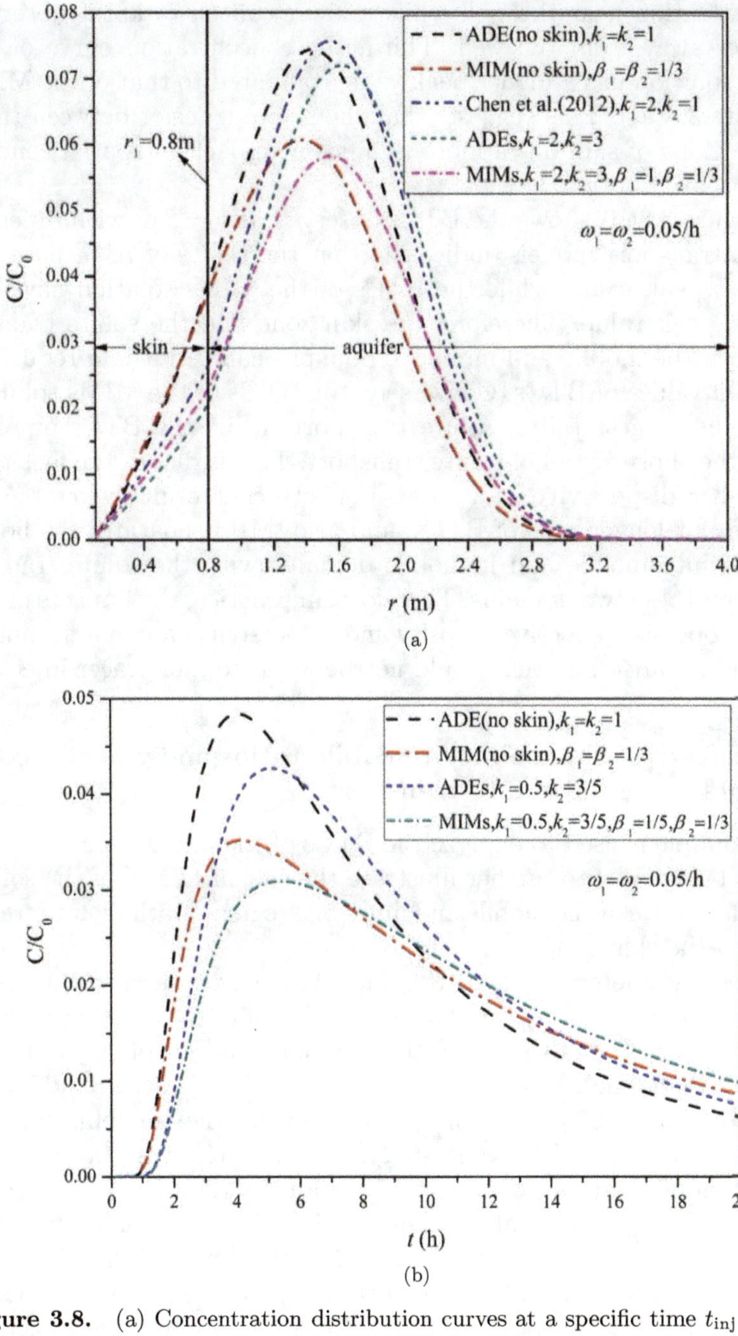

Figure 3.8. (a) Concentration distribution curves at a specific time $t_{\text{inj}} = 3\,\text{h}$ for different radial transport models. (b) BTCs for different radial transport models at $r = 2\,\text{m}$.

concentration near the well while a lower concentration is observed further away from the well. Third, the concentration curve of the ADE solution has a higher peak value compared to that of the MIMs (Figure 3.8(a)). This suggests that the solute transfer between these two regions of skin has a more significant impact on spatial concentration distribution.

Figure 3.8(b) shows the BTCs observed at $r = 2\,\mathrm{m}$ with different radial transport models, indicating that the BTCs of ADE have the highest peak value, while the BTCs of the MIMs solution have the lowest peak value. Therefore, the skin zone and the solute transfer between the mobile and immobile domains can result in a reduction in peak values of BTCs. Conversely, the BTCs of the MIMs solution have the longest tail of solute transport, while the BTCs of ADE have the shortest tail of solute transport. This is due to the fact that a smaller dispersivity or a larger porosity can result in lower peak values but longer tails of BTCs, and also, MIM separates the liquid regime into mobile and immobile domains with the solute transfer between these two domains. Therefore, in practice, the impacts of the skin properties (effective porosity and dispersivity) and mass transfer around an injection well should not be neglected for tracer injection tests.

(d) Effects of the mobile/immobile ratios and transfer coefficients on radial solute transport

An example is used to analyze the BTCs of the mobile domain using the MIMs model to further illustrate the salient effects of the solute transfer between the mobile and immobile regions on the solute transport in the skin zone.

The parameters of the MIMs model are given as $r_1 = 1\,\mathrm{m}, \alpha_1 = 0.25\,\mathrm{m}, \alpha_2 = 0.5\,\mathrm{m}, \theta_{2m} = 0.3, \theta_{2im} = 0.1$, and $\omega_1 = \omega_2 = 0.05/\mathrm{h}$, respectively. In addition, for the computations involving different mobile/immobile ratios, θ_{1m} and θ_{1im} will change accordingly with the constraint of $\theta_{1m} + \theta_{1im} = 0.4$. Hence, the immobile/mobile ratios $\beta_1 = 1, 3$ and 7, respectively, and a special case (an extreme case where there is no immobile domain in the skin zone) where $\theta_{1m} = 0.4, \theta_{1im} = 0$, is also considered in this section as a reference. A sampling point of $r = 2\,\mathrm{m}$ is picked in the aquifer formation zone for the discussion of BTCs. It is worthwhile to note that some related studies used fixed dispersivities (α_1 and α_2) to analyze the transport process (Dronfield and Silliman, 1993; Zhou *et al.*, 2017). As shown in Eq. (3.67), the relationship between D_1 and α_1 can be simplified as $D_1 = \alpha_1 v_1 = \alpha_1 Q/(2\pi r B\theta_{1m})$ while neglecting the molecular diffusion coefficient.

In Figure 3.9, the BTCs in radial flow are plotted with different mobile/immobile ratios β_1 in the case of constant dispersivity and total porosity. It is obvious that the BTCs are moderately sensitive to the β_1 in the skin zone, and a smaller β_1 means a smaller mobile porosity in the skin zone. The results also show that a smaller mobile porosity in the skin zone for the MIMs model leads to a higher concentration at the early injection stage but a lower concentration at the late stage; the same effects of mobile porosity on the BTCs can be found for the ADEs model as shown in Figure 3.5(b). In addition, it is notable that a smaller mobile porosity in the skin zone for the MIMs model results in a lower peak value of BTCs, while that of the ADEs model results in a higher peak value of BTCs. This highlights the different characteristics between the transports of the

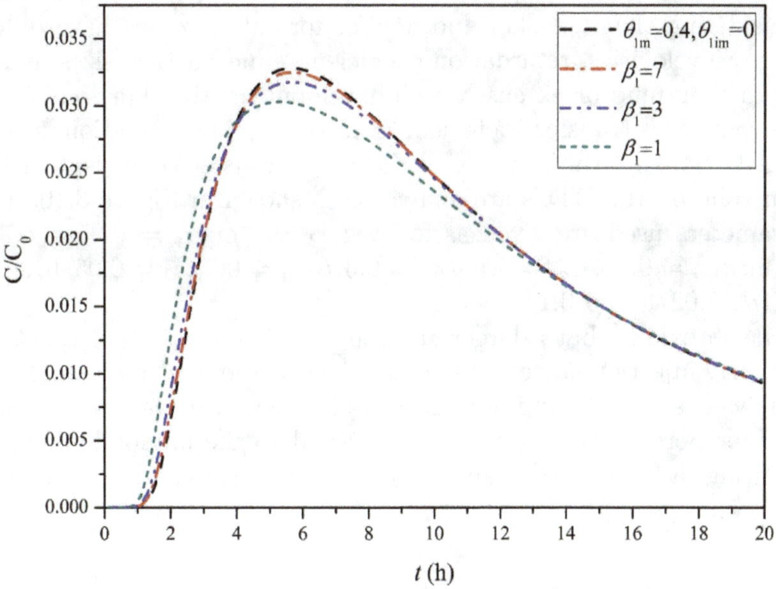

Figure 3.9. BTCs for different mobile/immobile ratios β_1 at $r = 2\,\text{m}$ under the condition of a fixed the total porosity with $\beta_1 = 1, 3,$ and 7, and a special case of $\theta_{1m} = 0.4, \theta_{1im} = 0$ is considered as a reference.

ADEs and MIMs models, which can be elaborated as follows. First, a smaller mobile porosity of the skin zone means a faster pore velocity, according to Eq. (3.69), resulting in faster solute transport in the skin zone. Therefore, the BTCs of the ADEs and MIMs models have higher concentrations at the early injection stage. Second, from time scale calculation, the mass transfer for MIMs model does not have enough time to reach equilibrium at the early injection stage. Therefore, the BTCs of the MIMs model have higher concentrations at the early injection stage. However, at the late injection stage, the mass transfer for the MIMs model has enough time to reach equilibrium. Therefore, a smaller β_1 means a larger immobile domain, which also signifies that a larger percentage of mass can be transferred to the immobile domain, resulting in lower values of the BTCs at the late injection stage.

In addition, the mass transfer coefficient has a significant influence on solute transport in the mobile domain. Several investigations have shown that the mass transfer coefficient is dependent on the pore velocity, retardation coefficient, and particle size (Abgaze and Sharma, 2015; Brusseau, 1992; Harmon *et al.*, 1992; Maraqa, 2001). Considering that the skin and aquifer formation zones have different pore velocities, retardation coefficients, and particle sizes due to the drilling mud or extensive well development, the skin zone has a different mass transfer coefficient from the aquifer formation zone.

Additionally, the impacts of the mass transfer coefficient of the skin zone on the BTCs are analyzed, as shown in Figure 3.10. The parameters used are given as follows: $r_1 = 1\,\text{m}, \alpha_1 = 0.25\,\text{m}, \alpha_2 = 0.5\,\text{m}, \theta_{1m} = 0.4, \theta_{1im} = 0.1, \theta_{2m} = 0.3, \theta_{2im} = 0.1, \omega_2 = 0.05/\text{h}, \omega_1 = 0.04/\text{h}, 0.02/\text{h}$ and $0.1/\text{h}$.

It is obvious that a larger ω_1 results in lower values of BTCs at the early injection stage but higher values at the late stage and lower peak values of BTCs in Figure 3.10. This is because a larger ω_1 means a larger percentage of mass is transferred to the immobile domain, resulting in lower BTC values at the early injection stage and lower peak values.

3.3.1.6 *Analytical solutions of Wang and Zhan (2013)*

In this section, we will focus on radial reactive solute transport in an aquifer-aquitard system. Aquitards consist of fine sediments such

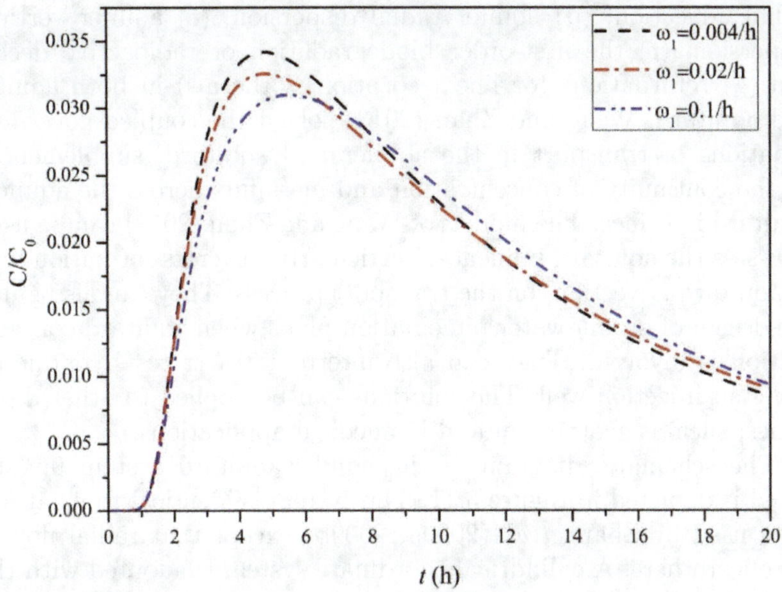

Figure 3.10. BTCs for different mass transfer coefficients at $r = 2\,\text{m}$ with $\omega_2 = 0.05/\text{h}, \omega_1 = 0.04/\text{h}, 0.02/h$, and $0.1/h$, respectively.

as clay and silt in low-energy depositional environments, and they are natural hydrogeological barriers to protect regional aquifers from contamination (Neuzil, 1986; Wang *et al.*, 2020; Yan *et al.*, 2001). Aquitards are widespread throughout the world, and their thickness could vary greatly (Jiao *et al.*, 2010; Ortiz-Zamora and Ortega-Guerrero, 2010). Aquitards have considerable total porosities and a strong capability of absorbing ions, thus are of great importance in the contaminant transport process (Lin and Puls, 2000). Once the contaminants enter an aquitard, they cannot be extracted easily due to the high porosity, low permeability, and the binding force between the contaminant ions and the negatively charged clay surface (Barbour *et al.*, 2012), thus become a long-term source of contamination for the adjacent aquifers (Hendry *et al.*, 2004; Liu and Ball, 2002).

Wang and Zhan (2013) investigated the radial reactive transport near an injecting well in an aquifer-aquitard system, considering the following processes. They are (a) aquitard advection perpendicularly to the aquifer-aquitard interface, (b) aquitard dispersion including both mechanical dispersion and molecular diffusion, (c) aquifer radial advection, (d) aquifer radial dispersion, (e) aquifer vertical dispersion, (f) the first-order biodegradation or radioactive decay, and (g) retardations (or linear sorption isotherms) in both aquifer and aquitard. Wang and Zhan (2013) solved the coupled governing equations of transport in the aquifer and aquitard, supplemented by the continuity of concentration and mass flux across the aquifer-aquitard interface. The models of Wang and Zhan (2013) can be used to assess the aquitard influence, particularly the cross-formation flow or aquitard advection, on the transport process. They can also guide the design of groundwater remediation plans when aquitard contamination is involved. They can also interpret tracer tests conducted using an injection well. The solutions can be applied to other disciplines, such as geothermal and biomedical applications.

The schematic diagram of the aquifer-aquitard system investigated is depicted in Figure 3.11. The system is identical to what has been used by Zhan *et al.* (2009a, 2009b) except that radial flow is of concern here. A cylindrical coordinate system is adopted with the

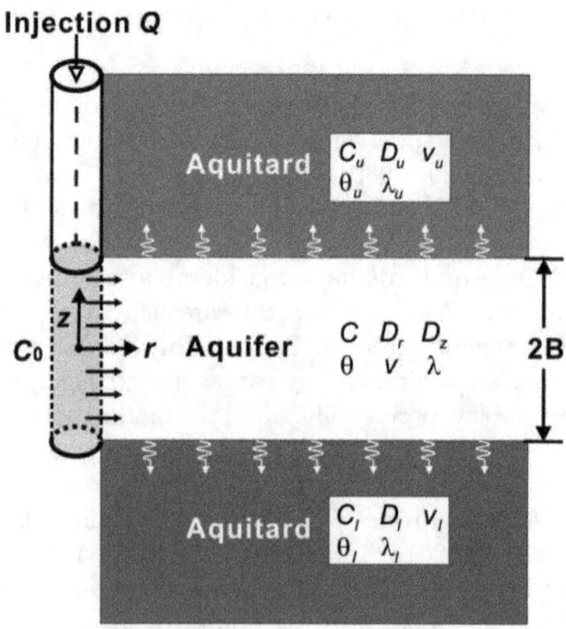

Figure 3.11. Schematic diagram of radial reactive dispersion in an aquifer-aquitard system.

r-axis horizontal and toward infinity from the well, and the z-axis vertical upward. The origin is at the center of the well and located in the plane of symmetry of the aquifer.

Several assumptions used in this section are briefly explained. First, a constant-rate injection well fully penetrates a homogeneous and horizontally isotropic aquifer to generate a steady-state radial flow field. Second, the aquifer-aquitard boundaries are horizontal and extend to infinity. The aquifer has a constant thickness. Third, the aquitards are homogeneous and sufficiently thick in respect to transport, i.e., the aquitard thickness will not affect the transport. Fourth, the injection well maintains a constant concentration (the first-kind boundary). The fourth assumption can be relaxed to accommodate other possible scenarios. For instance, a constant solute flux instead of a constant concentration can be used at the injection well.

(a) Mathematical model

Based on the assumptions mentioned above, the mathematical model for this problem can be described as follows. In the aquifer, the mathematical model is

$$R\frac{\partial C}{\partial t} = \frac{1}{r}\frac{\partial}{\partial r}\left(rD_r\frac{\partial C}{\partial r}\right) + D_z\frac{\partial^2 C}{\partial z^2} - v\frac{\partial C}{\partial r} - \lambda RC \quad (3.122)$$

$$C(r, z, t = 0) = 0 \quad (3.123)$$

$$C(r \to \infty, z, t) = 0 \quad (3.124)$$

$$C(r \to r_w, z, t) = C_0 \quad (3.125)$$

where C is resident concentration in the aquifer [Kg/m^3]; C_0 is a constant concentration [Kg/m^3] at $r \to r_w$; t is time [s]; r is radial distance from the center of the well [m]; r_w is radius of the well [m]; D_r and D_z are radial and vertical dispersion coefficients in the aquifer, respectively [m^2/s]; z represents vertical distance [m]; v is radial groundwater flow velocity in the aquifer [m/s]; λ is a constant rate for the first-order biodegradation, or radioactive decay, or the first-order reaction rate in the aquifer [1/s]; $R = 1+\rho_b K_d/\theta$, is known as the retardation factor of the aquifer [–], K_d is the equilibrium distribution coefficient [m^3/Kg], ρ_b is the bulk density of the aquifer material [Kg/m^3], and θ is porosity in the aquifer [–].

For the upper aquitard, the mathematical model is

$$R_u \frac{\partial C_u}{\partial t} = D_u \frac{\partial^2 C_u}{\partial z^2} - v_u \frac{\partial C_u}{\partial z} - \lambda_u R_u C_u \tag{3.126}$$

$$C_u(r, z = B, t) = C(r, z = B, t) \tag{3.127}$$

$$\left[\theta_u D_u \frac{\partial C_u(r,z,t)}{\partial z} - \theta_u v_u C_u(r,z,t)\right]\Bigg|_{z=B} = \theta D_z \frac{\partial C(r,z,t)}{\partial z}\Bigg|_{z=B} \tag{3.128}$$

$$C_u(r, z \to \infty, t) = 0 \tag{3.129}$$

$$C_u(r, z, t = 0) = 0 \tag{3.130}$$

where the subscript "u" refers to the parameter in the upper aquitard hereinafter; C_u is resident concentration [Kg/m^3]; D_u is vertical dispersion coefficient [m^2/s]; v_u is leakage flow velocity [m/s]; λ_u is the constant rate for radioactive decay or the first-order reaction rate [1/s]; θ_u is porosity [-]; R_u is the retardation factors [-], all referring to the upper aquitard. B is half vertical thickness of the aquifer [m].

For the lower aquitard, the mathematical model is

$$R_l \frac{\partial C_l}{\partial t} = D_l \frac{\partial^2 C_l}{\partial z^2} - v_l \frac{\partial C_l}{\partial z} - \lambda_l R_l C_l \tag{3.131}$$

$$C_l(r, z = -B, t) = C(r, z = -B, t) \tag{3.132}$$

$$\left[\theta_l D_l \frac{\partial C_l(r,z,t)}{\partial z} - \theta_l v_l C_l(r,z,t)\right]\Bigg|_{z=-B} = \theta D_z \frac{\partial C(r,z,t)}{\partial z}\Bigg|_{z=-B} \tag{3.133}$$

$$C_l(r, z \to -\infty, t) = 0 \tag{3.134}$$

$$C_l(r, z, t = 0) = 0 \tag{3.135}$$

where the subscript l refers to the parameter in the lower aquitard hereinafter; $C_l, D_l, v_l, \lambda_l, \theta_l$, and R_l are defined the same way as their counterparts for the upper aquitard. Dispersion coefficients can be calculated by the following equations:

$$D_r = \alpha_r |v| + D_{0a}, \quad D_z = \alpha_z |v| + D_{0a}, \tag{3.136}$$

$$D_u = \alpha_u |v_u| + D_{0u}, \quad D_l = \alpha_l |v_l| + D_{0l}, \tag{3.137}$$

where D_{0a}, D_{0u}, and D_{0l} are the effective diffusion coefficients in the aquifer, the upper aquitard, and the lower aquitard, respectively [m^2/s]; $D_{0a} = \tau_a D_0, D_{0u} = \tau_u D_0, D_{0l} = \tau_l D_0$, where τ_a, τ_u, and τ_l are the tortuosity values (between 0 and 1) in the aquifer, the upper aquitard, and the lower aquitard, respectively; D_0 is the free water molecular diffusion coefficient of the solute; α_r and α_z are the radial and vertical dispersivity values in the aquifer, respectively [m]; α_u and α_l are the vertical dispersivity values in the upper and lower aquitards, respectively [m]; $v = Q/(4\pi B\theta r)$, where Q is the injection rate (positive) [m^3/s]. Be aware that B used here is the half aquifer thickness, so "4π" is used in above expression of v. Some investigators may define B as the whole aquifer thickness and then "2π" instead of "4π" should be used for calculating v (Chen, 1985). The effective diffusion is expected to be much smaller than the mechanical dispersion in the aquifer, thus is ignored. However, the effective diffusion in either the upper or lower aquitard must be considered because they could be the dominating transport mechanism there, particularly when the aquitard advection is extremely small. For the case without any aquitard advection, the effective diffusion becomes the only transport process for the aquitard of concern.

(b) Dimensionless transform

To investigate the solute transport in an aquifer-aquitard system, the dimensionless variables are defined as

$$C_D = \frac{C}{C_0}, \quad C_{lD} = \frac{C_l}{C_0}, \quad C_{uD} = \frac{C_u}{C_0}, \quad M_{lD} = \frac{M_l}{C_0\alpha_r^2 b},$$

$$M_{uD} = \frac{M_u}{C_0\alpha_r^2 b}, \quad r_D = \frac{r}{\alpha_r}, \quad r_{wD} = \frac{r_w}{\alpha_r}, \quad t_D = \frac{\beta}{\alpha_r^2 R}t, \quad z_D = \frac{z}{B},$$

$$\gamma_u = \frac{v_u}{|v_u|}\frac{\eta_u}{\varphi_u}, \quad \gamma_l = \frac{v_l}{|v_l|}\frac{\eta_l}{\varphi_l}, \quad \delta_u = \frac{\theta\alpha_z\beta}{\theta_u\alpha_r\alpha_u|v_u|}, \quad \delta_l = \frac{\theta\alpha_z\beta}{\theta_l\alpha_r\alpha_l|v_l|},$$

$$\eta_u = \frac{v_u{\alpha_r}^2 R}{\beta B R_u}, \quad \eta_l = \frac{v_l{\alpha_r}^2 R}{\beta B R_l}, \quad \lambda_D = \frac{\lambda R{\alpha_r}^2}{\beta}, \quad \lambda_{uD} = \frac{\lambda_u{\alpha_r}^2 R}{\beta}$$

$$\lambda_{lD} = \frac{\lambda_l{\alpha_r}^2 R}{\beta}, \quad \varphi_u = \frac{|v_u|\alpha_u{\alpha_r}^2 R}{\beta B^2 R_u}, \quad \varphi_l = \frac{|v_l|\alpha_l{\alpha_r}^2 R}{\beta B^2 R_l},$$

$$\chi = \frac{\alpha_r\alpha_z}{B^2}, \quad \beta = Q/(2\pi B\theta) \tag{3.138}$$

Based on the dimensionless variables defined in Eqs. (3.122)–(3.138) can be rewritten in the dimensionless formats as follows. The mathematic model of the aquifer can be expressed as

$$\frac{\partial C_D}{\partial t_D} = \frac{1}{2r_D}\frac{\partial^2 C_D}{\partial r_D^2} + \frac{\chi}{2r_D}\frac{\partial^2 C_D}{\partial z_D^2} - \frac{1}{2r_D}\frac{\partial C_D}{\partial r_D} - \lambda_D C_D \quad (3.139)$$

$$C_D(r_D, z_D, t_D = 0) = 0 \quad (3.140)$$

$$C_D(r_D \to \infty, z_D, t_D) = 0 \quad (3.141)$$

$$C_D(r_D \to r_{wD}, z_D, t_D) = 1 \quad (3.142)$$

The mathematic model of the upper aquitard can be rewritten as

$$\frac{\partial C_{uD}}{\partial t_D} = \varphi_u \frac{\partial^2 C_{uD}}{\partial z_D^2} - \eta_u \frac{\partial C_{uD}}{\partial z_D} - \lambda_{uD} C_{uD} \quad (3.143)$$

$$C_{uD}(r_D, z_D = 1, t_D) = C_D(r_D, z_D = 1, t_D) \quad (3.144)$$

$$\left[\frac{\partial C_{uD}(r_D, z_D, t_D)}{\partial z_D} - \gamma_u C_{uD}(r_D, z_D, t_D)\right]\Bigg|_{z_D=1} = \frac{\delta_u}{r_D}\frac{\partial C_D(r_D, z_D, t_D)}{\partial z_D}\Bigg|_{z_D=1} \quad (3.145)$$

$$C_{uD}(r_D, z_D \to \infty, t_D) = 0 \quad (3.146)$$

$$C_{uD}(r_D, z_D, t_D = 0) = 0 \quad (3.147)$$

The mathematic model of the lower aquitard can be rewritten as

$$\frac{\partial C_{lD}}{\partial t_D} = \varphi_l \frac{\partial^2 C_{lD}}{\partial z_D^2} - \eta_l \frac{\partial C_{lD}}{\partial z_D} - \lambda_{lD} C_{lD}, \quad (3.148)$$

$$C_{lD}(r_D, z_D = -1, t_D) = C_D(r_D, z_D = -1, t_D) \quad (3.149)$$

$$\left[\frac{\partial C_{lD}(r_D, z_D, t_D)}{\partial z_D} - \gamma_l C_{lD}(r_D, z_D, t_D)\right]\Bigg|_{z_D=-1} = \frac{\delta_l}{r_D}\frac{\partial C_D(r_D, z_D, t_D)}{\partial z_D}\Bigg|_{z_D=-1} \quad (3.150)$$

$$C_{lD}(r_D, z_D \to \infty, t_D) = 0 \quad (3.151)$$

$$C_{lD}(r_D, z_D, t_D = 0) = 0 \quad (3.152)$$

where the subscript D refers to the dimensionless terms hereinafter; $\lambda_D, \lambda_{uD},$ and λ_{lD} are the dimensionless first-order reaction rates in the aquifer, upper and lower aquitards, respectively; φ_u and φ_l are the dimensionless dispersions in the upper and lower aquitard, respectively; η_u and η_l are the dimensionless velocity values in the upper and lower aquitards, respectively.

(c) Analytical solutions

Applying the Laplace transform to above equation groups, one has for the aquifer:

$$s\overline{C}_D = \frac{1}{2r_D}\frac{\partial^2 \overline{C}_D}{\partial r_D^2} + \frac{\chi}{2r_D}\frac{\partial^2 \overline{C}_D}{\partial z_D^2} - \frac{1}{2r_D}\frac{\partial \overline{C}_D}{\partial r_D} - \lambda_D \overline{C}_D \quad (3.153)$$

$$\overline{C}_D(r_D \to \infty, z_D, s) = 0 \quad (3.154)$$

$$\overline{C}_D(r_D \to r_{wD}, z_D, s) = \frac{1}{s} \quad (3.155)$$

For the upper aquitard:

$$s\overline{C}_{uD} = \varphi_u \frac{d^2 \overline{C}_{uD}}{dz_D^2} - \eta_u \frac{d\overline{C}_{uD}}{dz_D} - \lambda_{uD}\overline{C}_{uD} \quad (3.156)$$

$$\overline{C}_{uD}(r_D, z_D = 1, s) = \overline{C}_D(r_D, z_D = 1, s) \quad (3.157)$$

$$\left[\frac{d\overline{C}_{uD}(r_D, z_D, s)}{dz_D} - \gamma_u \overline{C}_{uD}(r_D, z_D, s)\right]\Bigg|_{z_D=1} = \frac{\delta_u}{r_D}\frac{d\overline{C}_D(r_D, z_D, s)}{dz_D}\Bigg|_{z_D=1} \quad (3.158)$$

$$\overline{C}_{uD}(r_D, z_D \to \infty, s) = 0 \quad (3.159)$$

For the lower aquitard:

$$s\overline{C}_{lD} = \varphi_l \frac{d^2 \overline{C}_{lD}}{dz_D^2} - \eta_l \frac{d\overline{C}_{lD}}{dz_D} - \lambda_{lD}\overline{C}_{lD} \quad (3.160)$$

$$\overline{C}_{lD}(r_D, z_D = -1, s) = \overline{C}_D(r_D, z_D = -1, s) \quad (3.161)$$

$$\left[\frac{d\overline{C}_{lD}(r_D, z_D, s)}{dz_D} - \gamma_l \overline{C}_{lD}(r_D, z_D, s)\right]\Bigg|_{z_D=-1}$$

$$= \frac{\delta_l}{r_D} \frac{d\overline{C}_D(r_D, z_D, s)}{dz_D}\bigg|_{z_D=-1} \tag{3.162}$$

$$\overline{C}_{lD}(r_D, z_D \to \infty, s) = 0 \tag{3.163}$$

In the Laplace domain, the governing equation is still a partial differential equation, and it is not easy to obtain the solutions directly. The z-axis passes through a lower aquitard, an aquifer, and an upper aquitard. Similar approaches have been employed by Zhan *et al.* (2009a, 2009b) for transport in an aquifer-aquitard system with the unilateral flow, the finite Fourier transform along the z-axis will be employed to separate the x and z components for the solution. The concentration in Laplace domain $\overline{C}_D$ can be written as

$$\overline{C}_D = \sum_{n=0}^{\infty} A_n \overline{C}_{rD}(r_D, n, s)\cos(\omega_n z_D + \mu_n) \tag{3.164}$$

where all terms have been defined and explained in the main text and will not be repeated hereinafter. Substituting Eq. (3.164) into Eq. (3.153), one has

$$\frac{d^2\overline{C}_{rD}}{dr_D^2} - \frac{d\overline{C}_{rD}}{dr_D} - \left[2r_D(\lambda_D + s) + \chi\omega_n^2\right]\overline{C}_{rD} = 0 \tag{3.165}$$

Equation (3.165) becomes an ordinary differential equation, and the general solution is

$$\overline{C}_{rD} = \exp\left(\frac{r_D}{2}\right)\sqrt{\xi_n}\left\{C_1 I_{\frac{1}{3}}\left[\frac{2}{3}\sqrt{2(\lambda_D + s)}\xi_n^{\frac{3}{2}}\right]\right.$$

$$\left. + C_2 K_{\frac{1}{3}}\left[\frac{2}{3}\sqrt{2(\lambda_D + s)}\xi_n^{\frac{3}{2}}\right]\right\} \tag{3.166}$$

$$\xi_n = r_D + \frac{4\chi{\omega_n}^2 + 1}{8(\lambda_D + s)} \tag{3.167}$$

where $I_{1/3}$ and $K_{1/3}$ are modified Bessel functions of the first kind and second kind, 1/3 order, respectively, and N_1 and N_2 are undetermined constants.

Substituting Eq. (3.166) into Eq. (3.164), the analytical solution in the Laplace domain becomes:

$$\overline{C}_D = \sum_{n=0}^{\infty} A_n \exp\left(\frac{r_D}{2}\right)\sqrt{\xi_n}\left\{C_1 I_{\frac{1}{3}}\left[\frac{2}{3}\sqrt{2(\lambda_D+s)}\xi_n^{3/2}\right]\right.$$
$$\left.+ C_2 K_{1/3}\left[\frac{2}{3}\sqrt{2(\lambda_D+s)}\xi_n^{3/2}\right]\right\}\cos\left(\omega_n z_D + \mu_n\right) \quad (3.168)$$

Considering the boundary condition at positive infinity along the r-axis, Eq. (3.154), one has

$$N_1 = 0 \quad (3.169)$$

Since both C_2 and A_n are the unknown constants, C_2 can be incorporated into A_n, and Eq. (3.168) becomes:

$$\overline{C}_D = \exp\left(\frac{r_D}{2}\right)\sum_{n=0}^{\infty} A_n \xi_n^{1/2} K_{1/3}\left[\frac{2}{3}\sqrt{2(\lambda_D+s)}\xi_n^{3/2}\right]\cos(\omega_n z_D + \mu_n) \quad (3.170)$$

Considering the boundary condition at the injection well Eq. (3.155) yields

$$\sum_{n=0}^{\infty} A_n \xi_n^{\frac{1}{2}}(r_{wD}) K_{1/3}\left[\frac{2}{3}\sqrt{2(\lambda_D+s)}\xi_n^{\frac{3}{2}}(r_{wD})\right]\cos(\omega_n z_D + \mu_n)$$
$$= \frac{1}{s}\exp\left(-\frac{r_{wD}}{2}\right) \quad (3.171)$$

Conducting the inverse Fourier transform of Eq. (3.171) results in

$$A_n = \frac{4}{s}\frac{F_1(\omega_n,\mu_n)}{F_2(\omega_n,\mu_n)}\exp\left(-\frac{r_{wD}}{2}\right)$$
$$\times\left\{\xi_n^{\frac{1}{2}}(r_{wD}) K_{\frac{1}{3}}\left[\frac{2}{3}\sqrt{2(\lambda_D+s)}\xi_n^{\frac{3}{2}}(r_{wD})\right]\right\}^{-1} \quad (3.172)$$

where $F_1(\omega_n,\mu_n) = \sin(\omega_n)\cos(\mu_n)$, $F_2(\omega_n,\mu_n) = 2\omega_n + \sin(2\omega_n)\cos(2\mu_n)$. If the values of the ω_n and μ_n are known, the constant A_n can be calculated easily. Substituting Eq. (3.172) into Eq. (3.170), the solution of the solute resident concentration in the aquifer is

$$\overline{C}_D = \exp\left(\frac{r_D - r_{wD}}{2}\right)\sum_{n=0}^{\infty}\frac{4}{s}\frac{\sin(\omega_n)\cos(\mu_n)}{2\omega_n + \sin(2\omega_n)\cos(2\mu_n)}$$

$$\times\cos(\omega_n z_D + \mu_n)\frac{\xi_n^{1/2}(r_D)}{\xi_n^{1/2}(r_{wD})}\frac{K_{1/3}\left(\frac{2}{3}\sqrt{2(\lambda_D + s)}\xi_n^{3/2}(r_D)\right)}{K_{1/3}\left(\frac{2}{3}\sqrt{2(\lambda_D + s)}\xi_n^{3/2}(r_{wD})\right)} \tag{3.173}$$

Governing equations in the aquitards in the Laplace domain are second-order linear ordinary differential equations, and their solutions are

$$\overline{C}_{uD} = \overline{C}_D(r_D, z_D = 1, s)$$

$$\times\exp\left\{(z_D - 1)\left[\frac{\eta_u}{2\varphi_u} - \frac{1}{2}\sqrt{\left(\frac{\eta_u}{\varphi_u}\right)^2 + 4\frac{\lambda_{uD} + s}{\varphi_u}}\right]\right\} \tag{3.174}$$

$$\overline{C}_{lD} = \overline{C}_D(r_D, z_D = -1, s)$$

$$\times\exp\left\{(z_D + 1)\left[\frac{\eta_l}{2\varphi_l} + \frac{1}{2}\sqrt{\left(\frac{\eta_l}{\varphi_l}\right)^2 + 4\frac{\lambda_{lD} + s}{\varphi_l}}\right]\right\} \tag{3.175}$$

Substituting Eqs. (3.173)–(3.175) into Eqs. (3.158) and (3.162) leads to:

$$\omega_n\tan(\omega_n + \mu_n) = \frac{2r_D(E + \gamma_u)}{\delta_u} \tag{3.176}$$

$$\omega_n\tan(\omega_n - \mu_n) = \frac{2r_D(F - \gamma_l)}{\delta_l} \tag{3.177}$$

where

$$E = \frac{1}{2}\sqrt{\left(\frac{\eta_u}{\varphi_u}\right)^2 + 4\frac{\lambda_{uD} + s}{\varphi_u}} - \frac{\eta_u}{2\varphi_u} \tag{3.178}$$

$$F = \frac{1}{2}\sqrt{\left(\frac{\eta_l}{\varphi_l}\right)^2 + 4\frac{\lambda_{lD} + s}{\varphi_l}} + \frac{\eta_l}{2\varphi_l} \tag{3.179}$$

After expanding the functions $\tan(\omega_n + \mu_n)$ and $\tan(\omega_n - \mu_n)$, one has

$$\left(\omega_n^2 - \frac{4r_D(E+\gamma_u)}{\delta_u}\frac{r_D(F-\gamma_l)}{\delta_l}\right)\tan(2\omega_n) = 2\omega_n\left(\frac{r_D(E+\gamma_u)}{\delta_u} + \frac{r_D(F-\gamma_l)}{\delta_l}\right) \quad (3.180)$$

$$\left(\omega_n^2 + \frac{4r_D(E+\gamma_u)}{\delta_u}\frac{r_D(F-\gamma_l)}{\delta_l}\right)\tan(2\mu_n) = 2\omega_n\left(\frac{r_D(E+\gamma_u)}{\delta_u} - \frac{r_D(F-\gamma_l)}{\delta_l}\right) \quad (3.181)$$

The equation groups of (3.180) and (3.181) deserve some notation. The solution domain of ω_n (or μ_n) is separated into an infinite series of sub-domains with a period of $\pi/2$, divided by an infinite series of singularity points at $\omega_n = \pm\frac{\pi}{4}, \pm\frac{3\pi}{4}, \pm\frac{5\pi}{4}, \ldots$. Zhan *et al.* (2009a) pointed out that there was only one root of ω_n existing within each sub-domain except the one sub-domain containing $\omega_n = \sqrt{\frac{4r_D(E+\gamma_u)}{\delta_u}\frac{r_D(F-\gamma_l)}{\delta_l}}$, where two roots of ω_n existed. The value of ω_n and μ_n can be obtained by Eqs. (3.180) and (3.181). Detailed discussion of techniques for solving similar equations as Eqs. (3.180) and (3.181) was illustrated in previous works (Zhan *et al.*, 2009a, 2009b).

It is difficult, if not impossible, to conduct analytical inverse Laplace transforms of Eqs. (3.173)–(3.175) to obtain the resident concentrations in the real-time domain, so the numerical inverse Laplace method is used.

An interesting question that we try to answer is this How much mass has been transported to the upper or lower aquitard, and how fast is such a transport process? To answer this question, we calculate the mass stored in the upper or lower aquitard, denoted as M_u and M_l, respectively, in the following:

$$M_u = \int_{r_w}^{\infty} 2\pi r dr \int_{B}^{\infty} C_u dz \quad (3.182)$$

$$M_l = \int_{r_w}^{\infty} 2\pi r dr \int_{-\infty}^{-B} C_l dz \quad (3.183)$$

Converting M_u and M_l into their dimensionless forms defined in Eq. (3.138), and subsequently conducting the Laplace transform, one has

$$\overline{M}_{uD} = \int_{r_{wD}}^{\infty} 2\pi r_D dr_D \int_{1}^{\infty} \overline{C}_D(r_D, z_D = 1, s)\exp\{E(1 - z_D)\}dz_D \tag{3.184}$$

$$\overline{M}_{lD} = \int_{r_{wD}}^{\infty} 2\pi r_D dr_D \int_{-\infty}^{-1} \overline{C}_D(r_D, z_D = -1, s)\exp\{F(1 + z_D)\}dz_D \tag{3.185}$$

where $\overline{M}_{uD}$ and $\overline{M}_{lD}$ represent the dimensionless forms of M_u and M_l in the Laplace domain, respectively. Carrying out the integrations of Eqs. (3.184) and (3.185) along the z-axis leads to:

$$\overline{M}_{uD} = \frac{2\pi}{E}\sum_{n=0}^{\infty} W \int_{r_{wD}}^{\infty} r_D \exp\left(\frac{r_D - r_{wD}}{2}\right)\xi_n^{\frac{1}{2}}(r_D)K_{\frac{1}{3}} \times \left[\frac{2}{3}\sqrt{2(\lambda_D + s)}\xi_n^{\frac{3}{2}}(r_D)\right] dr_D \tag{3.186}$$

$$\overline{M}_{lD} = \frac{2\pi}{F}\sum_{n=0}^{\infty} P \int_{r_{wD}}^{\infty} r_D \exp\left(\frac{r_D - r_{wD}}{2}\right)\xi_n^{\frac{1}{2}}(r_D)K_{\frac{1}{3}} \times \left[\frac{2}{3}\sqrt{2(\lambda_D + s)}\xi_n^{\frac{3}{2}}(r_D)\right] dr_D \tag{3.187}$$

where

$$W = \frac{\frac{4}{s}\frac{F_1(\omega_n,\mu_n)}{F_2(\omega_n,\mu_n)}\cos(\omega_n + \mu_n)}{\xi_n^{\frac{1}{2}}(r_{wD})K_{\frac{1}{3}}\left[\frac{2}{3}\sqrt{2(\lambda_D + s)}\xi_n^{\frac{3}{2}}(r_{wD})\right]} \quad \text{and}$$

$$P = \frac{\frac{4}{s}\frac{F_1(\omega_n,\mu_n)}{F_2(\omega_n,\mu_n)}\cos(\omega_n - \mu_n)}{\xi_n^{\frac{1}{2}}(r_{wD})K_{\frac{1}{3}}\left[\frac{2}{3}\sqrt{2(\lambda_D + s)}\xi_n^{\frac{3}{2}}(r_{wD})\right]}.$$

To obtain M_{uD} and M_{lD}, one needs to conduct the numerical integral and the inverse Laplace transform using the Stehfest method.

(d) Analytical model for the constant flux boundary condition at the well

The solution derived above is for the constant-concentration boundary condition (see Eq. (3.125)). Although constant concentration has been employed broadly in studying solute transport, the constant flux boundary condition (third-type or Robin boundary condition) may be a physically robust alternative because it maintains the mass balance across the boundary. Solutions derived with the constant concentration and constant flux are expected to be very close to each other except at the very beginning of the transport, where a steep concentration gradient (sharp interface) exists near the well. Nevertheless, for the completion of the study, we also provide the solutions for the constant flux boundary condition in the following with the detailed derivation listed in the supplementary materials. If a constant-flux boundary is employed there, one should replace Eq. (3.125) with the following equation:

$$\left(vC - D_r\frac{\partial C}{\partial r}\right)_{r\to r_w} = \frac{QC_0}{4\pi r_w B} \tag{3.188}$$

Based on the definition of the dimensionless parameters in Eq. (3.138), the constant-flux boundary condition Eq. (3.188) becomes:

$$\left(C_D - \frac{\partial C_D}{\partial r_D}\right)_{r_D\to r_{wD}} = 1 \tag{3.189}$$

Conducting Laplace transform to Eq. (3.189) and substituting it into Eq. (3.170), one has

$$\begin{aligned}\sum_{n=0}^{\infty} & A_n\cos(\omega_n z_D + \mu_n)\exp\left(\frac{r_{wD}}{2}\right)\xi_n^{\frac{1}{2}}(r_{wD}) \\ & \times\left\{\frac{1}{2}K_{\frac{1}{3}}(P') - \frac{1}{2}\xi_n^{-1}(r_{wD})K_{\frac{1}{3}}(P') - \frac{1}{2}\sqrt{2(\lambda_D+s)}\xi_n^{\frac{1}{2}}(r_{wD})\right. \\ & \left.\times\left[K_{-\frac{2}{3}}(P') + K_{\frac{4}{3}}(P')\right]\right\} = \frac{1}{s}\end{aligned} \tag{3.190}$$

where $P' = \frac{2}{3}\sqrt{2(\lambda_D+s)}\xi_n^{3/2}(r_{wD})$.

Conducting the inverse Fourier transform of Eq. (3.190) results in

$$A_n = \frac{4}{s}\frac{F_1(\omega_n,\mu_n)}{F_2(\omega_n,\mu_n)}\exp\left(-\frac{r_{wD}}{2}\right)\xi_n^{-\frac{1}{2}}(r_{wD})$$
$$\times\left\{\frac{1}{2}K_{\frac{1}{3}}(P') - \frac{1}{2}\xi_n^{-1}(r_{wD})K_{\frac{1}{3}}(P') - \frac{1}{2}\sqrt{2(\lambda_D+s)}\xi_n^{\frac{1}{2}}(r_{wD})\right.$$
$$\left.\times\left[K_{-\frac{2}{3}}(P') + K_{\frac{4}{3}}(P')\right]\right\}^{-1} \tag{3.191}$$

Thus, the analytical model for the constant flux boundary condition at the well is

$$\overline{C}_D = \exp\left(\frac{r_D - r_{wD}}{2}\right)\sum_{n=0}^{\infty}\frac{4}{s}\frac{F_1(\omega_n,\mu_n)}{F_2(\omega_n,\mu_n)}\cos(\omega_n z_D + \mu_n)\frac{\xi_n^{\frac{1}{2}}(r_D)}{\xi_n^{\frac{1}{2}}(r_{wD})}$$
$$\times\frac{K_{1/3}\left[\frac{2}{3}\sqrt{2(\lambda_D+s)}\xi_n^{3/2}(r_D)\right]}{\frac{1}{2}K_{1/3}(P') - \frac{1}{2}\xi_n^{-1}(r_{wD})K_{1/3}(P') - \sqrt{\frac{\lambda_D+s}{2}}\xi_n^{\frac{1}{2}}(r_{wD})\left[K_{-2/3}(P') + K_{4/3}(P')\right]} \tag{3.192}$$

where $P' = \frac{2}{3}\sqrt{2(\lambda_D+s)}\xi_n^{\frac{3}{2}}(r_{wD})$.

Similarly, the analytical solution of the solute transport in the aquitards can be derived easily, and the solutions in the upper and lower aquitards can also be expressed as Eqs. (3.174) and (3.175), respectively, provided that $\overline{C}_D$ from Eq. (3.192) is used. The main difference in the derivation between Eqs. (3.173) and (3.192) is the boundary condition at the injection well.

(e) Average volumetric sink/source approximation solution

The average volumetric sink/source approximation method (or the so-called AA method) is one of the commonly used approaches to derive the analytical solution for the solute transport problems in a fracture-matrix or aquifer-aquitard system. This approach treats the mass flux across the aquifer-aquitard interface as an averaged volumetric sink/source term in the governing equation of transport in the aquifer, and it ignores the variation of concentration over the aquifer thickness (Chen, 1985; Tang *et al.*, 1981; Tang and Aral, 1992). Chen (1985) derived an analytical solution of radial dispersion in an aquifer-aquitard system based on the AA method, excluding the aquitard leakage (advection) (Chen, 1985). The solution of Chen (1985) is named the AA solution from now on. In the following, we will expand the AA solution by including the aquitard leakage, and this solution will be called the expanded AA solution from now on. By including aquitard leakage, the expanded AA model can be described by the following governing equations:

$$R\frac{\partial C}{\partial t} = \frac{1}{r}\frac{\partial}{\partial r}\left(rD_r\frac{\partial C}{\partial r}\right) - v\frac{\partial C}{\partial r} - \lambda RC - \left(\frac{\theta_u v_u}{2\theta B}C_u - \frac{\theta_u D_u}{2\theta B}\frac{\partial C_u}{\partial z}\right)\Bigg|_{z=B} + \left(\frac{\theta_l v_l}{2\theta B}C_l - \frac{\theta_l D_l}{2\theta B}\frac{\partial C_l}{\partial z}\right)\Bigg|_{z=-B} \tag{3.193}$$

$$R_u\frac{\partial C_u}{\partial t} = D_u\frac{\partial^2 C_u}{\partial z^2} - v_u\frac{\partial C_u}{\partial z} - \lambda_u R_u C_u \tag{3.194}$$

$$R_l\frac{\partial C_l}{\partial t} = D_l\frac{\partial^2 C_l}{\partial z^2} - v_l\frac{\partial C_l}{\partial z} - \lambda_l R_l C_l \tag{3.195}$$

Equations (3.193)–(3.195), with the initial and boundary conditions mentioned in the section on the mathematical model, compose the mathematical model of the expanded AA solution.

Using the dimensionless variables defined in Eqs. (3.138), (3.193)–(3.195) can be rewritten with the dimensionless format as

$$\frac{\partial C_D}{\partial t_D} = \frac{1}{2r_D}\frac{\partial^2 C_D}{\partial r_D^2} - \frac{1}{2r_D}\frac{\partial C_D}{\partial r_D} - \lambda_D C_D - \left(R_u C_{uD} - T_u\frac{\partial C_{uD}}{\partial z_D}\right)\Bigg|_{z_D=1} + \left(R_l C_{lD} - T_l\frac{\partial C_{lD}}{\partial z_D}\right)\Bigg|_{z_D=-1} \tag{3.196}$$

$$\frac{\partial C_{uD}}{\partial t_D} = \phi_u\frac{\partial^2 C_{uD}}{\partial z_D^2} - \eta_u\frac{\partial C_{uD}}{\partial z_D} - \lambda_{uD} C_{uD} \tag{3.197}$$

$$\frac{\partial C_{lD}}{\partial t_D} = \phi_l\frac{\partial^2 C_{lD}}{\partial z_D^2} - \eta_l\frac{\partial C_{lD}}{\partial z} - \lambda_{lD} C_{lD} \tag{3.198}$$

where $T_u = \frac{\chi R}{2\delta_u}, T_l = \frac{\chi R}{2\delta_l}, R_u = \frac{\eta_u \chi R}{2\delta_u \varphi_u}, R_l = \frac{\eta_l \chi R}{2\delta_l \varphi_l}$.

Using Laplace transform, Eqs. (3.196)–(3.198) become:

$$\frac{1}{2r_D}\frac{\partial^2\overline{C}_D}{\partial r_D^2} - \frac{1}{2r_D}\frac{\partial\overline{C}_D}{\partial r_D} - (\lambda_D + s)\overline{C}_D - \left(R_u\overline{C}_{uD} - T_u\frac{\partial\overline{C}_{uD}}{\partial z_D}\right)\Bigg|_{z_D=1} + \left(R_l\overline{C}_{lD} - T_l\frac{\partial\overline{C}_{lD}}{\partial z_D}\right)\Bigg|_{z_D=-1} = 0, \tag{3.199}$$

$$\varphi_u\frac{\partial^2\overline{C}_{uD}}{\partial z_D^2} - \eta_u\frac{\partial\overline{C}_{uD}}{\partial z_D} - (\lambda_{uD} + s)\overline{C}_{uD} = 0 \tag{3.200}$$

$$\varphi_l\frac{\partial^2\overline{C}_{lD}}{\partial z_D^2} - \eta_l\frac{\partial\overline{C}_{lD}}{\partial z_D} - (\lambda_{lD} + s)\overline{C}_{lD} = 0 \tag{3.201}$$

The general solution of Eq. (3.200) is

$$\overline{C}_{uD} = \exp\left(\frac{\eta_u z_D}{2\phi_u}\right)\left[C_1\exp\left(\frac{z_D\varphi_u}{2}\right) + C_2\exp\left(\frac{-z_D\varphi_u}{2}\right)\right] \tag{3.202}$$

$$\phi_u = \sqrt{\left(\frac{\eta_u}{\varphi_u}\right)^2 + \frac{4(\lambda_{uD} + s)}{\varphi_u}} \tag{3.203}$$

where C_1 and C_2 are the unknown constants. Considering the boundary conditions, Eqs. (3.157) and (3.159), the values of both C_1 and C_2 can be determined:

$$C_1 = 0 \tag{3.204}$$

$$C_2 = \frac{\overline{C}_D}{\exp\left(-\frac{\varphi_u}{2}\right)\exp\left(\frac{\eta_u}{2\phi_u}\right)} \tag{3.205}$$

Therefore, the concentration in the Laplace domain is

$$\overline{C}_{uD} = \overline{C}_D\exp\left[(1 - z_D)\left(\frac{\phi_u}{2} - \frac{\eta_u}{2\varphi_u}\right)\right]. \tag{3.206}$$

Similarly, the solution of Eq. (3.201) considering the boundary conditions can be derived:

$$\overline{C}_{lD} = \overline{C}_D\exp\left[(1 + z_D)\left(\frac{\phi_l}{2} + \frac{\eta_l}{2\varphi_l}\right)\right]. \tag{3.207}$$

$$\phi_l = \sqrt{\left(\frac{\eta_l}{\varphi_l}\right)^2 + \frac{4(\lambda_{lD} + s)}{\varphi_l}}, \tag{3.208}$$

Substituting Eqs. (3.206) and (3.207) into Eq. (3.199) yields:

$$\frac{\partial^2 \overline{C}_D}{\partial r_D^2} - \frac{\partial \overline{C}_D}{\partial r_D} - 2r_D A \overline{C}_D = 0 \tag{3.209}$$

where

$$A = \lambda_D + s + R_u - R_l + T_u\left(\frac{\phi_u}{2} - \frac{\eta_u}{2\varphi_u}\right) + T_l\left(\frac{\phi_l}{2} + \frac{\eta_l}{2\varphi_l}\right).$$

The general solution of Eq. (3.209) is

$$\begin{aligned}\overline{C}_D = \exp\left(\frac{r_D}{2}\right)\sqrt{\xi(r_D)}\Big\{C_1 I_{\frac{1}{3}}\left[\frac{2}{3}\sqrt{2A}\xi^{\frac{3}{2}}(r_D)\right]\\ + C_2 K_{\frac{1}{3}}\left[\frac{2}{3}\sqrt{2A}\xi^{\frac{3}{2}}(r_D)^{\frac{3}{2}}\right]\Big\}\end{aligned} \tag{3.210}$$

$$\xi(r_D) = r_D + \frac{1}{8A} \tag{3.211}$$

where C_1 and C_2 are the unknown constants to be determined by the boundary conditions.

Based on Eqs. (3.154) and (3.155), the values of C_1 and C_2 can be figured out, and the analytical solution of Eqs. (3.193)–(3.195) in the Laplace domain as follows:

$$\overline{C}_D = \frac{1}{s}\exp\left(\frac{r_D - r_{wD}}{2}\right)\sqrt{\frac{r_D + \frac{1}{8A}}{r_{wD} + \frac{1}{8A}}}\frac{K_{1/3}\left[\frac{2}{3}\sqrt{2A}\xi^{3/2}(r_D)\right]}{K_{1/3}\left[\frac{2}{3}\sqrt{2A}\xi^{3/2}(r_{wD})\right]} \tag{3.212}$$

$$\overline{C}_{uD} = \overline{C}_D \exp\left\{(1 - z_D)\left[\frac{1}{2}\sqrt{\left(\frac{\eta_u}{\varphi_u}\right)^2 + \frac{4(\lambda_{uD} + s)}{\varphi_u}} - \frac{\eta_u}{2\varphi_u}\right]\right\} \tag{3.213}$$

$$\overline{C}_{lD} = \overline{C}_D \exp\left\{(1 + z_D)\left[\frac{1}{2}\sqrt{\left(\frac{\eta_l}{\varphi_l}\right)^2 + \frac{4(\lambda_{lD} + s)}{\varphi_l}} + \frac{\eta_l}{2\varphi_l}\right]\right\} \tag{3.214}$$

where

$$A = \lambda_D + s + \frac{\eta_u \chi R}{2\delta_u \varphi_u} - \frac{\eta_l \chi R}{2\delta_l \varphi_l} + \frac{\chi R}{4\delta_u}\left[\sqrt{\left(\frac{\eta_u}{\varphi_u}\right)^2 + \frac{4(\lambda_{uD}+s)}{\varphi_u}} - \frac{\eta_u}{\varphi_u}\right]$$
$$+ \frac{\chi R}{4\delta_l}\left[\sqrt{\left(\frac{\eta_l}{\varphi_l}\right)^2 + \frac{4(\lambda_{lD}+s)}{\varphi_l}} + \frac{\eta_l}{\varphi_l}\right].$$

When vertical velocity in the aquitard is set as 0, the solutions of Eqs. (3.212)–(3.214) for $z_D > 0$ become the previous AA solution.

Substituting Eqs. (3.212)–(3.214) into Eqs. (3.182) and (3.183) gives the dimensionless mass in the upper and lower aquitards in the Laplace domain as

$$\overline{M}_{uD} = \frac{1}{\frac{1}{2}\sqrt{\left(\frac{\eta_u}{\varphi_u}\right)^2 + \frac{4(\lambda_{uD}+s)}{\varphi_u}} - \frac{\eta_u}{2\varphi_u}} \int_{r_{wD}}^{\infty} 2\pi r_D \overline{C}_D dr_D \tag{3.215}$$

$$\overline{M}_{lD} = \frac{1}{\frac{1}{2}\sqrt{\left(\frac{\eta_l}{\varphi_l}\right)^2 + \frac{4(\lambda_{lD}+s)}{\varphi_l}} + \frac{\eta_l}{2\varphi_l}} \int_{r_{wD}}^{\infty} 2\pi r_D \overline{C}_D dr_D \tag{3.216}$$

Similarly, the solution of mass in the real-time domain can be obtained using the numerical integral method and the Stehfest method for Laplace inversion of Eqs. (3.215) and (3.216).

(f) Comparison of concentration distribution between the new solution and the expanded AA solution

Figure 3.12 shows the comparison of the solution of Wang and Zhan (2013) and the expanded AA solution for the representative dimensionless parameter inputs of $r_{wD} = 0.1$, $\delta_u = 3183$, $\delta_l = 3183$, $\eta_u = 0.09$, $\eta_l = 0.09$, $\lambda_D = 4.71 \times 10^{-6}$, $\lambda_{uD} = 4.71 \times 10^{-7}$, $\lambda_{lD} = 4.71 \times 10^{-7}$, $\varphi_u = 0.0019$, $\varphi_l = 0.0019$, $\chi = 0.2$. In this figure, the new solution and the expanded AA solution include identical upward advection in both the upper and lower aquitards ($v_u = v_l = 0.1$m/day), whereas the AA solution considers a case of $v_u = v_l = 0$. Some observations are notable. First, the AA solution is symmetrical, and is significantly different from the other two solutions. This is because the AA solution considers the transverse

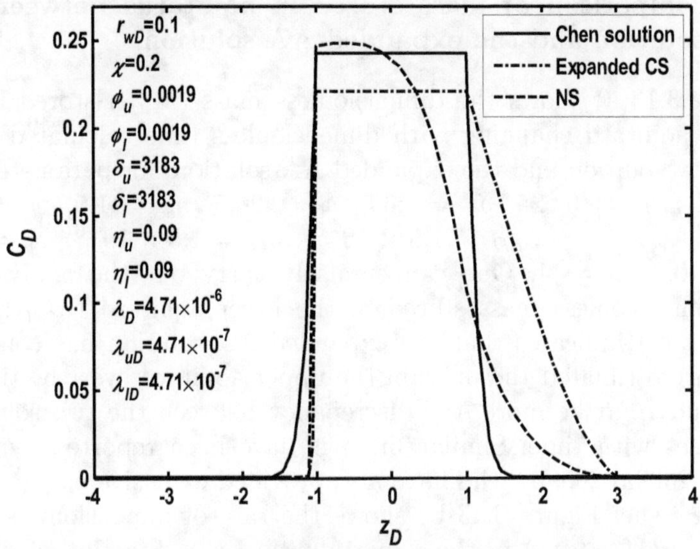

Figure 3.12. Concentration distribution of the new solution and the AA solutions along z_D for $r_D = 0.75$ and $t_D = 1.5$. "NS" represents the solution of Eqs. (3.173)–(3.175), and "Expanded CS" refers to the expanded Chen solution.

diffusion in the aquitard but ignores the advection in the aquitard. Second, the concentration distribution of the AA solution and the expanded AA solution is uniform in the aquifer, while the maximum concentration of the new solution is located at the lower aquifer-aquifer interface because of the upward advection velocities used for both aquitards in this figure. Third, the AA solution and the expanded AA solution overestimate the concentration at the upper aquifer-aquitard interface while they underestimate the concentration at the lower aquifer-aquitard interface. This is because concentrations obtained by the AA solution and the expanded AA solution in the aquifer represent the averaged values over the aquifer thickness, whereas the new solution is capable of showing the detailed concentration variation over the aquifer thickness. Fourth, the penetrating depths of the new solution and the expanded AA solution are almost the same. This observation is similar to the finding of Zhan *et al.* (2009a) for solute transport in an aquifer-aquitard system with unilateral flow.

(g) Comparison of mass stored in aquitards between the new solution and the expanded AA solution

Figure 3.13(a) shows the dimensionless mass (M_{uD}) stored in the upper aquitard changing with dimensionless time calculated using the new solution and the expanded AA solution, for parameter values of $r_{wD} = 0.025$, $\delta_u = 26.5$, $\delta_l = 26.5$, $\eta_u = 1.5$, $\eta_l = 1.5$, $\lambda_D = 1.5 \times 10^{-5}$, $\lambda_{uD} = 1.5 \times 10^{-6}$, $\lambda_{lD} = 1.5 \times 10^{-6}$, $\varphi_u = 0.9$, $\varphi_l = 0.9$, $\chi = 4$. One can similarly carry out the analysis on the dimensionless mass stored in the lower aquitard (M_{uD}). Figure 3.13(a) indicates that the expanded AA solution has considerably overestimated the mass in the upper aquitard over the time of the study. Furthermore, the discrepancy between the two solutions increases with time. Similar findings have been reported by Zhan *et al.* (2009a, 2009b) who have also provided extensive explanations on the issue. Figure 3.13(b) shows the rate of dimensionless mass change (dM_{uD}/dt_D) in the upper aquitard as a function of dimensionless time, calculated with the new solution and the expanded AA solution.

A few interesting observations can be made. First, the dM_{uD}/dt_D value calculated from the new solution increases with t_D sub-linearly (with a decreasing slope) and eventually approaches a constant after t_D greater than 3.0. Second, the dM_{uD}/dt_D value calculated from the expanded AA solution is universally greater than its counterpart calculated from the new solution at any given t_D; it also increases sub-linearly with t_D but with a greater slope than that of the new solution at the given t_D. Third, the dM_{uD}/dt_D value calculated from the expanded AA solution is still steadily increasing with t_D over the range of (0, 6.0) and it will take a much longer time to eventually reach its asymptotic limit than the case of the new solution. From a groundwater remediation perspective, utilization of the AA solution or the expanded AA solution may lead to much longer design time and higher costs for cleaning up the aquifer-aquitard system.

3.3.2 *The two-stage SWPP analytical models*

Single-well push-pull (SWPP) test is a popular technique to characterize the *in situ* geological formations and to remedy the polluted aquifer by a series of biogeochemical reactions (Istok, 2012;

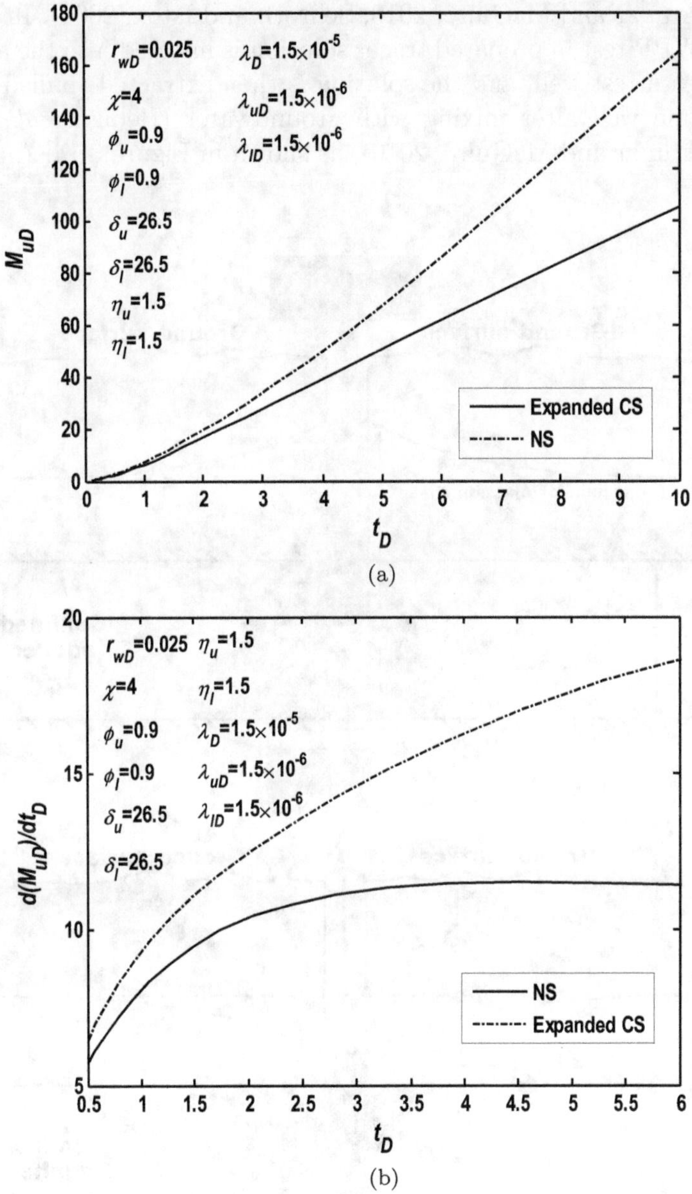

Figure 3.13. (a) Dimensionless mass along the z-axis transported to the upper aquitard (M_u) with t_D. (b) The slope of dimensionless mass transported to the upper aquitard with t_D.

Phanikumar and McGuire, 2010; Schroth and Istok, 2006). In a typical SWPP test, a prepared tracer solution is injected into the aquifer through a test well, and the solution is then extracted (pulled) from the same well after mixing with groundwater (Hebig *et al.*, 2015; Phanikumar and McGuire, 2010), as shown in Figure 3.14. A proper

(a)

(b)

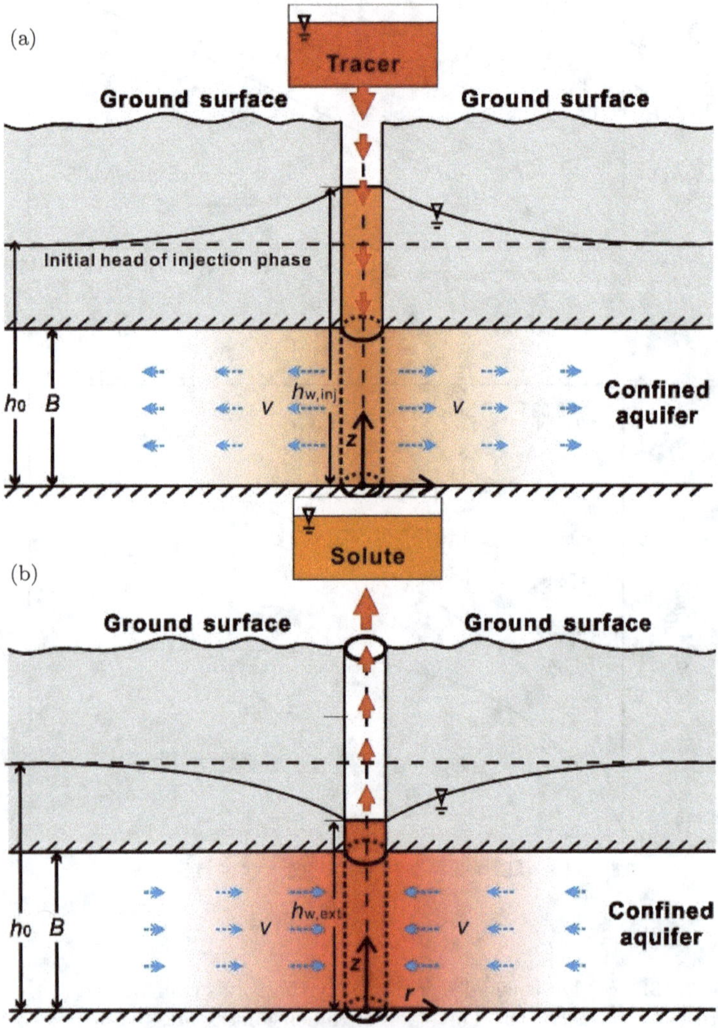

Figure 3.14. Schematic diagram of the SWPP test. (a) In the injection phase. (b) In the extraction phase.

model could best fit the observed BTCs in the extraction phase, allowing for the determination of aquifer parameters.

When characterizing *in situ* aquifer parameters for reactive transport, like dispersivity, biochemical reaction rates, and so on, a SWPP test has some irreplaceable advantages over its competitive test methods such as inter-well test, natural gradient test, etc. For instance, the SWPP test is easier to perform, has a lower cost, and is more efficient than other field tracer tests (Chen *et al.*, 2007; Giese and Powers, 2002; Moench, 1989; Wang and Zhan, 2019). As values of those reactive transport parameters usually could not be measured directly, one must use a properly crafted mathematical model to inversely estimate the values of parameters by best fitting the BTCs. The success or failure of such a parameter estimation exercise largely relies on the credibility of the chosen mathematical model, which must faithfully honor the relevant physical processes associated with SWPP tests.

As for the analytical solutions of the SWPP test, they have been widely used for applications due to the high efficiency and great accuracy of the solutions, like the model of Gelhar and Collins (1971) for a fully penetrating well, the model of Schroth and Istok (2005) for a point source/sink well, and the model of Huang *et al.* (2010) for a partially penetrating well, assuming that the advection, the dispersion and the first-order reaction were involved in the transport processes. Haggerty *et al.* (1998) and Snodgrass and Kitanidis (1998) presented a simplified method based on a well-mixed reactor to estimate the first-order and zero-order reaction rates without involving complex numerical modeling. Schroth and Istok (2006) provided two alternative models. One of them was a plug-flow model, and the other was a variably mixed reactor model. Schroth *et al.* (2001) presented a simplified method for estimating retardation factors, based on the model of Gelhar and Collins (1971). Istok *et al.* (2001) extended the models of Haggerty *et al.* (1998) and Snodgrass and Kitanidis (1998) to estimate the Michaelis-Menten kinetic parameters, which were used to describe the microbial respiration in the aquifer.

A careful review of existing mathematical models of SWPP tests shows that one issue has been grossly overlooked: the wellbore mixing effect (Gelhar and Collins, 1971; Istok *et al.*, 1997; Schroth and Istok, 2005; Schroth *et al.*, 2001). When ignoring the mixing effect in the injection phase of the SWPP test, the concentration of the tracer in the wellbore is constant and equal to the concentration of the injected tracer. Without the mixing effect in the extraction phase of the test, the computed BTC at the well represents the value between the wellbore and aquifer (e.g., the well screen), not the wellbore inside. The locations for sampling the groundwater or where the sensors are placed are somewhere inside the wellbore, not in the well screen. Recalling that all the information collected from SWPP tests comes from the wellbore, thus negligence of such a wellbore mixing effect will profoundly affect the accuracy of parameter estimation.

Wang *et al.* (2018) pointed out that the present commonly used three-dimensional (3D) numerical models of the SWPP test could cause significant errors in describing the mixing effect in confined aquifers. A major deficit of those models is the assumption that the wellbore water volume was constant, even when the wellbore mixing effect is taken into consideration, which is certainly erroneous, as the wellbore water volume changes quite dramatically at different phases of the SWPP test. Wang *et al.* (2018) derived a semi-analytical solution in the Laplace domain considering the wellbore mixing effect and numerically inversed the Laplace domain solution to obtain the real-time domain solution. Shi *et al.* (2020) developed new mathematical models with specific consideration of the wellbore mixing effect on the basis of the approximate solution of the SWPP test by Gelhar and Collins (1971) for a two-phase (injection and extraction) SWPP test, the simplified method to estimate the retardation factor by Schroth *et al.* (2001), and the approximate solution by Schroth and Istok (2005). A point to note is that Gelhar and Collins (1971) and Schroth *et al.* (2001) target different parameters to estimate. The approximate solution of Gelhar and Collins (1971) was derived by solving the ADE models, and it was widely used to inversely estimate the dispersivity. The simplified model of Schroth *et al.* (2001) was used to estimate the retardation factor, while it needed both observed BTCs by a conservative tracer SWPP test and a reactive tracer SWPP test. The difference between the models between Gelhar and Collins (1971) and Schroth and Istok (2005) is the well penetration. The former is for a fully penetrating well, and the latter is for a point-source well. The resting and chasing phases that are used in some SWPP tests are omitted in this section for the sake of developing simplified analytical solutions. Recently, Shi *et al.* (2023) developed a new general model considering the mixing effect, skin effect, scale effect, aquitard effect, and media heterogeneity (in which the solute transport is described in a mobile-immobile framework).

In this section, four analytical models of the two-stage SWPP test for estimating aquifer parameters are introduced. The first analytical model is developed by Huang *et al.* (2010), which is an exact analytical solution to 2D equations describing the governing processes acting on a dissolved compound during a modified push-pull test (advection, longitudinal and transverse dispersion, first-order decay, and rate-limited sorption/partitioning in steady, divergent, and convergent flow fields). As mentioned before, the model of Huang *et al.* (2010) ignored the wellbore mixing effect. Thus, the second analytical model, the SWPP test model, including a mixing effect in the wellbore, is developed by Wang *et al.* (2018). Two types of boundary conditions at the wellbore were considered: the RCC and the flux concentration continuity (FCC). Since the complexity of the mathematical expression of the second model, it is more difficult to generalize it in practice. Alternatively, Shi *et al.* (2020) proposed the third model, a simplified model, to estimate the retardation factor for a SWPP test with a reactive tracer, where the dispersivity had to be determined first using a SWPP test with a conservative tracer. The last model is a physically based new model and the associated analytical solutions in the Laplace domain, which was developed by Shi *et al.* (2023). Note that although all four models are two-stage SWPP test models, the last model differs from the first three in that the last model is a SWPP test model based on the MIM model and considers only the injection and chasing processes.

3.3.2.1 *Analytical solutions of Huang et al. (2010)*

The model of Huang *et al.* (2010) accounts for advection, longitudinal and transverse (vertical) dispersion, first-order degradation, and rate-limited interphase partitioning during nonuniform, steady radial flow in a homogeneous, confined aquifer. Constant concentration and the third-type (flux) boundary conditions are employed at the well screen. Figure 3.15 depicts the modeled scenario where in a fully screened well of length M, tracer is released from z_b to z_t.

In the model of Huang *et al.* (2010), flow is assumed to be in the radial direction only, with mechanical dispersion occurring in the

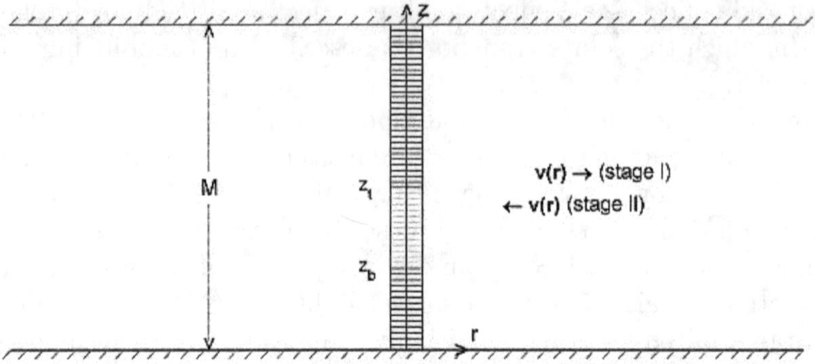

Figure 3.15. Conceptual model and coordinate system.

longitudinal (radial) and transverse (vertical) directions. Analytical solutions are obtained using Laplace and Finite Fourier transforms. The solution is derived assuming a two-stage SWPP test. During the injection phase, the flow is divergent as water is injected into the confined aquifer, and the tracer is released over a portion of the fully penetrating well screen. In the extraction phase, the flow is convergent as the well extracts water from the aquifer and tracer solutions are recovered through the screened section of the well.

(a) Mathematical model

The following equations describe the advective/dispersive transport of a reacting and sorbing compound in a radially symmetric, horizontal flow field:

$$r\left(\theta\frac{\partial C}{\partial t}+\rho_b\frac{\partial S}{\partial t}\right)=a_L\theta v\frac{\partial^2 C}{\partial r^2}\mp\theta v\frac{\partial C}{\partial r}+a_T\theta v\frac{\partial^2 C}{\partial z^2}-r\theta\lambda C-r\rho_b\lambda S \tag{3.217}$$

$$\frac{\partial S}{\partial t}=-\alpha(S-k_dC)-\lambda S \tag{3.218}$$

where C is the compound concentration in the liquid phase, S is the concentration in the solid phase, is the aquifer porosity, ρ_b is the aquifer bulk density, $q = Q/2\pi M$ where Q is well flow-rate, M is the confined aquifer thickness, a_L and a_T are the longitudinal and transverse dispersivity, respectively, λ is a first- order decay constant, α is the non- equilibrium sorption first- order mass transfer rate coefficient, and k_d is the adsorption partitioning coefficient.

Equations (3.217) and (3.218) are solved according to the following initial and boundary conditions during the phases of injection and extraction of the two-stage SWPP test:

In the injection phase:

$$C(r,z,0) = S(r,z,0) = 0 \tag{3.219a}$$

$$-\sigma a_L \frac{\partial C(r,z,t)}{\partial r} + C(r,z,t)$$

$$= \begin{cases} C_0, & r = r_w,\ z_b \le z \le z_t \\ 0, & r = r_w,\ 0 \le z < z_b,\ z_t < z \le M \end{cases} \tag{3.219b}$$

$$C(r \to \infty, z, 0) = 0, \quad 0 \le z \le M \tag{3.219c}$$

$$\frac{\partial C(r,z,t)}{\partial z} = 0, \quad r_w \le r < \infty, \quad z = 0, M \tag{3.219d}$$

In the extraction phase:

$$C(r,z,0) = C_I(r,z,t_1), \quad S(r,z,0) = S_I(r,z,t_1) \tag{3.220a}$$

$$\frac{\partial C(r,z,t)}{\partial z} = 0, \quad r = r_w, \quad 0 \le z \le M \tag{3.220b}$$

$$C(r \to \infty, z, 0) = 0, \quad 0 \le z \le M \tag{3.220c}$$

$$\frac{\partial C(r,z,t)}{\partial z} = 0, \quad r_w \le r < \infty, \quad z = 0, \quad M \tag{3.220d}$$

where C_0 is the concentration of the injected tracer, r_w is the well radius, z_b and z_t are the bottom and top elevation of the tracer injection segment of the well, respectively, σ is a control flag with values 1 or 0 for constant flux or constant concentration boundary conditions, respectively, and $C_I(r,z,t_1)$ and $S_I(r,z,t_1)$ are the initial distributions of liquid and solid phase compound concentrations, respectively, following the completion of the injection phase at the time t_1. Note from Eq. (3.219b) that although the aquifer thickness is M and the well is screened over the entire aquifer thickness, tracer injection is only over the vertical interval between z_t and z_b (where $z_t - z_b \le M$).

(b) Analytical solution in the injection phase

Solution of Eqs. (3.217) and (3.218) can be derived using Laplace transform and Finite Fourier Cosine transform methods.

Applying Laplace transforms with respect to $t(\hat{C})$ and Finite Fourier Cosine transforms with respect to $z(\overline{C})$, Eqs. (3.217) and (3.218) could be rewritten:

$$a_L \frac{d^2\widehat{\overline{C}}}{dr^2} - \frac{d\widehat{\overline{C}}}{dr} - [rf(s) + N(n)]\widehat{\overline{C}} = 0 \tag{3.221}$$

where

$$f(s) = \frac{1}{q}\left[\theta(s+\lambda) + \frac{\alpha k_d \rho_b (s+\lambda)}{s+\lambda+\alpha}\right] N(n) = a_T \frac{n^2\pi^2}{M^2} \tag{3.222}$$

and s and n are the Laplace and Finite Fourier Cosine transform parameters, respectively; $N(n)$ is a coefficient.

Letting $\widehat{\overline{C}} = y(x)e^{\frac{r}{2a_L}}$ and $x = a_L^{-1/3}\left(\frac{1}{4a_L} + rf + N\right)f^{-2/3}$, Eq. (3.221) takes the form of the well-known Airy equation:

$$\frac{d^2y}{dx^2} - xy = 0 \tag{3.223}$$

which has a general solution Abramowitz and Stegun (1968):

$$y = c_1 A_i(x) + c_2 B_i(x) \tag{3.224a}$$

or

$$\widehat{\overline{C}} = c_1 e^{\frac{r}{2a_L}} A_i(x) + c_2 e^{\frac{r}{2a_L}} B_i(x) \tag{3.224b}$$

where $A_i(\cdot)$ and $B_i(\cdot)$ are the Airy functions of the first and second kinds, respectively.

During the injection phase, the Finite Fourier Cosine transform the inner boundary condition (Eq. (3.219b)):

$$-\sigma a_L \frac{\partial \widehat{\overline{C}}}{\partial r} + \widehat{\overline{C}} = \frac{C_0}{p}\Lambda(n), \quad r = r_w \tag{3.225a}$$

where

$$\Lambda(n) = \frac{M}{n\pi}\left[\sin\left(\frac{n\pi z_t}{M}\right) - \sin\left(\frac{n\pi z_b}{M}\right)\right] \tag{3.225b}$$

Applying transformed boundary conditions (Eq. (3.219c)) and (Eq. (3.225a)), the constants in (Eq. (3.225a)) may be evaluated, resulting in

$$\widehat{\overline{C}}_1(r,n,s) = \frac{C_0}{p}\Lambda(n)e^{\frac{r-r_w}{2a_L}}\frac{A_i(x)}{\gamma_1 A_i(x_w) - \gamma_2 A_i'(x_w)} \tag{3.226}$$

The transformed concentration in the solid phase may likewise be written

$$\widehat{\overline{S}}_1(r,n,s) = \frac{\alpha k_d}{s+\lambda+\alpha}\widehat{\overline{C}}_1(r,n,s) \tag{3.227}$$

where the subscript 1 signifies the solutions are for the injection phase of the SWPP test and $\gamma_1 = 1-\sigma/2, \gamma_2 = \sigma a_L^{2/3} f^{1/3}$ and $x_w = a_L^{-1/3}\left(\frac{1}{4a_L} + r_w f + N\right) f^{-2/3}$.

To obtain the solution for the aqueous concentration in real time, setting $\sigma = 0$, Eq. (3.226) becomes:

$$\widehat{\overline{C}}_1(r,n,s) = C_0\Lambda(n)e^{\frac{r-r_w}{2a_L}}\overline{F}_{CC}(r,n,s) \tag{3.228}$$

where

$$\overline{F}_{CC}(r,n,s) = \frac{1}{s}\frac{A_i(x)}{A_i(x_w)} \tag{3.229}$$

Applying the complex inversion formula (Hildebrand, 1976), the inversion of Eq. (3.229) to the real-time domain may be written as

$$F_{CC}(r,n,t) = \frac{1}{2\pi i}\int_{\gamma-i\infty}^{\gamma+i\infty}\overline{F}_{CC}(r,n,p)e^{st}ds \tag{3.230}$$

Following the inversion methodology employed by Chen (1985) with the intermediate variable $f(s)$ substituted for the Laplace parameter (s) in the expression for x_w, the Bromwich contour for the complex inversion may consider multiple branch points to account for the rate-limited adsorption process. Along the imaginary axis of the s plane, s has two segments, respectively, on $[s_1, s_2]$ and $(-\infty, s_3]$, where, $s_1 = -1, s_2 = -(1+a)$ and $s_3 = -(\lambda+\alpha) - \alpha k_d \rho_b/\theta$, that make $f(s)$ negatively valued, which are the cuts linking the branch points, s_1, s_2 and s_3. As indicated by Abramowitz and Stegun (1968), the possible zeros of $A_i(x_w)$,which are the singular points of (3.230), are distributed in the range $(-\infty < x_w < 0)$. The interesting intervals are the cuts along $(s_2 < s < s_1)$ and $(-\infty < s < s_3)$. Based on this property, an improved Bromwich contour depicted in Figure 3.16 can be used to carry out the inversion.

An equivalent integration path with $(\gamma - i\infty < s < \gamma + i\infty)$ is composed of $(L_1+L_2+L_3+L_4+\Gamma)$. Thus, Eq. (3.230) can be written:

$$F_{CC}(r,n,t) = \sum_{i=0}^{3} \mathrm{Res}(s_i) + \frac{1}{2\pi i}\left(\int_{L_1+L_2+L_3+L_4+\Gamma}\right)\cdot F_{CC}(r,n,s)e^{st}ds \tag{3.231}$$

A unique nonzero residue is at $p_0 = 0$, which leads to:

$$\sum_{i=0}^{3} \mathrm{Res}(s_i) = \frac{A_i(x_0)}{A_i(x_{w0})} \tag{3.232}$$

where x_0 and x_{w0} are the specific values of x and x_w, in which the intermediate variable $f(0) = \frac{1}{q}\left\lfloor \theta\lambda + \frac{\alpha k_d \rho_b \lambda}{\lambda+\alpha} \right\rfloor$. Since the integrand

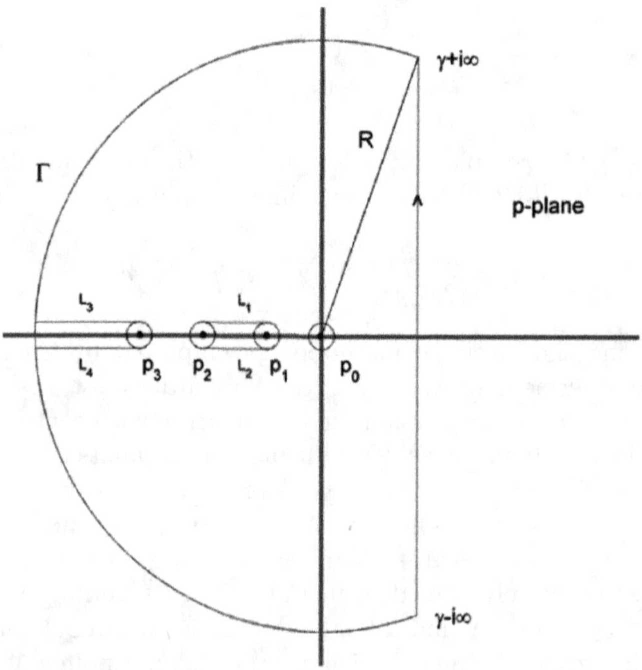

Figure 3.16. Modified Bromwich contour of integration for Laplace transform inversion.

approaches zero when R, the radius of the contour Γ, approaches infinity, there is no contribution along Γ. Along L_1 the integration is written as follows:

$$\frac{1}{2\pi i}\int_{L_1} \overline{F}_{CC}(r,n,s)e^{st}ds = -\frac{1}{\pi i}\int_{\xi_1}^{\xi_2} \frac{e^{-\xi^2 t}}{\xi}\frac{A_i(\eta e^{-2\pi i/3})}{A_i(\eta_w e^{-2\pi i/3})}d\xi \tag{3.233}$$

where

$$\eta = a_L^{-1/3}\left(\frac{1}{4a_L} - rh + N\right)h^{-2/3} \tag{3.234a}$$

$$\eta_w = a_L^{-1/3}\left(\frac{1}{4a_L} - r_w h + N\right)h^{-2/3} \tag{3.234b}$$

$$h = -\frac{1}{q}\left[\theta(\lambda - \xi^2) + \frac{\alpha k_d \rho_b(\lambda - \xi^2)}{\lambda + \alpha - \xi^2}\right] \tag{3.234c}$$

$$\xi_1 = \sqrt{\lambda} \tag{3.234d}$$

$$\xi_\nu = \sqrt{\lambda + \alpha} \tag{3.234e}$$

Following Abramowitz and Stegun (1968):

$$A_i(\eta e^{-2\pi i/3}) = \frac{1}{2}e^{-\pi i/3}[A_i(\eta) + iB_i(\eta)] \tag{3.235}$$

and the right-hand side of Eq. (3.233) takes the alternative form:

$$-\frac{1}{\pi i}\int_{\xi_1}^{\xi_2} \frac{e^{-\xi^2 t}}{\xi} \cdot \frac{A_i(\eta)A_i(\eta_w) + B_i(\eta)B_i(\eta_w) - i[A_i(\eta)B_i(\eta_w) - B_i(\eta)A_i(\eta_w)]}{A_i^2(\eta_w) + B_i^2(\eta_w)}d\xi \tag{3.236}$$

Along L_2, the integration is of the negative conjugate of Eq. (3.236). Combining L_1 and L_2:

$$\frac{1}{2\pi i}\int_{L_1+L_2} \overline{F}_{CC}(r,n,s)e^{st}ds$$
$$= -\frac{2}{\pi}\int_{\xi_1}^{\xi_2} \frac{e^{-\xi^2 t}}{\xi}\frac{A_i(\eta)B_i(\eta_w) - B_i(\eta)A_i(\eta_w)}{A_i^2(\eta_w) + B_i^2(\eta_w)}d\xi \tag{3.237}$$

The contribution from L_3 and L_4 can be derived in the same way. Substituting these path integrations and residuals into Eq. (3.231), the final inversion takes the form:

$$F_{CC}(r,n,t) = F_{CC0}(r) - \frac{2}{\pi}\left(\int_{\xi_1}^{\xi_2} \frac{e^{-\xi^2 t}}{\xi}\Psi(r,n,\xi)d\xi\right.$$
$$\left. + \int_{\xi_3}^{\infty} \frac{e^{-\xi^2 t}}{\xi}\Psi(r,n,\xi)d\xi\right) \tag{3.238}$$

where

$$F_{CC0}(r) = \frac{A_i(x_0)}{A_i(x_{w0})} \tag{3.239a}$$

$$\Psi(r,n,\xi) = \frac{A_i(\eta)B_i(\eta_w) - B_i(\eta)A_i(\eta_w)}{A_i^2(\eta_w) + B_i^2(\eta_w)} \tag{3.239b}$$

$$\xi_3 = \sqrt{\lambda + \alpha(1 + k_d\rho_b/\theta)} \tag{3.239c}$$

Substituting Eq. (3.238) into Eq. (3.228), the solution in the real-time domain can be obtained.

For a constant concentration boundary condition ($\sigma = 0$), Eq. (3.226) may be written in real time:

$$\hat{C}_1(r,n,t) = C_0\Lambda(n)e^{\frac{r-r_w}{2a_L}}F_{CC}(r,n,t) \tag{3.240}$$

where the expressions of $\Lambda(n)$ and $F_{CC}(r,n,t)$ are shown in Eqs. (3.225b) and (3.238), respectively.

As for a constant flux boundary condition ($\sigma = 1$), Eq. (3.226) may be written in real time:

$$\hat{\overline{C}}_1(r,n,s) = C_0\Lambda(n)e^{\frac{r-r_w}{2a_L}}\overline{F}_{CF}(r,n,s) \tag{3.241}$$

where

$$\overline{F}_{CF}(r,n,s) = \frac{1}{s}\frac{2A_i(x)}{A_i(x_w) - 2a_L^{2/3}f^{1/3}A_i'(x_w)} \tag{3.242}$$

Like the derivation of Eq. (3.238), the expression of $F_{CF}(r,n,t)$ can be obtained:

$$F_{CF}(r,n,t) = F_{CF0}(r,n) - \frac{4}{\pi}\left(\int_{\xi_1}^{\xi_2} \frac{e^{-\xi^2 t}}{\xi}\Omega(r,n,\xi)d\xi\right.$$
$$\left. + \int_{\xi_3}^{\infty} \frac{e^{-\xi^2 t}}{\xi}\Omega(r,n,\xi)d\xi\right) \tag{3.243}$$

where

$$F_{CF0}(r,n) = \frac{2A_i(x_0)}{A_i(x_{w0}) - 2a_L^{\frac{2}{3}} f_0^{\frac{1}{3}} A_i'(x_{w0})} \tag{3.244a}$$

$$\Omega(r,n,\xi) = \frac{A_i(\eta)w_1 - B_i(\eta)w_2}{w_1^2 + w_2^2} \tag{3.244b}$$

Substituting Eq. (3.243) into Eq. (3.241), the solution in the real-time domain is obtained:

$$\hat{C}_1(r,n,t) = C_0\Lambda(n)e^{\frac{r-r_w}{2a_L}} F_{CF}(r,n,t) \tag{3.245}$$

In the special case where a tracer is injected for a pulse length t_s less than t_1, the length of water injection during the injection stage, the boundary condition (Eq. (3.219b)) may be rewritten using the Heaviside step function ($H(\cdot)$):

$$-\sigma a_L \frac{\partial C}{\partial r} + C = \begin{cases} C_0 H(t_s - t), & r = r_w,\ z_b \le z \le z_t \\ 0, & r = r_w,\ 0 \le z < z_b,\ z_t < z \le M \end{cases} \tag{3.246}$$

Solutions for this slightly modified scenario are readily obtained using superposition:

$$\hat{C}_1(r,n,t) = C_0\Lambda(n)e^{\frac{r-r_w}{2a_L}} \left[F_{CC}(r,n,t) - F_{CC}(r,n,t-t_s)\right] \tag{3.247}$$

for the constant concentration boundary condition and

$$\hat{C}_1(r,n,t) = C_0\Lambda(n)e^{\frac{r-r_w}{2a_L}} \left[F_{CF}(r,n,t) - F_{CF}(r,n,t-t_s)\right] \tag{3.248}$$

for the constant flux boundary condition.

The solid phase concentration in real time can be obtained by inverting Eq. (3.227):

$$\hat{S}_1(r,n,t) = \alpha k_d \int_0^t e^{-(t-\tau)(\lambda+\alpha)}\hat{C}_1(r,n,\tau)d\tau$$

$$= C_0\alpha k_d\Lambda(n)e^{\frac{r-r_w}{2a_L}}F_{\text{conv}}(r,n,t) \quad (3.249)$$

where

$$F_{\text{conv}}(r,n,t) = F_{CC0}(r)\phi(t) - \frac{2}{\pi}\left(\int_{\xi_1}^{\xi_2}\frac{w(\xi,t)}{\xi}\Psi(r,n,\xi)d\xi\right)$$

$$+\int_{\xi_3}^{\infty}\frac{w(\xi,t)}{\xi}\Psi(r,n,\xi)d\xi \quad (3.250)$$

for a constant concentration boundary condition ($\sigma = 0$), and:

$$F_{\text{conv}}(r,n,t) = F_{CF0}(r)\phi(t) - \frac{4}{\pi}\left(\int_{\xi_1}^{\xi_2}\frac{w(\xi,t)}{\xi}\Omega(r,n,\xi)d\xi\right)$$

$$+\int_{\xi_3}^{\infty}\frac{w(\xi,t)}{\xi}\Omega(r,n,\xi)d\xi \quad (3.251)$$

for a constant flux boundary condition ($\sigma = 1$), and where $\phi(t) = \frac{1}{\lambda+\alpha}[1-e^{-(\lambda+\alpha)t}]$, and $w(\xi,t) = \frac{1}{\lambda+\alpha-\xi^2}[e^{-\xi_t^2} - e^{-(\lambda+\alpha)t}]$.

Finally, using the inverse Finite Fourier Cosine transform, the analytical solutions for concentrations in the aqueous and solid phase during stage 1 of the single- well push- pull test are obtained:

$$C_1(r,z,t) = \frac{1}{M}\hat{C}_1(r,0,t) + \frac{2}{M}\sum_{n=1}^{\infty}\hat{C}_1(r,n,t)\cos\left(\frac{n\pi z}{M}\right) \quad (3.252a)$$

$$S_1(r,z,t) = \frac{1}{M}\hat{S}_1(r,0,t) + \frac{2}{M}\sum_{n=1}^{\infty}\hat{S}_1(r,n,t)\cos\left(\frac{n\pi z}{M}\right) \quad (3.252b)$$

where $\hat{C}_1$ and $\hat{S}_1$ are as given in Eqs. (3.240), (3.245), and (3.249).

(c) Analytical solution in the extraction phase

Solving for the aqueous and solid phase concentrations during the extraction phase, Eq. (3.217) may be rewritten using Laplace and

Finite Fourier Cosine transforms (noting that the advection term is now positive, indicating convergent flow):

$$a_L \frac{d^2\widehat{\overline{C}}}{dr^2} + \frac{d\widehat{\overline{C}}}{dr} - [rf(s) + N(n)]\widehat{\overline{C}} = -g(r, n, s) \tag{3.253}$$

where $g(r, n, s)$ accounts for the nonzero initial conditions resulting from the injection phase tracer push:

$$g(r, n, s) = \frac{r}{q}\left[\theta \hat{C}_1(r, n, t_1) + \frac{\rho_b \alpha}{s + \lambda + \alpha}\hat{S}_1(r, n, t_1)\right] \tag{3.254}$$

The homogeneous form of Eq. (3.253) has two linear, independent solutions:

$$e^{-\frac{r}{2a_L}} A_i(x) \quad \text{and} \quad e^{-\frac{r}{2a_L}} B_i(x) \tag{3.255}$$

Green's function may be used to obtain the general solution to the heterogeneous form of Eq. (3.253):

$$\widehat{\overline{C}} = c_1 e^{-\frac{r}{2a_L}} A_i(x) + c_2 e^{-\frac{r}{2a_L}} B_i(x) + H(r, n, s) \tag{3.256}$$

where c_1 and c_2 are the integration constants and $H(r, n, s)$ is the specific solution of Eq. (3.253). Applying the boundary conditions (Eqs. (3.220b) and (3.220c)), the solution to Eq. (3.253) written in the transformed domains is

$$\widehat{\overline{C}}_2(r, n, s) = c_1(n, p) e^{-\frac{r}{2a_L}} A_i(x) + H(r, n, s) \tag{3.257}$$

where

$$c_1 = -\frac{\pi}{\gamma}\frac{B_i(x_w) - 2\gamma B_i'(x_w)}{A_i(x_w) - 2\gamma A_i'(x_w)} \int_{r_w}^{\infty} e^{\frac{r}{2a_L}} A_i(x) g(r, n, s) dr \tag{3.258a}$$

$$H = \frac{\pi}{\gamma} e^{-\frac{r}{2a_L}} \left[A_i(x) \int_{r_w}^{r} e^{\frac{r}{2a_L}} B_i(x) g(r, n, s) dr\right.$$
$$\left. + B_i(x) \int_{r}^{\infty} e^{\frac{r}{2a_L}} A_i(x) g(r, n, s) dr\right] \tag{3.258b}$$

$$\gamma = a_L^{2/3} f^{1/3} \tag{3.258c}$$

Given that SWPP tests employ solute concentrations extracted during the extraction of the SWPP test to obtain aquifer parameters, it is important to quantify the concentration at the extraction well $(r = r_w)$. For this condition, Eq. (3.257) simplifies to:

$$\widehat{\overline{C}}_2(r_w, n, s) = \frac{2e^{-\frac{r_w}{2a_L}}}{A_i(x_w) - 2\gamma A_i'(x_w)} \int_{r_w}^{\infty} e^{\frac{r}{2a_L}} A_i(x) g(r, n, s) dr \tag{3.259a}$$

$$\widehat{\overline{C}}_2(r_w, n, s) = \frac{e^{-\frac{r_w}{2a_L}}}{q} \int_{r_w}^{\infty} e^{\frac{r}{2a_L}} \overline{F}_2(r, n, s) r[\theta \hat{C}_1(r, n, t_1) + \frac{\rho_b \alpha}{s + \lambda + \alpha} \hat{S}_1(r, n, t_1)] dr \tag{3.259b}$$

where

$$\overline{F}_2(r, n, s) = \frac{2A_i(x)}{A_i(x_w) - 2\gamma A_i'(x_w)} \quad \text{or} \quad \cdot\overline{F}_2(r, n, s) = s\overline{F}_{CF}(r, n, s).$$

However, employing the Laplace transform of the derivative:

$$L[F'_{CF}(r, n, s)] = \overline{F}_{CF}(r, n, s) - F_{CF}(r, n, 0) \tag{3.260}$$

where $L[\cdot]$ indicates the Laplace transform and recognizing that $F_{CF}(r, n, 0) = 0$, the inversion of $\overline{F}_2(r, n, s)$ is

$$F_2(r, n, t) = F'_{CF}(r, n, t) = \frac{4}{\pi} \left(\int_{\xi_1}^{\xi_2} \xi e^{-\xi^2 t} \Omega(r, n, \xi) d\xi \right) + \int_{\xi_3}^{\infty} \xi e^{-\xi^2 t} \Omega(r, n, \xi) d\xi \tag{3.261}$$

Using the convolution principle:

$$\hat{C}_2(r_w, n, t) = \frac{e^{-\frac{r_w}{2a_L}}}{q} \int_{r_w}^{\infty} e^{\frac{r}{2a_L}} r[\theta F_2(r, n, t) \hat{C}_1(r, n, t_1) + \Phi(r, n, t) \hat{S}_1(r, n, t_1)] dr \tag{3.262}$$

where

$$\Phi(r, n, t) = \rho_b \alpha \int_0^t e^{-(t-\tau)(\lambda+\alpha)} F_2(r, n, \tau) d\tau \tag{3.263}$$

or

$$\Phi(r,n,t) = \frac{4\rho_b\alpha}{\pi}\left(\int_{\xi_1}^{\xi_2} \xi w(\xi,t)\Omega(r,n,\xi)d\xi\right) + \int_{\xi_3}^{\infty} \xi w(\xi,t)\Omega(r,n,\xi)d\xi \tag{3.264}$$

Finally, using the inverse Finite Fourier Cosine transform, the final analytical solution for the solute concentration at the extraction well during the extraction phase of the SWPP test is

$$C_2(r_w,z,t) = \frac{1}{M}\hat{C}_2(r_w,0,t) + \frac{2}{M}\sum_{n=1}^{\infty}\hat{C}_2(r_w,n,t)\cos\left(\frac{n\pi z}{M}\right) \tag{3.265}$$

where $\hat{C}_2$ is as given in Eq. (3.262).

Note that the concentration given by Eq. (3.265) is a function of the aqueous and solid phase concentration distribution at the completion of the injection phase (when $t = t_1$). Equation (3.265) provides an analytical solution that may be combined with optimization algorithms to obtain aquifer parameters from data obtained for SWPP tracer tests in confined aquifers with the injection compound released over only a portion of the well screen and rate-limited mass transfer between the aqueous and immobile (solid and non-aqueous phase liquid) phases.

3.3.2.2 *Analytical solutions of Wang et al. (2018)*

Unlike the model of Huang *et al.* (2010), in the mathematical model of Wang *et al.* (2018), the well is assumed to fully penetrate a homogeneous and horizontally isotropic aquifer with a constant thickness. A cylindrical coordinate system is adopted with the r-axis horizontal and toward infinity from the well and the z-axis vertically upward. The origin is at the center of the well and located in the plane of symmetry of the aquifer, as shown in Figures 3.14(a) and 3.14(b).

(a) Mathematical model

When advection, dispersion, and sorption are involved in the solute transport in the SWPP test, the governing equation for Fickian-type solute transport is (Gelhar and Collins, 1971; Wang *et al.*, 2017):

$$R\frac{\partial C}{\partial t} = \frac{1}{r}\frac{\partial}{\partial r}\left(r\alpha_r|v|\frac{\partial C}{\partial r}\right) - v\frac{\partial C}{\partial r} - \lambda RC, \quad r > r_w, \quad t > 0 \tag{3.266}$$

where C is used to represent the resident concentration [Kg/m^3]; t is time [s]; r is the radial distance from the center of the well [m]; r_w is well radius [m]; α_r is the pore scale dispersivity of aquifer [m]; v is radial flow velocity [m/s]; λ is the first-order reaction rate in the aquifer [1/s]; $R = 1 + \rho_b K_d/\theta$ is the retardation factor [–]; K_d is the equilibrium distribution coefficient [m^3/Kg]; ρ_b is bulk density of the aquifer material [Kg/m^3]; θ is porosity [–].

Assuming that the aquifer is initially free of tracer before the test, then the initial condition is

$$C(r, 0) = 0, \quad r > r_w \tag{3.267}$$

The laterally infinite boundary condition is

$$C(\infty, t) = 0, \quad t > 0. \tag{3.268}$$

For the FCC model, the boundary condition at the wellbore-aquifer interface is

$$\left[vC(r,t) - \alpha_r|v|\frac{\partial C(r,t)}{\partial r}\right]_{r=r_w} = [vC_{\text{inj}}(t)]_{r=r_w}, \quad 0 < t \leq t_{\text{inj}} \tag{3.269a}$$

$$\left[vC(r,t) - \alpha_r|v|\frac{\partial C(r,t)}{\partial r}\right]_{r=r_w} = [vC_{\text{ext}}(t)]_{r=r_w}, \quad t_{\text{inj}} < t \leq t_{\text{ext}} \tag{3.269b}$$

where the ending times of the injection and the extraction phases are t_{inj} and t_{ext}, respectively; C_{inj} and C_{ext} in Eqs. (3.269a) and (3.269b) represent the wellbore tracer concentrations at the injection and extraction phases, respectively.

For the RCC model, the boundary condition at the wellbore-aquifer interface is

$$C(r,t)_{r=r_w} = C_{\text{inj}}(t)|_{r=r_w}, \quad 0 < t \leq t_{\text{inj}} \tag{3.270a}$$

$$C(r,t)_{r=r_w} = C_{\text{ext}}(t)|_{r=r_w}, \quad t_{\text{inj}} < t \leq t_{\text{ext}} \tag{3.270b}$$

Interestingly, FCC and RCC perform essentially the same when α_r is 0. In previous studies, C_{inj} was assumed to be constant in Eqs. (3.269a) and (3.270a), and C_{ext} was assumed to be equal to $C(r,t)_{r=r_w}$ in Eq. (3.269b), as can be seen in Gelhar and Collins (1971), Huang *et al.* (2010), Schroth *et al.* (2001) and Chen *et al.* (2017). It is interesting to note that Eq. (3.269b) reduces to the Danckwerts boundary condition if the concentration continuity at the wellbore-aquifer interface is maintained at the extraction phase (Danckwerts, 1995):

$$\left[\frac{\partial C(r,t)}{\partial r}\right]\Bigg|_{r=r_w} = 0, \quad t_{\text{inj}} < t \leq t_{\text{ext}} \tag{3.271}$$

which has been widely used to describe the boundary condition of solute transport at the wellbore in the extraction phase of SWPP tests (Gelhar and Collins, 1971; Jung and Pruess, 2012; Schroth *et al.*, 2001).

However, the above mathematical model overlooks an important fact in the actual operation of an SWPP test: the wellbore mixing effect. In the following, we will revise the above mathematical model to incorporate the wellbore mixing effect.

At the early stage of the test, the tracer could be diluted after mixing with the tracer-free water in the wellbore. Based on the mass balance principle, the tracer concentration in the wellbore, considering the mixing effect, could be described as (Novakowski, 1992b, 1992c):

$$V_{w,\text{inj}}\frac{dC_{\text{inj}}(t)}{dt} = -\xi v(r_w)[C_{\text{inj}}(t) - C_0], \quad 0 < t \leq t_{\text{inj}} \tag{3.272}$$

where $V_{w,\text{inj}}$ is the constant wellbore water volume in the injection phase [m^3] and $V_{w,\text{inj}} = \pi r_w^2 h_{w,\text{inj}}$, in which $h_{w,\text{inj}}$ is the constant water level [m] in the wellbore due to a constant injection rate of Q_{inj} [m^3/s]; C_0 is the concentration of the prepared tracer [Kg/m^3]; $\xi = 2\pi r_w \theta B$ [m^2]; B is the aquifer thickness [m]. Equation (3.272) is briefly explained as follows. One can see that $\xi v(r_w) = \pi r^2 \theta v(r_w) = \pi r^2 q(r_w) = Q_{\text{inj}}$ is the total discharge from the test well into the aquifer, which should also be the rate of injection, where $q(r_w)$ is the Darcian velocity (or specific discharge; m/s) at the wellbore-aquifer interface. Therefore, the tracer mass flux into the wellbore is $Q_{\text{inj}}C_0$, and tracer mass flux leaving the wellbore and entering the aquifer is $Q_{\text{inj}}C_{\text{inj}}(t)$, which is represented by the right-hand side of Eq. (3.272), must equal the tracer mass accumulation rate in the wellbore, which is exactly the term on the left-hand side of Eq. (3.272).

Considering the water in the wellbore is initially free of the tracer, the initial condition becomes:

$$C_{\text{inj}}(t)|_{t=0} = 0 \tag{3.273}$$

The presumptions used in the above Eqs. (3.272) and (3.273) are briefly illustrated as follows. First, the tracer injected is completely mixed with water in the wellbore. This seems reasonable due to the extremely large hydraulic conductivity of the wellbore, since it could make the mixed processes completed almost instantaneously. Meanwhile, the complete mixing process could be achieved by installing an automatic mixer in the wellbore. Second, the water level in the wellbore is $h_{w,\text{inj}}$ before injecting the tracer solution into the well and maintains at that level for the rest of the injection phase.

Similarly, the concentration of the tracer in the extraction phase could be described as

$$V_{w,\text{ext}}\frac{dC_{\text{ext}}(t)}{dt} = -\xi\alpha_r|v|\left.\frac{\partial C(r,t)}{\partial r}\right|_{r=r_w}, \quad t_{\text{inj}} < t \le t_{\text{ext}} \tag{3.274}$$

where $V_{w,\text{ext}}$ is the volume of water in the wellbore in the extraction phase [m^3], and $V_{w,\text{ext}} = \pi r_w^2 h_{w,\text{ext}}$, in which $h_{w,\text{ext}}$ is the water level in the wellbore due to the flow rate Q_{ext} in the extraction phase. Equation (3.274) is the statement of tracer mass balance in the wellbore and can be explained as follows. First, the tracer entering the wellbore is fully mixed, and its concentration should be the same as $C(r_w, t)$ to satisfy the requirement of concentration continuity at the wellbore-aquifer interface:

$$C(r_w^+, t) = C(r_w^-, t), \quad t_{\text{inj}} < t \le t_{\text{ext}} \tag{3.275}$$

The tracer mass entering the wellbore per unit time ($\tilde{M}_{\text{in}}[1/(\text{Kg}\cdot\text{t})]$) is

$$\tilde{M}_{\text{in}} = 2\pi r_w B\left.\left[\theta|v|C(r,t) - \theta\alpha_r|v|\frac{\partial C(r,t)}{\partial r}\right]\right|_{r=r_w^+} \tag{3.276}$$

and notice that dispersive mass flux may not be 0.

To maintain a constant water volume in the wellbore, the flow rate from the aquifer into the wellbore should be the same as the rate of extraction Q_{ext}. Therefore, one has

$$Q_{\text{ext}} = 2\pi r_w B\theta v(r_w, t) \tag{3.277}$$

The tracer mass extracted from the wellbore per unit time ($\tilde{M}_{\text{out}}[1/(\text{Kg}\cdot\text{t})]$) should be:

$$\tilde{M}_{\text{out}} = Q_{\text{ext}} C(r_w, t) = [2\pi r_w B\theta |v| C(r, t)]|_{r=r_w} \tag{3.278}$$

The difference of tracer mass flux entering and leaving the wellbore $\Delta\tilde{M}[1/(\text{Kg}\cdot\text{t})]$ is

$$\Delta\tilde{M} = \tilde{M}_{\text{in}} - \tilde{M}_{\text{out}} = -\left.\left[2\pi r_w B\theta\alpha_r |v| \frac{\partial C(r,t)}{\partial r}\right]\right|_{r=r_w} \tag{3.279}$$

which is the right-hand side of Eq. (3.274), and according to the tracer mass balance requirement, it should equal the accumulation rate of tracer mass inside the wellbore, which is represented by the left-hand side of Eq. (3.274).

Equation (3.274) contains an assumption that the flow field is in a steady state during the extraction phase or that the hydraulic diffusion of the aquifer is infinite. The hydraulic diffusion refers to the ratio of hydraulic conductivity over specific storage of a confined aquifer. Although such an assumption has been widely used to describe solute transport in the SWPP test (Huang *et al.*, 2010; Jung and Pruess, 2012; Schroth and Istok, 2005), its validity may be acceptable only under certain circumstances. This is because the transient flow effect is inevitable when switching from the injection phase to the extraction phase. However, the time of such transition may be very short for the sufficiently large aquifer hydraulic diffusivities (Wang *et al.*, 2017). Therefore, we will test the accuracy of such a treatment by comparing the analytical solution under steady-state flow with the numerical model under transient flow.

The boundary conditions of this section, Eqs. (3.272)–(3.274), are the generalizations of the previously used boundary conditions by specifically including the wellbore mixing effect. For instance, when $r_w \to 0$ (or ignoring the wellbore mixing effect), Eq. (3.272) becomes:

$$C_{\text{inj}}(t) = C_0 \tag{3.280}$$

and Eq. (3.274) reduces to the so-called Danckwerts boundary condition:

$$\left.\frac{\partial C(r,t)}{\partial r}\right|_{r=r_w} = 0 \tag{3.281}$$

Equations (3.280) and (3.281) are the boundary conditions widely used in previous studies such as in Gelhar and Collins (1971), Huang *et al.* (2010), Schroth *et al.* (2001) and Chen *et al.* (2017).

The chemical reaction included in the governing equation focuses on the single species, for example, first-order biodegradation, radioactive decay, or the first-order reaction rate in the aquifer. As for the multispecies problems, the mixing processes among the species also needed to be considered (Paster *et al.*, 2014).

Similar to the treatment in previous studies (Gelhar and Collins, 1971; Huang *et al.*, 2010; Phanikumar and McGuire, 2010; Schroth *et al.*, 2001), the flow field is assumed to be steady state during the SWPP test, and the radial flow velocity is

$$v = \frac{Q}{2\pi B\theta r} = \frac{A}{r} \tag{3.282}$$

where $Q = \begin{cases} Q_{\text{inj}} & 0 < t \le t_{\text{inj}} \\ Q_{\text{ext}} & t_{\text{inj}} < t \le t_{\text{cha}} \end{cases}$; Q_{inj} and Q_{ext} are the flow rates of the well in the injection and extraction phases, respectively. Q_{inj} is positive, while Q_{ext} is negative.

As mentioned previously, such a steady state flow assumption might not always be valid during an SWPP test since the transient flow effect is inevitable at the start of the extraction phase of the test. For instance, for an injection–extraction (two-phase) SWPP test, it needs time for the flow to change direction from outward flow at the injection phase to inward flow at the extraction phase. How swiftly can such a change occur is mostly dependent on the hydraulic diffusivity of the aquifer. Wang *et al.* (2017) found that the accuracy of the steady-state flow assumption improved when the aquifer hydraulic diffusivity increased, as expected. The water level in the wellbore may affect the mixing effect, and such a water level could be calculated using the steady-state analytical solutions of (Papadopulos and Cooper Jr, 1967):

$$h_{w,\text{inj}} = h_0 - Q_{\text{inj}}/(4\pi T)F(u_w, S), \quad 0 < t \le t_{\text{inj}} \tag{3.283a}$$

$$h_{w,\text{ext}} = h_0 - Q_{\text{ext}}/(4\pi T)F(u_w, S), \quad t_{\text{inj}} < t \le t_{\text{ext}} \tag{3.283b}$$

where T is the transmissivity of the aquifer $[\text{m}^2/\text{t}]$; S is the storage of the aquifer [-]; $u_w = r_w^2 S/(4\pi T t)$; $F(u_w, S) = \frac{32S^2}{\pi^2} \int_0^\infty \frac{1-\exp(-\beta^2/(4u_w))}{\beta^2 \Delta\beta} d\beta$; $\Delta\beta = [\beta J_0(\beta) - 2SJ_1(\beta)]^2 + [\beta Y_0(\beta) - 2SY_1(\beta)]^2$; $J_0(\cdot)$ and $J(\cdot)$ are the first kind of Bessel functions with orders 0 and 1, respectively; $Y_0(\cdot)$ and $Y_1(\cdot)$ are the second kind of Bessel functions with orders 0 and 1, respectively.

(b) Dimensionless transform

One can define the following dimensionless parameters: $C_D = \frac{C}{C_0}, C_{\text{inj},D} = \frac{C_{\text{inj}}}{C_0}, C_{\text{ext},D} = \frac{C_{\text{ext}}}{C_0}, t_D = \frac{|A|}{\alpha_r^2 R} t, r_D = \frac{r}{\alpha_r}, r_{wD} = \frac{r_w}{\alpha_r}$, and $\lambda_D = \frac{\alpha_r^2 R}{|A|}\lambda$, where the subscript "$D$" represents the dimensionless parameter, $A = \frac{Q}{2\pi b\theta_m}$.

The dimensionless forms of the governing equation with initial and boundary conditions could be rewritten as

$$\frac{\partial C_D}{\partial t_D} = \frac{1}{r_D}\frac{\partial^2 C_D}{\partial r_D^2} - \frac{|A|}{A}\frac{1}{r_D}\frac{\partial C_D}{\partial r_D} - \lambda_D C_D, \quad t_D > 0 \tag{3.284}$$

$$C_D(r_D, t_D)|_{t_D=0} = 0, \quad r_D > r_{wD} \tag{3.285}$$

$$C_D(r_D, t_D)|_{r_D\to\infty} = 0, \quad t_D > 0 \tag{3.286}$$

$$\left[C_D - \frac{\partial C_D(r_D, t_D)}{\partial r_D}\right]\Bigg|_{r_D=r_{wD}} = C_{\text{inj},D}(t_D), \quad 0 < t_D \le t_{\text{inj},D} \tag{3.287a}$$

$$\left[C_D - \frac{\partial C_D(r_D, t_D)}{\partial r_D}\right]\Bigg|_{r_D=r_{wD}} = C_{\text{ext},D}(t_D), \quad t_{\text{inj},D} < t_D \le t_{\text{ext},D} \tag{3.287b}$$

$$C_D|_{r_D=r_{wD}} = C_{\text{inj},D}(t_D), \quad 0 < t_D \le t_{\text{inj},D} \tag{3.288a}$$

$$C_D|_{r_D=r_{wD}} = C_{\text{ext},D}(t_D), \quad t_{\text{inj},D} < t_D \le t_{\text{ext},D} \tag{3.288b}$$

where Eqs. (3.287a) and (3.287b) are the FCC boundary conditions at the wellbore, and Eqs. (3.288a) and (3.288b) are the RCC boundary conditions at the wellbore.

In the injection phase, the dimensionless forms of Eqs. (3.272) and (3.273) are

$$\beta_{\text{inj},D} r_{wD} \frac{dC_{\text{inj},D}(t_D)}{dt_D} = 1 - C_{\text{inj},D}(t_D), \quad 0 < t_D \leq t_{\text{inj},D} \tag{3.289a}$$

$$C_{\text{inj},D}(t)\big|_{t=0} = 0, \quad 0 < t_D \leq t_{\text{inj},D} \tag{3.289b}$$

where $\beta_{\text{inj},D} = \frac{V_{w,\text{inj}}}{\xi R \alpha_r}$.

In the extraction phase, the boundary condition of Eq. (3.274) is

$$\beta_{\text{ext},D} \frac{\partial C_D(r_D, t_D)}{\partial t_D}\bigg|_{r_D = r_{w,D}} = \frac{\partial C_D(r_D, t_D)}{\partial r_D}\bigg|_{r_D = r_{w,D}}, \quad t_{\text{inj},D} < t_D \leq t_{\text{ext},D} \tag{3.290}$$

where $\beta_{\text{ext},D} = -\frac{V_{w,\text{ext}} r_{wD}}{\xi R \alpha_r}$.

(c) Analytical solution with RCC model

In this model with a wellbore mixing effect, the mathematical model is solved using the Laplace transformation and Green's function technique.

(i) Solutions in the injection phase

Conducting Laplace transform on Eqs. (3.284) and (3.285), one has

$$\frac{\partial^2 \overline{C}_D}{\partial r_D^2} - \frac{\partial \overline{C}_D}{\partial r_D} - r_D (s + \lambda_D) \overline{C}_D = 0 \tag{3.291}$$

where the overbar represents the variable in the Laplace domain; s is the Laplace transform parameter with respect to t_D. The initial condition of Eq. (3.285) is included in Eq. (3.291). Since the flow velocity is positive, one has $|A|/A = 1$.

Conducting Laplace transform on Eqs. (3.288a) and (3.288b), the boundary condition of the wellbore is

$$\overline{C}_D\big|_{r_D = r_{wD}} = \frac{1}{s(\beta_{\text{inj},D} r_{wD} s + 1)} \tag{3.292}$$

Similarly, the boundary condition of the infinity in the Laplace domain is

$$\overline{C}_D(r_D, s)|_{r_D \to \infty} = 0 \tag{3.293}$$

Equations (3.291)–(3.293) compose a model of the second-order ordinary differential equation (ODE) with boundary conditions. The general solution of Eq. (3.291) is

$$\overline{C}_D(r_D, s) = a_0 \exp\left(\frac{y_{\text{inj}}}{2}\right) A_i(E^{1/3} y_{\text{inj}}) + b_0 \exp\left(\frac{y_{\text{inj}}}{2}\right) B_i(E^{1/3} y_{\text{inj}}), \quad 0 < t_D \leq t_{\text{inj},D} \tag{3.294}$$

where $y_{\text{inj}} = r_D + \frac{1}{4E}$, $E = s + \lambda_D$; a_0 and b_0 are constants which the boundary conditions could determine; $A_i(\cdot)$ and $B_i(\cdot)$ are the Airy functions of the first kind and second kind, respectively.

Substituting Eq. (3.294) into Eq. (3.293), one has

$$b_0 = 0 \tag{3.295}$$

Substituting Eqs. (3.292) and (3.294) into Eq. (3.288a), one has

$$a_0 \exp\left(\frac{y_{\text{inj},w}}{2}\right) A_i(E^{1/3} y_{\text{inj},w}) = \frac{1}{s(\beta_{\text{inj},D} r_{wD} s + 1)} \tag{3.296}$$

where $A_i'(\cdot)$ is the derivative of $A_i(\cdot)$, and $y_{\text{inj},w} = r_{wD} + \frac{1}{4E}$.

The value of a_0 could be determined from Eq. (3.296):

$$a_0 = \frac{1}{\exp\left(\frac{y_{\text{inj},w}}{2}\right) A_i\left(E^{1/3} y_{\text{inj},w}\right)} \frac{1}{s(\beta_{\text{inj},D} r_{wD} s + 1)} \tag{3.297}$$

Substituting Eqs. (3.297) and (3.295) into Eq. (3.294), one could obtain the Laplace-domain analytical solution of solute transport in the injection phase of the SWPP test.

(ii) Solutions in the extraction phase

Conducting Laplace transformation on Eq. (3.284) in the extraction phase, one has

$$\frac{\partial^2 \overline{C}_D}{\partial r_D^2} + \frac{\partial \overline{C}_D}{\partial r_D} - r_D(s + \lambda_D)\overline{C}_D + r_D C_D(r_D, t)|_{t=t_{\text{inj}}} = 0 \tag{3.298}$$

where $C_D(r_D,t)|_{t=t_{\text{inj}}}$ represents the initial concentration of the SWPP test in the extraction phase, which could be determined by Eq. (3.294).

The boundary condition of Eq. (3.290) in the Laplace domain becomes:

$$\beta_{\text{ext},D} s\overline{C}_D(r_D,s)|_{r_D=r_{wD}} - \beta_{\text{ext},D} C_D(r_D,t)|_{r_D=r_{wD},t=t_{\text{inj}}} = \left.\frac{\partial \overline{C}_D(r_D,s)}{\partial r_D}\right|_{r_D=r_{wD}} \tag{3.299}$$

Like the model of the SWPP test in the injection phase, Eqs. (3.293), (3.298) and (3.299) compose a model of the second-order ODE with boundary conditions. However, the governing equation is an inhomogeneous differential equation. In this section, we use the Green's function method to derive the analytical solution of Eq. (3.298).

Similar to Chen and Woodside (1988), Eq. (3.298) could be transferred into a self-adjoint form:

$$\frac{\partial^2 G}{\partial r_D^2} - \left(r_D\varsigma + \frac{1}{4}\right)G = -F(r_D) \tag{3.300}$$

where $G = \exp(r_D/2)\overline{C}_D$, $F(r_D) = \exp(r_D/2) r_D C_D(r_D,t_D)|_{t_D=t_{\text{inj},D}}$, $\varsigma = s + \lambda_D$.

The boundary conditions of Eqs. (3.286) and (3.299) could be rewritten as

$$G(r_D,s)|_{r_D\to\infty} = 0 \tag{3.301}$$

$$\left.\left[\left(\beta_{\text{ext},D} s + \frac{1}{2}\right)G - \frac{\partial G}{\partial r_D}\right]\right|_{r_D=r_{wD}} = \beta_{\text{ext},D}\exp(r_{wD}/2) C_D(r_D,t)|_{r_D=r_{wD},t=t_{\text{inj}}} \tag{3.302}$$

One could find that the boundary condition of Eq. (3.302) is inhomogeneous, and we need to make it homogeneous. Let $G = U(r_D) + V(r_D)$ and $V(r_D) = A + Br_D$, and substituting them into

Eqs. (3.301) and (3.302) yields:

$$U(r_D, s)|_{r_D\to\infty} = 0 \tag{3.303}$$

$$\left[\left(\beta_{\text{ext},D}s + \frac{1}{2}\right)U - \frac{\partial U}{\partial r_D}\right]\Bigg|_{r_D=r_{wD}} = 0 \tag{3.304}$$

where

$$A = -\frac{\beta_{\text{ext},D}\exp(r_{wD}/2)C_D(r_D,t)|_{r_D=r_{wD},t=t_{\text{inj}}}}{\left(\beta_{\text{ext},D}s+\frac{1}{2}\right)r_{wD} - 1 - \left(\beta_{\text{ext},D}s+\frac{1}{2}\right)r_D|_{r_D\to\infty}}r_D|_{r_D\to\infty},$$

$$B = \frac{\beta_{\text{ext},D}\exp(r_{wD}/2)C_D(r_D,t)|_{r_D=r_{wD},t=t_{\text{inj}}}}{\left(\beta_{\text{ext},D}s+\frac{1}{2}\right)r_{wD} - 1 - \left(\beta_{\text{ext},D}s+\frac{1}{2}\right)r_D|_{r_D\to\infty}}.$$

Defining a spatial operator: $L = -\frac{d^2}{dr_D^2} + \left(r_D\varsigma + \frac{1}{4}\right)$, one has

$$LG = L[U(r_D) + V(r_D)] = LU(r_D) + LV(r_D) = F(r_D) \tag{3.305}$$

and

$$LU(r_D) = F(r_D) - LV(r_D) \tag{3.306}$$

Let $f(r_D) = F(r_D) - LV(r_D)$, one has

$$\frac{\partial^2 U}{\partial r_D^2} - \left(r_D\varsigma + \frac{1}{4}\right)U = -f(r_D) \tag{3.307}$$

where

$$f(r_D) = \exp(r_D/2)r_D C_D(r_D,t)|_{t=t_{\text{inj}}} - \left(r_D\varsigma + \frac{1}{4}\right)(A + Br_D).$$

Right now, the model with inhomogeneous boundary conditions becomes a regular Sturm–Louisville problem. The general solution of Eqs. (3.303), (3.304), and (3.307) is

$$U(r_D, \varsigma; \chi) = \int_{r_{wD}}^{\infty} g(r_D, \varsigma; \chi) f(\chi) d\chi \tag{3.308}$$

where χ is a positive value varying between r_{wD} and ∞ (e.g., $r_{wD} \leq \chi \leq \infty$); $g(r_D, s, \chi)$ is the Green's function, and could be expressed as

$$g(r_D, \varsigma; \chi) = \begin{cases} g_1(r_D, \varsigma; \chi) = C_1 A_i\left(y_{\text{ext}}\right) + C_2 B_i\left(y_{\text{ext}}\right), & r_{wD} \leq y_{\text{ext}} < \chi \\ g_2(r_D, \varsigma; \chi) = C_3 A_i\left(y_{\text{ext}}\right) + C_4 B_i\left(y_{\text{ext}}\right), & \chi < y_{\text{ext}} < \infty \end{cases} \tag{3.309}$$

where $y_{\text{ext}} = \varsigma^{\frac{1}{3}}\left(r_D + \frac{1}{4\varsigma}\right)$; C_1, C_2, C_3 and C_4 are the coefficients which the following conditions could determine (Chen and Woodside, 1988):

(a) $g(r_D, \varsigma; \chi)$ satisfies the model of Eqs. (3.303), (3.304), and (3.307);
(b) $g_1(r_D, \varsigma; \chi) = g_2(r_D, \varsigma; \chi)$;
(c) $dg_2/dr_D|_{r_D=\chi^+} - dg_1/dr_D|_{r_D=\chi^-} = -1$.

Based on Eq. (3.303), one has

$$C_4 = 0 \tag{3.310}$$

Substituting Eq. (3.308) into Eq. (3.301), one has

$$\left.\left[\left(\beta_{\text{ext},D} s g_1 + \frac{1}{2} g_1 - \frac{\partial g_1}{\partial r_D}\right)\right]\right|_{r_D = r_{wD}} = 0 \tag{3.311}$$

which leads to:

$$C_1 = -C_2 W \tag{3.312}$$

where

$$W = \frac{\left(\beta_{\text{ext},D} s + \frac{1}{2}\right) B_i(y_{\text{ext},w}) - \varsigma^{\frac{1}{3}} B_i'(y_{\text{ext},w})}{\left(\beta_{\text{ext},D} s + \frac{1}{2}\right) A_i(y_{\text{ext},w}) - \varsigma^{\frac{1}{3}} A_i'(y_{\text{ext},w})}.$$

According to the condition of (b), one has

$$C_1 A_i\left(y_{\text{ext}}|_{r_D=\chi^+}\right) + C_2 B_i\left(y_{\text{ext}}|_{r_D=\chi^+}\right) = C_3 A_i\left(y_{\text{ext}}|_{r_D=\chi^-}\right) \tag{3.313}$$

Based on the condition of (c), one has

$$\left.\left[C_3 \varsigma^{\frac{1}{3}} A_i'\left(y_{\text{ext}}\right)\right]\right|_{r_D=\chi^+} - \left.\left[C_1 \varsigma^{\frac{1}{3}} A_i'\left(y_{\text{ext}}\right) + C_2 \varsigma^{\frac{1}{3}} B_i'\left(y_{\text{ext}}\right)\right]\right|_{r_D=\chi^-} = -1 \tag{3.314}$$

The values of C_1, C_2 and C_3 could be determined by Eqs. (3.311)–(3.314).

$$C_1 = -\frac{\pi A_i\left(y_{\text{ext}}|_{r_D=\div^+}\right)}{\varsigma^{\frac{1}{3}}}W, \quad C_2 = \frac{\pi A_i\left(y_{\text{ext}}|_{r_D=\chi^+}\right)}{\varsigma^{\frac{1}{3}}}, \quad \text{and}$$

$$C_3 = \frac{\pi A_i\left(y_{\text{ext}}|_{r_D=\chi^+}\right)}{\varsigma^{\frac{1}{3}}}\left[\frac{B_i\left(y_{\text{ext}}|_{r_D=\chi^+}\right)}{A_i\left(y_{\text{ext}}|_{r_D=\chi^+}\right)} - W\right].$$

(d) Analytical solution with FCC model

As for the FCC analytical solution, one could use the similar approach used for deriving the RCC solution.

(i) Solutions in the injection phase

Substituting Eq. (3.294) into Eq. (3.293), one has

$$b_0 = 0 \tag{3.315}$$

Substituting Eq. (3.294) into Eqs. (3.287a) and (3.292) yields:

$$a_0 = \frac{1}{\exp\left(\frac{y_{\text{inj},w}}{2}\right)\left[\frac{A_i(E^{1/3}y_{\text{inj},w})}{2} - E^{1/3}A_i'\left(E^{1/3}y_{\text{inj},w}\right)\right]} \times \frac{1}{s\left(\beta_{\text{inj},D}r_{wD}s+1\right)} \tag{3.316}$$

Substituting Eqs. (3.315) and (3.316) into Eq. (3.294), one could obtain the Laplace-domain analytical solutions of solute transport in the injection phase of the SWPP test.

(ii) Solutions in the extraction phase

As the wellbore boundary condition for the FCC model is the same as the one of the RCC model, the mathematic expression of the FCC solution is the same as the RCC solution in the extraction phase:

$$\overline{C}_D(r_D, s) = [U(r_D, \varsigma; \chi) + A + Br_D]\exp(-r_D/2) \tag{3.317a}$$

$$U(r_D, \varsigma; \chi) = \int_{r_{wD}}^{\infty} g(r_D, \varsigma; \chi)f(\chi)d\chi \tag{3.317b}$$

$$g(r_D, \varsigma; \chi) = \begin{cases} g_1(r_D, \varsigma; \chi) = C_1 A_i\left(y_{\text{ext}}\right) + C_2 B_i\left(y_{\text{ext}}\right), & r_{wD} \leq y_{\text{ext}} < \chi \\ g_2(r_D, \varsigma; \chi) = C_3 A_i\left(y_{\text{ext}}\right), & \chi < y_{\text{ext}} < \infty \end{cases} \tag{3.317c}$$

$$f(\chi) = \exp(\chi/2)\chi C_D(\chi, t)|_{t=t_{\text{inj}}} - \left(\chi\varsigma + \frac{1}{4}\right)(A + B\chi) \tag{3.317d}$$

where $B_i(\cdot)$ are the Airy functions of the second kind; $A_i'(\cdot)$ and $B_i'(\cdot)$ are the derivative of $A_i(\cdot)$ and $B_i(\cdot)$, respectively. χ is a positive value varying between r_{wD} and ∞ (e.g., $r_{wD} \leq \chi \leq \infty$) in Eq. (3.317c). $\varsigma = s + \lambda_D, y_{\text{ext}} = \varsigma^{\frac{1}{3}}\left(r_D + \frac{1}{4\varsigma}\right)$,

$$A = -\frac{\beta_{\text{ext},D} \exp(r_{wD}/2) C_{rD}(r_D, t)|_{r_D=r_{wD}, t=t_{\text{inj}}}}{\left(\beta_{\text{ext},D} s + \frac{1}{2}\right) r_{wD} - 1 - \left(\beta_{\text{ext},D} s + \frac{1}{2}\right) r_D|_{r_D \to \infty}} r_D|_{r_D \to \infty},$$

$$B = \frac{\beta_{\text{ext},D} \exp(r_{wD}/2) C_{rD}(r_D, t)|_{r_D=r_{wD}, t=t_{\text{inj}}}}{\left(\beta_{\text{ext},D} s + \frac{1}{2}\right) r_{wD} - 1 - \left(\beta_{\text{ext},D} s + \frac{1}{2}\right) r_D|_{r_D \to \infty}}$$

$$W = \frac{\left(\beta_{\text{ext},D} s + \frac{1}{2}\right) B_i(y_{\text{ext},w}) - \varsigma^{\frac{1}{3}} B_i'(y_{\text{ext},w})}{\left(\beta_{\text{ext},D} s + \frac{1}{2}\right) A_i(y_{\text{ext},w}) - \varsigma^{\frac{1}{3}} A_i'(y_{\text{ext},w})},$$

$$C_1 = -\frac{\pi A_i\left(y_{\text{ext}}|_{r_D=\chi^+}\right)}{\varsigma^{\frac{1}{3}}} W, \quad C_2 = \frac{\pi A_i\left(y_{\text{ext}}|_{r_D=\chi^+}\right)}{\varsigma^{\frac{1}{3}}},$$

$$C_3 = \frac{\pi A_i\left(y_{\text{ext}}|_{r_D=\chi^+}\right)}{\varsigma^{\frac{1}{3}}} \left[\frac{B_i\left(y_{\text{ext}}|_{r_D=\chi^+}\right)}{A_i\left(y_{\text{ext}}|_{r_D=\chi^+}\right)} - W\right].$$

3.3.2.3 *Analytical solutions of Shi et al. (2020)*

In this section, we will develop new mathematical models with specific consideration of the wellbore mixing effect on the basis of the approximate solution of the SWPP test by Gelhar and Collins (1971) for a two-stage (injection and extraction) SWPP test, the simplified method to estimate the retardation factor by Schroth *et al.* (2001), and the approximate solution by Schroth and Istok (2005). A point to note is that Gelhar and Collins (1971) and Schroth *et al.* (2001) target different parameters to estimate. The approximate solution of Gelhar and Collins (1971) was derived by solving the ADE models, and it was widely used to inversely estimate the dispersivity. The simplified model of Schroth *et al.* (2001) was used to estimate the retardation factor, while it needed both observed BTCs by a conservative tracer SWPP test and a reactive tracer SWPP test. The difference in the models between Gelhar and Collins (1971) and Schroth and Istok (2005) is the well penetration. The former is for a fully penetrating well, and the latter is for a point-source well. This section omits the resting and chasing phases that are used in some SWPP tests to develop simplified analytical solutions.

Like Figure 3.14, a schematic diagram of the SWPP test used in this section is shown in Figure 3.17. The schematic diagram of the point-source SWPP test can be seen in Schroth and Istok (2005). The mathematical model is the same as the ones shown in Section 3.3.2.2.

(a) Approximate solution of the SWPP test with the wellbore mixing effect

In this section, the approximate solution of Gelhar and Collins (1971) will be modified by considering the wellbore mixing effect.

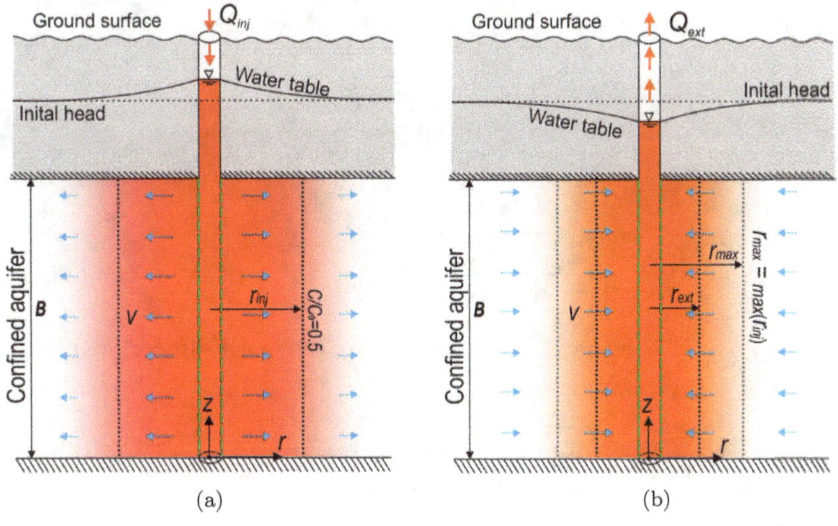

Figure 3.17. Schematic diagram of the SWPP test. (a) In the injection phase. (b) In the extraction phase.

The mathematical form of the new solution is the same as one of Gelhar and Collins (1971), but with different definitions of the involved variables.

In the injection phase, the solution is

$$\frac{C(r,t)}{C_0}=\frac{1}{2}\mathrm{erfc}\left[\frac{r^2-r_{\mathrm{inj}}^2(t)}{\sqrt{\frac{16}{3}\alpha_r\left(r_{\mathrm{inj}}^3(t)-r_w^3\right)}}\right],\quad 0<t\le t_{\mathrm{inj}},\quad r_w\le r \tag{3.318}$$

where $erfc(\cdot)$ is the complementary error function; $r_{\mathrm{inj}}(t)$ is the tracer frontal position at a specific time of t in the injection phase when the dispersive effect is ignored, as shown in Figure 3.14(a). Similar to Gelhar and Collins (1971) and Schroth *et al.* (2001), the tracer frontal position approximated by the location of the spatial concentration distribution where $\frac{C}{C_0}=0.5$, and it is

$$r_{\mathrm{inj}}(t)=\sqrt{\frac{Q_{\mathrm{inj}}t-V_{\mathrm{inj}}}{\pi BnR}+r_w^2},\quad 0<t\le t_{\mathrm{inj}} \tag{3.319}$$

In the extraction phase, the solution is

$$\frac{C(r,t)}{C_0}=\frac{1}{2}\mathrm{erfc}\left[\frac{r^2-r_{\mathrm{ext}}^2(t)}{\sqrt{\frac{16}{3}\alpha_r\left(2r_{\max}^3-r_{\mathrm{ext}}^3(t)-r_w^3\right)}}\right],\quad t>t_{\mathrm{inj}},\ r_w\le r \tag{3.320}$$

where $r_{\mathrm{ext}}(t)$ is the tracer frontal position in the extraction phase, and

$$r_{\mathrm{ext}}(t)=\sqrt{\frac{V_{\mathrm{ext}}-|Q_{\mathrm{ext}}|t}{\pi BnR}+r_{\max}^2},\quad t>t_{\mathrm{inj}} \tag{3.321}$$

where $r_{\max}$ is the maximum value of $r_{\mathrm{inj}}(t)$ in the injection phase, and it is

$$r_{\max}=\sqrt{\frac{Q_{\mathrm{inj}}t_{\mathrm{inj}}-V_{\mathrm{inj}}}{\pi BnR}+r_w^2},\quad t=t_{\mathrm{inj}} \tag{3.322}$$

One point to emphasize is that Eq. (3.320) represents the tracer concentration in the extraction phase in the aquifer, not in the wellbore. The tracer concentration in the wellbore is obtained by substituting Eq. (3.320) into Eq. (3.274):

$$C_{\text{ext}}(t) = C_{\text{ext}}(t=0) + \int_{t_{\text{inj}}}^{t} \frac{Q_{\text{ext}}\alpha_r}{V_{\text{ext}}} \frac{\partial}{\partial r} \times \left[\frac{1}{2}\text{erfc}\left(\frac{r^2 - r_{\text{ext}}^2(t)}{\sqrt{\frac{16}{3}\alpha_r \left(2r_{\text{max}}^3 - r_{\text{ext}}^3(t) - r_w^3\right)}} \right) \right] \Bigg|_{r=r_w} dt, \quad t > t_{\text{inj}} \tag{3.323}$$

As the complementary error function is included in Eq. (3.323), it is not a closed-form solution in a rigorous sense and requires numerical integration to complete the computation, which can be conducted straightforwardly.

As the new solutions are obtained by revising the solution of Gelhar and Collins (1971), it contains an assumption that the concentration in the wellbore at the end of the injection phase should be the same as the concentration of the prepared tracer injected into the wellbore, e.g., $C_{\text{ext}}(t=0) = C_0$.

Equations (3.318) and (3.320) are the new approximate solutions of the SWPP test proposed by considering the wellbore mixing effect. When the wellbore water volume is ignored (e.g., $V_{\text{inj}} = 0$ in Eqs. (3.319) and (3.322), and $V_{\text{ext}} = 0$ in Eq. (3.321)), the new solutions of Eqs. (3.318) and (3.320) reduce to the solutions of Gelhar and Collins (1971).

(b) Simplified model for retardation factor with the wellbore mixing effect

The retardation factor is an important parameter in the description of reactive transport in a porous medium. It was found that the retardation factor and dispersivity were not easy to be estimated simultaneously when best fitting the observed BTCs of reactive transport. Then, Schroth *et al.* (2001) proposed a simplified model to determine the retardation factor:

$$R = \frac{r_{\text{max},c}^2}{r_{\text{max},r}^2} \tag{3.324}$$

where $r_{\text{max},c}$ and $r_{\text{max},r}$ are maximum values of r_{inj} for a conservative tracer and a reactive tracer when the mixing effect is ignored, respectively. $r_{\max,c}$ is computed by Eq. (3.320) by setting $r_w = 0$ and $R = 1$ in the model of Schroth *et al.* (2001). $r_{\max,r}$ is determined by best fitting the BTCs of tracer concentration observed in the wellbore. Unfortunately, this model has not considered the mixing effect in the wellbore at all.

Considering the wellbore mixing effect, a modified method of Schroth *et al.* (2001) becomes:

$$R = \frac{r_{\max,c}^2 - r_w^2}{r_{\max,r}^2 - r_w^2} \tag{3.325}$$

Equations (3.324) and (3.325) demonstrate that one needs to conduct the SWPP test for both conservative tracer and reactive tracer of interest to determine the retardation factor. The procedure of determining the values of $r_{\max,c}$ and $r_{\max,r}$ are similar to that of Schroth *et al.* (2001) and Cassiani *et al.* (2005). First, $r_{\max,c}$, is calculated by Eq. (3.322) by setting $R = 1$. Second, as the values of α_r are constant for both conservative and reactive tracers, it could be obtained by best fitting the observed BTC of the conservative tracer with Eq. (3.320). Third, $r_{\max,r}$ is determined by best fitting the observed BTC of the reactive tracer with Eq. (3.320) as well. Finally, the values of R could be estimated by substituting $r_{\max,c}$ and $r_{\max,r}$ into Eq. (3.324) or Eq. (3.325).

If $r_w = 0$ in Eq. (3.325), the new model to estimate the retardation factor will become the model of Schroth *et al.* (2001) and (Cassiani *et al.*, 2005).

(c) Model extensions to a point source/sink

Schroth and Istok (2005) proposed an approximate solution for the SWPP test with a spherical flow field, which implied that the length of the well screen was much smaller than the aquifer thickness, and the wellbore could be taken as a point source/sink. The wellbore mixing effect was also not considered by Schroth and Istok (2005). Similar to the treatment of a fully penetrating well, the model of a point source/sink well with the wellbore mixing effect could be described as follows.

In the injection phase, one has

$$\frac{C(r,t)}{C_0} = \frac{1}{2}\text{erf}c\left[\frac{r^3 - r_{\text{inj}}^3(t)}{\left(\frac{36}{5}\alpha_r\left(r_{\text{inj}}^5(t) - r_w^5\right)\right)^{1/3}}\right], \quad 0 < t \le t_{\text{inj}} \tag{3.326}$$

where

$$r_{\text{inj}}(t) = \left(\frac{3}{4}\frac{Q_{\text{inj}}t - \pi r_w^2 h_{\text{inj}}}{\pi nR} + r_w^3\right)^{1/3}.$$

In the extraction phase, one has

$$\frac{C(r,t)}{C_0} = \frac{1}{2}\text{erf}c\left[\frac{r^3 - r_{\text{ext}}^3(t)}{\left(\frac{36}{5}\alpha_r\left(2r_{\max}^5 - r_{\text{ext}}^5(t) - r_w^5\right)\right)^{1/3}}\right], \quad 0 < t \le t_{\text{inj}} \tag{3.327}$$

where

$$r_{\max} = \left(\frac{3}{4}\frac{Q_{\text{inj}}t_{\text{inj}} - V_{\text{inj}}}{\pi nR} + r_w^3\right)^{1/3}, \quad \text{and}$$

$$r_{\text{ext}}(t) = \left(r_w^3 - \frac{3}{4}\frac{|Q_{\text{ext}}|\,t - V_{\text{ext}}}{\pi nR}\right)^{1/3}.$$

Similar to the treatment of Eq. (3.323), the concentration in the wellbore is obtained by substituting Eq. (3.327) into Eq. (3.274), and one has

$$C_{\text{ext}}(t) = C_{\text{ext}}(t=0) + \int_{t_{\text{inj}}}^{t}\frac{Q_{\text{ext}}\alpha_r}{V_{\text{ext}}}\frac{\partial}{\partial r}$$
$$\times\left(\frac{1}{2}\text{erf}c\left[\frac{r_w^3 - r_{\text{ext}}^3(t)}{\left(\frac{36}{5}\alpha_r\left(2r_{\max}^5 - r_{\text{ext}}^5(t) - r_w^5\right)\right)^{1/3}}\right]\right)dt,$$
$$t > t_{\text{inj}} \tag{3.328}$$

When $V_{\text{inj}} = V_{\text{ext}} = 0$, the new model of this section reduces to the model of Schroth and Istok (2005).

3.3.2.4 *Analytical solutions of Shi et al. (2023)*

As mentioned before, Shi *et al.* (2023) has extended the model of Li *et al.* (2020) and Hsieh and Yeh (2014). They developed a physically based new model and the associated analytical solutions in the Laplace domain by considering the mixing effect, skin effect, scale effect, aquitard effect, and media heterogeneity. This section will introduce the model of Shi *et al.* (2023).

There are some assumptions were included in the mathematical model of Shi *et al.* (2023): An aquifer is assumed to be confined, homogeneous, horizontally isotropic, with a constant thickness, and fully penetrated by a well from which the solute is injected. A cylindrical coordinate system is established with the r-axis horizontal and the z-axis vertically upward. The origin of the coordinate system is located at the intersection of the well center and the middle elevation of the aquifer. A schematic diagram of the problem is available in Figure 3.14.

(a) Mathematical model

Similar to Hsieh and Yeh (2014) and Li *et al.* (2020), a two-region (skin and formation) model of radial dispersion is employed to describe the skin effect. In the skin zone, the governing equations of radial dispersion are

$$\theta_{m1} R_{m1} \frac{\partial C_{m1}}{\partial t} = \frac{\theta_{m1}}{r} \frac{\partial}{\partial r}\left(r\alpha_1 |v_{a1}| \frac{\partial C_{m1}}{\partial r} \right) - \theta_{m1} v_{a1} \frac{\partial C_{m1}}{\partial r} - \omega_1 (C_{m1} - C_{im1}) - \theta_{m1}\mu_{m1} C_{m1}, \quad r_w \le r \le r_s \tag{3.329a}$$

$$\theta_{im1} R_{im1} \frac{\partial C_{im1}}{\partial t} = \omega_1 (C_{m1} - C_{im1}) - \theta_{im1}\mu_{im1} C_{im1}, \quad r_w \le r \le r_s \tag{3.329b}$$

$$\theta_{m2} R_{m2} \frac{\partial C_{m2}}{\partial t} = \frac{\theta_{m2}}{r} \frac{\partial}{\partial r}\left(r\alpha_2 |v_{a2}| \frac{\partial C_{m2}}{\partial r} \right) - \theta_{m2} v_{a2} \frac{\partial C_{m2}}{\partial r} - \omega_2 (C_{m2} - C_{im2}) - \theta_{m2}\mu_{m2} C_{m2}, \quad r > r_s \tag{3.329c}$$

$$\theta_{im2} R_{im2} \frac{\partial C_{im2}}{\partial t} = \omega_2 (C_{m2} - C_{im2}) - \theta_{im2}\mu_{im2} C_{im2}, \quad r > r_s \tag{3.329d}$$

where the subscripts "m" and "im" refer to parameters in the mobile and immobile domains, respectively; the subscripts "1" and "2" refer to parameters in the skin and formation regions, respectively; C_{m1} and C_{im1} are the mobile and immobile concentrations [Kg/m^3] of the skin zone, respectively; C_{m2} and C_{im2} are the mobile and immobile concentrations [Kg/m^3] of the formation zone, respectively; r is the radial distance [m] from the center of the well; r_w is the well radius; r_s is the radial distance [m] from the center of the well to the outer boundary of the skin zone; v_a is the average radial pore velocity [m/s] in the aquifer; $v_{a1} = \frac{u_{a1}}{\theta_{m1}}$; $v_{a2} = \frac{u_{a2}}{\theta_{m2}}$; u_{a1} and u_{a2} represent Darcian velocities [m/s] in the skin and formation zones, respectively; α_1 and α_2 represent the longitudinal dispersivities [m] in the skin and formation zones, respectively; μ_{m1}, μ_{im1}, μ_{m2} and μ_{im2} are reaction rates for the first-order reaction rate, or the first-order biodegradation, or the radioactive decay [1/s]; θ_{m1}, θ_{im1}, θ_{m2} and θ_{im2} are porosities; R_{m1}, R_{im1}, R_{m2} and R_{im2} are retardation factors [-]; ω_1 and ω_2 represent the first-order mass transfer coefficients [1/s] between the mobile and immobile dissolved phases in the skin and formation zones, respectively. One point to note is that the molecular diffusive effect is assumed to be negligible in the above governing equations.

Assuming that the skin and formation zone is initially free of solute, the initial conditions are

$$C_{m1}(r,t)|_{t=0} = C_{im1}(r,t)|_{t=0} = C_{m2}(r,t)|_{t=0} = C_{im2}(r,t)|_{t=0} = 0 \tag{3.330}$$

The outer boundary condition at an infinite distance is

$$C_{m2}(r,t)|_{r\to\infty} = C_{im2}(r,t)|_{r\to\infty} = 0 \tag{3.331}$$

Two types of models have been widely applied to the boundary condition at the well screen: the mass flux continuity (MFC) model and the RCC model. The RCC model is

$$[C_{m1}(r,t)]|_{r=r_w} = [C_{\text{inj}}(t)]|_{r=r_w}, \quad 0 < t \leq t_{\text{inj}} \tag{3.332a}$$

$$[C_{m1}(r,t)]|_{r=r_w} = [C_{\text{cha}}(t)]|_{r=r_w}, \quad t > t_{\text{inj}} \tag{3.332b}$$

and the MFC model is

$$\left.\left[C_{m1}(r,t) - \alpha_1 \frac{|v_{a1,\text{inj}}|}{v_{a1,\text{inj}}} \frac{\partial C_{m1}(r,t)}{\partial r}\right]\right|_{r=r_w} = [C_{\text{inj}}(t)]|_{r=r_w}, \quad 0 < t \leq t_{\text{inj}} \tag{3.333a}$$

$$\left[C_{m1}(r,t) - \alpha_1 \frac{|v_{a1,\text{cha}}|}{v_{a1,\text{cha}}} \frac{\partial C_{m1}(r,t)}{\partial r}\right]\Bigg|_{r=r_w} = [C_{\text{cha}}(t)]|_{r=r_w}, \quad t > t_{\text{inj}} \tag{3.333b}$$

where $v_{a1,\text{inj}}$ and $v_{a1,\text{cha}}$ refer to velocities in the injection and chasing phases, respectively.

It was demonstrated that the mass balance requirement could not be satisfied in the RCC model, while the resident concentration was not continuous in the MFC model of Wang *et al.* (2018). Many experimental studies demonstrated that the MFC model performed better than the RCC model of Novakowski (1992a). Therefore, the MFC model will be used to describe the boundary condition in the wellbore in this section. Comparing Eqs. (3.332) and (3.333), one may find that the main difference between these two models is whether the dispersivity is involved or not. Recently, Wang and Zhan (2019) pointed out that the conflicts between these two models could be resolved by a scale-dependent dispersivity, which was zero at the well screen and increased with the travel distance of the solute. This is because when the dispersivity is zero in Eqs. (3.333a) and (3.333b), the MFC model reduces to the RCC model.

When considering the mixing effect in the wellbore, one has

$$V_{w,\text{inj}} \frac{dC_{\text{inj}}}{dt} = -\xi v_{a1,\text{inj}}(r_w)[C_{\text{inj}}(t) - C_0], \quad 0 < t \le t_{\text{inj}} \tag{3.334}$$

$$V_{w,\text{cha}} \frac{dC_{\text{cha}}}{dt} = -\xi v_{a1,\text{cha}}(r_w)[C_{\text{cha}}(t)], \quad t > t_{\text{inj}} \tag{3.335}$$

where $V_{w,\text{inj}}$ is the volume [m^3] of water in the wellbore when $t \le t_{\text{inj}}$, and $V_{w,\text{inj}} = \pi r_w^2 h_{w,\text{inj}}$; $h_{w,\text{inj}}$ is the water level [m] in the wellbore when $t \le t_{\text{inj}}$; $\xi = 2\pi r_w \theta_{m1} B$; B is thickness [m] of aquifer; $V_{w,\text{cha}}$ is the volume [m^3] of water in the wellbore when $t > t_{\text{inj}}$, and $V_{w,\text{cha}} = \pi r_w^2 h_{w,\text{cha}}$; $h_{w,\text{cha}}$ is the water level [m] in the wellbore when $t > t_{\text{inj}}$; $v_{a1,\text{inj}}(r_w)$ is velocity at the well screen in the injection phase, and $v_{a1,\text{inj}}(r_w) = \frac{Q_{\text{inj}}}{2\pi B r_w \theta_{m1}}$; $v_{a1,\text{cha}}(r_w)$ is velocity at the well screen in the chasing phase, and it equals to $\frac{Q_{\text{cha}}}{2\pi B r_w \theta_{m1}}$; Q_{inj} and Q_{cha} are the well flow rates [m^3/s] in the injection and chasing phases, respectively. The mass balance for the well in Eqs. (3.334) and (3.335) is only relevant when velocity exceeds zero because it does not contain terms for possible diffusive losses.

The water level in the wellbore (e.g., $h_{w,\mathrm{inj}}$ and $h_{w,\mathrm{cha}}$) could be determined by solving the groundwater flow model. In the steady state, one has

$$Q_{\mathrm{inj}} = 2\pi r B K \frac{dh}{dr}, \quad 0 < t \le t_{\mathrm{inj}} \tag{3.336}$$

$$Q_{\mathrm{cha}} = 2\pi r B K \frac{dh}{dr}, \quad t > t_{\mathrm{inj}} \tag{3.337}$$

where K is the hydraulic conductivity [m/s], and $K = \begin{cases} K_1 \text{ when } r_w \le r < r_s \\ K_2 \text{ when } r_s \le r \end{cases}$; K_1 and K_2 are the hydraulic conductivities [m/s] of skin and formation zones, respectively.

By respectively conducting the integration on Eqs. (3.334) and (3.335) from r_w to r_s and from r_s to r_e, the water level in the wellbore could be obtained as follows:

$$h_{w,\mathrm{inj}} = h_0 + \frac{Q_{\mathrm{inj}}}{2\pi B K_1} \ln \frac{r_s}{r_w} + \frac{Q_{\mathrm{inj}}}{2\pi B K_2} ln \frac{r_e}{r_s}, \quad 0 < t \le t_{\mathrm{inj}} \tag{3.338}$$

$$h_{w,\mathrm{cha}} = h_0 + \frac{Q_{\mathrm{cha}}}{2\pi B K_1} \ln \frac{r_s}{r_w} + \frac{Q_{\mathrm{cha}}}{2\pi B K_2} ln \frac{r_e}{r_s}, \quad t > t_{\mathrm{inj}} \tag{3.339}$$

where r_e is the radial distance [m] from the center of the well to the outer boundary of the formation zone; h_0 is the hydraulic head [m] at r_e. One could find that a finite radius r_e is needed to keep h_w finite. It seems to contradict the boundary condition of the transport problems, which is at infinity, as shown in Eq. (3.331). The reason for such a "contradiction" could be explained as follows. The influence area is limited by the finite injection rate and the finite injection time of the well (from a plane view perspective), bounded by a circle with a radius r_e where the hydraulic head is almost constant, and the flow velocity is almost zero.

At the interface between the skin and formation zones, the concentration and dispersive flux must be continuous, and one has

$$C_{m1}(r_s, t) = C_{m2}(r_s, t), \quad t > 0 \tag{3.340}$$

$$\left[\alpha_1 |v_{a1}| \frac{\partial C_{m1}(r,t)}{\partial r} \right] \Bigg|_{r=r_s} = \left[\alpha_2 |v_{a2}| \frac{\partial C_{m2}(r,t)}{\partial r} \right] \Bigg|_{r=r_s}, \quad t > 0 \tag{3.341}$$

Since Darcy fluxes (advective solute fluxes) are continuous, dispersive fluxes must be continuous, which is Eq. (3.341).

Here, it is worthwhile to comment on the nature of using the MIM approach to describe transport in heterogeneous aquifers. First, it has been commonly observed that the aquifer heterogeneity renders the use of ADE invalid in many cases as ADE is developed and used primarily for homogeneous aquifers. ADE fails to explain the early breakthrough and long tailing phenomena that are frequently observed in transport in heterogeneous aquifers, as illustrated in the introduction. Second, a striking feature of a heterogeneous aquifer is that a sequence of mobile and less mobile regions co-exists. In contrast, a homogeneous aquifer may be simplified as a single (mobile) region. Ideally, suppose one knows precisely the spatial distribution of those mobile and less mobile regions and their associated flow and transport parameters. In that case, one can use a high-resolution numerical simulator to predict the flow and transport process precisely. Unfortunately, this is not feasible for most practical cases. Therefore, as an alternative, we have adopted the concept of the MIM approach in which two continuums consisting of a mobile domain and an immobile domain co-exist over the entire heterogeneous aquifer. Each of these two continua itself has uniform flow and transport parameters (such as porosity, retardation factor, etc.) for the sake of simplification. Furthermore, mass can transfer between these two continuums in a certain fashion, usually using the first-order rate-limited equation. Third, this alternative approach has successfully explained many phenomena that cannot be explained using ADE, for instance, the early breakthrough and long tailing issues. Later, the two-continuum MIM approach was expanded to multiple-continuum MIM or namely, the multi-rate MIM approach, to better capture the transport features in a heterogeneous aquifer (Elenius and Abriola, 2019; Guo *et al.*, 2019; van Genuchten and Wierenga, 1976). In summary, the MIM does not incorporate the spatial variation of flow and transport parameters that are mostly unknown. Instead, it is based on an alternative approach, using two or more interrelated continuum, and in each continuum, the flow and transport parameters remain uniform over space. To date, the validation of the MIM model has been tested by numerous experimental studies (Elenius and Abriola, 2019; Gao *et al.*, 2009b; Griffioen *et al.*, 1998).

(b) Analytical model with skin effect

Dimensionless forms of parameters used in the derivation of analytical solution are defined as

$$C_{m1D} = \frac{C_{m1}}{C_0}, \quad C_{im1D} = \frac{C_{im1}}{C_0}, \quad C_{m2D} = \frac{C_{m2}}{C_0}, \quad C_{im2D} = \frac{C_{im2}}{C_0},$$

$$C_{\text{inj},D} = \frac{C_{\text{inj}}}{C_0}, \quad C_{\text{cha},D} = \frac{C_{\text{cha}}}{C_0}, \quad t_D = \frac{|A|t}{\alpha_2^2 R_{m1}}, \quad t_{\text{inj},D} = \frac{|A|t_{\text{inj}}}{\alpha_2^2 R_{m1}},$$

$$r_D = \frac{r}{\alpha_2}, \quad r_{wD} = \frac{r_w}{\alpha_2}, \quad r_{sD} = \frac{r_s}{\alpha_2}, \quad r_{0D} = \frac{r_0}{\alpha_2}, \quad \mu_{m1D} = \frac{\alpha_2^2 \mu_{m1}}{A},$$

$$\mu_{im1D} = \frac{\alpha_2^2 R_{m1} \mu_{im1}}{R_{im1} A}, \quad \mu_{m2D} = \frac{\alpha_2^2 \mu_{m2} R_{m1}}{A R_{m2}}, \quad \mu_{im2D} = \frac{\alpha_2^2 R_{m1} \mu_{im2}}{R_{im2} A}$$

$$\text{and} \quad A = \frac{Q}{2\pi B \theta_{m1}}.$$

After the dimensionless transform, the governing equations become:

$$\frac{\partial C_{m1D}}{\partial t_D} = \frac{\lambda}{r_D} \frac{\partial^2 C_{m1D}}{\partial r_D^2} - \frac{1}{r_D} \frac{\partial C_{m1D}}{\partial r_D} - \varepsilon_{m1}(C_{m1D} - C_{im1D})$$
$$- \mu_{m1D} C_{m1D}, \quad r_{wD} < r_D \leq r_{sD} \tag{3.342a}$$

$$\frac{\partial C_{im1D}}{\partial t_D} = \varepsilon_{im1}(C_{m1D} - C_{im1D}) - \mu_{im1D} C_{im1D}, \quad r_{wD} < r_D \leq r_{sD} \tag{3.342b}$$

$$\frac{\partial C_{m2D}}{\partial t_D} = \frac{\eta}{r_D} \frac{\partial^2 C_{m2D}}{\partial r_D^2} - \frac{\eta}{r_D} \frac{\partial C_{m2D}}{\partial r_D} - \varepsilon_{m2}(C_{m2D} - C_{im2D})$$
$$- \mu_{m2D} C_{m2D}, \quad r_D > r_{sD} \tag{3.342c}$$

$$\frac{\partial C_{im2D}}{\partial t_D} = \varepsilon_{im2}(C_{m2D} - C_{im2D}) - \mu_{im2D} C_{im2D}, \quad r_D > r_{sD} \tag{3.342d}$$

where

$$\varepsilon_{m1} = \frac{\omega_1 \alpha_2^2}{A\theta_{m1}}, \quad \varepsilon_{im1} = \frac{\omega_1 \alpha_2^2 R_{m1}}{A\theta_{im1} R_{im1}}, \quad \varepsilon_{m2} = \frac{\omega_2 \alpha_2^2 R_{m1}}{A\theta_{m2} R_{m2}},$$

$$\varepsilon_{im2} = \frac{\omega_2 \alpha_2^2 R_{m1}}{A\theta_{im2} R_{im2}}, \quad \eta = \frac{\theta_{m1} R_{m1}}{\theta_{m2} R_{m2}} \quad \text{and} \quad \lambda = \frac{\alpha_1}{\alpha_2}.$$

As for the boundary conditions at the well screen, a Heaviside step function will be employed to combine them at the injection and chasing phases:

$$C_w(r_w,t) = C_{\text{inj}}[H(t) - H(t - t_{\text{inj}})] + C_{\text{cha}}H(t - t_{\text{inj}}), \quad t > 0 \tag{3.343}$$

where $H(t)$ is the Heaviside step function, $C_w(r_w,t)$ is concentration [Kg/m^3] in the wellbore.

The dimensionless initial conditions and dimensionless boundary conditions are

$$C_{m1D}(r_D,t_D)|_{t_D=0} = C_{im1D}(r_D,t_D)|_{t_D=0} = C_{m2D}(r_D,t_D)|_{t_D=0} = C_{im2D}(r_D,t_D)|_{t_D=0} = 0, \quad r_D > r_{wD} \tag{3.344a}$$

$$C_{m2D}(r_D,t_D)|_{r_D\to\infty} = C_{im2D}(r_D,t_D)|_{r_D\to\infty} = 0, \quad t_D > 0 \tag{3.344b}$$

$$\left.\left[C_{m1D}(r_D,t_D) - \lambda\frac{\partial C_{m1D}(r_D,t_D)}{\partial r_D}\right]\right|_{r_D=r_{wD}} = C_{\text{inj},D}(t_D), \quad 0 < t_D \le t_{\text{inj},D} \tag{3.345a}$$

$$\left.\left[C_{m1D}(r_D,t_D) - \lambda\frac{\partial C_{m1D}(r_D,t_D)}{\partial r_D}\right]\right|_{r_D=r_{wD}} = C_{\text{cha},D}(t_D), \quad t_D > t_{\text{inj},D} \tag{3.345b}$$

$$C_{\text{w}D}(r_{wD},t_D) = C_{\text{inj},D}[H(t_D) - H(t_D - t_{\text{inj},D})] + C_{\text{cha},D}H(t_D - t_{\text{inj},D}) \tag{3.345c}$$

The dimensionless forms of Eqs. (3.334) and (3.335) are

$$\beta_{\text{inj}}\frac{dC_{\text{inj},D}(t_D)}{dt_D} = 1 - C_{\text{inj},D}(t_D), \quad 0 < t_D \le t_{\text{inj},D} \tag{3.346a}$$

$$\beta_{\text{cha}}\frac{dC_{\text{cha},D}(t_D)}{dt_D} = -C_{\text{cha},D}(t_D), \quad t_D > t_{\text{inj},D} \tag{3.346b}$$

where

$$\beta_{\text{inj}} = \frac{V_{w,\text{inj}}r_{wD}}{\xi R_{m1}\alpha_2} \quad \text{and} \quad \beta_{\text{cha}} = \frac{V_{w,\text{cha}}r_{wD}}{\xi R_{m1}\alpha_2}.$$

The dimensionless forms of Eqs. (3.340) and (3.341) are

$$C_{m1D}(r_{sD}, t_D) = C_{m2D}(r_{sD}, t_D), \quad t_D > 0 \tag{3.347a}$$

$$\left[\lambda \frac{\partial C_{m1D}(r_D, t_D)}{\partial r_D}\right]\Big|_{r_D = r_{sD}} = \left[\frac{\partial C_{m2D}(r_D, t_D)}{\partial r_D}\right]\Big|_{r_D = r_{sD}}, \quad t_D > 0 \tag{3.347b}$$

Conducting Laplace transform to Eqs. (3.342a) and (3.342b), one has

$$s\overline{C}_{m1D} = \frac{\lambda}{r_D}\frac{\partial^2 \overline{C}_{m1D}}{\partial r_D^2} - \frac{1}{r_D}\frac{\partial \overline{C}_{m1D}}{\partial r_D} - \varepsilon_{m1}\left(\overline{C}_{m1D} - \overline{C}_{im1D}\right) - \mu_{m1D}\overline{C}_{m1D} \tag{3.348a}$$

$$s\overline{C}_{im1D} = \varepsilon_{im1}\left(\overline{C}_{m1D} - \overline{C}_{im1D}\right) - \mu_{im1D}\overline{C}_{im1D} \tag{3.348b}$$

Substituting Eq. (3.348b) into Eq. (3.348a), one has

$$\frac{\lambda}{r_D}\frac{\partial^2 \overline{C}_{m1D}}{\partial r_D^2} - \frac{1}{r_D}\frac{\partial \overline{C}_{m1D}}{\partial r_D} - E_1\overline{C}_{m1D} = 0, \quad r_{wD} < r_D \leq r_{sD} \tag{3.349}$$

where $E_1 = s + \varepsilon_{m1} + \mu_{m1D} - \frac{\varepsilon_{m1}\varepsilon_{im1}}{s+\varepsilon_{im1}+\mu_{im1D}}$.

Conducting Laplace transform to Eqs. (3.342c) and (3.342d), one has

$$s\overline{C}_{m2D} = \frac{\eta}{r_D}\frac{\partial^2 \overline{C}_{m2D}}{\partial r_D^2} - \frac{\eta}{r_D}\frac{\partial \overline{C}_{m2D}}{\partial r_D} - \varepsilon_{m2}\left(\overline{C}_{m2D} - \overline{C}_{im2D}\right) - \mu_{m2D}\overline{C}_{m2D} \tag{3.350a}$$

$$s\overline{C}_{im2D} = \varepsilon_{im2}\left(\overline{C}_{m2D} - \overline{C}_{im2D}\right) - \mu_{im2D}\overline{C}_{im2D} \tag{3.350b}$$

Substituting Eq. (3.350b) into Eq. (3.350a), one has

$$\frac{1}{r_D}\frac{\partial^2 \overline{C}_{m2D}}{\partial r_D^2} - \frac{1}{r_D}\frac{\partial \overline{C}_{m2D}}{\partial r_D} - E_2\overline{C}_{m2D} = 0, \quad r_D > r_{sD} \tag{3.351}$$

where $E_2 = \frac{1}{\eta}\left(s + \varepsilon_{m2} + \mu_{m2D} - \frac{\varepsilon_{m2}\varepsilon_{im2}}{s+\varepsilon_{im2}+\mu_{im2D}}\right)$.

The boundary conditions in the Laplace domain are

$$\left[\overline{C}_{m1D}(r_D,s)-\lambda\frac{\partial\overline{C}_{m1D}(r_D,s)}{\partial r_D}\right]\Bigg|_{r_D=r_{wD}}=\frac{1}{s(s\beta_{\text{inj}}+1)},\quad 0<t_D\le t_{\text{inj},D} \tag{3.352a}$$

$$\left[\overline{C}_{m1D}(r_D,s)-\lambda\frac{\partial\overline{C}_{m1D}(r_D,s)}{\partial r_D}\right]\Bigg|_{r_D=r_{wD}}=\frac{\beta_{\text{cha}}C_{\text{inj},D}(r_{wD},t_{\text{inj},D})}{(s\beta_{\text{cha}}+1)},\quad t_D>t_{\text{inj},D} \tag{3.352b}$$

$$\overline{C}_{m1D}(r_{sD},s)=\overline{C}_{m2D}(r_{sD},s),\quad t_D>0 \tag{3.352c}$$

$$\lambda\left[\frac{\partial\overline{C}_{m1D}(r_D,s)}{\partial r_D}\right]\Bigg|_{r_D=r_{sD}}=\left[\frac{\partial\overline{C}_{m2D}(r_D,s)}{\partial r_D}\right]\Bigg|_{r_D=r_{sD}},\quad t_D>0 \tag{3.352d}$$

$$\overline{C}_{m2D}(r_D,s)\big|_{r_D\to\infty}=0,\quad t_D>0 \tag{3.352e}$$

The general solutions of Eqs. (3.349) and (3.351) are respectively:

$$\overline{C}_{m1D}=N_1\text{exp}\left(\frac{r_D}{2\lambda}\right)A_i(y_1)+N_2\exp\left(\frac{r_D}{2\lambda}\right)B_i(y_1),\quad r_{wD}<r_D\le r_{sD} \tag{3.353a}$$

$$\overline{C}_{m2D}=N_3\text{exp}\left(\frac{r_D}{2}\right)A_i(y_2)+N_4\exp\left(\frac{r_D}{2}\right)B_i(y_2),\quad r_D>r_{sD} \tag{3.353b}$$

where $y_1=\left(\frac{E_1}{\lambda}\right)^{1/3}\left(r_D+\frac{1}{4\lambda E_1}\right)$; $y_2=(E_2)^{1/3}\left(r_D+\frac{1}{4E_2}\right)$; N_1, N_2, N_3 and N_4 are constants that the boundary conditions could determine. $A_i(\cdot)$ and $B_i(\cdot)$ are the Airy functions of the first kind and second kind, respectively.

Substituting Eq. (3.353b) into Eq. (3.352e), one has

$$N_4=0 \tag{3.354}$$

Substituting Eq. (3.353a) into Eq. (3.352a), one has

$$N_1 \exp\left(\frac{r_{wD}}{2\lambda}\right)\left[\frac{1}{2}A_i(y_w) - \lambda\left(\frac{E_1}{\lambda}\right)^{1/3} A_i'(y_w)\right]$$
$$+ N_2 \exp\left(\frac{r_{wD}}{2\lambda}\right)\left[\frac{1}{2}B_i(y_w) - \lambda\left(\frac{E_1}{\lambda}\right)^{1/3} \exp\left(\frac{r_{wD}}{2}\right) B_i'(y_w)\right] = F_1 \tag{3.355a}$$

where $y_w = \left(\frac{E_1}{\lambda}\right)^{1/3}\left(r_{wD} + \frac{1}{4\lambda E_1}\right)$, and $F_1 = \frac{1}{s(s\beta_{\text{inj}}+1)}$.

Substituting Eq. (3.353a) into Eq. (3.352b), one has

$$N_1 \exp\left(\frac{r_{wD}}{2\lambda}\right)\left[\frac{1}{2}A_i(y_w) - \lambda\left(\frac{E_1}{\lambda}\right)^{1/3} A_i'(y_w)\right]$$
$$+ N_2 \exp\left(\frac{r_{wD}}{2\lambda}\right)\left[\frac{1}{2}B_i(y_w) - \lambda\left(\frac{E_1}{\lambda}\right)^{1/3} \exp\left(\frac{r_{wD}}{2}\right) B_i'(y_w)\right] = F_2 \tag{3.355b}$$

where $F_2 = \frac{\beta_{\text{cha}} C_{\text{inj},D}(r_{wD}, t_{\text{inj},D})}{(s\beta_{\text{cha}}+1)}$.

Conducting Laplace transform on Eq. (3.345c), one has

$$\overline{C}_{wD}(r_{wD}, s) = C_{\text{inj},D}\frac{1-\exp(-t_{\text{inj},D}s)}{s}$$
$$+ C_{\text{cha},D}\frac{\exp(-t_{\text{inj},D}s)}{s}, \quad r_D = r_{wD} \tag{3.355c}$$

Thus, Eqs. (3.355a) and (3.355c) could be combined as the following equation:

$$N_1 \exp\left(\frac{r_{wD}}{2\lambda}\right)\left[\frac{1}{2}A_i(y_w) - \lambda\left(\frac{E_1}{\lambda}\right)^{1/3} A_i'(y_w)\right]$$
$$+ N_2 \exp\left(\frac{r_{wD}}{2\lambda}\right)\left[\frac{1}{2}B_i(y_w) - \lambda\left(\frac{E_1}{\lambda}\right)^{1/3} \exp\left(\frac{r_{wD}}{2}\right) B_i'(y_w)\right] = F \tag{3.355d}$$

where

$$F = C_{\text{inj},D}\frac{1-\exp(-t_{\text{inj},D}s)}{s} + C_{\text{cha},D}\frac{\exp(-t_{\text{inj},D}s)}{s}.$$

Substituting Eqs. (3.353a) and (3.353b) into Eq. (3.352c), one has

$$N_1\exp\left(\frac{r_{sD}}{2\lambda}\right)A_i(y_{1s}) + N_2\exp\left(\frac{r_{sD}}{2\lambda}\right)B_i(y_{1s}) = N_3\exp\left(\frac{r_{sD}}{2}\right)A_i(y_{2s}) \tag{3.356}$$

where $y_{1s} = \left(\frac{E_1}{\lambda}\right)^{1/3}\left(r_{sD} + \frac{1}{4\lambda E_1}\right)$; $y_{2s} = (E_2)^{1/3}\left(r_{sD} + \frac{1}{4E_2}\right)$.

Substituting Eqs. (3.353a) and (3.353b) into Eq. (3.352d) yields:

$$\begin{aligned} &N_1\exp\left(\frac{r_{sD}}{2\lambda}\right)\left[\frac{1}{2}A_i(y_{1s}) + \lambda\left(\frac{E_1}{\lambda}\right)^{1/3}A_i'(y_{1s})\right] \\ &\quad + N_2\exp\left(\frac{r_{sD}}{2\lambda}\right)\left[\frac{1}{2}B_i(y_{1s}) + \lambda\left(\frac{E_1}{\lambda}\right)^{1/3}B_i'(y_{1s})\right] \\ &\quad = N_3\exp\left(\frac{r_{sD}}{2}\right)\left[\frac{1}{2}A_i(y_{2s}) + (E_2)^{1/3}A_i'(y_{2s})\right] \end{aligned} \tag{3.357}$$

where $A_i'(\cdot)$ and $B_i'(\cdot)$ are the derivative of the Airy functions of the first kind and second kind, respectively.

The values of N_1, N_2, and N_3 could be determined by solving Eqs. (3.355d) and (3.357):

$$N_1 = \frac{F - H_2N_2}{H_1} \tag{3.358a}$$

$$N_2 = \frac{H_3H_8F - H_5H_6F}{H_1H_5H_7 + H_2H_3H_8 - H_2H_5H_6 - H_1H_4H_8} \tag{3.358b}$$

$$N_3 = \frac{H_3F}{H_1H_5} - \frac{H_2H_3N_2}{H_1H_5} + \frac{H_4N_2}{H_5} \tag{3.358c}$$

where

$$H_1 = \exp\left(\frac{r_{wD}}{2\lambda}\right)\left[\frac{1}{2}A_i(y_w) - \lambda\left(\frac{E_1}{\lambda}\right)^{1/3}A_i'(y_w)\right],$$

$$H_2 = \exp\left(\frac{r_{wD}}{2\lambda}\right)\left[\frac{1}{2}B_i(y_w) - \lambda\left(\frac{E_1}{\lambda}\right)^{1/3}\exp\left(\frac{r_{wD}}{2}\right)B_i'(y_w)\right],$$

$$H_3 = \exp\left(\frac{r_{sD}}{2\lambda}\right) A_i(y_{1s}), \quad H_4 = \exp\left(\frac{r_{sD}}{2\lambda}\right) B_i(y_{1s}),$$

$$H_5 = \exp\left(\frac{r_{sD}}{2}\right) A_i(y_{2s}),$$

$$H_6 = \exp\left(\frac{r_{sD}}{2\lambda}\right) \left[\frac{1}{2}A_i(y_{1s}) + \lambda\left(\frac{E_1}{\lambda}\right)^{1/3} A_i'(y_{1s})\right],$$

$$H_7 = \exp\left(\frac{r_{sD}}{2\lambda}\right) \left[\frac{1}{2}B_i(y_{1s}) + \lambda\left(\frac{E_1}{\lambda}\right)^{1/3} B_i'(y_{1s})\right],$$

$$H_8 = \exp\left(\frac{r_{sD}}{2}\right) \left[\frac{1}{2}A_i(y_{2s}) + (E_2)^{1/3} A_i'(y_{2s})\right],$$

and

$$F = C_{\text{inj},D}\frac{1-\exp(-t_{\text{inj},D}s)}{s} + C_{\text{cha},D}\frac{\exp(-t_{\text{inj},D}s)}{s}.$$

Substituting the expressions of N_1, N_2, N_3, and N_4 into Eqs. (3.353a)–(3.353b), one could get the solutions of Eq. (3.349) and Eq. (3.351).

Substituting Eqs. (3.353a) and (3.353b) into Eqs. (3.348b) and (3.350b), one could get the solutions of Eqs. (3.348b) and (3.350b):

$$\overline{C}_{im1D} = \frac{\varepsilon_{im1}}{s + \varepsilon_{im1} + \mu_{im1D}}\overline{C}_{m1D}, \quad r_{wD} \le r_D \le r_{sD} \tag{3.359a}$$

$$\overline{C}_{im2D} = \frac{\varepsilon_{im2}}{s + \varepsilon_{im2} + \mu_{im2D}}\overline{C}_{m2D}, \quad r_D > r_{sD} \tag{3.359b}$$

In the injection phase, the values of N_1 and N_2 are modified into N_1' and N_2' as follows:

$$N_1' = \frac{F_1 - H_2N_2'}{H_1'} \quad \text{and}$$

$$N_2' = \frac{H_3H_8F_1 - H_5H_6F_1}{H_1'H_5H_7 + H_2H_3H_8 - H_2H_5H_6 - H_1'H_4H_8}$$

where $H_1' = \exp\left(\frac{r_{wD}}{2\lambda}\right) A_i(y_w)$.

Substituting N_1' and N_2' into Eq. (3.353a), one has

$$\overline{C}_{\text{inj},D} = N_1'\exp\left(\frac{r_D}{2\lambda}\right) A_i(y_1) + N_2'\exp\left(\frac{r_D}{2\lambda}\right) B_i(y_1), \quad r_D = r_{wD} \tag{3.360}$$

In the chasing phase, the values of N_1 and N_2 are modified into N_1'' and N_2'' as follows:

$$N_1'' = \frac{F_2 - H_2 N_2''}{H_1'},$$

$$N_2'' = \frac{H_3 H_8 F_2 - H_5 H_6 F_2}{H_1' H_5 H_7 + H_2 H_3 H_8 - H_2 H_5 H_6 - H_1' H_4 H_8}$$

and substituting N_1'' and N_2'' into Eq. (3.353a), one has

$$\overline{C}_{\text{cha},D} = N_1''\exp\left(\frac{r_D}{2\lambda}\right) A_i(y_1) + N_2''\exp\left(\frac{r_D}{2\lambda}\right) B_i(y_1), \quad r_D = r_{wD} \tag{3.361}$$

Finally, the analytical solution is

$$\overline{C}_{m1D} = N_1\exp\left(\frac{r_D}{2\lambda}\right) A_i(y_1) + N_2\exp\left(\frac{r_D}{2\lambda}\right) B_i(y_1),$$
$$r_{wD} \le r_D \le r_{sD} \tag{3.362a}$$

$$\overline{C}_{im1D} = \frac{\varepsilon_{im1}}{s + \varepsilon_{im1} + \mu_{im1D}} \overline{C}_{m1D}, \quad r_{wD} \le r_D \le r_{sD} \tag{3.362b}$$

$$\overline{C}_{m2D} = N_3\exp\left(\frac{r_D}{2}\right) A_i(y_2), \quad r_D > r_{sD} \tag{3.363a}$$

$$\overline{C}_{im2D} = \frac{\varepsilon_{im2}}{s + \varepsilon_{im2} + \mu_{im2D}} \overline{C}_{m2D}, \quad r_D > r_{sD} \tag{3.363b}$$

where the expressions for N_1, N_2 and N_3 are listed in Eqs. (3.358a)–(3.358c).

This model considers the mixing effect, skin effect, and media heterogeneity (which is described using MIM) simultaneously, and the solute is injected into the well as a step source. This general solution encompasses many previous studies as special cases. For instance, when "$r_s \to \infty$", the skin effect is excluded; "$t_{\text{inj}} \to \infty$" implies that the solute is continuously injected into the well, while "$t_{\text{inj}} \to 0$" means that the solute is instantaneously injected into the well; "$\omega = 0$" implies that the MIM solution reduces to the ADE solution; "$V_{w,\text{inj}} = 0$" or "$r_w = 0$" shows that the mixing effect is excluded.

Therefore, the new solution reduces to the solutions of Hoopes and Harleman (1967), Gelhar and Collins (1971), Tang and Babu (1979),

Moench and Ogata (1981), Hsieh (1986), Tang and Peaceman (1987), and Philip (1994) when $r_s \to \infty$, $t_{\text{inj}} \to \infty$, $\omega = 0$, and $V_{w,\text{inj}} = 0$. The solution of Wang *et al.* (2018) is a special case of this section when $r_s \to \infty$, $t_{\text{inj}} \to \infty$, and $\omega = 0$.

(c) Analytical model with scale-dependent dispersivity

Due to the heterogeneities of the porous media, the dispersivity was found to be dependent on the travel distance of solute from the source, and such a phenomenon was first observed in the field scale experiment (Dagan, 1988; Gelhar *et al.*, 1992; Pickens and Grisak, 1981b). The field-scale effect (i.e., dispersivity growing with distance from the well) is usually considered to be a result of spatial heterogeneity at different scales in the aquifer. Subsequently, the scale-dependent dispersivity phenomenon was also found in controlled laboratory tests due to heterogeneities caused by the bridging effect and microstructures, although the sediments (as the porous media) are well sorted and carefully packed (Berkowitz *et al.*, 2000; Gao *et al.*, 2010; Silliman and Simpson, 1987; Wang *et al.*, 2019). For example, Silliman and Simpson (1987) found that the dispersivity continuously increased with distance, based on the experiments conducted in a 2.4 × 1.07 × 0.10 m sandbox. Berkowitz *et al.* (2000) obtained similar conclusions to Silliman and Simpson (1987) in the laboratory-controlled experiment. Wang and Zhan (2019) also concluded that the scale-dependent model performed better than the scale-independent model in interpreting observed BTCs of the laboratory-controlled experiment. To date, four types of functions have been widely used to describe scale-dependent dispersivity, including asymptotic, parabolic, exponential, and linear functions, as summarized by Pickens and Grisak (1981a). In this section, the model of the scale-independent dispersivity (e.g., Eqs. (3.362) and (3.363)) will be extended by considering the linear-asymptotic dispersivity model in the formation zone. As for the other types of scale-dependent functions, the analytical solutions could be derived using a similar approach. The formula of the linear distance-dependent dispersivity is

$$\alpha_2(r) = \begin{cases} kr, & r_s \le r \le r_0 \\ \alpha_0, & r > r_0 \end{cases} \tag{3.364}$$

where r_0 is the distance [m], where $\alpha_2(r_0) = \alpha_0$, k is a constant [–].

Substituting Eq. (3.364) into Eq. (3.329c), the dimensionless form of the governing equations becomes:

$$\frac{\partial C_{m2D}}{\partial t_D} = \frac{k\eta\partial^2 C_{m2D}}{\partial r_D^2} + \frac{k\eta - \eta}{r_D}\frac{\partial C_{m2D}}{\partial r_D} - \varepsilon_{m2}(C_{m2D} - C_{im2D}) - \mu_{m2D}C_{m2D}, \quad r_{sD} \le r_D \le r_{0D} \tag{3.365a}$$

$$\frac{\partial C_{m2D}}{\partial t_D} = \frac{\eta}{r_D}\frac{\partial^2 C_{m2D}}{\partial r_D^2} - \frac{\eta}{r_D}\frac{\partial C_{m2D}}{\partial r_D} - \varepsilon_{m2}(C_{m2D} - C_{im2D}) - \mu_{m2D}C_{m2D}, \quad r_D \ge r_{0D} \tag{3.365b}$$

Similarly, one could obtain the dimensionless initial conditions and dimensionless boundary conditions, the expressions of the dimensionless initial conditions and dimensionless boundary conditions are the same with Eqs. (3.344a) and (3.347b), except that:

$$\left[\lambda\frac{\partial C_{m1D}(r_D, t_D)}{\partial r_D}\right]\Big|_{r_D = r_{sD}} = \left[k\frac{\partial C_{m2D}(r_D, t_D)}{\partial r_D}\right]\Big|_{r_D = r_{sD}} \tag{3.366}$$

In the formation zone, we could obtain the boundary condition at $r_D = r_{0D}$:

$$kr_D\frac{\partial C_{m2D}(r_D, t_D)}{\partial r_D} = \frac{\partial C_{m2D}(r_D, t_D)}{\partial r_D}, \quad r_D = r_{0D} \tag{3.367}$$

Then, conducting Laplace transform to Eqs. (3.365a) and (3.365b), one has

$$s\overline{C}_{m2D} = k\eta\frac{\partial^2 \overline{C}_{m2D}}{\partial r_D^2} + \frac{k\eta - \eta}{r_D}\frac{\partial \overline{C}_{m2D}}{\partial r_D} - \varepsilon_{m2}\left(\overline{C}_{m2D} - \overline{C}_{im2D}\right) - \mu_{m2D}\overline{C}_{m2D}, \quad r_{sD} \le r_D \le r_{0D} \tag{3.368a}$$

$$s\overline{C}_{m2D} = \frac{\eta}{r_D}\frac{\partial^2 \overline{C}_{m2D}}{\partial r_D^2} - \frac{\eta}{r_D}\frac{\partial \overline{C}_{m2D}}{\partial r_D} - \varepsilon_{m2}\left(\overline{C}_{m2D} - \overline{C}_{im2D}\right) - \mu_{m2D}\overline{C}_{m2D}, \quad r_D \ge r_{0D} \tag{3.368b}$$

Substituting Eq. (3.350b) into Eqs. (3.368a) and (3.368b), one has

$$\frac{\partial^2 \overline{C}_{m2D}}{\partial r_D^2} + \frac{n}{r_D}\frac{\partial \overline{C}_{m2D}}{\partial r_D} - \varepsilon_1^2\overline{C}_{m2D} = 0, \quad r_{sD} \le r_D \le r_{0D} \tag{3.369a}$$

$$\frac{1}{r_D}\frac{\partial^2 \overline{C}_{m2D}}{\partial r_D^2} - \frac{1}{r_D}\frac{\partial \overline{C}_{m2D}}{\partial r_D} - \varepsilon_1^2 \overline{C}_{m2D} = 0, \quad r_D > r_{0D} \tag{3.369b}$$

where $n = 1 - \frac{1}{k}$ and $\varepsilon_1 = \sqrt{\frac{E_2}{k\eta}}$.

Like Eqs. (3.352a)–(3.352e) and (3.366), the boundary conditions at wellbore and infinity in the Laplace domain are

$$\left[\overline{C}_{m1D}(r_D, s) - \lambda \frac{\partial \overline{C}_{m1D}(r_D, s)}{\partial r_D}\right]\Bigg|_{r_D = r_{wD}} = \frac{1}{s\,(s\beta_{\text{inj}} + 1)} \tag{3.370a}$$

$$\left[\overline{C}_{m1D}(r_D, s) - \lambda \frac{\partial \overline{C}_{m1D}(r_D, s)}{\partial r_D}\right]\Bigg|_{r_D = r_{wD}} = \frac{\beta_{\text{cha}} C_{\text{inj},D}(r_{wD}, t_{\text{inj},D})}{(s\beta_{\text{cha}} + 1)} \tag{3.370b}$$

$$\overline{C}_{m1D}(r_{sD}, s) = \overline{C}_{m2D}(r_{sD}, s) \tag{3.370c}$$

$$\left[\lambda \frac{\partial \overline{C}_{m1D}(r_D, s)}{\partial r_D}\right]\Bigg|_{r_D = r_{sD}} = \left[k \frac{\partial \overline{C}_{m2D}(r_D, s)}{\partial r_D}\right]\Bigg|_{r_D = r_{sD}} \tag{3.370d}$$

$$\left[k r_D \frac{\partial \overline{C}_{m2D}(r_D, s)}{\partial r_D}\right]\Bigg|_{r_D = r_{0D}} = \left[\frac{\partial \overline{C}_{m2D}(r_D, s)}{\partial r_D}\right]\Bigg|_{r_D = r_{0D}} \tag{3.370e}$$

$$\overline{C}_{m1D}(r_{0D}, s) = \overline{C}_{m2D}(r_{0D}, s) \tag{3.370f}$$

$$\overline{C}_{m2D}(r_D, s)\big|_{r_D \to \infty} = 0 \tag{3.370g}$$

The general solutions of Eqs. (3.349) and (3.366) are respectively:

$$\overline{C}_{m1D} = \mathcal{T}_1 \exp\left(\frac{r_D}{2\lambda}\right) A_i(y_1) + \mathcal{T}_2 \exp\left(\frac{r_D}{2\lambda}\right) B_i(y_1), \quad r_{wD} < r_D \le r_{sD} \tag{3.371a}$$

$$\overline{C}_{m2D} = \mathcal{T}_3 r_D^m K_m(\varepsilon_1 r_D) + \mathcal{T}_4 r_D^m I_m(\varepsilon_1 r_D), \quad r_{sD} \le r_D \le r_{0D} \tag{3.371b}$$

$$\overline{C}_{m2D} = \mathcal{T}_5 \exp\left(\frac{r_D}{2}\right) A_i(y_3) + \mathcal{T}_6 \exp\left(\frac{r_D}{2}\right) B_i(y_3), \quad r_D > r_{0D} \tag{3.371c}$$

where $m = \frac{1}{2k}$; $y_3 = (\varepsilon_1)^{1/3}\left(r_D + \frac{1}{4\varepsilon_1}\right)$; $\mathcal{T}_1, \mathcal{T}_2, \mathcal{T}_3, \mathcal{T}_4, \mathcal{T}_5$ and $\mathcal{T}_6$ are constants which the boundary conditions could determine; I_m and K_m are the mth-order modified Bessel functions of the first kind and second kind, respectively.

Substituting Eq. (3.371c) into Eq. (3.370g), one has

$$\mathcal{T}_6 = 0 \tag{3.372}$$

Substituting Eq. (3.371a) into Eq. (3.370a), one has

$$\mathcal{T}_1 \exp\left(\frac{r_{wD}}{2\lambda}\right)\left[\frac{1}{2}A_i(y_w) - \lambda\left(\frac{E_1}{\lambda}\right)^{1/3} A_i'(y_w)\right] + \mathcal{T}_2 \exp\left(\frac{r_{wD}}{2\lambda}\right)\left[\frac{1}{2}B_i(y_w) - \lambda\left(\frac{E_1}{\lambda}\right)^{1/3} \exp\left(\frac{r_{wD}}{2}\right) B_i'(y_w)\right] = F_1 \tag{3.373}$$

Substituting Eq. (3.371a) into Eq. (3.370b), one has

$$\mathcal{T}_1 \exp\left(\frac{r_{wD}}{2\lambda}\right)\left[\frac{1}{2}A_i(y_w) - \lambda\left(\frac{E_1}{\lambda}\right)^{1/3} A_i'(y_w)\right] + \mathcal{T}_2 \exp\left(\frac{r_{wD}}{2\lambda}\right)\left[\frac{1}{2}B_i(y_w) - \lambda\left(\frac{E_1}{\lambda}\right)^{1/3} \exp\left(\frac{r_{wD}}{2}\right) B_i'(y_w)\right] = F_2 \tag{3.374a}$$

Like the treatment of Eq. (3.355d), Eqs. (3.373)–(3.374a) could be combined as the following equation:

$$\mathcal{T}_1 \exp\left(\frac{r_{wD}}{2\lambda}\right)\left[\frac{1}{2}A_i(y_w) - \lambda\left(\frac{E_1}{\lambda}\right)^{1/3} A_i'(y_w)\right] + \mathcal{T}_2 \exp\left(\frac{r_{wD}}{2\lambda}\right)\left[\frac{1}{2}B_i(y_w) - \lambda\left(\frac{E_1}{\lambda}\right)^{1/3} \exp\left(\frac{r_{wD}}{2}\right) B_i'(y_w)\right] = F \tag{3.374b}$$

where $F = C_{\text{inj},D}\frac{1-\exp(-t_{\text{inj},D}s)}{s} + C_{\text{cha},D}\frac{\exp(-t_{\text{inj},D}s)}{s}$.

Substituting Eqs. (3.371a) and (3.371b) into Eq. (3.370c), one has

$$\mathcal{T}_1\exp\left(\frac{r_{sD}}{2\lambda}\right) A_i(y_{1s}) + \mathcal{T}_2\exp\left(\frac{r_{sD}}{2\lambda}\right) B_i(y_{1s}) = \mathcal{T}_3 r_{sD}^m K_m(\varepsilon_1 r_{sD}) + \mathcal{T}_4 r_D^m I_m(\varepsilon_1 r_D) \tag{3.375}$$

Substituting Eqs. (3.371a) and (3.371b) into Eq. (3.370d) yields:

$$\mathcal{T}_1 \exp\left(\frac{r_{sD}}{2\lambda}\right)\left[\frac{1}{2}A_i(y_{1s}) + \lambda\left(\frac{E_1}{\lambda}\right)^{1/3} A_i'(y_{1s})\right]$$

$$+ \mathcal{T}_2 \exp\left(\frac{r_{sD}}{2\lambda}\right)\left[\frac{1}{2}B_i(y_{1s}) + \lambda\left(\frac{E_1}{\lambda}\right)^{1/3} B_i'(y_{1s})\right]$$

$$= -\mathcal{T}_3 k\varepsilon_1 r_{sD}^{m+1} K_{m-1}(\varepsilon_1 r_{sD}) + \mathcal{T}_4 k$$

$$\times \left\{m r_{sD}^{m-1} I_m(\varepsilon_1 r_D) + 0.5\varepsilon_1 r_{sD}^m \left[I_{m-1}(\varepsilon_1 r_D) + I_{m+1}(\varepsilon_1 r_D)\right]\right\} \quad (3.376)$$

where $K_{m-1}(\cdot)$ is the derivative of the mth-order modified Bessel function of the second kind, $I_{m-1}(\cdot)$ and $I_{m+1}(\cdot)$ are the derivatives of the mth-order modified Bessel function of the first kind.

Substituting Eqs. (3.371b) and (3.371c) into Eqs. (3.370e) and (3.370f) yields:

$$-\mathcal{T}_3 k\varepsilon_1 r_{0D}^{m+2} K_{m-1}(\varepsilon_1 r_{0D}) + \mathcal{T}_4 k\left\{m r_{0D}^m I_m(\varepsilon_1 r_{0D})\right.$$

$$\left. + 0.5\varepsilon_1 r_{0D}^{m+1}\left[I_{m-1}(\varepsilon_1 r_{0D}) + I_{m+1}(\varepsilon_1 r_{0D})\right]\right\}$$

$$= \mathcal{T}_5\left[0.5\exp\left(\frac{r_D}{2}\right)A_i(y_4) + \varepsilon_1^{1/3}\exp\left(\frac{r_D}{2}\right)A_i'(y_4)\right] \quad (3.377)$$

$$\mathcal{T}_3 r_{0D}^m K_m(\varepsilon_1 r_{0D}) + \mathcal{T}_4 r_{0D}^m I_m(\varepsilon_1 r_{0D}) = \mathcal{T}_5 \exp\left(\frac{r_{0D}}{2}\right)A_i(y_4) \quad (3.378)$$

where $y_4 = (\varepsilon_1)^{1/3}\left(r_{0D} + \frac{1}{4\varepsilon_1}\right)$.

The values of $\mathcal{T}_1, \mathcal{T}_2, \mathcal{T}_3, \mathcal{T}_4, \mathcal{T}_5$ and $\mathcal{T}_6$ could be determined by solving Eqs. (3.370)–(3.376), one has

$$\mathcal{T}_1 = \frac{F - W_2\mathcal{T}_2}{W_1} \quad (3.379a)$$

$$\mathcal{T}_2 = \frac{W_1W_5}{W_1W_4 - W_2W_3}\mathcal{T}_3 + \frac{W_1W_6}{W_1W_4 - W_2W_3}\mathcal{T}_4 - \frac{W_3F}{W_1W_4 - W_2W_3} \quad (3.379b)$$

$$\mathcal{T}_3 = \frac{W_{13}W_{15} - W_{12}W_{16}}{W_{11}W_{16} - W_{13}W_{14}}\mathcal{T}_4 \quad (3.379c)$$

$$\mathcal{T}_4 = \frac{W_3F(W_1W_8 - W_2W_7) - W_7F(W_1W_4 - W_2W_3)}{(W_1W_5\Theta + W_1W_6)(W_1W_8 - W_2W_7) \\ -(W_1W_9\Theta - W_1W_{10})(W_1W_4 - W_2W_3)} \tag{3.379d}$$

$$\mathcal{T}_5 = \frac{W_{14}}{W_{16}}\mathcal{T}_3 + \frac{W_{15}}{W_{16}}\mathcal{T}_4 \tag{3.379e}$$

where

$$W_1 = \exp\left(\frac{r_{wD}}{2\lambda}\right)\left[\frac{1}{2}A_i(y_w) - \lambda\left(\frac{E_1}{\lambda}\right)^{1/3} A_i'(y_w)\right],$$

$$W_2 = \exp\left(\frac{r_{wD}}{2\lambda}\right)\left[\frac{1}{2}B_i(y_w) - \lambda\left(\frac{E_1}{\lambda}\right)^{1/3} \exp\left(\frac{r_{wD}}{2}\right) B_i'(y_w)\right],$$

$$W_3 = \exp\left(\frac{r_{sD}}{2\lambda}\right) A_i(y_{1s}), \exp\left(\frac{r_{sD}}{2\lambda}\right) B_i(y_{1s}),$$

$$W_5 = r_{sD}^m K_m(\varepsilon_1 r_{sD}), W_6 = r_D^m I_m(\varepsilon_1 r_D),$$

$$W_7 = \exp\left(\frac{r_{sD}}{2\lambda}\right)\left[\frac{1}{2}A_i(y_{1s}) + \lambda\left(\frac{E_1}{\lambda}\right)^{1/3} A_i'(y_{1s})\right],$$

$$W_8 = \exp\left(\frac{r_{sD}}{2\lambda}\right)\left[\frac{1}{2}B_i(y_{1s}) + \lambda\left(\frac{E_1}{\lambda}\right)^{1/3} B_i'(y_{1s})\right],$$

$$W_9 = -k\varepsilon_1 r_{sD}^{m+1} K_{m-1}(\varepsilon_1 r_{sD}),$$

$$W_{10} = k\left\{m r_{sD}^{m-1} I_m(\varepsilon_1 r_D) + 0.5\varepsilon_1 r_{sD}^m \left[I_{m-1}(\varepsilon_1 r_D) + I_{m+1}(\varepsilon_1 r_D)\right]\right\},$$

$$W_{11} = -k\varepsilon_1 r_{0D}^{m+2} K_{m-1}(\varepsilon_1 r_{0D}),$$

$$W_{12} = k\{m r_{0D}^m I_m(\varepsilon_1 r_{0D}) + 0.5\varepsilon_1 r_{0D}^{m+1}\left[I_{m-1}(\varepsilon_1 r_{0D}) \\ + I_{m+1}(\varepsilon_1 r_{0D})\right]\},$$

$$W_{13} = 0.5\exp\left(\frac{r_D}{2}\right) A_i(y_4) + \varepsilon_1^{1/3}\exp\left(\frac{r_D}{2}\right) A_i'(y_4),$$

$$W_{14} = r_{0D}^m K_m(\varepsilon_1 r_{0D}), \quad W_{15} = r_{0D}^m I_m(\varepsilon_1 r_{0D}),$$

$$W_{16} = \exp\left(\frac{r_{0D}}{2}\right) A_i(y_4) \quad \text{and} \quad \Theta = \frac{W_{13}W_{15} - W_{12}W_{16}}{W_{11}W_{16} - W_{13}W_{14}}$$

In the injection phase, the values of $\mathcal{T}_1$ and $\mathcal{T}_2$ are modified into $\mathcal{T}_1'$ and $\mathcal{T}_2'$ as follows:

$$\mathcal{T}_1' = \frac{F_1 - W_2'\mathcal{T}_2'}{W_1'}, \quad \mathcal{T}_2' = \frac{W_1'W_5}{W_1'W_4 - W_2'W_3}$$
$$\mathcal{T}_3' + \frac{W_1'W_6}{W_1'W_4 - W_2'W_3}\mathcal{T}_4' - \frac{W_3F_1}{W_1'W_4 - W_2'W_3} \tag{3.380}$$

where

$$W_1' = \exp\left(\frac{r_{wD}}{2\lambda}\right) A_i(y_w), \quad W_2' = \exp\left(\frac{r_{wD}}{2\lambda}\right) B_i(y_w),$$
$$\mathcal{T}_3' = \frac{W_{13}W_{15} - W_{12}W_{16}}{W_{11}W_{16} - W_{13}W_{14}}\mathcal{T}_4', \quad \text{and}$$
$$\mathcal{T}_4' = \frac{W_3F_1(W_1'W_8 - W_2'W_7) - W_7F_1(W_1'W_4 - W_2'W_3)}{\begin{array}{c}(W_1'W_5\Theta + W_1'W_6)(W_1'W_8 - W_2'W_7)\\ -(W_1'W_9\Theta - W_1'W_{10})(W_1'W_4 - W_2'W_3)\end{array}}.$$

Substituting $\mathcal{T}_1'$ and $\mathcal{T}_2'$ into Eq. (3.371a), one has

$$\overline{C}_{\text{inj},D}(r_{wD}, \text{s}) = \mathcal{T}_1'\exp\left(\frac{r_D}{2\lambda}\right) A_i(y_1) + \mathcal{T}_2'\exp\left(\frac{r_D}{2\lambda}\right) B_i(y_1) \tag{3.381}$$

In the chasing phase, the values of $\mathcal{T}_1$ and $\mathcal{T}_2$ are changed into $\mathcal{T}_1''$ and $\mathcal{T}_2''$ as follows:

$$\mathcal{T}_1'' = \frac{F_2 - W_2'\mathcal{T}_2''}{W_1'}, \quad \mathcal{T}_2'' = \frac{W_1'W_5}{W_1'W_4 - W_2'W_3}\mathcal{T}_3''$$
$$+ \frac{W_1'W_6}{W_1'W_4 - W_2'W_3}\mathcal{T}_4'' - \frac{W_3F_2}{W_1'W_4 - W_2'W_3} \tag{3.382}$$

where

$$\mathcal{T}_3'' = \frac{W_{13}W_{15} - W_{12}W_{16}}{W_{11}W_{16} - W_{13}W_{14}}\mathcal{T}_4'',$$
$$\mathcal{T}_4'' = \frac{W_3F_2(W_1'W_8 - W_2'W_7) - W_7F_2(W_1'W_4 - W_2'W_3)}{\begin{array}{c}(W_1'W_5\Theta + W_1'W_6)(W_1'W_8 - W_2'W_7)\\ -(W_1'W_9\Theta - W_1'W_{10})(W_1'W_4 - W_2'W_3)\end{array}}.$$

Substituting $\mathcal{T}_1''$ and $\mathcal{T}_2''$ into Eq. (3.374a), one has

$$\overline{C}_{\text{cha},D}(r_{wD},\text{s}) = \mathcal{T}_1''\exp\left(\frac{r_D}{2\lambda}\right) A_i(y_1) + \mathcal{T}_2''\exp\left(\frac{r_D}{2\lambda}\right) B_i(y_1) \tag{3.383}$$

Finally, the modified solutions with scale-dependent dispersivity are

$$\overline{C}_{m1D} = \mathcal{T}_1\exp\left(\frac{r_D}{2\lambda}\right)A_i(y_1) + \mathcal{T}_2\exp\left(\frac{r_D}{2\lambda}\right) B_i(y_1), \quad r_{wD} \le r_D \le r_{sD} \tag{3.384a}$$

$$\overline{C}_{im1D} = \frac{\varepsilon_{im1}}{s + \varepsilon_{im1} + \mu_{im1D}}\overline{C}_{m1D}, \quad r_{wD} \le r_D \le r_{sD} \tag{3.384b}$$

$$\overline{C}_{m2D} = \mathcal{T}_3 r_D^m K_m(\varepsilon_1 r_D) + \mathcal{T}_4 r_D^m I_m(\varepsilon_1 r_D), \quad r_{sD} \le r_D \le r_{0D} \tag{3.385a}$$

$$\overline{C}_{m2D} = \mathcal{T}_5\exp\left(\frac{r_D}{2}\right) A_i(y_3) + \mathcal{T}_6\exp\left(\frac{r_D}{2}\right) B_i(y_3), \quad r_D > r_{0D} \tag{3.385b}$$

$$\overline{C}_{im2D} = \frac{\varepsilon_{im2}}{s + \varepsilon_{im2} + \mu_{im2D}}\overline{C}_{m2D}, \quad r_D > r_{sD} \tag{3.385c}$$

where $m = \frac{1}{2k}$; $K_m(\cdot)$ is mth-order modified Bessel function of the second kind, $I_m(\cdot)$ is mth-order modified Bessel function of the first kind; the expressions for $\mathcal{T}_1, \mathcal{T}_2, \mathcal{T}_3, \mathcal{T}_4, \mathcal{T}_5$ and $\mathcal{T}_6$ are listed in Eqs. (3.379a)–(3.379e); $y_3 = (\varepsilon_1)^{1/3}\left(r_D + \frac{1}{4\varepsilon_1}\right)$; $y_4 = (\varepsilon_1)^{1/3}\left(r_{0D} + \frac{1}{4\varepsilon_1}\right)$; $C_{\text{inj},D}$ and $C_{\text{cha},D}$ could be determined by Eq. (3.381) and Eq. (3.383).

Substituting Eq. (3.364) into the dispersivity coefficient (D_α), one has

$$D_\alpha = \alpha_2|v_{a1}| + D_0 = \begin{cases} \dfrac{krQ}{2\pi r B\theta_{m1}} + D_0 = \dfrac{kQ}{2\pi B\theta_{m1}} + D_0, & r_s \le r \le r_0 \\ \dfrac{\alpha_0 Q}{2\pi r B\theta_{m2}} + D_0, & r > r_0 \end{cases} \tag{3.386}$$

where D_0 is the molecular diffusion coefficient [m^2/s]. A few interesting features are notable here. First, because of the unique feature of a divergent flow field in which the velocity is inversely proportional to the radial distance, and the use of a dispersivity function that is proportional to the radial distance when $r \leq r_0$, the dispersion coefficient in Eq. (3.386) becomes constant. However, one must be aware that if other types of dispersivity equations are used (such as exponential and parabolic functions), the dispersion coefficient in Eq. (3.386) will depend on the radial distance from the well. Second, even when D_α becomes constant for a linear dispersivity function when $r \leq r_0$, the mechanical dispersion is still dominant since the value of D_0 is generally much smaller than the mechanical dispersion term of $\frac{kQ}{2\pi B\theta_{m1}}$. For instance, the diffusion coefficient in water ranges from 1×10^{-9} to 2×10^{-9} m^2/s, and it is much smaller in the porous media (Freeze and Cherry, 1979). When $k = 0.01, Q = 0.1$ m^3/s, $B =$ 1 m, $\theta_{m1} = 0.3$, one has $\frac{kQ}{2\pi B\theta_{m1}} = 5.3 \times 10^{-4}$ m^2/s. Therefore, it is reasonable to ignore the molecular diffusion effect when $r \leq r_0$. The values of $\frac{kQ}{2\pi B\theta_{m1}}$ is dependent on k. The chosen value of $k = 0.01$ is from experimental studies, for instance, $k = 0.018$ in Chen *et al.* (2007).

(d) Analytical model in an aquifer-aquitard system

Regardless of Eqs. (3.362) and (3.363) or Eqs. (3.384) and (3.385), the aquifer is assumed to be completely isolated from the underlying and overlying aquitards (strictly confined), which might not be true in real applications. As stated before (Zhan *et al.*, 2009e, 2009f), it is nearly impossible to maintain a strictly confined condition in terms of transport. That is because if a solute in the aquifer is in contact with the upper or lower aquitard, molecular diffusion will always drive the solute from high concentration aquifer into the solute-free aquitard, even if the cross-formation flow in the aquitard does not exist. In fact, such diffusion-driven transport of solute into the aquitard and the subsequent back diffusion (from aquitard to aquifer when the aquifer solute concentration drops below the solute concentration in the aquitards) is responsible for many long tails in aquifer BTCs. The importance of aquitard in regulating solute transport has indeed been recognized by a number of investigators, such as Chen (1985, 1986, 1991), Yates (1988), Novakowski (1992a), Wang and Zhan (2013), and Zhou *et al.* (2017).

In this section, the solutions of Eqs. (3.362) and (3.363) will be extended considering underlying and overlying aquitards.

The governing equations are revised as

$$\theta_{m1}R_{m1}\frac{\partial C_{m1}}{\partial t} = \frac{\theta_{m1}}{r}\frac{\partial}{\partial r}\left(r\alpha_1|v_{a1}|\frac{\partial C_{m1}}{\partial r}\right)$$

$$-\theta_{m1}v_{a1}\frac{\partial C_{m1}}{\partial r} - \omega_1(C_{m1} - C_{im1}) - \theta_m\mu_{m1}C_{m1}$$

$$-\left.\frac{\theta_{um}D_u}{2b}\frac{\partial C_{um}}{\partial z}\right|_{z=b} + \left.\frac{\theta_{lm}D_l}{2b}\frac{\partial C_{lm}}{\partial z}\right|_{z=-b}, \quad r_w \le r \le r_s \tag{3.387a}$$

$$\theta_{im1}R_{im1}\frac{\partial C_{im1}}{\partial t} = \omega_1(C_{m1} - C_{im1}) - \theta_{im1}\mu_{im1}C_{im1}, \quad r_w \le r \le r_s \tag{3.387b}$$

$$\theta_{m2}R_{m2}\frac{\partial C_{m2}}{\partial t} = \frac{\theta_{m2}}{r}\frac{\partial}{\partial r}\left(r\alpha_2|v_{a2}|\frac{\partial C_{m2}}{\partial r}\right)$$

$$-\theta_{m2}v_{a2}\frac{\partial C_{m2}}{\partial r} - \omega_2(C_{m2} - C_{im2})$$

$$-\theta_{m2}\mu_{m2}C_{m2} - \left.\frac{\theta_{um}D_u}{2b}\frac{\partial C_{um}}{\partial z}\right|_{z=b} + \left.\frac{\theta_{lm}D_l}{2b}\frac{\partial C_{lm}}{\partial z}\right|_{z=-b}, \quad r > r_s \tag{3.387c}$$

$$\theta_{im2}R_{im2}\frac{\partial C_{im2}}{\partial t} = \omega_2(C_{m2} - C_{im2}) - \theta_{im2}\mu_{im2}C_{im2}, \quad r > r_s \tag{3.387d}$$

$$\theta_{um}R_{um}\frac{\partial C_{um}}{\partial t} = \theta_{um}D_u\frac{\partial^2 C_{um}}{\partial z^2}$$

$$-\omega_u(C_{um} - C_{\mathrm{uim}}) - \theta_{um}\mu_{um}C_{um}, \quad z \ge b \tag{3.388a}$$

$$\theta_{\mathrm{uim}}R_{\mathrm{uim}}\frac{\partial C_{\mathrm{uim}}}{\partial t} = \omega_u(C_{um} - C_{\mathrm{uim}}) - \theta_{\mathrm{uim}}\mu_{\mathrm{uim}}C_{\mathrm{uim}}, \quad z \ge b \tag{3.388b}$$

$$\theta_{lm}R_{lm}\frac{\partial C_{lm}}{\partial t} = \theta_{lm}D_l\frac{\partial^2 C_{lm}}{\partial z^2} - \omega_l(C_{lm} - C_{\mathrm{lim}})$$

$$-\theta_{lm}\mu_{lm}C_{lm}, \quad z \le -b \tag{3.389a}$$

$$\theta_{\mathrm{lim}}R_{\mathrm{lim}}\frac{\partial C_{\mathrm{lim}}}{\partial t} = \omega_l(C_{lm} - C_{\mathrm{lim}}) - \theta_{\mathrm{lim}}\mu_{\mathrm{lim}}C_{\mathrm{lim}}, \quad z \le -b \tag{3.389b}$$

where the subscripts "u" and "l" refer to the parameters in the upper and lower aquitard, respectively; C_{um} and C_{uim} are the mobile and immobile concentrations [Kg/m^3] of the upper aquitard, respectively; C_{lm} and C_{lim} are the mobile and immobile concentrations [Kg/m^3] of the upper aquitard, respectively; z represents the vertical distance [m]; D_u and D_l are the vertical dispersion coefficients [m^3/s] of the upper and lower aquitards, respectively; u_1 and u_2 represent Darcian velocity [m/s] of the skin and formation zones, respectively; $\mu_{um}, \mu_{\text{uim}}, \mu_{lm}$ and μ_{lim} are reaction rates for first-order biodegradation, or radioactive decay, or the first-order reaction rate [1/s], respectively; $\theta_{um}, \theta_{\text{uim}}, \theta_{lm}$ and θ_{lim} are the mobile and immobile porosities [-], respectively; $R_{um} = 1 + \frac{\rho_b K_d}{\theta_{um}}$ and $R_{\text{uim}} = 1 + \frac{\rho_b K_d}{\theta_{\text{uim}}}$ are regarded as retardation factors [-] for the mobile and immobile regions of the upper aquitard, respectively; $R_{lm} = 1 + \frac{\rho_b K_d}{\theta_{lm}}$ and $R_{\text{lim}} = 1 + \frac{\rho_b K_d}{\theta_{\text{lim}}}$ could be regarded as retardation factors [-] for the mobile and immobile regions in the lower aquitard, respectively; ω_u and ω_l are the first-order mass transfer coefficients [1/s] of the upper aquitard and lower aquitard, respectively.

Subject to the following initial and boundary conditions:

$$\begin{aligned} C_{m1}(r,t)|_{t=0} &= C_{im1}(r,t)|_{t=0} = C_{m2}(r,t)|_{t=0} \\ &= C_{im2}(r,t)|_{t=0} = C_{um}(r,z,t)|_{t=0} = C_{\text{uim}}(r,z,t)|_{t=0} \\ &= C_{lm}(r,z,t)|_{t=0} = C_{\text{lim}}(r,z,t)|_{t=0} = 0, \quad r \geq r_w \end{aligned} \tag{3.390}$$

$$\begin{aligned} C_{m2}(r,t)|_{r\to\infty} &= C_{im2}(r,t)|_{r\to\infty} = C_{um}(r,z,t)|_{r\to\infty} \\ &= C_{\text{uim}}(r,z,t)|_{r\to\infty} = C_{lm}(r,z,t)|_{r\to-\infty} \\ &= C_{\text{lim}}(r,z,t)|_{r\to-\infty} = 0, \quad r \geq r_w \end{aligned} \tag{3.391}$$

$$C_{m1}(r,t) = C_{um}(r,z=b,t), \quad r_w \leq r \leq r_s \tag{3.392a}$$

$$C_{m2}(r,t) = C_{um}(r,z=b,t), \quad r > r_s \tag{3.392b}$$

$$C_{m1}(r,t) = C_{lm}(r,z=-b,t), \quad r_w \leq r \leq r_s \tag{3.393a}$$

$$C_{m2}(r,t) = C_{lm}(r,z=-b,t), \quad r > r_s \tag{3.393b}$$

The flux concentration continuity is applied in the wellbore's boundary condition, which is the same as Eqs. (3.333a) and (3.333b). The wellbore mixing effect is also taken into account in the inner boundary condition, which is the same as Eqs. (3.334) and (3.335).

The dimensionless parameters are defined as

$$C_{umD}=\frac{C_{um}}{C_0},\quad C_{\text{uim}D}=\frac{C_{\text{uim}}}{C_0},\quad C_{lmD}=\frac{C_{lm}}{C_0},\quad C_{\text{lim}D}=\frac{C_{\text{lim}}}{C_0},$$

$$\mu_{umD}=\frac{\alpha_2^2R_m\mu_m}{R_{um}A},\quad \mu_{\text{uim}D}=\frac{\alpha_2^2R_m\mu_{im}}{R_{um}A},\quad \mu_{lmD}=\frac{\alpha_2^2R_m\mu_m}{AR_{lm}},$$

$$\text{and}\quad \mu_{\text{lim}D}=\frac{\alpha_2^2R_m\mu_{im}}{R_{lm}A}.$$

The dimensionless forms of the governing equation could be rewritten as

$$\begin{aligned}\frac{\partial C_{m1D}}{\partial t_D}&=\frac{\lambda}{r_D}\frac{\partial^2C_{m1D}}{\partial r_D^2}-\frac{1}{r_D}\frac{\partial C_{m1D}}{\partial r_D}\\&\quad-\varepsilon_{m1}(C_{m1D}-C_{im1D})-\mu_{m1D}C_{m1D}\\&\quad-\frac{\theta_{um}\alpha_2^2D_u}{2A\theta_{m1}b^2}\frac{\partial C_{umD}}{\partial z_D}\bigg|_{z=1}+\frac{\theta_{lm}\alpha_2^2D_l}{2Ab^2\theta_{m1}}\frac{\partial C_{lmD}}{\partial z_D}\bigg|_{z=-1},\\&\quad r_{wD}\le r_D\le r_{sD}\end{aligned}\tag{3.394a}$$

$$\frac{\partial C_{im1D}}{\partial t_D}=\varepsilon_{im1}(C_{m1D}-C_{im1D})-\mu_{im1D}C_{im1D},\quad r_{wD}\le r_D\le r_{sD}\tag{3.394b}$$

$$\begin{aligned}\frac{\partial C_{m2D}}{\partial t_D}&=\frac{1}{r_D}\frac{\partial^2C_{m2D}}{\partial r_D^2}-\frac{1}{r_D}\frac{\partial C_{m2D}}{\partial r_D}-\varepsilon_{m2}(C_{m2D}-C_{im2D})\\&\quad-\mu_{m2D}C_{m2D}-\frac{\theta_{um}\alpha_2^2D_u}{2A\theta_{m2}\text{b}^2}\frac{\partial C_{umD}}{\partial z_D}\bigg|_{z=1}\\&\quad+\frac{\theta_{lm}\alpha_2^2D_l}{2Ab^2\theta_{m2}}\frac{\partial C_{lmD}}{\partial z_D}\bigg|_{z=-1},\quad r_D>r_{sD}\end{aligned}\tag{3.394c}$$

$$\frac{\partial C_{im2D}}{\partial t_D}=\varepsilon_{im2}(C_{m2D}-C_{im2D})-\mu_{im2D}C_{im2D},\quad r_D>r_{sD}\tag{3.394d}$$

$$\begin{aligned}\frac{\partial C_{umD}}{\partial t_D}&=\frac{R_{m1}\alpha_2^2D_u}{Ab^2R_{um}}\frac{\partial^2C_{umD}}{\partial z_D^2}\\&\quad-\varepsilon_{um}(C_{umD}-C_{\text{uim}D})-\mu_{umD}C_{umD},\quad z_D\ge1,t_D>0\end{aligned}\tag{3.395a}$$

$$\frac{\partial C_{\text{uim}D}}{\partial t_D} = \varepsilon_{\text{uim}}(C_{umD} - C_{\text{uim}D}) - \mu_{\text{uim}D}C_{\text{uim}D}, \quad z_D \geq 1, t_D > 0 \tag{3.395b}$$

$$\frac{\partial C_{lmD}}{\partial t_D} = \frac{R_{m1}\alpha_2^2 D_l}{Ab^2 R_{lm}} \frac{\partial^2 C_{lmD}}{\partial z_D^2} - \varepsilon_{lm}(C_{lmD} - C_{\text{lim}D}) - \mu_{lmD}C_{lmD}, \quad z_D \leq -1, t_D > 0 \tag{3.396a}$$

$$\frac{\partial C_{\text{lim}D}}{\partial t_D} = \varepsilon_{\text{lim}}(C_{lmD} - C_{\text{lim}D}) - \mu_{\text{lim}D}C_{\text{lim}D}, \quad z_D \leq -1, t_D > 0 \tag{3.396b}$$

where

$$\varepsilon_{m1} = \frac{\omega_1\alpha_2^2}{A\theta_{m1}}, \quad \varepsilon_{im1} = \frac{\omega_1\alpha_2^2 R_{m1}}{A\theta_{im1}R_{im1}}, \quad \varepsilon_{m2} = \frac{\omega_2\alpha_2^2}{A\theta_{m2}},$$

$$\varepsilon_{im2} = \frac{\omega_2\alpha_2^2 R_{m1}}{A\theta_{im2}R_{im2}}, \quad \varepsilon_{um} = \frac{\omega_u\alpha_2^2 R_{m1}}{A\theta_{um}R_{um}}, \quad \varepsilon_{\text{uim}} = \frac{\omega_u\alpha_2^2 R_{m1}}{A\theta_{\text{uim}}R_{\text{uim}}},$$

$$\varepsilon_{lm} = \frac{\omega_l\alpha_2^2 R_{m1}}{A\theta_{lm}R_{lm}}, \quad \varepsilon_{\text{lim}} = \frac{\omega_l\alpha_2^2 R_{m1}}{A\theta_{\text{lim}}R_{\text{lim}}}.$$

Substituting the dimensionless parameters into Eqs. (3.390)–(3.393), one has

$$\begin{aligned} C_{m1D}(r_D, t_D)|_{t_D=0} &= C_{im1D}(r_D, t_D)|_{t_D=0} = C_{m2D}(r_D, t_D)|_{t_D=0} \\ &= C_{im2D}(r_D, t_D)|_{t_D=0} = .C_{umD}(r_D, z_D, t_D)|_{t_D=0} \\ &= C_{\text{uim}D}(r_D, z_D, t_D)|_{t_D=0} = C_{lmD}(r_D, z_D, t_D)|_{t_D=0} \\ &= C_{\text{lim}D}(r_D, z_D, t_D)|_{t_D=0} = 0 \end{aligned} \tag{3.397}$$

$$\begin{aligned} C_{m2D}(r_D, t_D)|_{r_D\to\infty} &= C_{im2D}(r_D, t_D)|_{r_D\to\infty} \\ &= C_{umD}(r_D, z_D, t_D)|_{z_D\to\infty} = C_{\text{uim}D}(r_D, z_D, t_D)|_{z_D\to\infty} \\ &= C_{lmD}(r_D, z_D, t)|_{z_D\to-\infty} = C_{\text{lim}D}(r_D, z_D, t_D)|_{z_D\to-\infty} = 0 \end{aligned} \tag{3.398}$$

$$C_{m1D}(r_D, t_D) = C_{umD}\left(r_D, z_D = 1, t_D\right), \quad r_{wD} \leq r_D \leq r_{sD} \tag{3.399a}$$

$$C_{m2D}(r_D, t_D) = C_{umD}\left(r_D, z_D = 1, t_D\right), \quad r_{wD} \leq r_D \leq r_{sD} \tag{3.399b}$$

$$C_{m2D}(r_D,t_D)=C_{lmD}\left(r_D,z_D=-1,t_D\right),\quad r_D>r_{sD} \tag{3.399c}$$

$$C_{m2D}(r_D,t_D)=C_{lmD}\left(r_D,z_D=-1,t_D\right),\quad r_D>r_{sD} \tag{3.399d}$$

Conducting Laplace transform to Eqs. (3.395a) and (3.395b), one has

$$s\overline{C}_{umD}=\frac{R_{m1}\alpha_2^2 D_u}{AB^2R_{um}}\frac{\partial^2\overline{C}_{umD}}{\partial z_D^2} - \left(\varepsilon_{um}+\mu_{umD}\right)\overline{C}_{umD}+\varepsilon_{um}\overline{C}_{\text{uim}D},\quad z_D\geq 1 \tag{3.400a}$$

$$s\overline{C}_{\text{uim}D}=\varepsilon_{\text{uim}}\left(\overline{C}_{umD}-\overline{C}_{\text{uim}D}\right)-\mu_{\text{uim}D}\overline{C}_{\text{uim}D},\quad z_D\geq 1 \tag{3.400b}$$

Substituting Eq. (3.400b) into Eq. (3.400a), Eq. (3.400a) could be rewritten as

$$\frac{R_{m1}\alpha_2^2 D_u}{Ab^2R_{um}}\frac{\partial^2\overline{C}_{umD}}{\partial z_D^2} - \left(s+\varepsilon_{um}+\mu_{umD}-\frac{\varepsilon_{um}\varepsilon_{\text{uim}}}{s+\mu_{\text{uim}D}+\varepsilon_{\text{uim}}}\right)\overline{C}_{umD}=0,\quad z_D\geq 1 \tag{3.401}$$

Similarly, Eqs. (3.396a) and (3.396b) become:

$$\frac{R_{m1}\alpha_2^2 D_l}{Ab^2R_{lm}}\frac{\partial^2\overline{C}_{lmD}}{\partial z_D^2} - \left(s+\varepsilon_{lm}+\mu_{lmD}-\frac{\varepsilon_{lm}\varepsilon_{\text{lim}}}{s+\mu_{\text{lim}D}+\varepsilon_{\text{lim}}}\right)\overline{C}_{lmD}=0,\quad z_D\leq -1 \tag{3.402}$$

Equations (3.397)–(3.399d) and (3.401)–(3.402) compose a model of the second-order ordinary differential equation (ODE) with boundary conditions, and the general solution of Eq. (3.401) is

$$\overline{C}_{umD}=A_1e^{a_1z_D}+B_1e^{a_2z_D},\quad z_D\geq 1 \tag{3.403a}$$

Similarly, the general solution of Eq. (3.402) is

$$\overline{C}_{lmD}=A_2e^{b_1z_D}+B_2e^{b_2z_D},\quad z_D\leq -1 \tag{3.403b}$$

where

$$a_1 = \sqrt{s + \varepsilon_{um} + \mu_{umD} - \frac{\varepsilon_{um}\varepsilon_{\text{uim}}}{s + \mu_{\text{uim}D} + \varepsilon_{\text{uim}}}},$$

$$a_2 = -\sqrt{s + \varepsilon_{um} + \mu_{umD} - \frac{\varepsilon_{um}\varepsilon_{\text{uim}}}{s + \mu_{\text{uim}D} + \varepsilon_{\text{uim}}}},$$

$$b_1 = \sqrt{s + \varepsilon_{lm} + \mu_{lmD} - \frac{\varepsilon_{lm}\varepsilon_{\text{lim}}}{s + \mu_{\text{lim}D} + \varepsilon_{\text{lim}}}} \quad \text{and}$$

$$b_2 = -\sqrt{s + \varepsilon_{lm} + \mu_{lmD} - \frac{\varepsilon_{lm}\varepsilon_{\text{lim}}}{s + \mu_{\text{lim}D} + \varepsilon_{\text{lim}}}}.$$

Substituting Eqs. (3.403a) and (3.403b) into Eqs. (3.399a)–(3.399d), one has

$$\overline{C}_{umD} = B_1 e^{a_2 z_D}, \quad z_D \geq 1 \tag{3.404a}$$

$$\overline{C}_{lmD} = A_2 e^{b_1 z_D}, \quad z_D \leq -1 \tag{3.404b}$$

where $B_1 = \overline{C}_{mD}\exp(-a_2)$, and $A_2 = \overline{C}_{mD}\exp(b_1)$.

Thus, the solutions of the aquitards are

$$\overline{C}_{umD} = \overline{C}_{m1D}\exp(a_2 z_D - a_2), \quad r_{wD} \leq r_D \leq r_{sD} \tag{3.405a}$$

$$\overline{C}_{umD} = \overline{C}_{m2D}\exp(a_2 z_D - a_2), \quad r_D > r_{sD} \tag{3.405b}$$

$$\overline{C}_{\text{uim}D} = \frac{\varepsilon_{\text{uim}}}{s + \varepsilon_{\text{uim}} + \mu_{\text{uim}D}}\overline{C}_{umD}, \quad r_D \geq r_{wD} \tag{3.405c}$$

$$\overline{C}_{lmD} = \overline{C}_{m1D}\exp(b_1 z_D + b_1), \quad r_{wD} \leq r_D \leq r_{sD} \tag{3.406a}$$

$$\overline{C}_{lmD} = \overline{C}_{m2D}\exp(b_1 z_D + b_1), \quad r_D > r_{sD} \tag{3.406b}$$

$$\overline{C}_{\text{lim}D} = \frac{\varepsilon_{\text{lim}}}{s + \varepsilon_{\text{lim}} + \mu_{\text{lim}D}}\overline{C}_{lmD}, \quad r_D \geq r_{wD} \tag{3.406c}$$

The dimensionless forms of Eqs. (3.334) and (3.335) become:

$$\left[C_{m1D} - \lambda\frac{\partial C_{m1D}(r_D, t_D)}{\partial r_D}\right]\Bigg|_{r=r_{wD}} = C_{\text{inj},D}(t_D), \quad 0 < t_D \leq t_{\text{inj},D} \tag{3.407a}$$

$$\left[C_{m1D} - \lambda\frac{\partial C_{m1D}(r_D, t_D)}{\partial r_D}\right]\Bigg|_{r=r_{wD}} = C_{\text{cha},D}(t_D), \quad t_D > t_{\text{inj},D} \tag{3.407b}$$

$$\beta_{\text{inj}}\frac{dC_{\text{inj},D}(t_D)}{dt_D} = 1 - C_{\text{inj},D}(t_D), \quad 0 < t_D \le t_{\text{inj},D} \tag{3.408a}$$

$$\beta_{\text{cha}}\frac{dC_{\text{cha},D}(t_D)}{dt_D} = -C_{\text{cha},D}(t_D), \quad t_D > t_{\text{inj},D} \tag{3.408b}$$

where $\beta_{\text{inj}} = \frac{V_{w,\text{inj}}r_{wD}}{\xi R_m \alpha_2}$ and $\beta_{\text{cha}} = \frac{V_{w,\text{cha}}r_{wD}}{\xi R_m \alpha_2}$.

After applying Laplace transform to Eqs. (3.394a) and (3.394b), one has

$$\begin{aligned} s\overline{C}_{m1D} = {} & \frac{\lambda}{r_D}\frac{\partial^2 \overline{C}_{m1D}}{\partial r_D^2} - \frac{1}{r_D}\frac{\partial \overline{C}_{m1D}}{\partial r_D} \\ & - \varepsilon_{m2}\left(\overline{C}_{m1D} - \overline{C}_{im1D}\right) - \mu_{m1D}\overline{C}_{m1D} \\ & - \frac{\theta_{um}\alpha_2^2 D_u}{2A\theta_{m1}b^2}\frac{\partial \overline{C}_{umD}}{\partial z_D}\bigg|_{z=1} + \frac{\theta_{lm}\alpha_2^2 D_l}{2Ab^2\theta_{m1}}\frac{\partial \overline{C}_{lmD}}{\partial z_D}\bigg|_{z=-1}, \\ & r_{wD} \le r_D \le r_{sD} \end{aligned} \tag{3.409a}$$

$$s\overline{C}_{im1D} = \varepsilon_{im2}\left(\overline{C}_{m1D} - \overline{C}_{im1D}\right) - \mu_{im1D}\overline{C}_{im1D}, \quad r_{wD} \le r_D \le r_{sD} \tag{3.409b}$$

Substituting Eq. (3.409b) into Eq. (3.409a), Eq. (3.409a) could be rewritten as

$$\begin{aligned} & \frac{\lambda}{r_D}\frac{\partial^2 \overline{C}_{m1D}}{\partial r_D^2} - \frac{1}{r_D}\frac{\partial \overline{C}_{m1D}}{\partial r_D} - \left(\varepsilon_{m1} + \mu_{m1D} - \frac{\varepsilon_{m1}\varepsilon_{im1}}{s + \mu_{im1D} + \varepsilon_{im1}}\right) \\ & \overline{C}_{m1D} - \frac{\theta_{um}\alpha_2^2 D_u}{2A\theta_{m1}b^2}\frac{\partial \overline{C}_{umD}}{\partial z_D}\bigg|_{z=1} + \frac{\theta_{lm}\alpha_2^2 D_l}{2Ab^2\theta_{m1}}\frac{\partial \overline{C}_{lmD}}{\partial z_D}\bigg|_{z=-1}, \\ & r_{wD} \le r_D \le r_{sD} \end{aligned} \tag{3.410}$$

After applying Laplace transform to Eqs. (3.394c) and (3.394d), the following equations would be obtained:

$$\begin{aligned} s\overline{C}_{m2D} = {} & \frac{\eta}{r_D}\frac{\partial^2 \overline{C}_{m2D}}{\partial r_D^2} - \frac{\eta}{r_D}\frac{\partial \overline{C}_{m2D}}{\partial r_D} - \varepsilon_{m2}(C_{m2D} - C_{im2D}) \\ & - \frac{\theta_{um}\alpha_2^2 D_u}{2A\theta_{m2}b^2}\frac{\partial \overline{C}_{umD}}{\partial z_D}\bigg|_{z=1} + \frac{\theta_{lm}\alpha_2^2 D_l}{2Ab^2\theta_{m2}}\frac{\partial \overline{C}_{lmD}}{\partial z_D}\bigg|_{z=-1}, \\ & r_D > r_{sD} \end{aligned} \tag{3.411a}$$

$$s\overline{C}_{im2D} = \varepsilon_{im2}\left(\overline{C}_{m2D} - \overline{C}_{im2D}\right) - \mu_{im2D}\overline{C}_{im2D}, \quad r_D > r_{sD} \tag{3.411b}$$

Substituting Eq. (3.411b) into Eq. (3.411a), one has

$$s\overline{C}_{m2D} = \frac{\eta}{r_D}\frac{\partial^2 \overline{C}_{m2D}}{\partial r_D^2} - \frac{\eta}{r_D}\frac{\partial \overline{C}_{m2D}}{\partial r_D} - \left(\varepsilon_{m2} + \mu_{m2D} - \frac{\varepsilon_{m2}\varepsilon_{im2}}{s + \mu_{im2D} + \varepsilon_{im2}}\right) \overline{C}_{m2D} - \frac{\theta_{um}\alpha_2^2 D_u}{2A\theta_{m2}b^2}\frac{\partial \overline{C}_{umD}}{\partial z_D}\bigg|_{z=1} + \frac{\theta_{lm}\alpha_2^2 D_l}{2Ab^2\theta_{m2}}\frac{\partial \overline{C}_{lmD}}{\partial z_D}\bigg|_{z=-1},$$

$$r_D > r_{sD} \tag{3.412}$$

Substituting Eqs. (3.405a) and (3.406a) into Eq. (3.410), one has

$$\frac{\lambda}{r_D}\frac{\partial^2 \overline{C}_{m1D}}{\partial r_D^2} - \frac{1}{r_D}\frac{\partial \overline{C}_{m1D}}{\partial r_D} - E_3\overline{C}_{m1D} = 0, \quad r_{wD} \le r_D \le r_{sD} \tag{3.413a}$$

Substituting Eqs. (3.405b) and (3.406b) into Eq. (3.412), one has

$$\frac{1}{r_D}\frac{\partial^2 \overline{C}_{m2D}}{\partial r_D^2} - \frac{1}{r_D}\frac{\partial \overline{C}_{m2D}}{\partial r_D} - E_4\overline{C}_{m2D} = 0, \quad r_D > r_{sD} \tag{3.413b}$$

where

$$E_3 = s + \varepsilon_{m1} + \mu_{m1D} - \frac{\varepsilon_{m1}\varepsilon_{im1}}{s + \mu_{im1D} + \varepsilon_{im1}} - \frac{a_2\theta_{um}\alpha_2^2 D_u}{2A\theta_{m1}b^2} + \frac{b_1\theta_{lm}\alpha_2^2 D_l}{2Ab^2\theta_{m1}},$$

and

$$E_4 = \frac{1}{\eta}\left(s + \varepsilon_{m2} + \mu_{m2D} - \frac{\varepsilon_{m2}\varepsilon_{im2}}{s + \mu_{im2D} + \varepsilon_{im2}} - \frac{a_2\theta_{um}\alpha_2^2 D_u}{2A\theta_{m2}b^2} + \frac{b_1\theta_{lm}\alpha_2^2 D_l}{2Ab^2\theta_{m2}}\right).$$

The boundary conditions of the wellbore and infinity in the Laplace domain are

$$\left[\overline{C}_{m1D} - \lambda\frac{\partial \overline{C}_{m1D}(r_D, s)}{\partial r_D}\right]\bigg|_{r_D = r_{wD}} = \frac{1}{s\left(s\beta_{\text{inj}} + 1\right)}, \quad 0 < t_D \le t_{\text{inj},D} \tag{3.414a}$$

$$\left[\overline{C}_{m1D} - \lambda \frac{\partial \overline{C}_{m1D}(r_D, s)}{\partial r_D}\right]\Bigg|_{r_D = r_{wD}} = \frac{\beta_{\text{cha}} C_{\text{inj},D}\left(r_{wD}, t_{\text{inj},D}\right)}{\left(s\beta_{\text{cha}} + 1\right)},$$

$$t_D > t_{\text{inj},D} \tag{3.414b}$$

$$\overline{C}_{m1D}(r_{sD}, s) = \overline{C}_{m2D}(r_{sD}, s) \tag{3.414c}$$

$$\lambda \frac{\partial \overline{C}_{m1D}(r_D, s)}{\partial r_D}\Bigg|_{r_D = r_{sD}} = \frac{\partial \overline{C}_{m2D}(r_D, s)}{\partial r_D}\Bigg|_{r_D = r_{sD}} \tag{3.414d}$$

$$\overline{C}_{m2D}(r_D, s)\big|_{r_D \to \infty} = 0 \tag{3.414e}$$

Therefore, the general solutions of Eqs. (3.414a) and (3.414b) are

$$\overline{C}_{m1D} = T_1 \exp\left(\frac{r_D}{2\lambda}\right) A_i\left(\varphi_1\right) + T_2 \exp\left(\frac{r_D}{2\lambda}\right) B_i\left(\varphi_1\right),$$

$$r_{wD} \le r_D \le r_{sD} \tag{3.415a}$$

$$\overline{C}_{m2D} = T_3 \exp\left(\frac{r_D}{2}\right) A_i\left(\varphi_2\right) + T_4 \exp\left(\frac{r_D}{2}\right) B_i\left(\varphi_2\right), \quad r_D > r_{sD} \tag{3.415b}$$

where $\varphi_1 = \left(\frac{E_3}{\lambda}\right)^{1/3}\left(r_D + \frac{1}{4\lambda E_3}\right), \varphi_2 = {E_4}^{1/3}\left(r_D + \frac{1}{4E_4}\right), T_1, T_2, T_3$ and T_4 are constants that the boundary conditions could determine.

Substituting Eq. (3.415b) into Eq. (3.414e), one has

$$T_4 = 0 \tag{3.416}$$

Substituting Eq. (3.415a) into Eq. (3.414a), one has

$$T_1 \exp\left(\frac{r_{wD}}{2\lambda}\right)\left[\frac{1}{2} A_i\left(\varphi_w\right) - \lambda\left(\frac{E_3}{\lambda}\right)^{1/3} A_i'\left(\varphi_w\right)\right] + T_2 \exp\left(\frac{r_{wD}}{2\lambda}\right)\left[\frac{1}{2} B_i\left(\varphi_w\right) - \lambda\left(\frac{E_3}{\lambda}\right)^{1/3} B_i'\left(\varphi_w\right)\right] = F_1 \tag{3.417a}$$

Substituting Eq. (3.415a) into Eq. (3.414b), one has

$$T_1 \exp\left(\frac{r_{wD}}{2\lambda}\right)\left[\frac{1}{2} A_i\left(\varphi_w\right) - \lambda\left(\frac{E_4}{\lambda}\right)^{1/3} A_i'\left(\varphi_w\right)\right] + T_2 \exp\left(\frac{r_{wD}}{2\lambda}\right)\left[\frac{1}{2} B_i\left(\varphi_w\right) - \lambda\left(\frac{E_4}{\lambda}\right)^{1/3} B_i'\left(\varphi_w\right)\right] = F_2 \tag{3.417b}$$

where $\varphi_w = \left(\frac{E_3}{\lambda}\right)^{1/3}\left(r_{wD} + \frac{1}{4\lambda E_3}\right)$, $F_1 = \frac{1}{s(s\beta_{\text{inj}}+1)}$ and $F_2 = \frac{\beta_{\text{cha}} C_{\text{inj},D}(r_{wD}, t_{\text{inj},D})}{(s\beta_{\text{cha}}+1)}$.

Substituting Eqs. (3.415a) and (3.415b) into Eq. (3.414c), one has

$$T_1 \exp\left(\frac{r_{sD}}{2\lambda}\right) A_i(\varphi_{1s}) + T_2 \exp\left(\frac{r_{sD}}{2\lambda}\right) B_i(\varphi_{1s}) = T_3 \exp\left(\frac{r_{sD}}{2}\right) A_i(\varphi_{2s}) \tag{3.418}$$

where $\varphi_{1s} = \left(\frac{E_3}{\lambda}\right)^{1/3}\left(r_{sD} + \frac{1}{4\lambda E_3}\right)$, $\varphi_{2s} = {E_4}^{1/3}\left(r_D + \frac{1}{4E_4}\right)$.

Substituting Eqs. (3.415a) and (3.415b) into Eq. (3.414d), one has

$$\begin{aligned} T_1 \exp\left(\frac{r_{sD}}{2\lambda}\right) &\left[\frac{1}{2} A_i(\varphi_{1s}) + \lambda\left(\frac{E_3}{\lambda}\right)^{1/3} A_i'(\varphi_{1s})\right] \\ &+ T_2 \exp\left(\frac{r_{sD}}{2\lambda}\right)\left[\frac{1}{2} B_i(\varphi_{1s}) + \lambda\left(\frac{E_3}{\lambda}\right)^{1/3} B_i'(\varphi_{1s})\right] \\ &= T_3 \exp\left(\frac{r_{sD}}{2}\right)\left[\frac{1}{2} A_i(\varphi_{2s}) + {E_4}^{1/3} A_i'(\varphi_{2s})\right] \end{aligned} \tag{3.419}$$

The values of T_1, T_2, and T_3 could be determined by solving Eqs. (3.417a)–(3.419):

$$T_1 = \frac{F - G_2 T_2}{G_1} \tag{3.420a}$$

$$T_2 = \frac{G_3 G_8 F - G_5 G_6 F}{G_1 G_5 G_7 + G_2 G_3 G_8 - G_2 G_5 G_6 - G_1 G_4 G_8} \tag{3.420b}$$

$$T_3 = \frac{G_3 F}{G_1 G_5} - \frac{G_2 G_3 T_2}{G_1 G_5} + \frac{G_4 T_2}{G_5} \tag{3.420c}$$

where

$$G_1 = \exp\left(\frac{r_{wD}}{2\lambda}\right)\left[\frac{1}{2} A_i(\varphi_w) - \lambda\left(\frac{E_3}{\lambda}\right)^{1/3} A_i'(\varphi_w)\right],$$

$$G_2 = \exp\left(\frac{r_{wD}}{2\lambda}\right)\left[\frac{1}{2} B_i(\varphi_w) - \lambda\left(\frac{E_3}{\lambda}\right)^{1/3} B_i'(\varphi_w)\right],$$

$$G_3 = \exp\left(\frac{r_{sD}}{2\lambda}\right) A_i(\varphi_{1s}), \quad G_4 = \exp\left(\frac{r_{sD}}{2\lambda}\right) B_i(\varphi_{1s}),$$

$$G_5 = \exp\left(\frac{r_{sD}}{2}\right) A_i(\varphi_{2s}),$$

$$G_6 = \exp\left(\frac{r_{sD}}{2\lambda}\right)\left[\frac{1}{2}A_i(\varphi_{1s}) + \lambda\left(\frac{E_3}{\lambda}\right)^{1/3} A_i'(\varphi_{1s})\right],$$

$$G_7 = \exp\left(\frac{r_{sD}}{2\lambda}\right)\left[\frac{1}{2}B_i(\varphi_{1s}) + \lambda\left(\frac{E_3}{\lambda}\right)^{1/3} B_i'(\varphi_{1s})\right],$$

$$G_8 = \exp\left(\frac{r_{sD}}{2}\right)\left[\frac{1}{2}A_i(\varphi_{2s}) + E_4{}^{1/3} A_i'(\varphi_{2s})\right],$$

and $F = C_{\text{inj},D}\frac{1-\exp(-t_{\text{inj},D}s)}{s} + C_{\text{cha},D}\frac{\exp(-t_{\text{inj},D}s)}{s}$.

In the injection phase, the values of T_1 and T_2 are modified into T_1' and T_2' as follows:

$$T_1' = \frac{F_1 - G_2' T_2'}{G_1'} \quad \text{and}$$

$$T_2' = \frac{G_3 G_8 F_1 - G_5 G_6 F_1}{G_1' G_5 G_7 + G_2' G_3 G_8 - G_2' G_5 G_6 - G_1' G_4 G_8} \tag{3.421}$$

where $G_1' = \frac{1}{2}\exp\left(\frac{r_{wD}}{2\lambda}\right) A_i(\varphi_w)$ and $G_2' = \frac{1}{2}\exp\left(\frac{r_{wD}}{2\lambda}\right) B_i(\varphi_w)$.

Substituting T_1' and T_2' into Eq. (3.415a), one has

$$\overline{C}_{\text{inj},D}(r_{wD}, \text{s}) = T_1'\exp\left(\frac{r_{wD}}{2\lambda}\right) A_i(\varphi_w) + T_2'\exp\left(\frac{r_{wD}}{2\lambda}\right) B_i(\varphi_w) \tag{3.422}$$

In the chasing phase, the values of T_1 and T_2 are modified into T_1' and T_2' as follows:

$$T_1'' = \frac{F_2 - G_2' T_2'}{G_1'} \quad \text{and}$$

$$T_2'' = \frac{G_3 G_8 F_2 - G_5 G_6 F_2}{G_1' G_5 G_7 + G_2' G_3 G_8 - G_2' G_5 G_6 - G_1' G_4 G_8} \tag{3.423}$$

Substituting T_1'' and T_2'' into Eq. (3.415a), one has

$$\overline{C}_{\text{cha},D}(r_{wD}, \text{s}) = T_1''\exp\left(\frac{r_{wD}}{2\lambda}\right) A_i(\varphi_w) + T_2''\exp\left(\frac{r_{wD}}{2\lambda}\right) B_i(\varphi_w) \tag{3.424}$$

Conducting Laplace transform on Eq. (3.345c), one has

$$\overline{C}_{wD}(r_{wD}, s) = C_{\text{inj},D}\frac{1-\exp(-t_{\text{inj},D}s)}{s} + C_{\text{cha},D}\frac{\exp(-t_{\text{inj},D}s)}{s} \tag{3.425}$$

where $C_{\text{inj},D}$ and $C_{\text{cha},D}$ could be determined by Eqs. (3.422)–(3.424).

Substituting Eqs. (3.415a) and (3.415b) into Eqs. (3.417a)–(3.419) and (3.425), one has

$$\overline{C}_{m1D} = T_1, \exp\left(\frac{r_D}{2\lambda}\right) A_i(\varphi_1) + T_2 \exp\left(\frac{r_D}{2\lambda}\right) B_i(\varphi_1), \quad r_{wD} < r_D \leq r_{sD} \tag{3.426a}$$

$$\overline{C}_{m2D} = T_3 \exp\left(\frac{r_D}{2}\right) A_i(\varphi_2), \quad r_D > r_{sD} \tag{3.426b}$$

3.3.2.5 *Robustness of assumptions*

The values of $V_{w,\text{inj}} = 0$ and $V_{w,\text{ext}} = 0$ represent no mixing effects in the wellbore for both injection and extraction phases; thus, the new model reduces to the case of a fully penetrating well as used by Huang *et al.* (2010). A main innovation of Wang *et al.* (2018) and Shi *et al.* (2020) is to develop a new model of SWPP tests with the wellbore mixing effect based on two assumptions, which is reflected in Eqs. (3.272)–(3.274), whose robustness should be checked for in applications. This can be done by comparing the theoretically developed equations with experimental data from an actual field site or laboratory study. Although there is a great deal of SWPP tests in the publications, the information of $h_{w,\text{inj}}$, $h_{w,\text{ext}}$, and well radius is not available yet. Alternatively, a high-resolution numerical simulation will be employed for comparison.

Besides the two assumptions mentioned above, the models of Wang *et al.* (2018) and Shi *et al.* (2020) include other assumptions: homogeneous aquifer and Fickian-type solute transport, which have been widely used in previous studies (Chen *et al.*, 2017; Gelhar and Collins, 1971; Schroth and Istok, 2005; Schroth *et al.*, 2001). Several concerns of these assumptions will be explained in the following section.

(a) First assumption: A model of the complete mixing process in the wellbore

The first assumption of the analytical solution of Wang *et al.* (2018) and Shi *et al.* (2020) is that the tracer injected into the wellbore is completely mixed with the tracer-free water in the wellbore prior to the start of the injection. Due to the extremely large hydraulic conductivity of a wellbore (if the wellbore is also treated as an equivalent porous media), such a treatment seems reasonable. In this section, we will employ a numerical approach to test the accuracy of such an assumption. MODFLOW/MT3DMS, developed by the US Geological Survey, is a 3D finite difference model to simulate groundwater flow (MODFLOW) and reactive solute transport (MT3DMS) in the aquifer. To date, the MODFLOW/MT3DMS package has been widely used to replicate the SWPP test. In this section, GMS 7.0 (which includes MODFLOW/MT3DMS) will be employed for the calculation.

The mixing effect has been automatically considered in the traditional well package of MODFLOW/MT3DMS. In setting the information of well in GMS 7.0, the boundary condition of the ADE should be "well" in the "PointSource/Sink BC" package of MT3DMS rather than "Specified conc". Numerical solutions of the ADE model may suffer from nonnegligible numerical error, especially for advection-dominating problems. Looking at the velocity distribution around the well, one could conclude that advection is the greatest at the wellbore-aquifer interface and decreases with radial distance away from the well. The numerical error could be decreased to a negligible level by increasing the grid resolution of the numerical solution. A nonuniform grid system will be employed, and the aquifer dimensions are 0.6 m in thickness, 100 m in length, and 100 m in width. To increase the accuracy of the numerical solution, a non-uniform grid system is used. The aquifer is discretized horizontally into 173 rows and 173 columns and vertically into 1 layer. The dimensions of the well block (cell representing well) are $0.02\,\text{m} \times 0.02\,\text{m} \times 0.6\,\text{m}$, which is located at the center of the aquifer, as shown in Figure 3.18. The flow field is assumed to be in a steady state. The hydraulic conductivity is 1 m/day, and the specific storage is 0. The initial head of the aquifer is 0.6 m. Four lateral boundary conditions are constant head with 0.6 m, as shown in Figure 3.18.

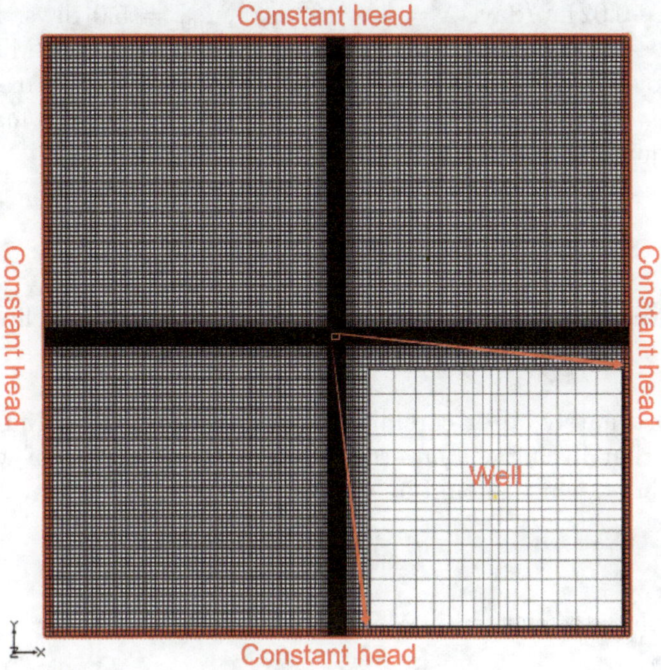

Figure 3.18. Grid system of MODFLOW/MT3DMS. The yellow dot represents the well, and the red lines represent the constant boundary.

The concentration at four lateral boundaries is 0 during the whole SWPP test process, and the influence of the aquifer dimension and the boundary conditions could be negligible.

MT3DMS contains five approaches when solving the ADE model: a third-order total variation diminishing scheme (ULTIMATE), a standard finite difference method, a particle tracking-based method of characteristics (MOC), a modified MOC (MMOC), and a hybrid MOC/MMOC (HMOC). It was found that each technique had its strengths and limitations (Zheng and Wang, 1999). In this section, we will check the performance of these methods for modeling solute transport associated with SWPP tests. Two sets of radial dispersivities are used for comparison: $\alpha_r = 0.5\,\text{m}$ and $\alpha_r = 0.1\,\text{m}$, where the smaller dispersivity case is regarded as closer to the advection-dominating transport scenario, and the other one is regarded as closer to the dispersion-dominating transport scenario. The other parameters used in this section are $Q_{\text{inj}} = 0.02\,\text{m}^3/\text{day}$,

$Q_{\text{ext}} = -0.02\,\text{m}^3/\text{day}$, $\lambda = 1 \times 10^{-7}\,\text{s}^{-1}$, $t_{\text{inj}} = 5.0$ days, $R = 1$, $\theta = 0.3$, and $B = 0.6\,\text{m}$. Since the cells of MODFLOW/MT3DMS are cuboid instead of cylinders that are preferred to approximate a wellbore, the well radius could be calculated through the mass balance requirement at the wellbore cell:

$$r_w = \sqrt{\frac{\Delta x \Delta y}{\pi}} \tag{3.427}$$

where Δx and Δy are the horizontal dimensions of the wellbore cell in MODFLOW/MT3DMS and $\Delta x = \Delta y = 0.02$ m is used in this section.

Unfortunately, after carefully examining the code of MT3DMS, we find that it contains an error related to the volume of water in the wellbore, which should be computed by

$$V_{w,\text{inj}} = \pi r_w^2 h_w \tag{3.428}$$

but is computed instead in MT3DMS by

$$V_{w,\text{inj}} = \pi r_w^2 B \tag{3.429}$$

where h_w is the vertical height of water in the wellbore, which is time-dependent for the transient flow model and B is the aquifer thickness. Therefore, Eq. (3.429) only works when the height of the water column in the wellbore is the same as the aquifer thickness. This is not true for most cases; thus, we recommend the use of Eq. (3.428) instead of Eq. (3.429) in MT3DMS for dealing with a wellbore. This correction is very important whenever the wellbore mixing effect is a concern.

Figures 3.19(a) and 3.19(b) show the comparison between the solutions of this section and MOFLOW/MT3DMS, where $h_0 = B$ and $V_{w,\text{inj}} = V_{w,\text{ext}} = \pi r_w^2 B$. The special choice of $V_{w,\text{inj}}$ and $V_{w,\text{ext}}$ here to eliminate the errors caused by the use of Eq. (3.429) in MT3DMS, although the values of $V_{w,\text{inj}}$ and V_w, ext may not be the same for actual problems. Since Figures 3.19(a) and 3.19(b) have too many curves, we also plot them in semi-log diagrams for better examination, as shown in Figures 3.20(a) and 3.20(b), 3.21 and 3.22.

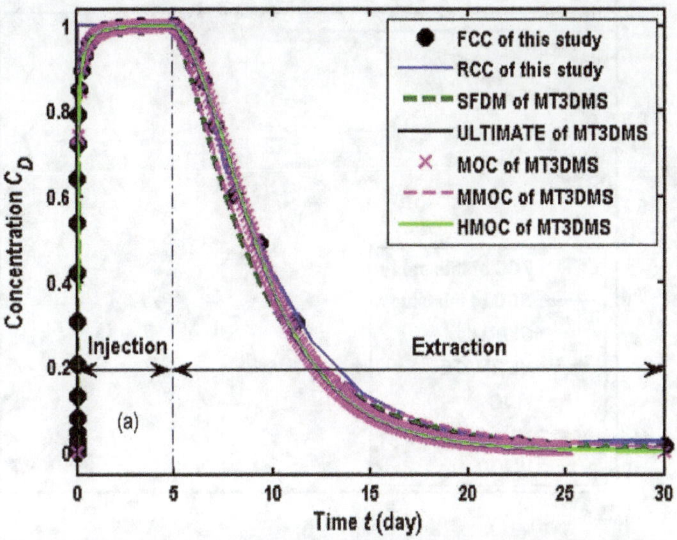

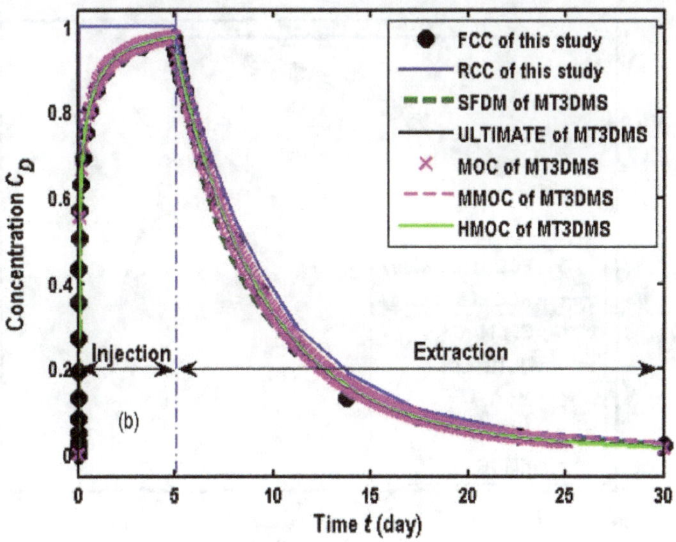

Figure 3.19. Comparison of BTCs at the wellbore between the solutions of this section and of MODFLOW/MT3DMS. (a) $\alpha_r = 0.5\,\mathrm{m}$. (b) $\alpha_r = 0.1\,\mathrm{m}$. FCC = flux concentration continuity; RCC = resident concentration continuity; SFDM = standard finite difference method; MOC = method of characteristics; MMOC = modified MOC; HMOC = hybrid MOC/MMOC.

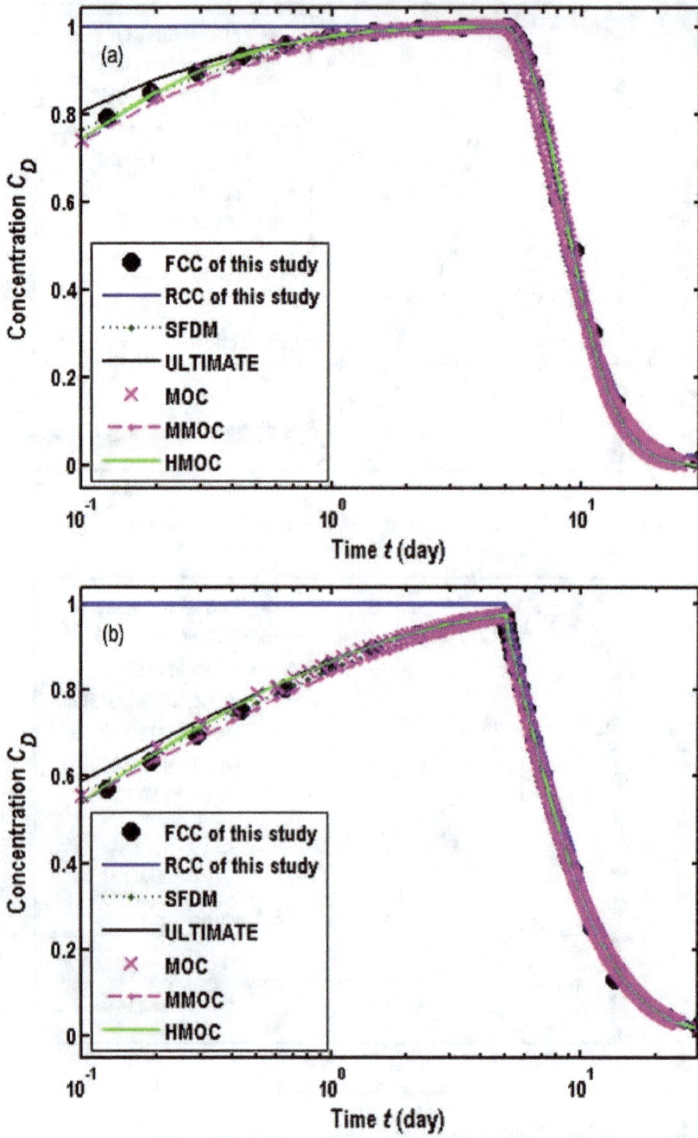

Figure 3.20. BTCs of Figures 3.19(a) and 3.19(b) presented in a semi-log diagram.

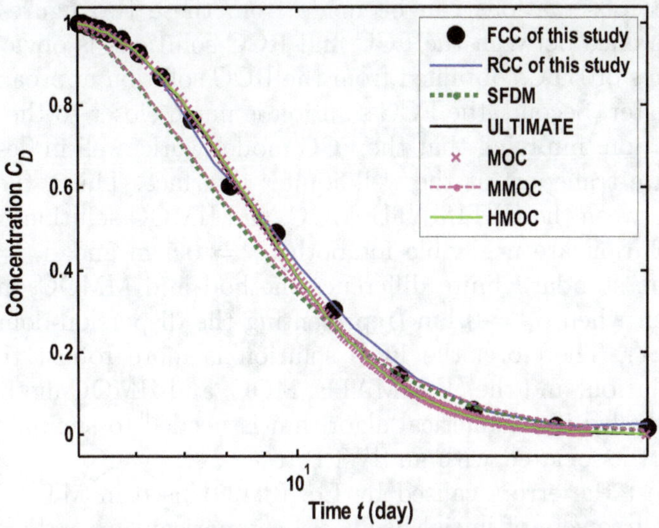

Figure 3.21. BTCs of Figure 3.19(a) presented in a semi-log diagram in the extraction phase.

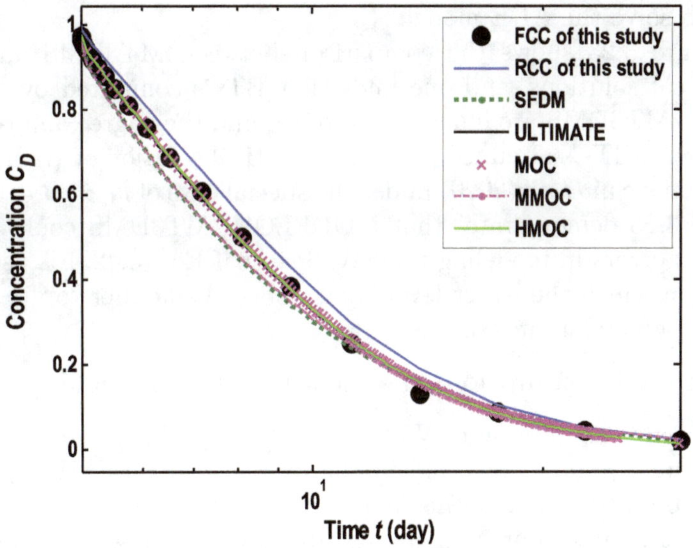

Figure 3.22. BTCs of Figure 3.19(b) presented in a semi-log diagram in the extraction phase.

Three observations can be made from these two figures. First, the difference between the FCC and RCC solutions is obvious, and the values of BTCs obtained from the RCC solution approach unity much faster. Second, the FCC solution is much closer to the numerical solution, implying that the FCC model works well in describing the solute transport at the well-aquifer interface. Third, the differences between the ULTIMATE, MOC, and HMOC solutions and the FCC solution are negligible for both $\alpha_r = 0.5$ m and $\alpha_r = 0.1\,\text{m}$, while the standard finite difference method and MMOC only perform well when $\alpha_r = 0.1\,\text{m}$ (representing the dispersion-dominating transport). Therefore, the FCC solution is more robust than the RCC solution, and the ULTIMATE, MOC, and HMOC methods are recommended if a numerical algorithm is needed to simulate solute transport associated with an SWPP test.

To test the errors caused by Eq. (3.429) used in MT3DMS, we choose three sets of initial heads for comparison: $h_0 = B = 0.6\,\text{m}$, $h_0 = 6.1\,\text{m}$ and $h_0 = 15.0\,\text{m}$. A greater h_0 represents a larger volume of water in the wellbore, leading to a greater mixing effect. The other parameters are the same as those in Figure 3.19(a), and the MOC is used to solve the ADE model.

Figure 3.23 shows the comparison between MT3DMS and the analytical solutions, and one finds that BTCs computed by MODFLOW/MT3DMS are independent of h_0, and they agree almost perfectly with BTCs obtained with the analytical solution of Wang *et al.* (2018) and Shi *et al.* (2020) under the special case of $h_0 = B = 0.6$ m. Figure 3.23 demonstrates that MODFLOW/MT3DMS could result in great errors in modeling the SWPP test if Eq. (3.429) is used for a case in which the water level is far above the aquifer top (e.g., h_0 is much greater than B).

(b) Second assumption: the steady-state flow field

The second assumption of Wang *et al.* (2018) and Shi *et al.* (2020) is the steady-state flow field in the extraction phase. Wang *et al.* (2017) pointed out that this assumption worked well for cases with sufficiently large aquifer hydraulic diffusivities. However, they did not consider the wellbore mixing effect; thus, it is still unclear about the impact of the steady state flow assumption on the SWPP model when the mixing effect is included. This issue will be addressed in the following.

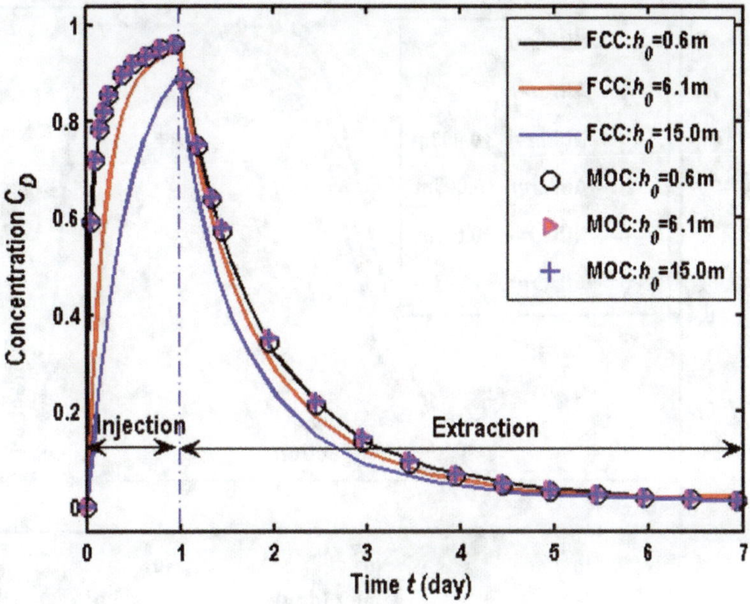

Figure 3.23. Comparison of BTCs at the wellbore for different initial heads of the aquifer. FCC = flux concentration continuity; MOC = method of characteristics.

To accommodate the transient flow effect into solute transport, MODFLOW/MT3DMS may be applicable. However, the present form of MODFLOW/MT3DMS has an error (Eq. (3.429)) for modeling the SWPP test, so it is not selected for the following analysis. Instead, we will revise the numerical model of Wang *et al.* (2017) by considering both the transient flow field and mixing effect to test the steady-state flow assumption. The mathematical model of the groundwater flow is the same as the one used by Wang *et al.* (2017), while the model of solute transport is the same as the one used in the models of Wang *et al.* (2018) and Shi *et al.* (2020).

Figure 3.24 shows the comparison of the analytical and numerical solutions for the different locations, where the flow model in numerical solutions is transient. The parameters used in this case are $K = 10.0\,\text{m/day}$, $S_s = 1 \times 10^{-4}\,\text{m}^{-1}$, $h_0 = 15\,\text{m}$, $\alpha_r = 0.1\,\text{m}$, $Q_{\text{inj}} = 0.02\,\text{m}^3/\text{day}$, $Q_{\text{ext}} = -0.02\,\text{m}^3/\text{day}$, $\lambda = 1 \times 10^{-7}\,\text{s}^{-1}$, $t_{\text{inj}} = 1.0$ days, $R = 1$, $\theta = 0.3$, and $B = 0.6\,\text{m}$. The numerical solution always agrees with the analytical solution well.

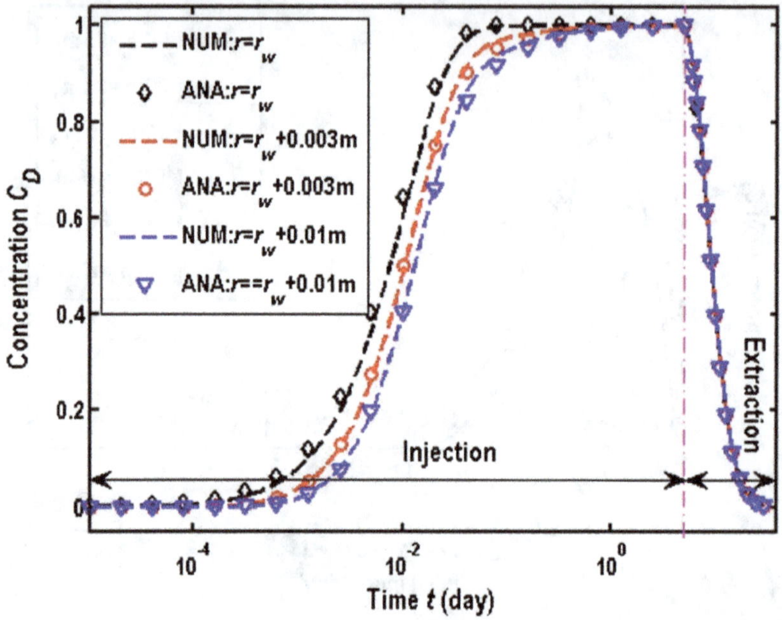

Figure 3.24. Comparison of BTCs at the wellbore between the analytical and numerical solutions of Wang *et al.* (2018). "NUM" represents the numerical solutions of the transient flow model, and "ANA" represents the analytical solutions of Wang *et al.* (2018).

(c) Assumption of the homogeneous aquifer

One of the fundamental assumptions included in the models of Wang *et al.* (2018) and Shi *et al.* (2020) is that the aquifer is assumed to be homogeneous. Although it has been widely used in previous studies and was found to perform well in many cases (Chen *et al.*, 2017; Gelhar and Collins, 1971; Schroth and Istok, 2005; Schroth *et al.*, 2001), numerous studies demonstrated that it is an upscaling treatment and an approximation for the reality (Hansen *et al.*, 2016; Molinari *et al.*, 2015; Pedretti *et al.*, 2013, 2014). Recently, Hansen *et al.* (2016) pointed out that the path reversibility and the radial flow field in the SWPP test, resulting from the homogeneous aquifer, were idealizations for reality. Molinari *et al.* (2015) found that highly conductive materials (like gravel channels) could produce preferential channels through which chemicals moved rapidly, and the assumption of a homogeneous aquifer could cause unacceptable errors in interpreting the convergent flow tracer tests. Therefore, it is necessary to test the influence of heterogeneity on the SWPP test. In this section, two types of aquifers are introduced for the comparison: A single confined aquifer representing the horizontal heterogeneity and a stratified aquifer representing the vertical heterogeneity.

As for a single confined aquifer, the autocorrelated multi-Gaussian and isotropic hydraulic conductivity fields are generated by stochastic modeling using the "gstat" package in the S environments (Pebesma, 2004). Four cases are considered, including the variance of the natural logarithm of hydraulic conductivity $\sigma^2_{\mathrm{LnK}} = 0.1$, 1, 2, and 4, respectively, as shown in Figure 3.25. The other parameters

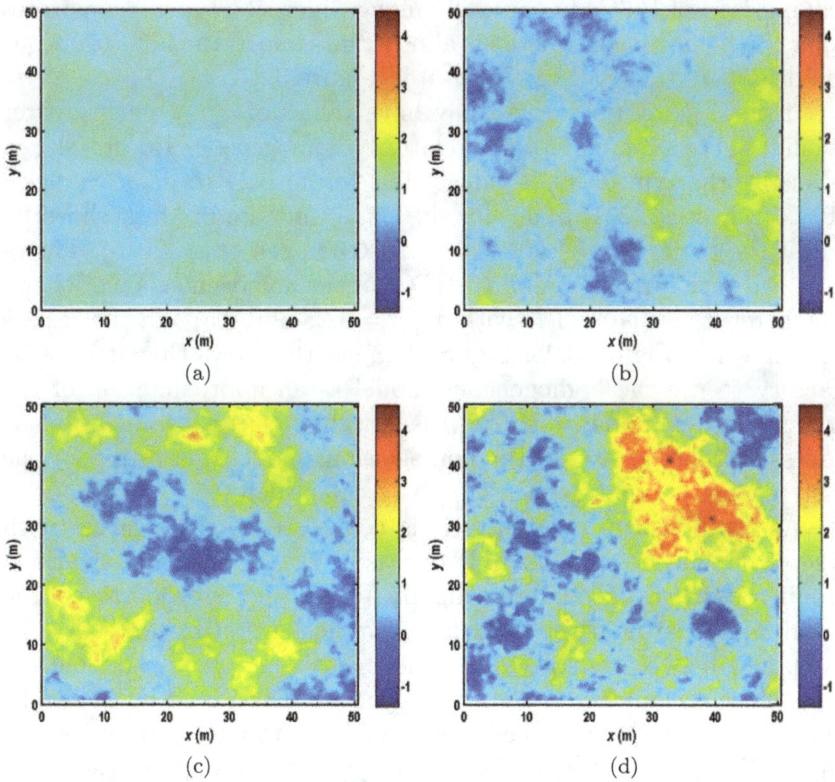

Figure 3.25. Heat map showing $\log_{10} K$ for a single realization. (a) $\sigma^2_{\mathrm{LnK}} = 0.1$; (b) $\sigma^2_{\mathrm{LnK}} = 1$; (c) $\sigma^2_{\mathrm{LnK}} = 2$; and (d) $\sigma^2_{\mathrm{LnK}} = 4$.

involved in the simulations are the same as the ones used in Hansen *et al.* (2016), such as the following: The correlation length is 4 m for the exponential semi-variogram; a geometric mean hydraulic conductivity is 10^{-4} m/s; the aquifer dimensions are 50 m in length, 50 m in width, and 10 m in depth; $Q_{\text{inj}} = Q_{\text{ext}} = 1$ L/s; and $t_{\text{inj}} = 40$ h. MODFLOW/MT3DMS conducts numerical simulations of the groundwater flow and solute transport.

Figure 3.26 shows the travel path of the particles and the contour map of the hydraulic head. The observations from this figure are consistent with the conclusions of Hansen *et al.* (2016) that the radial flow field of the SWPP test is asymmetric for the heterogeneous aquifer. Meanwhile, a greater σ^2_{LnK} could result in a more irregular shape of the contour map and the travel pathway. However, the comparison of BTCs in the well among different σ^2_{LnK} demonstrates that the homogeneous model is almost the same as the heterogeneous model when $\sigma^2_{\text{LnK}} \le 0.1$, as shown in Figure 3.27.

In fact, many field cases may have values of σ^2_{LnK} much greater than 0.1. For example, for the relatively homogeneous Borden site in Canada, the typical σ^2_{LnK} values range from 0.24 to 0.38 (Sudicky, 1986; Yeh *et al.*, 2015), and for the Cape Cod site in Massachusetts, USA, the typical σ^2_{LnK} values are 0.20–0.32 (Yeh *et al.*, 2015). Therefore, we have further compared BTCs of the homogeneous model and the heterogeneous models with $\sigma^2_{\text{LnK}} = 0.25$ and 0.5, and the results are shown in Figure 3.28. Figure 3.28 clearly shows that it is satisfactory to use the homogeneous model as an approximation of the heterogeneous model when σ^2_{LnK} is less than or equal to 0.25, but it is less satisfactory to use the homogeneous model as a proxy for the heterogeneous aquifer when σ^2_{LnK} is greater than 0.5.

As for a stratified aquifer, it is composed of three layers with the same thickness of 0.2 m. The reason for choosing the stratified aquifer is that the SWPP test mainly focuses on the properties of the local aquifer around the well, for example, several meters from the well, and it is reasonable to assume that the aquifer is horizontally homogeneous but vertically heterogeneous. Denoting the hydraulic conductivities of the three layers from the top to bottom as K_1, K_2, and K_3, respectively, three cases are considered, and each case contains two sets of parameters, as shown in Table 3.6. Greater difference in the hydraulic conductivity among the layers means greater vertical heterogeneity of the aquifer. The layer with great hydraulic

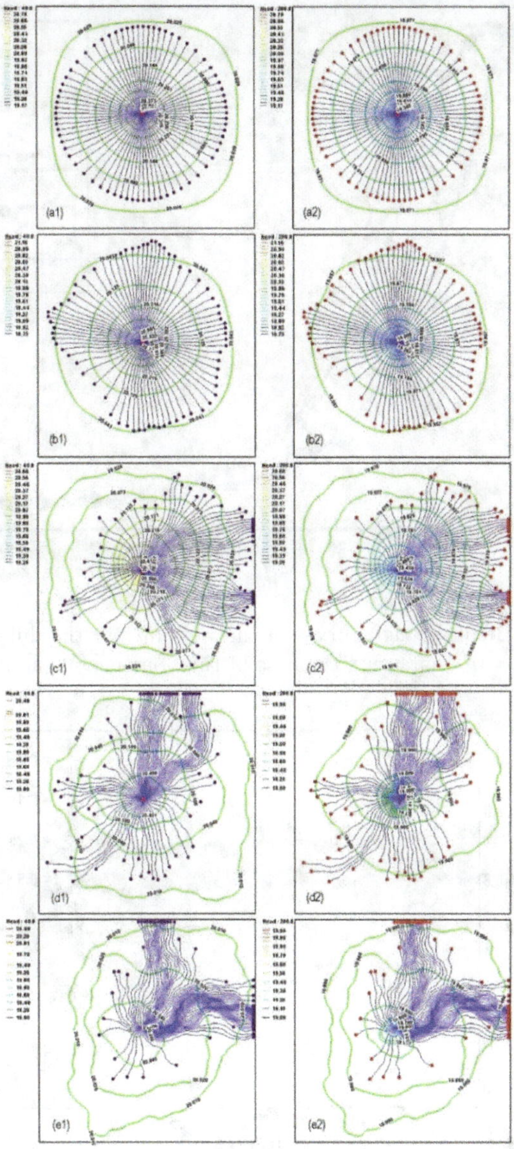

Figure 3.26. Contour map of hydraulic head and particle path. The red and purple dots represent the starting and ending points of the particle path, respectively. In a(1) to e(1), the starting points are in the well; in a(2) to e(2), the ending points are in the well. In (a1) and (a2), $\sigma^2_{\mathrm{LnK}} = 4$ representing the homogenous aquifer. In (b1) and (b2), $\sigma^2_{\mathrm{LnK}} = 0.1$. In (c1) and (c2), $\sigma^2_{\mathrm{LnK}} = 1$. In (d1) and (d2), $\sigma^2_{\mathrm{LnK}} = 2$. In (e1) and (e2), $\sigma^2_{\mathrm{LnK}} = 0$. In the first column, the 40 starting points of the particle are in the wellbore.

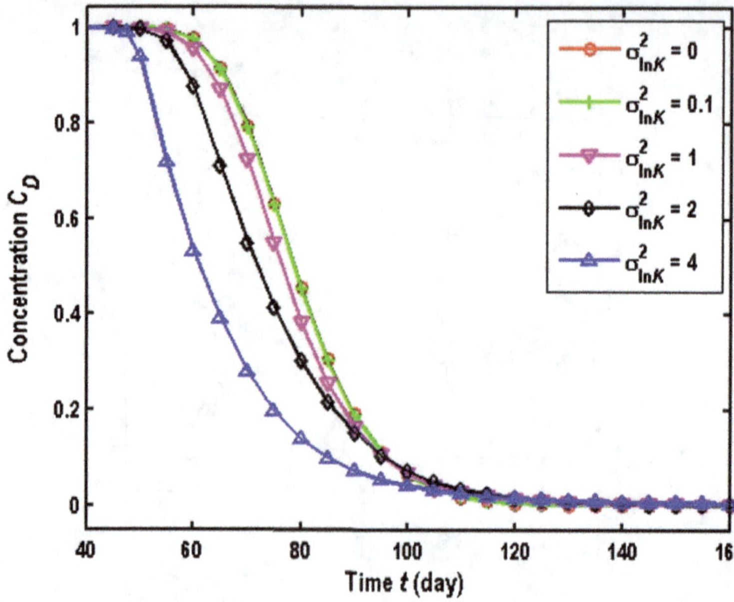

Figure 3.27. Breakthrough curves in the wellbore for the different σ^2_{LnK}. The legend of "$\sigma^2_{\mathrm{LnK}} = 0$" represents the case of the homogenous aquifer.

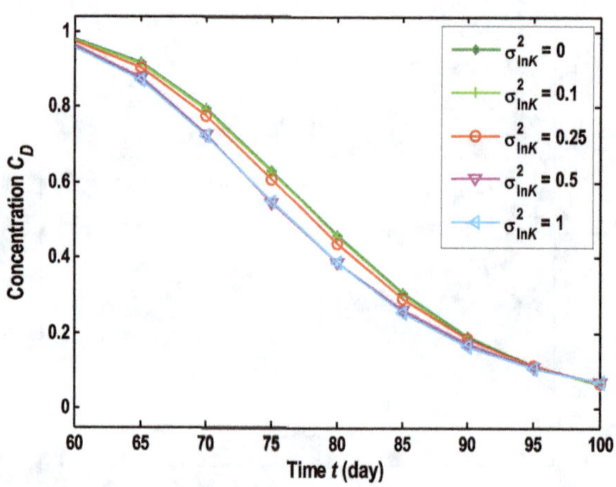

Figure 3.28. BTCs for different σ^2_{lnK} in the wellbore of the SWPP test.

Table 3.6. Hydraulic conductivities of the three-layer model and the estimated values by the homogeneous model.

	Actual values of three-layer model			Estimated value by homogeneous model
Cases	K_1[m/day]	K_2[m/day]	K_3[m/day]	K[m/day]
Case 1	30	20	10	22
	30	10	20	22
Case 2	50	25	10	31.2
	50	10	25	31.2
Case 3	50	25	1	28.5
	50	1	25	28.5

conductivity represents the preferential channel. The other parameters are the same as the ones used in Figure 3.19. A very large conductivity is set for the wellbore cells of the finite difference grid system in the numerical simulation (e.g., $1.0 \times 105\,\text{m/day}$), representing negligible head losses with the wellbore.

In this case, the transient models are considered for both groundwater flow and solute transport. As for the flow field, we use the homogeneous model to best fit the hydraulic heads in the wellbore produced by the three-layer models, as shown in Table 3.6 and Figure 3.29. The difference in the hydraulic heads between the two types of models seems negligible. The estimated hydraulic conductivity represents the average hydraulic conductivity of the heterogeneous aquifer. In each case, the hydraulic heads of the three-layer models are almost the same for the two sets of parameters, implying that the vertical order of the layers with different hydraulic conductivities may not influence the hydraulic head in the wellbore.

Figure 3.30 shows the comparison of the BTCs between the homogeneous model based on the estimated hydraulic conductivity and the three-layer model. One may find that the difference in BTCs is obvious between the two types of models, and the difference becomes larger for the greater vertically heterogeneous aquifer. Different from the flow field, BTCs are also dependent on the vertical order of the layers for the stratified aquifer. Therefore, the homogeneous SWPP model might be a good approximation for the vertically less heterogeneous aquifer. For example, the variations of hydraulic conductivities among layers are smaller than 50% in this case.

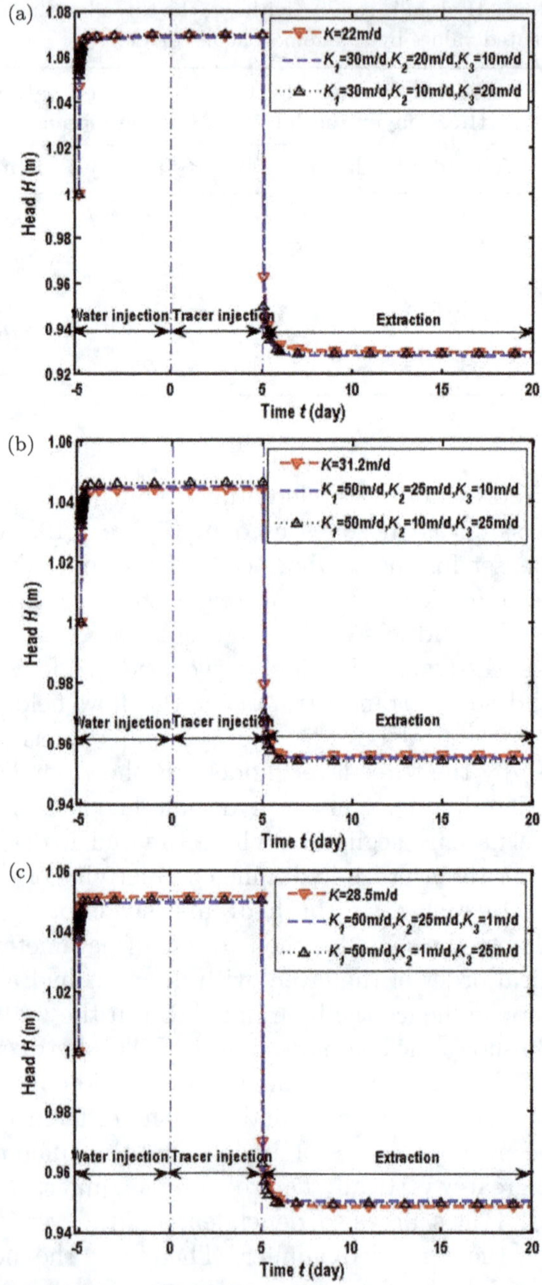

Figure 3.29. Fitness of hydraulic head in the wellbore of the three-layer models by the homogeneous model.

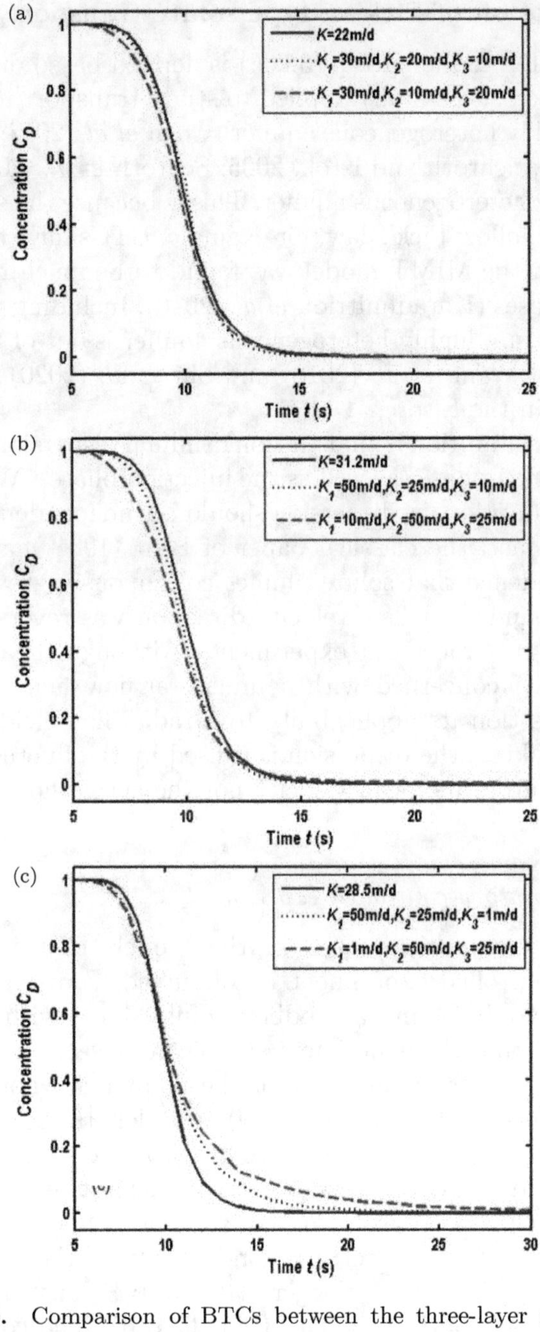

Figure 3.30. Comparison of BTCs between the three-layer model and the homogeneous model.

(d) Assumption of Fickian-type solute transport

The radial ADE model of Eq. (3.266) is derived based on Fick's law, which has been successfully applied to study transport in a homogeneous or weakly heterogeneous aquifer (Chen *et al.*, 2017; Gelhar and Collins, 1971; Schroth and Istok, 2005; Schroth *et al.*, 2001) but not for the highly heterogeneous aquifer. This is because the solute transport may not follow Fick's law, for example, BTCs have a heavy tail. Alternatively, the MIMT model was found to be much more powerful for such cases (Karadimitriou *et al.*, 2016). Including the wellbore mixing effect in a highly heterogeneous aquifer is out of the scope of the models of Wang *et al.* (2018) and Shi *et al.* (2020) and will be investigated in the future.

Another assumption is that the longitudinal dispersion is included in the extraction phase (as well as the injection phase). We think this is a valid assumption, as dispersion should be independent of the flow direction. In fact, the classical paper of Bear (1961) have provided abundant evidence that solute plumes continuously grow with time when the groundwater flow velocity direction was reversed multiple times during the tracer test experiments. Although the experiments of Bear (1961) concerned with a unilateral flow field, there is no reason to question its applicability to a radial flow field. As fundamentally speaking, the dispersion is caused by the fluctuation of the pore velocity from its mean velocity not the actual flow direction of the mean flow.

3.3.2.6 *Laboratory-controlled experiment*

To test the assumptions used in the models of this section, a laboratory-controlled experiment is conducted. Four constant head reservoirs laterally bound a sandbox aquifer filled with well-sorted medium sand, and the experiment setup can be seen in Figure 3.31. The top and bottom boundaries of the aquifer are nonpermeable. The dimensions of the sandbox are 80 cm in length, 80 cm in width, and 30 cm in height. An injection/extraction well fully penetrates the aquifer and is located at the center of the aquifer with a radius of 2.5 cm. NaCl is used as a conservative tracer, and the concentration is obtained by the electrical conductivity of the solute, which is measured by Solinst Levelogger (LTC 5M; Solinst Levelogger is a registered trademark of Solinst Canada Ltd). The concentration

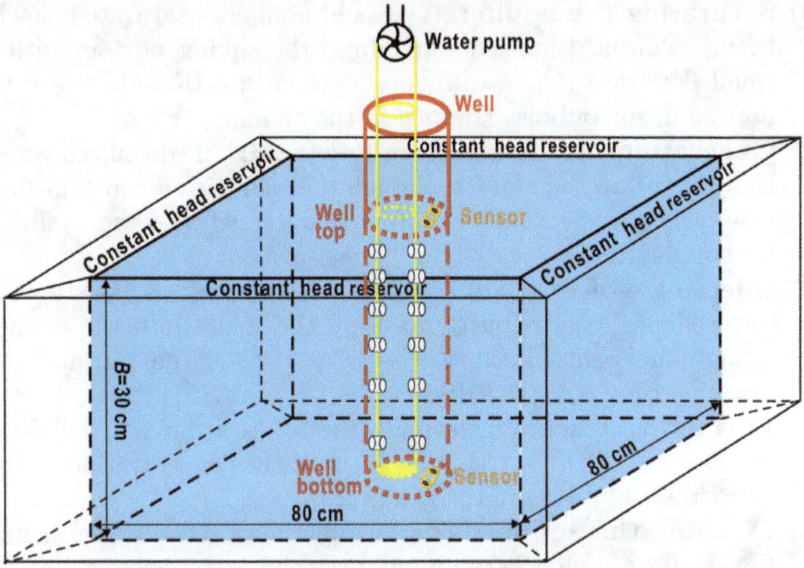

Figure 3.31. Schematic diagram of the experiment setup. The blue region represents the confined aquifer. The red cylinder represents the wellbore, and the perforated cylinder is the well screen.

is observed at two locations in the wellbore: well top and well bottom. If the BTCs are the same for both points, it implies that the assumption of the fully mixing effect is reasonable. The parameters used in this experiment are $Q_{\text{inj}} = 34.5\,\text{cm}^3/\text{s}$, $Q_{\text{ext}} = -34.5\,\text{cm}^3/\text{s}$, $t_{\text{inj}} = 210\,\text{s}$, $\theta = 0.38$, $B = 0.3\,\text{m}$, $h_0 = 54.8\,\text{cm}$, $h_w = 58.7\,\text{cm}$, $r_w = 2.5\,\text{cm}$, and $h_{w,\text{ext}} = 50.9\,\text{cm}$.

The procedure of the experiment is

(a) **Filling the sandbox of the aquifer:** First, the well-sorted medium sands are washed using deionized water and are completely mixed. Second, the box is packed with clean sand slowly from the bottom until the aquifer thickness is up to 30 cm. Finally, a Polymethyl methacrylate (PMMA) seal is put over the sand aquifer to generate a confined condition. Two concentration sensors (embedded in the Solinst Levelogger) are installed at the top and bottom of the well screen inside the wellbore but outside the perforated PVC pipe (yellow pipe as shown in Figure 3.31), respectively. The usage of the PVC pipe will be explained shortly.

(b) **Saturating the aquifer:** The sand aquifer is saturated slowly by injection of deionized water from the aquifer bottom with a small flow rate ($Q_{\text{inj}} = 0.1\,\text{cm}^3/\text{s}$) to ensure that there are no noticeable air bubbles trapped in the aquifer.
(c) **Generating the steady-state flow field:** The deionized water is injected into the aquifer through the well with a constant flow rate of $Q_{\text{inj}} = 34.5\,\text{cm}^3/\text{s}$, until the water level of the wellbore becomes stable.
(d) **Injecting tracer (NaCl):** When water level is constant in the wellbore, continuously injecting the solute into the aquifer though the well. The flow rate is $Q_{\text{inj}} = 34.5\,\text{cm}^3/\text{s}$, and the injection time is $t_{\text{inj}} = 210\,\text{s}$.
(e) **Extracting tracer:** Extracting the solute from the well with the flow rate $Q_{\text{ext}} = -34.5\,\text{cm}^3/\text{s}$, until the concentration of the extraction solute is approaching zero.
(f) **Cleaning the aquifer:** After completing a set of experiments, the sandbox aquifer is thoroughly cleaned using deionized water until the NaCl tracer concentration is below the detectable level.
(g) **Repeating the experiment:** The above steps (c)–(f) are repeated three times to minimize the operational errors of the experiments.

As the radius of the wellbore is very small in the experiment, it is difficult to install a mixer inside the wellbore. Here, we have invented an innovative experimental procedure to make sure that the injection tracer is completely (and instantaneously) mixed with the original water in the wellbore. To serve this purpose, a perforated PVC pipe (yellow pipe as shown in Figure 3.31) with a smaller radius of 0.6 cm than the radius (2.5 cm) of the wellbore (red pipe) is connected with the water pump. The length of the yellow pipe is the same as the red one.

The perforated holes are identical with the same radius of 0.15 cm. After trying multiple times, we found that the distribution of the holes in the perforated PVC pipe is better to be non-uniformly distributed over the depth to achieve a complete and instantaneous mixing effect. Specifically, five layers of holes are located at $z =$ 25 cm, 23 cm, 19 cm, 13 cm, and 5 cm, respectively in the perforated PVC pipe. At each layer, four holes are uniformly distributed over its circumference.

It is important to point out that the bottom of the yellow perforated PVC pipe must be sealed completely, as shown in Figure 3.31. If the bottom of the perforated PVC pipe is not sealed, this will result in an incomplete mixing of tracer in the wellbore, which is detected through two concentration sensors (embedded in the Solinst Leveloggers) installed near the top and bottom of the wellbore. On the contrary, if the bottom of the perforated PVC pipe is sealed completely, hydrodynamics inside the PVC pipe (particularly near the bottom of the PVC pipe) is altered, and two concentration sensors near the top and bottom of the wellbore show identical concentrations at any given moment, indicating complete and instantaneous mixing over the entire length of the wellbore.

Figure 3.32 shows the fitness of the observed BTCs using the new FCC model of Wang *et al.* (2018), where the estimated dispersivity is $\alpha_r = 0.2\,\text{cm}$ and $\lambda = 1 \times 10^{-7}\,\text{s}^{-1}$. To test if the fitting is sensitive to the value of dispersivity, we also include the BTC of $\alpha_r = 0.2\,\text{cm}$ and $\lambda = 1 \times 10^{-7}\,\text{s}^{-1}$ in Figure 3.32. It is evident from this figure that

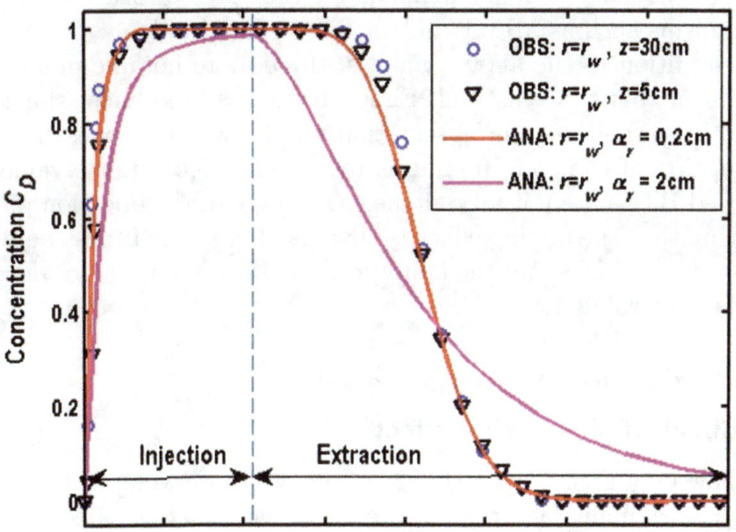

Figure 3.32. Fitness of observed BTCs by the new analytical solution of Wang *et al.* (2018). "OBS" represents the observed BTC, and "ANA" represents the analytical solution.

the BTC is sensitive to the value of dispersivity as the two curves of $\alpha_r = 0.2\,\text{cm}$ and $\alpha_r = 2\,\text{cm}$ differ distinctly from each other. This finding is somewhat different from Hansen *et al.* (2016), which reported that essentially no dispersion other than minor pore scale dispersion would be visible under the push-pull condition. A careful inspection of Hansen *et al.* (2016) revealed that the MIMT rate used in their analysis (which is similar to our first-order reaction constant) was $\lambda = 10\,\text{h}^{-1}$ or $\lambda = 6.9 \times 10^{-3}\,\text{s}^{-1}$, which is more than 4 orders of magnitude greater than the first-order reaction rate of the model of Wang *et al.* (2018) ($\lambda = 1 \times 10^{-7}\,\text{s}^{-1}$). This implies that the mass transfer between the mobile and immobile domains in Hansen *et al.* (2016) plays a dominating role in the transport process and overshadows the dispersion. That is probably why the result of Hansen *et al.* (2016) is insensitive to the dispersion. In contrast, the model of Wang *et al.* (2018) does not include MIMT, but instead, includes a very weak first-order reaction process (with $\lambda = 1 \times 10^{-7}\,\text{s}^{-1}$), so the sensitivity of our solution to dispersion is clearly evident. In summary, as far as we can see, whether the push-pull process is sensitive to dispersion or not depends on the rates of other transport mechanisms such as MIMT and/or first-order reaction.

In addition to the above analysis, three more notable points can be seen in this figure as well. First, the BTCs are almost the same for both top and bottom observation points, which implies that the assumption of complete mixing is reasonable. Second, the values of observed BTCs are not unit at the early stage of the injection phase, implying that the mixing effect is obvious. Third, the fitness between the observed BTCs and the analytical solution is very good, demonstrating the robustness of the new model.

3.3.2.7 *Parameter sensitivity analysis*

(a) Effect of the mixing effect

As shown in Eqs. (3.272)–(3.274), the wellbore volume ($V_{w,\text{inj}}$ or $V_{w,\text{ext}}$) is a main factor associated with the mixing effect. When $V_{w,\text{inj}} = 0$ and $V_{w,\text{ext}} = 0$, the new solution will reduce to the solution of Huang *et al.* (2010) for a fully penetrating well without the mixing effect. However, the values of $V_{w,\text{inj}}$ and $V_{w,\text{ext}}$ are usually not small for actual SWPP tests, especially for deep confined aquifer cases in which the water column in the wellbore could be considerable.

In this section, the effect of r_w on BTCs obtained at the test well will be investigated.

Figure 3.33 shows BTCs for different r_w, where a greater well radius represents a greater wellbore volume. The solutions of Huang *et al.* (2010) are also included for the purpose of comparison. The parameters used in this case are $\alpha_r = 2.5\,\text{cm}$, $Q_{\text{inj}} = 45\,\text{cm}^3/\text{s}$, $Q_{\text{ext}} = -45\,\text{cm}^3/\text{s}$, $t_{\text{inj}} = 1000\,\text{s}$, $\lambda = 1 \times 10^{-7}\,\text{s}^{-1}$, $R = 1$, $\theta = 0.38$, $B = 0.8\,\text{m}$, $h_0 = 95\,\text{cm}$, $h_{w,\text{inj}} = 100\,\text{cm}$, and $h_{w,\text{ext}} = 90\,\text{cm}$. Several interesting observations could be made from this figure. First, the influence of r_w on BTCs in the wellbore is not negligible. Apparently, a larger r_w results in a smaller concentration of BTCs in the injection phase but a larger concentration of BTCs in the extraction phase. This is because a larger r_w represents a greater $V_{w,\text{inj}}$ or $V_{w,\text{ext}}$ thus reflects a stronger mixing effect in the wellbore. Specifically, a larger Vw, inj means that there is a larger volume of tracer-free water in the wellbore prior to the injection, which can cause greater dilution of the tracer solution injected into the wellbore. Second, the solution of Wang *et al.* (2018) is close to the solution of Huang *et al.* (2010)

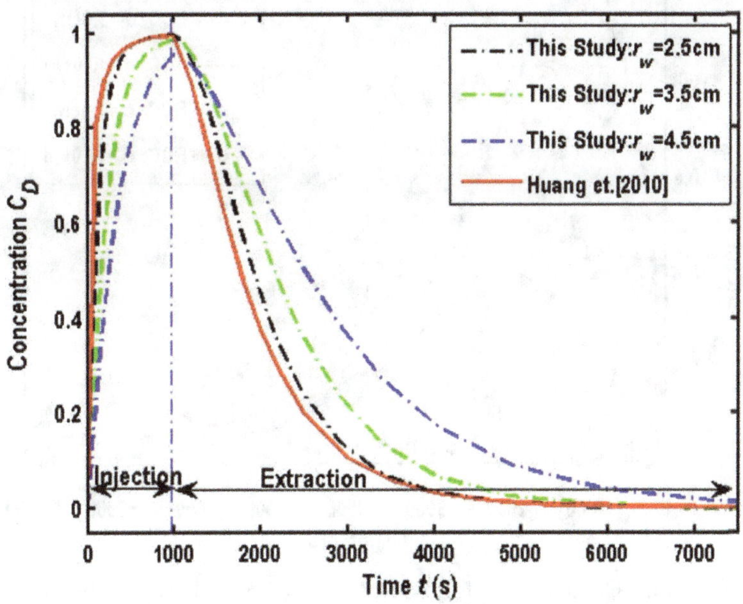

Figure 3.33. BTCs at the wellbore for the different well radius.

for a smaller well radius. This is because the model of Huang *et al.* (2010) does not consider well mixing effect, and it only works when r_w is equal to 0.

(b) Comparison between the FCC and RCC solutions

The FCC and RCC models have been widely used to describe the solute transport at the wellbore. Due to its simple expression, the RCC model is preferred in the analytical treatment (Chen *et al.*, 2017; Gelhar and Collins, 1971; Schroth and Istok, 2005; Schroth *et al.*, 2001); however, such a model is often being criticized for not rigorously satisfying the mass balance requirement. In the model of Wang *et al.* (2018), the analytical solutions of both FCC and RCC models are derived. Figure 3.34 shows the comparison of the FCC and RCC solutions for different α_r. The parameters used in this case are $Q_{\text{inj}} = 45\,\text{cm}^3/\text{s}$, $Q_{\text{ext}} = -45\,\text{cm}^3/\text{s}$, $t_{\text{inj}} = 1000\,\text{s}$, $\lambda = 1 \times 10^{-7}\,\text{s}^{-1}$, $R = 1$, $\theta = 0.38$, $B = 0.8\,\text{m}$, $h_0 = 95\,\text{cm}$, $h_{w,\text{inj}} = 100\,\text{cm}$, and $h_{w,\text{ext}} = 90\,\text{cm}$.

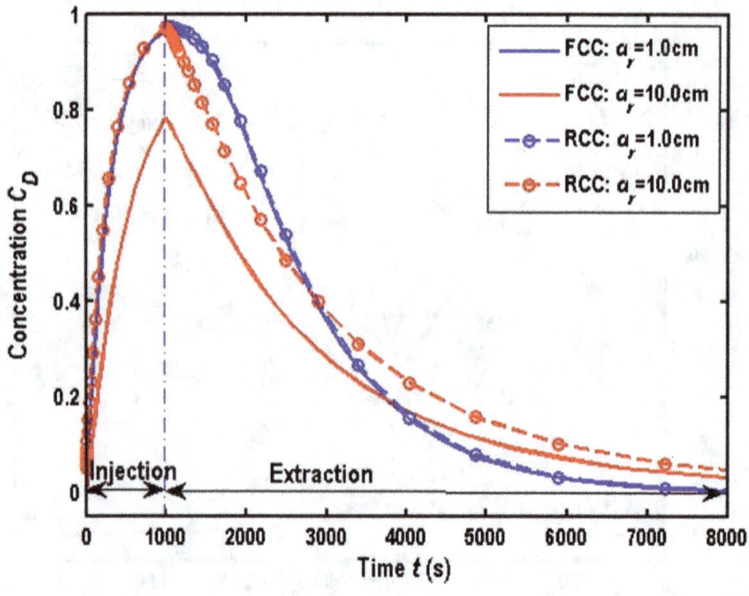

Figure 3.34. Comparison of BTCs at the wellbore between the RCC and FCC solutions. FCC = flux concentration continuity; MOC = method of characteristics.

One may find that the FCC and RCC solutions become the same for smaller values of α_r in the injection phase, for example, $\alpha_r = 1\,\text{cm}$ for this case, and the difference between them becomes larger with increasing α_r. This observation could be explained by comparing Eqs. (3.269a), (3.269b), (3.270a), and (3.270b). One may find that the difference between the FCC and RCC models is in the dispersive terms of Eqs. (3.269a) and (3.269b). A smaller α_r could decrease the difference between the FCC and RCC models, and they become the same when $\alpha_r = 0\,\text{cm}$.

3.3.3 *The multi-stage SWPP analytical models*

Section 3.3.2 presented four different two-stage SWPP test analytical solutions developed by Huang *et al.* (2010), Wang *et al.* (2018), Shi *et al.* (2020, 2023), respectively. Unlike these four analytical models, a complete SWPP test might contain four principal phases: (1) injection of a prepared solution (tracer) into a targeted aquifer, (2) injection of a chaser, (3) rest period, and (4) extraction of the mixture solution (Hebig *et al.*, 2015; Phanikumar and McGuire, 2010a), as shown in Figure 3.35. The chaser is used to push the prepared solution out of the borehole into the aquifer, in which the chaser is usually the groundwater collected from the same aquifer. The second and third phases are optional, but Hebig *et al.* (2015) pointed out that the chaser was recommended and presented a model to determine the optimal chaser volume.

In this section, two analytical models of the four-stage SWPP test for estimating aquifer parameters are introduced. The first analytical model is developed by Chen *et al.* (2017), which investigates the difference between local and nonlocal transport BTCs in SWPP tests using the mobile-immobile model with a first-order mass transfer rate for the four-stage SWPP test. A general implicit Euler method was also proposed by Chen *et al.* (2017) to solve the coupled fractional-in-time-and-space model (FTS) and to match the four-stage SWPP test data with arbitrary boundary conditions in a radial coordinate system. The second model was developed by Wang *et al.* (2020), who extended the analytical model of Chen *et al.* (2017) by considering the mixing effect and the aquitard effect under the MIM framework.

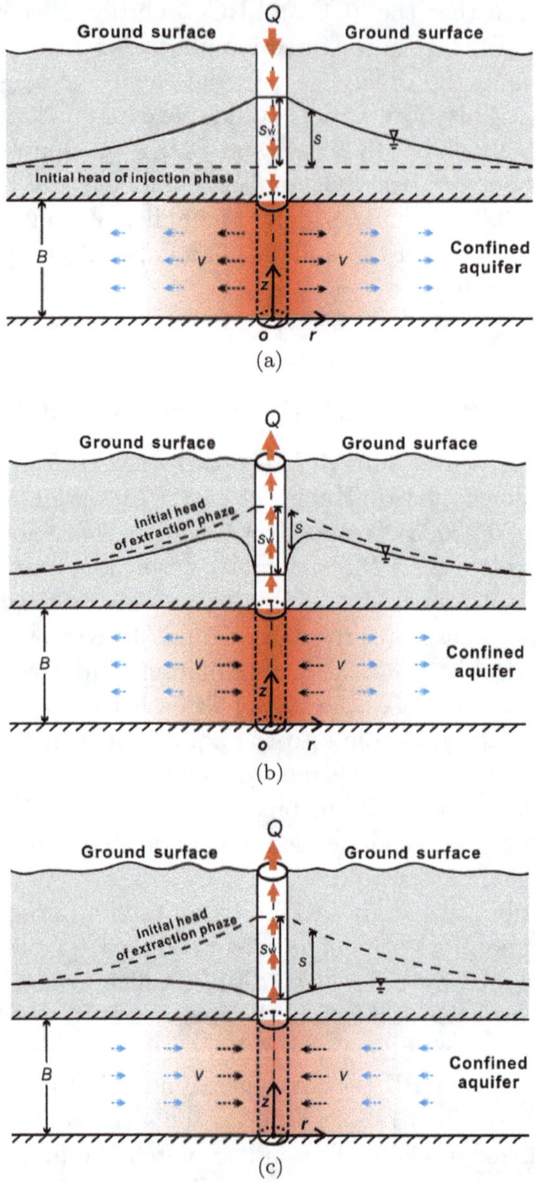

Figure 3.35. The schematic diagram of the multi-stage SWPP test. (a) In the injection phase, (b) at the beginning of the extraction phase, and (c) at the late stage of the extraction phase.

3.3.3.1 *Analytical solutions of Chen et al. (2017)*

In this section, we introduce an analytical solution developed by Chen *et al.* (2017) to the mobile-immobile model with a first-order mass transfer rate for the four-stage SWPP test to investigate the difference between local and nonlocal transport BTCs in SWPP tests. A general implicit Euler method was proposed to solve the coupled fractional-in-time-and-space model (FTS) and to match the four-stage SWPP test data with arbitrary boundary conditions in a radial coordinate system. This model can be used to predict solute distribution in mobile and immobile zones at different stages.

The assumptions made in the approach of Chen *et al.* (2017) are as follows: (1) the well is fully penetrating, and the injection rate and pumping rate are both constant but not necessarily the same; (2) the aquifer is horizontally isotropic with a uniform thickness, and extends sufficiently far from the well so the outer boundary is not a model constraint; (3) the upper and lower boundaries of the aquifer are impermeable to flow and transport; (4) the regional flow is neglected and flow is only driven by injecting (or pumping) during the experimental period; (5) transverse dispersion is ignored and only radial dispersion is considered; and (6) flow is in a steady state during injection, chasing, and pumping stages.

(a) Mathematical model

Based on the assumptions mentioned above, the mass balance equation of mobile zone transport in a radial flow system without a source and sink term is

$$\frac{\partial C_m}{\partial t} + \beta \frac{\partial C_{im}}{\partial t} = -\frac{1}{r}\frac{\partial}{\partial r}[rq(r,t)] \tag{3.430}$$

where C_m and C_{im} are the solute concentrations [Kg/m^3] in the mobile and immobile zones, respectively; t is the transport time [s]; r is the radial distance to the center of an injection or pumping well [m]; $\beta = \theta_{im}/\theta_m$ is the capacity ratio (or immobile-mobile ratio); θ_{im} and θ_m are the porosities in the immobile and mobile zones, respectively; $q(r,t)$ is the mass flux term. The mass flux term of Fickian transport can be represented as the sum of advective flux and dispersive flux:

$$q(r,t) = v(r)C_m - D(r)\frac{\partial C_m}{\partial r} \tag{3.431}$$

where v_r is the advective velocity [m/s] and $D(r)$ is the dispersion coefficient [m^2/s]. $v(r)$ and $D(r)$ are both distance-dependent and given by $v(r) = Q/(2\pi b\theta_m r)$ and $D(r) = aQ/(2\pi b\theta_m r)$, respectively, where Q is the injection or pumping rate [m^3/s], b is the aquifer thickness [m], and a is the radial dispersivity [m]. Note that the fluid flow field is assumed to be in a steady state. For simplification, the set of parameters $Q/(2\pi b\theta_m)$ is replaced by κ_v from now on.

The governing equation describing an SWPP test with a first-order rate-limited mobile-immobile mass transfer is thus obtained by substituting Eq. (3.431) into Eq. (3.430), one has

$$\frac{\partial C_m}{\partial t} + \beta\frac{\partial C_{im}}{\partial t} = L(t)C_m \tag{3.432}$$

$$\theta_{im}\frac{\partial C_{im}}{\partial t} = \omega(C_m - C_{im}) \tag{3.433}$$

where

$$L(t)\begin{cases} -\dfrac{\kappa_v}{r}\dfrac{\partial}{\partial r} + \dfrac{a\kappa_v}{r}\dfrac{\partial^2}{\partial r^2}, & 0 \le t < t_{\text{inj}} \\ -\dfrac{\kappa_v}{r}\dfrac{\partial}{\partial r} + \dfrac{a\kappa_v}{r}\dfrac{\partial^2}{\partial r^2}, & t_{\text{inj}} \le t < t_{\text{cha}} \\ 0, & t_{\text{cha}} \le t < t_{\text{rest}} \\ \dfrac{\kappa_v}{r}\dfrac{\partial}{\partial r} + \dfrac{a\kappa_v}{r}\dfrac{\partial^2}{\partial r^2}, & t_{\text{rest}} \le t < t_{\text{pump}} \end{cases} \tag{3.434}$$

Here $L(t)$ is a temporally dependent spatial operator; t_{inj}, t_{cha}, t_{rest}, and t_{pump} are the ending times of injection, chasing, resting, and pumping stages, respectively, for a complete SWPP test. Note that the ADE for the injection and chasing stages are the same if one assumes that the injection rates of these two stages remain identical. This assumption is valid, and it is actually a normal setting in SWPP tests (Le Borgne and Gouze, 2008). The difference between these two stages is reflected in the boundary conditions at the wellbore, which will be discussed later. Actually, water chasing does not add new solutes into the aquifer, and the objective of injecting chasing fluid is to push the tracer farther from the injection well to obtain more information on subsurface properties while pumping back. If the chasing stage is not included in the field experiments, the part of $L(t)$ between t_{inj} and t_{cha} can be omitted.

For non-Fickian transport, more specifically, nonlocal transport in space, the mass flux term is given by

$$q(r,t) = v(r)C_m - D(r)\left[p\frac{\partial^{\alpha-1}C_m}{\partial r^{\alpha-1}} + (1-p)\frac{\partial^{\alpha-1}C_m}{\partial(-r)^{\alpha-1}}\right] \tag{3.435}$$

where $\partial^{\alpha-1}C_m/\partial r^{\alpha-1}$ and $\partial^{\alpha-1}C_m/\partial(-r)^{\alpha-1}$ are the Riemann–Liouville fractional derivatives; p is the skewness indicating the proportion of preferential path in the direction of flow and $p = 1$ and $p = 0$ represent maximally positive skewness and maximally negative skewness, respectively; α is the fractional order in space satisfying $1 < \alpha \leq 2$, which describes the degree of nonlocal transport in space; $D(r) = a_{FS}\kappa_v/r$ and a_{FS} is fractional dispersivity [$\mathrm{m}^{\alpha-1}$] (Benson *et al.*, 2004). Note that a_{FS} is not the same as the dispersivity in classical ADE because its unit is a fraction of the unit length. Inserting Eq. (3.435) into Eq. (3.432), the fractional-in-space (FS) model describing SWPP tests is obtained as

$$\frac{\partial C_m}{\partial t} + \beta\frac{\partial C_{im}}{\partial t} = L_{FS}(t)C_m \tag{3.436}$$

where

$$L_{FS}(t)\begin{cases} -\dfrac{\kappa_v}{r}\dfrac{\partial}{\partial r} + \dfrac{a_{FS}\kappa_V}{r}\dfrac{\partial^\alpha}{\partial r^\alpha}, & 0 \leq t < t_{\text{inj}} \\ -\dfrac{\kappa_v}{r}\dfrac{\partial}{\partial r} + \dfrac{a_{FS}\kappa_V}{r}\dfrac{\partial^\alpha}{\partial r^\alpha}, & t_{\text{inj}} \leq t < t_{\text{cba}} \\ 0, & t_{\text{cha}} \leq t < t_{\text{rest}} \\ \dfrac{\kappa_v}{r}\dfrac{\partial}{\partial r} + \dfrac{a_{FS}\kappa_V}{r}\dfrac{\partial^\alpha}{\partial(-r)^\alpha}, & t_{\text{rest}} \leq t < t_{\text{pump}} \end{cases} \tag{3.437}$$

Note that the skewness p is set to be 1 in the injection and chasing stages because the distribution of tracer concentration skews to the far side along the flow direction, and p is set to be 0 in the pumping stage as the flow is reversed and the distribution skews toward the wellbore. The temporal effect of non-Fickian transport is incorporated into the term $\beta \partial C_{im}/\partial t$. For first-order rate-limited mass transfer between the mobile and immobile zones, C_{im} and $\partial C_{im}/\partial t$ are given as (Schumer *et al.*, 2003):

$$C_{im} = \omega e^{-\omega t} * C_m + C_{im}(r, 0)e^{-\omega t} \tag{3.438}$$

$$\frac{\partial C_{im}}{\partial t} = \frac{\partial}{\partial t}[f(t) * C_m] + \frac{1}{\omega}\frac{\partial}{\partial t}[f(t)C_{im}(r, 0)] \tag{3.439}$$

where $f(t) = \omega e^{-\omega t}$ is the memory function, which is the probability density of solutes entering the immobile zone at $t = 0$ and remaining there at the time t (Carrera *et al.*, 1998); "$*$" is the convolution sign; ω is the first-order mass transfer rate [1/s]; $C_m(r, 0)$ and $C_{im}(r, 0)$ are the initial concentrations in the mobile and immobile zones, respectively. For SWPP tests, $C_{im}(r, 0)$ is zero, so the relation between the mobile and immobile concentrations for SWPP tests can be simplified as

$$C_{im} = f(t) * C_m \tag{3.440}$$

$$\frac{\partial C_{im}}{\partial t} = \frac{\partial}{\partial t}[f(t) * C_m] \tag{3.441}$$

For first-order rate-limited mass transfer, the memory function follows an exponential form (Haggerty *et al.*, 2000). However, heavy tailing BTCs are often observed in field tracer tests, and they are attributed to the power law memory function, $f(t) = \tau_\gamma t^{-\gamma}/\Gamma(1-\gamma)$, where γ is the power law index, $\Gamma(x)$ denotes the gamma function and τ_γ is a scale factor for the power law that is always set to be 1 [$s^{\gamma-1}$].

Note that a cutoff time is introduced to the memory function in some recent studies to capture the transition of BTC at a late time from a power law to an exponential trend, which is observed for solute transport in streams (Aubeneau *et al.*, 2014; Meerschaert *et al.*, 2008; Zhang *et al.*, 2015). However, the late time transition is rarely observed in SWPP tests to the best of our knowledge, so the cutoff time is not considered in this section. Substituting Eq. (3.441) with the power law memory function into Eq. (3.436) and then using the relation:

$$\frac{\partial^\gamma C_m}{\partial t^\gamma} = \frac{\partial}{\partial t}\left[\frac{\tau_\gamma t^{-\gamma}}{\Gamma(1-\gamma)} * C_m\right] \tag{3.442}$$

A time-dependent fractional model is developed to interpret anomalous transport behavior. This equation considers a time-dependent power law memory function and a time-dependent fractional advection-dispersion operator. One can get the fractional-in-time-and-space (FTS) model:

$$\frac{\partial C_m}{\partial t} + \beta\tau_{\gamma_{\text{inj}}}\frac{\partial^{\gamma_{\text{inj}}} C_m}{\partial t^{\gamma_{\text{inj}}}} = -\frac{\kappa_V}{r}\frac{\partial C_m}{\partial t} + \frac{a_{FS}\kappa_V}{r}\frac{\partial^\alpha C_m}{\partial r^\alpha}, \quad 0 \le t < t_{\text{inj}} \tag{3.443a}$$

$$\frac{\partial C_m}{\partial t} + \beta\tau_{\gamma_{\text{cha}}}\frac{\partial^{\gamma_{\text{cha}}} C_m}{\partial t^{\gamma_{\text{cha}}}} = -\frac{\kappa_V}{r}\frac{\partial C_m}{\partial t} + \frac{a_{FS}\kappa_V}{r}\frac{\partial^\alpha C_m}{\partial r^\alpha}, \quad t_{\text{inj}} \le t < t_{\text{cha}} \tag{3.443b}$$

$$\frac{\partial C_m}{\partial t} + \beta\tau_{\gamma_{\text{rest}}}\frac{\partial^{\gamma_{\text{rest}}} C_m}{\partial t^{\gamma_{\text{rest}}}} = 0, \quad t_{\text{cha}} \le t < t_{\text{rest}} \tag{3.443c}$$

$$\frac{\partial C_m}{\partial t} + \beta\tau_{\gamma_{\text{pump}}}\frac{\partial^{\gamma_{\text{pump}}} C_m}{\partial t^{\gamma_{\text{pump}}}} = \frac{\kappa_V}{r}\frac{\partial C_m}{\partial r} + \frac{a_{rs}\kappa_V}{r}\frac{\partial^\alpha C_m}{\partial(-r)^\alpha}, \quad t_{\text{rest}} \le t < t_{\text{pump}} \tag{3.443d}$$

where $\partial^\gamma C_m/\partial t^\gamma$ with $\gamma = \{\gamma_{\text{inj}}, \gamma_{\text{cha}}, \gamma_{\text{rest}}, \gamma_{\text{pump}}\}$ is the Riemann-Liouville fractional derivative; $\gamma_{\text{inj}}, \gamma_{\text{cha}}, \gamma_{\text{rest}}$ and γ_{pump} are the time-fractional indices for the injection, chasing, resting, and pumping stages, respectively, satisfying $0 < \gamma_{\text{inj}}, \gamma_{\text{cha}}, \gamma_{\text{rest}}, \gamma_{\text{pump}} < 1$; $\tau_{\gamma_{\text{inj}}}$, $\tau_{\gamma_{\text{cha}}}$, $\tau_{\gamma_{\text{rest}}}$ and $\tau_{\gamma_{\text{pump}}}$ are the scale factors for the corresponding fractional indices. The reason that the time-fractional index is stage-dependent is that the retention time distributions for stages after injection are modified with particles pre-existing in the immobile zone. The temporally dependent spatial operator remains the same as that of the FS model. The FTS model (Eq. (3.443)) considers both space and time-fractional derivatives, which account for the nonlocal transport and power law retention time of solutes in the immobile zone, respectively. Note that $\gamma = 1$ is equivalent to adding a retardation factor of $(1 + \beta\tau_\gamma)$ for C_m. This is because the power law memory function $f(t) = \tau_\gamma\delta(t)$ when $\gamma \to 1$ and convolution of $\partial C_m/\partial t$ with $\tau_\gamma\delta(t)$ is $\tau_\gamma\partial C_m/\partial t$. The fractional-in-time (FT) model can be produced simply by letting the fractional order of space equal to 2. The immobile concentration $C_{im}(r,t)$ during the SWPP test can be calculated by employing Eq. (3.440). The total concentration $C_{\text{tot}}(r,t)$ that is defined as $C_{\text{tot}}(r,t) = \theta_m C_m + \theta_{im} C_{im}$ is obtainable as well.

The boundary conditions for a four-stage SWPP test without mixing effect are

$$C_m(r_w,t) = C_{\text{inj}}, \quad 0 < t < t_{\text{inj}} \tag{3.444}$$

$$C_m(r_w,t) = 0, \quad t_{\text{inj}} < t < t_{\text{cha}} \tag{3.445}$$

$$\frac{\partial C_m(r_w,t)}{\partial t} = 0, \quad t_{\text{rest}} < t < t_{\text{pump}} \tag{3.446}$$

$$C_m(r_e,t) = 0, \quad 0 < t < t_{\text{pump}} \tag{3.447}$$

$$C_m(r,0) = C_{im}(r,0) = 0, \quad r_w < r < r_e \tag{3.448}$$

Note that the boundary condition for the resting stage is not specified because this stage only involves a temporal derivative, and the advection-dispersion operator is not included. Thus, spatial derivatives are not a concern. Therefore, the boundary condition, which is needed when the governing equation involves spatial derivative(s), is irrelevant for the resting stage.

A solver is developed to solve Eq. (3.443) using an implicit finite-difference method. The implicit Euler method is proven to be unconditionally stable if the shifted Grunwald formula represents the Riemann–Liouville derivative. Details of the numerical scheme to solve the fractional models are elaborated below.

(b) Implicit Euler method for the fractional models

In the injection phase, the time fractional Riemann–Liouville derivative $\partial^{\gamma_{\text{inj}}} c_m / \partial t^{\gamma_{\text{inj}}}$ is defined as

$$\frac{\partial^{\gamma_{\text{inj}}} c_m(r,t)}{\partial t^{\gamma_{\text{inj}}}} = \frac{1}{\Gamma(1-\gamma_{\text{inj}})} \frac{\partial}{\partial t} \int_0^t \frac{c_m(r,\eta)}{(t-\eta)^{\gamma_{\text{inj}}}} d\eta \tag{3.449}$$

The spatial fractional Riemann–Liouville derivative $\frac{\partial^\alpha c_m}{\partial r^\alpha}$ is defined as

$$\frac{\partial^\alpha c_m(r,t)}{\partial r^\alpha} = \frac{1}{\Gamma(2-\alpha)} \frac{\partial^2}{\partial r^2} \int_0^r \frac{c_m(\xi,t)}{(r-\xi)^{\alpha-1}} d\xi \tag{3.450}$$

Note that the time-fractional Riemann–Liouville derivative $\partial^{\gamma_{\text{inj}}} c_m / \partial t^{\gamma_{\text{inj}}}$ is identical to the time-fractional Caputo derivative:

$$\frac{\partial^{\gamma_{\text{inj}}} c_m(r,t)}{\partial t^{\gamma_{\text{inj}}}} = \frac{1}{\Gamma(1-\gamma_{\text{inj}})} \int_0^t (t-\eta)^{-\gamma_{\text{inj}}} \frac{\partial c_m(r,\eta)}{\partial \eta} d\eta \tag{3.451}$$

with the initial condition that $c_m(r, t = 0) = 0$. It is preferred to study the fractional differential equation with the Caputo derivative because its properties facilitate numerical solutions.

The discretization scheme for time is $t_k = k\Delta t$ ($k = 0, 1, 2, \ldots, t_{\text{pump}}/\Delta$) where Δt is a uniform time step and $t_{\text{pump}}/\Delta t$ is the total number of time steps. t_k satisfies $0 \leq t_k \leq t_{\text{pump}}$. Note that $k = 0$ indicates the initial concentration. For space discretization, the distance of the grid i to the wellbore is $r_i = i\Delta r$ ($i = 1, 2, \ldots, m$) where Δr is the uniform radial step and m is the total number of grids. r_i satisfies $r_w \leq r_i \leq r_e$.

The finite-difference approximation to the Caputo derivative Eq. (3.451) is

$$\begin{aligned}
\frac{\partial^{\gamma_{\text{inj}}} c_m(r_i, t_{k+1})}{\partial t^{\gamma_{\text{inj}}}} &= \frac{1}{\Gamma(1-\gamma_{\text{inj}})} \sum_{j=0}^{k} \frac{c_m(r_i, t_{j+1}) - c_m(r_i, t_j)}{\Delta t} \int_{j\Delta t}^{(j+1)\Delta t} \\
&\quad \times \frac{d\xi}{(t_{k+1}-\xi)^{\gamma_{\text{inj}}}} + O(\Delta t) = \frac{\Delta t^{1-\gamma_{\text{inj}}}}{\Gamma(2-\gamma_{\text{inj}})} \\
&\quad \times \sum_{j=0}^{k} \frac{c_m(r_i, t_{k+1-j}) - c_m(r_i, t_{k-j})}{\Delta t} \\
&\quad \times [(j+1)^{1-\gamma_{\text{inj}}} - j^{1-\gamma_{\text{inj}}}] + O(\Delta t)
\end{aligned} \tag{3.452}$$

For the Riemann–Liouville derivative, a shifted Grünwald formula is used, which guarantees that the implicit Euler method is unconditionally stable:

$$\frac{\partial^{\alpha} c_m(r_i, t_{k+1})}{\partial r^{\alpha}} = \frac{1}{\Delta r^{\alpha}} \sum_{j=0}^{i+1} g_j c_m(r_i - (j-1)\Delta r, t_{k+1}) + O(\Delta t + \Delta r) \tag{3.453}$$

where $g_j = \frac{\Gamma(j-\alpha)}{[\Gamma(-\alpha)\Gamma(j+1)]}$, $j = 1, 2, 3, \ldots$.

Substituting Eqs. (3.452) and (3.453) into Eq. (3.443) for the injection stage and then discretizing the first derivative terms using the implicit finite-difference method, one gets:

$$\frac{c_m(r_i,t_{k+1})-c_m(r_i,t_k)}{\Delta t}+\frac{\beta\tau_{\gamma_{\rm inj}}\Delta t^{1-\gamma_{\rm inj}}}{\Gamma(2-\gamma_{\rm inj})}$$
$$\times\sum_{j=0}^{k}\frac{c_m(r_i,t_{k+1-j})-c_m(r_i,t_{k-j})}{\Delta t}[(j+1)^{1-\gamma_{\rm inj}}-j^{1-\gamma_{\rm inj}}]$$
$$=-b(r_i)\frac{c_m(r_i,t_{k+1})-c_m(r_{i-1},t_{k+1})}{\Delta r}+\frac{a(r_i)}{\Delta r^{\alpha}}\sum_{j=0}^{i+1}g_jc_m$$
$$\times(r_i-(j-1)\Delta r,t_{k+1}),\quad 0\le k<\frac{t_{\rm inj}}{\Delta t} \tag{3.454}$$

where $b(r_i)=\frac{k_\nu}{r_i}$ · and · $a(r_i)=\frac{\alpha_Lk_\nu}{r_i}$. Denote $c_m(r_i,t_k)=c_i^k$, $a_i=a(r_i)$, $\cdot b_i=b(r_i)$, $e_j=(j+1)^{1-\gamma_{\rm inj}}-j^{1-\gamma_{\rm inj}}$, $u_i=a_i\Delta t\Delta r^{-\alpha}$, $w_i=b_i\Delta t\Delta r^{-1}$, $z=\frac{\beta\tau_{\gamma_{\rm inj}}\Delta t^{1-\gamma_{\rm inj}}}{\Gamma(2-\gamma_{\rm inj})}$, and Eq. (3.454) can be rewritten as

$$c_i^{k+1}-c_i^k+z\sum_{j=0}^{k}e_j(c_i^{k+1-j}-c_i^{k-j})$$
$$=-w_i(c_i^{k+1}-c_{i-1}^{k+1})+u_i\sum_{j=0}^{i+1}g_jc_{i-j+1}^{k+1},\quad 0\le k<\frac{t_{\rm inj}}{\Delta t} \tag{3.455}$$

Rearranging Eq. (3.455), one has

$$-u_ig_0c_{i+1}^{k+1}+(w_i-u_ig_1+ze_0+1)c_i^{k+1}-(w_i+u_ig_2)c_{i-1}^{k+1}$$
$$-u_i\sum_{j=3}^{i+1}g_jc_{i-j+1}^{k+1}=c_i^k+z\sum_{j=0}^{k-1}(e_j-e_{j+1})c_i^{k-j}+ze_kc_i^0,$$
$$0\le k<\frac{t_{\rm inj}}{\Delta t} \tag{3.456}$$

The matrix form of Eq. (3.456) is

$$\begin{cases}A_{\rm inj}U^1=(ze_0-ze_1+1)U^0+H_{\rm inj},\quad k=0\\ A_{\rm inj}U^{k+1}=U^k+z\sum_{j=0}^{k-1}(e_j-e_{j+1})U^{k-j}\\ \qquad +ze_kU^0+H_{\rm inj},\quad 0<k<\frac{t_{\rm inj}}{\Delta t}\end{cases} \tag{3.457}$$

where $U^k = (c_1^k, c_2^k, \ldots, c_m^k)^T$, $H_{\text{inj}} = ((w_1 + u_1 g_2)c_{\text{inj}}, \overbrace{0, \ldots, 0}^{m-1})^T$ and coefficient matrix A_{inj} is

$$A_{\text{inj}(i,j)} = \begin{cases} -u_i g_{i-j+1}, & j < i-1 \\ -w_i - u_i g_2, & j = i-1 \\ w_i - u_i g_1 + z e_0 + 1, & j = i \\ -u_i g_0, & j = i+1 \\ 0, & j > i+1 \end{cases} \tag{3.458}$$

In the chasing phase, the discretized equation for the chasing stage is the same as Eq. (3.456). Note that the term $\sum_{j=1}^{k-1}(e_j - e_{j+1})c_i^{k-j}$ contains all the past concentration information at grid i, including the whole injection stage and the chasing stage before time step $k+1$. Therefore, the matrix form should be modified from Eq. (3.457) accordingly:

$$A_{\text{cha}} U^{k+1} = U^k + z \sum_{j=0}^{k-1} (e_j - e_{j+1}) U^{k-j} + z e_k U^0, \quad k = \frac{t_{\text{inj}}}{\Delta t} \tag{3.459a}$$

$$A_{\text{cha}} U^{k+1} = U^k + z \sum_{j=k-t_{\text{inj}}/\Delta t}^{k-1} (e_j - e_{j+1}) U^{k-j} + z \sum_{j=0}^{k-t_{\text{inj}}/\Delta t - 1} \times (e_j - e_{j+1}) U^{k-j} + z e_k U^0, \quad \frac{t_{\text{inj}}}{\Delta t} < k < \frac{t_{\text{cha}}}{\Delta t} \tag{3.459b}$$

where $e_j = (j+1)^{1-\gamma_{\text{cha}}} - j^{1-\gamma_{\text{cha}}}$, $z = \beta \tau_{\gamma_{\text{cha}}} \Delta t^{1-\gamma_{\text{cha}}} / \Gamma(2-\gamma_{\text{cha}})$, $A_{\text{cha}} = A_{\text{inj}}$ and $\frac{t_{\text{cha}}}{\Delta t}$ is the total time steps at the end of the chasing stage. Note that the term H_{inj} is omitted here because the concentration of chasing fluid is zero.

In the resting phase, the Eq. (3.443) is discretized similarly:

$$\frac{c_m(r_i, t_{k+1}) - c_m(r_i, t_k)}{\Delta t} + \frac{\Delta t^{1-\gamma}}{\Gamma(2-\gamma)} \sum_{j=0}^{k} \frac{c_m(r_i, t_{k+1-j}) - c_m(r_i, t_{k-j})}{\Delta t} \times [(j+1)^{1-\gamma} - j^{1-\gamma}] = 0, \quad \frac{t_{\text{cha}}}{\Delta t} \le k < \frac{t_{\text{rest}}}{\Delta t} \tag{3.460}$$

where $\frac{t_{\text{rest}}}{\Delta t}$ is the total time steps at the end of the resting stage. Following the conventions of Eq. (3.455) and rearranging Eq. (3.460), one has

$$(ze_0+1)c_i^{k+1}=c_i^k+z\sum_{j=0}^{k-1}(e_j-e_{j+1})c_i^{k-j}+ze_kc_i^0,\ \frac{t_{\text{cha}}}{\Delta t}\le k<\frac{t_{\text{rest}}}{\Delta t} \tag{3.461}$$

The matrix form of Eq. (3.461) is

$$\begin{aligned}A_{\text{rest}}U^{k+1}&=U^k+z\sum_{j=k-t_{\text{inj}}/\Delta t}^{k-1}(e_j-e_{j+1})U^{k-j}\\&\quad+z\sum_{j=k-t_{\text{cha}}/\Delta t}^{k-t_{\text{inj}}/\Delta t-1}(e_j-e_{j+1})U^{k-j}+ze_kU^0,\quad k=\frac{t_{\text{cha}}}{\Delta t}\end{aligned} \tag{3.462a}$$

$$\begin{aligned}A_{\text{rest}}U^{k+1}&=U^k+z\sum_{j=k-t_{\text{inj}}/\Delta t}^{k-1}(e_j-e_{j+1})U_{\text{inj}}^{k-j}\\&\quad+z\sum_{j=k-t_{\text{cha}}/\Delta t}^{k-t_{\text{inj}}/\Delta t-1}(e_j-e_{j+1})U^{k-j}\\&\quad+z\sum_{j=0}^{k-t_{\text{cha}}/\Delta t-1}(e_j-e_{j+1})U^{k-j}+ze_kU^0,\\&\quad\frac{t_{\text{cha}}}{\Delta t}<k<\frac{t_{\text{rest}}}{\Delta t}\end{aligned} \tag{3.462b}$$

where $e_j=(j+1)^{1-\gamma_{\text{rest}}}-j^{1-\gamma_{\text{rest}}}$, $z=\beta\tau_{\gamma_{\text{rest}}}\Delta t^{1-\gamma_{\text{rest}}}/\Gamma(2-\gamma_{\text{rest}})$ and A_{rest} is

$$A_{\text{rest}(i,j)}=\begin{cases}0, & j<i-1\\0, & j=i-1\\z_ie_0+1, & j=i\\0, & j=i+1\\0, & j>i+1\end{cases} \tag{3.463}$$

In the extraction phase, the spatial fractional Riemann–Liouville derivative $\frac{\partial^\alpha c_m}{\partial(-r)^\alpha}$ is defined as

$$\frac{\partial^\alpha c_m(r,t)}{\partial(-r)^\alpha} = \frac{1}{\Gamma(2-\alpha)}\frac{\partial^2}{\partial r^2}\int_r^{r_e}\frac{c_m(\xi,t)}{(r-\xi)^{\alpha-1}}d\xi \tag{3.464}$$

Using the shifted Grünwald formula, one has

$$\frac{\partial^\alpha c_m(r_i,t_{k+1})}{\partial(-r)^\alpha} = \frac{1}{\Delta r^\alpha}\sum_{j=0}^{m-i+1} g_j c_m(r_i+(j-1)\Delta r, t_{k+1}) + O(\Delta t+\Delta r) \tag{3.465}$$

and one can get the discretized form of Eq. (3.443) for the extraction phase:

$$\begin{aligned}&\frac{c_m(r_i,t_{k+1})-c_m(r_i,t_k)}{\Delta t}\\&+\frac{\Delta t^{1-\gamma}}{\Gamma(2-\gamma)}\sum_{j=0}^{k}\frac{c_m(r_i,t_{k+1-j})-c_m(r_i,t_{k-j})}{\Delta t}\\&\times[(j+1)^{1-\gamma}-j^{1-\gamma}] = b(r_i)\frac{c_m(r_i,t_{k+1})-c_m(r_{i-1},t_{k+1})}{\Delta r}\\&+\frac{a(r_i)}{\Delta r^\alpha}\sum_{j=0}^{m-i+1} g_j c_m(r_i+(j-1)\Delta r, t_{k+1}),\ \frac{t_{\text{rest}}}{\Delta t}\le k<\frac{t_{\text{pump}}}{\Delta t}\end{aligned} \tag{3.466}$$

where $\frac{t_{\text{pump}}}{\Delta t}$ is the total time steps at the end of the pumping stage. Following the conventions of Eq. (3.455) and rearranging Eq. (3.464), one gets:

$$\begin{aligned}&-u_i\sum_{j=2}^{m-i+1} g_j c_{i+j-1}^{k+1} + (w_i+ze_0+1-u_ig_1)c_i^{k+1} - (w_i+u_ig_0)c_{i-1}^{k+1}\\&= c_i^k + z\sum_{j=0}^{k-1}(e_j-e_{j+1})c_i^{k-j} + ze_kc_i^0,\quad \frac{t_{\text{rest}}}{\Delta t}\le k<\frac{t_{\text{pump}}}{\Delta t}\end{aligned} \tag{3.467}$$

The matrix form of equation (3.465) is

$$A_{\text{pump}}U^{k+1} = U^k + z\sum_{j=k-t_{\text{inj}}/\Delta t}^{k-1}(e_j - e_{j+1})U_{\text{inj}}^{k-j} + z\sum_{j=k-t_{\text{cha}}/\Delta t}^{k-t_{\text{inj}}/\Delta t-1}(e_j - e_{j+1})U^{k-j} + z\sum_{j=0}^{k-t_{\text{cha}}/\Delta t-1}(e_j - e_{j+1})U^{k-j} + ze_kU^0, \quad k = \frac{t_{\text{rest}}}{\Delta t} \tag{3.468a}$$

$$A_{\text{pump}}U^{k+1} = U^k + z\sum_{j=k-t_{\text{inj}}/\Delta t}^{k-1}(e_j - e_{j+1})U^{k-j} + z\sum_{j=k-t_{\text{cha}}/\Delta t}^{k-t_{\text{inj}}/\Delta t-1}(e_j - e_{j+1})U^{k-j} + z\sum_{j=k-t_{\text{rest}}/\Delta t}^{k-t_{\text{rest}}/\Delta t-1}(e_j - e_{j+1})U^{k-j} + z\sum_{j=0}^{k-t_{\text{pump}}/\Delta t-1}(e_j - e_{j+1})U^{k-j} + ze_kU^0, \times \frac{t_{\text{rest}}}{\Delta t} < k < \frac{t_{\text{pump}}}{\Delta t} \tag{3.468b}$$

where $e_j = (j+1)^{1-\gamma_{\text{pump}}} - j^{1-\gamma_{\text{pump}}}$, $z = \beta\tau_{\gamma_{\text{pump}}}/\Delta t^{1-\gamma_{\text{pump}}}/\Gamma(2-\gamma_{\text{pump}})$ and A_{pump} is

$$A_{\text{pump}(i,j)} = \begin{cases} 0, & j < i-1 \\ -w_i - u_i g_0, & j = i-1 \\ w_i + ze_0 + 1 - u_i g_1, & j = i \\ -u_i g_{j-i+1}, & j > i \end{cases} \tag{3.469}$$

For the FT model, only the Caputo derivative is discretized, as shown above, and the Fickian flux term is approximated using the traditional implicit finite-difference method. The FS model is solved similarly.

The analytical solution can serve as a benchmark to test the numerical solver of the fractional model with an integer fractional index. Note that this calibration only holds for the FS model with $\alpha = 1$. Suppose time fraction is considered, such as in the FT or FTS model. In that case, the semi-analytical solution is not equal to that of the fractional model with gamma equals 1 because the analytical model assumes the memory function has an exponential form. In contrast, the FT and FTS models assume the memory function is a Dirac delta function $\delta(t)$. The basic scheme for the derivation of the analytical solution is to obtain the analytical solution for each stage sequentially and to use the final distribution of a previous stage as the initial condition for the current stage. This operation is only valid for the analytical model because no memory effect is incorporated for the exponential memory function. For the fractional model with the power law memory function, this operation will miss all the past concentration information at each grid point and is thus incorrect.

(c) Analytical solution in the injection and chasing phases

In this section, the analytical solution assuming the first-order rate-limited mass transfer between the mobile and immobile zones is proposed for a four-stage SWPP test. The governing equation of transport in the mobile zone during the injection and chasing stages is

$$\frac{\partial C_m}{\partial t} + \beta \frac{\partial C_{im}}{\partial t} = -\frac{\kappa_\nu}{r}\frac{\partial C_m}{\partial r} + \frac{a\kappa_\nu}{r}\frac{\partial^2 C_{im}}{\partial r^2} \tag{3.470}$$

The immobile zone transport still satisfies Eq. (3.433). The outer boundary condition and initial conditions are already given in Eqs. (3.447) and (3.448), respectively. The wellbore boundary condition is modified to

$$C_m(r_w, t) = C_{\text{inj}}[H(t) - H(t - t_{\text{inj}})], \quad 0 < t < t_{\text{cha}} \tag{3.471}$$

where $H(t)$ is the Heaviside step function. The dimensionless form of the governing equation in the mobile zone (Eq. (3.470)) considering Eq. (3.433) is

$$\frac{\partial C_{mD}}{\partial \tau} = -\frac{1}{\rho}\frac{\partial C_{mD}}{\partial \rho} + \frac{1}{\rho}\frac{\partial^2 C_{mD}}{\partial \rho^2} - \varepsilon_m(C_{mD} - C_{imD}) \tag{3.472}$$

where $C_{mD} = c_m/c_{\text{inj}}$, $C_{imD} = c_{im}/c_{\text{inj}}$, $\rho = r/a$, $\varepsilon_m = \omega a^2/\theta_m \kappa_v$, $\tau = \kappa_v t/a^2$.

The dimensionless form of the immobile zone equation is

$$\frac{\partial C_{imD}}{\partial \tau} = \varepsilon_{im}(C_{mD} - C_{imD}) \tag{3.473}$$

where $\varepsilon_m = \omega a^2/\theta_{im} \kappa_v$.

Conducting Laplace transform to Eq. (3.472), one can get the mobile concentration in the Laplace domain

$$\frac{1}{\rho}\frac{\partial \overline{C}_{mD}}{\partial \rho^2} - \frac{1}{\rho}\frac{\overline{C}_{mD}}{\partial \rho} - \varepsilon_m(\overline{C}_{mD} - \overline{C}_{imD}) - s\overline{C}_{mD} = 0 \tag{3.474}$$

The dimensionless immobile concentration in the Laplace domain $\overline{C}_{miD}$ is

$$\overline{C}_{imD}(\rho, s) = \frac{\varepsilon_{im}\overline{C}_{mD}(\rho, s)}{s + \varepsilon_{im}} \tag{3.475}$$

The immobile concentration in the Laplace domain is given in Eq. (3.475). Substituting Eq. (3.475) into Eq. (3.474), one obtains:

$$\frac{1}{\rho}\frac{\partial \overline{C}_{mD}}{\partial \rho^2} - \frac{1}{\rho}\frac{\overline{C}_{mD}}{\partial \rho} - A_1\overline{C}_{mD} = 0 \tag{3.476}$$

where

$$A_1 = \varepsilon_m + s - \frac{\varepsilon_m \varepsilon_{im}}{s + \varepsilon_{im}} \tag{3.477}$$

The boundary conditions (Eqs. (3.444) and (3.447)) in Laplace domain are respectively:

$$\overline{C}_{mD}(\rho_w, s) = \frac{1 - \exp(-\tau_{\text{inj}} s)}{s} \tag{3.478}$$

$$\overline{C}_{mD}(\rho_e, s) = 0 \tag{3.479}$$

where ρ_e is the dimensionless outer radius.

Note that Eq. (3.476) is a second-order ordinary differential equation (ODE) with variable coefficients. The general solution of Eq. (3.476) is

$$\overline{C}_{mD} = N_1 \exp\left(\frac{y}{2}\right) A_i(A_1^{1/3} y) + N_2 \exp\left(\frac{y}{2}\right) B_i(A_1^{1/3} y) \tag{3.480}$$

where $y = \rho + \frac{1}{4A_1}$, N_1 and N_2 are constants. As $B_i(x)$ diverges when $\rho \to \infty$, N_2 must be zero. Substituting the boundary conditions Eqs. (3.478) and (3.479) into Eq. (3.480), N_1 is obtained:

$$N_1 = \frac{1 - \exp(-\tau_{\text{inj}} s)}{s} \frac{\exp(-\frac{y_w}{2})}{A_i(A_1^{1/3} y_w)} \tag{3.481}$$

where $y_w = \rho_w + \frac{1}{4A_1}$, $\rho_w = r_w/a$. Substituting Eq. (3.481) and $N_2 = 0$ into Eq. (3.480), the analytical solution for injection and chasing stages can be obtained in the Laplace domain

$$\overline{C}_{mD} = \frac{1 - \exp(-\tau_{\text{inj}} s)}{s} \exp\left(\frac{y - y_w}{2}\right) \frac{A_i(A_1^{1/3} y)}{A_i(A_1^{1/3} y_w)} \tag{3.482}$$

Combining Eqs. (3.475) and (3.482), one has

$$\overline{C}_{imD}(\rho, s) = \frac{\varepsilon_{im}[1 - \exp(-\tau_{\text{inj}} s)]}{(s + \varepsilon_{im}) s} \exp\left(\frac{y - y_w}{2}\right) \frac{A_i(A_1^{1/3} y)}{A_i(A_1^{1/3} y_w)} \tag{3.483}$$

(d) Analytical solution in the resting phase

The governing equation of the mobile zone during the resting stage is

$$\frac{\partial C_m}{\partial t} + \beta \frac{\partial C_{im}}{\partial t} = 0 \tag{3.484}$$

and the immobile zone equation is the same as Eq. (3.433). The initial condition for C_m is

$$C_m(\rho, \tau_{\text{rest}} \to \tau_{\text{cha}}) = C_m(\rho, \tau_{\text{cha}}) \tag{3.485}$$

To solve the initial value problem, one more condition is required. Two assumptions are made here. The first assumption is that the total solute mass of the mobile and immobile zones is constant from the beginning to the end of the resting stage (Eq. (3.486)). The second assumption is that the mobile and immobile concentrations are the same in the resting stage when resting time goes to infinity (Eq. (3.487)):

$$C_{mD}(\rho,\tau_{\text{cha}})\theta_m + C_{imD}(\rho,\tau_{\text{cha}})\theta_{im}$$
$$= C_{mD}(\rho,\tau_\infty)\theta_m + C_{imD}(\rho,\tau_\infty)\theta_{im} \tag{3.486}$$
$$C_{mD}(\rho,\tau_\infty) = C_{imD}(\rho,\tau_\infty) \tag{3.487}$$

The dimensionless mobile zone equation in the resting stage is

$$\frac{\partial C_{mD}}{\partial \tau} = -\varepsilon_m(C_{mD} - C_{imD}) \tag{3.488}$$

Rearranging Eq. (3.488), one has

$$C_{imD} = C_{mD} + \frac{1}{\varepsilon_m}\frac{\partial C_{mD}}{\partial \tau} \tag{3.489}$$

Substituting Eq. (3.489) into Eq. (3.473), one can get:

$$\frac{\partial^2 C_{mD}}{\partial \tau^2} + (\varepsilon_m + \varepsilon_{im})\frac{\partial C_{mD}}{\partial \tau} = 0 \tag{3.490}$$

Equation (3.490) is a second-order ODE with constant coefficients. The general solution is

$$C_{mD} = N_3 \exp(l_1\tau) + N_4 \exp(l_2\tau) \tag{3.491}$$

where l_1 and l_2 are the roots of the following characteristic equation:

$$l^2 + (\varepsilon_m + \varepsilon_{im})l = 0 \tag{3.492}$$

The solution of Eq. (3.492) is

$$l_1 = 0 \tag{3.493}$$
$$l_2 = -(\varepsilon_m + \varepsilon_{im}) \tag{3.494}$$

The initial condition in dimensionless form is

$$C_{mD}(\rho,\tau_{\text{rest}} \to \tau_{\text{cha}}) = C_{mD}(\rho,\tau_{\text{cha}}) \tag{3.495}$$

Coefficients N_3 and N_4 are calculated by substituting Eqs. (3.493)–(3.495) and Eqs. (3.486)–(3.487) into Eq. (3.491), then the mobile concentration in the resting stage is obtained:

$$C_{mD}(\rho,\tau) = (N_4 - N_3)\exp[-(\varepsilon_m + \varepsilon_{im})\tau] + N_3 \tag{3.496}$$

where

$$N_3 = \frac{C_{mD}(\rho, \tau_{\text{cha}})\theta_m + C_{imD}(\rho, \tau_{\text{cha}})\theta_{im}}{\theta_m + \theta_{im}} \tag{3.497}$$

$$N_4 = C_{mD}(\rho, \tau_{\text{cha}}) \tag{3.498}$$

and the dimensionless $C_{imD}(\rho, \tau)$ of the resting stage is

$$C_{imD}(\rho, \tau) = -\frac{\varepsilon_{im}}{\varepsilon_m} C_{imD}(\rho, \tau) + \left(\frac{\varepsilon_{im}}{\varepsilon_m} + 1\right) C_1 \tag{3.499}$$

(e) Analytical solution in the extraction phase

In the extraction phase, the direction of advective flux is reversed. Thus, the governing equation of the mobile zone for this stage is modified accordingly:

$$\frac{\partial C_m}{\partial t} + \beta \frac{\partial C_{im}}{\partial t} = \frac{\kappa_v}{r} \frac{\partial C_m}{\partial r} + \frac{a\kappa_v}{r} \frac{\partial^2 C_m}{\partial r^2} \tag{3.500}$$

The initial conditions for the mobile and immobile zones are, respectively:

$$C_m(\rho, \tau_{\text{pump}} \to \tau_{\text{rest}}) = C_m(\rho, \tau_{\text{rest}}) \tag{3.501}$$

$$C_{im}(\rho, \tau_{\text{pump}} \to \tau_{\text{rest}}) = C_{im}(\rho, \tau_{\text{rest}}) \tag{3.502}$$

The dimensionless mobile and immobile stage equations in the Laplace domain are, respectively:

$$\frac{1}{\rho} \frac{\partial \overline{C}_{mD}}{\partial \rho^2} + \frac{1}{\rho} \frac{\overline{C}_{mD}}{\partial \rho} - \varepsilon_m(\overline{C}_{mD} - \overline{C}_{imD})$$
$$-s\overline{C}_{mD} + C_{mD}(\rho, \tau_{\text{rest}}) = 0 \tag{3.503}$$

$$s\overline{C}_{imD} - C_{imD}(\rho, \tau_{\text{rest}}) = \varepsilon_{im}(\overline{C}_{mD} - \overline{C}_{imD}) \tag{3.504}$$

Substituting $\overline{C}_{imD}$ obtained from Eq. (3.504) into Eq. (3.503), one can get:

$$\frac{1}{\rho} \frac{\partial^2 \overline{C}_{mD}}{\partial \rho^2} + \frac{1}{\rho} \frac{\partial \overline{C}_{mD}}{\partial \rho} - A_3 \overline{C}_{mD} + B_3(\rho) = 0 \tag{3.505}$$

Chen and Woodside (1988) investigated the pumping problem without an immobile zone in the aquifer and proposed a closed-form analytical solution. An analytical solution with an immobile zone is proposed based on their work. The mobile concentration in the Laplace domain $\overline{C}_{mD}(\rho, s)$ is

$$\overline{C}_{mD}(\rho, s) = \pi e^{-\frac{\rho}{2}} \left\{ \int_{\rho_w}^{\rho} H_1 F(\lambda) d\lambda + \int_{\rho}^{\infty} H_2 F(\lambda) d\lambda - \int_{\rho_w}^{\infty} H_3 F(\lambda) d\lambda \right\} \tag{3.506}$$

where

$$F(\lambda) = \lambda e^{\frac{\lambda}{2}} B_3(\lambda) \tag{3.507a}$$

$$H_1 = \frac{A_i(z) B_i(u)}{A_3^{1/3}} \tag{3.507b}$$

$$H_2 = \frac{A_i(u) B_i(z)}{A_3^{1/3}} \tag{3.507c}$$

$$H_3 = \frac{A_i(z) A_i(u)}{A_3^{1/3}} \frac{A_3^{1/3} B_i'(z_w) - \frac{1}{2} B_i(z_w)}{A_3^{1/3} A_i'(z_w) - \frac{1}{2} A_i(z_w)} \tag{3.507d}$$

$$z = A_3^{1/3} \left(\rho + \frac{1}{4A_3} \right) \tag{3.507e}$$

$$z_w = A_3^{1/3} \left(\rho_w + \frac{1}{4A_3} \right) \tag{3.507f}$$

$$u = A_3^{1/3} \left(\lambda + \frac{1}{4A_3} \right) \tag{3.507g}$$

$$A_3 = \varepsilon_m + s - \frac{\varepsilon_m \varepsilon_{im}}{s + \varepsilon_{im}} \tag{3.507h}$$

$$B_3(\rho) = \frac{\varepsilon_m C_{imD}(\rho, \tau_{\text{rest}})}{s + \varepsilon_{im}} + C_{mD}(\rho, \tau_{\text{rest}}) \tag{3.507i}$$

The immobile concentration in Laplace domain $\overline{C}_{imD}(\rho, s)$ is

$$\overline{C}_{imD}(\rho, s) = \frac{\varepsilon_{im} \overline{C}_{mD}(\rho, s) + C_{imD}(\rho, \tau_{\text{rest}})}{s + \varepsilon_{im}} \tag{3.508}$$

Both $\overline{C}_{mD}(\rho, s)$ and $\overline{C}_{imD}(\rho, s)$ are inversed to the real domain using the de Hoog or Stehfest methods.

3.3.3.2 *Analytical solutions of Wang et al. (2020)*

Based on the model of Chen *et al.* (2017), Wang *et al.* (2020) developed a new model of SWPP tests considering the mixing effect and the aquitard effect under the MIM framework. A cylindrical coordinate system is employed in this section, and the origin is located at the well center, as shown in Figure 3.36(c). The z-axis and the r-axis are vertical and horizontal, respectively. Figure 3.36 is a schematic diagram of the model investigated in this section.

The concept of homogeneity here deserves clarification. Even though the homogeneity assumption is commonly used in developing analytical and numerical models of subsurface flow and transport, one should be aware that a rigorous sense of homogeneity probably never exists in a real-world setting (unless the media are composed of idealized glass balls, as in some laboratory experiments). Therefore, the homogeneity concept here should be envisaged as a media whose hydraulic parameters vary within relatively narrow ranges or the so-called weak heterogeneity. The Borden site of Canada (Sudicky, 1988) is one example of weak aquifer heterogeneity. Wang *et al.* (2018)

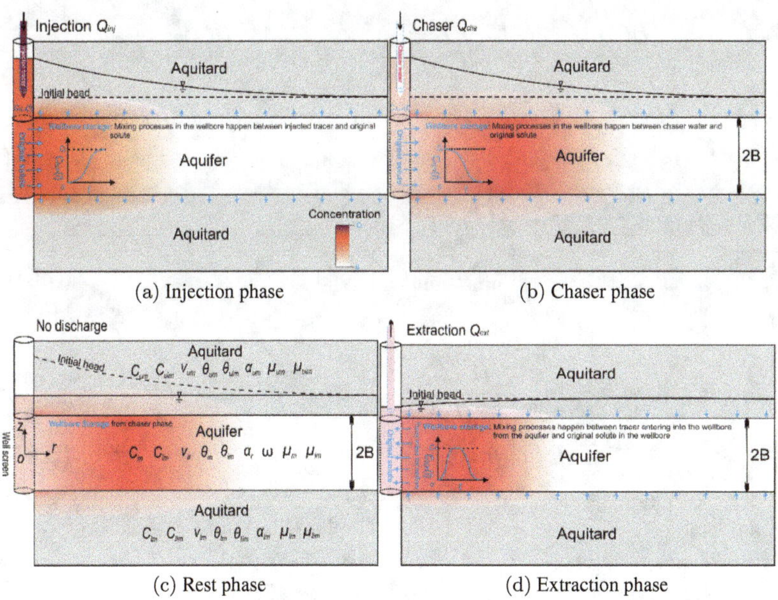

(a) Injection phase (b) Chaser phase

(c) Rest phase (d) Extraction phase

Figure 3.36. The schematic diagram of the SWPP test in an aquifer-aquitard system.

employed a stochastic modeling technique to test the assumption of homogeneity associated with the SWPP test and found that such an assumption could be used to approximate a heterogeneous aquifer when the variance of spatial hydraulic conductivity was small.

(a) Mathematical model

Considering the advective effect, dispersive effect and first-order chemical reaction in describing solute transport under the MIM framework, the governing equations of the SWPP test are

$$\theta_m R_m \frac{\partial C_m}{\partial t} = \frac{\theta_m}{r}\frac{\partial}{\partial r}\left(rD_r\frac{\partial C_m}{\partial r}\right) - \theta_m v_a \frac{\partial C_m}{\partial r} - \omega_a(C_m - C_{im})$$

$$-\theta_m\mu_m C_m - \left(\frac{\theta_{um}v_{um}}{2B}C_{um} - \frac{\theta_{um}D_u}{2B}\frac{\partial C_{um}}{\partial z}\right)\Bigg|_{z=B}$$

$$+\left(-\frac{\theta_{lm}v_{lm}}{2B}C_{lm} - \frac{\theta_{lm}D_l}{2B}\frac{\partial C_{lm}}{\partial z}\right)\Bigg|_{z=-B} \quad (3.509a)$$

$$\theta_{im}R_{im}\frac{\partial C_{im}}{\partial t} = \omega_a(C_m - C_{im}) - \theta_{im}\mu_{im}C_{im} \quad (3.509b)$$

$$\theta_{um}R_{um}\frac{\partial C_{um}}{\partial t} = \theta_{um}D_u\frac{\partial^2 C_{um}}{\partial z^2} - \theta_{um}v_{um}\frac{\partial C_{um}}{\partial z}$$

$$-\omega_u(C_{um} - C_{\text{uim}}) - \theta_{um}\mu_{um}C_{um}, \quad z \geq B \quad (3.510a)$$

$$\theta_{\text{uim}}R_{\text{uim}}\frac{\partial C_{\text{uim}}}{\partial t} = \omega_u(C_{um} - C_{\text{uim}})$$

$$-\theta_{\text{uim}}\mu_{\text{uim}}C_{\text{uim}}, \quad z \geq B \quad (3.510b)$$

$$\theta_{lm}R_{lm}\frac{\partial C_{lm}}{\partial t} = \theta_{lm}D_l\frac{\partial^2 C_{lm}}{\partial z^2} - \theta_{lm}v_{lm}\frac{\partial C_{lm}}{\partial z} - \omega_l(C_{lm} - C_{\text{lim}})$$

$$-\theta_{lm}\mu_{lm}C_{lm}, \quad z \leq -B \quad (3.511a)$$

$$\theta_{\text{lim}}R_{\text{lim}}\frac{\partial C_{\text{lim}}}{\partial t} = \omega_l(C_{lm} - C_{\text{lim}}) - \theta_{\text{lim}}\mu_{\text{lim}}C_{\text{lim}}, \quad z \leq -B \quad (3.511b)$$

where subscripts "u" and "l" refer to parameters in the upper and lower aquitards, respectively; subscripts "m" and "im" refer to parameters in the mobile and immobile domains, respectively; C_m and C_{im} are the concentrations [Kg/m^3] of the aquifer; C_{um}, C_{uim}, C_{lm} and C_{lim} are concentrations [Kg/m^3] of the aquitards; t is the time [s]; B is half of the aquifer thickness [m]; r is the radial distance [m]; z represents the vertical distance [m]; r_w is the well radius [m]; D_r is aquifer dispersion coefficient [m^2/s]; D_u and D_l are vertical dispersion coefficients [m^2/s] of the upper aquitard and lower aquitard, respectively; v_a represents the average velocity [m/s] in the aquifer and $v_a = \frac{u_a}{\theta_m}$; u_a is Darcian velocity [m/s]; v_{um} and v_{lm} are vertical velocities [m/s] in the aquitards; μ_m, μ_{im}, μ_{um}, μ_{uim}, μ_{lm} and μ_{lim} are reaction rates; θ_m, θ_{im}, θ_{um}, θ_{uim}, θ_{lm} and θ_{lim} are the porosities [-]; $R_m = 1 + \frac{\rho_b K_d}{\theta_m}$, $R_{im} = 1 + \frac{\rho_b K_d}{\theta_{im}}$, $R_{um} = 1 + \frac{\rho_b K_d}{\theta_{um}}$, $R_{\mathrm{uim}} = 1 + \frac{\rho_b K_d}{\theta_{\mathrm{uim}}}$, $R_{lm} = 1 + \frac{\rho_b K_d}{\theta_{lm}}$ and $R_{\mathrm{lim}} = 1 + \frac{\rho_b K_d}{\theta_{\mathrm{lim}}}$ are the retardation factors [dimensionless]; K_d is the equilibrium distribution coefficient [m^3/Kg]; ρ_b is the bulk density [Kg/m^3]; ω_a, ω_u and ω_l are the first-order mass transfer coefficients [1/s].

The symbol of the advection term is positive in the extraction phase in the above equations, while it is negative before that. The dispersions are assumed to be linearly changing with the flow velocity, and one has

$$D_r = \alpha_r|v_r| + D_r^* \tag{3.512a}$$

$$D_u = \alpha_u|v_u| + D_u^* \tag{3.512b}$$

$$D_l = \alpha_l|v_l| + D_l^* \tag{3.512c}$$

where α_r, α_u and α_l are dispersivities [m] of the aquifer, upper aquitard, and lower aquitard, respectively; D_r^*, D_u^* and D_l^* are the diffusion coefficients [m^2/s].

Initial conditions are

$$\begin{aligned} C_m(r,t)|_{t=0} &= C_{im}(r,t)|_{t=0} = C_{um}(r,z,t)|_{t=0} = C_{\mathrm{uim}}(r,z,t)|_{t=0} \\ &= C_{lm}(r,z,t)|_{t=0} = C_{\mathrm{lim}}(r,z,t)|_{t=0} = 0, \quad r \geq r_w \end{aligned} \tag{3.513}$$

The boundary conditions at infinity are

$$\begin{aligned} C_m(r,t)|_{r\to\infty} &= C_{im}(r,t)|_{r\to\infty} = C_{um}(r,z,t)|_{z\to\infty} \\ &= C_{\mathrm{uim}}(r,z,t)|_{z\to\infty} = C_{lm}(r,z,t)|_{z\to-\infty} \\ &= C_{\mathrm{lim}}(r,z,t)|_{z\to-\infty} = 0, r \geq r_w \end{aligned} \tag{3.514}$$

Due to the concentration continuity at the aquifer-aquitard interface, one has

$$C_m(r,t) = C_{um}(r, z = B, t) \tag{3.515a}$$

$$C_m(r,t) = C_{lm}(r, z = -B, t) \tag{3.515b}$$

The flux concentration continuity (FCC) is applied on the surface of the wellbore, and one has

$$\left[v_a C_m(r,t) - \alpha_r |v_a| \frac{\partial C_m(r,t)}{\partial r}\right]\bigg|_{r=r_w} = \left[v_a C_{\text{inj},m}(t)\right]\big|_{r=r_w}, \quad 0 < t \le t_{\text{inj}} \tag{3.516}$$

$$\left[v_a C_m(r,t) - \alpha_r |v_a| \frac{\partial C_m(r,t)}{\partial r}\right]\bigg|_{r=r_w} = \left[v_a C_{\text{cha},m}(t)\right]\big|_{r=r_w}, \quad t_{\text{inj}} < t \le t_{\text{cha}} \tag{3.517}$$

$$[C_m(r,t)]|_{r=r_w} = C_{\text{res},m}(r_w, t), \quad t_{\text{cha}} < t \le t_{\text{res}} \tag{3.518}$$

$$\left[v_a C_m(r,t) - \alpha_r |v_a| \frac{\partial C_m(r,t)}{\partial r}\right]\bigg|_{r=r_w} = \left[v_a C_{\text{ext},m}(t)\right]\big|_{r=r_w}, \quad t_{\text{res}} < t \le t_{\text{ext}} \tag{3.519}$$

where t_{inj}, t_{cha}, t_{res} and t_{ext} are the end moments [s] of the injection phase, the chaser phase, the rest phase and the extraction phase, respectively; $C_{\text{inj},m}(t)$, $C_{\text{cha},m}(t)$, $C_{\text{res},m}(t)$ and $C_{\text{ext},m}(t)$ represent the wellbore concentrations [Kg/m^3] of tracer in the injection phase, the chaser phase, the rest phase and the extraction phase, respectively. Equations (3.516)–(3.519) indicate that the flux continuity across the interface between the well and the formation is only considered for the mobile continuum (or mobile domain).

The variation of the concentration with the mixing effect in the injection phase could be described by (Wang *et al.*, 2018):

$$V_{w,\text{inj}} \frac{dC_{\text{inj},m}}{dt} = -\xi v_a(r_w)[C_{\text{inj},m}(t) - C_0], \quad 0 < t \le t_{\text{inj}} \tag{3.520a}$$

$$C_{\text{inj},m}(t)\big|_{t=0} = 0, \quad 0 < t \le t_{\text{inj}} \tag{3.520b}$$

$$V_{w,\text{inj}} = \pi r_w^2 h_{w,\text{inj}} \tag{3.520c}$$

$$\xi = 2\pi r_w \theta_m 2B \tag{3.520d}$$

where $h_{w,\text{inj}}$ is the wellbore water depth [m] in the injection phase, C_0 is concentration [Kg/m^3] of prepared tracer.

As for the chaser phase, the models describing the concentration variation in the wellbore could be obtained using the mass balance principle:

$$V_{w,\text{cha}} \frac{dC_{\text{cha},m}}{dt} = -\xi v_a(r_w)[C_{\text{cha},m}(t)], \quad t_{\text{inj}} < t \leq t_{\text{cha}} \tag{3.521a}$$

$$C_{\text{cha},m}(t)|_{t=t_{\text{inj}}} = C_{\text{inj},m}(t)|_{t=t_{\text{inj}}}, \quad t_{\text{inj}} < t \leq t_{\text{cha}} \tag{3.521b}$$

$$V_{w,\text{cha}} = \pi r_w^2 h_{w,\text{cha}} \tag{3.521c}$$

where $h_{w,\text{cha}}$ is the wellbore water depth [m] in the chaser phase.

In the extraction phase, the boundary condition is (Wang *et al.*, 2018):

$$V_{w,\text{ext}} \left.\frac{dC_{\text{ext},m}}{dt}\right|_{r=r_w} = -\xi \alpha_r v_a(r_w) \left.\frac{dC_{\text{ext},m}}{dt}\right|_{r=r_w}, \quad t_{\text{res}} < t \leq t_{\text{ext}} \tag{3.522a}$$

$$C_{\text{ext},m}(t)|_{t=t_{\text{res}}} = C_{\text{res},m}(t)|_{t=t_{\text{res}}}, \quad t_{\text{res}} < t \leq t_{\text{ext}} \tag{3.522b}$$

$$V_{w,\text{ext}} = \pi r_w^2 h_{w,\text{ext}} \tag{3.522c}$$

where $h_{w,\text{ext}}$ is the wellbore water depth [m] in the extraction phase.

The flow problem must be solved first before investigating the transport problem of the SWPP test. The velocity involved in the advection and dispersion terms of Eqs. (3.509a) and (3.509b) is

$$v_a(r_w) = \frac{Q}{4\pi r_w B \theta_m}, \quad r \geq r_w \tag{3.523}$$

where Q is the pumping rate [m^3/s], and it is negative for injection and positive for pumping. The use of Eq. (3.523) implies that quasi-steady state flow can be established very quickly near the injection/pumping well. Thus, the flow velocity becomes independent of time. This approximation is generally acceptable given the very limited spatial range of influence of most SWPP tests.

The water levels in the wellbore in Eqs. (3.520)–(3.522) could be calculated by the models of Moench (1985):

$$h_w = \lim_{t\to\infty} \{\mathcal{L}^{-1}[\overline{h}_w(p)]\} \tag{3.524}$$

where p is Laplace transform variable; $\mathcal{L}^{-1}$ represents the inverse Laplace transform; the over bar represents the Laplace-domain variable, and

$$\overline{h}_w(p) = h_0 - \frac{Q}{8\pi KB}\frac{2[K_0(x) + xS_wK_1(x)]}{p\{pW_D[K_0(x) + xS_wK_1(x)] + xK_1(x)\}} \tag{3.525}$$

$$W_D = \frac{1}{4BS_a} \tag{3.526}$$

$$x = \frac{(p + \overline{q})}{2} \tag{3.527}$$

$$\overline{q} = (\gamma')^2 m'\coth(m') + (\gamma'')^2 m''\coth(m'') \tag{3.528}$$

$$m' = \frac{\left(\frac{S_uB_up}{S_aB}\right)^{1/2}}{\gamma'} \tag{3.529}$$

$$m'' = \frac{\left(\frac{S_lB_lp}{S_aB}\right)^{1/2}}{\gamma''} \tag{3.530}$$

$$\gamma' = r_w\left(\frac{K_u}{2K_aBB_u}\right)^{1/2} \tag{3.531}$$

$$\gamma'' = r_w\left(\frac{K_l}{2K_aBB_l}\right)^{1/2} \tag{3.532}$$

where K_u and K_l are hydraulic conductivities [m/s]; S_u and S_l are specific storages [1/s]; S_w is the wellbore skin factor [dimensionless]; B_u and B_l are thicknesses [m]; $K_0(\cdot)$ and $K_1(\cdot)$ are the modified Bessel functions.

(b) Dimensionless transform

To reduce the complexity in analyzing the influence of input parameters on the output, the dimensionless parameters are introduced as follows:

$$C_{mD} = \frac{C_m}{C_0}, \quad C_{imD} = \frac{C_{im}}{C_0}, \quad C_{\text{inj},mD} = \frac{C_{\text{inj},m}}{C_0},$$

$$C_{\text{inj},imD} = \frac{C_{\text{inj},im}}{C_0}, \quad C_{\text{cha},mD} = \frac{C_{\text{cha},m}}{C_0}, \quad C_{\text{cha},imD} = \frac{C_{\text{cha},im}}{C_0},$$

$$C_{\text{res},mD} = \frac{C_{\text{res},m}}{C_0}, \quad C_{\text{res},imD} = \frac{C_{\text{res},im}}{C_0}, \quad C_{\text{ext},mD} = \frac{C_{\text{ext},m}}{C_0},$$

$$C_{\text{ext},imD} = \frac{C_{\text{ext},im}}{C_0}, \quad C_{umD} = \frac{C_{um}}{C_0}, \quad C_{\text{uim}D} = \frac{C_{\text{uim}}}{C_0},$$

$$C_{lmD} = \frac{C_{lm}}{C_0}, \quad C_{\text{lim}D} = \frac{C_{\text{lim}}}{C_0}, \quad t_D = \frac{|A|}{\alpha_r^2 R_m} t, \quad r_D = \frac{r}{\alpha_r},$$

$$r_{wD} = \frac{r_w}{\alpha_r}, \quad z_D = \frac{z}{B}, \quad \mu_{mD} = \frac{\alpha_r^2 \mu_m}{A}, \quad \mu_{imD} = \frac{\alpha_r^2 R_m \mu_{im}}{R_{im} A},$$

$$\mu_{umD} = \frac{\alpha_r^2 \mu_{um}}{A}, \quad \mu_{\text{uim}D} = \frac{\alpha_r^2 R_m \mu_{\text{uim}}}{R_{im} A}, \quad \mu_{lmD} = \frac{\alpha_r^2 \mu_{lm}}{A} \quad \text{and}$$

$$\mu_{\text{lim}D} = \frac{\alpha_r^2 R_m \mu_{\text{lim}}}{R_{im} A}, \quad \text{and} \quad A = \frac{Q}{4\pi B \theta_m} \tag{3.533}$$

By substituting these dimensionless parameters into the governing equations, one could obtain the dimensionless model of the SWPP test:

$$\begin{aligned} \frac{\partial C_{mD}}{\partial t_D} &= \frac{1}{r_D}\frac{\partial^2 C_{mD}}{\partial r_D^2} - \frac{1}{r_D}\frac{\partial C_{mD}}{\partial r_D} - \varepsilon_m (C_{mD} - C_{imD}) - \mu_{mD} C_{mD} \\ &\quad - \left(\frac{\theta_{um}\alpha_r^2 v_{um}}{2A\theta_m B} C_{umD} - \frac{\theta_{um}\alpha_r^2 D_u}{2A\theta_m B^2}\frac{\partial C_{umD}}{\partial z_D} \right)\Bigg|_{z_D=1} \\ &\quad + \left(\frac{\theta_{lm}\alpha_r^2 v_{lm}}{2AB\theta_m} C_{lmD} - \frac{\theta_{lm}\alpha_r^2 D_l}{2AB^2\theta_m}\frac{\partial C_{lmD}}{\partial z_D} \right)\Bigg|_{z_D=-1}, \\ &\quad r_D \geq r_{wD} \end{aligned} \tag{3.534a}$$

$$\frac{\partial C_{imD}}{\partial t_D} = \varepsilon_{im}(C_{mD} - C_{imD}) - \mu_{imD} C_{imD}, \quad r_D \geq r_{wD} \tag{3.534b}$$

$$\begin{aligned} \frac{\partial C_{umD}}{\partial t_D} &= \frac{R_m \alpha_r^2 D_u}{AB^2 R_{um}}\frac{\partial^2 C_{umD}}{\partial z_D^2} - \frac{R_m v_{um}\alpha_r^2}{ABR_{um}}\frac{\partial C_{um}D}{\partial z_D} \\ &\quad - \varepsilon_{um}(C_{umD} - C_{\text{uim}D}) - \mu_{umD} C_{umD}, \quad z_D \geq 1 \end{aligned} \tag{3.535a}$$

$$\frac{\partial C_{\text{uim}D}}{\partial t_D} = \varepsilon_{\text{uim}}(C_{umD} - C_{\text{uim}D}) - \mu_{\text{uim}D} C_{\text{uim}D}, \quad z_D \geq 1 \tag{3.535b}$$

$$\frac{\partial C_{lmD}}{\partial t_D} = \frac{R_m \alpha_r^2 D_l}{AB^2 R_{lm}} \frac{\partial^2 C_{lmD}}{\partial z_D^2} + \frac{R_m v_{lm} \alpha_r^2}{ABR_{lm}} \frac{\partial C_{lmD}}{\partial z_D} - \varepsilon_{lm}(C_{lmD} - C_{\text{lim}D}) - \mu_{lmD} C_{lmD}, \quad z_D \leq -1 \tag{3.536a}$$

$$\frac{\partial C_{\text{lim}D}}{\partial t_D} = \varepsilon_{\text{lim}}(C_{lmD} - C_{\text{lim}D}) - \mu_{\text{lim}D} C_{\text{lim}D}, \quad z_D \leq -1 \tag{3.536b}$$

where

$$\varepsilon_m = \frac{\omega_a \alpha_r^2}{A\theta_m}, \quad \varepsilon_{im} = \frac{\omega_a \alpha_r^2 R_m}{A\theta_m R_{im}}, \quad \varepsilon_{um} = \frac{\omega_u \alpha_r^2 R_m}{A\theta_{um} R_{um}},$$

$$\varepsilon_{\text{uim}} = \frac{\omega_u \alpha_r^2 R_m}{A\theta_{um} R_{\text{uim}}}, \quad \varepsilon_{lm} = \frac{\omega_l \alpha_r^2 R_m}{A\theta_{lm} R_{lm}}, \quad \varepsilon_{\text{lim}} = \frac{\omega_l \alpha_r^2 R_m}{A\theta_{lm} R_{\text{lim}}}.$$

Substituting the dimensionless parameters into Eqs. (3.513) and (3.514), one could obtain the dimensionless boundary conditions and dimensionless initial conditions for the injection phase:

$$\begin{aligned} C_{mD}(r_D, t_D)|_{t_D=0} &= C_{imD}(r_D, t_D)|_{t_D=0} = C_{umD}(r_D, z_D, t_D)|_{t_D=0} \\ &= C_{\text{uim}D}(r_D, z_D, t_D)|_{t_D=0} \\ &= C_{lmD}(r_D, z_D, t_D)|_{t_D=0} \\ &= C_{\text{lim}D}(r_D, z_D, t_D)|_{t_D=0} = 0 \end{aligned} \tag{3.537}$$

$$\begin{aligned} C_{mD}(r_D, t_D)|_{r_D\to\infty} &= C_{imD}(r_D, t_D)|_{r_D\to\infty} \\ &= C_{umD}(r_D, z_D, t_D)|_{z_D\to\infty} \\ &= C_{\text{uim}D}(r_D, z_D, t_D)|_{z_D\to\infty} \\ &= C_{lmD}(r_D, z_D, t)|_{z_D\to-\infty} \\ &= C_{\text{lim}D}(r_D, z_D, t_D)|_{z_D\to-\infty} = 0 \end{aligned} \tag{3.538}$$

$$C_{mD}(r_D, t_D) = C_{umD}(r_D, z_D = 1, t_D) \tag{3.539a}$$

$$C_{mD}(r_D, t_D) = C_{lmD}(r_D, z_D = -1, t_D) \tag{3.539b}$$

The analytical solution will be derived using the Laplace transform method and Green's functions method, and detailed information is provided in the following sections.

(c) Analytical solution in the injection phase

Conducting Laplace transform to Eqs. (3.535a) and (3.535b), one has

$$s\overline{C}_{umD} = \frac{R_m\alpha_r^2 D_u}{AB^2 R_{um}}\frac{\partial^2 \overline{C}_{umD}}{\partial z_D^2} - \frac{R_m v_{um}\alpha_r^2}{ABR_{um}}\frac{\partial \overline{C}_{umD}}{\partial z_D} - (\varepsilon_{um} + \mu_{umD})\overline{C}_{umD} + \varepsilon_{um}\overline{C}_{\text{uim}D}, \quad z_D \geq 1 \tag{3.540a}$$

$$s\overline{C}_{\text{uim}D} = \varepsilon_{\text{uim}}(\overline{C}_{umD} - \overline{C}_{\text{uim}D}) - \mu_{\text{uim}D}\overline{C}_{\text{uim}D}, \quad z_D \geq 1 \tag{3.540b}$$

Substituting Eq. (3.540b) into Eq. (3.540a) will lead to:

$$s\overline{C}_{umD} = \frac{R_m\alpha_r^2 D_u}{AB^2 R_{um}}\frac{\partial^2 \overline{C}_{umD}}{\partial z_D^2} - \frac{R_m v_{um}\alpha_r^2}{ABR_{um}}\frac{\partial \overline{C}_{umD}}{\partial z_D} - \left(\varepsilon_{um} + \mu_{umD} - \frac{\varepsilon_{um}\varepsilon_{\text{uim}}}{s + \mu_{\text{uim}D} + \varepsilon_{\text{uim}}}\right)\overline{C}_{umD}, \quad z_D \geq 1 \tag{3.541}$$

Similarly, Eqs. (3.536a) and (3.536b) become:

$$s\overline{C}_{lmD} = \frac{R_m\alpha_r^2 D_l}{AB^2 R_{lm}}\frac{\partial^2 \overline{C}_{lmD}}{\partial z_D^2} + \frac{R_m v_{lm}\alpha_r^2}{ABR_{lm}}\frac{\partial \overline{C}_{lmD}}{\partial z_D} - (\varepsilon_{lm} + \mu_{lmD})\overline{C}_{lmD} + \varepsilon_{lm}\overline{C}_{\text{lim}D}, \quad z_D \leq -1 \tag{3.542a}$$

$$s\overline{C}_{\text{lim}D} = \varepsilon_{\text{lim}}(\overline{C}_{lmD} - \overline{C}_{\text{lim}D}) - \mu_{\text{lim}D}\overline{C}_{\text{lim}D}, \quad z_D \leq -1 \tag{3.542b}$$

Substituting Eq. (3.542b) into Eq. (3.542a) results in

$$s\overline{C}_{lmD} = \frac{R_m\alpha_r^2 D_l}{AB^2 R_{lm}}\frac{\partial^2 \overline{C}_{lmD}}{\partial z_D^2} + \frac{R_m v_{lm}\alpha_r^2}{ABR_{lm}}\frac{\partial \overline{C}_{lmD}}{\partial z_D} - \left(\varepsilon_{lm} + \mu_{lmD} - \frac{\varepsilon_{lm}\varepsilon_{\text{lim}}}{s + \mu_{\text{lim}D} + \varepsilon_{\text{lim}}}\right)\overline{C}_{lmD}, \quad z_D \leq -1 \tag{3.543}$$

Equations (3.538), (3.539a), (3.539b), and (3.541) compose a model of the second-order ordinary differential equation (ODE) with boundary conditions, the general solution of Eq. (3.541) is

$$\overline{C}_{umD} = A_1 \exp(b_1 z_D) + B_1 \exp(b_2 z_D) \tag{3.544a}$$

Similarly, the general solution of Eq. (3.543) is

$$\overline{C}_{lmD} = A_2 \exp(b_1 z_D) + B_2 \exp(b_2 z_D) \tag{3.544b}$$

where

$$a_1 = \frac{\frac{R_m v_{um} \alpha_r^2}{ABR_{um}} + \sqrt{\begin{array}{l}\left(\frac{R_m v_{um} \alpha_r^2}{ABR_{um}}\right)^2 + 4\frac{R_m \alpha_r^2 D_u}{AB^2 R_{um}} \\ \times \left(s + \varepsilon_{um} + \mu_{umD} - \frac{\varepsilon_{um}\varepsilon_{\text{uim}}}{s+\mu_{\text{uim}D}+\varepsilon_{\text{uim}}}\right)\end{array}}}{2\frac{R_m \alpha_r^2 D_u}{AB^2 R_{um}}},$$

$$a_2 = \frac{\frac{R_m v_{um} \alpha_r^2}{ABR_{um}} - \sqrt{\begin{array}{l}\left(\frac{R_m v_{um} \alpha_r^2}{ABR_{um}}\right)^2 + 4\frac{R_m \alpha_r^2 D_u}{AB^2 R_{um}} \\ \times \left(s + \varepsilon_{um} + \mu_{umD} - \frac{\varepsilon_{um}\varepsilon_{\text{uim}}}{s+\mu_{\text{uim}D}+\varepsilon_{\text{uim}}}\right)\end{array}}}{2\frac{R_m \alpha_r^2 D_u}{AB^2 R_{um}}},$$

$$b_1 = \frac{-\frac{R_m v_{lm} \alpha_r^2}{ABR_{lm}} + \sqrt{\begin{array}{l}\left(\frac{R_m v_{lm} \alpha_r^2}{ABR_{lm}}\right)^2 + 4\frac{R_m \alpha_r^2 D_l}{AB^2 R_{lm}} \\ \times \left(s + \varepsilon_{lm} + \mu_{lmD} - \frac{\varepsilon_{lm}\varepsilon_{\text{lim}}}{s+\mu_{\text{lim}D}+\varepsilon_{\text{lim}}}\right)\end{array}}}{2\frac{R_m \alpha_r^2 D_l}{AB^2 R_{lm}}} \quad \text{and}$$

$$b_2 = \frac{-\frac{R_m v_{lm} \alpha_r^2}{ABR_{lm}} - \sqrt{\begin{array}{l}\left(\frac{R_m v_{lm} \alpha_r^2}{ABR_{lm}}\right)^2 + 4\frac{R_m \alpha_r^2 D_l}{AB^2 R_{lm}} \\ \times \left(s + \varepsilon_{lm} + \mu_{lmD} - \frac{\varepsilon_{lm}\varepsilon_{\text{lim}}}{s+\mu_{\text{lim}D}+\varepsilon_{\text{lim}}}\right)\end{array}}}{2\frac{R_m \alpha_r^2 D_l}{AB^2 R_{lm}}}.$$

Substituting Eqs. (3.544a) and (3.544b) into Eqs. (3.538) and (3.539b) leads to:

$$\overline{C}_{umD} = B_1 \exp(a_2 z_D) \tag{3.545a}$$

$$\overline{C}_{lmD} = A_2 \exp(b_1 z_D) \tag{3.545b}$$

where $B_1 = \overline{C}_{mD} \exp(-a_2)$, $B_2 = 0$, $A_1 = 0$ and $A_2 = \overline{C}_{mD} \exp(b_1)$.

Thus, we could obtain the solutions for the aquitards as

$$\overline{C}_{umD} = \overline{C}_{mD}\exp(a_2 z_D - a_2) \tag{3.546a}$$

$$\overline{C}_{\text{uim}D} = \frac{\varepsilon_{\text{uim}}}{s + \varepsilon_{\text{uim}} + \mu_{\text{uim}D}}\overline{C}_{umD} \tag{3.546b}$$

$$\overline{C}_{lmD} = \overline{C}_{mD}\exp(b_1 z_D + b_1) \tag{3.547a}$$

$$\overline{C}_{\text{lim}D} = \frac{\varepsilon_{\text{lim}}}{s + \varepsilon_{\text{lim}} + \mu_{\text{lim}D}}\overline{C}_{lmD} \tag{3.547b}$$

In the injection phase, the dimensional boundary conditions Eq. (3.516) and Eqs. (3.520a) and (3.520b) are transformed into their dimensionless forms:

$$\left[C_{mD} - \frac{\partial C_{mD}(r_D, t_D)}{\partial r_D}\right]\Bigg|_{r=r_{wD}} = C_{\text{inj},mD}(t_D), \quad 0 < t_D \le t_{\text{inj},D} \tag{3.548}$$

$$\beta_{\text{inj}}\frac{dC_{\text{inj},mD}(t_D)}{dt_D} = 1 - C_{\text{inj},mD}(t_D), \quad 0 < t_D \le t_{\text{inj},D} \tag{3.549a}$$

$$C_{\text{inj},mD}(t_D = 0) = 0 \tag{3.549b}$$

where $\beta_{\text{inj}} = \frac{V_{w,\text{inj}} r_{wD}}{\xi R_m \alpha_r}$.

Conducting Laplace transform to Eqs. (3.534a)–(3.534b), one has

$$\begin{aligned} s\overline{C}_{mD} = {} & \frac{1}{r_D}\frac{\partial^2 \overline{C}_{mD}}{\partial r_D^2} - \frac{1}{r_D}\frac{\partial \overline{C}_{mD}}{\partial r_D} - (\varepsilon_m + \mu_{mD})\overline{C}_{mD} + \varepsilon_m \overline{C}_{imD} \\ & - \left(\frac{\theta_{um}\alpha_r^2 v_{um}}{2A\theta_m B}\overline{C}_{umD} - \frac{\theta_{um}\alpha_r^2 D_u}{2A\theta_m B^2}\frac{\partial \overline{C}_{umD}}{\partial z_D}\right)\Bigg|_{z_D=1} \\ & + \left(\frac{\theta_{lm}\alpha_r^2 v_{lm}}{2A\theta_m B}\overline{C}_{lmD} - \frac{\theta_{lm}\alpha_r^2 D_l}{2AB^2\theta_m}\frac{\partial \overline{C}_{lmD}}{\partial z_D}\right)\Bigg|_{z_D=-1}, \quad r_D \ge r_{wD} \end{aligned} \tag{3.550a}$$

$$\overline{C}_{imD} = \frac{\varepsilon_{im}}{(s + \mu_{imD} + \varepsilon_{im})}\overline{C}_{mD}, \quad r_D \ge r_{wD} \tag{3.550b}$$

Substituting Eqs. (3.546a), (3.547a), and (3.550b) into Eq. (3.550a), one has

$$\frac{1}{r_D}\frac{\partial^2 \overline{C}_{mD}}{\partial r_D^2} - \frac{1}{r_D}\frac{\partial \overline{C}_{mD}}{\partial r_D} - E\overline{C}_{mD} = 0 \tag{3.551}$$

where

$$E = s + \varepsilon_m + \mu_{mD} - \frac{\varepsilon_m \varepsilon_{im}}{s + \mu_{imD} + \varepsilon_{im}} + \frac{\theta_{um}\alpha_r^2 v_{um}}{2A\theta_m B} - \frac{\theta_{lm}\alpha_r^2 v_{lm}}{2AB\theta_m}$$
$$- \frac{a_2\theta_{um}\alpha_r^2 D_u}{2A\theta_m B^2} + \frac{b_1\theta_{lm}\alpha_r^2 D_l}{2AB^2\theta_m}.$$

The boundary conditions of the wellbore and infinity in the Laplace domain are

$$\left[\overline{C}_{mD} - \frac{\partial \overline{C}_{mD}(r_D, s)}{\partial r_D}\right]\Bigg|_{r=r_{wD}} = \overline{C}_{\text{inj},mD}(s) \quad (3.552\text{a})$$

$$\overline{C}_{mD}(r_D, s)\big|_{r_D\to\infty} = 0 \quad (3.552\text{b})$$

Conducting Laplace transform on Eqs. (3.549a)–(3.549b), one has

$$\overline{C}_{\text{inj},mD}(r_w, s) = \frac{1}{s(s\beta_{\text{inj}} + 1)} \quad (3.553)$$

Equations (3.551), (3.552a)–(3.552b), and (3.553) compose a model of the second-order ordinary differential equation (ODE) with boundary conditions. The general solution of Eq. (3.551) is

$$\overline{C}_{mD}(r_D, s) = \phi_1 \exp\left(\frac{y_{\text{inj}}}{2}\right) A_i\left(E^{1/3} y_{\text{inj}}\right)$$
$$+ \phi_2 \exp\left(\frac{y_{\text{inj}}}{2}\right) B_i(E^{1/3} y_{\text{inj}}) \quad (3.554)$$

where $y_{\text{inj}} = r_D + \frac{1}{4E}$, $y_{\text{inj},w} = r_{wD} + \frac{1}{4E}$; ϕ_1 and ϕ_2 are constants which the boundary conditions could determine; $A_i(\cdot)$ and $B_i(\cdot)$ are the Airy functions of the first kind and second kind, respectively. As $B_i(r_D)$ diverges when $r_D \to \infty$, ϕ_2 must be zero.

Substituting Eqs. (3.554), (3.553) and $\phi_2 = 0$ into Eq. (3.552a), the value of ϕ_1 is

$$\phi_1 = \frac{1}{s(s\beta_{\text{inj}} + 1)} \frac{1}{\exp(\frac{y_{\text{inj},w}}{2})[\frac{A_i(E^{1/3} y_{\text{inj},w})}{2} - E^{1/3} A_i'(E^{1/3} y_{\text{inj}})]} \quad (3.555)$$

where $A_i'(\cdot)$ is the derivative of the Airy function.

Substituting Eq. (3.555) and $\phi_2 = 0$ into Eqs. (3.554) and (3.550b), one could obtain the Laplace-domain analytical solution of solute transport in the injection phase of the SWPP test.

(d) Analytical solution in the chasing phase

For the chaser phase, conducting Laplace transform on Eqs. (3.535a) and (3.535b), one has

$$\frac{R_m\alpha_r^2 D_u}{AB^2 R_{um}}\frac{\partial^2 \overline{C}_{umD}}{\partial z_D^2} - \frac{R_m v_{um}\alpha_r^2}{ABR_{um}}\frac{\partial \overline{C}_{umD}}{\partial z_D} - (s+\varepsilon_{um}+\mu_{umD})\overline{C}_{umD}$$
$$+\varepsilon_{um}\overline{C}_{\text{uim}D} + C_{umD}(r_D, z_D, t_{\text{inj},D}) = 0, \quad z_D \geq 1, \tag{3.556a}$$
$$s\overline{C}_{\text{uim}D} - C_{\text{uim}D}(r_D, z_D, t_{\text{inj},D})$$
$$= \varepsilon_{\text{uim}}(\overline{C}_{umD} - \overline{C}_{\text{uim}D}) - \mu_{\text{uim}D}\overline{C}_{\text{uim}D} \tag{3.556b}$$

Equation (3.556b) could be rewritten as

$$\overline{C}_{\text{uim}D} = \frac{\varepsilon_{\text{uim}}}{s+\varepsilon_{\text{uim}}+\mu_{\text{uim}D}}\overline{C}_{umD} + \frac{C_{\text{uim}D}(r_D, z_D, t_{\text{inj},D})}{s+\varepsilon_{\text{uim}}+\mu_{\text{uim}D}} \tag{3.556c}$$

Substituting Eq. (3.556c) into Eq. (3.556a), one has

$$\frac{R_m\alpha_r^2 D_u}{AB^2 R_{um}}\frac{\partial^2 \overline{C}_{umD}}{\partial z_D^2} - \frac{R_m v_{um}\alpha_r^2}{ABR_{um}}\frac{\partial \overline{C}_{umD}}{\partial z_D}$$
$$-\left(s+\varepsilon_{um}+\mu_{umD} - \frac{\varepsilon_{um}\varepsilon_{\text{uim}}}{s+\varepsilon_{\text{uim}}+\mu_{\text{uim}D}}\right)\overline{C}_{umD}$$
$$+C_{umD}(r_D, z_D, t_{\text{inj},D}) + \frac{\varepsilon_{um}C_{\text{uim}D}(r_D, z_D, t_{\text{inj},D})}{s+\varepsilon_{\text{uim}}+\mu_{\text{uim}D}} = 0.\ z_D \geq 1 \tag{3.557}$$

Similarly, Eqs. (3.536a) and (3.536b) become:

$$\frac{R_m\alpha_r^2 D_l}{AB^2 R_{lm}}\frac{\partial^2 \overline{C}_{lmD}}{\partial z_D^2} + \frac{R_m v_{lm}\alpha_r^2}{ABR_{lm}}\frac{\partial \overline{C}_{lmD}}{\partial z_D} - (s+\varepsilon_{lm}+\mu_{lmD})\overline{C}_{lmD}$$
$$+\varepsilon_{lm}\overline{C}_{\text{lim}D} + C_{lmD}(r_D, z_D, t_{\text{inj},D}) = 0, \quad z_D \leq -1 \tag{3.558a}$$
$$s\overline{C}_{\text{lim}D} - C_{\text{lim}D}(r_D, z_D, t_{\text{inj},D})$$
$$= \varepsilon_{\text{lim}}(\overline{C}_{lmD} - \overline{C}_{\text{lim}D}) - \mu_{\text{lim}D}C_{\text{lim}D} \tag{3.558b}$$

Equation (3.556b) could be rewritten as

$$\overline{C}_{\text{lim}D} = \frac{\varepsilon_{\text{lim}}}{s + \varepsilon_{\text{lim}} + \mu_{\text{lim}D}} \overline{C}_{lmD} + \frac{C_{\text{lim}D}(r_D, z_D, t_{\text{inj},D})}{s + \varepsilon_{\text{lim}} + \mu_{\text{lim}D}} \quad (3.558c)$$

Substituting Eq. (3.558c) into Eq. (3.558a), one has

$$\frac{R_m \alpha_r^2 D_l}{AB^2 R_{lm}} \frac{\partial^2 \overline{C}_{lmD}}{\partial z_D^2} + \frac{R_m v_{lm} \alpha_r^2}{ABR_{lm}} \frac{\partial \overline{C}_{lmD}}{\partial z_D}$$
$$- \left(s + \varepsilon_{lm} + \mu_{lmD} - \frac{\varepsilon_{lm}\varepsilon_{\text{lim}}}{s + \varepsilon_{\text{lim}} + \mu_{\text{lim}D}} \right) \overline{C}_{lmD}$$
$$+ C_{lmD}(r_D, z_D, t_{\text{inj},D}) + \frac{\varepsilon_{lm} C_{\text{lim}D}(r_D, z_D, t_{\text{inj},D})}{s + \varepsilon_{\text{lim}} + \mu_{\text{lim}D}} = 0. \; z_D \leq -1 \quad (3.559)$$

where $C_{umD}(r_D, z_D, t_{\text{inj},D})$ and $C_{\text{uim}D}(r_D, z_D, t_{\text{inj},D})$ are respectively the mobile and immobile concentrations [Kg/m^3] of the upper aquitard at the end of the injection phase, $C_{lmD}(r_D, z_D, t_{\text{inj},D})$ and $C_{\text{lim}D}(r_D, z_D, t_{\text{inj},D})$ are respectively the mobile and immobile concentrations [Kg/m^3] of the lower aquitard at the end of the injection phase. In this section, we use the Green's function method to derive the analytical solution of Eqs. (3.557) and (3.559).

Notice that the boundary condition of Eq. (3.539a) is inhomogeneous; thus, we need to homogenize it first. Letting $\overline{C}_{umD} = \hbar(z_D) + \jmath_1 + \jmath_2 z_D$, and substituting them into Eqs. (3.538) and (3.539a) yields:

$$[\hbar(z_D)]|_{z_D \to \infty} = 0 \quad (3.560a)$$

$$[\hbar(z_D)]|_{z_D = 1} = 0 \quad (3.560b)$$

where $\jmath_1 = -\jmath_2 z_{eD}$ and $\jmath_2 = \frac{\overline{C}_{mD}(r_D, s)}{1 - z_{eD}}$.

Defining the spatial operator:

$$L_u = - \left[\frac{R_m \alpha_r^2 D_u}{AB^2 R_{um}} \frac{d^2}{dz_D^2} - \frac{R_m v_{um} \alpha_r^2}{ABR_{um}} \frac{d}{dz_D} - E_u \right],$$

one has

$$L_u \overline{C}_{umD} = L_u[\hbar(z_D) + \jmath_1] = F_u(z_D) \quad (3.561)$$

Let $f_u(z_D) = F_u(z_D) - L_u[\mathcal{s}_1 + \mathcal{s}_2 z_D]$, one has

$$\frac{R_m \alpha_r^2 D_u}{AB^2 R_{um}} \frac{d^2 \mathcal{k}}{dz_D^2} - \frac{R_m v_{um} \alpha_r^2}{ABR_{um}} \frac{d\mathcal{k}}{dz_D} - E_u \mathcal{k} = -f_u(z_D) \tag{3.562}$$

where

$$E_u = s + \varepsilon_{um} + \mu_{umD} - \frac{\varepsilon_{um}\varepsilon_{\text{uim}}}{s + \varepsilon_{\text{uim}} + \mu_{\text{uim}D}},$$

$$F_u(z_D) = C_{umD}(r_D, z_D, t_{\text{inj},D}) + \frac{\varepsilon_{um} C_{\text{uim}D}(r_D, z_D, t_{\text{inj},D})}{s + \varepsilon_{\text{uim}} + \mu_{\text{uim}D}} \quad \text{and}$$

$$f_u(z_D) = C_{umD}(r_D, z_D, t_{\text{inj},D}) + \frac{\varepsilon_{um} C_{\text{uim}D}(r_D, z_D, t_{\text{inj},D})}{s + \varepsilon_{\text{uim}} + \mu_{\text{uim}D}}$$
$$- \frac{R_m v_{um} \alpha_r^2}{ABR_{um}} \mathcal{s}_2 - E_u(\mathcal{s}_1 + \mathcal{s}_2 z_D).$$

The general solution of Eq. (3.557) is

$$\overline{C}_{umD} = \int_1^\infty g_u(z_D, E_u; \eta_u) f_u(\eta_u) d\eta_u + \frac{z_D - z_{eD}}{1 - z_{eD}} \overline{C}_{mD}(r_D, s),$$
$$z_D \geq 1 \tag{3.563}$$

where $f_u(\eta_u) = C_{umD}(r_D, \eta_u, t_{\text{inj},D}) + \frac{\varepsilon_{um} C_{\text{uim}D}(r_D, \eta_u, t_{\text{inj},D})}{s + \varepsilon_{\text{uim}} + \mu_{\text{uim}D}} - \frac{R_m v_{um} \alpha_r^2}{ABR_{um}} \mathcal{s}_2 - E_u(\mathcal{s}_1 + \mathcal{s}_2 \eta_u)$, η_u is a positive value varying between 1 and ∞ (e.g., $1 \leq \eta_u \leq \infty$); $g_u(z_D, E_u; \eta_u)$ is the Green's function, and could be expressed as

$$g_u(z_D, E_u; \eta_u) = \begin{cases} g_{u1}(z_D, E_u; \eta_u) = N_1 \exp(a_1 z_D) + N_2 \exp(a_2 z_D) \\ \qquad 1 \leq z_D < \eta_u \\ g_{u2}(z_D, E_u; \eta_u) = N_3 \exp(a_1 z_D) + N_4 \exp(a_2 z_D) \\ \qquad \eta_u \leq z_D < \infty \end{cases} \tag{3.564}$$

where N_1, N_2, N_3 and N_4 are coefficients to be determined using the following conditions (Chen and Woodside, 1988):

$g_u(z_D, E_u; \eta_u)$ satisfies the model of Eqs. (3.560a), (3.560b) and (3.562);

$$g_{u1}(z_D, E_u; \eta_u) = g_{u2}(z_D, E_u; \eta_u);$$
$$\left.\frac{dg_{u2}}{dz_D}\right|_{z_D = \eta_u^+} - \left.\frac{dg_{u1}}{dz_D}\right|_{z_D = \eta_u^-} = -\frac{AB^2 R_{um}}{R_m \alpha_r^2 D_u};$$

Substituting Eq. (3.564) into Eq. (3.560a), one has

$$N_3 = 0 \tag{3.565}$$

Substituting Eq. (3.564) into Eq. (3.560b), one has

$$N_1 \exp(a_1) + N_2 \exp(a_2) = 0 \tag{3.566a}$$

According to Eq. (3.566a), one has

$$N_1 = -N_2 \exp(a_2 - a_1) \tag{3.566b}$$

According to the above condition of b), one has

$$N_1 \exp(a_1\eta_u) + N_2 \exp(a_2\eta_u) = N_4 \exp(a_2\eta_u) \tag{3.567}$$

According to the above condition of c), one has

$$N_4 a_2 \exp(a_2\eta_u) - [N_1 a_1 \exp(a_1\eta_u) + N_2 a_2 \exp(a_2\eta_u)] = -\frac{AB^2 R_{um}}{R_m \alpha_r^2 D_u} \tag{3.568}$$

In the chaser phase, the values of N_1, N_2, N_3 and N_4 could be determined by Eqs. (3.566a)–(3.568), namely:

$$N_1 = -N_2 \exp(a_2 - a_1),$$

$$N_2 = \frac{-AB^2 R_{um}}{R_m \alpha_r^2 D_u [(a_1 - a_2)\exp(a_2 - a_1)\exp(a_1\eta_u)]}, \quad N_3 = 0 \quad \text{and}$$

$$N_4 = N_2 - N_2 \exp(a_2 - a_1)\exp(a_1\eta_u - a_2\eta_u).$$

As for the analytical solution of the lower aquitard, one could use a similar approach as that used for deriving the analytical solution of the upper aquitard, and the general solution of Eq. (3.559) could be described as

$$\overline{C}_{lmD} = \int_{-1}^{-\infty} g_l(z_D, E_l; \eta_l) f_l(\eta_l) d\eta_l + \frac{z_{eD} + z_D}{z_{eD} - 1} \overline{C}_{mD}(r_D, z_D, s),$$

$$z_D \leq -1 \tag{3.569a}$$

$$g_l(z_D, E_l; \eta_l) = \begin{cases} g_{l1}(z_D, E_l; \eta_l) = M_1 \exp(b_1 z_D) + M_2 \exp(b_2 z_D) & -1 \le z_D < \eta_l \\ g_{l2}(z_D, E_l; \eta_l) = M_3 \exp(b_1 z_D) + M_4 \exp(b_2 z_D) & \eta_l \le z_D < -\infty \end{cases} \tag{3.569b}$$

$$f_l(\eta_l) = C_{lmD}(r_D, \eta_l, t_{\text{inj},D}) + \frac{\varepsilon_{lm} C_{\text{lim}D}(r_D, \eta_l, t_{\text{inj},D})}{s + \varepsilon_{\text{lim}} + \mu_{\text{lim}D}} + \frac{R_m v_{lm} \alpha_r^2}{ABR_{lm}} \frac{\overline{C}_{mD}}{z_{eD} - 1} - \overline{C}_{mD} E_l \frac{z_{eD} + \eta_l}{z_{eD} - 1} \tag{3.569c}$$

where η_l is a negative value varying between -1 and $-\infty$ (e.g., $-1 \le \eta_l \le -\infty$); $g_l(z_D, E_l; \eta_l)$ is the Green's function, $E_l = s + \varepsilon_{lm} + \mu_{lmD} - \frac{\varepsilon_{lm}\varepsilon_{\text{lim}}}{s+\varepsilon_{\text{lim}}+\mu_{\text{lim}D}}$, and the values of M_1, M_2, M_3 and M_4 could be described as $M_1 = -M_2 \exp(b_1 - b_2), M_2 = \frac{-AB^2 R_{lm}}{R_m \alpha_r^2 D_l [\exp(b_2\eta_l - b_1\eta_l) - b_2 \exp(b_2 \eta_l)]}$, $M_3 = M_2 \exp(b_2\eta_l - b_1\eta_l) - M_2 \exp(b_1 - b_2)$, $M_4 = 0$, and the values of a_1, a_2, b_1, and b_2 are the same as used in the injection phase.

In the chaser phase, the dimensional boundary conditions Eqs. (3.521a) and (3.521b) are transformed into dimensionless forms as

$$\beta_{\text{cha},D} \left. \frac{\partial C_{mD}(r_D, t_D)}{\partial t_D} \right|_{r_D = r_{wD}} = C_{mD}(r_D, t_D), \quad t_{\text{inj},D} < t_D \le t_{\text{cha},D} \tag{3.570a}$$

$$C_{\text{cha},mD}(r_D, t_D)|_{t_D = t_{\text{inj},D}} = C_{\text{inj},mD}(r_D, t_D)|_{t_D = t_{\text{inj},D}}, \quad t_{\text{inj},D} < t_D \le t_{\text{cha},D} \tag{3.570b}$$

where $\beta_{\text{cha},D} = -\frac{V_{w,\text{cha}} r_{wD}}{\xi R_m \alpha_r}$.

Conducting Laplace transform on Eqs. (3.534a) and (3.534b) in the chaser phase, one has

$$s\overline{C}_{mD} - C_{mD}(r_D, t_{\text{inj},D}) = \frac{1}{r_D} \frac{\partial^2 \overline{C}_{mD}}{\partial r_D^2} - \frac{1}{r_D} \frac{\partial \overline{C}_{mD}}{\partial r_D} - (\varepsilon_m + \mu_{mD}) \overline{C}_{mD} + \varepsilon_m \overline{C}_{imD} - \left. \left(\frac{\theta_{um} \alpha_r^2 v_{um}}{2A\theta_m B} \overline{C}_{umD} - \frac{\theta_{um} \alpha_r^2 D_u}{2A\theta_m B^2} \frac{\partial \overline{C}_{umD}}{\partial z_D} \right) \right|_{z_D = 1}$$

$$+\left(\frac{\theta_{lm}\alpha_r^2 v_{lm}}{2A\theta_m B}\overline{C}_{lmD} - \frac{\theta_{lm}\alpha_r^2 D_l}{2AB^2\theta_m}\frac{\partial \overline{C}_{lmD}}{\partial z_D}\right)\bigg|_{z_D=-1},$$

$$r_D \geq r_{wD} \tag{3.571a}$$

$$\overline{C}_{imD} = \frac{\varepsilon_{im}}{(s+\mu_{imD}+\varepsilon_{im})}\overline{C}_{mD} + \frac{C_{imD}(r_D, t_{\text{inj},D})}{(s+\mu_{imD}+\varepsilon_{im})},$$

$$r_D \geq r_{wD} \tag{3.571b}$$

where $C_{mD}(r_D, t_{\text{inj},D})$ and $C_{imD}(r_D, t_{\text{inj},D})$ are respectively the mobile and immobile concentrations [ML^{-3}] of the aquifer at the end of the injection phase, which could be calculated by Eqs. (3.554) and (3.550b).

After substituting Eqs. (3.563), (3.569a)–(3.569c) and (3.571b) into Eq. (3.571a), one has

$$\frac{1}{r_D}\frac{\partial^2 \overline{C}_{mD}}{\partial r_D^2} - \frac{1}{r_D}\frac{\partial \overline{C}_{mD}}{\partial r_D} - E_a\overline{C}_{mD} + F = 0, \quad r_D \geq r_{wD} \tag{3.572}$$

where

$$E_a = s + \varepsilon_m + \mu_{mD} - \frac{\varepsilon_m\varepsilon_{im}}{s+\mu_{imD}+\varepsilon_{im}} + \frac{\theta_{um}\alpha_r^2 v_{um}}{2A\theta_m B} - \frac{\theta_{lm}\alpha_r^2 v_{lm}}{2AB^2\theta_m}$$

$$- \frac{1}{1-z_{eD}}\frac{\theta_{um}\alpha_r^2 D_u}{2A\theta_m B^2} + \frac{1}{z_{eD}-1}\frac{\theta_{lm}\alpha_r^2 D_l}{2AB^2\theta_m} \quad \text{and}$$

$$F = C_{mD}(r_D, t_{\text{inj},D}) + \frac{\varepsilon_m C_{imD}(r_D, t_{\text{inj}})}{s+\mu_{imD}+\varepsilon_{im}}.$$

The boundary conditions of Eqs. (3.570a) and (3.570b) in Laplace domain becomes:

$$\overline{C}_{\text{cha},mD}(r_{wD}, s) = \frac{\beta_{\text{cha},D}}{s\beta_{\text{cha},D}+1}C_{\text{inj},mD}(r_D, t_D)|_{t_D=t_{\text{inj},D}} \tag{3.573}$$

The boundary conditions of the wellbore and infinity in the Laplace domain are

$$\left[\overline{C}_{mD} - \frac{\partial \overline{C}_{mD}(r_D, s)}{\partial r_D}\right]\bigg|_{r=r_{wD}} = \frac{\beta_{\text{cha},D}}{s\beta_{\text{cha},D}+1}C_{\text{inj},mD}(r_D, t_D)|_{t_D=t_{\text{inj},D}} \tag{3.574a}$$

$$\overline{C}_{\text{cha},mD}(r_{wD}, s)\big|_{r_D\to\infty} = 0 \tag{3.574b}$$

Like the model of the SWPP test in the injection phase, Eqs. (3.572) and (3.573)–(3.574b) compose a model of the second-order ordinary differential equation (ODE) with boundary conditions. However, the governing equation is an inhomogeneous differential equation. In this section, we use the Green's function method to derive the analytical solution of Eq. (3.572).

Notice that the boundary condition of Eq. (3.574a) is inhomogeneous, and we need to homogenize it first. Assigning $\overline{C}_{mD} = \Psi(r_D)+\delta_1+\delta_2 r_D$, and substituting it into Eqs. (3.574a) and (3.574b) yields:

$$\left[\Psi(r_D, s) - \frac{\partial\Psi(r_D, s)}{\partial r_D}\right]\Bigg|_{r=r_{wD}} = 0 \tag{3.575a}$$

$$\Psi(r_D, s)|_{r_D\to\infty} = 0 \tag{3.575b}$$

where

$$\delta_1 = -\frac{\beta_{\text{cha},D}}{s\beta_{\text{cha},D}+1}\frac{r_D|_{r_D\to\infty}}{(r_{wD} - r_D|_{r_D\to\infty} - 1)}C_{\text{inj},mD}(r_D, t_D)|_{t_D=t_{\text{inj},D}}$$

and

$$\delta_2 = \frac{\beta_{\text{cha},D}}{s\beta_{\text{cha},D}+1}\frac{1}{(r_{wD} - r_D|_{r_D\to\infty} - 1)}C_{\text{inj},mD}(r_D, t_D)|_{t_D=t_{\text{inj},D}}.$$

Defining a spatial operator: $L = -\left[\frac{d^2}{dr_D^2} - \frac{d}{dr_D} - r_D E_a\right]$, one has

$$L\overline{C}_{mD} = L[\Psi(r_D) + \delta_1 + \delta_2 r_D] = F r_D \tag{3.576}$$

Let $\varphi(r_D) = Fr_D - L(\delta_1 + \delta_2 r_D)$, one has

$$\frac{\partial^2\Psi}{\partial r_D^2} - \frac{\partial\Psi}{\partial r_D} - r_D E_a \Psi = -\varphi(r_D) \tag{3.577}$$

where $\varphi(r_D) = Fr_D - [\delta_2 + r_D E_a(\delta_1 + \delta_2 r_D)]$.

The general solution of Eqs. (3.575a)–(3.577) is

$$\Psi(r_D, E_a; \eta) = \int_{r_{wD}}^{\infty} g(r_D, E_a; \eta)\varphi(\eta)d\eta. \tag{3.578}$$

where η is a positive value varying between r_{wD} and ∞ (e.g., $r_{wD} \le \eta \le \infty$); $g(r_D, E_a; \eta)$ is the Green's function, and could be expressed as

$$g(r_D, E_a; \eta) = \begin{cases} g_1(r_D, E_a; \eta) = \mathcal{T}_1 \exp(\frac{y_{\text{cha}}}{2}) A_i \left(E_a{}^{\frac{1}{3}} y_{\text{cha}}\right) \\ +\mathcal{T}_2 \exp\left(\frac{y_{\text{cha}}}{2}\right) B_i \left(E_a{}^{\frac{1}{3}} y_{\text{cha}}\right) r_{wD} \le y_{\text{cha}} \le \eta \\ g_2(r_D, E_a; \eta) = \mathcal{T}_3 \exp(\frac{y_{\text{cha}}}{2}) A_i \left(E_a{}^{\frac{1}{3}} y_{\text{cha}}\right) \\ +\mathcal{T}_4 \exp\left(\frac{y_{\text{cha}}}{2}\right) B_i \left(E_a{}^{\frac{1}{3}} y_{\text{cha}}\right) \eta \le y_{\text{cha}} \le \infty \end{cases} \tag{3.579}$$

where $\varphi(\eta) = F\eta - [\delta_2 + \eta E_a(\delta_1 + \delta_2 \eta)]$, $y_{\text{cha}} = r_D + \frac{1}{4E_a}$. As $B_i(r_D)$ diverges when $r_D \to \infty$, $\mathcal{T}_4$ must be zero. Substituting Eq. (3.578) into Eq. (3.575a), one has

$$\left[g_1 - \frac{\partial g_1}{\partial r_D}\right]\Bigg|_{r_D = r_{wD}} = 0 \tag{3.580}$$

According to Eq. (3.580), one has

$$\mathcal{T}_1 = -\mathcal{T}_2 X. \tag{3.581}$$

where

$$X = \frac{\frac{1}{2} B_i \left(E_a{}^{1/3} y_{\text{cha},w}\right) - E_a{}^{1/3} B_i' \left(E_a{}^{1/3} y_{\text{cha},w}\right)}{\frac{1}{2} A_i \left(E_a{}^{1/3} y_{\text{cha},w}\right) - E_a{}^{1/3} A_i' \left(E_a{}^{1/3} y_{\text{cha},w}\right)}$$

and $y_{\text{cha},w} = r_{wD} + \frac{1}{4E_a}$.

According to the above condition of (b), one has

$$\mathcal{T}_1 A_i \left(E_a{}^{\frac{1}{3}} y_{\text{cha}}|_{r_D = \eta^+}\right) + \mathcal{T}_2 B_i \left(E_a{}^{\frac{1}{3}} y_{\text{cha}}|_{r_D = \eta^+}\right) = \mathcal{T}_3 A_i \left(E_a{}^{1/3} y_{\text{cha}}|_{r_D = \eta^-}\right) \tag{3.582}$$

According to the above condition of (c), one has

$$\left[\frac{1}{2}\mathcal{T}_3 \exp\left(\frac{y_{\text{cha}}}{2}\right) A_i \left(E_a{}^{\frac{1}{3}} y_{\text{cha}}\right) + E_a{}^{\frac{1}{3}} \mathcal{T}_3 \exp\left(\frac{y_{\text{cha}}}{2}\right) A_i' \times \left(E_a{}^{\frac{1}{3}} y_{\text{cha}}\right)\right]\Bigg|_{r_D = \eta^-} - \left[\frac{1}{2}\mathcal{T}_1 \exp\left(\frac{y_{\text{cha}}}{2}\right) A_i \left(E_a{}^{\frac{1}{3}} y_{\text{cha}}\right) + E_a{}^{\frac{1}{3}} \mathcal{T}_1 \exp\left(\frac{y_{\text{cha}}}{2}\right) A_i' \left(E_a{}^{\frac{1}{3}} y_{\text{cha}}\right)\right]\Bigg|_{r_D = \eta^+}$$

$$
\begin{aligned}
&-\left[\frac{1}{2}\mathcal{T}_2\exp\left(\frac{y_{\mathrm{cha}}}{2}\right)B_i\left(E_a{}^{\frac{1}{3}}y_{\mathrm{cha}}\right)\right.\\
&\left.\left.+E_a{}^{\frac{1}{3}}\mathcal{T}_2\exp\left(\frac{y_{\mathrm{cha}}}{2}\right)B_i'\left(E_a{}^{\frac{1}{3}}y_{\mathrm{cha}}\right)\right]\right|_{r_D=\eta^+}=-1 \qquad (3.583)
\end{aligned}
$$

For the solution in the chaser phase, the values of $\mathcal{T}_1, \mathcal{T}_2$,$\mathcal{T}_3$, and $\mathcal{T}_4$ could be determined by Eqs. (3.581)–(3.583), namely:

$$
\mathcal{T}_1=-\frac{\pi A_i(y_{\mathrm{ext}}|_{r_D=\eta^+})}{E_a{}^{1/3}}X, \mathcal{T}_2=\frac{\pi A_i(y_{\mathrm{ext}}|_{r_D=\eta^+})}{E_a{}^{1/3}},
$$

$$
\mathcal{T}_3=\frac{\pi A_i(y_{\mathrm{ext}}|_{r_D=\eta^+})}{E_a{}^{1/3}}\left[\frac{B_i(y_{\mathrm{ext}}|_{r_D=\eta^+})}{A_i(y_{\mathrm{ext}}|_{r_D=\eta^+})}-X\right] \quad \text{and} \quad \mathcal{T}_4=0.
$$

(e) Analytical solution in the resting phase

In the rest phase, the flow velocity becomes zero, and the advection and dispersion terms drop out of the governing equations. After conducting Laplace transform on Eqs. (3.535a) and (3.535b), the following equations would be obtained:

$$
\begin{aligned}
&(s+\varepsilon_{um}+\mu_{umD})\overline{C}_{umD}\\
&\quad -\varepsilon_{um}\overline{C}_{\mathrm{uim}D}-C_{umD}(r_D,z_D,t_{\mathrm{cha},D})=0,\ z_D\geq 1 \qquad (3.584a)
\end{aligned}
$$

$$
\begin{aligned}
\overline{C}_{\mathrm{uim}D}&=\frac{\varepsilon_{\mathrm{uim}}}{s+\varepsilon_{\mathrm{uim}}+\mu_{umD}}\overline{C}_{umD}+\frac{C_{\mathrm{uim}D}(r_D,z_D,t_{\mathrm{cha},D})}{s+\varepsilon_{\mathrm{uim}}+\mu_{umD}},\\
&\quad z_D\geq 1 \qquad (3.584b)
\end{aligned}
$$

Substituting Eq. (3.584b) into Eq. (3.584a), one has

$$
\begin{aligned}
&\left(s+\varepsilon_{um}+\mu_{umD}-\frac{\varepsilon_{um}\varepsilon_{\mathrm{uim}}}{s+\varepsilon_{\mathrm{uim}}+\mu_{\mathrm{uim}D}}\right)\overline{C}_{umD}\\
&\quad -C_{umD}(r_D,z_D,t_{\mathrm{cha},D})-\frac{\varepsilon_{um}C_{\mathrm{uim}D}(r_D,z_D,t_{\mathrm{cha},D})}{s+\varepsilon_{\mathrm{uim}}+\mu_{\mathrm{uim}D}}=0,\quad z_D\geq 1
\end{aligned}
$$

(3.585)

Similarly, Eqs. (3.536a) and (3.536b) become:

$$(s+\varepsilon_{lm}+\mu_{lmD})\overline{C}_{lmD}-\varepsilon_{lm}\overline{C}_{\text{lim}D} - C_{lmD}(r_D,z_D,t_{\text{cha},D})=0, \quad z_D\le -1 \tag{3.586a}$$

$$\overline{C}_{\text{lim}D}=\frac{\varepsilon_{\text{lim}}}{s+\varepsilon_{\text{lim}}+\mu_{lmD}}\overline{C}_{lmD}+\frac{C_{\text{lim}D}(r_D,z_D,t_{\text{cha},D})}{s+\varepsilon_{\text{lim}}+\mu_{lmD}}, \quad z_D\le -1 \tag{3.586b}$$

Substituting Eq. (3.586b) into Eq. (3.586a), one has

$$\left(s+\varepsilon_{lm}+\mu_{lmD}-\frac{\varepsilon_{lm}\varepsilon_{\text{lim}}}{s+\varepsilon_{\text{lim}}+\mu_{\text{lim}D}}\right)\overline{C}_{lmD} - C_{lmD}(r_D,z_D,t_{\text{cha},D})-\frac{\varepsilon_{lm}C_{\text{lim}D}(r_D,z_D,t_{\text{cha},D})}{s+\varepsilon_{\text{lim}}+\mu_{\text{lim}D}}=0, \ z_D\le -1 \tag{3.587}$$

According to Eqs. (3.585) and (3.587), one has

$$\overline{C}_{umD}=\frac{C_{umD}(r_D,z_D,t_{\text{cha},D})+\frac{\varepsilon_{um}C_{\text{uim}D}(r_D,z_D,t_{\text{cha},D})}{s+\varepsilon_{\text{uim}}+\mu_{\text{uim}D}}}{\left(s+\varepsilon_{um}+\mu_{umD}-\frac{\varepsilon_{um}\varepsilon_{\text{uim}}}{s+\varepsilon_{\text{uim}}+\mu_{\text{uim}D}}\right)}, \quad z_D\ge 1 \tag{3.588a}$$

$$\overline{C}_{lmD}=\frac{C_{lmD}(r_D,z_D,t_{\text{cha},D})+\frac{\varepsilon_{lm}C_{\text{lim}D}(r_D,z_D,t_{\text{cha},D})}{s+\varepsilon_{\text{lim}}+\mu_{\text{lim}D}}}{\left(s+\varepsilon_{lm}+\mu_{lmD}-\frac{\varepsilon_{lm}\varepsilon_{\text{lim}}}{s+\varepsilon_{\text{lim}}+\mu_{\text{lim}D}}\right)}, \quad z_D\le -1 \tag{3.588b}$$

where $C_{umD}(r_D,z_D,t_{\text{cha},D})$ and $C_{\text{uim}D}(r_D,z_D,t_{\text{cha},D})$ are respectively the mobile and immobile concentrations [Kg/m^3] of the upper aquitard at the end of the chaser phase, $C_{lmD}(r_D,z_D,t_{\text{cha},D})$ and $C_{\text{lim}D}(r_D,z_D,t_{\text{cha},D})$ are respectively the mobile and immobile concentrations [Kg/m^3] of the lower aquitard at the end of the chaser phase.

Similarly, the dimensionless governing equation of the mobile zone during the rest phase is

$$\frac{\partial C_{mD}}{\partial t_D}=-\varepsilon_m(C_{mD}-C_{imD})-\mu_{mD}C_{mD}, \quad r_D\ge r_{wD} \tag{3.589a}$$

$$\frac{\partial C_{imD}}{\partial t_D} = \varepsilon_{im}(C_{mD} - C_{imD}) - \mu_{imD}C_{imD}, \quad r_D \geq r_{wD} \quad (3.589b)$$

Conducting Laplace transform to Eqs. (3.589a) and (3.589b) for the rest phase, one has

$$s\overline{C}_{mD} - C_{mD}(r_D, t_{\text{cha},D}) = -\varepsilon_m\left(\overline{C}_{mD} - \overline{C}_{imD}\right) - \mu_{mD}\overline{C}_{mD}, \quad r_D \geq r_{wD} \quad (3.590a)$$

$$s\overline{C}_{imD} - C_{imD}(r_D, t_{\text{cha},D}) = \varepsilon_{im}\left(\overline{C}_{mD} - \overline{C}_{imD}\right) - \mu_{imD}\overline{C}_{imD}, \quad r_D \geq r_{wD} \quad (3.590b)$$

According to Eqs. (3.590a) and (3.590b), one has

$$\overline{C}_{mD} = \frac{C_{mD}(r_D, t_{\text{cha},D}) + \frac{\varepsilon_m C_{imD}(r_D, t_{\text{cha},D})}{(s+\mu_{imD}+\varepsilon_{im})}}{\left[s + \varepsilon_m + \mu_{mD} - \frac{\varepsilon_m\varepsilon_{im}}{(s+\mu_{imD}+\varepsilon_{im})}\right]} \quad (3.591a)$$

$$\overline{C}_{imD} = \frac{C_{imD}(r_D, t_{\text{cha},D})}{(s + \mu_{imD} + \varepsilon_{im})} + \frac{\varepsilon_{im}\overline{C}_{mD}}{(s + \mu_{imD} + \varepsilon_{im})} \quad (3.591b)$$

(f) Analytical solution in the extraction phase

Contrary to the injection and chaser phases, the direction of advective flux is reversed in the extraction stage, Eqs. (3.535a) and (3.536a) are modified as

$$\frac{\partial C_{umD}}{\partial t_D} = \frac{R_m\alpha_r^2 D_u}{AB^2 R_{um}}\frac{\partial^2 C_{umD}}{\partial z_D^2} + \frac{R_m v_{um}\alpha_r^2}{ABR_{um}}\frac{\partial C_{um}D}{\partial z_D} - \varepsilon_{um}\left(C_{umD} - C_{\text{uim}D}\right) - \mu_{umD}C_{umD}, \quad z_D \geq 1 \quad (3.592a)$$

$$\frac{\partial C_{lmD}}{\partial t_D} = \frac{R_m\alpha_r^2 D_l}{AB^2 R_{lm}}\frac{\partial^2 C_{lmD}}{\partial z_D^2} - \frac{R_m v_{lm}\alpha_r^2}{ABR_{lm}}\frac{\partial C_{lmD}}{\partial z_D} - \varepsilon_{lm}\left(C_{lmD} - C_{\text{lim}D}\right) - \mu_{lmD}C_{lmD}, \quad z_D \leq -1 \quad (3.592b)$$

Conducting Laplace transform on Eqs. (3.535b) and (3.592a), one has

$$s\overline{C}_{umD} - C_{umD}(r_D, z_D, t_{\text{res},D}) = \frac{R_m \alpha_r^2 D_u}{AB^2 R_{um}} \frac{\partial^2 \overline{C}_{umD}}{\partial z_D^2} + \frac{R_m v_{um} \alpha_r^2}{ABR_{um}} \frac{\partial \overline{C}_{umD}}{\partial z_D} - \varepsilon_{um}\left(\overline{C}_{umD} - \overline{C}_{\text{uim}D}\right) - \mu_{umD}\overline{C}_{umD}, \quad z_D \geq 1 \tag{3.593a}$$

$$\overline{C}_{\text{uim}D} = \frac{\varepsilon_{\text{uim}}\overline{C}_{umD}}{s + \varepsilon_{\text{uim}} + \mu_{\text{uim}D}} + \frac{C_{\text{uim}D}(r_D, z_D, t_{\text{res},D})}{s + \varepsilon_{\text{uim}} + \mu_{\text{uim}D}}, \quad z_D \geq 1 \tag{3.593b}$$

Substituting Eqs. (3.593b) into Eq. (3.593a), one can have:

$$\frac{R_m \alpha_r^2 D_u}{AB^2 R_{um}} \frac{\partial^2 \overline{C}_{umD}}{\partial z_D^2} + \frac{R_m v_{um} \alpha_r^2}{ABR_{um}} \frac{\partial \overline{C}_{umD}}{\partial z_D} - \left(s + \varepsilon_{um} + \mu_{umD} - \frac{\varepsilon_{um}\varepsilon_{\text{uim}}}{s + \varepsilon_{\text{uim}} + \mu_{\text{uim}D}}\right)\overline{C}_{umD} + C_{umD}(r_D, z_D, t_{\text{res},D}) + \frac{\varepsilon_{um} C_{\text{uim}D}(r_D, z_D, t_{\text{res},D})}{s + \varepsilon_{\text{uim}} + \mu_{\text{uim}D}} = 0, \ z_D \geq 1 \tag{3.594}$$

Similarly, conducting Laplace transform on Eqs. (3.536b) and (3.592b), one has

$$s\overline{C}_{lmD} - C_{lmD}(r_D, z_D, t_{\text{res},D}) = \frac{R_m \alpha_r^2 D_l}{AB^2 R_{lm}} \frac{\partial^2 \overline{C}_{lmD}}{\partial z_D^2} - \frac{R_m v_{lm} \alpha_r^2}{ABR_{lm}} \frac{\partial \overline{C}_{lmD}}{\partial z_D} - \varepsilon_{lm}(\overline{C}_{lmD} - \overline{C}_{\text{lim}D}) - \mu_{lmD}\overline{C}_{lmD}, \quad z_D \leq -1 \tag{3.595a}$$

$$\overline{C}_{\text{lim}D} = \frac{\varepsilon_{\text{lim}}\overline{C}_{lmD}}{s + \varepsilon_{\text{lim}} + \mu_{\text{lim}D}} + \frac{C_{\text{lim}D}(r_D, z_D, t_{\text{res},D})}{s + \varepsilon_{\text{lim}} + \mu_{\text{lim}D}}, \quad z_D \leq -1 \tag{3.595b}$$

Substituting Eqs. (3.595b) into Eq. (3.595a), one has

$$\frac{R_m \alpha_r^2 D_l}{AB^2 R_{lm}} \frac{\partial^2 \overline{C}_{lmD}}{\partial z_D^2} - \frac{R_m v_{lm} \alpha_r^2}{ABR_{lm}} \frac{\partial \overline{C}_{lmD}}{\partial z_D} - \left(s + \varepsilon_{lm} + \mu_{lmD} - \frac{\varepsilon_{lm}\varepsilon_{\mathrm{lim}}}{s + \varepsilon_{\mathrm{lim}} + \mu_{\mathrm{lim}D}} \right) \overline{C}_{lmD} + C_{lmD}(r_D, z_D, t_{\mathrm{res},D}) + \frac{\varepsilon_{lm} C_{\mathrm{lim}D}(r_D, z_D, t_{\mathrm{res},D})}{s + \varepsilon_{\mathrm{lim}} + \mu_{\mathrm{lim}D}} = 0, \ z_D \le -1 \tag{3.596}$$

where $C_{umD}(r_D, z_D, t_{\mathrm{res},D})$ and $C_{\mathrm{uim}D}(r_D, z_D, t_{\mathrm{res},D})$ are respectively the mobile and immobile concentrations [Kg/m^3] of the upper aquitard at the end of the rest phase, $C_{lmD}(r_D, z_D, t_{\mathrm{res},D})$ and $C_{\mathrm{lim}D}(r_D, z_D, t_{\mathrm{res},D})$ are respectively the mobile and immobile concentrations [Kg/m^3] of the lower aquitard at the end of the rest phase.

One could use a similar approach of obtaining the analytical solution of aquitards in the chaser phase to derive the solution of aquitards in the extraction phase. The general solution of (3.594) is

$$\overline{C}_{umD} = \int_1^\infty g_u(z_D, E_u; \mathscr{b}_u) f_u(\mathscr{b}_u) d\mathscr{b}_u + \frac{z_D - z_{eD}}{1 - z_{eD}} \overline{C}_{mD}(r_D, s), \quad z_D \ge 1 \tag{3.597a}$$

$$g_u(z_D, E_u; \mathscr{b}_u) = \begin{cases} g_{u1}(z_D, E_u; \mathscr{b}_u) = H_1 \exp(m_1 z_D) + H_2 \exp(m_2 z_D) & 1 \le z_D < \mathscr{b}_u \\ g_{u2}(z_D, E_u; \mathscr{b}_u) = H_3 \exp(m_1 z_D) + H_4 \exp(m_2 z_D) & \mathscr{b}_u \le z_D < \infty \end{cases} \tag{3.597b}$$

$$f_u(\mathscr{b}_u) = C_{umD}(r_D, \mathscr{b}_u, t_{\mathrm{res},D}) + \frac{\varepsilon_{um} C_{\mathrm{uim}D}(r_D, \mathscr{b}_u, t_{\mathrm{res},D})}{s + \varepsilon_{\mathrm{uim}} + \mu_{\mathrm{uim}D}} + \frac{R_m v_{um} \alpha_r^2}{ABR_{um}} \frac{\overline{C}_{mD}(r_D, s)}{1 - z_{eD}} - \frac{\mathscr{b}_u - z_{eD}}{1 - z_{eD}} E_u \overline{C}_{mD}(r_D, s) \tag{3.597c}$$

The general solution of Eq. (3.596) could be described as

$$\overline{C}_{lmD} = \int_{-1}^{-\infty} g_l(z_D, E_l; \mathscr{b}_l) f_l(\mathscr{b}_l) d\mathscr{b}_l + \frac{z_D + z_{eD}}{z_{eD} - 1} \overline{C}_{mD}(r_D, s), \quad z_D \leq -1 \tag{3.598a}$$

$$g_l(z_D, E_l; \mathscr{b}_l) = \begin{cases} g_{l1}(z_D, E_l; \mathscr{b}_l) = I_1 \exp(n_1 z_D) + I_2 \exp(n_2 z_D) - 1 \leq z_D < \mathscr{b}_l \\ g_{l2}(z_D, E_l; \mathscr{b}_l) = I_3 \exp(n_1 z_D) + I_4 \exp(n_2 z_D) \mathscr{b}_l \leq z_D < -\infty \end{cases} \tag{3.598b}$$

$$f_l(\mathscr{b}_l) = C_{mD}(r_D, \mathscr{b}_l, t_{\text{res},D}) + \frac{\varepsilon_{lm} C_{\text{lim}D}(r_D, \mathscr{b}_l, t_{\text{res},D})}{s + \varepsilon_{\text{lim}} + \mu_{\text{lim}D}} - \frac{R_m v_{lm} \alpha_r^2}{ABR_{lm}} \frac{\overline{C}_{mD}(r_D, s)}{z_{eD} - 1} - \frac{\mathscr{b}_l + z_{eD}}{z_{eD} - 1} E_l \overline{C}_{mD}(r_D, s) \tag{3.598c}$$

where $\mathscr{b}_u$ is a positive value varying between 1 and ∞; $\mathscr{b}_l$ is a negative value varying between -1 and $-\infty$; $g_u(z_D, E_u; b_u)$ and $g_l(z_D, E_l; \mathscr{b}_l)$ are the Green's functions, $H_1 \sim H_4$ and $I_1 \sim I_4$ are constants which could be determined by the boundary conditions and conditions of (a)–(c), the values of $H_1 \sim H_4$ and $I_1 \sim I_4$ are as follows:

$$H_1 = -H_2 \exp(m_2 - m_1),$$

$$H_2 = \frac{-AR_{um} B^2}{R_m \alpha_r^2 D_u [(m_1 - m_2) \exp(m_2 - m_1) \exp(m_1 \mathscr{b}_u)]}, \quad H_3 = 0,$$

$$H_4 = H_2 - H_2 \exp(m_2 - m_1) \exp(m_1 \mathscr{b}_u - m_2 \mathscr{b}_u),$$

$$I_1 = -I_2 \exp(n_1 - n_2), \quad I_2 = \frac{-AB^2 R_{lm}}{R_m \alpha_r^2 D_l [\exp(n_2 \mathscr{b}_l - n_1 \mathscr{b}_l) - n_2 \exp(n_2 \mathscr{b}_l)]},$$

$$I_3 = I_2 \exp(n_2 \mathscr{b}_l - n_1 \mathscr{b}_l) - I_2 \exp(n_1 - n_2), \quad I_4 = 0,$$

$$m_1 = \frac{-\frac{R_m v_{um} \alpha_r^2}{ABR_{um}} + \sqrt{\left(\frac{R_m v_{um} \alpha_r^2}{ABR_{um}}\right)^2 + 4 \frac{R_m \alpha_r^2 D_u}{AB^2 R_{um}} \left(s + \varepsilon_{um} + \mu_{umD} - \frac{\varepsilon_{um} \varepsilon_{\text{uim}}}{s + \mu_{\text{uim}D} + \varepsilon_{\text{uim}}}\right)}}{2 \frac{R_m \alpha_r^2 D_u}{AB^2 R_{um}}},$$

$$m_2 = \frac{-\frac{R_m v_{um}\alpha_r^2}{ABR_{um}} - \sqrt{\left(\frac{R_m v_{um}\alpha_r^2}{ABR_{um}}\right)^2 + 4\frac{R_m\alpha_r^2 D_u}{AB^2 R_{um}}\left(s + \varepsilon_{um} + \mu_{umD} - \frac{\varepsilon_{um}\varepsilon_{\mathrm{uim}}}{s+\mu_{\mathrm{uim}D}+\varepsilon_{\mathrm{uim}}}\right)}}{2\frac{R_m\alpha_r^2 D_u}{AB^2 R_{um}}},$$

$$n_1 = \frac{\frac{R_m v_{lm}\alpha_r^2}{ABR_{lm}} + \sqrt{\left(\frac{R_m v_{lm}\alpha_r^2}{ABR_{lm}}\right)^2 + 4\frac{R_m\alpha_r^2 D_l}{AB^2 R_{lm}}\left(s + \varepsilon_{lm} + \mu_{lmD} - \frac{\varepsilon_{lm}\varepsilon_{\mathrm{lim}}}{s+\mu_{\mathrm{lim}D}+\varepsilon_{\mathrm{lim}}}\right)}}{2\frac{R_m\alpha_r^2 D_l}{AB^2 R_{lm}}} \quad \text{and}$$

$$n_2 = \frac{\frac{R_m v_{lm}\alpha_r^2}{ABR_{lm}} - \sqrt{\left(\frac{R_m v_{lm}\alpha_r^2}{ABR_{lm}}\right)^2 + 4\frac{R_m\alpha_r^2 D_l}{AB^2 R_{lm}}\left(s + \varepsilon_{lm} + \mu_{lmD} - \frac{\varepsilon_{lm}\varepsilon_{\mathrm{lim}}}{s+\mu_{\mathrm{lim}D}+\varepsilon_{\mathrm{lim}}}\right)}}{2\frac{R_m\alpha_r^2 D_l}{AB^2 R_{lm}}}.$$

Similarly, contrary to the injection and chaser phases, the direction of advective flux is reversed in the extraction stage, and Eq. (3.534a) is modified as

$$\begin{aligned}\frac{\partial C_{mD}}{\partial t_D} &= \frac{1}{r_D}\frac{\partial^2 C_{mD}}{\partial r_D^2} + \frac{1}{r_D}\frac{\partial C_{mD}}{\partial r_D} - \varepsilon_m(C_{mD} - C_{imD}) - \mu_{mD}C_{mD} \\ &\quad - \left(-\frac{\theta_{um}\alpha_r^2 v_{um}}{2A\theta_m B}C_{umD} - \frac{\theta_{um}\alpha_r^2 D_u}{2A\theta_m B}\frac{\partial C_{umD}}{\partial z_D}\right)\bigg|_{z=1} \\ &\quad + \left(-\frac{\theta_{lm}\alpha_r^2 v_{lm}}{2AB^2\theta_m}C_{lmD} - \frac{\theta_{lm}\alpha_r^2 D_l}{2AB^2\theta_m}\frac{\partial C_{lmD}}{\partial z_D}\right)\bigg|_{z=-1}, \quad r_D \geq r_{wD}\end{aligned} \tag{3.599}$$

In the extraction phase, the dimensional boundary conditions Eqs. (3.522a) and (3.522b) are transformed to the dimensionless format:

$$\beta_{\mathrm{ext},D}\frac{\partial C_{mD}(r_D,t_D)}{\partial t_D}\bigg|_{r_D=r_{wD}} = \frac{\partial C_{mD}(r_D,t_D)}{\partial r_D}\bigg|_{r_D=r_{wD}}, \quad t_{\mathrm{res},D} < t_D \leq t_{\mathrm{ext},D} \tag{3.600a}$$

$$C_{mD}(r_D,t_D)|_{t_D=t_{\mathrm{res},D}} = C_{\mathrm{res},mD}(r_D,t_D)|_{t_D=t_{\mathrm{res},D}} \tag{3.600b}$$

where $\beta_{\mathrm{ext},D} = -\frac{V_{w,\mathrm{ext}}r_{wD}}{\xi R_m\alpha_r}$.

Conducting Laplace transform on Eqs. (3.591) and (3.534b) in the extraction phase, one has

$$\begin{aligned}
&s\overline{C}_{mD} - C_{mD}(r_D, t_{\text{res}}) \\
&\quad = \frac{1}{r_D}\frac{\partial^2 \overline{C}_{mD}}{\partial r_D^2} + \frac{1}{r_D}\frac{\partial \overline{C}_{mD}}{\partial r_D} - (\varepsilon_m + \mu_{mD})\overline{C}_{mD} + \varepsilon_m \overline{C}_{imD} \\
&\qquad - \left(-\frac{\theta_{um}\alpha_r^2 v_{um}\overline{C}_{umD}}{2A\theta_m b} - \frac{\theta_{um}\alpha_r^2 D_u}{2A\theta_m b}\frac{\partial \overline{C}_{umD}}{\partial z_D}\right)\Bigg|_{z_D=1} \\
&\qquad - \left(\frac{\theta_{lm}\alpha_r^2 v_{lm}\overline{C}_{lmD}}{2Ab^2\theta_m} + \frac{\theta_{lm}\alpha_r^2 D_l}{2Ab^2\theta_m}\frac{\partial \overline{C}_{lmD}}{\partial z_D}\right)\Bigg|_{z_D=-1}, \\
&r_D \geq r_{wD}
\end{aligned} \tag{3.601a}$$

$$\overline{C}_{imD} = \frac{\varepsilon_{im}}{(s + \mu_{imD} + \varepsilon_{im})}\overline{C}_{mD} + \frac{C_{imD}(r_D, t_{\text{res}})}{s + \mu_{imD} + \varepsilon_{im}}, \quad r_D \geq r_{wD} \tag{3.601b}$$

After substituting Eqs. (3.597a)–(3.598c) and Eq. (3.601b) into Eq. (3.601a), one has

$$\frac{\partial^2 \overline{C}_{mD}}{\partial r_D^2} + \frac{\partial \overline{C}_{mD}}{\partial r_D} - r_D \zeta \overline{C}_{mD} + r_D \Lambda = 0 \tag{3.602}$$

where

$$\begin{aligned}
\Lambda &= C_{mD}(r_D, t_{\text{res}}) + \frac{\varepsilon_m C_{imD}(r_D, t_{\text{res}})}{s + \mu_{imD} + \varepsilon_{im}}; \\
\zeta &= s + \varepsilon_m + \mu_{mD} - \frac{\varepsilon_{im}\varepsilon_m}{s + \mu_{imD} + \varepsilon_{im}} \\
&\quad - \frac{\theta_{um}\alpha_r^2 v_{um}}{2A\theta_m B} + \frac{\theta_{lm}\alpha_r^2 v_{lm}}{2AB^2\theta_m} - \frac{1}{1 - z_{eD}}\frac{\theta_{um}\alpha_r^2 D_u}{2A\theta_m b} \\
&\quad + \frac{1}{z_{eD} - 1}\frac{\theta_{lm}\alpha_r^2 D_l}{2Ab^2\theta_m}, \quad C_{imD}(r_D, t_{\text{res}}) \text{ and } C_{mD}(r_D, t_{\text{res}})
\end{aligned}$$

represent the initial concentrations in the immobile and mobile domains of the SWPP test in the rest phase.

The boundary condition of Eqs. (3.600a) and (3.600b) in the Laplace domain becomes:

$$s\beta_{\text{ext},D}\overline{C}_{mD}(r_D,s)\big|_{r_D=r_{wD}} - \beta_{\text{ext},D}C_{\text{res},m}(r_D,t_D)\big|_{t_D=t_{\text{res},D}}$$
$$= \left.\frac{\partial\overline{C}_{mD}(r_D,s)}{\partial r_D}\right|_{r_D=r_{wD}} \quad (3.603)$$

Like the model of the SWPP test in the injection phase, Eqs. (3.538), (3.594), and (3.603) compose a model of the second-order ordinary differential equation (ODE) with boundary conditions. However, the governing equation is an inhomogeneous differential equation. In this section, we use the Green's function method to derive the analytical solution of Eq. (3.602).

Similar to Chen and Woodside (1988), Eq. (3.602) could be transferred into a self-adjoint form:

$$\frac{\partial^2 G}{\partial r_D^2} - \left(r_D\zeta + \frac{1}{4}\right)G = -\ell\left(r_D\right) \quad (3.604)$$

where $G = \exp(r_D/2)\overline{C}_{mD}$ and $\ell(r_D) = \exp(r_D/2)r_D\Lambda$.

The boundary conditions of Eqs. (3.538) and (3.603) could be rewritten

$$G(r_D,s)|_{r_D=\infty} = 0 \quad (3.605a)$$

$$\left.\left[\left(s\beta_{\text{ext},D} + \frac{1}{2}\right)G - \frac{\partial G}{\partial r_D}\right]\right|_{r_D=r_{wD}} = \beta_{\text{ext},D}\exp(r_{wD}/2)C_{mD}$$
$$\times\left(r_{wD},t_{\text{res},D}\right) \quad (3.605b)$$

One could find that the boundary condition of Eq. (3.605b) is inhomogeneous, and we need to homogenize it first. Assigning $G = U(r_D) + V(r_D)$ and $V(r_D) = \sigma_1 + \sigma_2 r_D$, and substituting them into Eqs. (3.605a) and (3.605b) yields:

$$U(r_D,s)|_{r_D=\infty} = 0 \quad (3.606a)$$

$$\left.\left[\left(s\beta_{\text{ext},D} + \frac{1}{2}\right)U - \frac{\partial U}{\partial r_D}\right]\right|_{r_D=r_{wD}} = 0 \quad (3.606b)$$

where

$$\sigma_1 = -\frac{\beta_{\text{ext},D}\exp(r_{wD}/2)C_{mD}(r_{wD}, t_{\text{res},D})}{\left(s\beta_{\text{ext},D}+\frac{1}{2}\right)r_{wD} - 1 - \left(s\beta_{\text{ext},D}+\frac{1}{2}\right)r_D|_{r_D\to\infty}} r_D|_{r_D\to\infty},$$

$$\sigma_2 = \frac{\beta_{\text{ext},D}\exp(r_{wD}/2)C_{mD}(r_{wD}, t_{\text{res},D})}{\left(s\beta_{\text{ext},D}+\frac{1}{2}\right)r_{wD} - 1 - \left(s\beta_{\text{ext},D}+\frac{1}{2}\right)r_D|_{r_D\to\infty}}.$$

After defining a spatial operator: $\mathrm{L} = -\frac{d^2}{dr_D^2} + \left(r_D\zeta + \frac{1}{4}\right)$, one has

$$LG = LU(r_D) + LV(r_D) = \ell(r_D) \tag{3.607}$$

and

$$LU(r_D) = \ell(r_D) - LV(r_D) \tag{3.608}$$

Let $f(r_D) = \ell(r_D) - LV(r_D)$, one has

$$\frac{\partial^2 U}{\partial r_D^2} - \left(r_D\zeta + \frac{1}{4}\right)U = -f(r_D) \tag{3.609}$$

where $f(r_D) = \exp(r_D/2)r_D\Lambda - \left(r_D\zeta + \frac{1}{4}\right)(\sigma_1 + \sigma_2 r_D)$.

Right now, the model with an inhomogeneous boundary condition becomes a regular Sturm–Louisville problem. The general solution of Eqs. (3.606a), (3.606b), and (3.609) is

$$U(r_D, \zeta; \varepsilon) = \int_{r_{wD}}^{\infty} g(r_D, \zeta; \varepsilon) f(\varepsilon) d\varepsilon \tag{3.610}$$

where ε is a positive value varying between r_{wD} and ∞ (e.g., $r_{wD} \leq \varepsilon \leq \infty$); $g(r_D, \zeta; \varepsilon)$ is the Green's function, and could be expressed as

$$g(r_D, \zeta; \varepsilon) = \begin{cases} g_1(r_D, \zeta; \varepsilon) = P_1 A_i(y_{\text{ext}}) + P_2 B_i(y_{\text{ext}}) r_{wD} \\ \qquad \leq y_{\text{ext}} \leq \varepsilon \\ g_2(r_D, \zeta; \varepsilon) = P_3 A_i(y_{\text{ext}}) + P_4 B_i(y_{\text{ext}}) \varepsilon \\ \qquad \leq y_{\text{ext}} \leq \infty \end{cases} \tag{3.611}$$

where $f(\varepsilon) = \exp(\varepsilon/2)\varepsilon\Lambda - \left(\varepsilon\zeta + \frac{1}{4}\right)(\sigma_1 + \sigma_2\varepsilon)$, $y_{\text{ext}} = \zeta^{1/3}\left(r_D + \frac{1}{4\zeta}\right)$ P_1, P_2, P_3 and P_4 are coefficients to be determined. As $B_i(r_D)$ diverges when $r_D \to \infty$, P_4 has to be zero.

Substituting Eq. (3.611) into Eq. (3.606b), one has

$$\left[\left(s\beta_{\text{ext},D} + \frac{1}{2}\right) g_1 - \frac{\partial g_1}{\partial r_D}\right]\Bigg|_{r_D=r_{wD}} = 0 \tag{3.612}$$

which leads to

$$P_1 = -P_2 W \tag{3.613}$$

where

$$W = \frac{\left(s\beta_{\text{ext},D} + \frac{1}{2}\right) B_i(y_{\text{ext},w}) - \zeta^{1/3} B_i'(y_{\text{ext},w})}{\left(s\beta_{\text{ext},D} + \frac{1}{2}\right) A_i(y_{\text{ext},w}) - \zeta^{1/3} A_i'(y_{\text{ext},w})},$$

$$y_{\text{ext},w} = \zeta^{1/3}\left(r_{wD} + \frac{1}{4\zeta}\right).$$

According to the properties of Green's function, one has

$$P_1 A_i(y_{\text{ext}}|_{r_D=\varepsilon^+}) + P_2 B_i(y_{\text{ext}}|_{r_D=\varepsilon^+}) = P_3 A_i(y_{\text{ext}}|_{r_D=\varepsilon^-}) \tag{3.614}$$

$$[P_3 \zeta^{1/3} A_i'(y_{\text{ext}})]_{r_D=\varepsilon^-} - [P_1 \zeta^{\frac{1}{3}} A_i'(y_{\text{ext}}) + P_2 \zeta^{\frac{1}{3}} B_i'(y_{\text{ext}})]_{r_D=\varepsilon^+} = -1 \tag{3.615}$$

The values of P_1, P_2 and P_3 could be determined by Eqs. (3.602)–(3.604), namely:

$$P_1 = -\frac{\pi A_i(y_{\text{ext}}|_{r_D=\varepsilon^+})}{\zeta^{1/3}} W, \quad P_2 = \frac{\pi A_i(y_{\text{ext}}|_{r_D=\varepsilon^+})}{\zeta^{1/3}}, \quad \text{and}$$

$$P_3 = \frac{\pi A_i(y_{\text{ext}}|_{r_D=\varepsilon^+})}{\zeta^{1/3}}\left[\frac{B_i(y_{\text{ext}}|_{r_D=\varepsilon^+})}{A_i(y_{\text{ext}}|_{r_D=\varepsilon^+})} - W\right].$$

In sum, the analytical solutions of SWPP tests, including both wellbore storage and the aquitard effect under the MIM framework, are as follows:

In the injection phase of the SWPP test, the solutions in the Laplace domain are

$$\overline{C}_{mD}(r_D, s) = \phi_1 \exp\left(\frac{y_{\text{inj}}}{2}\right) A_i\left(E^{1/3} y_{\text{inj}}\right), \quad r_D \geq r_{wD} \tag{3.616a}$$

$$\overline{C}_{imD} = \frac{\varepsilon_{im}}{(s + \mu_{imD} + \varepsilon_{im})} \overline{C}_{mD}, \quad r_D \geq r_{wD} \tag{3.616b}$$

$$\overline{C}_{umD} = \overline{C}_{mD} \exp(a_2 z_D - a_2), \quad z_D \geq 1 \tag{3.616c}$$

$$\overline{C}_{\text{uim}D} = \frac{\varepsilon_{\text{uim}}}{s + \varepsilon_{\text{uim}} + \mu_{\text{uim}D}} \overline{C}_{umD}, \quad z_D \geq 1 \tag{3.616d}$$

$$\overline{C}_{lmD} = \overline{C}_{mD} \exp(b_1 z_D + b_1), \quad z_D \leq -1 \tag{3.616e}$$

$$\overline{C}_{\text{lim}D} = \frac{\varepsilon_{\text{lim}}}{s + \varepsilon_{\text{lim}} + \mu_{\text{lim}D}} \overline{C}_{lmD}, \quad z_D \leq -1 \tag{3.616f}$$

where s represents the Laplace transform parameter for t_D; $A_i(\cdot)$ is the Airy function $A_i'(\cdot)$ is the derivative of the Airy function; the expressions for a_2, b_1, E, y_{inj}, $y_{\text{inj},w}$, ε_m, ε_{im}, ε_{um}, ε_{uim}, ε_{lm}, ε_{lim}, β_{inj}, and ϕ_1 are listed in Table 3.7.

In the chaser phase, the solutions of the SWPP test in Laplace domain are

$$\overline{C}_{mD} = \Psi(r_D) + \delta_1 + \delta_2 r_D, \quad r_D \geq r_{wD} \tag{3.617a}$$

$$\overline{C}_{imD} = \frac{\varepsilon_{im}}{(s + \mu_{imD} + \varepsilon_{im})} \overline{C}_{mD} + \frac{C_{imD}(r_D, t_{\text{inj},D})}{(s + \mu_{imD} + \varepsilon_{im})},$$
$$r_D \geq r_{wD} \tag{3.617b}$$

$$\Psi(r_D, E_a; \eta) = \int_{r_{wD}}^{\infty} g(r_D, E_a; \eta)\, \varphi(\eta) d\eta, \quad r_D \geq r_{wD} \tag{3.617c}$$

$$\overline{C}_{umD} = \int_{1}^{\infty} g_u(z_D, E_u; \eta_u) f_u(\eta_u) d\eta_u + \frac{z_D - z_{eD}}{1 - z_{eD}}$$
$$\times \overline{C}_{mD}(r_D, s), \quad z_D \geq 1 \tag{3.617d}$$

$$\overline{C}_{\text{uim}D} = \frac{\varepsilon_{\text{uim}}}{s + \varepsilon_{\text{uim}} + \mu_{\text{uim}D}} \overline{C}_{umD} + \frac{C_{\text{uim}D}(r_D, z_D, t_{\text{inj},D})}{s + \varepsilon_{\text{uim}} + \mu_{\text{uim}D}},$$
$$z_D \geq 1 \tag{3.617e}$$

$$\overline{C}_{lmD} = \int_{-1}^{-\infty} g_l(z_D, E_l; \eta_l) f_l(\eta_l) d\eta_l + \frac{z_{eD} + z_D}{z_{eD} - 1}$$
$$\times \overline{C}_{mD}(r_D, z_D, s), \quad z_D \leq -1 \tag{3.617f}$$

$$\overline{C}_{\text{lim}D} = \frac{\varepsilon_{\text{lim}}}{s + \varepsilon_{\text{lim}} + \mu_{\text{lim}D}} \overline{C}_{lmD} + \frac{C_{\text{lim}D}(r_D, z_D, t_{\text{inj},D})}{s + \varepsilon_{\text{lim}} + \mu_{\text{lim}D}},$$
$$z_D \leq -1 \tag{3.617g}$$

Table 3.7. Expressions of the coefficients in the solutions expressed in Eqs. (3.616a)–(3.616f).

a_2	$\dfrac{\dfrac{R_m v_{um}\alpha_r^2}{ABR_{um}} - \sqrt{\left(\dfrac{R_m v_{um}\alpha_r^2}{ABR_{um}}\right)^2 + 4\dfrac{R_m\alpha_r^2 D_u}{AB^2R_{um}}\left(s+\varepsilon_{um}+\mu_{umD}-\frac{\varepsilon_{um}\varepsilon_{\text{uim}}}{s+\mu_{\text{uim}D}+\varepsilon_{\text{uim}}}\right)}}{2\dfrac{R_m\alpha_r^2 D_u}{AB^2R_{um}}}$
b_1	$\dfrac{-\dfrac{R_m v_{lm}\alpha_r^2}{ABR_{lm}} + \sqrt{\left(\dfrac{R_m v_{lm}\alpha_r^2}{ABR_{lm}}\right)^2 + 4\dfrac{R_m\alpha_r^2 D_l}{AB^2R_{lm}}\left(s+\varepsilon_{lm}+\mu_{lmD}-\frac{\varepsilon_{lm}\varepsilon_{\text{lim}}}{s+\mu_{\text{lim}D}+\varepsilon_{\text{lim}}}\right)}}{2\dfrac{R_m\alpha_r^2 D_l}{AB^2R_{lm}}}$
E	$s+\varepsilon_m+\mu_{mD}-\dfrac{\varepsilon_m\varepsilon_{im}}{s+\mu_{imD}+\varepsilon_{im}}+\dfrac{\theta_{um}\alpha_r^2 v_{um}}{2A\theta_m B}-\dfrac{\theta_{lm}\alpha_r^2 v_{lm}}{2A\theta_m B}$ $-\dfrac{a_2\theta_{um}\alpha_r^2 D_u}{2A\theta_m B^2}+\dfrac{b_1\theta_{lm}\alpha_r^2 D_l}{2AB^2\theta_m}$
y_{inj}	$r_D+\dfrac{1}{4E}$
$y_{\text{inj},w}$	$r_{wD}+\dfrac{1}{4E}$
ε_m	$\dfrac{\omega_a\alpha_r^2}{A\theta_m}$
ε_{im}	$\dfrac{\omega_a\alpha_r^2}{A\theta_{im}}$
ε_{um}	$\dfrac{\omega_u\alpha_r^2 R_m}{A\theta_{um}R_{um}}$
ε_{uim}	$\dfrac{\omega_l\alpha_r^2 R_m}{A\theta_{um}R_{\text{uim}}}$
β_{inj}	$\dfrac{V_{w,\text{inj}}r_{wD}}{\xi R_m\alpha_r}$
ξ	$4\pi r_w\theta B$
ϕ_1	$\dfrac{1}{s(s\beta_{\text{inj}}+1)}\dfrac{1}{\exp\left(\frac{y_{\text{inj},w}}{2}\right)\left[\frac{A_i\left(E^{1/3}y_{\text{inj},w}\right)}{2}-E^{1/3}A_i'\left(E^{1/3}y_{\text{inj}}\right)\right]}$

where η varies between r_{wD} and ∞, e.g., $r_{wD} \leq \eta \leq \infty$; η_u varies between 1 and ∞; η_l varies between -1 and $-\infty$; $C_{mD}(r_D, t_{\text{inj},D})$ and $C_{imD}(r_D, t_{\text{inj},D})$ are the concentrations [Kg/m^3] of the aquifer at the end of injection stage, which could be calculated by Eqs. (3.616a) and (3.616b) after applying the inverse Laplace transform, $C_{umD}(r_D, z_D, t_{\text{inj},D})$ and $C_{\text{uim}D}(r_D, z_D, t_{\text{cha},D})$ represent the concentrations [Kg/m^3] of the upper aquitard at the end of the injection phase, which could be calculated by Eqs. (3.616c) and (3.616d) after applying the inverse Laplace transform, $C_{lmD}(r_D, z_D, t_{\text{inj},D})$ and $C_{\text{lim}D}(r_D, z_D, t_{\text{inj},D})$ are the concentrations [Kg/m^3] of the lower aquitard at the end of the injection phase, which could be calculated by Eq. (3.616e) and (3.616f) after applying the inverse Laplace transform, $g(r_D, E_a; \eta), g_u(z_D, E_u; \eta_u)$ and $g_l(z_D, E_l; \eta_l)$ are the Green's functions; the expressions for $g(r_D, E_a; \eta), g_u(z_D, E_u; \eta_u)$, $g_l(z_D, E_l; \eta_l)$, δ_1, δ_2, $\jmath_1$, $\jmath_2$, E_a, E_u, $E_l, y_{\text{cha}}, y_{\text{cha},w}, F, \varphi(r_D), f_u(\eta_u)$, $X, M_1, M_2, M_3, M_4, N_1, N_2, N_3, N_4, \mathcal{T}_1, \mathcal{T}_2, \mathcal{T}_3, \mathcal{T}_4$ and $\beta_{\text{cha},D}$ are listed in Table 3.8.

For the rest phase, the solutions of the SWPP test in the Laplace domain are

$$\overline{C}_{mD} = \frac{C_{mD}(r_D, t_{\text{cha},D}) + \frac{\varepsilon_m C_{imD}(r_D, t_{\text{cha},D})}{(s+\mu_{imD}+\varepsilon_{im})}}{\left(s + \varepsilon_m + \mu_{mD} - \frac{\varepsilon_m \varepsilon_{im}}{s+\mu_{imD}+\varepsilon_{im}}\right)}, \quad r_D \geq r_{wD} \tag{3.618a}$$

$$\overline{C}_{imD} = \frac{C_{imD}(r_D, t_{\text{cha},D})}{(s + \mu_{imD} + \varepsilon_{im})} + \frac{\varepsilon_{im} \overline{C}_{mD}}{(s + \mu_{imD} + \varepsilon_{im})}, \quad r_D \geq r_{wD} \tag{3.618b}$$

$$\overline{C}_{umD} = \frac{C_{umD}(r_D, z_D, t_{\text{cha},D}) + \frac{\varepsilon_{um} C_{\text{uim}D}(r_D, z_D, t_{\text{cha},D})}{s+\varepsilon_{\text{uim}}+\mu_{\text{uim}D}}}{\left(s + \varepsilon_{um} + \mu_{umD} - \frac{\varepsilon_{um} \varepsilon_{\text{uim}}}{s+\varepsilon_{\text{uim}}+\mu_{\text{uim}D}}\right)}, \quad z_D \geq 1 \tag{3.618c}$$

$$\overline{C}_{\text{uim}D} = \frac{\varepsilon_{\text{uim}}}{s + \varepsilon_{\text{uim}} + \mu_{umD}} \overline{C}_{umD} + \frac{C_{\text{uim}D}(r_D, z_D, t_{\text{cha},D})}{s + \varepsilon_{\text{uim}} + \mu_{umD}}, \quad z_D \geq 1 \tag{3.618d}$$

$$\overline{C}_{lmD} = \frac{C_{lmD}(r_D, z_D, t_{\text{cha},D}) + \frac{\varepsilon_{lm} C_{\text{lim}D}(r_D, z_D, t_{\text{cha},D})}{s+\varepsilon_{\text{lim}}+\mu_{\text{lim}D}}}{\left(s + \varepsilon_{lm} + \mu_{lmD} - \frac{\varepsilon_{lm} \varepsilon_{\text{lim}}}{s+\varepsilon_{\text{lim}}+\mu_{\text{lim}D}}\right)}, \quad z_D \leq -1 \tag{3.618e}$$

Table 3.8. Expressions of the coefficients in the solutions expressed in Eqs. (3.617a)–(3.617g).

δ_1	$-\dfrac{\beta_{\text{cha},D}}{s\beta_{\text{cha},D}+1}\dfrac{r_D\|_{r_D\to\infty}}{(r_{wD}-r_D\|_{r_D\to\infty}-1)}C_{\text{inj},mD}(r_D,t_D)\|_{t_D=t_{\text{inj},D}}$
δ_2	$\dfrac{\beta_{\text{cha},D}}{s\beta_{\text{cha},D}+1}\dfrac{1}{(r_{wD}-r_D\|_{r_D\to\infty}-1)}C_{\text{inj},mD}(r_D,t_D)\|_{t_D=t_{\text{inj},D}}$
$\jmath_1$	$-\jmath_2 z_{eD}$
$\jmath_2$	$\dfrac{\overline{C}_{mD}(r_D,s)}{1-z_{eD}}$
$\beta_{\text{cha},D}$	$-\dfrac{V_{w,\text{cha}}r_{wD}}{\xi R_m\alpha_r}$
E_a	$s+\varepsilon_m+\mu_{mD}-\dfrac{\varepsilon_m\varepsilon_{im}}{s+\mu_{imD}+\varepsilon_{im}}+\dfrac{\theta_{um}\alpha_r^2 v_{um}}{2A\theta_m B}-\dfrac{\theta_{lm}\alpha_r^2 v_{lm}}{2AB^2\theta_m}$ $-\dfrac{1}{1-z_{eD}}\dfrac{\theta_{um}\alpha_r^2 D_u}{2A\theta_m B^2}+\dfrac{1}{z_{eD}-1}\dfrac{\theta_{lm}\alpha_r^2 D_l}{2AB^2\theta_m}$
E_u	$s+\varepsilon_{um}+\mu_{umD}-\dfrac{\varepsilon_{um}\varepsilon_{\text{uim}}}{s+\varepsilon_{\text{uim}}+\mu_{\text{uim}D}}$
E_l	$s+\varepsilon_{lm}+\mu_{lmD}-\dfrac{\varepsilon_{lm}\varepsilon_{\text{lim}}}{s+\varepsilon_{\text{lim}}+\mu_{\text{lim}D}}$
F	$C_{mD}(r_D,t_{\text{inj},D})+\dfrac{\varepsilon_m C_{imD}(r_D,t_{\text{inj}})}{s+\mu_{imD}+\varepsilon_{im}}$
$\varphi(\eta)$	$F\eta-[\delta_2+\eta E_a(\delta_1+\delta_2\eta)]$
$f_u(\eta_u)$	$C_{umD}(r_D,\eta_u,t_{\text{inj},D})+\dfrac{\varepsilon_{um}C_{\text{uim}D}(r_D,\eta_u,t_{\text{inj},D})}{s+\varepsilon_{\text{uim}}+\mu_{\text{uim}D}}-\dfrac{R_m v_{um}\alpha_r^2}{ABR_{um}}\jmath_2$ $-E_u(\jmath_1+\jmath_2\eta_u)$
$f_l(\eta_l)$	$C_{lmD}(r_D,\eta_l,t_{\text{inj},D})+\dfrac{\varepsilon_{lm}C_{\text{lim}D}(r_D,\eta_l,t_{\text{inj},D})}{s+\varepsilon_{\text{lim}}+\mu_{\text{lim}D}}+\dfrac{R_m v_{lm}\alpha_r^2}{ABR_{lm}}\dfrac{\overline{C}_{mD}}{z_{eD}-1}$ $-\overline{C}_{mD}E_l\dfrac{z_{eD}+\eta_l}{z_{eD}-1}$
$g(r_D,E_a;\eta)$	$g_1(r_D,E_a;\eta)=\mathcal{T}_1\exp(\frac{y_{\text{cha}}}{2})A_i(E_a^{\frac{1}{3}}y_{\text{cha}})+\mathcal{T}_2\exp(\frac{y_{\text{cha}}}{2})B_i(E_a^{\frac{1}{3}}y_{\text{cha}})$ $r_{wD}\le y_{\text{cha}}\le\eta$ $g_2(r_D,E_a;\eta)=\mathcal{T}_3\exp\left(\dfrac{y_{\text{cha}}}{2}\right)A_i(E_a^{\frac{1}{3}}y_{\text{cha}})+\mathcal{T}_4\exp\left(\dfrac{y_{\text{cha}}}{2}\right)B_i(E_a^{\frac{1}{3}}y_{\text{cha}})$ $\eta\le y_{\text{cha}}\le\infty$
$g_u(z_D,E_u;\eta_u)$	$g_{u1}(z_D,E_u;\eta_u)=N_1\exp(a_1z_D)+N_2\exp(a_2z_D)$ $1\le z_D<\eta_u$ $g_{u2}(z_D,E_u;\eta_u)=N_3\exp(a_1z_D)+N_4\exp(a_2z_D)$ $\eta_u\le z_D<\infty$

(*Continued*)

Table 3.8. *(Continued)*

$g_l(z_D, E_l; \eta_l)$	$g_{u1}(z_D, E_l; \eta_l) = M_1 \exp(b_1 z_D) + M_2 \exp(b_2 z_D) \quad -1 \le z_D < \eta_l$ $g_{u2}(z_D, E_l; \eta_l) = M_3 \exp(b_1 z_D) + M_4 \exp(b_2 z_D) \quad \eta_l \le z_D < -\infty$
M_1	$-M_2 \exp(b_1 - b_2)$
M_2	$\dfrac{-AB^2 R_{lm}}{R_m \alpha_r^2 D_l \left[\exp(b_2\eta_l - b_1\eta_l) - b_2 \exp(b_2\eta_l)\right]}$
M_3	$M_2 \exp(b_2\eta_l - b_1\eta_l) - M_2 \exp(b_1 - b_2)$
M_4	0
N_1	$-N_2 \exp(a_2 - a_1)$
N_2	$\dfrac{-AB^2 R_{um}}{R_m \alpha_r^2 D_u [(a_1 - a_2)\exp(a_2 - a_1)\exp(a_1\eta_u)]}$
N_3	0
N_4	$N_2 - N_2 \exp(a_2 - a_1)\exp(a_1\eta_u - a_2\eta_u)$
X	$\dfrac{\frac{1}{2} B_i({E_a}^{1/3} y_{\text{cha},w}) - {E_a}^{1/3} B_i'({E_a}^{1/3} y_{\text{cha},w})}{\frac{1}{2} A_i({E_a}^{1/3} y_{\text{cha},w}) - {E_a}^{1/3} A_i'({E_a}^{1/3} y_{\text{cha},w})}$
$\mathcal{J}_1$	$-\dfrac{\pi A_i(y_{\text{ext}}\vert_{r_D=\eta^+})}{E^{1/3}} X$
$\mathcal{J}_2$	$\dfrac{\pi A_i(y_{\text{ext}}\vert_{r_D=\eta^+})}{{E_a}^{1/3}}$
$\mathcal{J}_3$	$\dfrac{\pi A_i(y_{\text{ext}}\vert_{r_D=\eta^+})}{{E_a}^{1/3}} \left[\dfrac{B_i(y_{\text{ext}}\vert_{r_D=\eta^+})}{A_i(y_{\text{ext}}\vert_{r_D=\eta^+})} - X\right]$
$\mathcal{J}_4$	0
y_{cha}	$r_D + \dfrac{1}{4E_a}$
$y_{\text{cha},w}$	$r_{wD} + \dfrac{1}{4E_a}$

$$\overline{C}_{\text{lim}D} = \frac{\varepsilon_{\text{lim}}}{s + \varepsilon_{\text{lim}} + \mu_{lmD}} \overline{C}_{lmD} + \frac{C_{\text{lim}D}(r_D, z_D, t_{\text{cha},D})}{s + \varepsilon_{\text{lim}} + \mu_{lmD}}, \quad z_D \le -1 \tag{3.618f}$$

where $C_{mD}(r_D, t_{\text{cha},D})$ and $C_{imD}(r_D, t_{\text{cha},D})$ are the concentrations [Kg/m^3] of the aquifer at the end of the chaser phase, which could be calculated by Eqs. (3.617a) and (3.617b) after applying the inverse Laplace transform, $C_{umD}(r_D, z_D, t_{\text{cha},D})$ and $C_{\text{uim}D}(r_D, z_D, t_{\text{cha},D})$ are the concentrations [Kg/m^3] of the upper aquitard at the end of the chaser phase, which could be computed by Eqs. (3.617d) and (3.617e) after applying the inverse

Laplace transform, $C_{lmD}(r_D, z_D, t_{\text{cha},D})$ and $C_{\text{lim}D}(r_D, z_D, t_{\text{cha},D})$ are the concentrations [Kg/m^3] of the lower aquitard at the end of the chaser phase, which could be calculated by Eqs. (3.617f) and (3.617g) after applying the inverse Laplace transform.

As for the extraction phase of the SWPP test, the solutions in Laplace domain are

$$\overline{C}_{mD}(r_D, s) = \exp(-r_D/2)[U(r_D, \zeta; \varepsilon) + \sigma_1 + \sigma_2 r_D],$$
$$r_D \geq r_{wD} \tag{3.619a}$$

$$\overline{C}_{imD} = \frac{\varepsilon_{im}}{(s + \mu_{imD} + \varepsilon_{im})}\overline{C}_{mD} + \frac{C_{imD}(r_D, \text{res})}{s + \mu_{imD} + \varepsilon_{im}},$$
$$r_D \geq r_{wD} \tag{3.619b}$$

$$U(r_D, \zeta; \varepsilon) = \int_{r_{wD}}^{\infty} g(r_D, \zeta; \varepsilon) f(\varepsilon) d\varepsilon \tag{3.619c}$$

$$\overline{C}_{umD} = \int_1^{\infty} g_u(z_D, E_u; \vartheta_u) f_u(\vartheta_u) d\vartheta_u + \frac{z_D - z_{eD}}{1 - z_{eD}}\overline{C}_{mD}(r_D, s),$$
$$z_D \geq 1 \tag{3.619d}$$

$$\overline{C}_{\text{uim}D} = \frac{\varepsilon_{\text{uim}}\overline{C}_{umD}}{s + \varepsilon_{\text{uim}} + \mu_{\text{uim}D}} + \frac{C_{\text{uim}D}(r_D, z_D, t_{\text{res},D})}{s + \varepsilon_{\text{uim}} + \mu_{\text{uim}D}}, \quad z_D \geq 1 \tag{3.619e}$$

$$\overline{C}_{lmD} = \int_{-1}^{-\infty} g_l(z_D, E_l; \vartheta_l)\ f_l(\vartheta_l) d\vartheta_l + \frac{z_D + z_{eD}}{z_{eD} - 1}\overline{C}_{mD}(r_D, s),$$
$$z_D \leq -1 \tag{3.619f}$$

$$\overline{C}_{\text{lim}D} = \frac{\varepsilon_{\text{lim}}\overline{C}_{lmD}}{s + \varepsilon_{\text{lim}} + \mu_{\text{lim}D}} + \frac{C_{\text{lim}D}(r_D, z_D, t_{\text{res},D})}{s + \varepsilon_{\text{lim}} + \mu_{\text{lim}D}}, \quad z_D \leq -1 \tag{3.619g}$$

where $C_{mD}(r_D, t_{\text{res},D})$ and $C_{imD}(r_D, t_{\text{res},D})$ are the concentrations [Kg/m^3] of the aquifer at the end of the rest phase, which could be calculated by Eqs. (3.618a) and (3.618b) after applying the inverse Laplace transform, $C_{umD}(r_D, z_D, t_{\text{res},D})$ and $C_{\text{uim}D}(r_D, z_D, t_{\text{res},D})$ are the concentrations [Kg/m^3] of the upper aquitard at the end of the rest phase, which could be calculated by Eqs. (3.618c) and (3.618d) after applying the inverse Laplace transform, $C_{lmD}(r_D, z_D, t_{\text{res},D})$

and $C_{\text{lim}D}(r_D, z_D, t_{\text{res},D})$ are the concentrations [Kg/m^3] of the lower aquitard at the end of the rest phase, which could be calculated by Eqs. (3.618e) and (3.618f) after applying the inverse Laplace transform; ϑ_u varies between 1 and ∞; ϑ_l varies between -1 and $-\infty$; ε varies between r_{wD} and ∞ (e.g., $r_{wD} \le \varepsilon \le \infty$); $g(r_D, \zeta; \varepsilon)$, $g_u(z_D, E_u; \vartheta_u)$ and $g_l(z_D, E_l; \vartheta_l)$ are the Green's functions; the expressions for $g(r_D, \zeta; \varepsilon)$, $g_u(z_D, E_u; \vartheta_u)$, $g_l(z_D, E_l; \vartheta_l)$, $\sigma_1, \sigma_2, \Lambda, \zeta, f(\varepsilon), f_u(\vartheta_u), f_l(\vartheta_l), H_1 \sim H_4, I_1 \sim I_4$, $m_1 \sim m_2$, $n_1 \sim n_2$, $P_1 \sim P_4$, W, y_{ext}, $y_{\text{ext},w}$ and $\beta_{\text{ext},D}$ are listed in Table 3.9.

Because the analytical solutions in Laplace domain are too complex, it seems impossible to transform it into the real time domain analytically. Alternatively, a numerical method will be used for the invers Laplace transform. Here, the Stehfest method will be applied to conduct the inverse Laplace transform, since it performed well for radial-dispersion problems (Wang *et al.*, 2018).

3.3.3.3 *Assumptions included in the SWPP test model of Wang et al. (2020)*

The new SWPP test model of Wang *et al.* (2020) is a generalization of several previous studies; for instance, the solution of Wang *et al.* (2020) reduces to the solution of Gelhar and Collins (1971) when $\omega_a = \omega_u = \omega_l = D_u = D_l = v_{um} = v_{lm} = V_{w,\text{inj}} = V_{w,\text{cha}} = V_{w,\text{ext}} = t_{\text{cha}} = t_{\text{res}} = 0$, and Wang *et al.* (2018) when $\omega_a = \omega_u = \omega_l = D_u = D_l = v_{um} = v_{lm} = t_{\text{cha}} = t_{\text{res}} = 0$. "$t_{\text{cha}} = t_{\text{res}} = 0$" represents the four-phase SWPP test becomes the two-phase SWPP test, where the chaser and rest phases are excluded.

When $\omega_u = \omega_l = D_u = D_l = v_{um} = v_{lm} = V_{w,\text{inj}} = V_{w,\text{cha}} = V_{w,\text{ext}} = 0$, the solution of Wang *et al.* (2020) reduces to the solution of Chen *et al.* (2017). Figure 3.37 shows the comparison of BTCs between them, and the parameters used in such a comparison are $R_m = R_{im} = R_{um} = R_{\text{uim}} = R_{lm} = R_{\text{lim}} = 1$, $\theta_{um} = \theta_{lm} = 0.1$, $\alpha_r = \alpha_u = \alpha_l = 0.1\text{m}$, $\mu_m = \mu_{im} = \mu_{um} = \mu_{\text{uim}} = \mu_{lm} = \mu_{\text{lim}} = 10^{-6}\text{d}^{-1}$, $r_w = 0.2\text{m}$, $Q_{\text{inj}} = 2.5\,\text{m}^3/\text{d}$, $Q_{\text{cha}} = 2.5\,\text{m}^3/\text{d}$, $Q_{\text{res}} = 0\,\text{m}^3/\text{d}$, $Q_{\text{ext}} = -2.5\,\text{m}^3/\text{d}$, $t_{\text{inj}} = 100$ day, $t_{\text{cha}} = 50$ day, $t_{\text{res}} = 40$ day, $B = 5\,\text{m}$, $\theta_m = 0.3$, $\theta_{im} = 0.15$, $\theta_{\text{uim}} = \theta_{\text{lim}} = 0.1$, and $\omega = 0.001\text{d}^{-1}$. Parameters of "$h_{w,\text{inj}} = h_{w,\text{cha}} = h_{w,\text{res}} = h_{w,\text{ext}} = 0$" represent $V_{w,\text{inj}} = 0$, $V_{w,\text{cha}} = 0$ and $V_{w,\text{ext}} = 0$, and implies that wellbore storage is neglected. The values of $v_{um} = v_{lm} = 0\,\text{m/d}$ mean that aquitards are neglected. As shown

Table 3.9. Expressions of the coefficients in the solutions expressed in Eqs. (3.619a)–(3.619g).

Λ	$C_{mD}(r_D, t_{\text{res}}) + \dfrac{\varepsilon_m C_{imD}(r_D, t_{\text{res}})}{s + \mu_{imD} + \varepsilon_{im}}$
$\beta_{\text{ext},D}$	$-\dfrac{V_{w,\text{ext}} r_{wD}}{\xi R_m \alpha_r}$
ζ	$s + \varepsilon_m + \mu_{mD} - \dfrac{\varepsilon_{im}\varepsilon_m}{s + \mu_{imD} + \varepsilon_{im}} - \dfrac{\theta_{um}\alpha_r^2 v_{um}}{2A\theta_m B} + \dfrac{\theta_{lm}\alpha_r^2 v_{lm}}{2AB^2\theta_m}$ $-\dfrac{1}{1 - z_{eD}} \dfrac{\theta_{um}\alpha_r^2 D_u}{2A\theta_m b} + \dfrac{1}{z_{eD} - 1} \dfrac{\theta_{lm}\alpha_r^2 D_l}{2Ab^2\theta_m}$
$f(\varepsilon)$	$\exp(\varepsilon/2)\varepsilon\Lambda - \left(\varepsilon\zeta + \dfrac{1}{4}\right)(\sigma_1 + \sigma_2\varepsilon)$
$f_u(\mathscr{b}_u)$	$C_{umD}(r_D, \mathscr{b}_u, t_{\text{res},D}) + \dfrac{\varepsilon_{um} C_{\text{uim}D}(r_D, \mathscr{b}_u, t_{\text{res},D})}{s + \varepsilon_{\text{uim}} + \mu_{\text{uim}D}}$ $+\dfrac{R_m v_{um}\alpha_r^2}{ABR_{um}} \dfrac{\overline{C}_{mD}(r_D, s)}{1 - z_{eD}} - \dfrac{\mathscr{b}_u - z_{eD}}{1 - z_{eD}} E_u \overline{C}_{mD}(r_D, s)$
$f_l(\mathscr{b}_l)$	$C_{mD}(r_D, \mathscr{b}_l, t_{\text{res},D}) + \dfrac{\varepsilon_{lm} C_{\text{lim}D}(r_D, \mathscr{b}_l, t_{\text{res},D})}{s + \varepsilon_{\text{lim}} + \mu_{\text{lim}D}} - \dfrac{R_m v_{lm}\alpha_r^2}{ABR_{lm}}$ $\times \dfrac{\overline{C}_{mD}(r_D, s)}{z_{eD} - 1} - \dfrac{\mathscr{b}_l + z_{eD}}{z_{eD} - 1} E_l \overline{C}_{mD}(r_D, s)$
$g(r_D, \zeta; \varepsilon)$	$g_1(r_D, \zeta; \varepsilon) = P_1 A_i(y_{\text{ext}}) + P_2 B_i(y_{\text{ext}})$ $r_{wD} \le y_{\text{ext}} < \varepsilon$ $g_2(r_D, \zeta; \varepsilon) = P_3 A_i(y_{\text{ext}}) + P_4 B_i(y_{\text{ext}})$ $\varepsilon \le y_{\text{ext}} < \infty$
$g_u(z_D, E_u; \mathscr{b}_u)$	$g_{u1}(z_D, E_u; \mathscr{b}_u) = H_1 \exp(m_1 z_D) + H_2 \exp(m_2 z_D)$ $1 \le z_D < \mathscr{b}_u$ $g_{u2}(z_D, E_u; \mathscr{b}_u) = H_3 \exp(m_1 z_D) + H_4 \exp(m_2 z_D)$ $\mathscr{b}_u \le z_D < \infty$
$g_l(z_D, E_l; \mathscr{b}_l)$	$g_{l1}(z_D, E_l; \mathscr{b}_l) = I_1 \exp(n_1 z_D) + I_2 \exp(n_2 z_D)$ $-1 \le z_D < \mathscr{b}_l$ $g_{l2}(z_D, E_l; \mathscr{b}_l) = I_3 \exp(n_1 z_D) + I_4 \exp(n_2 z_D)$ $\mathscr{b}_l \le z_D < -\infty$
H_1	$-H_2 \exp(m_2 - m_1)$
H_2	$\dfrac{-AR_{um}B^2}{R_m\alpha_r^2 D_u[(m_1 - m_2)\exp(m_2 - m_1)\exp(m_1\mathscr{b}_u)]}$
H_3	0
H_4	$H_2 - H_2 \exp(m_2 - m_1)\exp(m_1\mathscr{b}_u - m_2\mathscr{b}_u)$
I_1	$-I_2 \exp(n_1 - n_2)$
I_2	$\dfrac{-AB^2 R_{lm}}{R_m\alpha_r^2 D_l[\exp(n_2\mathscr{b}_l - n_1\mathscr{b}_l) - n_2\exp(n_2\mathscr{b}_l)]}$
I_3	$I_2 \exp(n_2\mathscr{b}_l - n_1\mathscr{b}_l) - I_2 \exp(n_1 - n_2)$
I_4	0

(Continued)

Table 3.9. (*Continued*)

m_1	$\dfrac{-\dfrac{R_m v_{um}\alpha_r^2}{ABR_{um}} + \sqrt{\left(\dfrac{R_m v_{um}\alpha_r^2}{ABR_{um}}\right)^2 + 4\dfrac{R_m\alpha_r^2 D_u}{AB^2R_{um}}\left(s+\varepsilon_{um}+\mu_{umD}-\dfrac{\varepsilon_{um}\varepsilon_{\mathrm{uim}}}{s+\mu_{\mathrm{uim}D}+\varepsilon_{\mathrm{uim}}}\right)}}{2\dfrac{R_m\alpha_r^2 D_u}{AB^2R_{um}}}$
m_2	$\dfrac{-\dfrac{R_m v_{um}\alpha_r^2}{ABR_{um}} - \sqrt{\left(\dfrac{R_m v_{um}\alpha_r^2}{ABR_{um}}\right)^2 + 4\dfrac{R_m\alpha_r^2 D_u}{AB^2R_{um}}\left(s+\varepsilon_{um}+\mu_{umD}-\dfrac{\varepsilon_{um}\varepsilon_{\mathrm{uim}}}{s+\mu_{\mathrm{uim}D}+\varepsilon_{\mathrm{uim}}}\right)}}{2\dfrac{R_m\alpha_r^2 D_u}{AB^2R_{um}}}$
n_1	$\dfrac{\dfrac{R_m v_{lm}\alpha_r^2}{ABR_{lm}} + \sqrt{\left(\dfrac{R_m v_{lm}\alpha_r^2}{ABR_{lm}}\right)^2 + 4\dfrac{R_m\alpha_r^2 D_l}{AB^2R_{lm}}\left(s+\varepsilon_{lm}+\mu_{lmD}-\dfrac{\varepsilon_{lm}\varepsilon_{\mathrm{lim}}}{s+\mu_{\mathrm{lim}D}+\varepsilon_{\mathrm{lim}}}\right)}}{2\dfrac{R_m\alpha_r^2 D_l}{AB^2R_{lm}}}$
n_2	$\dfrac{\dfrac{R_m v_{lm}\alpha_r^2}{ABR_{lm}} - \sqrt{\left(\dfrac{R_m v_{lm}\alpha_r^2}{ABR_{lm}}\right)^2 + 4\dfrac{R_m\alpha_r^2 D_l}{AB^2R_{lm}}\left(s+\varepsilon_{lm}+\mu_{lmD}-\dfrac{\varepsilon_{lm}\varepsilon_{\mathrm{lim}}}{s+\mu_{\mathrm{lim}D}+\varepsilon_{\mathrm{lim}}}\right)}}{2\dfrac{R_m\alpha_r^2 D_l}{AB^2R_{lm}}}$
P_1	$-\dfrac{\pi A_i\left(y_{\mathrm{ext}}\vert_{r_D=\varepsilon^+}\right)}{\zeta^{1/3}}W$
P_2	$\dfrac{\pi A_i\left(y_{\mathrm{ext}}\vert_{r_D=\varepsilon^+}\right)}{\zeta^{1/3}}$
P_3	$\dfrac{\pi A_i\left(y_{\mathrm{ext}}\vert_{r_D=\varepsilon^+}\right)}{\zeta^{1/3}}\left[\dfrac{B_i\left(y_{\mathrm{ext}}\vert_{r_D=\varepsilon^+}\right)}{A_i\left(y_{\mathrm{ext}}\vert_{r_D=\varepsilon^+}\right)} - W\right]$
P_4	0
W	$\dfrac{\left(s\beta_{\mathrm{ext},D}+\frac{1}{2}\right)B_i\left(y_{\mathrm{ext},w}\right)-\zeta^{1/3}B_i'\left(y_{\mathrm{ext},w}\right)}{\left(s\beta_{\mathrm{ext},D}+\frac{1}{2}\right)A_i\left(y_{\mathrm{ext},w}\right)-\zeta^{1/3}A_i'\left(y_{\mathrm{ext},w}\right)}$
y_{ext}	$\zeta^{1/3}\left(r_D+\frac{1}{4\zeta}\right)$
$y_{\mathrm{ext},w}$	$\zeta^{1/3}\left(r_{wD}+\frac{1}{4\zeta}\right)$
σ_1	$-\dfrac{\beta_{\mathrm{ext},D}\exp(r_{wD}/2)C_{mD}(r_{wD},t_{\mathrm{res},D})}{\left(s\beta_{\mathrm{ext},D}+\frac{1}{2}\right)r_{wD}-1-\left(s\beta_{\mathrm{ext},D}+\frac{1}{2}\right)r_D\vert_{r_D\to\infty}}r_D\vert_{r_D\to\infty}$
σ_2	$\dfrac{\beta_{\mathrm{ext},D}\exp(r_{wD}/2)C_{mD}(r_{wD},t_{\mathrm{res},D})}{\left(s\beta_{\mathrm{ext},D}+\frac{1}{2}\right)r_{wD}-1-\left(s\beta_{\mathrm{ext},D}+\frac{1}{2}\right)r_D\vert_{r_D\to\infty}}$

in Figure 3.37, both solutions agree well for the mobile and immobile domains.

The new solutions of Wang *et al.* (2020) consider the fact that all values of ω_a, ω_u, ω_l, D_u, D_l, v_{um}, v_{lm}, $V_{w,\mathrm{inj}}$, $V_{w,\mathrm{cha}}$, and $V_{w,\mathrm{ext}}$ are nonzero in reality. In addition, two assumptions remain. First, the groundwater flow is horizontal in the aquifer and is vertical in the aquitard. This treatment relies on the basis that the permeability of

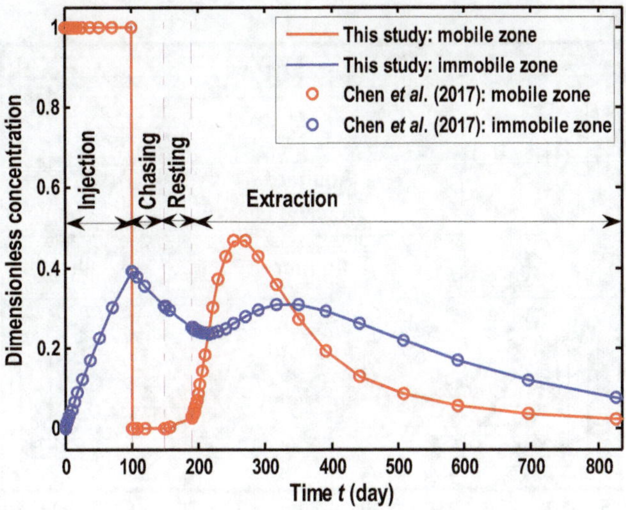

Figure 3.37. Comparison of BTCs at the well screen computed by the solution of Wang *et al.* (2020) and Chen *et al.* (2017).

the aquitard is smaller than the permeability of the aquifer (Moench, 1985). Second, the model is simplified for the solute transport. For example, only vertical dispersion and advection effects are considered in the aquitard, and only radial dispersion and advection effects are considered in the aquifer.

To test the three assumptions mentioned above, a 3D finite-element method with the help of COMSOL Multiphysics will be used to solve the 3D model. The grid mesh of the aquifer-aquitard system in the numerical modeling can be seen in Figure 3.38. In this modeling, the finite thickness of the aquitard is used to approximate its infinite thickness and the finite radial length of the aquifer is used to approximate its infinite radial length. Such treatment works well when the tracer has not approached the boundary.

In this section, four sets of aquitard hydraulic conductivities are employed, such as $K_u = K_l = 0.1\ K_a, K_u = K_l = 0.02\ K_a, K_u = K_l = 0.01\ K_a$, and $K_u = K_l = 0.001\ K_a$. A point to note is that the extreme case of $K_u = K_l = 0.1\ K_a$. The data used here are only for examining the robustness of comparison, while the real values of K_u and K_l are usually much lower than $0.1\ K_a$. In other words, the other three cases mentioned above are more likely to occur in real applications.

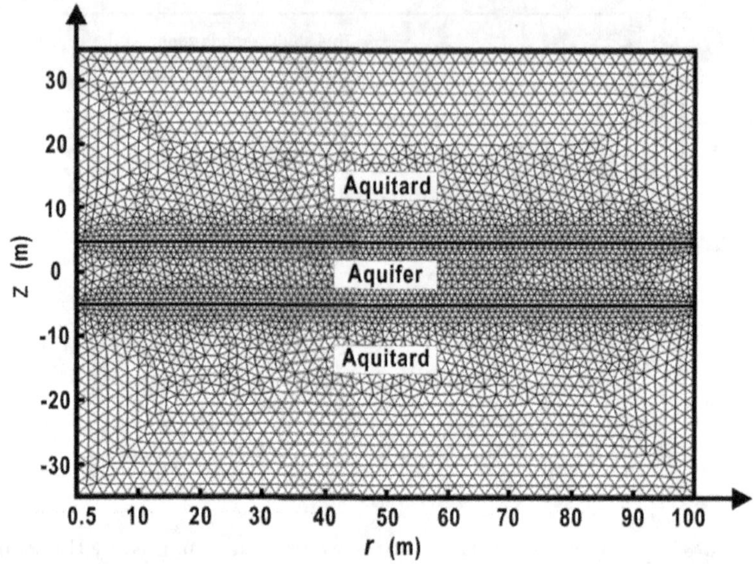

Figure 3.38. The grid mesh of the aquifer-aquitard system used in the Galerkin finite element program using COMSOL Multiphysics.

The initial drawdown and the initial concentration are 0 for aquifers and aquitards. The hydraulic parameters are $K_a = 0.1\,\text{m/day}$, $S_a = S_u = S_l = 10^{-4}\,\text{m}^{-1}$, and the other parameters are $R_m = R_{im} = R_{um} = R_{\text{uim}} = R_{lm} = R_{\text{lim}} = 1$, $\theta_{um} = \theta_{lm} = 0.1$, $\alpha_r = 2.5\,\text{m}$, $\alpha_u = \alpha_l = 0.5\,\text{m}$, $\mu_m = \mu_{im} = \mu_{um} = \mu_{\text{uim}} = \mu_{lm} = \mu_{\text{lim}} = 10^{-7}\,\text{s}^{-1}$, $r_w = 0.5\,\text{m}$, $Q_{\text{inj}} = Q_{\text{cha}} = 50\,\text{m}^3/\text{day}$, $Q_{\text{res}} = 0\,\text{m}^3/\text{day}$, $Q_{\text{ext}} = -50\,\text{m}^3/\text{d}$, $t_{\text{inj}} = 250\,\text{day}$, $t_{\text{cha}} = 50\,\text{day}$, $t_{\text{res}} = 50\,\text{day}$, $B = 10\,\text{m}$, $\theta_m = 0.25$, $\theta_{im} = 0.05$, and $\omega = 0.01\,\text{day}^{-1}$. The comparison of concentration between the analytical of Wang *et al.* (2020) and numerical solutions is shown in Figures 3.39 and 3.40.

The following discussion will focus on the two assumptions mentioned above. Figures 3.39(a)–3.39(c) represent the snapshots of concentration distributions in the aquifer along the r-axis at different times. One may conclude that curves with smaller K_u and K_l values are closer to the analytical solution. This is because aquitards with smaller K_u and K_l (when K_a remains constant) could make flow closer to the horizontal direction (or parallel with the aquitard-aquifer interface) in the aquifer and closer to the vertical direction (or perpendicular with the aquitard-aquifer interface) in the aquitard,

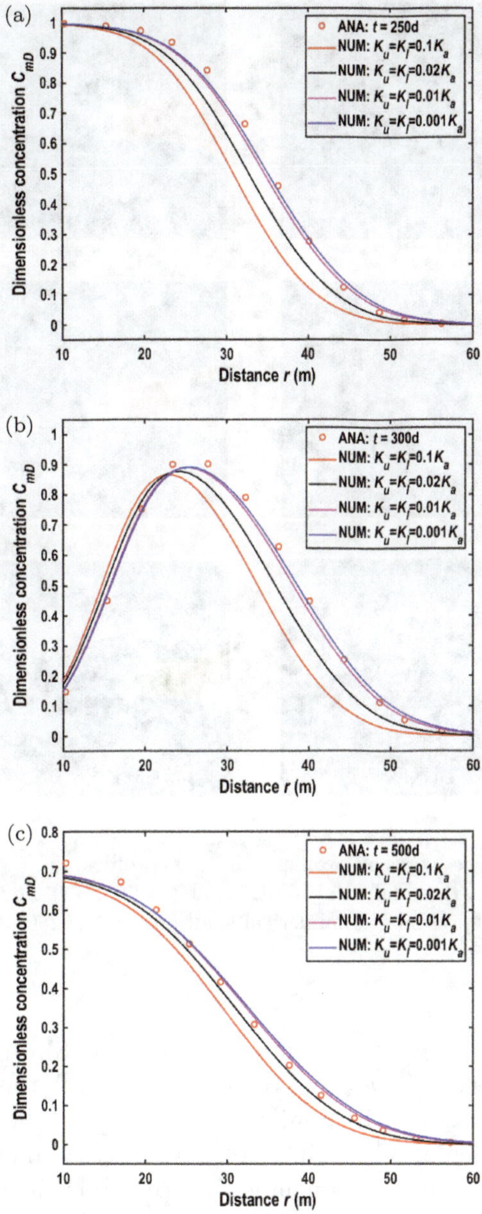

Figure 3.39. Comparison of the concentration distribution between the analytical and numerical solutions along the r-axis at $z = 0$ m. "ANA" and "NUM" represent the analysis of Wang *et al.* (2020) and numerical solutions, respectively. (a) At the end of the injection phase: $t = 250$ days; (b) At the end of the chasing phase: $t = 300$ days; (c) In the extraction phase: $t = 500$ days.

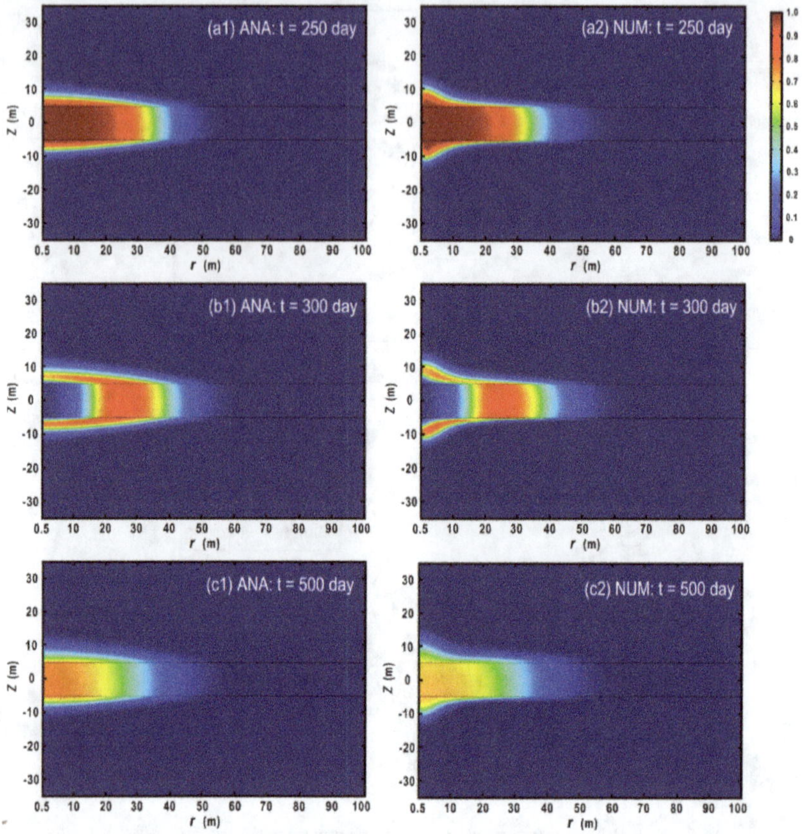

Figure 3.40. The vertical profiles (the r–z profiles) of the concentrations. (a1)–(c1) represent the analytical solutions at $t = 250$, 300, and 500 day, respectively. (a2)–(c2) represent the numerical solutions at $t = 250$, 300, and 500 day, respectively.

according to the law of refraction (Fetter, 2018). In other words, when the values of K_u/K_a and K_l/K_a approach 0, the flow direction becomes horizontal in the aquifer and vertical in the aquitard, and then the numerical model reduces to the analytical model. Therefore, from this figure, one may conclude that the above-mentioned second assumption in Section 3.3 works well in the aquifer when K_u/K_a and K_l/K_a are smaller than 0.01.

Figure 3.40 shows the comparison of the analytical and numerical solutions for aquitards. Figures 3.40(a1)–3.40(c1) represent the snapshots of concentration distributions obtained from the analytical solution of Wang *et al.* (2020) at different times, and Figures 3.40(a2)–3.40(c2) represent the snapshots of concentration distributions obtained from the numerical solution. One may find that the contour maps obtained from both solutions are almost the same in the aquifer but very different in the aquitards. Therefore, the above-mentioned second assumption is generally unacceptable in describing solute transport in the aquitard in the SWPP test, but works well when the aquifer is of the primary concern.

3.3.3.4 *Parameter sensitivity analysis*

(a) A global sensitivity analysis

From the analytical solutions of Eqs. (3.616)–(3.619), one may find that BTCs are affected by several parameters, like α_u, v_{um}, θ_{um}, ω, α_r, θ_m and V_w. As α_l, v_{lm}, θ_{lm} have a similar effect on the results with α_u, v_{um}, θ_{um}, they have been excluded in the following analysis. In this section, a global sensitivity analysis is conducted using the model of Morris (1991), which is a one-step-at-a-time method. Morris (1991) employed μ_k and σ_k to represent the importance of the input parameters on the output concentration, and they could be computed by (Lin *et al.*, 2019; Morris, 1991):

$$\mu_k = \sum_{l=1}^{M} (|EE_k^l|/M), \quad k = 1, 2, \ldots, N \tag{3.620a}$$

$$\sigma_k = \sqrt{\frac{1}{M}\sum_{l=1}^{M} (EE_k^l - \mu_k)}, \quad k = 1, 2, \ldots, N \tag{3.620b}$$

where M is the total sampling number, assuming that the range of parameter value is divided to M intervals; N is the total parameter number of interest, and it is 7 in this section; k is the kth parameter. In this section, $M = 50$; $EE_k^l = \frac{C_{mD}(P_1,P_2,\ldots,P_k+l\Delta,\ldots,P_N)-C_{mD}(P_1,P_2,\ldots,P_k,\ldots,P_N)}{l\Delta}$, where P_i is the random value of the ith parameter in the range of $(P_{i,0}, P_{i,lim})$; $P_{i,0}$ and $P_{i,lim}$ are the smallest and largest values of P_i, as shown in Table 3.10; Δ is a small increment defined as $1/(M-1)$.

Table 3.10. Parameter range used in the sensitivity analysis.

Parameter	Unit	Range
α_u	m	0.05–0.50
α_r	m	0.50–1.00
v_{um}	m/d	0–0.01
θ_{um}	—	0–0.2
ω	1/s	0.0001–0.001
θ_m	—	0.20–0.40
V_w	m^3	0.10–500

Note: "—" represents a dimensionless unit.

A larger μ_k means a higher sensitive effect of the kth parameter on the output and a larger σ_k represents that the kth parameter has a greater interaction effect with others. Figures 3.41(a) and 3.41(b) represent the variation of μ_k and σ_k with time in the wellbore, respectively. The values of μ_k are greater for α_r and θ_m than for the others, as shown in Figures 3.41(a), indicating that the influence of θ_m and α_r on the results is more obvious than others. However, the values of σ_k is large for α_u, θ_{um}, α_r, θ_m, and V_w, demonstrating that these parameters interact strongly with others; namely, their influence on results cannot be ignored.

(b) Effect of the aquitard

In this section, we try to figure out how the aquitards will affect BTCs of the SWPP tests. Since the porosity is an important factor of concern, three sets of porosity values are used for the aquitards: $\theta_{um} = \theta_{lm} = 0$, 0.1, and 0.25. The other parameters are from the case in Figure 3.40.

Figure 3.42 shows the difference between the models with and without aquitards for different flow velocities in the aquitard. The case of $\theta_{um} = \theta_{lm} = 0$ represents the model without the aquitard. The difference is not obvious at the beginning of the extraction phase, while such a difference is obvious at a later time. Meanwhile, the smaller aquitard porosity makes the value of BTCs in the aquifer greater at a given time. When the aquitard is ignored, the values of BTCs are the greatest. Therefore, the aquitard effect on transport in the aquifer is quite obvious and should not be ignored in general.

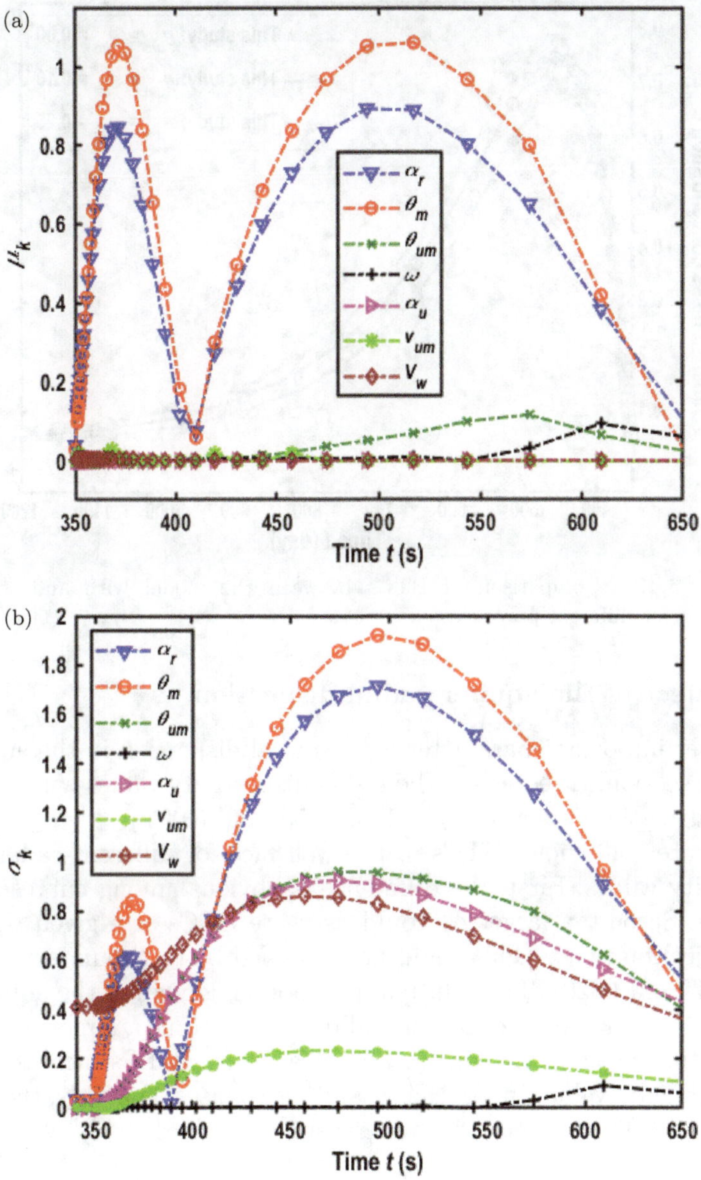

Figure 3.41. Sensitivity analysis. (a) Variation of μ_k with time in the wellbore. (b) Variation σ_k with time in the wellbore.

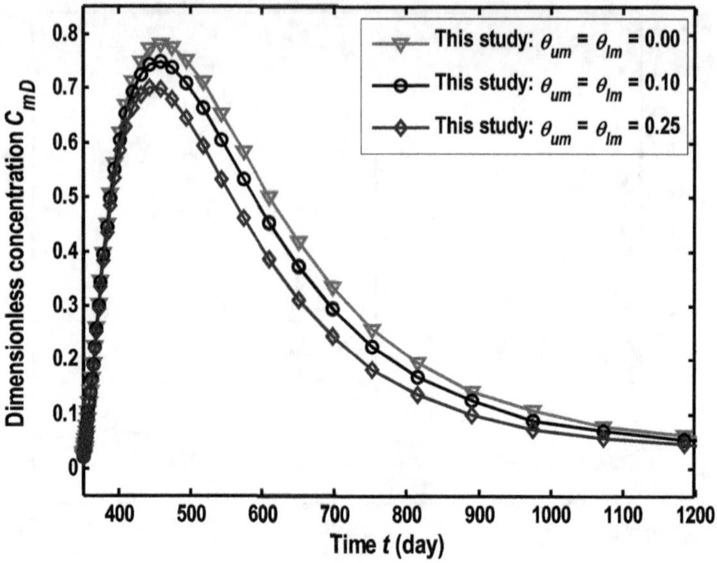

Figure 3.42. Comparison of BTCs between the model with and without aquitards for different porosities.

(c) Effect of the aquifer radial dispersion

Another important parameter is the radial dispersion in the aquifer. In this section, three sets of the radial dispersivity values will be used to analyze the influence: $\alpha_r = 1.25$, 2.50, and 5.00 m.

Figure 3.43 shows BTCs in the well face for different radial dispersivity values. First, the difference is obvious among curves in all phases. Second, a larger α_r could decrease BTCs at a given time of the injection phase. This could be explained by the boundary condition of Eq. (3.520). The solute in the mobile domain of the aquifer is transported by both advection and dispersion, thus a larger α_r could lower the values of C_m in the well face. Third, BTCs increase with increasing α_r values in the chaser and rest phases. Fourthly, the peak values of BTCs decrease with increasing α_r values.

3.3.3.5 *Comparison among the two models using the field SWPP test*

To test the performance of the new model, the field data reported in Chen *et al.* (2017) will be employed. Specifically, the experimental

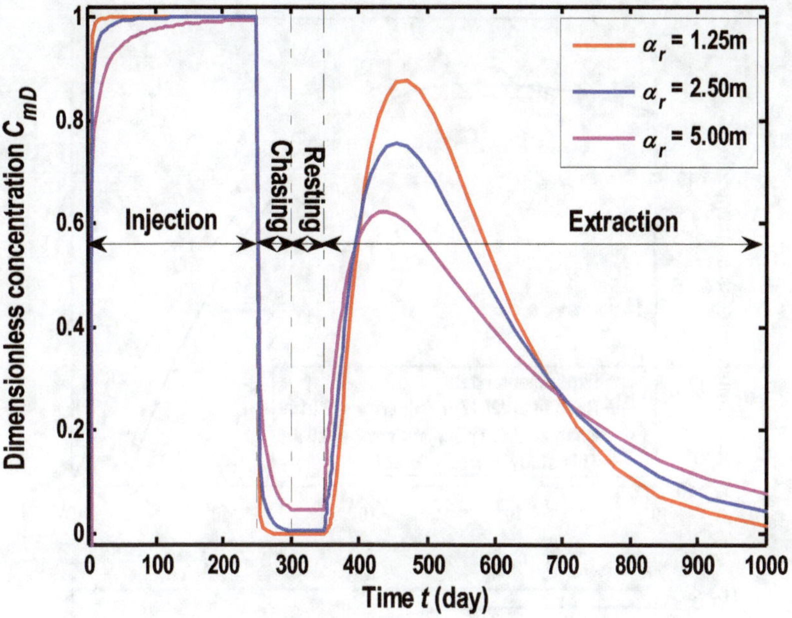

Figure 3.43. BTCs in the wellbore for different α_r.

data conducted in the borehole TW3 will be analyzed. The reason for choosing this dataset is that this borehole penetrated several layers, and it had been interpreted by Chen *et al.* (2017) before (using a model without considering the aquitard effect and the wellbore storage). The physical parameters of the SWPP test are $r_w = 0.1$ m, $Q_{\text{inj}} = Q_{\text{cha}} = 7.78$ L/min, $Q_{\text{res}} = 0$ L/min, $Q_{\text{ext}} = 12$ L/min, $t_{\text{inj}} = 180$ min, $t_{\text{cha}} = 26.74$ min, $t_{\text{res}} = 10080$ min, $B = 4$ m. The other information on experimental data can be seen in the references of Assayag *et al.* (2009) and Yang *et al.* (2014).

Figure 3.44(a) shows the fitness of the computed and observed BTCs. The estimated parameters are $\theta_{um} = 0.05$, $\theta_{lm} = 0.0$, $\theta_m = 0.1$, $\theta_{im} = 0.068$, $\alpha_r = 0.5$ m, $\alpha_u = 0.35$ m, $\alpha_l = 0.0$m, $R_m = R_{im} = R_{um} = R_{\text{uim}} = R_{lm} = R_{\text{lim}} = 1$, $\mu_m = \mu_{im} = \mu_{um} = \mu_{\text{uim}} = \mu_{lm} = \mu_{\text{lim}} = 10^{-7}$1/s, and $\omega = 0.001\,\text{day}^{-1}$, and $h_{w,\text{inj}} = h_{w,\text{cha}} = 32$ m, $h_{w,\text{res}} = 30$ m, $h_{w,\text{ext}} = 28$ m. Apparently, the fitness of the new solution is better than the model of Chen *et al.* (2017). As for the error between the observed and computed BTCs, the new solution is also smaller than that of Chen *et al.* (2017) as

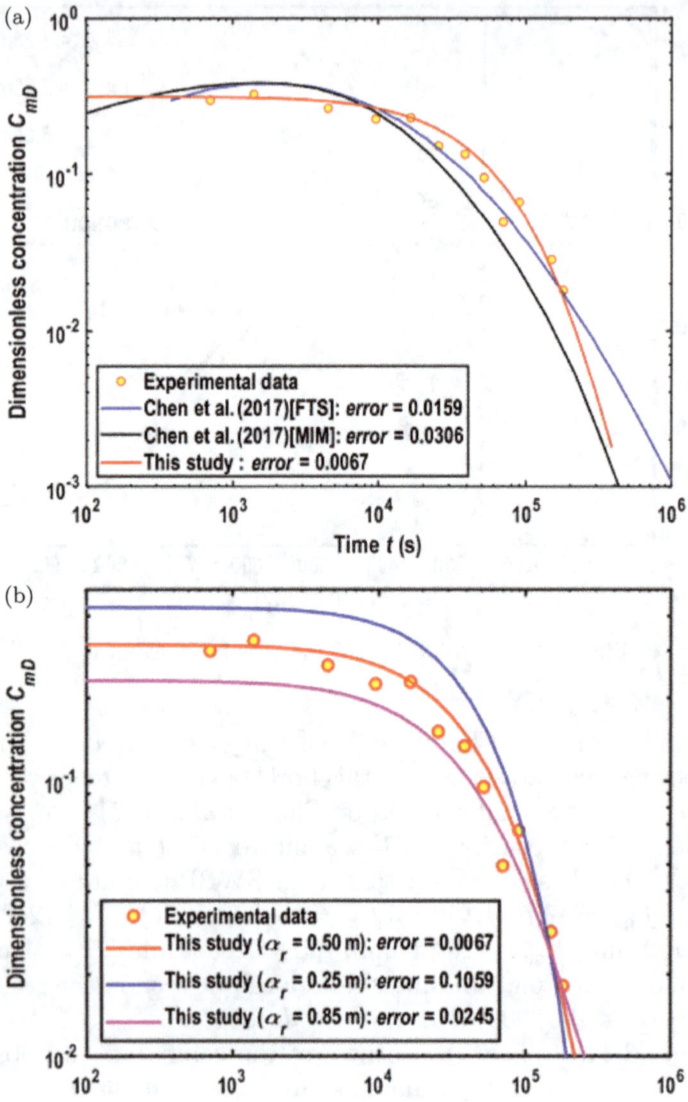

Figure 3.44. Fitness of observed BTC. (a) Fitness of the observed data by different models; (b) influence of the dispersivity of the aquifer on BTCs.

well, where the error is defined as

$$\text{Error} = \sum_{i=1}^{N} (C_{\text{OBS}} - C_{\text{COM}})^2 \tag{3.621}$$

where C_{OBS} and C_{COM} are the observed and computed concentrations, respectively, and N is the number of sampling points.

How accurate these parameters estimated by best fitting the observed data are in the representative of the real aquifer will be discussed as follows. The values of the retardation factor and reaction rate demonstrate that the chemical reaction and sorption are weak for the tracer of KBr in the SWPP test. It is not surprising since KBr is commonly treated as a "conservative" tracer. The porosity of the real aquifer ranges from 0.01 to 0.1, according to the well log analysis (Yang *et al.*, 2014), where the estimated values are located. The estimated porosity represents the average values of the aquifer and aquitards. The estimated dispersivity of the aquifer is 0.7134 m by Chen *et al.* (2017), which is similar to ours. The values of water level in the test could be observed directly; however, these data are not available, and they have to be estimated in the model of Wang *et al.* (2020). To evaluate the uncertainty in the estimated parameters, the sensitivity of the dispersivity on BTCs is analyzed, as shown in Figure 3.44(b). One may conclude that the estimated values of the model of Wang *et al.* (2020) seem to be representative of reality since $Error$ is smallest for $\alpha_r = 0.5\,\text{m}$.

3.4 Numerical Models of Radial Tracer Test

3.4.1 *3D reactive transport around the well*

Given the intricate nature of reactive transport in a wellbore-aquifer system, the expertise of hydrogeologists, environmental engineers, and researchers is crucial. Their understanding of the physical and biochemical processes of solute, combined with observations during experimental studies, has been instrumental in the development and application of numerical modeling techniques. (Ma *et al.*, 2011; McMillan *et al.*, 2014). The previous numerical models, which treated the wellbore as a source/sink, did not adequately describe the wellbore storage of solute transport in the confined aquifer. They failed to consider the actual water level in the wellbore, instead using the aquifer thickness to handle the wellbore storage, which is incorrect. This oversight underscores the urgent need for improved numerical models that do not ignore the wellbore mixing effect.

According to the treatment of solute transport in the wellbore, the mathematical models could be classified into two types. The first type of mathematical model considers the wellbore as an inner boundary condition of reactive transport in the aquifer (named the IBC model), and the IBC model is usually used in analytical solutions. The second type of mathematical model, the SS model, treats the well as a source or sink and is commonly used in numerical solutions due to the complexity of hydrogeological conditions, such as heterogeneity and transiency of the flow field. We acknowledge and appreciate the challenging environments in which our audience operates.

The SS model of the tracer test is composed of two parts: solute transport in the wellbore and the aquifer. In the confined aquifer, the wellbore is open to the air, hence making it unconfined. After a careful literature review, we found that previous numerical solutions of 2D and 3D SS models of tracer test in the Cartesian coordinate system did not properly deal with the mixing processes between the original water and tracers entering into the operational wellbore in a confined aquifer, as they did not consider the actual water level in the wellbore. It, instead, used the aquifer thickness to tackle the wellbore storage, which is incorrect.

The objectives of this section include the revisit of the treatment of wellbore storage in mathematical models of reactive transport in a wellbore-aquifer system, the revision of the numerical solution of the SS model describing the wellbore mixing processes between solutes in the wellbore during the solute transport, and the validation of the revised model, and the influence of wellbore storage on tests involving an injection or an extraction well (Gan and Wang, 2024).

3.4.1.1 *Revisit of models for reactive transport in a wellbore-aquifer system*

(a) The IBC model of solute transport

When the wellbore is considered as an inner boundary condition, the wellbore-aquifer system reduces the aquifer system, as the concentration variation of solute in the wellbore is not included in the governing equation (Veling, 2012; Wang and Zhan, 2013), e.g.,

$$\frac{\partial(\theta C^k)}{\partial t} = L_{DSP}(C^k) + L_{ADV}(C^k) + L_{SSM}(C^k) + L_{RCT}(\theta C^k), \quad t > 0 \tag{3.622}$$

$$C^k(r, z, t)|_{r=r_w} = f(z, t), \quad t > 0 \tag{3.623a}$$

or

$$\left[v_r C^k(r, z, t) - \alpha_r |v_r| \frac{\partial C^k(r, z, t)}{\partial r}\right]\Bigg|_{r=r_w} = v_r f(z, t), \quad t > 0 \tag{3.623b}$$

where C^k is dissolved concentration of species k [Kg/m^3]; k is a positive integral to account for the number of species [dimensionless]; t is time [s]; r is the radial distance from the wellbore [m]; z is vertical distance; θ is the porosity of the porous medium [dimensionless]; r_w is the wellbore radius [m]; α_r and v_r are the radial dispersivity [m^2/s] and radial flow velocity [m/s], respectively; $L_{DSP}(\cdot)$, $L_{ADV}(\cdot)$, $L_{SSM}(\cdot)$, and $L_{RCT}(\cdot)$ are the operators for dispersion, advection, other sink/sources excluding the discharge/recharge in the wellbore, and chemical reaction terms, respectively; $f(t)$ represents the concentration variations of solute in the wellbore, which is a function of time.

Equation (3.622) is the multi-species governing equation of reactive transport. Equations (3.623a) and (3.623b) are the inner boundary conditions, representing the RCC and the flux concentration continuity at the well-aquifer interface, respectively, while Eq. (3.623b) is recommended since it could keep the mass balance for solute transport in the aquifer. The difference between them can be seen in Schwartz *et al.* (1999) and Novakowski (1992a).

This type of model is generally established in the radial coordinate system, such as (Wang and Zhan, 2013). This is because the flow field is radial when only one well exists, and the regional flow (or ambient flow) is negligibly small. The advantage of the radial coordinate system is that it could simplify the mathematical models from two dimensions into one dimension (Chen, 1985, 2010; Chen *et al.*, 2011b, 2012; Novakowski, 1992a) or from three dimensions into two dimensions (Chen, 1985, 2010; Chen *et al.*, 2012; Novakowski, 1992a), for which elegant analytical models may be developed. As for the 2D radial transport, the operators of $L_{DSP}(\cdot)$, $L_{ADV}(\cdot)$, $L_{SSM}(\cdot)$, and $L_{RCT}(\cdot)$ are

$$L_{DSP}(C^k) = \frac{1}{r}\frac{\partial}{\partial r}\left(rD_{rr}\frac{\partial C^k}{\partial r} + rD_{rz}\frac{\partial C^k}{\partial z}\right) + \frac{\partial}{\partial z}\left(D_{zr}\frac{\partial C^k}{\partial z} + D_{zz}\frac{\partial C^k}{\partial z}\right) \tag{3.624}$$

$$L_{ADV}(C^k) = v_r\frac{\partial C^k}{\partial r} + v_z\frac{\partial C^k}{\partial z} \tag{3.625}$$

$$L_{SSM}(C^k) = q_s C^k \tag{3.626}$$

$$L_{RCT}(\theta C^k) = \sum R_n \tag{3.627}$$

where D_{rr}, D_{rz}, D_{zr} and D_{zz} are the four components of the dispersion coefficient tensor [m^2/s], respectively; $\sum R_n$ is chemical reaction term [Kg/(m$^3 \cdot$ s)]; q_s is the volumetric flow rate per unit volume, which does not include the pumpage of the wellbore [1/s]; v_z is vertical flow [m/s].

However, these types of models have two shortcomings. First, the flow field may not be radial in realistic aquifer settings, for instance, when more than one well exists, or the regional flow could be ignored. Second, Eq. (3.623a) or Eq. (3.623b) is used to describe the transport at the well screen, and the concentration in the wellbore is required, e.g., $f(z, t)$, which may be unknown in reality. For simplicity, many studies have assumed that $f(z, t)$ equals the concentration of the injected solute inside the wellbore (Chen, 1985, 2010; Phanikumar and McGuire, 2010; Yeh and Chang, 2013):

$$f(z, t) = C^k_{\text{inj}} \tag{3.628}$$

where C^k_{inj} is the concentration of species k in the injected solute [Kg/m^3]. This is not true since Eq. (3.628) does not consider the mixing processes between original water and tracers entering the operational wellbore and overestimates the concentration values in the wellbore. Novakowski (1992a) presented a well model considering the wellbore storage for different scenarios based on the mass balance principle, while the flow field was assumed to be in a steady state, and the mixing processes were assumed to be instantaneously completed.

(b) The SS model of solute transport

Because of the limitations of models where the wellbore is treated as an inner boundary condition, the popular way is to take the wellbore as a source/sink term in the governing equation of reactive transport in the numerical modeling. The governing equation of multi-species reactive transport in the wellbore-aquifer system becomes (Konikow and Grove, 1977; Zheng and Wang, 1999):

$$\begin{aligned}\frac{\partial(\theta C^k)}{\partial t} &= L_{DSP}(C^k) + L_{ADV}(C^k) + L_{SSM}(C^k) \\ &\quad + L_{RCT}(\theta C^k) + q_w C_w^k, \quad t > 0 \end{aligned} \tag{3.629}$$

where q_w is volumetric flow rate per unit volume of aquifer [m/s], and it is positive for injection (when the well acts as a source) and negative for extraction (when the well acts as a sink); C_w^k is the concentration of species k [Kg/m^3], and it is equal to C^k in case of extraction ($q_s < 0$); the operators of $L_{DSP}(\cdot)$ and $L_{ADV}(\cdot)$ are different from ones defined in Eq. (3.622), for instance:

$$L_{DSP}(C^k) = \frac{\partial}{\partial x_i}\left(\theta D_{ij}\frac{\partial C^k}{\partial x_j}\right) \tag{3.630}$$

$$L_{ADV}(C^k) = \frac{\partial}{\partial x_i}(\theta v_i C^k) \tag{3.631}$$

where x_i is distance [m] along the respective Cartesian coordinate axis, $i = 1$, 2, and 3, representing the x-, y-, and z-axes, respectively; D_{ij} is hydrodynamic dispersion coefficient tensor [m^2/s]; v_i is flow velocity. The definitions of $L_{SSM}(\cdot)$ and $L_{RCT}(\cdot)$ are the same as the ones in Eq. (3.622). The boundary of Eq. (3.625) is set far away from the well, where the concentration is equal to the background value.

Different from the first approach in Eqs. (3.622)–(3.628), only the values of q_w and C_w^k are required, which are generally available. Equations (3.629)–(3.631) have been widely employed for solute transport modeling in many software packages, like MODFLOW/MT3DMS (Zheng and Wang, 1999), FEFLOW (Trefry and Muffels, 2007),

TOUGH2 (Pruess *et al.*, 1999). The finite-difference scheme of reactive transport in the wellbore-aquifer system is (Konikow and Grove, 1977; Zheng and Wang, 1999):

$$MA_{\text{Cell}} = NMF_x + NMF_y + NMF_z + NMFSS_{\text{Cell}} + NMFR_{\text{Cell}} \tag{3.632}$$

where

$$NMF_x = -\frac{\partial}{\partial x}(C^k \theta v_x^* \Delta y \Delta z)\Delta x \tag{3.633a}$$

$$NMF_y = -\frac{\partial}{\partial y}(C^k \theta v_y^* \Delta x \Delta z)\Delta y \tag{3.633b}$$

$$NMF_z = -\frac{\partial}{\partial z}(C^k \theta v_z^* \Delta x \Delta y)\Delta z \tag{3.633c}$$

$$NMFSS_{\text{Cell}} = W_s C_s^k \Delta x \Delta y \Delta z \tag{3.633d}$$

$$NMFR_{\text{Cell}} = \theta \Delta x \Delta y \Delta z \sum R_k \tag{3.633e}$$

$$MA_{\text{Cell}} = \frac{\partial}{\partial t}(C^k \theta \Delta x \Delta y \Delta z) \tag{3.633f}$$

where v_x^*, v_y^*, and v_z^* are instantaneous mass velocities [m/s] along the x, y, and z axes, respectively; W_s is volumetric flux per volume of porous medium [m/s]; C_s^k is the concentration [Kg/m^3] of the solute in the source or sink fluid; Δx, Δy, and Δz are the dimensions [m] of cells along the x, y, and z axes, respectively; MA is mass accumulation rate [Kg/s]; NMF is net mass flux [Kg/s]; $NMFSS$ is net mass flux by source and sink [Kg/s]; $NMFR$ is net mass flux by reactions [Kg/s]. When $\Delta x = \Delta y$ and $\Delta x \Delta y = \pi r_w^2$ in the wellbore cell, W_s is injection or extraction rate per volume, and C_s^k is the concentration of the injection or extraction solute.

Equations (3.632)–(3.623) are used in the WELL package of MT3DMS for modeling reactive transport in the wellbore-aquifer system, which is suitable for the one-cell wellbore model, not for the multi-node well (MNW) model. The MNW model refers to the case when the wellbore is vertically discretized into several cells (e.g., Cell 1, Cell 2, Cell 3, and Cell 4, as shown in Figure 3.45). In the one-cell wellbore model, the intra-borehole flow is ignored. As for the MNW model, the intra-borehole flow is considered, and there is a special package was developed for both groundwater flow and solute

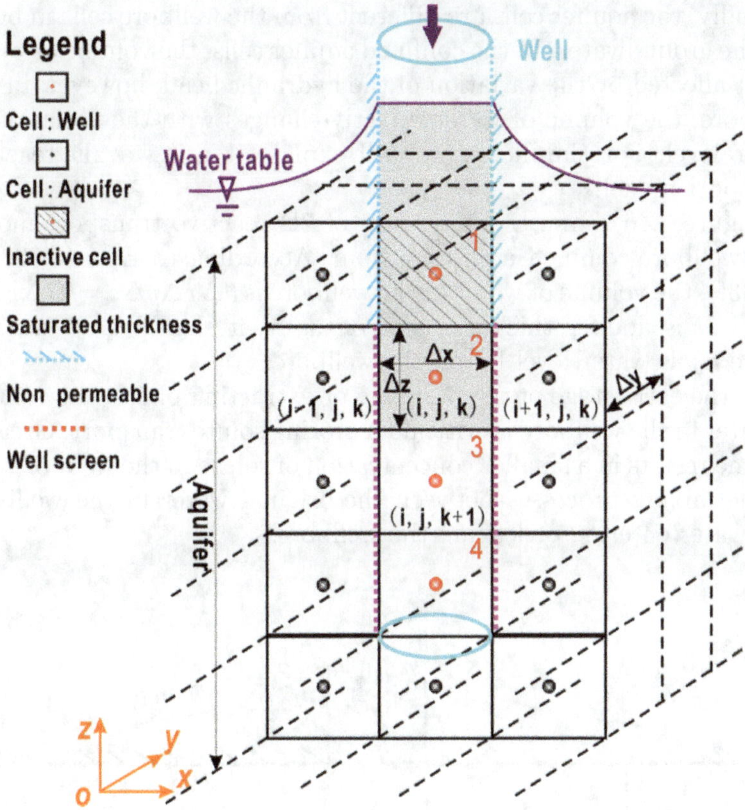

Figure 3.45. Schematic diagram of the grid system in a numerical simulation of a partially penetrating well. Black lines represent the discretization of the aquifer, including the wellbore in the aquifer (e.g., Cell 1, Cell 2, Cell 3, and Cell 4). The part of the wellbore located above the aquifer is not included in the grid system.

transport based on MODFLOW/MT3DMS, named the MNW package (Konikow and Hornberger, 2006; Konikow *et al.*, 2009).

(c) Weaknesses of the SS models

Equations (3.633a)–(3.633f) demonstrate that the weakness of the second type of model is that the wellbore is treated as a part of the aquifer, resulting in the following two problems. First, the porosity of the wellbore is unity, but it is assumed to be the same as the porosity of the surrounding aquifer. Second, the term of $\theta \Delta x \Delta y \Delta z$ represents the volume of the water in the cells of the grid system in Figure 3.45, regardless of aquifer and wellbore. It is constant.

Actually, the aquifer cells are different from the wellbore cells in bearing the groundwater. In the confined aquifer cells, the volume of water is not affected by the variation of the hydraulic head; however, in the wellbore, the volume of water directly changes with the variation of water level. For simplicity, we will explain it using a 2D reactive transport model.

Figure 3.46 shows a grid system of 2D reactive transport model in a wellbore-confined aquifer system. According to Eqs. (3.633a)–(3.633f), the volume of water in the wellbore is $\theta \Delta x \Delta y \Delta z = B \Delta x \Delta y$, where B is aquifer thickness [m]. Actually, it should be $h_w \Delta x \Delta y$, where h_w is water level [m] in the wellbore.

In the early stage of the pumping or extracting phase, the volume of water in the wellbore is critical for storing solute transport. Greater volume results in a smaller concentration of solute in the wellbore due to the mixing processes between the original water in the wellbore and water entering or leaving the wellbore.

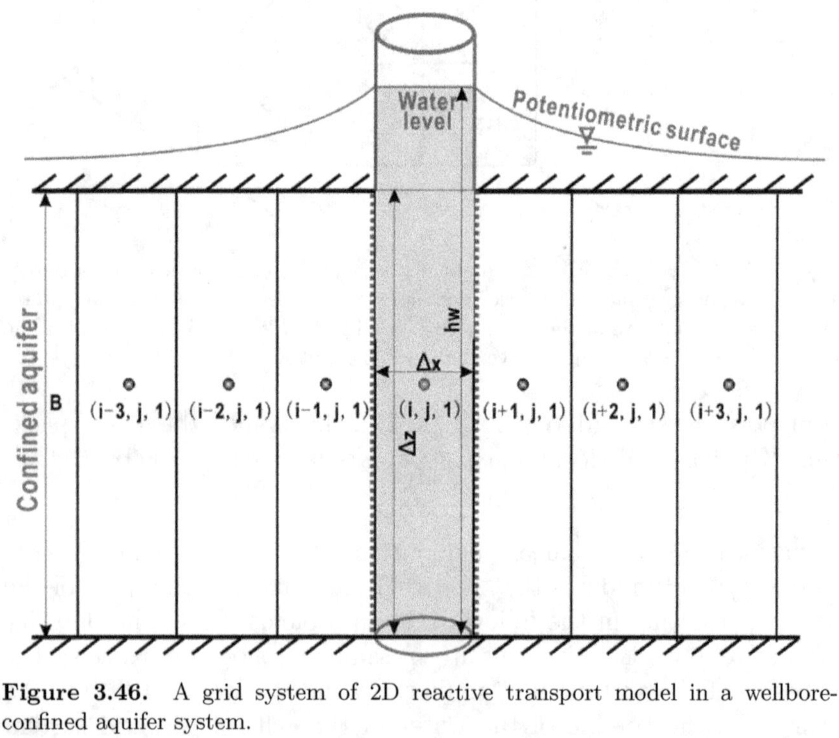

Figure 3.46. A grid system of 2D reactive transport model in a wellbore-confined aquifer system.

3.4.1.2 *New model treating wellbore storage*

(a) Revision of Eqs. (3.629)–(3.633)

Equations (3.629)–(3.633) are called the "previous models" from now on and will be revised by considering the water level variation in a wellbore. Including the wellbore cells in the numerical simulation of flow in a well-aquifer system imposes new challenges. For instance, the simulated aquifer is confined, whereas the simulated open wellbore is unconfined. The wellbore may include permeable screen sections and impermeable casings, as shown in Figure 3.45. Therefore, how to treat the wellbore cells in the numerical models needs to be clarified.

Figure 3.45 shows the grid system for the general case in the numerical simulation. The well is discretized into several cells, e.g., Cell 1, Cell 2, Cell 3, and Cell 4, and such a well is named a multi-node well. When the well is discretized into one cell, a multi-node well reduces to a one-cell well. Cell 2, Cell 3, and Cell 4 in Figure 3.45 represent the permeable screen, which could be treated as point sources/sinks in the model. Cell 1 in Figure 3.45 represents the impermeable casing, which is the uppermost cell above the screen inside the wellbore.

As for Cell 1, the lateral boundary is impermeable, which implies that it can only exchange water with Cell 2. Therefore, Cell 1, wellbore above Cell 1, and Cell 2 could be combined into one cell, e.g., a revised Cell 2. The volume of water in this revised Cell 2 is the summation of water in Cell 1, water in the wellbore above Cell 1, and water in the original Cell 2. Namely, the volume of water in this revised cell is $\Delta x \Delta y B_{\text{Cell2},w}$, where $B_{\text{Cell2},w} = h_w - z_{\text{Cell2,bot}}$; $z_{\text{Cell2,bot}}$ is the vertical coordinate of the bottom of Cell 2; h_w is the water level of the wellbore. For a confined aquifer, one has $B_{\text{Cell2},w} > \Delta z_{\text{Cell2},w}$, where $\Delta z_{\text{Cell2},w}$ represents the vertical dimension of Cell 2. The validity of such treatment will be investigated below.

Therefore, the mass balance for the revised Cell 2 should be:

$$MA_{\text{Cell2}} = \frac{\partial}{\partial t}(C^k \Delta x \Delta y B_{\text{Cell2},w}) \tag{3.634}$$

and Eq. (3.633d) becomes:

$$NMFSS_{\text{Cell}} = W_s C_s^k \Delta x \Delta y B_{\text{Cell2},w} \tag{3.635}$$

Since the porosity of the revised Cell 2 is unity, $NMFR_{\text{Cell}}$ in Eq. (3.632) becomes:

$$NMFR_{\text{Cell}} = \Delta x \Delta y B_{\text{Cell2},w} \sum R_k \tag{3.636}$$

The other terms of the revised Cell 2 in Eq. (3.632) are the same as their counterparts in Eqs. (3.633a)–(3.633f). As for other wellbore cells, the mass balance equations are the same as ones in Eq. (3.632), except that porosity is set as unity. As for aquifer cells, the mass balance equations are the same as Eq. (3.632).

Similar to MODFLOW/MT3DMS, the finite-difference method will be employed to solve Eq. (3.629). The code of MT3DMS will be revised to accommodate the above special treatments of the wellbore cells (particularly the revised Cell 2 in Figure 3.46) in this section. MODFLOW computes the flow field. The changes to the original MT3DMS code are explained as follows.

(b) Changes to the original MT3DMS code

MT3DMS is a 3D numerical model used to describe the multispecies solute transport in both confined and unconfined aquifer (Zheng and Wang, 1999). As mentioned above, the traditional model may cause errors in modeling solute transport in a well-confined aquifer system. The main reason is that water in an open wellbore is unconfined but is treated as confined, which is incorrect. Consequently, the thickness of water in the wellbore is assumed to be constant and the same as the aquifer thickness, which is not true. In this section, the MT3DMS code is revised to eliminate this error for numerical simulation.

The original computer program of MT3DMS consists of a main program and many modules, such as ADV (Advection Package), DSP (Dispersion Package), SSM (Sink & Source Mixing Package), RCT (Chemical Reaction Package), BTN (Basic Transport Package), FMI (Flow Model Interface Package), GCG (Generalized Conjugate Gradient Package), and so on. The ADV, DSP, SSM, and RCT packages are used to solve advective, dispersive, sink & source, and chemical terms in the governing equation of the solute transport, respectively. As Eqs. (3.634), (3.635), and (3.636) are respectively formulated in the BTN, SSM and RCT packages, the changes of the code mainly happen in these two modules. Detailed information on the changes can be seen in the revised MT3DMS code.

In the BTN module of the original code of MT3MDS version 4, "RETA(N, ICOMP)" in the computing code means retardation factor. As for Cell 2 in Figure 3.45, the corresponding code is revised by

$$\text{RETA(N, ICOMP)} = \text{RETA(N, ICOMP)}*B_{\text{Cell2},w}/\text{DZ(N)} \tag{3.637}$$

where $B_{\text{Cell2},w}$ is computed by MODFLOW, and DZ(N) is the cell's vertical thickness. When the water level in Cell 2 is equal to the cell's vertical thickness, the code (3.637) reduces the original one.

In the SSM module of the original code of MT3MDS version 4, "RHS(N)" in the computing code represents Eq. (3.635). As for Cell 2, the corresponding code is revised by

$$\text{RHS(N)} = \text{RHS(N)} - \text{QSS*CTMP*DELR(J)*DELC(I)}*B_{\text{Cell2},w} \tag{3.638}$$

where QSS is the pumping rate, CTMP is concentration, DELR(J) and DELC(I) are the horizontal dimensions of the cell. When $B_{\text{Cell2},w}$ = DH(J, I, K), is the vertical thickness of the cell; the code (3.638) reduces the original one.

Similarly, the code DELR(J)*DELC(I)* DH(J, I, K) in the RCT module is revised into DELR(J)*DELC(I)*$B_{\text{Cell2},w}$ for Cell 2 in Figure 3.45.

Equations used in the revised Cell 2 could similarly treat solute transport in a one-cell wellbore (when the well is discretized into a cell).

(c) Transport models in the wellbore

In actual applications, the flow field is complex for either an injection well or an extraction well, as shown in the laboratory-controlled experiment of Schwartz *et al.* (1999), due to turbulent flow caused by the injection/extraction apparatus (usually a pipe) with a smaller diameter than that of the wellbore itself. Different from transport in porous media, the mechanism of solute transport in the wellbore is like transport in surface water bodies (e.g., rivers). Therefore, diffusion effect and advection are considered for solute transport in the wellbore, while mechanical dispersion is absent (because there are no porous media inside the wellbore). In this section, the MNW model (Konikow and Hornberger, 2006) is used to describe the groundwater flow and solute transport in the wellbore, which is based on MODFLOW/MT3DMS.

3.4.1.3 *Accuracy of the new model*

To test the accuracy of the new model of this section, a proven analytical solution will be desirable to serve as the benchmark of comparison. Unfortunately, it seems not easy to derive a general-purpose analytical solution that can accommodate many realistic field conditions, such as flow transiency, etc. It is also necessary to test the new model with an analytical solution considering the actual well construction, such as the skin effect. However, the available analytical solutions of the two-region model have not considered the wellbore storage. For instance, Chen *et al.* (2012) assumed that $f(z,t)$ in Eq. (3.623a) was constant and independent of location and time and was equal to the concentration of the injected solute. Therefore, we will employ the analytical solution of an injection test by Novakowski (1992a), who considered the wellbore storage.

Figures 3.47(a) and 3.47(b) show the comparison of the BTCs by the analytical and numerical methods in the wellbore, where the vertical axis represents the relative (or normalized) concentration C/C_0, and C_0 is the constant concentration of the injected solute. The legend of "ANA" represents the analytical solutions by Novakowski (1992a).

The parameters used in this case are as follows: The aquifer dimensions are $100\,\mathrm{m} \times 100\,\mathrm{m} \times 6\,\mathrm{m}$; the horizontal hydraulic conductivity is 10 m/day; the horizontal anisotropy is 1.0, where the horizontal anisotropy is the ratio between the two horizontal principal components of the hydraulic conductivity; the injection flow rate is $20\,\mathrm{m}^3/\mathrm{day}$; the porosity is 0.3; the longitudinal dispersivity is 0.5 m; the ratio of horizontal transverse dispersivity to longitudinal dispersivity is 0.1; the ratio of vertical transverse dispersivity to longitudinal dispersivity is 0.01. As the well radius is always constant, three sets of initial conditions of the hydraulic head are employed to test the influence of water level on the wellbore storage: $h_0 = 6\,\mathrm{m}$, $h_0 = 30\,\mathrm{m}$, and $h_0 = 60\,\mathrm{m}$. A greater initial head implies a greater water level in the wellbore. Since the depth of the wellbore might be greater than 100 m, sometimes 1000 m for a deep confined aquifer, the maximum value of 60 m for h_0 is not unusual for the initial hydraulic head. As the flow is assumed to be steady state, the information of the specific yield and the specific storage is not needed. The spatial discretization is $\Delta x = 0.4\,\mathrm{m}$, $\Delta y = 0.4\,\mathrm{m}$, and $\Delta z = 6\,\mathrm{m}$. The aquifer

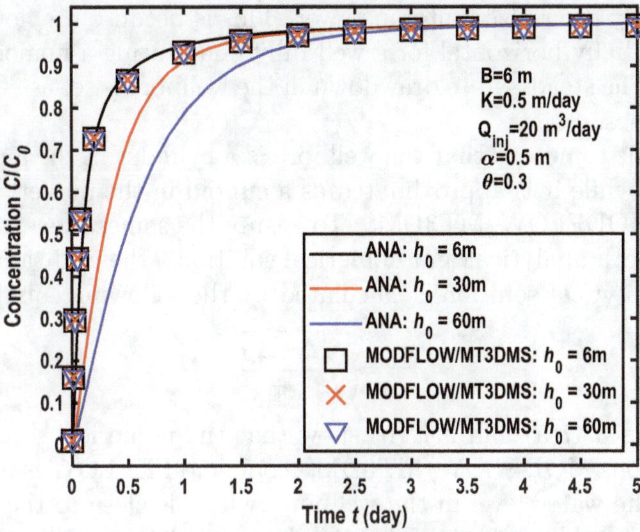

(a) MODFLOW/MT3DMS

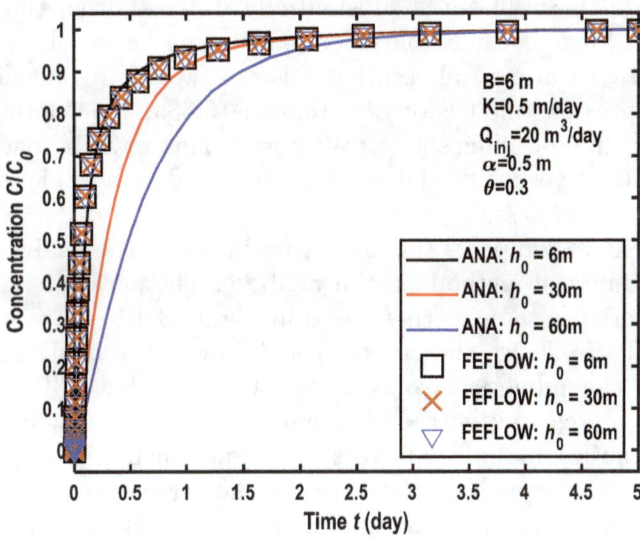

(b) FEFLOW

Figure 3.47. Comparison between BTCs based on analytical and numerical methods in the wellbore under steady-state flow conditions. ANA: Analytical solutions.

is vertically discretized into one layer. This is because the flow direction is radially horizontal for a well fully penetrating a homogeneous aquifer. The steady-state drawdown in the wellbore is set as −0.346 m for all cases.

A point to note is that the wellbore is a cylinder in the analytical solution, while it is approximated as a cuboid in the numerical solution by MODFLOW/MT3DMS. To ensure the same water volume is used in both analytical and numerical solutions, the well radius (r_w) of the analytical solution is calculated by the following equation:

$$r_w = \sqrt{\frac{\Delta x \Delta y}{\pi}} \tag{3.639}$$

Figures 3.47(a) and 3.47(b) show that the numerical solution by the previous MODFLOW/MT3DMS code and FEFLOW is independent of the water level in the wellbore, which is close to the analytical solution when the initial water level inside the wellbore is 6 m (the same as the aquifer thickness). However, when the initial water level inside the wellbore is substantially different from the aquifer thickness of 6 m, considerable discrepancies can be seen between the analytical and numerical solutions. These two figures demonstrate that the previous models of Eqs. (3.629)–(3.633) may cause significant errors in describing solute transport around a wellbore when the initial water level inside the wellbore is considerably different from the aquifer thickness.

Figure 3.48 compares the BTCs produced by the analytical and revised numerical methods in the wellbore. The parameters used in this case are the same as those used in Figure 3.47.

This figure shows the comparison between the analytical solution and the numerical solution by the revised MT3DMS code, and they match well with each other, with some minor discrepancies (but noticeable). Two factors may cause such minor discrepancies. First, the vertical surface area of the screen in the analytical solution (cylinder) is different from that in the numerical solution (cuboid) when the volume of the cuboid well is equal to the volume of the cylinder well. For instance, based on the setting of this section ($r_w = 0.226$ m, $B = 6$ m, $\Delta x = 0.4$ m), the surface area of a cylinder is $2\pi r_w B = 8.51\,\text{m}^2$, while the vertical surface area of a cuboid is $4\Delta x B = 9.60\,\text{m}^2$. Such a difference in the surface area of the screen may generate a minor discrepancy between the

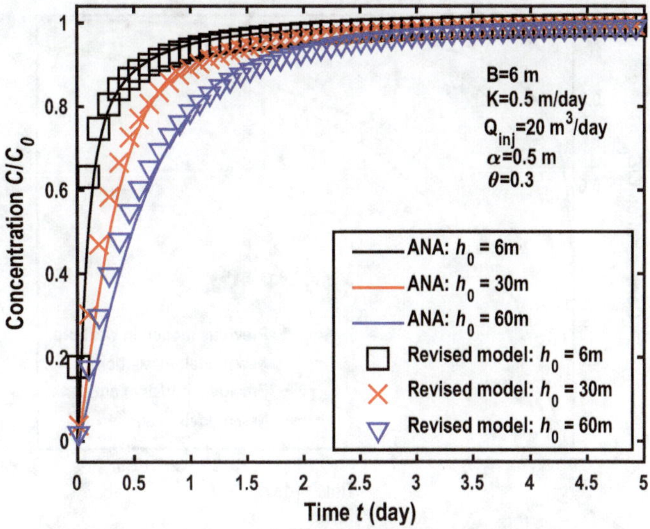

Figure 3.48. Comparison between BTCs based on analytical and revised numerical methods in the wellbore under steady-state flow conditions. ANA: Analytical solutions.

analytical and numerical solutions. Second, numerical error (like numerical dispersion) may not be eliminated in the finite-difference solution.

3.4.1.4 *Discussions*

In aquifer remediation practices, both injection and extraction wells have been widely employed (Anderson *et al.*, 2015; El-Kadi, 1988). Although the influence of the wellbore storage has been investigated for an injection well test case in Section 3.4.1.3, an important assumption should not be overlooked: the flow is under steady-state conditions, which might not always be satisfied in actual applications. In this section, the wellbore storage of solute transport around both injection and extraction wells under transient flow conditions will be investigated to see how the flow transiency will affect the results obtained from the steady-state assumption.

(a) The injection test in a homogeneous aquifer

Figure 3.49 shows the comparison of BTCs between the previous models of Eqs. (3.629)–(3.633) and new models of this section

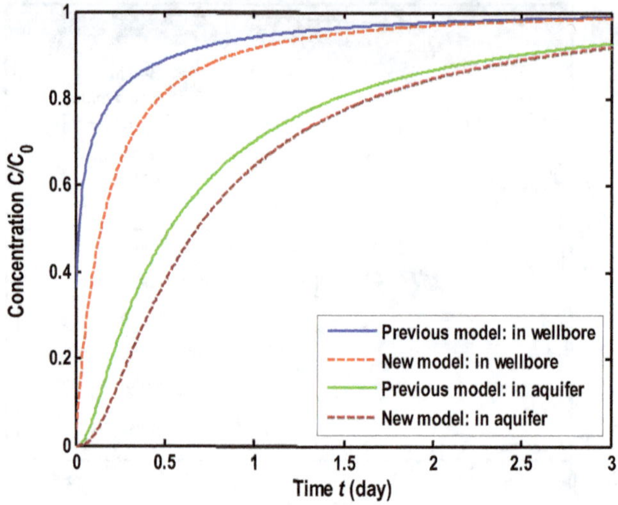

Figure 3.49. Comparison of BTCs for the injection well test.

Eqs. (3.634)–(3.636) at different vertical locations. The value of specific storage is $0.00001\,\text{m}^{-1}$, and the initial water table is 60 m. The parameters used are the same as the ones in Figure 3.47. The legend of "in aquifer" represents the case that the observed BTCs are 3 m away from the wellbore.

This figure demonstrates that the values of BTCs computed by the previous models are smaller than those of the new model, and the difference between them is obvious in both the aquifer and wellbore. It also demonstrates that errors associated with Eqs. (3.629)–(3.633) are not negligible for the case with an injection well.

(b) The extraction test in a homogeneous aquifer

Figure 3.50 shows the comparison of BTCs between the previous models and the revised models around an extraction well. The legend of "in aquifer" represents the case that the observed BTCs are 3 m away from the wellbore. The flow rate is $-20\,\text{m}^3/\text{day}$, where the negative sign represents extraction. The initial solute plume is a cuboid. Horizontally, the extraction well is located at the center of the plume cuboid, and the side faces of the plume are 5 m away from the well center. The other parameters are the same as the ones in Figure 3.49. This figure shows that BTCs computed by the new model are above ones by the previous model, due to the wellbore storage considered in the new model. The difference of BTCs between

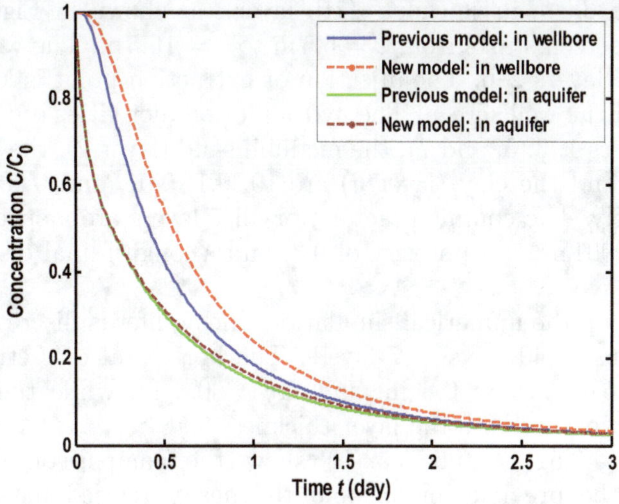

Figure 3.50. Comparison of BTCs for the extraction well test.

them is obvious in the wellbore, but not in the aquifer. One may find that the wellbore storage is generally not significant for solute transport in the aquifer at locations not too close to the wellbore, e.g., 3 m in this section. The reason is that the solute mainly moves toward the well due to the convergent flow field, and the wellbore storage on the aquifer could be ignored.

Therefore, this figure demonstrates that the errors caused by Eqs. (3.629)–(3.633) are generally negligible for reactive transport in the aquifer at locations not too close to the wellbore (such as 3 m), but those errors are not negligible for the wellbore.

(c) The injection test in a homogeneous aquifer

Generally, aquifer homogeneity is a simplification of reality. In this section, the aquifer is assumed to be vertically heterogeneous. The vertical heterogeneity is manifested in multiple layers (i.e., a multi-aquifer system). The dimensions of the multi-aquifer system are assumed to be 32 m in length, 32 m in width, and 20 m in height. It is composed of 7 layers from top to bottom: A coarse sand layer with 3.5 m in thickness (1st layer), a medium sand layer with 2 m in thickness (2nd layer), another coarse sand layer with 3.5 m in thickness (3rd layer), a fine sand layer with 2 m in thickness (4th layer), one more coarse sand layer with 3.5 m in thickness (5th layer), a clay layer with 2 m in thickness (6th layer), and finally a coarse sand layer with 3.5 m in thickness (7th layer), as shown in Figure 3.51. The well screen starts from $z = 5.5\,\text{m}$ to $z = 16.5\,\text{m}$. The well screen is open in layers 2–6. The injection or extraction point is located at the top of the well screen. The hydraulic conductivities of the coarse sand (layers 1, 3, 5, and 7), the medium sand (layer 2), the fine sand (layer 4) and the clay (layer 6) are 10, 0.1, 0.01, and 0.001 m/day, respectively. The common factors for all 7 layers are a specific storage of $0.0001\,\text{m}^{-1}$, a porosity of 0.3, and a longitudinal dispersivity of 0.5 m.

To set up the numerical simulation, the aquifer is discretized into 40 columns × 40 rows × 7 layers. The horizontal discretization is uniform, e.g., $\Delta x = 0.8$ m, and $\Delta y = 0.8$ m, while the vertical discretization matches the layer thickness.

Figures 3.52(a) and 3.52(b) show the comparison of BTCs between the previous model and the newly revised models. The injection well is located at the aquifer center with a flow rate of $20\,\text{m}^3/\text{day}$. Legends of "$z = 15\,\text{m}$", "$z = 12.75\,\text{m}$", "$z = 10\,\text{m}$",

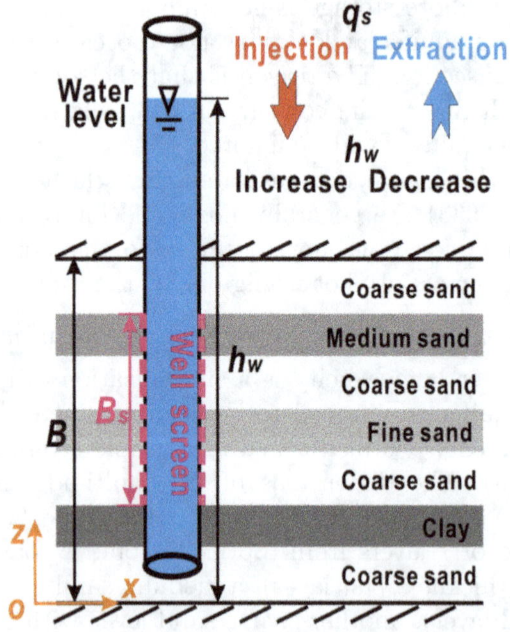

Figure 3.51. Schematic diagram of the well-aquifer system. The well partially penetrates the stratified aquifer.

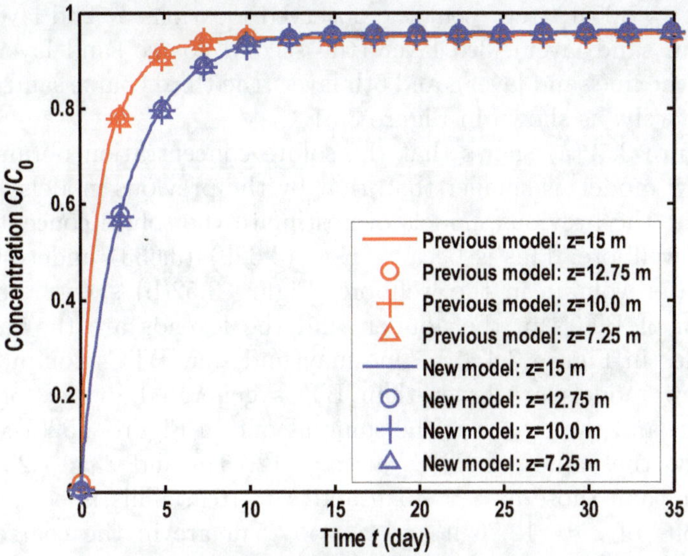

(a) BTC in the wellbore

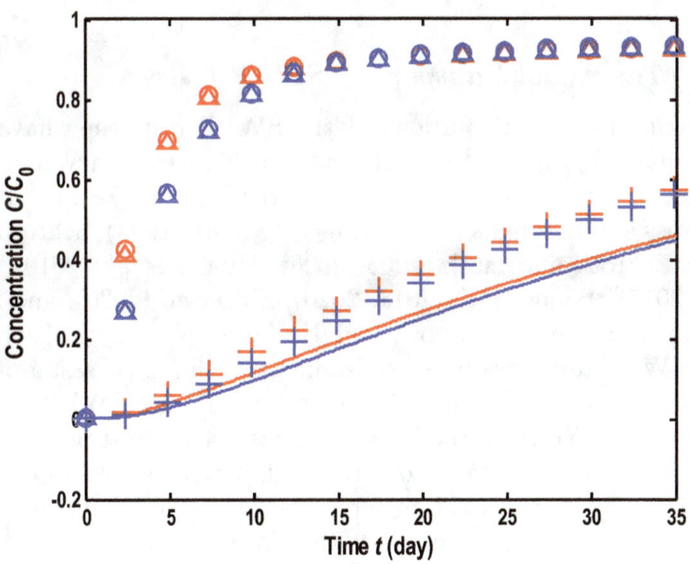

(b) BTC in the aquifer, 1.2 m away from the well.

Figure 3.52. Comparison of BTCs for the two-well test under transient flow conditions.

and "$z = 7.25\,\text{m}$" represent the observed locations at 2nd layer (the medium sand layer), 3rd layer (the second coarse sand layer), 4th layer (the fine sand layer), and 5th layer (the third coarse sand layer), respectively, as shown in Figure 3.51.

Figure 3.52(a) shows that the solute concentration obtained by the new models is smaller than that by the previous models, implying that the previous models overestimate the solute concentration in the wellbore. This is because Eqs. (3.629)–(3.633) underestimate the water volume in the wellbore. Figure 3.52(b) shows the comparison of BTCs in the aquifer, and the legends are the same as the ones in Figure 3.47(a). One may find that BTCs computed by the new models are lower than BTCs generated by the previous models, and the reason is the same as one in Figure 3.52(a). Additionally, the values of BTCs at $z = 12.75\,\text{m}$ and $z = 7.25\,\text{m}$ are greater than those at $z = 15\,\text{m}$ and $z = 10\,\text{m}$. This is because the locations of $z = 12.75\,\text{m}$ and $z = 7.25\,\text{m}$ are in the coarse sand layer, whose higher permeability makes solute transport much easier and faster.

3.4.2 *The radial transport of SWPP tests*

As for the analytical solutions of the SWPP test, they have been widely used for applications, due to the high efficiency and great accuracy of the solutions, like the model of Gelhar and Collins (1971), Snodgrass and Kitanidis (1998), Haggerty *et al.* (1998), Schroth *et al.* (2001), Schroth and Istok (2005, 2006), Huang *et al.* (2010), Chen *et al.* (2017), Wang *et al.* (2018, 2020), Shi *et al.* (2020), and so on. However, the mentioned-above analytical or semi-analytical solutions of the SWPP test were based on some over-simplified assumptions.

First, transient flow effect is inevitable during a SWPP test. For instance, when starting the injection phase of the test, groundwater flow must experience an initial transient period before approaching a quasi-steady state (if the injection period is long enough). When switching from the chaser phase to the rest phase, i.e., shutting down the well, a transient flow effect will also be activated. When changing from the rest phase to the extraction phase, one will once again activate the transient flow effect. Mathias (2010) investigated the steady-state flow assumption on the injected plume for radial transport, and found that such an assumption could result in an overestimation of the plume front. As for the SWPP test, such treatment may also create errors in parameter estimations.

Second, flow is assumed to be horizontal, which may be true for a nearly fully penetrating well. When the length of the injection/extraction region was relatively small compared to both aquifer thickness and the spatial dimension of the tracer plume at the end of the injection phase, the flow field in the SWPP test might become spherical (Schroth and Istok, 2005). In other words, vertical flow components become so great that they cannot be ignored.

Third, numerical models consider wellbore storage effects with respect to groundwater flow but not solute concentrations. The wellbore storage for solute transport refers to the variation of the solute injected in the wellbore during the processes of the test. Obviously, the assumption of ignoring the wellbore storage is not reasonable for solute transport. For instance, excluding the wellbore storage may overestimate the concentration in the wellbore in the injection phase since the finite volume of water in the wellbore could dilute the solute concentration in the early stage before the pre-test water inside the wellbore was completely flushed out of the borehole into the aquifer. Similarly, such a treatment could underestimate the concentration in the wellbore in the early stage of the chase phase due to the high concentration of solute in the wellbore at the end of the injection phase. In general, the chase concentration is usually set as zero. However, under some circumstances, investigators may use a non-zero concentration for the chase phase. For example, Phanikumar and McGuire (2010) used 10 mg/L for Cl^- and 2 mg/L for SO_4^{2-} in their chase solutions. Therefore, the concentration at the well screen may not be zero in the early stage of the rest phase when the chase concentration was not zero.

Actually, the assumptions mentioned above used in the analytical and semi-analytical solutions can be relaxed in the numerical models, such as MODFLOW/MT3DMS (Harbaugh *et al.*, 2000; Zheng and Wang, 1999), FEFLOW (Diersch, 2014), SUTRA (Voss, 1984), STOMP (Nichols *et al.*, 1997), and so on. For instance, Phanikumar and McGuire (2010) used a finite-difference method to model solute transport in a SWPP test including all four phases: injection, chaser, rest, and extraction. Such a model was capable of dealing with a number of species involving several reaction kinetics of chemical reactions. Hellerich *et al.* (2003) established a centrally weighted

Crank–Nicholson finite-difference scheme of the SWPP test. Istok *et al.* (1999), Schroth *et al.* (2000), Cassiani *et al.* (2005), and Davis *et al.* (2005) employed a STOMP (Subsurface Transport Over Multiple Phases) code (White and Oostrom, 1996) to solve ADE of the SWPP test. STOMP is a fully implicit volume-integrated finite-difference simulator for modeling 1D, 2D, and 3D groundwater flow and transport (White and Oostrom, 1996).

In this section, the models of Wang *et al.* (2017) and Wang and Zhan (2019) are introduced. Wang *et al.* (2017) investigated the influences of a finite hydraulic diffusivity on the results and found it might be significant, since both advective and dispersive transport were related to the flow velocity. Actually, their model still contained some assumptions, e.g., considering wellbore storage effects with respect to groundwater flow, but not solute concentrations. Wang and Zhan (2019) developed multi-species reactive transport associated with SWPP tests with a better conceptual model that acknowledges the realistic circumstances that have been either overlooked or overly simplified in previous investigations. First, their model employed a more realistic finite hydraulic diffusivity instead of an infinite hydraulic diffusivity to describe the flow field. Second, their model proposed a better way to handle the boundary condition of transport at the wellbore by consider the wellbore storage effect for both groundwater and solute transport during the SWPP tests. The influence of the assumptions used in previous studies on the reactive transport in the SWPP test were investigated based on the model of Wang and Zhan (2019).

3.4.2.1 *SWPP test in transient Forchheimer flow field*

A schematic diagram of the four-phase SWPP test investigated is depicted in Figure 3.35. A cylindrical coordinate system is adopted with the r-axis horizontal and toward infinity from the well, and the z-axis vertical upward. The origin is at the center of the well and located in the plane of symmetry of the aquifer. The well fully penetrates a homogeneous and horizontally isotropic aquifer with a constant thickness, of which the boundary is horizontal and extends to infinity.

(a) Mathematical model of solute transport in the SWPP test

Assuming that advection, dispersion, and sorption are involved in the transport processes, the governing equation of solute transport in a saturated and homogeneous porous aquifer around the well is

$$R\frac{\partial C}{\partial t} = \frac{1}{r}\frac{\partial}{\partial r}\left(rD_L\frac{\partial C}{\partial r}\right) - v_a\frac{\partial C}{\partial r} + \kappa RC, \quad r \geq r_w \tag{3.640}$$

where C is resident concentration of the aquifer [Kg/m^3]; t is time [s]; r is radial distance from the center of the well [m]; $v_a = v/\theta$ is average radial pore velocity [m/s], v is radial specific discharge or Darcian velocity [m/s], and θ represents porosity [dimensionless]; κ is a constant rate for first-order biodegradation, or radioactive decay, or the first-order reaction rate in the aquifer [1/s]; $R = 1+\rho_b K_d/\theta$, is known as the retardation factor of the aquifer [dimensionless]; K_d is the equilibrium distribution coefficient [m^3/Kg]; ρ_b is the bulk density of the aquifer material [Kg/m^3]; D_L represents the radial dispersion coefficient of the aquifer [m^2/s], and it is composed of mechanical dispersion and molecule diffusion:

$$D_L = \alpha_L v_a + D_0 \tag{3.641}$$

where α_L is the radial dispersivity [m]; D_0 is the effective diffusion coefficient in the aquifer [m^2/s]. The initial condition is

$$C(r, t = 0) = C_0(r) \tag{3.642}$$

where $C_0(r)$ is a given function of the radial distance.

The infinite boundary condition is

$$\frac{\partial C(r \to \infty, t)}{\partial r} = 0 \tag{3.643}$$

Assuming the ending times of the injection, chaser, rest, and extraction periods are t_{inj}, t_{cha}, t_{res}, and t_{ext}, respectively, the boundary conditions at the wellbore are

$$\left(v_a C - D_L\frac{\partial C}{\partial r}\right)\bigg|_{r\to r_w^-} = (v_a C_0^{\text{inj}})_{r\to r_w^+}, \quad 0 < t \leq t_{\text{inj}} \tag{3.644a}$$

$$\left(v_a C - D_L\frac{\partial C}{\partial r}\right)\bigg|_{r\to r_w^-} = (v_a C_0^{\text{cha}})_{r\to r_w^+}, \quad t_{\text{inj}} < t \leq t_{\text{inj}} + t_{\text{chs}} \tag{3.644b}$$

$$\left(v_a C - D_L \frac{\partial C}{\partial r}\right)\bigg|_{r\to r_w^-} = 0, \quad t_{\text{inj}} + t_{\text{chs}} < t \le t_{\text{inj}} + t_{\text{chs}} + t_{\text{res}} \tag{3.644c}$$

and

$$\frac{\partial C}{\partial r}\bigg|_{r\to r_w^-} = 0, \quad t_{\text{inj}} + t_{\text{chs}} + t_{\text{res}} < t \le t_{\text{inj}} + t_{\text{chs}} + t_{\text{res}} + t_{\text{ext}} \tag{3.644d}$$

in which C_0^{inj} and C_0^{cha} in Eqs. (3.644a) and (3.644b) represents the concentrations of tracer in the injection and chaser periods, respectively. When the original groundwater is injected into the aquifer during the chaser period, C_0^{cha} becomes 0.

(b) Mathematical model of groundwater flow in the SWPP

In nearly all previous studies, flow of the SWPP test was assumed to be steady state (Gelhar and Collins, 1971; Huang *et al.*, 2010; Phanikumar and McGuire, 2010; Schroth and Istok, 2005), and it is

$$v_a = Q/(2\pi B\theta r) \tag{3.645}$$

where Q is the flow rate of the well [m^3/s], and it is positive for injection and negative for pumping; B is the thickness of the aquifer [m]. During the rest period, Q is zero.

In this section, we employ the Forchheimer law to calculate the radial flow velocity:

$$v + \beta v|v| = K_\beta \frac{\partial s}{\partial r} \tag{3.646}$$

where β is a non-Darcian factor [s/m]; K_β can be regarded as an apparent hydraulic conductivity [m/s] of the aquifer (Mathias, 2010; Wang *et al.*, 2014); s is drawdown [m]. When $\beta = 0$, K_β become the hydraulic conductivity, and Eq. (3.646) reduces to Darcy's law.

The drawdown in Eq. (3.646) can be obtained by solving the following mass balance equation:

$$\frac{\partial v(r,t)}{\partial r} + \frac{v(r,t)}{r} = S_s \frac{\partial s(r,t)}{\partial t} \tag{3.647}$$

where S_s is the specific storage of aquifer [1/s]. The initial condition is

$$s(r, t = 0) = 0. \tag{3.648}$$

The infinity boundary condition is

$$\frac{\partial s(r \to \infty, t)}{\partial r} = 0 \tag{3.649}$$

The boundary conditions at the wellbore for all four phases are

$$2\pi r B v(r,t)|r \to r_w - \pi r_w^2 \frac{ds_w(t)}{dt} = Q, \tag{3.650}$$

where s_w is the drawdown inside the wellbore; r_w is the well radius. The second term on the left-hand side of Eq. (3.650) represents the wellbore storage.

The flow and transport models are coupled through the Darcian velocity, which will be firstly calculated by Eqs. (3.646)–(3.650). By setting $S_S = 0$, the transient flow model reduces to the steady-state model.

(c) Numerical solution of the SWPP test

Due to the non-linearity of the flow problem, it seems impossible to obtain analytical solutions of Eqs. (3.640) and (3.647). In this section, a finite-difference method will be called in for solving such problems. To minimize errors of the numerical solution, simulations will be conducted with a non-uniform grid system (Mathias *et al.*, 2008; Wang *et al.*, 2014; Wen *et al.*, 2009), in which the spatial steps are smaller near the wellbore and become progressively greater away from the wellbore. Assuming that the spatial domain $[r_w, r_e]$ is discretized into N nodes, excluding the inner (r_w) and outer (r_e) boundary nodes, the radial distance of node r_i is

$$r_i = (r_{i-1/2} + r_{i+1/2})/2, \quad i = 1, 2, \ldots, N \tag{3.651}$$

where $r_{i+1/2}$ is calculated as follows:

$$\log_{10}(r_{i+1/2}) = \log_{10}(r_{wD}) + i\left[\frac{\log_{10}(r_{eD}) - \log_{10}(r_{wD})}{N}\right], \quad i = 1, 2, \ldots, N \tag{3.652}$$

The distance of the infinity boundary condition is approximated by a relatively large distance r_e, at which no tracer approaches (Mathias *et al.*, 2008; Wang *et al.*, 2014; Wen *et al.*, 2009). Equations (3.651)–(3.652) represent a space domain discretized logarithmically.

After the spatial discretization, Eq. (3.640) becomes:

$$\frac{\partial C}{\partial t} = \frac{r_{i-1/2}(\alpha_L v_{a,i} + D_0)\frac{C_i - C_{i-1}}{r_i - r_{i-1}} - r_{i+1/2}(\alpha_L v_{a,i+1} + D_0)\frac{C_{i+1}-C_i}{r_{i+1}-r_i}}{Rr_i(r_{i+1/2} - r_{i-1/2})}$$

$$- \frac{v_{a,i}}{R}\frac{C_{i+1/2} - C_{i-1/2}}{r_{i+1/2} - r_{i-1/2}} + \kappa C_i \tag{3.653}$$

where the values of i in Eq. (3.653) are $2, 3, \ldots, N-1$; $v_{a,i}$ and C_i are the flow velocity v_a and concentration C at node i, respectively. Equations (3.643), (3.644), and (3.653) consist of a set of N ordinary differential functions with time, which will be solved by a stiff integrator ODE15s, which is available in any standard version of MATLAB (Ashino *et al.*, 2000; Shampine and Reichelt, 1997; Shampine *et al.*, 1999).

One of the main disadvantages of the finite-difference solution of ADE is that it often suffers from numerical errors (like numerical oscillation and numerical dispersion), which may become non-negligible for advection-dominating transport. Many studies have shown that such errors could be reduced by refining spatial and temporal discretization. A benchmark solution is needed to test the grid system used in the numerical solution.

As for the finite-difference solution of ADE, one of the main disadvantages is that it often suffers from numerical errors (like numerical oscillation and numerical dispersion), which may become non-negligible for advection-dominating transport. Many studies showed that such errors could be reduced by refining the spatial and/or temporal discretization. In order to test the grid system used in the numerical solution, a benchmark solution is needed.

Unfortunately, benchmark solutions of the SWPP test are not available to date, but some semi-analytical solutions are available for single-well tests involving only a single phase (injection) (Moench and Ogata, 1981b). Alternatively, the accuracy of the finite-difference solution could be tested by comparing BTCs of the numerical and semi-analytical methods for the injection phase. In this regard, the semi-analytical solution of Moench and Ogata (1981) for a single-well single-phase test serves as a benchmark, whose expression in Laplace domain is

$$\overline{C} = \frac{C_0^{\text{inj}}}{s} \exp\left(\frac{r - r_w}{2\alpha_L}\right) \frac{A_i[s^{-2/3}(sr/\alpha_L + 0.25)]}{A_i[s^{-2/3}(sr_w/\alpha_L + 0.25)]} \tag{3.654}$$

where $\overline{C}$ is the concentration in Laplace domain; s is the Laplace transform variable; A_i is the Airy function. Equation (3.654) assumed that molecular diffusion coefficient D_0 is negligible, which is usually true for the injection phase. It seems difficult to conduct analytical inverse Laplace transform of Eq. (3.654) to obtain the concentration in the real-time domain. Instead, Moench and Ogata (1981) found that the Stehfest method (Stehfest, 1970a, 1970b) worked very well for numerical inverse Laplace transform of the problem here. The Stehfest method contains a parameter, N_{Stehfest}, which is an even positive integer representing the optimal number of summation terms, and it is chosen to be 12 based on our numerical exercises and the recommendation of Moench and Ogata (1981).

Figure 3.53 shows comparison between the semi-analytical and numerical solutions, in which the groundwater flow is steady state. The parameters used in Figure 3.53 are as follows: $B = 1\,\text{m}$, $r_w = 0.1\,\text{m}$, $\alpha_L = 1\,\text{m}$, $\theta = 0.1\,\text{m}$, and $Q = 50\,\text{m}^3/\text{day}$. The results show that BTCs by the semi-analytical and numerical methods coincide. Assuming that the well flow rate in the SWPP test is smaller than or equal to $50\,\text{m}^3/\text{day}$, one may conclude that the numerical solution of the SWPP test is accurate and reliable.

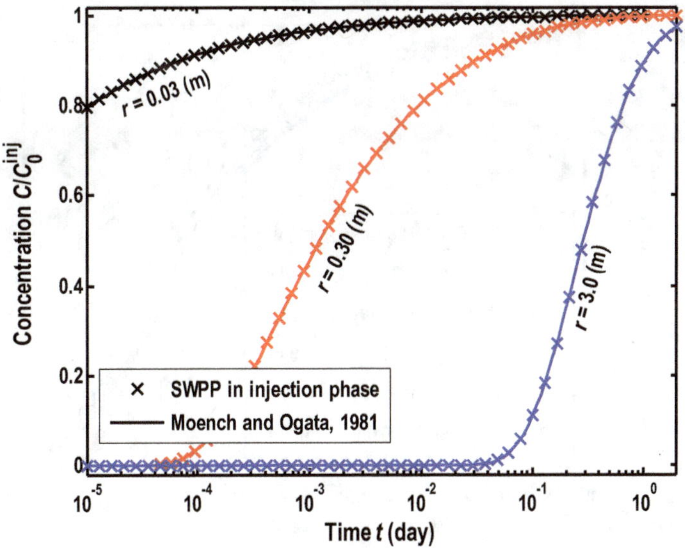

Figure 3.53. Comparison of the semi-analytical solution and the numerical solution in the injection phase of SWPP.

Assuming that the well flow rate in the SWPP test is smaller than or equal to $50\,\text{m}^3/\text{day}$, one may conclude that the numerical solution of the SWPP test is accurate and reliable. Additionally, Huang *et al.* (2010) presented an exact analytical solution of the SWPP test, considering partial well penetration, advection, longitudinal and transverse dispersions, first-order decay, and rate-limited sorption/partitioning in steady flow fields. Huang *et al.* (2010) only considered the injection and extraction phases. Figure 3.54 shows the comparison between the concentration of this section and Huang *et al.* (2010) in the well for different α_r of 1.0, 5, and 10 m, and the other parameters are the same with ones in Figure 3.53. For the purpose of comparison, the general model of this section is applied for a special case with steady-state flow and two phases (injection and extraction). Obviously, the numerical solution of this section is sufficiently accurate for such a special case.

(d) Effect of flow field on solute transport in the SWPP test

Figure 3.55 shows BTCs at the well during the extraction phase for different values of the specific storage. The BTC of the steady-state flow field is included as reference, since most previous SWPP test models contained an assumption that flows were steady state during

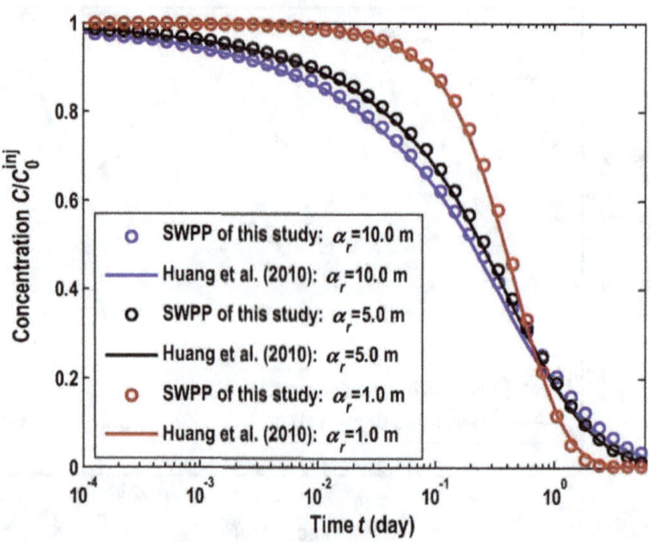

Figure 3.54. Comparison of the analytical and numerical solutions of SWPP.

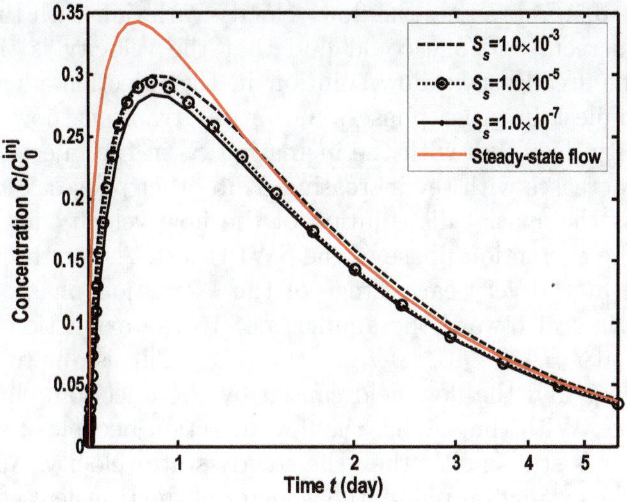

Figure 3.55. BTCs at the well during the extraction phase for different values of S_s.

all phases. As presented in Batu (1998), the ranges of S_s for the loose sand, dense sand and dense sandy gravel are 4.9×10^{-4} 1/m to 1.0×10^{-3} 1/m, 1.3×10^{-3} 1/m to 2.0×10^{-3} 1/m and 4.9×10^{-5} 1/m to 1.0×10^{-4} 1/m, respectively. Therefore, $\mathrm{S_s} = 1.0 \times 10^{-3}$ 1/m, 1.0×10^{-5} 1/m, and 1.0×10^{-7} 1/m are chosen for representing different types of aquifers. The other parameters are set as $K_\beta = 1$ m/day, $\beta = 0.000001$ day/m, $B = 1$ m, $\theta = 0.2$, $\alpha_r = 10$ m, $r_w = 0.1$ m, $t_{\text{inj}} = 0.5$ day, $t_{\text{cha}} = 0.1$ day, $t_{\text{res}} = 0.1$ day, $t_{\text{ext}} = 5$ day, $Q_{\text{inj}} = 50\,\text{m}^3/\text{day}$, $Q_{\text{cha}} = 50\,\text{m}^3/\text{day}$, and $Q_{\text{ext}} = -50\,\text{m}^3/\text{day}$, where Q_{inj}, Q_{cha}, and Q_{ext} denote the rates of injection, chaser, and extraction, respectively.

The legend of "Steady-state flow" in Figure 3.55 represents that flow is steady state in the injection, chaser, and pumping phases, while the others assume that flow is transient in all phases of the SWPP test. Obviously, the difference between solution of the transient flow and steady-state flow is not negligible, and such a difference increases with the increasing S_s. The concentration of the steady-state solution is greater than others at the beginning. This can be explained by the flow velocity field around the well, since it is a main factor affecting the solute transport in the SWPP test, as shown in Eqs. (3.640) and (3.641).

Figure 3.56 represents the flow velocity variation with time along the well screen, and one could find that the velocity is the greatest for the steady-state flow solution in the injection phase, while is the smallest in other phases. As for the transient flow solution, the velocity decreases with the increasing S_s in the injection phase, while it increases with the increasing S_s in other phases. Figure 3.57 represents the spatial distribution of the flow velocity for different time in the extraction phase of the SWPP test. As for the transient flow field at the very early stage of the extraction phase, the flow direction is still toward the aquifer, not to the extraction well, for instance, at $t = 1.0 \times 10^{-3} + t_{\text{inj}} + t_{\text{chs}} + t_{\text{res}}$. This is due to the wellbore storage and the flow field caused by the injection, chaser, and rest phases. With time going, the flow direction becomes toward the well, but it is still smaller than the steady-state velocity. At the late stage of the extraction phase, the velocity of the transient state is up to the velocity of steady state. In a word, the absolute flow velocity of the transient flow model is smaller than that of the steady-state model; even the flow direction is opposite to ones of the transient flow at the early stage of the extraction phase. Therefore, the solute

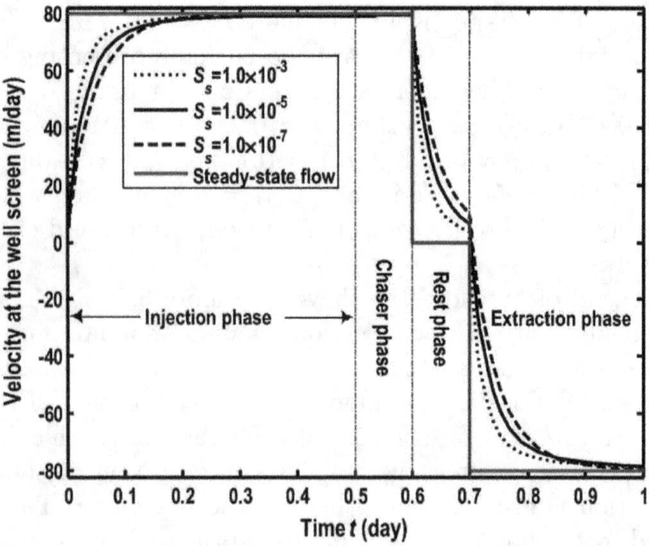

Figure 3.56. The variation of the flow velocity at the wellbore with time for different S_s.

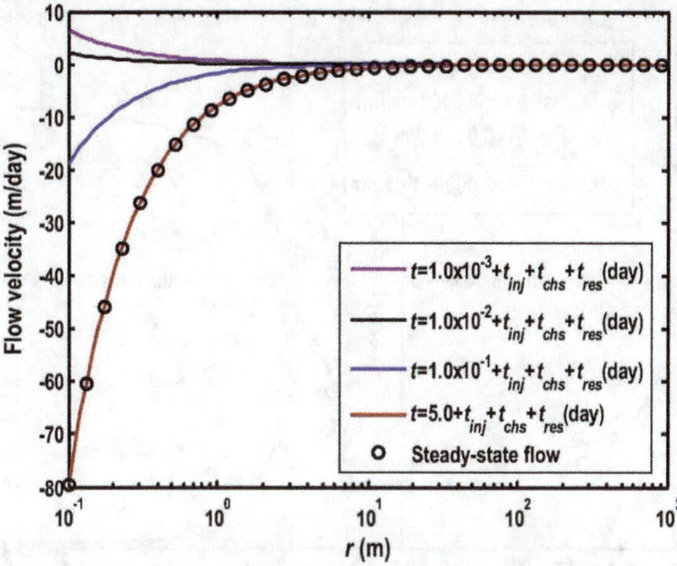

Figure 3.57. Spatial distribution of the flow velocity during the extraction phase for different times when $S_s = 1.0 \times 10^{-5}$ 1/m.

moves slower toward the well for the transient flow model than for the steady-state model.

Another interesting observation in Figure 3.54 is that the concentration is 0 at $t = t_{\text{inj}} + t_{\text{chs}} + t_{\text{res}}$, and this is because the operation of the chaser pushes the solute away from the well into the aquifer.

The groundwater flow may not follow Darcy's law when the velocity is sufficiently high, which may happen around the pumping/injecting well. The Forchheimer law reduces Darcy's law when the velocity becomes very small, which may happen at distances far from the well. The inertial force coefficient (β) is of primary concern for the non-Darcian flow described by the Forchheimer equation. Referring to Wang *et al.* (2014), the values of β are chosen as 1.0×10^{-2} day/m and 1.0×10^{-6} day/m. The other parameters are set as $K_\beta = 1\,\text{m/day}$, $\beta = 0.000001$ day/m, $B = 1\,\text{m}$, $\theta = 0.2$, $\alpha_r = 10\,\text{m}$, $r_w = 0.1\,\text{m}$, $t_{\text{inj}} = 0.5$ day, $t_{\text{cha}} = 0.1$ day, $t_{\text{res}} = 0.1$ day, $t_{\text{ext}} = 5$ day, $Q_{\text{inj}} = 50\,\text{m}^3/\text{day}$, $Q_{\text{cha}} = 50\,\text{m}^3/\text{day}$, and $Q_{\text{ext}} = -50\,\text{m}^3/\text{day}$. Figure 3.58 shows the time drawdown and time velocity at the wellbore, and one can find the differences among the drawdowns of different β values are obvious, but the differences

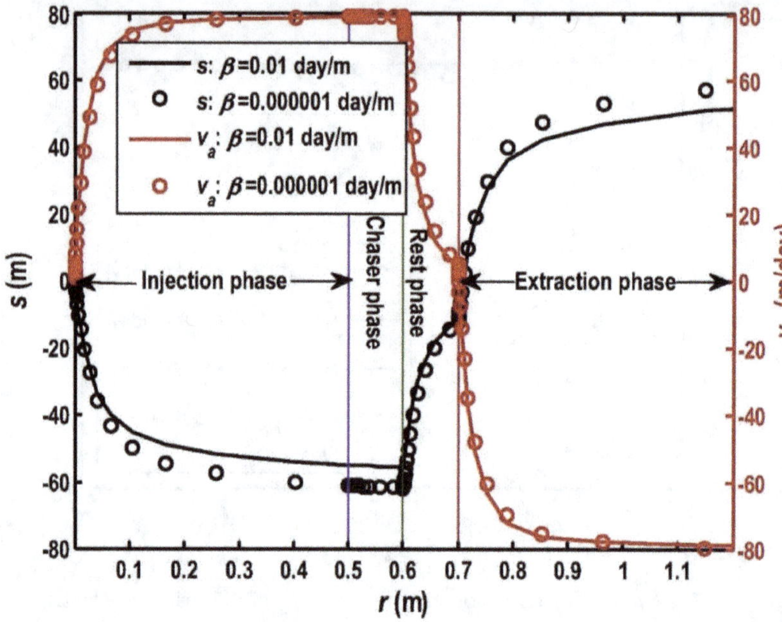

Figure 3.58. The time-drawdown and time-velocity at the wellbore for different β.

among the velocity curves are negligible. As solute transport depends on the velocity, not on the drawdown, therefore the effect of β might be much less significant to solute transport near the well.

(e) Effect of the dispersivity on solute transport in the SWPP test

Figure 3.59 shows the effect of α_r on the concentration in the extraction phase, in which "Transient flow" assumes that the flow is transient in all phases of SWPP test. In addition to the observations which can be seen in Figure 3.58, one could find that the differences among the solutions for transient and steady-state flows increase with the decreasing α_r at late stage. The peak value increases with the decreasing α_r, while the time for the concentration approaching the peak value decreases with the decreasing α_r. This is because a smaller α_r represents a smaller dispersive effect, thus a relatively larger advective effect.

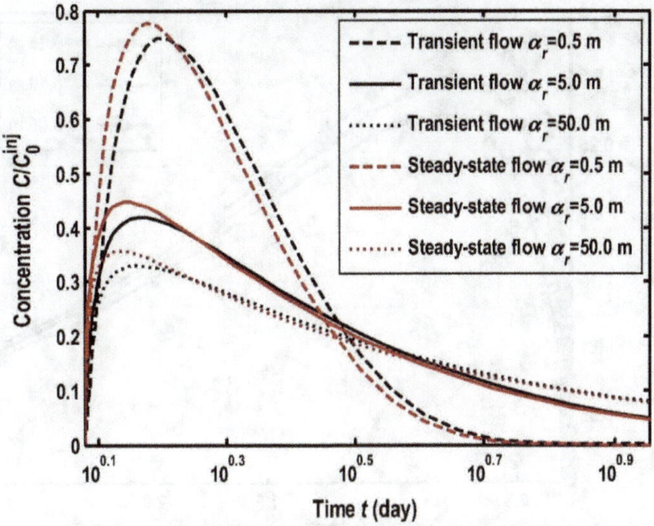

Figure 3.59. Comparison between BTCs at the well during the extraction phase under the transient and the steady-state flow conditions for different α_r.

(f) Effect of wellbore storage on solute transport in the SWPP test

The SWPP test usually only involves a smaller aquifer volume surrounding the well, and thus it is best suited for small scale parameter estimations. For such small-scale problems, the wellbore storage might be an important factor affecting BTCs in the wellbore during the extraction phase. From Eq. (3.650), one could also see that the radius of the well is a main parameter influencing the wellbore storage. Figure 3.60 shows BTCs for different r_w. The parameters used in this case are the same with ones in Figure 3.54, where $S_s = 1.0 \times 10^{-3}$ 1/m. The differences among BTCs of different r_w are obvious, and a larger r_w could result in a smaller concentration in the well. These observations demonstrate that the wellbore storage usually cannot be ignored for interpreting the SWPP test results.

(g) Parameter estimations

As discussed in the above sections, the difference between the solutions of the SWPP test under transient and steady-state flows is

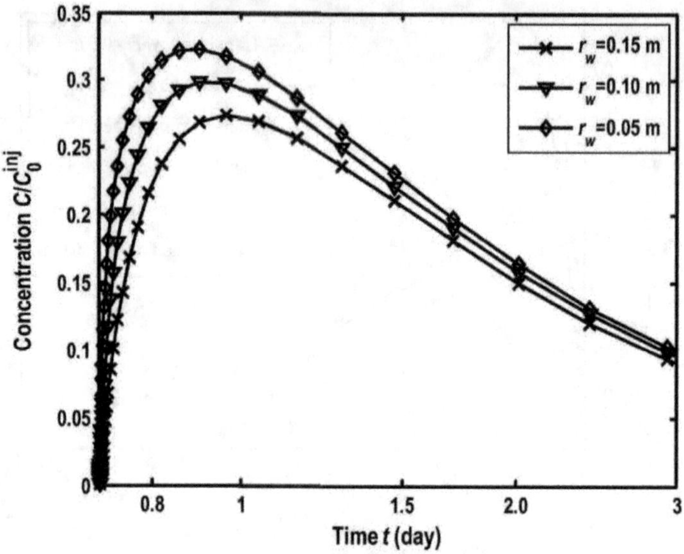

Figure 3.60. BTCs at the wellbore during the extraction phase for different r_w.

obvious. From a practical standpoint, hydrogeologists are interested in the accuracy of parameter estimation based on these two models, i.e., the transient flow and steady-state flow models. In this section, the results calculated by the transient flow model are taken as the benchmarks, and BTCs obtained from such benchmarks at the well will be fitted by the steady-state model. By comparing the parameters used in the transient flow model and estimated by the steady-state flow model, one could check the discrepancy between these two models. For instance, if the difference in the parameters between them is negligible, one could conclude that both models are acceptable; otherwise, the steady-state flow model may generate non-negligible errors and should not be used.

Figure 3.61 shows BTCs that are well-fitted by the steady-state model for different S_s. The parameters are set as $K_\beta = 1\,\text{m/day}$, $\beta = 0.000001$ day/m, $B = 1\,\text{m}$, $\theta = 0.2$, $\alpha_r = 10\,\text{m}$, $r_w = 0.1\,\text{m}$, $t_{\text{inj}} = 0.5$ day, $t_{\text{cha}} = 0.1$ day, $t_{\text{res}} = 0.1$ day, $t_{\text{ext}} = 5$ day, $Q_{\text{inj}} = 50\,\text{m}^3/\text{day}$, $Q_{\text{cha}} = 50\,\text{m}^3/\text{day}$, and $Q_{\text{ext}} = -50\,\text{m}^3/\text{day}$. The numerical solution of the SWPP test under the transient flow condition in all phases creates the circle lines. The dashed lines are created by the model under the steady-state flow condition throughout the

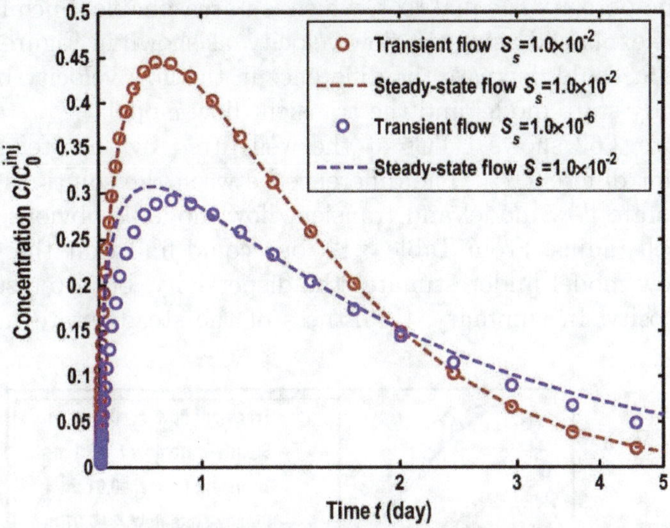

Figure 3.61. BTCs at the wellbore fitted by the steady-state flow model for different S_s.

Table 3.11. The parameter estimated by the solution of the SWPP test under the steady-state flow for different S_S.

Parameters	Real values used in the transient flow model	Estimated by the steady-state flow model	
		$S_S = 0.01$	$S_S = 0.00001$
α_r (m)	$\alpha_r = 10$	$0.5\alpha_r$	$0.9\alpha_r$
θ (dimensionless)	$\theta = 0.2$	0.41θ	5.0θ

SWPP test's operation. Using the trial-and-error method, we find that it is difficult to fit BTCs at the late stage of the SWPP test.

Table 3.11 shows the parameter estimated by the steady-state model. One could find that the fitness of $S_s = 1.0 \times 10^{-2}$ 1/m is good, and both dispersivity and porosity estimated by the steady-state model are smaller than their real values used in the transient flow model for smaller S_s as shown in Table 3.11. In other words, the parameters are underestimated by the steady-state model. As for the case of $S_s = 1.0 \times 10^{-6}$ 1/m, the fitness is very bad, and the dispersivity is also under-estimated by the steady-state model, while the porosity is equal to 1, which is unreasonable. Such findings could be explained using the flow velocity, as shown in Figure 3.55, a smaller S_s could decrease the difference in the flow velocity between the steady-state model and the transient flow model.

Figure 3.62 shows BTCs at the well fitted by the steady-state model for different r_w. The difference between the solutions by the steady-state flow model and transient flow model is obvious for the large well radius. From Table 3.12, one could find that the steady-state flow model underestimates the dispersivity, but overestimates the porosity. In summary, the fitness of the steady-state model is

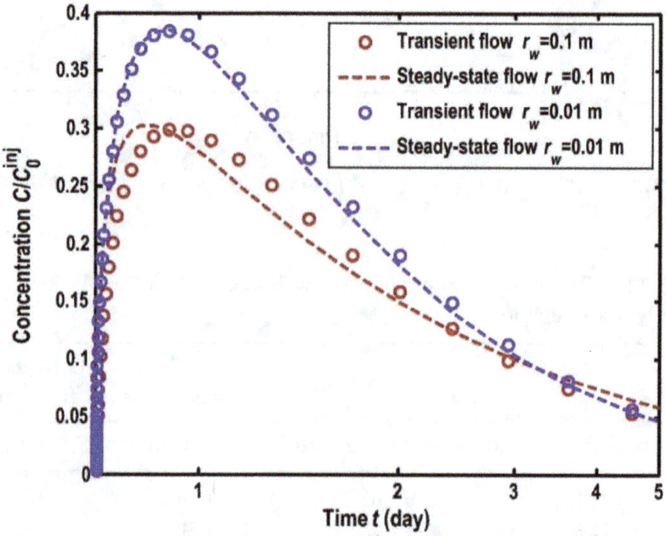

Figure 3.62. BTCs at the wellbore fitted by the steady-state flow model for different r_w.

Table 3.12. The parameter estimated by the solution of the SWPP test under the steady-state flow for different r_w.

Parameters	Real values used in the transient flow model	Estimated by the steady-state flow model $r_w = 0.1$	Estimated by the steady-state flow model $r_w = 0.01$
α_r (m)	$\alpha_r = 10$	$1.1\alpha_r$	$0.8\alpha_r$
θ (dimensionless)	$\theta = 0.2$	4.5θ	1.8θ

good for the smaller S_s and the smaller r_w, however, the difference the parameters between the real and the estimated values is still very large.

3.4.2.2 *A SWPP test with wellbore storages*

A complete SWPP test includes injection, chase, rest, and extraction phases, where the second and third phases are not necessary but are recommended to increase the duration of reaction. Due to the complex mechanics of biogeochemical reactions, aquifer properties, and so on, previous mathematical or numerical models contain some assumptions which may over simplify the actual physics, for instance, the hydraulic diffusivity of the aquifer is infinite. The Robin or the third-type boundary condition was often used in previous studies at the well screen in the injection, chase, and rest phases by ignoring the mixing effect of the volume of water in the wellbore (namely, wellbore storage). In this section, we presented a multi-species reactive SWPP model considering the wellbore storage for both groundwater flow and solute transport, and a finite aquifer hydraulic diffusivity. The models of wellbore storage for both solute transports are derived based on the mass balance. The Danckwerts boundary condition instead of the Robin condition is employed for solute transport across the well screen in the rest phase. The well fully penetrates a confined aquifer with a constant thickness. The aquifer is homogeneous, and the influence of the regional flow is ignored.

(a) Revisit of the previous model

The general form of the governing equation for a multi-species reactive SWPP test is

$$\frac{\partial C_i}{\partial t} + \frac{\rho_b}{\theta}\frac{\partial S_i}{\partial t} = -\sum_{j=1}^{N-1}[\mathcal{H}(t-t_j^*) - \mathcal{H}(t-t_{j+1}^*)]\lambda_j C_i^{n_j} \pm F_j \quad (3.655)$$

where C_i is the aqueous phase concentration of the ith reactive solute [Kg/m^3], S_i is the solid phase concentration of the ith reactive solute [Kg/m^3], t is the time in the SWPP test [s], ρ_b is the bulk density, θ is the porosity, $\mathcal{H}$ is the Heaviside step function, λ_j and n_j are the constant, N is the number of the segment, t_j^* and t_{j+1}^* are the times at two ends of the segment j, and F_j Monod/Michaelis-Menten kinetics. For simplicity, we only present the reactive processes of the chemicals as described by Eq. (3.655), while the expressions of the transport (e.g., dispersion, diffusion, and advection) could be seen in Phanikumar and McGuire (2010). They used it to describe biogeochemical reactive transport of an arbitrary number of species including Monod/Michaelis-Menten kinetics, and the sorption models could be isotherm (Freundlich, Langmuir and linear sorption), one-site kinetic and two-site kinetic. As for the studies of Gelhar and Collins (1971), Snodgrass and Kitanidis (1998), Haggerty *et al.* (1998), Schroth *et al.* (2001), Schroth and Istok (2005, 2006), Huang *et al.* (2010), Chen *et al.* (2017), Wang *et al.* (2018, 2020), Shi *et al.* (2020) and so on, the governing equation is a special case of Eq. (3.655).

As mentioned above, several assumptions may be debatable in previous studies and could be the source of errors for the actual applications. Firstly, the transport model is composed of a set of ADEs built on the basis of flow velocity which is assumed to be time-independent:

$$v_r = \frac{Q}{2\pi r B\theta}, \quad r \geq r_w \tag{3.656}$$

where r_w is the well radius [m]; r is the radius distance from the center of the well [m]; θ represents the porosity [dimensionless]; B is the aquifer thickness [m]; Q is the flow rate of the well [m^3/s]; $v_r = u_r/\theta$ is the average radial pore velocity [m/s] and u_r is the radial Darcian velocities [m/s]. Equation (3.656) implies that the hydraulic diffusivity of the aquifer is infinite, thus the flow velocity is independent of time. Meanwhile, the wellbore storage is negligible or the well radius r_w is assumed to be infinitesimal in formulating Eq. (3.656).

The second assumption of the model is the boundary condition of the well screen in the rest phase of the SWPP test, in which a Robin condition (or a third-type condition) is employed to describe the aqueous solute transport (Chen *et al.*, 2017; Gelhar and Collins, 1971; Phanikumar and McGuire, 2010):

$$\left.\left(v_r C - D_r \frac{\partial C}{\partial r}\right)\right|_{r \to r_w} = 0, \quad t_{\text{inj}} + t_{\text{cha}} < t \leq t_{\text{inj}} + t_{\text{cha}} + t_{\text{res}} \tag{3.657}$$

where t_{inj}, t_{cha}, t_{res}, and t_{ext}, represent the durations [s] of the injection, chase, rest, and extraction phases, respectively; C is resident concentration of the aqueous phase [Kg/m^3] to represent C_i in Eq. (1); D_r is the dispersion coefficient [m^2/s], which is

$$D_r = \alpha_r v_r + D_0 \tag{3.658}$$

in which α_r is the radial dispersivity [m]; D_0 is the effective diffusion coefficient in the aquifer [m^2/s].

Third, a constant solute concentration in the wellbore is applied in the injection and chase phases without considering the solute diluted effect in the wellbore (Chen *et al.*, 2017; Gelhar and Collins, 1971; Phanikumar and McGuire, 2010):

$$\left(v_r C - D_r \frac{\partial C}{\partial r}\right)\Bigg|_{r\to r_w} = v_r C_0^{\text{inj}}, \quad 0 < t \le t_{\text{inj}} \tag{3.659a}$$

or

$$C|_{r\to r_w} = C_0^{\text{inj}}, \quad 0 < t \le t_{\text{inj}} \tag{3.659b}$$

$$\left(v_r C - D_r \frac{\partial C}{\partial r}\right)\Bigg|_{r\to r_w} = v_r C_0^{\text{cha}}, \quad t_{\text{inj}} < t \le t_{\text{inj}} + t_{\text{cha}} \tag{3.660a}$$

or

$$C|_{r\to r_w} = C_0^{\text{cha}}, \quad t_{\text{inj}} < t \le t_{\text{inj}} + t_{\text{cha}} \tag{3.660b}$$

where C_0^{inj} and C_0^{cha} represent the solute concentrations [Kg/m^3] injected into the wellbore during the injection and chase phases, respectively. A detailed discussion of mentioned-above assumptions can be seen in Phanikumar and McGuire (2010).

Fourth, the hydraulic diffusivity is assumed to be infinite when estimating reaction rates; for instance, one of the simplest models of such reactions may be the first-order reaction:

$$\frac{\partial C}{\partial t} = -\lambda C \tag{3.661}$$

where λ is the first-order reaction rate constant. Beside the first-order reaction, Eq. (3.661) could be used to describe the first-order biodegradation and radioactive decay. Haggerty *et al.* (1998) presented a simplified method to estimate λ for the SWPP test:

$$\ln\left(\frac{C_{\text{rec}}(t^*)}{C_{\text{tra}}(t^*)}\right) = \ln\left(\frac{1-\exp(-\lambda t_{\text{inj}})}{\lambda t_{\text{inj}}}\right), \tag{3.662}$$

where t^* is the time since the end of injection; $C_{\text{rec}}(t^*)$ is the reactant concentration; $C_{\text{tra}}(t^*)$ is the concentration of a conservative tracer. To obtain the value of λ, the reactant and the conservative tracer should be fully mixed and injected into the aquifer simultaneously to conduct the SWPP test. After measuring the data of $C_{\text{rec}}(t^*)$ and $C_{\text{tra}}(t^*)$ in the extraction phase, one could fit the data of $\ln\left(\frac{C_{\text{rec}}(t^*)}{C_{\text{tra}}(t^*)}\right) \sim t^*$ using a linear function and the slope of t^* is the estimation of λ. Snodgrass and Kitanidis (1998) derived a similar model for estimating λ:

$$\ln\left(\frac{C_{\text{rea}}(t^*)}{C_{\text{tra}}(t^*)}\right) = \ln\left(\frac{C_{\text{rec}}^0}{C_{\text{tra}}^0}\right) - \lambda t^* \tag{3.663}$$

Comparing Eq. (3.662) with Eq. (3.663), one could find that the difference is the first terms on the right sides of equations, while λ is the slope for both Eqs. (3.662) and (3.663). Although the accuracy of both models has been tested by a number of investigators, previous studies on reactive transport were based on an assumption that the aquifer hydraulic diffusivity was infinite (e.g., Eq. (3.655) of Reinhard *et al.* (1997), and Eq. (3.656) of Haggerty *et al.* (1998)).

Actually, the assumptions of Eqs. (3.656)–(3.663) are debatable for the actual applications, and may cause errors in modelling the solute transport in the SWPP test. The second and third assumptions relate to the wellbore storage of the solute transport in the SWPP test. In the following section, the new models will be proposed to investigate the potential errors when these assumptions are involved.

(b) A revised model with a finite hydraulic diffusivity

As for the first assumption mentioned above, Wang *et al.* (2017) demonstrated that it might result in non-negligible errors in parameter estimation, particularly for the estimation of dispersivity. A minor point to note is that the model of Wang *et al.* (2017) mainly focused on conservative solute transport, rather than reactive transport. Nevertheless, the pore velocity of transient flow is calculated by Darcy's law:

$$v_r = \frac{K_r}{\theta}\frac{\partial s}{\partial r} \tag{3.664}$$

where K_r is the radial hydraulic conductivity [m/s]; s is drawdown [m], which could be obtained by solving the following mass balance equation with the proper initial and boundary conditions:

$$\frac{\partial v_r}{\partial r}+\frac{v_r}{r}=\frac{S_s}{\theta}\frac{\partial s(r,t)}{\partial r}, \quad r\geq r_w \tag{3.665}$$

$$s(r,t)|_{t=0}=0 \tag{3.666}$$

$$v_r|_{r\to\infty}=0 \tag{3.667}$$

$$(2\pi B v_r)|_{r\to r_w}-\frac{\pi r_w^2}{\theta}\frac{ds_w(t)}{dt}=Q \tag{3.668}$$

where S_s is the specific storage of aquifer [1/s]; s_w is the drawdown inside the wellbore [m].

As for the second assumption in the rest phase, as shown in Eq. (3.657), it implies that the concentration of solute is zero in the wellbore. This assumption works when the chase concentration is zero, and the prepared solution is completely pushed out of the borehole into the aquifer at the end of the chase phase. However, the chase concentration might not be zero; for instance, nature water contains Cl^- in itself when sodium chloride is taken as the tracer (Phanikumar and McGuire, 2010). Consequently, the concentration in the early stage of the rest phase, which is close to the concentration at the end of the chase phase, is not zero. This is because the water level in the wellbore is greater than the hydraulic head in the surrounding aquifer due to the wellbore storage, resulting in a positive flux from the wellbore into the aquifer. Correspondingly, when the chase concentration is not zero or the prepared solution in the injection phase is not completely pushed out of the wellbore, the concentration in the wellbore may not be zero in the early stage of the rest phase.

In this section, the Danckwerts condition for transport at the well screen in the rest period (Danckwerts, 1953) was employed:

$$\left.\frac{\partial C}{\partial r}\right|_{r\to r_w}=0, \quad t_{\text{inj}}+t_{\text{cha}}<t\leq t_{\text{inj}}+t_{\text{cha}}+t_{\text{res}} \tag{3.669}$$

Actually, Eq. (3.669) acknowledges the continuity of concentration and continuity of mass flux simultaneously across the well screen, namely, $C|_{r\to r_w^-}=C|_{r\to r_w^+}$ and $(v_rC)|_{r\to r_w^-}=\left(v_rC-D_r\frac{\partial C}{\partial r}\right)\big|_{r\to r_w^+}$, where $-$ and $+$ signs in the subscript of r_w represent approaching the well screen from inside the well and outside the well, respectively.

The third assumption mentioned above seems not reasonable at the early stage of the injection and chase phases, because the concentration of the injected solute will be affected by the finite volume of water in the wellbore. Take the chase phase as an example: it is impossible to immediately reduce the solute concentration inside the wellbore from a certain level during the tracer injection phase to zero when switching to the chase phase, even when the solute concentration in the chase phase is zero. This is because the wellbore with a finite radius contains a certain finite mass of solute at the moment of switching from injection of a tracer to injection of a chaser. Therefore, it will take some time to completely flush out the residual tracer inside the wellbore after the start of the chase phase, and a larger wellbore will take a longer time to flush out the residual tracer inside the wellbore. This means that the concentration at the wellbore/aquifer interface will not drop to zero immediately after the start of the chase phase. Instead, it will take a finite period of time to gradually approach zero during the chase phase. Similarly, the boundary condition of the well screen in the injection phase might not be appropriate in previous studies if the wellbore storage effect is of a concern. Therefore, the value of solute concentration inside the wellbore should be smaller or equal to C_0^{inj} in the injection phase and greater or equal to C_0^{cha} in the chase phase.

Here, we will develop a new approach to take care of the concentration in the wellbore in the injection and chase phases based on the mass balance principle, i.e.,

$$\Delta m = C_0^{\text{inj}} Q\Delta t = C_w^{t+\Delta t}(V^t + Q\Delta t) - C_w^t V^t, \quad 0 < t \le t_{\text{inj}} \tag{3.670}$$

$$\Delta m = C_0^{\text{cha}} Q\Delta t = C_w^{t+\Delta t}(V^t + Q\Delta t) - C_w^t V^t, \quad t_{\text{inj}} < t \le t_{\text{inj}} + t_{\text{cha}} \tag{3.671}$$

where Δm represents the mass entering into the well during time interval Δt; C_w^t and $C_w^{t+\Delta t}$ represent the solute concentrations in the wellbore at the time t and $t + \Delta t$, respectively; V^t represents the volume of water in the wellbore at the time t. The initial values of C_w^t and V^t at the injection phase are

$$C_w^t|_{t=0} = 0 \quad (3.672)$$

$$V^t|_{t=0} = \pi r_w^2 (H_w|_{t=0}) \quad (3.673)$$

where H_w represents the water depth of the wellbore.

In the chase phase, one has

$$C_w^t|_{t=t_{\text{inj}}^-} = C_w^t|_{t=t_{\text{inj}}^+} \quad (3.674)$$

$$V^t|_{t=t_{\text{inj}}^+} = \pi r_w^2 (H_w|_{t=t_{\text{inj}}^+}) \quad (3.675)$$

where the $-$ and $+$ signs in the subscripts of Eqs. (3.674)–(3.675) hereinafter represent approaching the limit from left and right sides of t_{inj}, respectively.

(c) Capability of the new SWPP model of Wang and Zhan (2019)

Different from the model of Wang *et al.* (2017), the multi-species reactive transport models are used to describe the nonlinear biogeochemical reactive processes considering wellbore effects not only for groundwater flow but also for solute concentrations. The model of Wang and Zhan (2019) is an extension of Phanikumar and McGuire (2010) that ignored the wellbore storage for both groundwater flow and solute transport, and assumed that the aquifer hydraulic diffusivity was infinite. The Danckwerts condition rather than the Robin condition is applied at the well screen in the rest phase of this section. Therefore, the model of Wang and Zhan (2019) is more powerful in describing an arbitrary number of species and user-defined reaction rate expressions, including Monod/Michaelis-Menten kinetics.

(d) Numerical solution of the SWPP test

To minimize numerical errors and to increase computational efficiency, we employ a non-uniform grid system for simulations (Wang *et al.*, 2014), which is

$$r_i = \frac{r_{i-1/2} + r_{i+1/2}}{2}, \quad i = 1, 2, 3, \ldots, N_r \quad (3.676)$$

where N_r represents the number of nodes in the discretization of the spatial domain $[r_w, r_e]$, r_w and r_e respectively represent the distances of inner and outer boundary nodes; r_i is the radial distance of node; $r_{i+1/2}$ is calculated as follows:

$$\log_{10}(r_{i+1/2}) = \log_{10}(r_w) + i\left[\frac{\log_{10}(r_e) - \log_{10}(r_w)}{N}\right], \quad i = 0, 1, 2, \ldots, N_r \tag{3.677}$$

The value of $r_{i-1/2}$ can be calculated using a similar method. Equations (3.676)–(3.677) represent a space domain discretized logarithmically, and the spatial steps are smaller near the wellbore and become progressively greater away from it.

Similarly, we logarithmically discretize the temporal domain

$$t_i = \frac{t_{i-1/2} + t_{i+1/2}}{2}, \quad i = 1, 2, 3, \ldots, M \tag{3.678}$$

where M represents the number of nodes in the discretization of the temporal domain; t_i is the time of the node i; $t_{i+1/2}$ is calculated as follows in the injection phase:

$$\log_{10}(t_{i+1/2}) = \log_{10}(t_0) + i\left[\frac{\log_{10}(t_{\text{inj}}) - \log_{10}(t_0)}{M}\right], \quad i = 1, 2, 3, \ldots, M \tag{3.679}$$

where t_0 is a very small positive value representing the first-time step, such as $t_0 = 1.0 \times 10^{-7}$ h.

As for the chase, one has

$$t_{i+1/2} = 10^{\log_{10}(t_0)+i\left[\frac{\log_{10}(t_{\text{cha}})-\log_{10}(t_0)}{M}\right]} + t_{\text{inj}}, \quad i = 1, 2, 3, \ldots, M \tag{3.680}$$

Similarly, in the rest phase, one has

$$t_{i+1/2} = 10^{\log_{10}(t_0)+i\left[\frac{\log_{10}(t_{\text{res}})-\log_{10}(t_0)}{M}\right]} + t_{\text{inj}} + t_{\text{cha}}, \quad i = 1, 2, 3, \ldots, M \tag{3.681}$$

In the extraction phase, one has

$$t_{i+1/2} = 10^{\log_{10}(t_0)+i\left[\frac{\log_{10}(t_{\text{ext}})-\log_{10}(t_0)}{M}\right]} + t_{\text{inj}} + t_{\text{cha}} + t_{\text{res}},$$
$$i = 1, 2, 3, \ldots, M \quad (3.682)$$

Before using the new model in this section, it is necessary to evaluate the numerical errors (like artificial oscillation and numerical dispersion) of the solution. Unfortunately, the benchmark analytical solutions of the SWPP test with a finite hydraulic diffusivity are not available up to date. Alternatively, the accuracy of the finite-difference solution could be tested by comparing it with the numerical solution of Wang *et al.* (2017) which was proven to be accurate and robust.

Figure 3.63 shows the comparison of BTCs between the solutions of Wang *et al.* (2017) and of this section, where the parameters used are similar to Figure 3 of Phanikumar and McGuire (2010): $B = 8\,\text{m}$, $r_w = 0.052\,\text{m}$, $\alpha_r = 1\,\text{m}$, $\theta = 0.38\,\text{m}$, $D_0 = 0\,\text{m}^2/\text{h}$, $t_{\text{inj}} = 94.32\,\text{h}$, $t_{\text{cha}} = 0\,\text{h}$, $t_{\text{res}} = 0$ hour, $t_{\text{ext}} = 405.6\,\text{h}$, injection flow rate

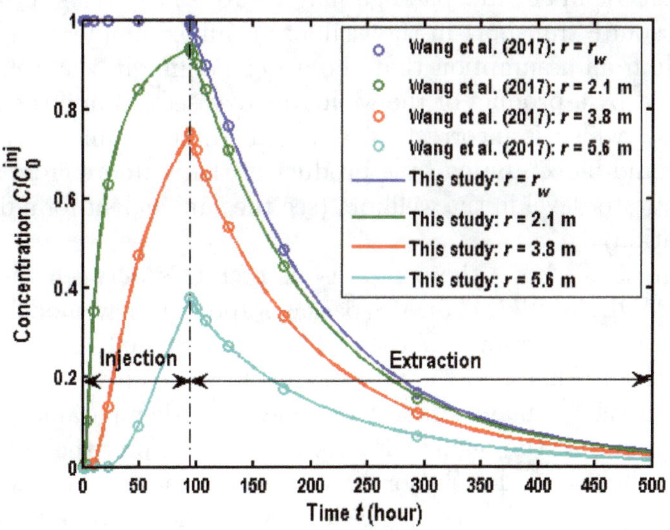

Figure 3.63. Comparison of BTCs between the solutions of Wang *et al.* (2017) and of this model, where C_0^{inj} represents the concentration of the prepared solute in the injection phase.

$Q_{\text{inj}} = 0.1\,\text{m}^3/\text{hour}$, and extraction flow rate $Q_{\text{ext}} = -0.11\,\text{m}^3/\text{h}$. It shows a small oscillation in the numerical solutions, which might be caused by numerical errors.

By comparing the solution of this model with Wang *et al.* (2017), one may conclude that the solution of this section appears to be accurate and reliable since the mean square error between two solutions is smaller than 0.01 for all cases in Figure 3.63. In the wellbore ($r = r_w$), the concertation is equal to C_0^{inj}, as shown in Figure 3.63. This is due to the boundary condition of the wellbore, e.g.,

$$C|_{r \to r_w} = C_0^{\text{inj}}, \quad 0 < t \le t_{\text{inj}} \tag{3.683}$$

In the aquifer, the values of BTCs increase with the decreasing distance from the wellbore.

It is also necessary to test the accuracy of the new models against the numerical software packages. Since the code of the original MODFLOW/MT3DMS package is open source and could be downloaded freely from the website of the United States Geological Survey, it is preferred by many modelers and is selected as the basis of comparison in this section. Unfortunately, such an open-source MODFLOW/MT3DMS package may create some errors in describing the solute transport in the wellbore-confined aquifer. The errors come from an assumption that the water volume in the wellbore is computed by a product of the wellbore cross section and the aquifer thickness, which is incorrect. The actual water volume in the wellbore should be computed by a product of the wellbore cross section and the water level in the wellbore (see the supplement for a detailed explanation).

Figure 3.63 shows the comparisons of BTCs between the open-source MODFLOW/MT3DMS package and the new model of this section. The water level of the wellbore is assumed to be equal to the aquifer thickness in the new models for the purpose of comparison, although it may not be true, and the other parameters used are the same as the ones in Figure 3.63. Therefore, the agreement between the two models demonstrates the accuracy of the new model. Figure 3.64 shows that the concentration in the wellbore is not unit in the injection phase, and this is because the new model considers the wellbore storage for both groundwater flow and solute transport.

It is worthwhile pointing out that an advanced version of MODFLOW/MT3DMS, namely MODFLOW-SURFACT, includes

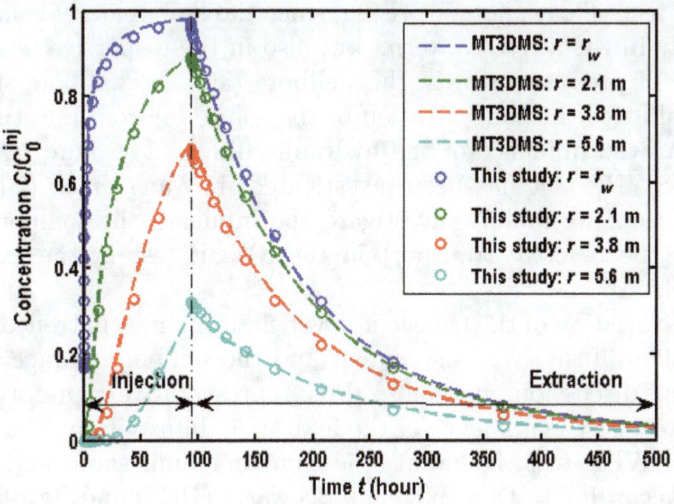

Figure 3.64. Comparison of BTCs between the solutions of MODFLOW/MT3DMS and the model of Wang and Zhan (2019).

a fracture-well package (FWL4 and FWL5) to overcome the problems in the original open-source MODFLOW well package. The FWL4 and FWL5 packages calculate the water volume using simulated heads, not aquifer thicknesses (see the MODFLOW-SURFACT manual, Vol I, Sect. 3.2, Eq. (24) for details). FEFLOW also has a similar package, referred to as a discrete feature to simulate a pumping/extraction well, if one chooses to do so. Additionally, with a FEFLOW model, the model mesh can be highly discretized to accurately represent well dimensions using a subset of elements (in centers). The modeler can assign a porosity of the unit for those elements representing the wells, rather than assuming the same porosity of the surrounding materials. In the future, we will conduct a comprehensive comparative investigation of the method proposed in this section and those of MODFLOW-SURFACT and FEFLOW for understanding the effects of well mixing and wellbore storage for both flow and transport processes involving an aquifer–well system.

(e) Effect of wellbore storage on the SWPP test in the transient flow field

Revisiting the assumptions used in previous studies, one may find that the flow field and the wellbore storage are key factors for the

SWPP test. This is not surprising, since the flow velocity is included not only in the advective term, but also in the dispersive term. The volume of pre-test water in the wellbore (i.e., the wellbore storage) may influence the concentration of the solute injected into the wellbore. As the influence of the hydraulic diffusivity solute transport in the SWPP test has been investigated in Wang *et al.* (2017), in this section, we mainly investigate the influence the wellbore storage on the reactive transport in the SWPP test in the transient flow field.

The variation of the transient flow field is mainly controlled by the hydraulic diffusivity of the aquifer and the wellbore storage. In the following discussion, we choose three representative types of porous media to test the influence of the hydraulic diffusivity on the results of the SWPP test, including fine sand, medium sand, and coarse sand. According to Domenico and Schwartz (1990) and Batu (1998), one could obtain the values of the hydraulic diffusivity for above mentioned three types of media: $4.17 \times 10\,\mathrm{m}^2/\mathrm{h}$ (with $K_r = 4.17 \times 10^{-3}\,\mathrm{m/h}$ and $S_S = 1.0\times10^{-4}\,\mathrm{m}^{-1}$) for the fine sand, $4.17\times102\,\mathrm{m}^2/\mathrm{h}$ (with $K_r = 4.17 \times 10^{-2}\,\mathrm{m/h}$ and $S_S = 1.0 \times 10^{-4}\,\mathrm{m}^{-1}$) for the medium sand, and $4.17 \times 104\,\mathrm{m}^2/\mathrm{h}$ (with $K_r = 4.17 \times 10^{-1}\,\mathrm{m/h}$ and $S_S = 1.0 \times 10^{-5}\,\mathrm{m}^{-1}$) for the coarse sand. Generally, the hydraulic diffusivity of the aquifer correlates to the grain size of the media, and the value is smaller for the smaller grain size, e.g., fine sand.

The parameters related to the solute transport mainly come from the studies of Phanikumar and McGuire (2010), who interpreted the field experimental data of the SWPP test conducted by Phanikumar and McGuire (2010). Except for parameters specifically mentioned otherwise, the default values used in the following section are $C_0^{\mathrm{inj}} = 100\,\mathrm{mg/L}$, $C_0^{\mathrm{cha}} = 10\,\mathrm{mg/L}$, $B = 0.1\,\mathrm{m}$, $r_w = 0.0125\,\mathrm{m}$, $\alpha_r = 0.01\,\mathrm{m}$, $\theta = 0.33$, $D_0 = 0\,\mathrm{m}^2/\mathrm{h}$, $t_{\mathrm{inj}} = 0.6\,\mathrm{h}$, $t_{\mathrm{cha}} = 0.067\,\mathrm{h}$, $t_{\mathrm{res}} = 0.0333\,\mathrm{h}$, $t_{\mathrm{ext}} = 3.6\,\mathrm{h}$, $Q_{\mathrm{inj}} = 0.0333\,\mathrm{m}^3/\mathrm{hour}$, $Q_{\mathrm{cha}} = 0.0255\,\mathrm{m}^3/\mathrm{h}$, and $Q_{\mathrm{ext}} = 0.0333\,\mathrm{m}^3/\mathrm{h}$, which can be found in Figure 5 of Phanikumar and McGuire (2010).

Figures 3.65(a) and 3.65(b) show the comparison of BTCs between the Robin and Danckwerts conditions at the wellbore for different porous media, where $C_0^{\mathrm{cha}} = 10\,\mathrm{mg/L}$ in Figure 3.65(a) and $C_0^{\mathrm{cha}} = 0.0\,\mathrm{mg/L}$ in Figure 3.65(b). For the purpose of comparison, the boundary conditions at the wellbore in the injection and chase phases are still described by Eqs. (3.659) and (3.660).

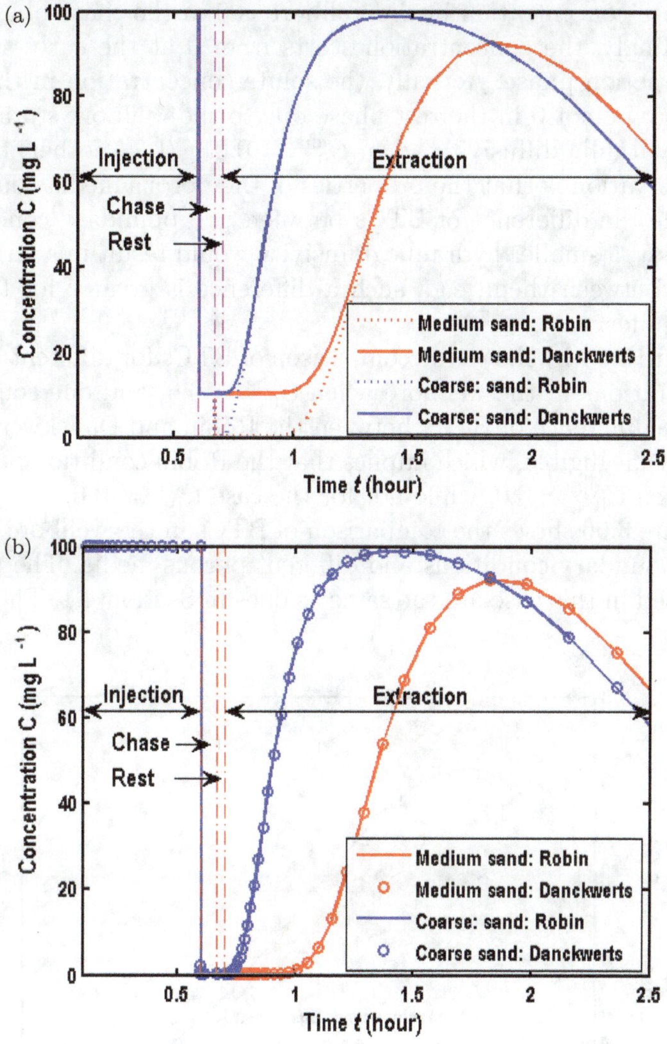

Figure 3.65. Comparison of BTCs in the wellbore between the Robin and Danckwerts conditions: (a) $C_0^{\text{cha}} = 10.0\,\text{mg/L}$; (b) $C_0^{\text{cha}} = 0.0$.

Figure 3.65(a) shows that the difference of BTCs between two boundary conditions is significant at the early stage of the extraction phase when $C_0^{\text{cha}} = 0.01\,\text{g/L}$, and BTCs of the Danckwerts condition are above BTCs of the Robin condition. With time going, such a difference becomes negligible. As for the curves of the Robin condition, the solute concentration in the wellbore is 0 in the chase phase, correspondingly, the concentration starts from 0 at the early stage of the extraction phase. Actually, the solute concentration in the wellbore may be not 0 in the rest phase due to the wellbore storage and finite hydraulic diffusivity when $C_0^{\text{cha}} = 0.01\,\text{g/L}$. Another interesting observation is that the properties of the porous media could also influence the difference of BTCs between two boundary conditions. Obviously, a smaller hydraulic diffusivity would result in a larger difference between them, e.g., such a difference is greater for the fine sand aquifer.

Figure 3.65(b) shows the comparison of BTCs for different boundary conditions in the wellbore when $C_0^{\text{cha}} = 0.0$, and one could find that the difference of BTCs between the Robin and Danckwerts conditions is negligible, which implies that the Robin condition performs well when $C_0^{\text{cha}} = 0.0$, while not for the case $C_0^{\text{cha}} \neq 0.0$.

Figure 3.66 shows the comparison of BTCs in the wellbore for different boundary conditions and different porous media. The parameters used in this case are the same as ones in Section 4.1. The initial

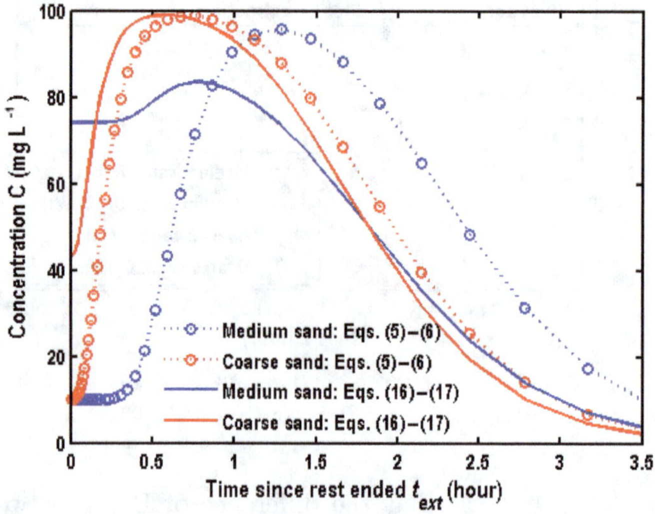

Figure 3.66. The BTCs in the wellbore for the different boundary conditions at the wellbore in the injection and chase phases.

head is 1 m. The boundary condition of the wellbore in the rest phase is described by the Danckwerts condition.

Two interesting observations can be seen. Firstly, the difference of BTCs between the two boundary conditions at the wellbore is obvious, and such a difference is larger for the medium sand than for the coarse sand, implying that it increases with the decreasing hydraulic diffusivity. Secondly, the values of BTCs obtained from Eqs. (3.670) and (3.671) are greater at the early stage of the extraction phase, while the peak values of BTC are smaller. In another word, the model of Eqs. (3.659) and (3.660) may underestimate the concentration in the early stage of the extraction phase, while overestimating the peak values of BTCs.

These observations can be explained as follows. The model of Eqs. (3.659) and (3.660) assumes that the volume of water in the wellbore is negligible, and the concentration in the wellbore is close to 10.0 mg/L in the rest phase, due to $C_0^{\mathrm{cha}} = 10.0\,\mathrm{mg/L}$. As for the model of Eqs. (3.670) and (3.671), the volume of water in the wellbore is non-negligible and could dilute the concentration in the injection phase; namely, the solute concentration in the wellbore could not immediately rises to C_0^{inj} at the early stage of the injection phase, thus resulting in smaller peak values of BTCs. Similarly, the concentration in the wellbore could not immediately reduce to C_0^{cha} at the early stage of the chase phase, which makes the concentration larger at the early stage of the extraction phase based on the model of Eqs. (3.670) and (3.671).

(f) Uniqueness of estimated parameters

In many previous studies, the parameters related to solute transport were estimated only by best fitting the observed BTCs in the wellbore by the simplified models ignoring the influence of the flow field on reactive transport. Actually, BTCs in the wellbore might be also affected by hydraulic parameters, such as hydraulic conductivity, specific storage, and porosity. In this section, the new reactive model of the SWPP test is established in the transient flow field, and it could be used to estimate the values of hydraulic conductivity, specific storage, porosity, dispersivity, and reaction rates by best fitting the observed by BTCs in the wellbore, e.g., the values could be obtained by minimizing the sum of absolute differences between the observed and calculated BTCs in the wellbore:

$$\mathcal{F} = \sum_{i=1}^{o} \left| C_{\mathrm{CAL}}(K_r, S_s, \alpha_r, \lambda_j, t, r_w)\big|_{t=t_i} - C_{\mathrm{OBS}}(t, r_w)|_{t=t_i} \right| \tag{3.684}$$

where $C_{\mathrm{CAL}}(K_r, S_s, \alpha_r, \lambda_j, t, r_w)\big|_{t=t_i}$ and $C_{\mathrm{OBS}}(t, r_w)|_{t=t_i}$ represent the concentration calculated by the new models of the SWPP test and the observations at $t = t_i$, respectively; o is the number of the observed data. Equation (3.684) shows that there are four parameters needed to be estimated, and it is necessary to test the uniqueness of solutions.

Although Wang *et al.* (2017) investigated the influences of the hydraulic diffusivity on the results, their model still contained some assumptions. For instance, the wellbore storage effect of solute transport was ignored, and the reaction was assumed to be in first order (linear).

In this section, two approaches are used to investigate the uniqueness of the estimated parameters. The first approach is sensitivity analysis. When the influence of the hydraulic parameters on the solute transport is not sensitive, the steady-state assumption could be used to approximate the flow field in the model of the SWPP test, and otherwise the transient flow field needed to be considered. The second approach is to best fit the BTCs of the transient-flow model by the steady-state-flow model. If the estimated parameters by the steady-state-flow model are nearly the same with the input parameters of the transient-flow model, the results could be thought to be unique.

McCuen (1985) proposed the sensitivity model of a dependent variable, which was normalized as (Kabala, 2001; Yang and Yeh, 2009):

$$\mathrm{SC}_{i,j} = I_j \frac{\partial C_i}{\partial I_j} \tag{3.685}$$

where $\mathrm{SC}_{i,j}$ is the sensitivity coefficient of the jth parameter I_j at the ith time; C_i is the concentration at the ith time. In this section, the differentiation of Eq. (3.686) will be approximated by a finite difference scheme:

$$\mathrm{SC}_{i,j} = I_j \frac{C_i(I_j + \Delta I_j) - C_i(I_j)}{\Delta I_j} \tag{3.686}$$

where ΔI_j is a small increment.

From the mathematical models of the groundwater flow, one may find that both hydraulic conductivity and specific storage could affect the flow field. Since greater hydraulic conductivity or smaller specific storage could shorten the time, it takes to approach the steady state, we will employ hydraulic diffusivity, which is the ratio of the two parameters, for the sensitivity analysis.

Figure 3.67 shows the sensitivity of the hydraulic diffusivity on BTCs, and one may find that it is not sensitive to the hydraulic diffusivity when the values of hydraulic diffusivity are sufficiently large. This might be because the time in approaching the steady state is very short when the hydraulic diffusivity values are sufficiently large (for instance, greater than $4.17 \times 10^2\ \mathrm{m^2h^{-1}}$), and the influence of the transient flow could be ignored. Therefore, the

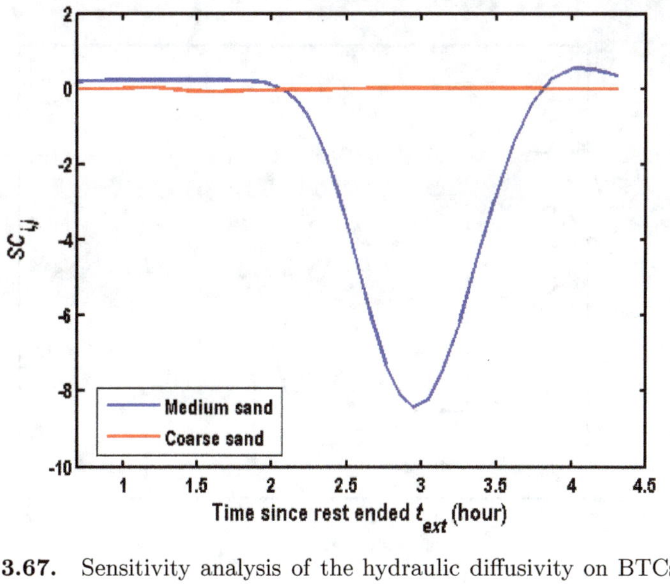

Figure 3.67. Sensitivity analysis of the hydraulic diffusivity on BTCs in the extraction phase.

steady-state assumption could be used to approximate the flow field in the SWPP test when the hydraulic diffusivity is greater than 4.17×10^2 m^{2}h^{-1}. Otherwise, the steady-state assumption is not recommended. Figure 3.68 shows that the BTCs in the wellbore are sensitive to both dispersivity and porosity.

Besides the sensitivity analysis, the uniqueness analysis is also important for parameter estimation, which is used to check whether there exist two or more sets of parameters for the same BTCs. Similar to the treatment in previous studies, we first use the transient model of this section to reproduce BTCs based on a set of given input parameters and then estimate the values of parameters by best fitting such BTCs. If the values of the input parameters are different from the estimated parameter when the fitness is very good, one could conclude that the solution is not unique, and the parameters estimated from Eq. (3.684) may not be reliable.

The new model of this section has four physical parameters: hydraulic conductivity, specific storage, dispersivity, and porosity, and one chemical parameter (reaction rate). Wang *et al.* (2017) investigated the uniqueness of solutions for the flow field, and the results showed that BTCs of the SWPP test were not unique for the

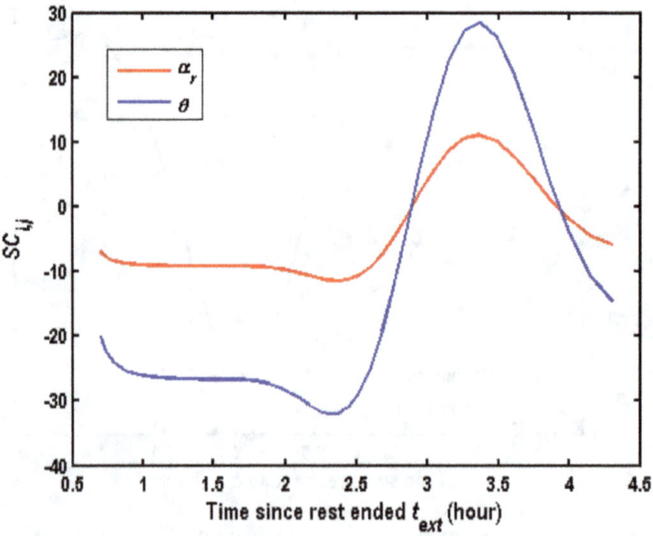

Figure 3.68. Sensitivity analysis of dispersivity and porosity on BTCs in the extraction phase.

flow-related parameters. For instance, BTCs with a steady-state flow field were almost the same as BTCs with a transient flow field, as shown in Figures 3.61 and 3.62. It implies that one may not inversely determine the hydraulic parameters of a flow field only by best fitting the observed BTCs in the wellbore, and additional aquifer tests are required to supplement the SWPP test to determine the flow-related parameters. However, Wang *et al.* (2017) did not investigate the uniqueness of porosity and dispersion when the hydraulic parameters were given, which will be discussed in this section.

Figure 3.69 shows a comparison of BTCs for different dispersivities and porosities but for the same hydraulic parameters, and one can see that the curves of $\alpha_r = 0.01\,\text{m}$ and $\theta = 0.33$ are almost the same as the curves of $\alpha_r = 0.006\,\text{m}$ and $\theta = 0.9$. Therefore, Eq. (3.684) may not be used to determine α_r and θ simultaneously. Fortunately, the porosity could be measured in the laboratory from core samples or determined by the SWPP test with drift flow. When the values of K_r, S_S, and θ are given, the dispersivity could be determined uniquely by Eq. (3.684).

In summary, it seems impossible to determine all parameters (K_r, S_S, α_r, and θ) simultaneously by only best fitting the observed

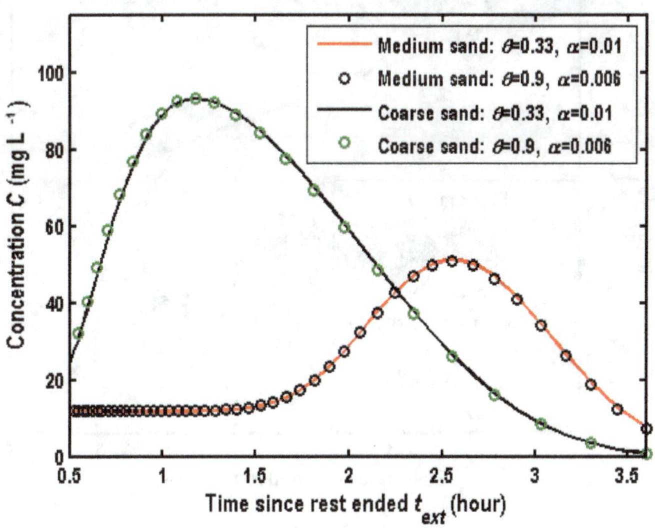

Figure 3.69. Comparison of BTCs for different dispersivities and porosities but for the same hydraulic parameters.

BTCs in the wellbore of the SWPP test using Eq. (3.684). Therefore, before determining the parameters related to the solute transport (α_r and θ), the hydraulic parameters (K_r and S_S) needed to be estimated by supplementary aquifer tests, or by best fitting the pressure data measured during the SWPP test, e.g., the pumping phase. The value of α_r could be determined by Eq. (3.684) when the porosity is given.

(g) Chemical parameter estimation

The models estimating the reaction rate are based on several assumptions in previous studies, e.g., Eqs. (3.662) and (3.663). To test the applicability of those equations, we will use the model of this section to reproduce the data of $\ln(C_{\text{rec}}/C_{\text{tra}}) \sim t^*$ based on a set of given parameters, and then using Eqs. (3.662) and (3.663) (which is based on an infinite hydraulic diffusivity presumption) to estimate λ (denoted as $\widetilde{\lambda}$) by best fitting $\ln(C_{\text{rec}}/C_{\text{tra}}) \sim t^*$. Two species involved in this case are Cl^{-1} and SO_4^{-2}, in which C_{tra} and C_{rec} represent the concentrations of Cl^{-1} and SO_4^{-2}, respectively. Figure 3.70 shows the fitness of the simulated $\ln(C_{\text{rec}}/C_{\text{tra}}) \sim t^*$ in the wellbore using a linear function, with the detailed information shown in Table 3.13. Two sets of λ are employed in the discussions for the

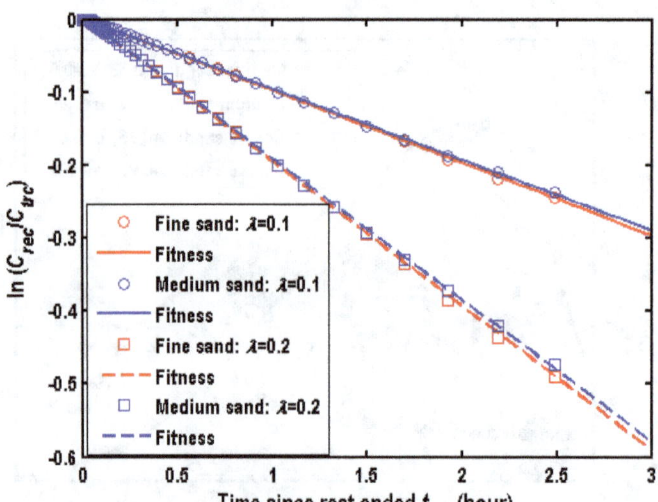

Figure 3.70. Fitness of $\ln(C_{\text{rec}}/C_{\text{tra}}) \sim t^*$ produced by the numerical solution of Wang and Zhan (2019) with the first-order reaction in the different porous media.

Table 3.13. Reaction parameters estimated by linear functions.

K (m day^{-1})	S_S (m^{-1})	λ (h)	$\tilde{\lambda}$ (h)	Intercept of linear function	$\ln\left(\frac{1-\exp(-\lambda t_{\text{inj}})}{\lambda t_{\text{inj}}}\right)$	$\ln\left[\frac{C^0_{\text{rec}}}{C^0_{\text{trc}}}\right]$
0.1	0.0001	0.1	0.0991	0.0017	−0.0299	0
1	0.0001	0.1	0.097	0.0016	−0.0299	0
0.1	0.0001	0.2	0.1981	0.0034	−0.0594	0
1	0.0001	0.2	0.1939	0.0031	−0.0594	0

reactant, e.g., $\lambda = 0.1$ and $0.2\,\text{h}^{-1}$. One may conclude that the simplified models of Eqs. (3.662) and (3.663) with an infinite hydraulic diffusivity perform well in the estimation of λ for reactive transport under the finite hydraulic diffusivity condition.

This simplified model of Eq. (3.663) has been widely used to estimate λ, due to the advantages that λ could be determined directly by best fitting the observed $\ln(C_{\text{rec}}/C_{\text{tra}}) \sim t^*$ curves, without knowing the aquifer properties, such as porosity, dispersivity, and hydraulic diffusivity. However, this model is proposed based on the first-order reaction, which is a linear reaction, as shown in Eq. (3.661). Whether this approach still works for nonlinear reactions is still questionable, and it will be investigated in the following section.

Assuming that the extraction time since the rest phase ended could be divided into $N-1$ segments, Phanikumar and McGuire (2010) employed the Heaviside unit step function to describe a type of nonlinear biogeochemical reaction:

$$\frac{\partial C}{\partial t} = -\sum_{j=1}^{N-1} [\mathcal{H}(t - t_j^*) - \mathcal{H}(t - t_{j+1}^*)]\lambda_j C_i^{n_j} \tag{3.687}$$

where λ_j is the reaction constant in the temporal segment j, and the Heaviside step function $\mathcal{H}(\cdot)$ is

$$\mathcal{H}(t - t_j^*) - \mathcal{H}\left(t - t_{j+1}^*\right) = \begin{cases} 0 & \text{if } t < t_j^* \\ 1 & \text{if } t_j^* < t < t_{j+1}^* \\ 0 & \text{if } t_{j+1}^* < t \end{cases} \tag{3.688}$$

Equation (3.686) is a series of piece-wise linear ($n_j = 1$) or nonlinear ($n_j \neq 1$) functions, which are an extension of Eq. (3.683).

To test the influence of the hydraulic diffusivity on the accuracy of this approach in estimating λ_j for the nonlinear reactions, the model of this section is used to reproduce the data of $\ln(C_{\mathrm{rec}}/C_{\mathrm{tra}}) \sim t^*$ with a set of specific λ_j, n_j and t_j^* for three types of porous media. Figures 3.71 and 3.72 represent the computed $\ln(C_{\mathrm{rec}}/C_{\mathrm{tra}}) \sim t^*$ based on the model of the chemical reactions described by the piece-wise linear and the nonlinear functions, respectively. The values of λ_j and t_j^* of Figure 3.71 are obtained by fitting the observation data using a piecewise linear function (e.g., $n_j = 1$) proposed by Phanikumar and McGuire (2010). The circle represents the experiments data observed by McGuire *et al.* (2002). The parameters related to the chemical reactions in Figure 3.72 are from Phanikumar and McGuire (2010) by fitting the observation data using a nonlinear function: $\lambda_j = 0.25$, $n_j = 0.25$, $N = j = 1$. Comparing Figures 3.71 and 3.72, one may find that the influence of the hydraulic diffusivity on the computed $\ln(C_{\mathrm{rec}}/C_{\mathrm{tra}}) \sim t^*$ is negligible for the chemical reaction described by the piecewise linear function, which is similar to the first-order reaction as shown in Figure 3.70. However, the influence of the hydraulic diffusivity on the relationship of $\ln(C_{\mathrm{rec}}/C_{\mathrm{tra}}) \sim t^*$

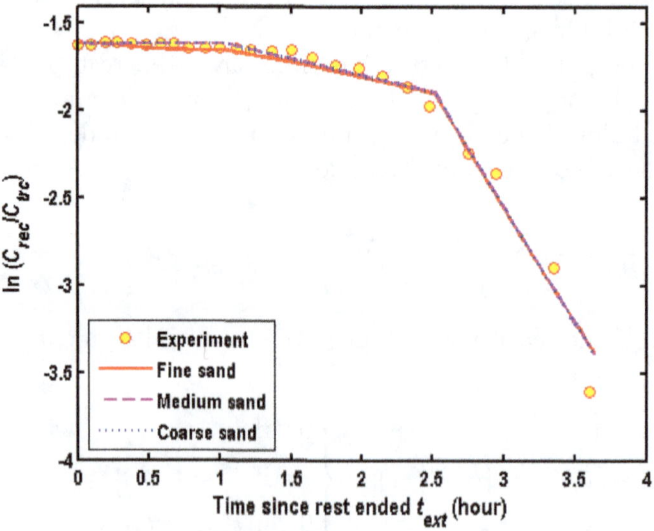

Figure 3.71. Computed $\ln(C_{\mathrm{rec}}/C_{\mathrm{tra}}) \sim t^*$ by the model of Wang and Zhan (2019) using a piece-wise linear function to describe the nonlinear chemical reactions.

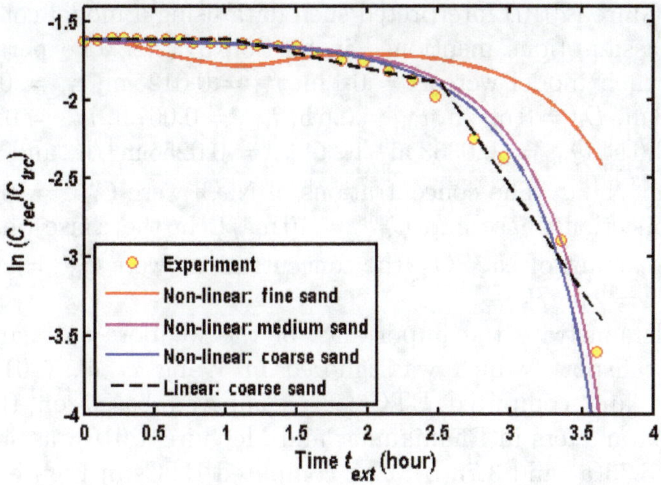

Figure 3.72. Computed $\ln(C_{\text{rec}}/C_{\text{tra}}) \sim t^*$ by the model of Wang and Zhan (2019) using a nonlinear function to describe the nonlinear chemical reactions.

cannot be ignored if one attempts to use nonlinear functions to describe such a chemical reaction. The difference between the curves of different porous media is obvious in Figure 3.72. The agreement between the observed and computed data is well for the medium and coarse sands, but not for the fine sand in Figure 3.72. This is because the hydraulic diffusivity values of the medium and coarse sands are larger than that of the fine sand, thus are close to the assumption of an infinite hydraulic diffusivity used in Phanikumar and McGuire (2010).

Therefore, one may conclude that λ_j, n_j and t_j^* could be determined by directly best fitting the observed $\ln(C_{\text{rec}}/C_{\text{tra}}) \sim t^*$ when the nonlinear reactions are described by the piece-wise linear functions, in a similar way to estimate the linear reaction rate by Eq. (3.661). However, such an approach may not work if one attempts to use nonlinear functions to describe such nonlinear reactions.

(h) Field applications

To test the model of this section, the field data of a SWPP test conducted in a single well by McGuire *et al.* (2002) will be employed. In this test, the prepared solution contains Na_2SO_4 (as a reactant) and NaCl (as a conservative tracer). The reactant and the tracer were well mixed and then injected into a targeted aquifer. Phanikumar and McGuire (2010) interpreted such data using a model containing several assumptions mentioned in Section 3.4.2.2. The parameters used in their model were $B = 0.1\,\text{m}$, $r_w = 0.0125\,\text{m}$, $\alpha_r = 0.001\,\text{m}$, $\theta = 0.33\,\text{m}$, $D_0 = 0\,\text{m}^2/\text{h}$, $t_{\text{inj}} = 0.6\,\text{h}$, $t_{\text{cha}} = 0.067\,\text{h}$, $t_{\text{res}} = 0.0333\,\text{h}$, $t_{\text{ext}} = 3.6\,\text{h}$, $Q_{\text{inj}} = 0.0333\,\text{m}^3/\text{h}$, $Q_{\text{cha}} = 0.0255\,\text{m}^3/\text{h}$, and $Q_{\text{ext}} = -0.011\,\text{m}^3/\text{hour}$. The concentrations of NaCl were $C_0^{\text{inj}} = 100\,\text{mg/L}$ in the injection phase and $C_0^{\text{cha}} = 10\,\text{mg/L}$ in the chase phase. As for the reactant of Na_2SO_4, the concentrations were $C_0^{\text{inj}} = 20\,\text{mg/L}$ and $C_0^{\text{cha}} = 2\,\text{mg/L}$.

To demonstrate the importance of the wellbore storage of the solute transport, which was ignored in Wang *et al.* (2017), the observed and computed BTCs are compared based on the estimated parameters in Phanikumar and McGuire (2010), as shown in Figures 3.73(a) and 3.73(b). The computed BTCs in Figure 3.73(a) and 3.73(b) are located at $r = r_w + 0.15\,\text{m}$ and $r = r_w$, respectively. The legend of "PPTEST" represents the solution of Phanikumar and McGuire (2010). The new model produces these and the others, ignoring the wellbore storage effect on the solute transport.

From Figures 3.73(a) and 3.73(b), several interesting observations could be made. First, the difference of BTCs among different porous media is obvious. BTCs of the coarse sand aquifer are close to the solution of "PPTEST", as shown in Figure 3.73(a). This is because the hydraulic diffusivity of the coarse sand aquifer is the largest, which is close to the assumption used in "PPTEST" that hydraulic diffusivity is infinite. Second, the wellbore concentration is $10\,\text{mg/L}$ at the early stage of the extraction phase for Cl^-. This is mainly due to the chosen boundary condition at the well screen. Third, the peak values of BTCs increase with the decreasing hydraulic diffusivity, and the arrival times of peak values increase with the decreasing hydraulic diffusivity. Such an observation is also found in Figures 3.65(a) and 3.65(b). Fourth, the configuration of BTCs in the aquifer (at $r = r_w + 0.15\,\text{m}$) computed by the model of this section shows that the concentration firstly decreases with time and then increases with time, as shown in Figure 3.73(a). This observation could be explained by the corresponding flow field, as shown in Figure 3.74. Looking at the flow velocity in the aquifer at $r = r_w + 0.15\,\text{m}$, one may find that the flow direction is still outward from the wellbore in the early stage of the extraction phase,

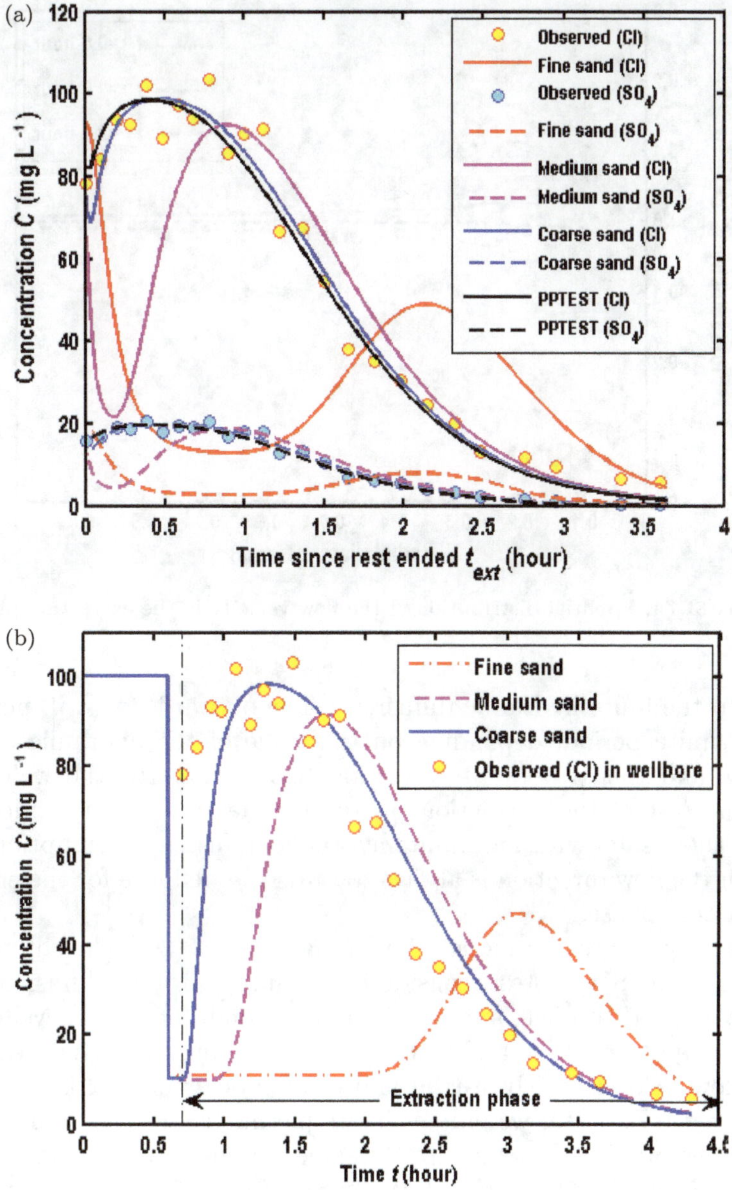

Figure 3.73. BTCs for the different porous media with a piece-wise linear function to describe chemical reactions: (a) Cl^{-1} and SO_4^{-2} in the aquifer at $r = r_w + 0.15\,\text{m}$, (b) Cl^{-1} in the wellbore.

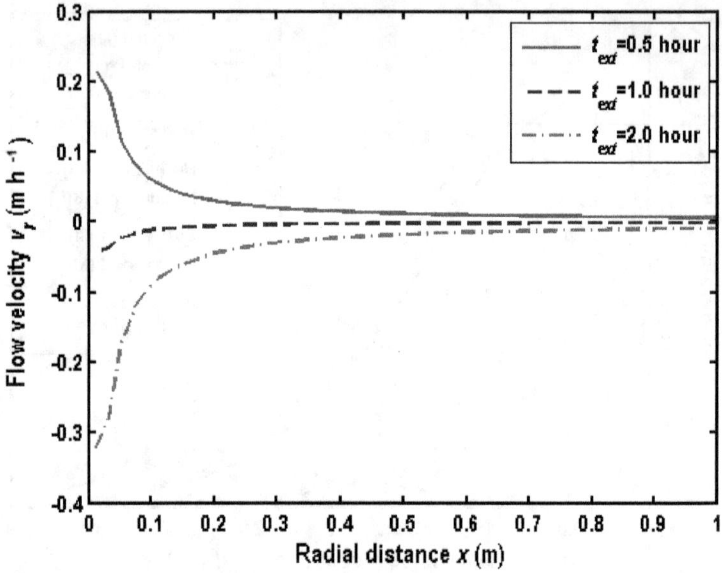

Figure 3.74. Spatial distribution of the flow velocity in the extraction phase.

due to the finite hydraulic diffusivity. The outward flow will persist for a finite period, depending on the value of the hydraulic diffusivity, and then reverse its direction to flow towards the wellbore for the rest of the extraction phase. This feature is very different from the results with an infinite hydraulic diffusivity assumption, in which the flow direction is always towards the wellbore for the entire extraction phase.

We try to use the model of Wang and Zhan (2019) to interpret BTCs of the SWPP test, considering a finite hydraulic diffusivity, a finite wellbore storage, and new boundary conditions of the wellbore at the injection, chase and rest phases. Assuming the initial head of the flow field is 1 m. In a trial and error process of best fitting the observed BTC data, we only estimate parameters of K_r, S_s and α_r, while the other parameters are the same as those used for producing Figure 5 of Phanikumar and McGuire (2010). Figure 3.75 demonstrates the fitness of the observed BTC data in the wellbore when $K_r = 1.0\,\text{m/h}$, $S_s = 1.0 \times 10^{-5}\,1/\text{m}$, and $\alpha_r = 0.015\,\text{m}$. Since the hydraulic diffusivity of this case greater than the hydraulic diffusivity of medium sand ($4.17 \times 10^2\,\text{m}^2/\text{h}$), the influence of the flow field

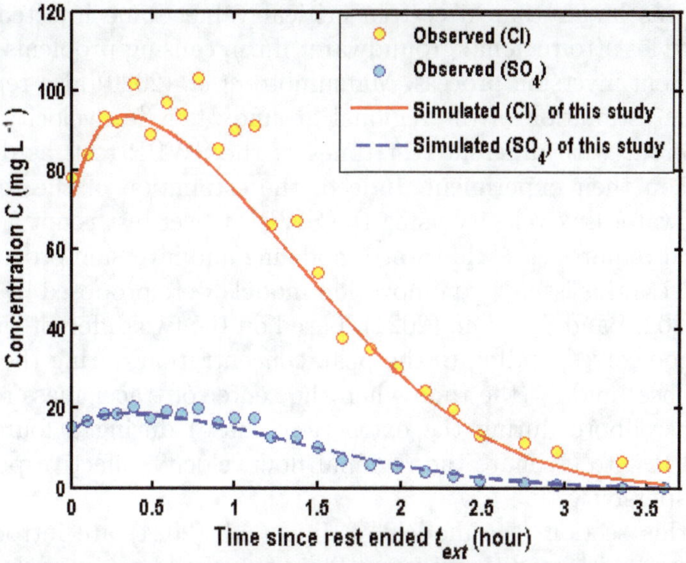

Figure 3.75. Fitness of the field SWPP test data by the new model of Wang and Zhan (2019).

could be negligible. Assuming the porosity is close to reality, the dispersivity is unique.

3.4.3 *The SWPP tests with regional groundwater flow*

The analytical solutions, as mentioned before, were all based on the assumption that the regional flow was negligible. When the regional groundwater flow is included, it is extremely challenging to obtain the analytical solution of the SWPP test. To serve this purpose, Leap and Kaplan (1988) and Hall *et al.* (1991) proposed a simplified model to estimate the regional flux velocity and effective porosity using a two-phase SWPP test, respectively. By taking the rest phase into account, Paradis *et al.* (2017, 2018) improved the above simplified models. In principle, the calculated regional groundwater flow velocity should be the same, regardless of the tracer test setup and selected models of parameter estimation. However, Johnsen and Whitson (2011) reported that when the duration of the injection phase was larger than a certain critical value, some injected tracers were lost (to regional groundwater flux), causing problems in the subsequent inversion process. Matsumoto *et al.* (2020) also reported that the estimation of the regional groundwater flux velocity could be different with different rest times of the SWPP test, as demonstrated in their experiment. Indeed, the estimation of the regional groundwater flux velocity using the SWPP tracer test is not a trivial case and requires careful forward modeling and inversion procedures. To address this issue, two innovation models were proposed by Wang *et al.* (2021) and Jin *et al.* (2024), based on the two different times t_p (the time corresponding to the peak concentration during the injection phase) and t_a (the time when the center of tracer mass returns to the wellbore during the extraction phase) during a four-phase SWPP test to estimate the regional flow velocity, effective porosity and dispersivity.

In this section, the models of Jin *et al.* (2024) are introduced. Compared with traditional curve-fitting methods, the models of Jin *et al.* (2024) require fewer observation data and do not require any optimization algorithms, thus avoiding the problem of falling into local optimum. Considering the similarity of the governing equations of heat transport and solute transport in the subsurface, this method may also be used to estimate the parameters of heat transport at laboratory and field scales (after proper conversion of solute transport parameters to heat transport parameters).

3.4.3.1 *Mathematical model*

For the conceptual model concerned in the model of Jin *et al.* (2024), a fully penetrating well is located at a homogeneous confined aquifer with a thickness of B and a uniform regional groundwater flow field presented over the entire period of test. The coordinate system is established as follows: the origin is located at the center of the test well, the x-axis points to the regional groundwater flow direction, and the y-axis is perpendicular to the regional groundwater flow direction, as shown in Figure 3.76. The length and width (m) of the aquifer can be set sufficiently far from the test well so that the lateral boundary conditions would not affect the result of the SWPP test. The mathematical model of the SWPP test will be described as follows.

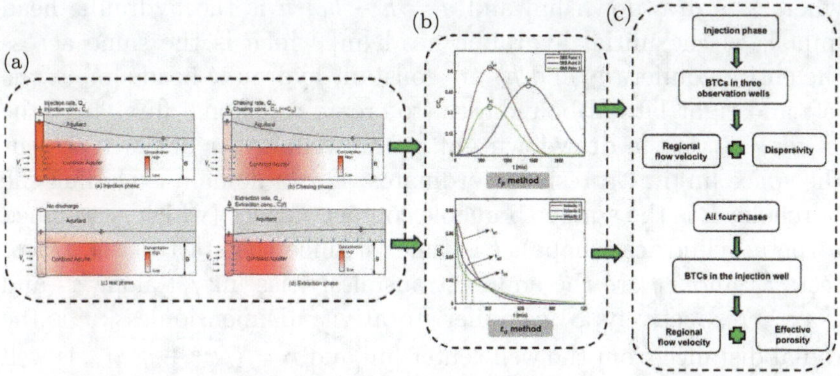

Figure 3.76. The schematic diagram of this section. (a) the conceptual diagram of the four-phase SWPP test with regional flow field; (b) the two types of time corresponding to BTCs; and (c) the procedure of parameter inversion based on t_p and t_a methods.

(a) Mathematical model of groundwater flow

Assuming that the principal directions of hydraulic conductivity are consistent with the directions of the coordinate axes, the governing equation of the transient 2D (in the xy plane) groundwater flow can be written as

$$S\frac{\partial s}{\partial t} = \frac{\partial}{\partial x}\left(T_x\frac{\partial s}{\partial x}\right) + \frac{\partial}{\partial y}\left(T_y\frac{\partial s}{\partial y}\right), \quad (x, y) \in D, \quad t > 0 \tag{3.689a}$$

$$s(x, y, t)|_{t=0} = 0 \tag{3.689b}$$

$$\oint_R Bnv_r(r, t)dr - \pi r_w^2\frac{ds_w}{dt} = -Q \tag{3.689c}$$

$$h(x, y, t)|_{x=-\frac{L}{2}} = h_1 \tag{3.689d}$$

$$h(x, y, t)|_{x=\frac{L}{2}} = h_2 \tag{3.689e}$$

$$\left.\frac{\partial s(x, y, t)}{\partial y}\right|_{y=-\frac{L}{2}} = 0 \tag{3.689f}$$

$$\left.\frac{\partial s(x, y, t)}{\partial y}\right|_{y=\frac{L}{2}} = 0 \tag{3.689g}$$

where s is drawdown [m] and $s = h - h_0$; h is the hydraulic head [m]; h_0 is the initial hydraulic head [m] which is the same across the entire aquifer, h_1 and h_2 are constant hydraulic heads [m] at the left and right lateral boundaries (to create a regional flux along the x-axis); s_w is the drawdown [m] in the wellbore; x and y represent the space [m] in Cartesian coordinates; D is a domain excluding the borehole; R is the aquifer-borehole contact surface (which is assumed to be a cylindrical shape); t is time [s] since the start of the tracer test; T_x and T_y are the aquifer transmissivities [m^2/s] along x- and y-axes, respectively; S is aquifer storativity [Dimensionless]; r is the radial distance from the well center [m] and $r = \sqrt{x^2 + y^2}$; r_w is well radius [m]; n is the effective porosity [Dimensionless]; v_i is the radial groundwater flux velocity [m/s] and $v_r = \sqrt{v_x^2 + v_y^2}$, in which v_x and v_y are the flux velocities [m/s] along the x- and y-axes, respectively; $x = \pm L/2$ and $y = \pm L/2$ are the four lateral boundaries in a square shape of aquifer, where L is the dimension along the x- or y-axes, and the value of L is sufficiently large that the lateral boundaries cause no disturbance to the SWPP test.

A complete SWPP test is composed of four phases, and one has

$$Q = \begin{cases} Q_{\mathrm{inj}}, & 0 < t \leq t_{\mathrm{inj}} \\ Q_{\mathrm{cha}}, & t_{\mathrm{inj}} < t \leq t_{\mathrm{inj}} + t_{\mathrm{cha}} \\ 0, & t_{\mathrm{inj}} + t_{\mathrm{cha}} < t \leq t_{\mathrm{inj}} + t_{\mathrm{cha}} + t_{\mathrm{res}} \\ Q_{\mathrm{ext}}, & t_{\mathrm{inj}} + t_{\mathrm{cha}} + t_{\mathrm{dri}} < t \leq t_{\mathrm{inj}} + t_{\mathrm{cha}} + t_{\mathrm{res}} + t_{\mathrm{ext}} \end{cases} \tag{3.690}$$

where Q_{inj}, Q_{cha} and Q_{ext} are the flow rates [m^3/s] of the well in the injection, chase, and extraction phases, respectively; t_{inj}, t_{cha}, t_{res}, and t_{ext} are the durations [s] of the injection, chase, rest, and extraction phases, respectively.

(b) Mathematical model of solute transport

Assuming that the solute transport is governed by ADE, which for a non-reactive solute without source/sink term can be written as

$$\frac{\partial C}{\partial t} = \frac{\partial}{\partial x}\left(D_{xx}\frac{\partial C}{\partial x} + D_{xy}\frac{\partial C}{\partial y}\right) + \frac{\partial}{\partial y}\left(D_{yx}\frac{\partial C}{\partial x} + D_{yy}\frac{\partial C}{\partial y}\right) - \frac{\partial}{\partial x}(Cv_x) - \frac{\partial}{\partial y}(Cv_y) \quad (x, y) \in D, \quad t > 0 \tag{3.691a}$$

$$C(x,y,t)|_{t=0} = 0 \tag{3.691b}$$

$$\oint_R \left[v_r C - D_r \frac{\partial C}{\partial r} \right] dr = 2\pi r_w v_r C_w \tag{3.691c}$$

$$\begin{aligned} C(x,y,t)|_{x=-\frac{L}{2}} &= C(x,y,t)|_{x=\frac{L}{2}} = C(x,y,t)|_{y=-\frac{L}{2}} \\ &= C(x,y,t)|_{y=\frac{L}{2}} = 0 \end{aligned} \tag{3.691d}$$

where C is solute concentration in the aquifer [Kg/m^3]; C_w is solute concentration inside the wellbore [Kg/m^3]; D_{xx}, D_{xy}, D_{yx}, and D_{yy} are the four components of a 2D dispersion coefficient [m^2/s] tensor in the xy plane, D_r is the radial dispersion coefficient [m^2/s]. The hydrodynamic dispersion coefficients are velocity-dependent tensor depicted as

$$D_{xx} = \frac{\alpha_L v_x^2}{|\overrightarrow{v_r}|} + \frac{\alpha_T v_y^2}{|\overrightarrow{v_r}|} + D_0 \tag{3.691e}$$

$$D_{xy} = D_{yx} = (\alpha_L - \alpha_T) \frac{v_x v_y}{|\overrightarrow{v_r}|} \tag{3.691f}$$

$$D_{yy} = \frac{\alpha_L v_y^2}{|\overrightarrow{v_r}|} + \frac{\alpha_T v_x^2}{|\overrightarrow{v_r}|} + D_0 \tag{3.691g}$$

where α_L and α_T are the longitudinal and transverse dispersivities [m], respectively, and the arrow above v_i indicates a vector; D_0 is the effective molecular diffusion coefficient in the aquifer [m^2/s]. The radial dispersion coefficient on the left side of Eq. (3.691c) is $D_r = \alpha_r v_r$, where α_r is the radial dispersivity and is assumed to be the same as the longitudinal dispersivity α_L [m]. The flow and transport models can be coupled through the groundwater flow velocity v_x and v_y:

$$v_x = K_x \frac{\partial s}{\partial x} \tag{3.692a}$$

$$v_y = K_y \frac{\partial s}{\partial y} \tag{3.692b}$$

where K_x and K_y are the hydraulic conductivities [m/s] along the x- and y-axes, respectively, and $K_x = \frac{T_x}{B}$, $K_y = \frac{T_y}{B}$.

3.4.3.2 *Numerical model*

When the regional groundwater flow is included, it is extremely challenging to find the analytical solution for the SWPP test. However, the mathematical models of Eqs. (3.689)–(3.691) could be solved by many software packages. COMSOL Multiphysics is more flexible in dealing with the aquifer-wellbore interface (due to its finite-element approach which is more effective in handling circular boundaries) and will be employed in this section. Considering that the numerical solution may contain errors, e.g., numerical dispersion and numerical oscillation, which are related to the discretization of the grid system, the semi-analytical solution of Suk *et al.* (2023) will be employed to benchmark the numerical simulation. In this section, the built-in PARDISO solver is adopted, and the relative tolerance is set to 1e-3. The temporal discretization is managed automatically by COMSOL Multiphysics based on the convergence rates of each system. Besides, the root means square error ($RMSE$) of the analytical versus numerical solutions of the normalized concentration is employed to guide the spatial discretization of the grid system. Figure 3.77 shows the

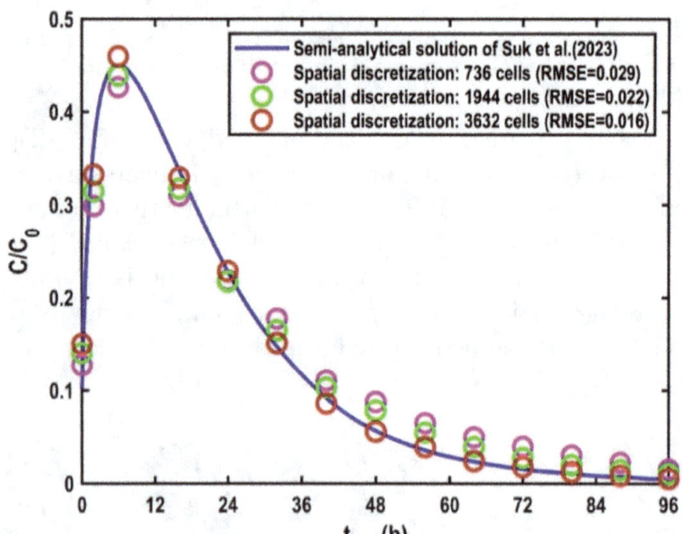

Figure 3.77. Comparison of the analytical solution and numerical simulation of COMSOL under different spatial discretization.

comparison of the analytical solution and numerical solution with different spatial discretization meshes. We find that the numerical error is negligible ($RMSE = 0.016$) when the number of elements in the grid system is 3632 for the problem of concern here. A further increase of the number of elements above 3632 does not noticeably improve the accuracy of the numerical modeling but requires greater computational time and memory, so the number of cells used in this investigation is set to 3632.

3.4.4 *Parameter estimation by the SWPP tests with regional flow*

3.4.4.1 *Estimation method with t_p*

The problem concerned in this section is the slug tracer injection process in a homogeneous and isotropic confined aquifer that is free of tracer initially. The slug tracer injection implies that the duration of the injection is much shorter than the time scales of other processes, Thus, the injection can be regarded as instantaneous. Assuming that the groundwater flow is along the x-axis with a uniform velocity and the injection of tracer does not change the state of groundwater flow appreciably, and the boundaries will not affect the results of the slug injection phase. Therefore, we consider that the process of tracer injection in the slug SWPP test can be approximatively portrayed by the analytical solution of Sauty (1980), which can be written as follows:

$$C(x,y,t) = \frac{M}{4\pi Bnt\sqrt{D_L D_T}} \exp\left[-\frac{(x-\bar{v}t)^2}{4D_L t}\right] \exp\left(-\frac{y^2}{4D_T t}\right) \tag{3.693}$$

where M is the total amount of solute contained in the slug [Kg]; n is the effective porosity (or the portion of porosity directly contributing to flow and transport); $\bar{v}$ is the average groundwater flow velocity [m/s] along the x-axis; $D_L = \alpha_L \bar{v} + D_0$ and $D_T = \alpha_T \bar{v} + D_0$ are the longitudinal and transverse dispersion coefficients, respectively [m^2/s]. During the slug injection phase, the tracer will pass through the porous medium as a pulse with a peak concentration at a downstream location. The peak time (t_p) corresponding to the peak value can be determined by the following equation (Liang *et al.*, 2018):

$$\frac{dln(C(x,y,t))}{dt} = 0 \tag{3.694}$$

For most practical problems, it is usually correct to neglect the effective molecular diffusion coefficient (D_0) unless the groundwater flow velocity ($\bar{v}$) is extremely small, such as in clays and silt. Substituting Eq. (3.693) into Eq. (3.694) and solving it yields:

$$t_p = -2\frac{\alpha_L}{\bar{v}} + \sqrt{4\left(\frac{\alpha_L}{\bar{v}}\right)^2 + \left(\frac{x}{\bar{v}}\right) + \frac{\alpha_L}{\alpha_T}\left(\frac{y}{\bar{v}}\right)^2} \tag{3.695}$$

It can be seen from Eq. (3.695) that t_p depends on both the x and y coordinates of the observation point. If the peak time at three observational points (x_1, y_1), (x_2, y_2) and (x_3, y_3) are t_{p1}, t_{p2} and t_{p3}, respectively, one can obtain the following formulas to evaluate v_p, α_L, and α_T simultaneously on the basis of Eq. (3.695):

$$\bar{v} = \sqrt{\frac{t_{p1}(x_3^2y_2^2 - x_2^2y_3^2) + t_{p2}(x_1^2y_3^2 - x_3^2y_1^2) + t_{p3}(x_2^2y_1^2 - x_1^2y_2^2)}{t_{p2}t_{p3}y_1^2(t_{p2} - t_{p3}) + t_{p1}t_{p3}y_2^2(t_{p3} - t_{p1}) + t_{p1}t_{p2}y_3^2(t_{p1} - t_{p2})}} \tag{3.696a}$$

$$\alpha_L = \frac{u^2(y_1^2t_{p3}^2 - y_3^2t_{p1}^2) + x_1^2y_3^2 - x_3^2y_1^2}{4v_p(t_{p1}y_3^2 - t_{p3}y_1^2)} \tag{3.696b}$$

$$\alpha_T = \frac{y_3^2\alpha_L}{t_{p3}^2{v_p}^2 + 4t_{p3}v_p\alpha_L - x_3^2} \tag{3.696c}$$

For the convenience of application, one can choose the following three locations $(x_1, 0)$, $(x_2, 0)$ and (x_1, y_1), and above equations can be simplified to:

$$\bar{v} = \sqrt{\frac{t_{p2}x_1^2 - t_{p1}x_2^2}{t_{p1}t_{p2}(t_{p1} - t_{p2})}} \tag{3.697a}$$

$$\alpha_L = \frac{1}{4}\left(\frac{x_1^2}{v_pt_{p1}} - v_pt_{p1}\right) \tag{3.697b}$$

$$\alpha_T = \frac{y_1^2\alpha_L}{t_{p3}^2{v_p}^2 + 4t_{p3}v_p\alpha_L - x_1^2} \tag{3.697c}$$

It should be noted that the estimation of α_T requires that at least one observation point must be located off the x direction ($y_1 \neq 0$).

3.4.4.2 *Estimation method with* t_a

The presumption used to calculate t_a is that displacement of the center of mass of the tracer, after completion of the injection, chase, rest, and extraction phases of a push-pull test is expected to be zero. This implies that all the injected mass of the tracer will be completely recovered from the aquifer. This presumption is valid under the following constraints: (1) the strength and duration of the injection and chase phases should be controlled within limits that the farthest traveled tracer particles are still within the capture zone of the extraction phase and (2) the injected tracer will not be trapped or absorbed in any porous media, especially in the less permeable zones (or the so-called immobile zones), thus allowing for a nearly 100% recovery of tracer. The displacement of the center of tracer mass is controlled by the superposition of two different flow fields: (1) the axisymmetric flow caused by injection/chase/extraction and (2) the unilateral flow caused by the regional flow. The displacement of the center of tracer mass during the injection/chase phases is always zero, due to the axisymmetric nature of flow near the well. Therefore, the displacement of the center of tracer mass can be solely attributed to the regional flow of groundwater and extraction phase, and written as

$$r_{a1} + r_{a2} + r_{a3} + r_{a4} + (-r_e) = 0 \tag{3.698}$$

where r_{a1}, r_{a2}, r_{a3}, and r_{a4} are the displacement [m] of the center of tracer mass due to regional groundwater flow during the injection, chase, rest, and extraction phases, respectively. r_e is the displacement [m] of the center of tracer mass due to extraction.

The steady-state unidirectional regional flow can be expressed by Darcy's law as

$$\bar{v} = -\frac{K_x \frac{dh}{dx}}{n} \tag{3.699a}$$

and in general, the flow velocity can be also expressed as

$$\bar{v} = \frac{\Delta r}{\Delta t} \tag{3.699b}$$

where $\bar{v}$ is the regional groundwater velocity [m/s]; Δr is the traveled distance induced by regional groundwater flow [m]; Δt is the elapsed time [s].

The volume of water injected or extracted from a confined aquifer within a constant flow rate by a fully penetrating well is given by

$$V = |Q|t \tag{3.700a}$$

where V is the injected or extracted volume [m^3] of water; Q is the constant injection or extraction rate [m^3/s].

Assuming the confined aquifer is homogeneous and isotropic and possesses no regional hydraulic gradient, then the volume in Eq. (3.700a) can be described as

$$V = \pi Bn(r^2 - r_w^2) \tag{3.700b}$$

where r is the average radius [m] of the moving tracer. Combining Eqs. (3.700a) and (3.700b) leads to:

$$r = \sqrt{\left(\frac{|Q|t}{\pi Bn} + r_w^2\right)} \tag{3.701}$$

which describes the leading or trailing average position of tracer particles within an expanding or contracting cylindrical volume as the tracer is injected or extracted from the aquifer.

Furthermore, the average displacement of tracer due to extraction occurs when one-half of the tracer mass has been recovered and is given by

$$M_h = \int_0^{t_a} C(t)dt \tag{3.702}$$

where M_h is one-half of the total extracted tracer mass (which is also the total injected tracer mass, if assuming 100% recovery) [Kg]; $C(t)$ is the concentration of the tracer as a function of time [Kg/m^3]; t_a is the time elapsed from the start of pumping until the center of mass of the tracer returns to the wellbore [s].

The components in Eq. (3.699b) can be substituted by their corresponding terms given in Eqs. (3.699b) and (3.701) as follows:

$$\bar{v}t_{\text{inj}} + \bar{v}t_{\text{cha}} + \bar{v}t_{\text{res}} + \bar{v}t_a - \sqrt{\left(\frac{|Q_{\text{ext}}|t_a}{\pi Bn} + r_w^2\right)} = 0, \tag{3.703}$$

Then Eq. (3.703) is rearranged to solve for the average flow velocity of groundwater:

$$\bar{v} = \frac{\sqrt{\left(\frac{|Q_{\text{ext}}|t_a}{\pi B n} + r_w^2\right)}}{(t_{\text{inj}} + t_{\text{cha}} + t_{\text{res}} + t_a)}, \tag{3.704a}$$

Equations (3.699a) and (3.704a) are solved jointly to obtain the equation for the effective porosity:

$$n = \frac{\pi B\left[{K_x}^2\left(\frac{dh}{dx}\right)^2(t_{\text{inj}} + t_{\text{cha}} + t_{\text{res}} + t_a)^2 - r_w^2\right]}{|Q_{\text{ext}}|t_a} \tag{3.704b}$$

It is worth pointing out that Eqs. (3.704a) and (3.704b) can be simplified to the model of Paradis *et al.* (2018) when ignoring the wellbore storage and the movement of the center of tracer mass during the chase phase, and to the model of Leap and Kaplan (1988) and Hall *et al.* (1991) when the movement of the center of tracer mass during the rest phase is also neglected.

Compared with the curve-fitting of BTCs which is a common practice for parameter estimation, the t_p and t_a methods have several advantages which are briefly outlined as follows. Firstly, the curve-fitting method requires many measured concentrations to capture the entire BTCs, leading to a high cost of data collection. In contrast, the t_p and t_a methods only require the identification of the time corresponding to the peak concentration and the center of mass returns to the wellbore, respectively. Therefore, the t_p and t_a methods provide direct parametric inversion methods in cases where complete BTCs are not available. Secondly, the curve-fitting method involves an optimized operation that sometimes falls into the local optimal estimates rather than the global ones. For example, both the EKF, UCODE, and PEST methods are unconstrained local search calibration methods with the disadvantage that the resulting parameter values may lie in the local minimum of the objective function, thus leading to false or even physically unrealistic parameter values.

3.4.4.3 *Parameter sensitivity analysis*

(a) The effects of the injection phase on the regional flow field

In parameter inversion process, the influence of the injection phase on the regional flow field has not been taken into account in the previous simplified models (Hall *et al.*, 1991; Leap and Kaplan, 1988; Li *et al.*, 2019; Paradis *et al.*, 2018). However, the influence of the injection phase on the regional flow field may be unavoidable in the whole SWPP test, which will cause errors in parameter estimation. In this section, a high-resolution numerical model is used to identify the influence of different injection conditions on the regional flow field, which could provide a theoretical basis for the subsequent parameter estimation models. The parameters used in this section are $B = 10\,\text{m}$, $S_s = 0.001$, $T_x = T_y = 80\,\text{m}^2/\text{d}$, $r_w = 0.1\,\text{m}$, $n = 0.3$, $\bar{v} = 2.0 \times 10^{-5}\,\text{m/s}$, $\alpha_L = 0.02\,\text{m}$, $\alpha_T = 0.002\,\text{m}$, $t_{\text{inj}} = 5$ min. These values are representatives of SWPP tests in field applications. Figure 3.78 shows the contours of the hydraulic head and streamline at the end of the injection phase for different Q_{inj}. One can find that with the increase of Q_{inj}, the streamlines near the wellbore are increasingly distorted.

To examine how much distortion to the regional flow field is caused by the well operation (injection/chase/extraction), we have defined a velocity relative deviation: $Dev = |(\bar{v} - v_i)/v_i|$, where $\bar{v}$ represents the undisturbed regional groundwater flow velocity and v_i represents the flow velocity at different locations at the end of the injection phase. Table 3.14 summarizes the Dev at different observation points at the end of the injection phase, and OBS1 to OBS5 represent the distances of the observation points from the wellbore along the x-axis being 1.0, 2.0, 5.0, 10.0, and 15.0 m, respectively. In all the cases, the closer to the wellbore, the greater the deviation of $\bar{v}$ and v_i. It is worth noting that the deviations at 2 m from the wellbore are 18.5% and 37.5% for $Q_{\text{inj}} = 0.5$ and $1.0\,\text{m}^3/\text{h}$, respectively, and the deviation is negligible as the distance from the wellbore increases. The effect of the injection phase on the regional flow field may be compensated during the extraction phase if using a similar amount of extraction rate as the injection rate because the injection and extracting fluid will move along the same paths (but in opposite directions). This point is similar to the influence of aquifer

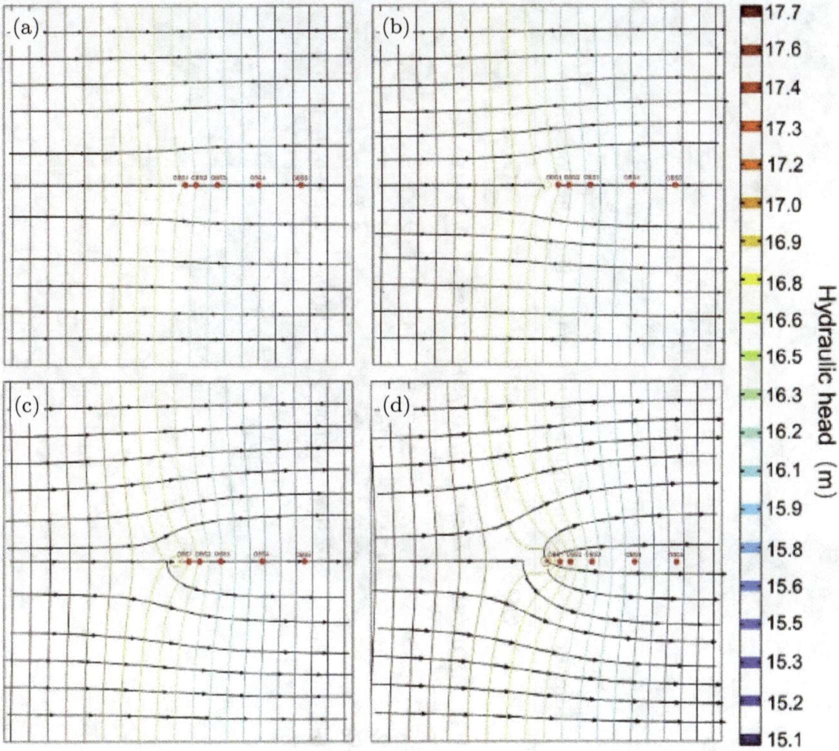

Figure 3.78. Contours of hydraulic head and streamline at the end of the injection phase for different injection rates. (a) $Q_{\text{inj}} = 0.5\,\text{m}^3/\text{h}$; (b) $Q_{\text{inj}} = 1.0\,\text{m}^3/\text{h}$; (c) $Q_{\text{inj}} = 2.0\,\text{m}^3/\text{h}$; (d) $Q_{\text{inj}} = 4.0\,\text{m}^3/\text{h}$.

heterogeneity on SWPP tests (Chen *et al.*, 2017; Gouze *et al.*, 2008). Therefore, we argue that the effect of the injection phase on the regional flow field is secondary when the injection rate is less than or equal to $1.0\,\text{m}^3/\text{h}$.

(b) Estimation of the regional groundwater flow velocity

The applicability of the parameter estimation method discussed above is illustrated as follows when the effective molecular diffusion is negligible (which is true for most practical problems). Firstly, one can generate BTCs based on known input parameters through the forward numerical model. Secondly, one identifies the t_p and t_a for a given location based on the generated BTCs. Thirdly, based on

Table 3.14. The differences in flow velocities at different observation points at the end of the injection phase.

Q_{inj} (m^3/h)	OBS1		OBS2		OBS3		OBS4		OBS5	
	v_i(m/s)	*Dev*	v_i(m/s)	*Dev*	v_i(m/s)	*Dev*	v_i(m/s)	*Dev*	v_i(m/s)	*Dev*
0.5	2.73e-5	36.5%	2.37e-5	18.5%	2.15e-5	7.5%	2.08e-5	4.0%	2.06e-5	3.0%
1.0	3.47e-5	73.5 %	2.75e-5	37.5%	2.31e-5	15.5%	2.16e-5	8.0%	2.12e-5	6.0%
2.0	4.93e-5	146.5%	3.47e-5	73.5%	2.62e-5	31.0%	2.33e-5	16.5%	2.25e-5	12.5%
4.0	7.90e-5	295.0%	4.97e-5	148.5%	3.24e-5	62.0%	2.66e-5	33.0%	2.49e-5	24.5%

the identified t_p and t_a, one can use Eqs. (3.697) and (3.704) to estimate the regional flow velocity, dispersivity and effective porosity. Fourthly, one can compare the estimated parameters with their actual counterparts to check the accuracy of the proposed inversion methods. Here, we define the following parameters of relative errors to reflect the difference between the calculated and the real values:

$$Error_{v_p} = |\hat{v}_p - \bar{v}|/\bar{v} \tag{3.705a}$$

$$Error_{\alpha_L} = |\widehat{\alpha}_L - \alpha_L|/\alpha_L \tag{3.705b}$$

$$Error_{\alpha_T} = |\widehat{\alpha}_T - \alpha_T|/\alpha_T \tag{3.705c}$$

$$Error_{v_a} = |\hat{v}_a - \bar{v}|/\bar{v} \tag{3.705d}$$

$$Error_n = |\hat{n} - n|/n \tag{3.705e}$$

where the hat sign $\wedge$ above the symbol represents the estimated values hereinafter, $\hat{v}_p$ and $\hat{v}_a$ represent the estimated regional groundwater flow velocities using the t_p and t_a methods, respectively. In the following analysis, the symbols and terminology are listed in Table 3.15.

It is important to note that the accuracy of the proposed parameter estimation methods relies heavily on the accurate determination of t_p and t_a. In theory, it is desirable to have an infinitely small-time interval of measurement so the t_p and t_a values can be precisely determined. However, given that such errors are unavoidable, one can reduce the errors by decreasing the time interval of measurement in practice. BTCs with the t_p and t_a methods are shown in Figures 3.79 and 3.80 under different regional groundwater flow velocities, e.g., $\bar{v} = 1.0 \times 10^{-5}$ m/s, 1.5×10^{-5} m/s, 2.0×10^{-5} m/s, 2.5×10^{-5} m/s, 3.0×10^{-5} m/s, and 3.5×10^{-5} m/s, respectively. In the numerical modeling, the regional groundwater flow velocity of $\bar{v}$ is computed by Darcy's law based on the computed hydraulic gradient.

To withdraw the injected tracer, two sets of extraction rates are employed, e.g., $Q_{\text{ext}} = -2.0\,\text{m}^3/\text{h}$ used in Figure 3.80(a) and $Q_{\text{ext}} = -4.0\,\text{m}^3/\text{h}$ used in Figure 3.80(b). The other parameters are $B = 10$ m, $S_s = 0.001$, $T_x = T_y = 80\,\text{m}^2/\text{d}$, $r_w = 0.1$ m, $n = 0.3$, $\alpha_L = 0.02$ m, $\alpha_T = 0.002$ m, $t_{\text{inj}} = 5$ min, $t_{\text{cha}} = 25$ min, $t_{\text{res}} = 420$ min and $t_{\text{ext}} = 1200$ min. BTCs with the t_p method are obtained at three observation points (0.4 m, 0 m), (0.8 m, 0 m) and (0.4 m, 0.15 m), and the peak time of three observation points can be

Table 3.15. The symbols and nomenclature used in this section.

Symbols	Nomenclature	Symbols	Nomenclature
B[m]	Thickness of the aquifer	v_x[m/s]	Groundwater flow velocity along the x-axis during the slug SWPP test
L[m]	Width of the aquifer	v_y[m/s]	Groundwater flow velocity along the y-axis during the slug SWPP test
r_w[m]	The Radius of test well	$\bar{v}$[m/s]	Undisturbed regional groundwater flow velocity
s[m]	Drawdown during the slug SWPP test	v_i[m/s]	Groundwater flow velocity at the end of injection phase
S[−]	Storativity of the aquifer	Dev[−]	The relative deviation of $\bar{v}$ and v_e
n[−]	Effective porosity of the aquifer	t_p[min]	The time corresponding to the peak value of BTCs during the injection phase
K[m/d]	Hydraulic conductivity of the aquifer	t_a[min]	The time corresponding to the mass center returns to the wellbore during the extraction phase
T[m^2/d]	Transmissivity of the aquifer	r_{a1}[m]	Displacement of the mass center due to regional groundwater flow during the injection phase
h_0[m]	Initial head of the aquifer	r_{a2}[m]	Displacement of the mass center due to regional groundwater flow during the chase phase
h_1[m]	Constant head of left boundary	r_{a3}[m]	Displacement of the mass center due to regional groundwater flow during the rest phase
h_2[m]	Constant head of right boundary	r_{a4}[m]	Displacement of the mass center due to regional groundwater flow during the extraction phase

α_L[m]	Longitudinal dispersivity of the aquifer	r_e[m]	Displacement of the center of tracer mass due to extraction
α_T[m]	Transverse dispersivity of the aquifer	M[kg]	Total amount of tracer injected during the injection phase
Q_{inj}[m^3/h]	Injection rate	M_h[kg]	One-half of the total extracted tracer mass
Q_{cha}[m^3/h]	Chasing rate	$\hat{v}_a$[m/s]	Estimated groundwater flow velocity based on the t_a method
Q_{ext}[m^3/h]	Extraction rate	$\hat{n}$[−]	Estimated aquifer porosity base on t_a method
t_{inj}[min]	Duration of the injection time	$\widehat{\alpha}_L$[m]	Estimated longitudinal dispersivity base on t_p method
t_{cha}[min]	Duration of the chase time	$\widehat{\alpha}_T$[m]	Estimated transverse dispersivity based on the t_p method
t_{res}[min]	Duration of the rest time	$\hat{v}_p$[m/s]	Estimated groundwater flow velocity base on t_p method
t_{ext}[min]	Duration of the extraction time	$\hat{v}_p$[m/s]	Estimated groundwater flow velocity based on the t_p method

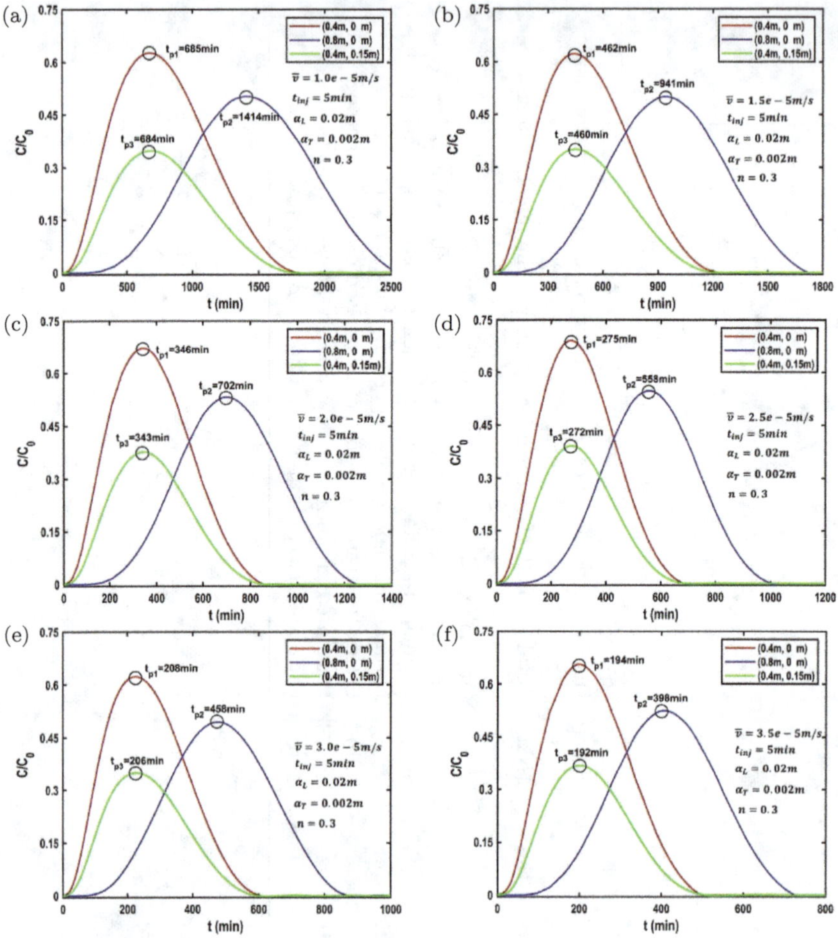

Figure 3.79. BTCs for three observation points based on t_p model with different regional groundwater velocities. (a) $v_a = 1.0 \times 10^{-5}$ m/s; (b) $v_a = 1.5 \times 10^{-5}$ m/s; (c) $v_a = 2.0 \times 10^{-5}$ m/s; (d) $v_a = 2.5 \times 10^{-5}$ m/s; (e) $v_a = 3.0 \times 10^{-5}$ m/s; and (f) $v_a = 3.5 \times 10^{-5}$ m/s.

obtained based on the obtained BTCs. For the t_a method, one can find that different regional groundwater flow velocities have great impacts on BTCs, and the tracer concentration is smaller at the early stage of the extraction phase with a greater regional groundwater flow velocity. Additionally, it is notable that a larger regional groundwater flow velocity will result in a longer tailing.

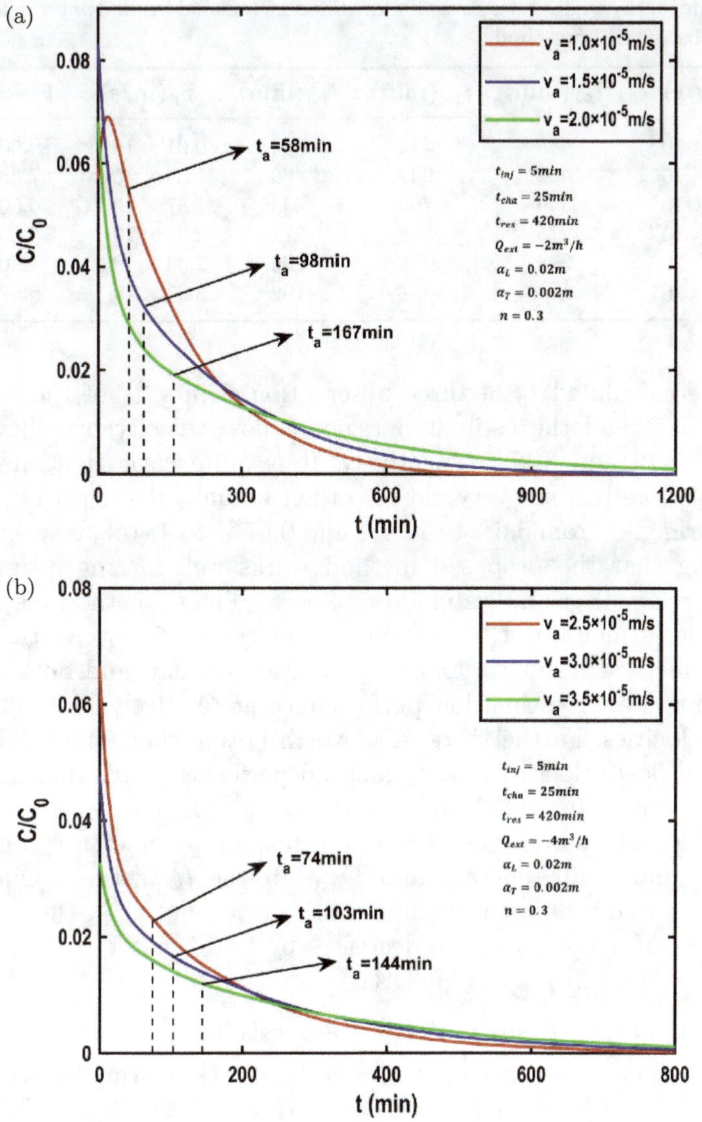

Figure 3.80. BTCs in the wellbore during the extraction phase based on t_a model with different regional groundwater velocities. (a) BTCs under relatively low regional groundwater velocities; (b) BTCs under relatively high regional groundwater velocities.

Table 3.16. Estimation results for different regional groundwater velocities based on t_p method.

$\bar{v}$(m/s)	t_{p1}(min)	t_{p2}(min)	t_{p3}(min)	$\hat{v}_p$(m/s)	Error$_{v_p}$
1.0×10^{-5}	685	1414	684	0.91×10^{-5}	9.00%
1.5×10^{-5}	462	941	460	1.39×10^{-5}	7.33%
2.0×10^{-5}	346	702	343	1.87×10^{-5}	6.50%
2.5×10^{-5}	275	558	272	2.36×10^{-5}	5.60%
3.0×10^{-5}	208	458	206	2.64×10^{-5}	12.0%
3.5×10^{-5}	194	398	192	3.27×10^{-5}	6.57%

The calculated t_p of three observation points, t_a of the extraction phase and the estimated regional flow velocity are shown in Tables 3.16 and 3.17, respectively. It is quite clear that the estimated $\hat{v}_p$ and $\hat{v}_a$ are very close to their actual values, with relative errors ranging from 5.6% to 12.0% and 0.57 % to 48.0%, respectively, showing that the proposed method works well for the estimation of the regional groundwater flow velocity. The t_p method results in an underestimation of the regional groundwater flow velocity, while the t_a method results in an overestimation of that, and both methods show better estimation performance at relatively high regional flow velocities. Furthermore, it is worth noting that under different regional flow velocities, the t_a method performs better than that of Leap and Kaplan (1988) and Paradis *et al.* (2018).

Table 3.18 shows the recovery rate and parameter estimation results under different t_{res} and Q_{ext} for the t_a method. One can find that under different combinations of t_{dri} and Q_{ext}, the relative error of parameter inversion decreases by 18.0% as the recovery rate increases from 51.61% to 90.64%.

(c) Estimation results of the dispersivity

To test the performance of the t_p method in estimating dispersivity, six sets of values are employed, e.g., $\alpha_L = 0.01$, 0.05, 0.1, 0.3, 1.0, and 2.0 m, and $\alpha_L = 10\alpha_T$, respectively. The other parameters are $B = 10$ m, $S_s = 0.001$, $T_x = T_y = 80\,\text{m}^2/\text{d}$, $r_w = 0.1$ m, $n = 0.3$, $\bar{v} = 3.5 \times 10^{-5}$ m/s, $t_{\text{inj}} = 5$ min.

Similarly, BTCs are obtained at three observation points (0.4 m, 0 m), (0.8 m, 0 m) and (0.4 m, 0.15 m), and t_p of three observation

Table 3.17. Estimation results for different regional groundwater velocities based on t_a method.

			$\hat{v}_a$(m/s)			Error_{v_a}		
$\bar{v}$(m/s)	t_a (min)	Q_{ext} (m^3/h)	Leap and Kaplan (1988)	Paradis *et al.* (2018)	This model	Leap and Kaplan (1988)	Paradis *et al.* (2018)	This model
1.0×10^{-5}	58	−2.0	1.58×10^{-5}	1.56×10^{-5}	1.48×10^{-5}	58.0%	56.0%	48.0%
1.5×10^{-5}	98	−2.0	1.90×10^{-5}	1.88×10^{-5}	1.79×10^{-5}	26.7%	25.3%	19.3%
2.0×10^{-5}	167	−2.0	2.18×10^{-5}	2.16×10^{-5}	2.07×10^{-5}	9.00%	8.00%	3.50%
2.5×10^{-5}	78	−4.0	3.11×10^{-5}	3.07×10^{-5}	2.89×10^{-5}	24.4%	22.8%	15.6%
3.0×10^{-5}	106	−4.0	3.39×10^{-5}	3.35×10^{-5}	3.16×10^{-5}	13.0%	11.7%	5.33%
3.5×10^{-5}	172	−4.0	3.74×10^{-5}	3.70×10^{-5}	3.52×10^{-5}	6.86%	5.72%	0.57%

Table 3.18. Recovery rate and parameter estimation results under different t_{res} and Q_{ext} for t_a method.

$\bar{v}$(m/s)	t_{res} (min)	Q_{ext} (m^3/h)	t_a (min)	$\widehat{v}_a$(m/s)	Recovery rate	$Error_{v_a}$
2.0×10^{-5}	480	−1.0	262	1.47×10^{-5}	69.67%	26.5%
2.0×10^{-5}	540	−1.0	276	1.38×10^{-5}	63.46%	31.0%
2.0×10^{-5}	600	−1.0	292	1.30×10^{-5}	57.34%	35.0%
2.0×10^{-5}	660	−1.0	311	1.23×10^{-5}	51.61%	38.5%
2.0×10^{-5}	720	−2.0	265	1.59×10^{-5}	90.64%	20.5%
2.0×10^{-5}	780	−2.0	308	1.56×10^{-5}	85.30%	22.0%
2.0×10^{-5}	840	−2.0	358	1.53×10^{-5}	79.93%	23.5%
2.0×10^{-5}	900	−2.0	414	1.50×10^{-5}	74.56%	25.0%

points can be obtained based on their BTCs, as shown in Figure 3.81. It can be seen that an increasing dispersivity value leads to increasingly asymmetric shapes of BTCs at the above three observation points, and decreasing peak concentrations and t_p values.

The calculated t_p values of three observation points and the estimation dispersivity values are shown in Table 3.19, respectively. It is obvious that the estimated $\widehat{\alpha}_L$ and $\widehat{\alpha}_T$ closely approximate their actual values with relative errors ranging from 6.0% to 28.6% and 7.0% to 37.0%, respectively. This demonstrates the effectiveness of the t_p method in dispersivity estimation, particularly when dealing with relatively small dispersivity values, such as when the longitudinal dispersivity is less than 0.3 m and the transverse dispersivity is less than 0.03 m. Additionally, the t_p method usually results in an underestimation of α_L and an overestimation of α_T.

(d) Estimation results of the effective porosity

Similarly, to test the performance of the t_a method in estimating effective porosity, six sets of values are employed, e.g., $n = 0.2$, 0.25, 0.3, 0.35, 0.4, and 0.45, respectively. The other parameters are $B = 10\,\text{m}$, $S_s = 0.001$, $T_x = T_y = 80\,\text{m}^2/\text{d}$, $r_w = 0.1\,\text{m}$, $\bar{v} = 2.0 \times 10^{-5}\,\text{m/s}$, $\alpha_L = 0.02\,\text{m}$, $\alpha_T = 0.002\,\text{m}$, $Q_{ext} - 2.0\,\text{m}^3/\text{h}$, $t_{inj} = 5$ min, $t_{cha} = 25$ min, $t_{res} = 360$ min, and $t_{ext} = 1200$ min. BTCs under different effective porosities during the extraction phase are shown in Figure 3.82. One can find that the tracer concentration is larger at the early stage of the extraction phase with a greater porosity, and a lower porosity will result in a longer tailing.

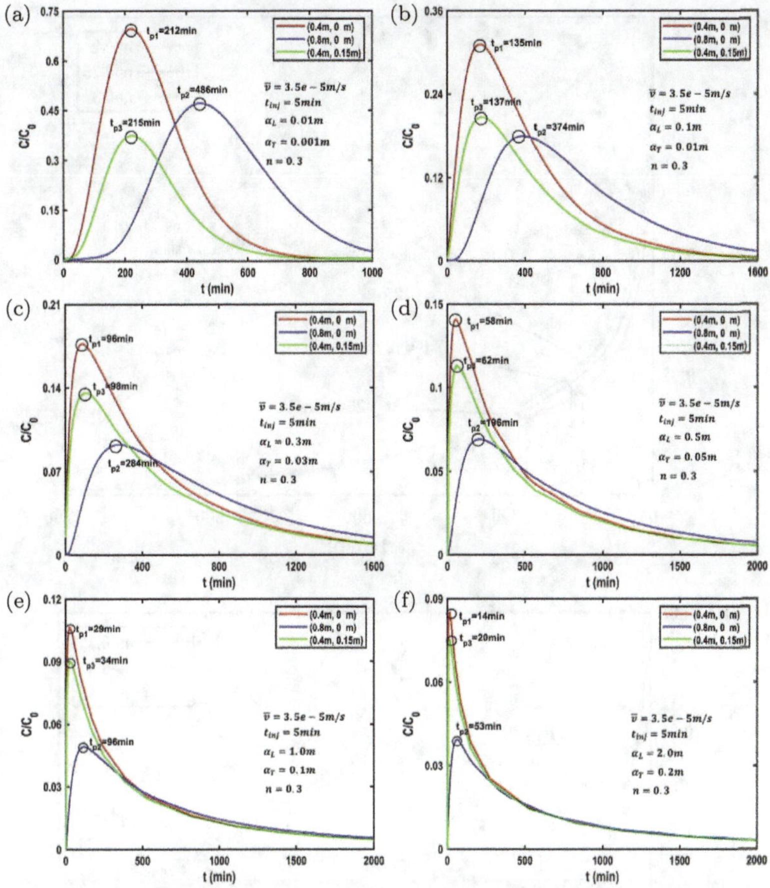

Figure 3.81. BTCs for three observation points based on t_p model with different dispersivities. (a) $\alpha_L = 0.01\,\text{m}$; (b) $\alpha_L = 0.1\,\text{m}$; (c) $\alpha_L = 0.3\,\text{m}$; (d) $\alpha_L = 0.5\,\text{m}$; (e) $\alpha_L = 1.0\,\text{m}$; and (f) $\alpha_L = 2.0\,\text{m}$.

Table 3.19. Estimation results for different dispersivities based on t_p method.

α_L (m)	α_T (m)	t_{p1} (min)	t_{p2} (min)	t_{p3} (min)	$\widehat{\alpha}_L$ (m)	$\widehat{\alpha}_T$(m)	Error_{α_L}	Error_{α_T}
0.01	0.001	212	486	215	0.0094	0.00093	6.00%	7.00%
0.1	0.01	138	376	142	0.0989	0.0114	11.0%	11.4%
0.3	0.03	85	286	89	0.223	0.035	25.7%	16.7%
0.5	0.05	56	198	60	0.372	0.069	28.6%	28.8%
1.0	0.1	30	98	36	0.727	0.137	27.3%	37.0%
2.0	0.2	15	55	21	1.475	0.250	26.3%	25.0%

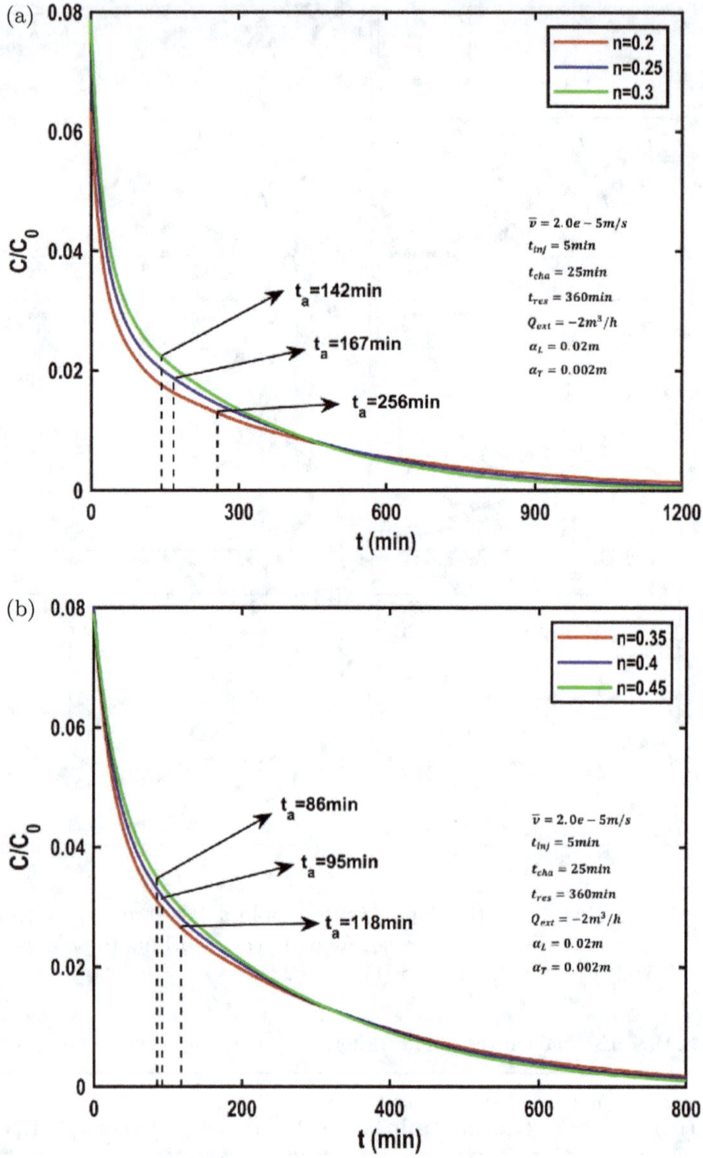

Figure 3.82. BTCs in the wellbore during the extraction phase based on t_a model with different porosities. (a) BTCs under relatively low porosities and (b) BTCs under relatively high porosities.

Table 3.20. Estimation results for different porosities based on t_a method.

			$\widehat{n}$			Error_n		
n	Q_{ext} (m^3/h)	t_a (h)	Hall *et al.* (1991)	Paradis *et al.* (2018)	This model	Hall *et al.* (1991)	Paradis *et al.* (2018)	This model
0.2	−2.0	258	0.181	0.184	0.199	9.50%	8.00%	0.50%
0.25	−2.0	172	0.201	0.205	0.224	19.6%	18.0%	10.4%
0.3	−2.0	145	0.215	0.219	0.241	28.3%	27.0%	19.7%
0.35	−2.0	118	0.237	0.242	0.267	32.6%	30.9%	23.7%
0.4	−2.0	95	0.266	0.272	0.302	33.5%	32.0%	24.5%
0.45	−2.0	86	0.283	0.289	0.322	37.1%	35.8%	28.4%

The calculated t_a during the extraction phase and the estimation results are shown in Table 3.20. Obviously, as the effective porosity increases, the calculated t_a values decrease instead. The estimated $\hat{n}$ value is very close to its actual value with the relative errors ranging from 0.5% to 28.4%, demonstrating that the t_a method also works well for estimation of the effective porosity. It should be noted that the t_a method usually results in an underestimation of effective porosity and the estimation error would increase with increasing effective porosity. Moreover, the t_a method performs better than Hall *et al.* (1991) and Paradis *et al.* (2018) under different effective porosities.

3.5 Circulation Well

3.5.1 *Background of circulation well*

A deep understanding of well hydraulics and the associated radial solute transport is crucial to enhanced groundwater remediation techniques because extraction of subsurface contaminants and delivery of reagent must be assisted by pumping well and extraction well, respectively. In some remediation techniques, a groundwater circulation well (GCW) system, which includes pumping and injection wells operated simultaneously, is employed to enhance the decontamination by reagent delivery (Knox *et al.*, 1997; Trötschler *et al.*, 2005), microorganism stimulation (Blanford *et al.*, 2001) as well as the extraction of harmful substances. Vertical circulation well (VCW), which is a common example of GCW, is a dual-screen single-well system installed with a built-in small pump to generate circulating water between the upper and lower chambers of the same well. The concept "dipole flow test" was first formally proposed to be a field-tested technology by Kabala (1993) for the determination of vertical permeability. Since then, extensive studies of dipole flow have been carried out through mathematical modeling, physical experiments and field site tests (Bodin *et al.*, 2012; Hvilshoj *et al.*, 2000; Johnson and Simon, 2007; Schaad, 1998; Simon *et al.*, 2000; Sutton *et al.*, 2000; Xiang and Kabala, 1997; Zlotnik and Ledder, 1996; Zlotnik *et al.*, 2007; Zlotnik and Zurbuchen, 1998). For example, Xiang and Kabala (1997) analyzed the effects of hydraulic conductivity and location of layers as well as chamber size on the estimated hydraulic conductivity anisotropy ratio of the layered aquifer; Sutton *et al.* (2000) employed a stream tube modeling approach to interpret tracer test with dipole-flow; Johnson and Simon (2007) employed *in situ* data of hydrogeological features and contaminant distribution in addition to chemical tracer tests to form a robust protocol for characterizing flow patterns around VCW.

Almost at the same time, VCW was also introduced for *in situ* aquifer remediation (Herrling and Stamm, 1993; Herrling *et al.*, 1991). Specifically, it was initially used in air stripping technology to remove volatile organic compounds (VOC) such as chlorinated and aromatic hydrocarbons (Allmon *et al.*, 1999; Office and Support, 1998). Soon afterward, VCW was coupled with other techniques to enhance the removal of more extensive volatile contaminants. For example, Knox *et al.* (1997) carried out surfactant-enhanced remediation of perchloroethylene through VCW in field sites. Blanford *et al.* (2001) conducted a pilot-scale test with cyclodextrin-enhanced VCW for the removal of trichloroethene. Trötschler *et al.* (2005) added hydrogen peroxide through VCW to enhance the natural attenuation of heterocyclic hydrocarbon mediated by microorganisms in both laboratory and field sites.

To date, it is still a customary research scheme to conduct remediation from laboratory scale to pilot scale and then to large field-scale application. As long as the laboratory experimental remediation performance is satisfactory, the next step is to prepare the *in situ* installation of the remediation apparatus. However, before the implementation of field operation, which is often costly and time-consuming, it is critical to thoroughly investigate the *in situ* hydrogeological conditions and then to seek a suit of optimal designing parameters, such as pumping/injection rate and location and length of each screen, etc. As for groundwater remediation strategies, particle tracking simulation is a popular technique and contributes mainly in two aspects: quantification of capture zone of extraction well (Cunningham *et al.*, 2004; Liao *et al.*, 2015; Mulligan and Ahlfeld, 1999; Warner *et al.*, 2006) and generation of advection-dominated solute plume (Cirpka and Kitanidis, 2001; Crane and Blunt, 1999; Elder *et al.*, 2002; Ginn *et al.*, 1995). The characterization of the capture zone depends on the streamline distribution, i.e., the particle trajectory, which indicates how a pump-and-treat system prevents contaminants from escaping from capture due to regional groundwater flow (Warner *et al.*, 2006). In complicated flow fields generated by ambient hydrologic stresses and multiple pump-and-treat facilities, rather than conventional modeling tools, the particle tracking simulation is used to analyze the contamination leakance (Liao *et al.*, 2015). The premise of advection-dominated solute transport is the allowable neglection of local dispersion and diffusion, which is particularly acceptable if the interacting species have strongly varying strengths of sorption and the chromatographic effects dominate the longitudinal mixture (Cirpka and Kitanidis, 2000; Oya and Valocchi, 1998). With the particle tracking technique to transform the coordinate, a multi-dimensional advection equation is transformed into a travel-time-based equation (1D), which can be solved using a single-point upstream weighting method (Aziz, 1979). Consequently, the concentration results in the travel-time domain can be projected to those in the spatial domain by grid volume flux averaging (Crane and Blunt, 1999).

Similarly, particle-tracking methods have been employed in VCW remediation studies. Two methods have been commonly used. One is the fourth-order Runge-Kutta method (Shampine *et al.*, 1975), as employed in the works of Philip and Walter (1992) and Xia *et al.* (2019). The other is the semi-analytical method of Pollock (1988), as applied in the study of Cirpka and Kitanidis (2001). The work by Philip and Walter (1992) and Xia *et al.* (2019) traced particle paths to characterize the capture zone in a VCW flow field while Cirpka and Kitanidis (2001) placed emphasis on the description of travel-time-based concentration during reactive transport through tandem VCWs. In addition, visualization of the VCW flow field alone by Xia *et al.* (2019) and Philip and Walter (1992) still have limitations in optimizing design parameters because the area of the capture zone is merely a qualitative representation. Overall, a more quantitative and user-friendly index is needed. In terms of computational efficiency, the time step Δt in the fourth-order Runge-Kutta algorithm must be adjusted as local velocity changes; otherwise, the algorithm becomes computationally too expensive for practical use. In the work by Cirpka and Kitanidis (2001), a series of reaction-dependent parameters as detailed as possible is required to implement the travel-time-based reactive transport, but the parameters involving the mass transfer and microbial activity are usually difficult to obtain in the preliminary field-testing stage. Ideally, it will be very helpful to develop an easy-to-use field guide that can provide the best design parameters with respect to given hydrogeological conditions based on rigorous physical principles and theoretical basis. As such, an accessible quantitative index that can be utilized to estimate the performance of each suit of design parameters will be very useful. It is noteworthy that the particle tracking method can provide the desired quantitative index, which involves the travel time of particles from the injection chamber to the extraction chamber or the recovery ratio of particles after a given time. In general, these critical issues highlighted above motivated us to advance the present particle-tracking method for the preliminary screening of a VCW remediation system, in particular, to introduce a highly practical index to measure circulation performance.

Despite the practicability in the preliminary stage, the travel time or recovery ratio of particles can only indicate the circulation efficiency of VCW. After the reactive reagent is employed, it is the spatiotemporal distribution of solute that can provide convincing evidence and validation for degradation efficiency, which requires an accurate characterization of multi-species reactive transport in a VCW system. Cirpka and Kitanidis (2001) simulated the spatiotemporal performance of bioremediation in a multi-VCW system, but their travel-time scheme is unable to consider the interactions of reagents transverse to the flow direction. Meanwhile, their neglect of local-scale dispersion may introduce errors if the primary reagent hardly absorbs. To our best knowledge, Chen *et al.* (2011a) might have been the first attempt to analytically characterize the spatiotemporal distribution of solute due to advection-dispersion in a VCW-induced flow field. They employed the Laplace transform finite difference (LTFD) method, which transforms the advection-dispersion governing equation into the Laplace domain, where a group of linear algebraic equations composed of coefficient matrix and concentration vectors at each node are solved. It should be noted that the reagent of Chen *et al.* (2010, 2011b) was conservative, making the LTFD method unable to characterize the reactive transport of multispecies. This is because, for microbially induced redox reactions, the reaction rates of electron acceptor (EA), electron donor (ED) and biomass depend closely on the concentrations of multi-species. As such, the elements of concentration vectors of Chen *et al.* (2010, 2011a) have to be mathematically related to the vector elements in other equation groups, making the linear algebraic equation group impractical to solve.

In terms of dealing with multi-species transport governing equations, conventional numerical methods such as finite difference (FD), finite volume and finite element method, which are implemented with Euler integration in time, are able to elaborate sorption, mass transfer and microbially mediated reaction while describing advection-dispersion process. Previous studies have employed FD-based numerical method to investigate single-species solute transport or multi-species reactive transport induced by pumping or injection well (Phanikumar and McGuire, 2010; Wang *et al.*, 2017; Zhu *et al.*, 2019), in which the purely radial flow (advection) direction makes the upstream or downstream node explicit in the FD scheme. Thus, one can use batch processing of either upstream (or upwind) or downstream (or downwind) differentiation to solve such a one-way governing equation at each node. Unlike the solute transport characterized by one-way differentiation, it is a concern for VCW-induced solute transport that the determination of upstream or downstream direction depends on the location of each node. This node-dependent upstream direction is common in a flow field impacted by two hydrologic stresses with opposite driving forces generating strongly bent streamlines, such as in the VCW or other pumping-and-injection systems. To our knowledge, the studies of solute transport with the FD method accounting for strongly bent streamlines are rarely reported; therefore, we can conclude that an effective method to characterize multispecies reactive transport in a VCW-induced flow field is not available yet.

For a specific VCW remediation project, hydrogeological condition, design parameters and injection mode are key indices for solute transport (Chen *et al.*, 2010, 2011a). The hydrogeological condition mainly refers to the hydraulic conductivity anisotropy ratio, which means the ratio of radial hydraulic conductivity to vertical one. The design parameters involve the injection/extraction rate, and the distance between injection and extraction screened intervals. The injection mode often indicates the time and frequency of reagent infusion. For more accurate optimization and prediction of VCW remediation projects, it is essential to understand the effects of these indices on bioremediation performance.

In summary, it is critical to develop effective and easy-to-use methods for designing and predicting VCW-induced remediation based on rigorous experimental principles and a theoretical basis. Zhu *et al.* (2020) has outlined two primary approaches to fulfill this objective. The first is to use a new particle-tracking method to delineate the effects of hydrogeological conditions and design parameters on the travel time and recovery ratio of particles and to provide recommendations for optimizing the design parameters in a preliminary VCW-induced remediation design. The second is to develop an innovative FD method, in which the upstream direction depends on the exact location of each node, to describe multispecies reactive transport in a VCW-induced flow field, with a view to determining the effects of key hydrogeological conditions, design parameters and injection mode on the bioremediation performance within the VCW system. The integration of the above-proposed methods can guide the prediction of remediation performance during the operation of VCW. In this section, the model of Zhu *et al.* (2020) is introduced.

3.5.2 *Mathematical model of groundwater flow in the GCW system*

The conceptual model of groundwater flow associated with a VCW system is shown in Figure 3.83. The assumptions are similar to those in the studies of Philip and Walter (1992) and Xia *et al.* (2019) as follows. (1) The aquifer is confined, homogeneous, vertically anisotropic and horizontally isotropic, and radially infinite with

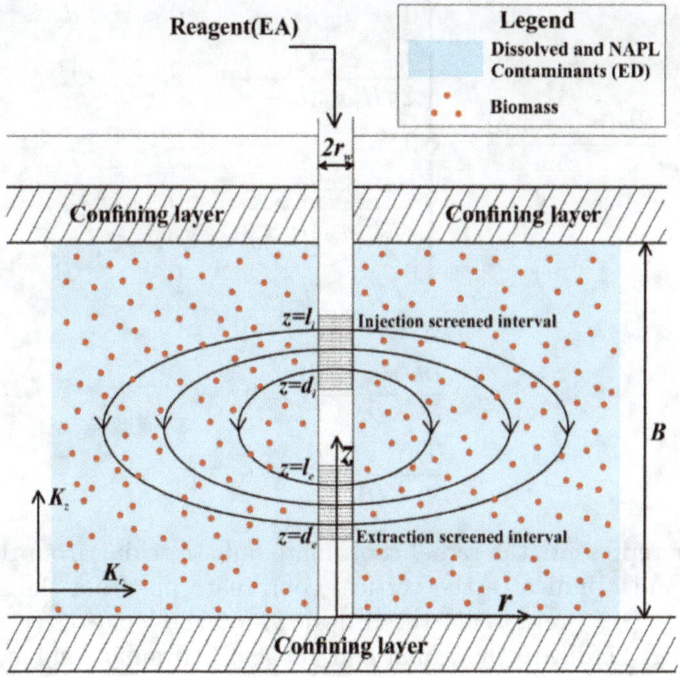

Figure 3.83. Schematic diagram of a VCW remediation system in a NAPL-contaminated confined aquifer. The light blue area near the well indicates both dissolved and NAPL contamination, the red dots represent biomass.

a finite constant thickness. (2) The ambient groundwater flow can be neglected compared to the VCW-induced flow. (3) The flow field is in a steady state. The double-screened single-well has a radius of r_w and a constant pumping/injection rate of Q. A circulating flow originates from the upper injection screened interval and terminates at the lower extraction screened interval. Without considering ambient groundwater flow, the VCW-induced flow field has axial symmetry. Therefore, the cylindrical coordinate system, as shown in Figure 3.83, is employed.

The governing equation, as well as the initial and boundary conditions can be described as (Sutton *et al.*, 2000; Xiang and Kabala, 1997):

$$K_r \frac{\partial^2 s(r,z)}{\partial r^2} + \frac{K_r}{r}\frac{\partial s(r,z)}{\partial r} + K_z \frac{\partial^2 s(r,z)}{\partial z^2} = 0 \qquad (3.706)$$

$$\frac{\partial s(r=r_w,z)}{\partial r}=\begin{cases}0 & 0<z<d_e\\ \dfrac{-Q}{2\pi K_r r_w(l_e-d_e)} & d_e<z<l_e\\ 0 & l_e<z<d_i\\ \dfrac{-Q}{2\pi K_r r_w(l_i-d_i)} & d_i<z<l_i\\ 0 & l_i<z<B\end{cases} \tag{3.707}$$

$$s(r\to+\infty,z)=0 \tag{3.708}$$

$$\frac{\partial s(r,z=0)}{\partial z}=0 \tag{3.709}$$

$$\frac{\partial s(r,z=B)}{\partial z}=0 \tag{3.710}$$

where r represents the radial coordinate [m] stretching from the well center and z indicates the vertical coordinate [m] from the aquifer bottom; $s(r,z)$ represents the drawdown at the location (r,z) [m]; K_r and K_z indicate the radial and vertical hydraulic conductivities [m/s], respectively; r_w is the well radius [m] and B is the aquifer thickness [m]; l_i and l_e are the top coordinates of the injection and extraction screened intervals [m], respectively; d_i and d_e are the bottom coordinates of the injection and extraction screened intervals [m], respectively. Equation (3.708) indicates that there is no drawdown far from the well. Equations (3.709) and (3.710) are the upper and lower no-flow boundary conditions, respectively.

Equations (3.706)–(3.710) have analytical solutions of drawdown distribution as demonstrated in previous studies (Chen *et al.*, 2010, 2011a; Sutton *et al.*, 2000; Zlotnik and Zurbuchen, 1998), where the specific discharge (or Darcy velocity) distribution can be calculated based on Darcy's law and the following expressions:

$$q_r(r,z)=\frac{Q}{\pi^2 r_w}\sum_1^{\infty}\left(\frac{-R_{ne}}{2\Delta_e}+\frac{R_{ni}}{2\Delta_i}\right)\frac{K_1\left(\frac{n\pi}{aB}r\right)}{K_1\left(\frac{n\pi}{aB}r_w\right)}\cos\left(\frac{n\pi z}{B}\right) \tag{3.711a}$$

$$q_z(r,z)=\frac{Q}{\pi^2 r_w}\sum_1^{\infty}\left(\frac{-R_{ne}}{2\Delta_e}+\frac{R_{ni}}{2\Delta_i}\right)\frac{K_0\left(\frac{n\pi}{aB}r\right)}{K_1\left(\frac{n\pi}{aB}r_w\right)}\sin\left(\frac{n\pi z}{B}\right) \tag{3.711b}$$

where $q_r(r,z)$ and $q_z(r,z)$ are the radial and vertical components of Darcy (average) velocity [m/s], respectively; $K_0(x)$ and $K_1(x)$ are the second kind of modified Bessel function with order zero and one, respectively. The other symbols used in the above equations are defined as $a = \sqrt{\frac{K_r}{K_z}}$; $2\Delta_e = l_e - d_e$; $2\Delta_i = l_i - d_i$; $R_{ne} = \frac{1}{n}\left[\sin\left(\frac{n\pi l_e}{B}\right) - \sin\left(\frac{n\pi d_e}{B}\right)\right]$; $R_{ni} = \frac{1}{n}\left[\sin\left(\frac{n\pi l_i}{B}\right) - \sin\left(\frac{n\pi d_i}{B}\right)\right]$. In the following section, the Darcy velocity is used for the particle-tracking computation.

Previous studies including (Cirpka and Kitanidis, 2001) neither introduced the application of Pollock (1988) method into VCW flow in detail nor applied travel time or recovery ratio of particles to estimate the circulation efficiency. We attempt to revisit the particle-tracking method of Pollock (1988) in a VCW-induced flow field when biochemical parameters, dispersivities, and *in situ* contamination conditions are unclear for the sake of preliminary VCW remediation design.

The method of Pollock (1988) characterized particle paths based on an FV model, and a schematic diagram of a particle path between 2D grids is shown in Figure 3.84. We first used Eqs. (3.711a) and (3.711b) to calculate the radial and vertical Darcy velocities at the midpoint of all interfaces between the grids, such as $q_r(i-1/2,j)$, $q_r(i+1/2,j)$, $q_z(i,j-1/2)$, and $q_z(i,j+1/2)$ between grid (i,j) and its circumjacent grids. Then, the Darcy velocity distribution inside the grid (i,j) obeys a simple linear interpolation scheme, and the velocity gradient with distance is expressed as (Pollock, 1988):

$$A_r(i,j) = \frac{q_r(i+1/2,j) - q_r(i-1/2,j)}{\Delta r} \tag{3.712a}$$

$$A_z(i,j) = \frac{q_z(i,j-1/2) - q_z(i,j+1/2)}{\Delta r} \tag{3.712b}$$

where $A_r(i,j)$ and $A_z(i,j)$ indicate the radial and vertical Darcy velocity gradients with distance [1/s], respectively; Δr and Δz are the radial and vertical spatial steps, respectively (see Figure 3.84); the grid (i,j) has four intergrid Darcy velocity components, in which $q_r(i-1/2,j)$ and $q_r(i+1/2,j)$ are radial inflow and outflow Darcy velocities, respectively, and $q_z(i,j+1/2)$ and $q_z(i,j-1/2)$ are vertical inflow and outflow Darcy velocities, respectively.

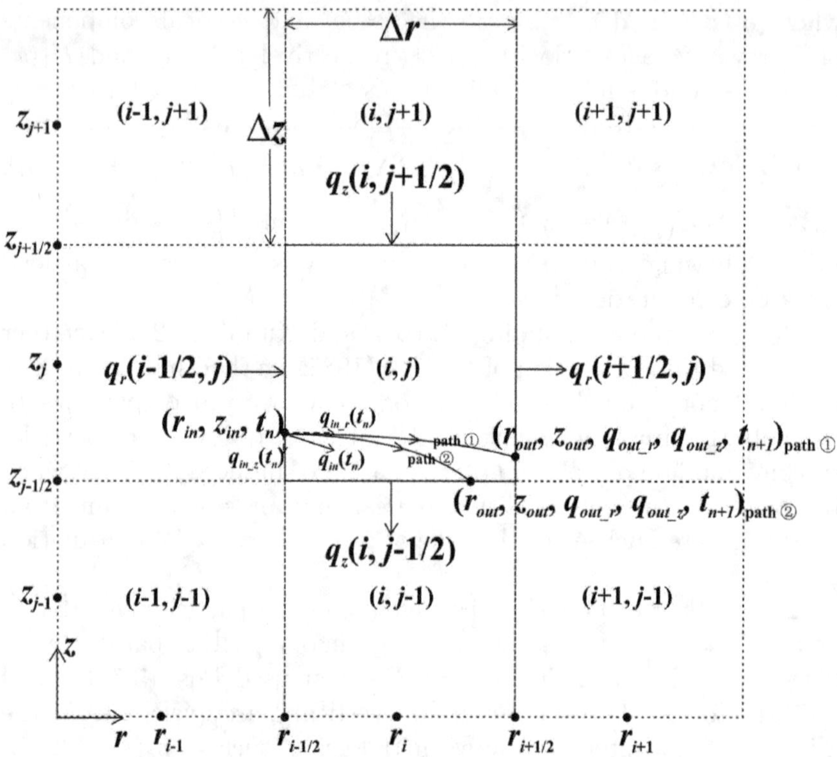

Figure 3.84. Schematic expression of the trajectory of a particle in a specific cell using the particle-tracking method of Pollock (1988).

Pollock (1988) derived expressions for the time spent traveling through the grid (i, j) in radial and vertical directions, respectively, as

$$\Delta t_r(i,j) = 1/A_r(i,j) \cdot \ln(q_r(i+1/2,j)/q_r(i-1/2,j)) \qquad (3.713a)$$

$$\Delta t_z(i,j) = 1/A_z(i,j) \cdot \ln(q_z(i,j-1/2)/q_z(i,j+1/2)) \qquad (3.713b)$$

The expressions for the radial and vertical Darcy velocity components at the exit of the grid (i, j) are respectively:

$$q_{\text{out}_r}(t_{n+1}) = q_{in\_r}(t_n) \cdot \exp(A_r(i,j) \cdot \Delta t) \qquad (3.714a)$$

$$q_{\text{out}_z}(t_{n+1}) = q_{in\_z}(t_n) \cdot \exp(A_z(i,j) \cdot \Delta t) \qquad (3.714b)$$

where the traveling duration Δt inside the grid (i, j) yields the radial and vertical exit position of a particle, respectively, as

$$r_{\text{out}}(t_{n+1}) = r_{\text{in}}(t_n) + 1/A_r(i,j) \cdot [q_{\text{out}\,r}(t_{n+1}) - q_{\text{in\_}r}(t_n)] \quad (3.715\text{a})$$

$$z_{\text{out}}(t_{n+1}) = z_{\text{in}}(t_n) + 1/A_z(i,j) \cdot [q_{\text{out}\,z}(t_{n+1}) - q_{\text{in\_}z}(t_n)] \quad (3.715\text{b})$$

where $r_{\text{in}}(t_n)$ and $z_{\text{in}}(t_n)$ are the radial and vertical coordinates of a particle at time point t_n, respectively, at which the particle is ready to enter the grid (i, j) with radial and vertical Darcian velocity components as $q_{in\_r}(t_n)$ and $q_{in\_z}(t_n)$, respectively. Note that Δt in Eqs. (3.714a) and (3.714b) is determined to be the minor value of $\Delta t_r(i, j)$ and $\Delta t_z(i, j)$ in Eqs. (3.713a) and (3.713b), respectively, which means that if $\Delta t_r(i, j) < \Delta t_z(i, j)$, the particle will choose path (1) to leave the grid (i, j), otherwise, it will choose path (2), as Figure 3.84 shows. Therefore, t_{n+1} is equal to $t_n + \Delta t$ and represents the time point at which the particle leaves the grid (i, j).

As Figure 3.85 shows, the number of particles starting circulation from the injection chamber is N_{inj}, which is 80 and large enough to

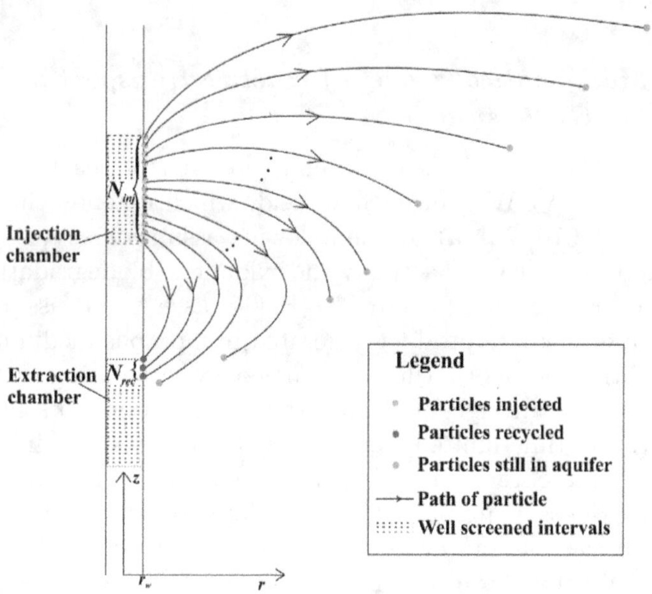

Figure 3.85. Schematic diagram of both the departure and arrival of a series of particles, N_{inj} and N_{rec} are the total number of particles departing from the injection chamber and arriving at the extraction chamber, respectively.

reveal statistical significance. The particles are distributed uniformly along the entire screened interval to cover and represent all streamtubes; thus, the initial vertical location of each particle can be easily determined. The initial velocity of each particle can be calculated by Eq. (3.707). Note that in the studies of a partially penetrating well, nonuniform flux boundary at the well face is often recognized (Cassiani and Kabala, 1998; Chang and Chen, 2002, 2003; Feng and Zhan, 2016; Mathias and Wen, 2015) and has also been applied (Zhu *et al.*, 2020). Considering that the nonuniform flux boundary has very limited effects on the flow field near well under steady state (Feng and Zhan, 2016), we employ a uniform flux boundary at the well face, as Eq. (3.707) indicates. We assume that as soon as $r_{\text{out}}(t_n+1)$ in Eq. (3.715a) equals zero, this particle has finished its travel and is recycled, then the number of particles recycled at a certain

$$P(t) = \frac{N_{\text{rec}}(t)}{N_{\text{inj}}} \tag{3.716}$$

The default parameters for the baseline simulation are listed in Table 3.21.

3.5.3 *Mathematical model of reactive transport in the GCW system*

As Figure 3.83 shows, we consider a problem of reactive reagent transport in a VCW-induced flow field, which extends the case of Chen *et al.* (2010, 2011a) by the following assumptions. (1) Though dissolved oxygen (DO) is actually the oxidant in biodegradation, this section regards hydrogen peroxide (H_2O_2) as EA and assumes DO to be an intermediate product generated in a porous medium by the decomposition of H_2O_2, which is carried by circulating water into the aquifer only through VCW, like the operation in Trötschler *et al.* (2005). (2) A time-limited injection of H_2O_2 does not inhibit the activity of microorganisms (Tindall *et al.*, 2005; Trötschler *et al.*, 2005). (3) ED is benzene, which is a common biodegradable aromatic hydrocarbon and has polluted the aquifer before the installation of VCW; they have both dissolved and nonaqueous-phase liquid (NAPL) phases and are assumed to uniformly fill the aquifer to a certain scale. (4) The Indigenous aerobic microorganism was cultivated to a considerable amount in advance by oxygen-saturated water like

Table 3.21. Parameters utilized in the baseline simulations of particle tracking and conservative solute transport.

Symbol	Definition	Input value	Unit
Q	Flow rate	0.01	m^3/min
B	Aquifer thickness	10	m
K_r	Horizontal hydraulic conductivity	$4 \times 10 - 4$	m/min
K_z	Vertical hydraulic conductivity	$4 \times 10 - 4$	m/min
K_r/K_z	Hydraulic conductivity anisotropy ratio	4	—
θ	Effective porosity of aquifer	0.3	—
r_w	Well radius	0.05	M
N_{inj}	Particles injected from upper chamber	80	—
d_e	Vertical coordinate of extraction screen bottom	2	M
l_e	Vertical coordinate of extraction screen top	4	m
d_i	Vertical coordinate of injection screen bottom	6	m
l_i	Vertical coordinate of injection screen top	8	m
α_L	Longitudinal dispersivity	0.2	m
α_T	Transverse dispersivity	0.05	M
t_0	Injection duration of conservative solute	100	min
M	Injected mass of conservative solute	10	kg
C^{ini}	Initial concentration of conservative solute	0	g/m^3
Re	Radius of influence	10	m
Δr	Radial space step	0.1	m
Δz	Vertical space step	0.1	m
Δt	Time step	5	s

Note: t_0 refers to injection duration of conservative solute and is different from the t_0 in Table 3.22, which indicates the injection duration of EA.

the pretreatment in Trötschler *et al.* (2005) and it has the same distribution as contaminants. (5) The effects of NAPL phase, biomass proliferation and microbially generated gas on hydraulic conductivity can be neglected. (6) In this VCW system, the water injected comes from the anoxic groundwater extracted from the confined aquifer; therefore, the DO brought into the system after cessation of H_2O_2 injection can be overlooked.

A point to note is that the advanced VCW remediation technologies nowadays involve much more than simple aerobic biodegradation of organic compounds. Transport of multiple species with simultaneous reactions among the species is a common phenomenon associated with many VCW remediation projects, which is also the primary concern of this investigation. Specifically, we choose aerobic biodegradation of benzene as an example and attempt to establish a universal methodology that can be applied as long as the species and parameters are confirmed in actual engineering practices of remediation involving a VCW.

An ADE considering reaction for 3D axisymmetric multispecies reactive transport can be described as

$$\begin{aligned}\frac{1}{r}\frac{\partial}{\partial r}\left\{r\left[D_{rr}(r,z)\frac{\partial C_i(r,z,t)}{\partial r}+D_{rz}(r,z)\frac{\partial C_i(r,z,t)}{\partial z}\right.\right.\\ \left.\left.-\,v_r(r,z)\cdot C_i(r,z,t)\right]\right\}+\frac{\partial}{\partial z}\left[D_{zr}(r,z)\frac{\partial C_i(r,z,t)}{\partial r}\right.\\ \left.+D_{zz}(r,z)\frac{\partial C_i(r,z,t)}{\partial z}-v_z(r,z)\cdot C_i(r,z,t)\right]+r_i\\ =R_i\frac{\partial C_i(r,z,t)}{\partial t}\end{aligned}\tag{3.717}$$

where the subscript i represents EA or mobile ED, that is, $C_{\mathrm{EA}}(r,z,t)$ and $C_{\mathrm{ED\_mob}}(r,z,t)$ indicate the concentrations [Kg/m^3] of EA and mobile ED at the time t and location (r,z), respectively. Ri here only refers to the retardation factor of ED for a linear isotherm and can be calculated as $R_{\mathrm{ED}}=1+\rho_{\mathrm{bulk}}K_d/\theta$, where ρ_{bulk} is the bulk density of porous media [Kg/m^3], K_d is the partitioning coefficient of ED [m^3/Kg], and θ is the effective porosity [-]. r_i is each species' source and sink term, and it is positive for the source and negative for the sink. Molecular diffusion is assumed to be negligible compared to mechanical dispersion. D_{ij} indicates a component of the dispersion coefficient tensor [m^2/s], which can be defined as (Bear, 2012):

$$D_{rr}(r,z)=\alpha_L\frac{v_r(r,z)^2}{|v(r,z)|}+\alpha_T\frac{v_z(r,z)^2}{|v(r,z)|}\tag{3.718a}$$

$$D_{rz}(r,z)=D_{zr}(r,z)=(\alpha_L-\alpha_T)\frac{v_r(r,z)\cdot v_z(r,z)}{|v(r,z)|}\tag{3.718b}$$

$$D_{zz}(r,z)=\alpha_L\frac{v_z(r,z)^2}{|v(r,z)|}+\alpha_T\frac{v_r(r,z)^2}{|v(r,z)|}\tag{3.718c}$$

where α_L and α_T are the longitudinal and transverse dispersivities [m], respectively. v_r and v_z are the radial and vertical flow velocities [m/s], respectively, and they can be obtained from Darcy velocities q_r and q_z in Eqs. (3.711a) and (3.711b) as $v_r = q_r/\theta$ and $v_z = q_z/\theta$, respectively. $|v(r,z)| = \sqrt{v_r(r,z)^2 + v_z(r,z)^2}$ is the magnitude of velocity.

The governing Eq. (3.717) is coupled with initial and boundary conditions as

$$C_i(r, z, t = 0) = C_i^{ini}, \quad i = \text{EA}, \quad \text{ED}, \quad \text{or bio} \tag{3.719}$$

$$C_i(r \to \infty, z, t = 0) = C_i^{ini}, \quad i = \text{EA}, \quad \text{ED}, \quad \text{or bio} \tag{3.720}$$

$$v_z(r, 0, t)C_i(r, 0, t) - D_{zz}(r, 0)\frac{\partial C_i(r, z = 0, t)}{\partial z}, \quad i = \text{EA or ED} \tag{3.721}$$

$$v_z(r, B, t)C_i(r, B, t) - D_{zz}(r, m)\frac{\partial C_i(r, z = B, t)}{\partial z} = 0, \quad i = \text{EA or ED} \tag{3.722}$$

$$v_z(r_w, z)C_{\text{EA}}(r_w, z, t) - D_{zz}(r_w, z)\frac{\partial C_{\text{EA}}(r = r_w, z, t)}{\partial z} = 0, \quad 0 < z < d_e \tag{3.723a}$$

$$\frac{\partial C_{\text{EA}}(r = r_w, z, t)}{\partial r} = 0, \quad d_e < z < l_e \tag{3.723b}$$

$$v_r(r_w, z)C_{\text{EA}}(r = r_w, z, t) - D_{rr}(r_w, z)\frac{\partial C_{\text{EA}}(r = r_w, z, t)}{\partial r} = 0, \quad l_e < z < d_i \tag{3.723c}$$

$$\frac{M_{\text{EA}}}{t_0} = 2\pi r_w(l_i - d_i)\theta\left[v_r(r_w, z)C_{\text{EA}}(r = r_w, z, t). - D_{rr}(r_w, z)\frac{\partial C_{\text{EA}}(r = r_w, z, t)}{\partial r}\right], \quad d_i < z < l_i \tag{3.723d}$$

$$v_r(r_w, z)C_{\text{EA}}(r = r_w, z, t) - D_{rr}(r_w, z)\frac{\partial C_{\text{EA}}(r = r_w, z, t)}{\partial r} = 0, \quad l_i < z < B \tag{3.723e}$$

where Eq. (3.719) describes the initial condition of each species, $C_{\text{EA}}^{\text{ini}}$ is the initial concentration of EAs and is assumed to be $0.01\,\text{g/m}^3$ after the preincubation of microorganisms. M_{EA}/t_0 and t_0 are the injected mass rate and injection duration of the reagent, respectively. Note that $C_{\text{ED}}^{\text{ini}}$ here should be classified into $C_{\text{EA\_mob}}^{\text{ini}}$ and $C_{\text{EA\_imm}}^{\text{ini}}$, which are the initial concentrations of mobile ED and immobile ED, respectively. The values of $C_{\text{EA\_mob}}^{\text{ini}}$ mob and $C_{\text{EA\_imm}}^{\text{ini}}$ are listed in Table 3.22, $C_{\text{EA\_mob}}^{\text{ini}}$ equals to the solubility of benzene C_{sol} due to

Table 3.22. Parameters utilized in the baseline simulation of reactive transport.

Symbol	Definition	Input value	Unit
t_0	Injection duration of EA	1	h
M_{EA}/t_0	Injection rate of H_2O_2 during one injection	8	kg/h
$C_{\text{EA}}^{\text{ini}}$	Initial concentration of EA	0.01	g/m^3
$C_{\text{ED\_mob}}^{\text{ini}}$	Initial concentration of dissolved ED	1800	g/m^3
$C_{\text{ED\_imm}}^{\text{ini}}$	Initial concentration of NAPL phase ED	0.5	g/kg solid particle
$C_{\text{bio}}^{\text{ini}}$	Initial concentration of mobile biomass	0.3	g dry biomass/m^3 porous media
C_{sol}	Solubility of ED	1800	g/m^3
ρ_{bulk}	Bulk density	1.8×10^6	g/m^3
μ_{max}	Maximum growth rate of aerobic biomass	12[a]	1/d
K_{DO}	Half saturation constants of DO	0.2[b]	g/m^3
K_{ED}	Half saturation constants of ED	1.0[c]	g/m^3
k_{dec}	Decay rate of biomass	0.1[d]	1/d
k_{NAPL}	Mass transfer rate from NAPL to mobile phase	1.5[e]	1/d
Y_{O_2}	Yield coefficient of DO	2.15[b]	—
Y_{ED}	Yield coefficient of ED	1.0c	—
$Y_{H_2O_2 \to O_2}$	Transform coefficient between H_2O_2 and O_2	0.485	—
K_d	Partitioning coefficient	1.3×10^{-3f}	m^3/g
ρ_s	Dry density of biomass	1500[f]	g/m^3

Sources: [a]Trigueros *et al.* (2010). [b]Kinzelbach *et al.* (1991). [c]Bekins *et al.* (1998). [d]Thullner *et al.* (2004). [e]Mobile *et al.* (2012). [f]Rockhold *et al.* (2005).

the existence of NAPL. $C_{\text{bio}}^{\text{ini}}$ in Table 3.22 is the initial concentration of biomass. In this section, the mobility of biomass is assumed to be negligible; thus, this section only accounts for the reactive source and sink of biomass in the multispecies system of concern.

In Eq. (3.717), r_{EA} and $r_{\text{ED\_mob}}$ are the source and sink terms of EA and mobile ED [Kg/(m^3 · s)], respectively. Similarly, we define r_{bio} as the source and sink term of biomass [Kg/(m^3 · s)]. In this three-species system, r_{EA} and r_{bio} are dependent on the aerobic redox reaction, $r_{\text{ED\_mob}}$ is determined by aerobic reaction and mass transfer between NAPL and dissolved phase. The expressions of the above three source and sink terms are written as follows:

$$r_{\text{bio}} = C_{\text{bio}} k_{gr} - C_{\text{bio}} k_{\text{dec}} \tag{3.724}$$

where C_{bio} is the concentration of functional biomass facilitating aerobic degradation (referred to total volume of porous media) [Kg/m^3]; k_{dec} is the decay rate of biomass [1/s]; k_{gr} is the growth rate of biomass [1/s], which can be expressed as

$$k_{gr} = \mu_{\max} \frac{C_{\text{EA}} Y_{H_2O_2 \to O_2}}{C_{\text{EA}} Y_{H_2O_2 \to O_2} + K_{DO}} \frac{C_{\text{ED\_mob}}}{C_{\text{ED\_mob}} + K_{\text{ED}}} F_{\text{lim}} \tag{3.725a}$$

where $\mu_{\max}$ is the maximum growth rate of aerobic organisms [1/s]; $Y_{H_2O_2 \to O_2}$ is the transform coefficient between H_2O_2 and O_2, which indicates the mass of O_2 generated by unit mass of H_2O_2. K_{DO} and K_{ED} are the half-saturation constants of DO and ED [Kg/ (m^3 · s)], respectively. F_{lim} describes the self-inhibition factor due to space limitation in pores and can be expressed as

$$F_{\text{lim}} = 1 - \frac{C_{\text{bio}}}{\rho_s * \theta} \tag{3.725b}$$

The expression of r_{EA} is

$$r_{\text{EA}} = -C_{\text{bio}} k_{gr} Y_{O_2} / Y_{H_2O_2 \to O_2} \tag{3.726}$$

where Y_{O_2} is the yield coefficient of DO in aerobic redox reaction [-]. As for ED, the NAPL release into dissolved phase should be considered and the expressions of $r_{\text{ED\_mob}}$ and $r_{\text{ED\_imm}}$ are, respectively,

$$r_{\text{ED\_imm}} = -k_{\text{NAPL}}(C_{\text{sol}} - C_{\text{ED\_mob}}) \frac{\theta}{\rho_{\text{bulk}}} \tag{3.727a}$$

$$r_{\text{ED\_mob}} = -C_{\text{bio}} k_{gr} Y_{ED} + k_{\text{NAPL}}(C_{\text{sol}} - C_{\text{ED\_mob}}) \tag{3.727b}$$

where k_{NAPL} is the mass transfer rate [1/s]; ρ_{bulk} is the bulk density of medium; C_{sol} is the solubility of ED [Kg/m^3]; Y_{ED} is the yield coefficient of ED [–].

3.5.4 *Numerical model of reactive transport in the GCW system*

The conventional FD method to describe radial solute transport, such as Wang *et al.* (2017) and Zhu *et al.* (2019), employed a spatial discretization on ADE as follows

$$\frac{\partial C_i}{\partial t} = \frac{r_{i-1/2}\alpha_r v_{a,i}\frac{C_{i-1}-C_i}{r_i-r_{i-1}} - r_{i+1/2}\alpha_r v_{a,i+1}\frac{C_i-C_{i+1}}{r_{i+1}-r_{i1}}}{r_i(r_{i+1/2}-r_{i-1/2})} + \frac{v_{a,i}(C_{i-1/2}-C_{i+1/2})}{r_{i+1/2}-r_{i-1/2}} \tag{3.728}$$

where i indicates the number of nodes; $i-1/2$ and $i+1/2$ indicate the midpoint of two adjacent points; α_r is the radial dispersivity. Equation (3.728) is generally solved by a stiff integrator ODE15s in MATLAB. However, this method is probably only fit for purely radial flow fields, which can be discretized into nodes with the same directions of inflow and outflow, that is, $v_{a,i}$ and $v_{a,i+1}$ have the same direction, which is toward the pumping well center or from the injection well center outward. However, in a VCW-induced flow field, there exist eight kinds of combinations of inflow and outflow directions as indicated in Figure 3.86, making Eq. (3.728) invalid to solve Eq. (3.717) numerically.

In this section, we propose a node-dependent finite difference (NDFD) method to solve Eq. (3.717). We adopt the same coordinate system for nodes rather than the grid in Figure 3.84. As Figure 3.86 indicates, $(r_i,\ z_j)$ represents the coordinate of the node (i,j). As the node ⑤ in Figure 3.86 represents, $v_r(r_{i-1/2}, z_j)$ and $v_r(r_{i+1/2}, z_j)$ are the two radial velocities on both sides of the node (i,j), and $v_z(r_i, z_{j-1/2})$ and $v_z(r_i, z_{j+1/2})$ are the two vertical velocities on both sides of the node (i,j). The red and blue arrows represent the positive (+) and negative (−) velocity directions, respectively. In our method, the expressions of spatial advection-dispersion are dependent on the locations of nodes, which have various combinations of

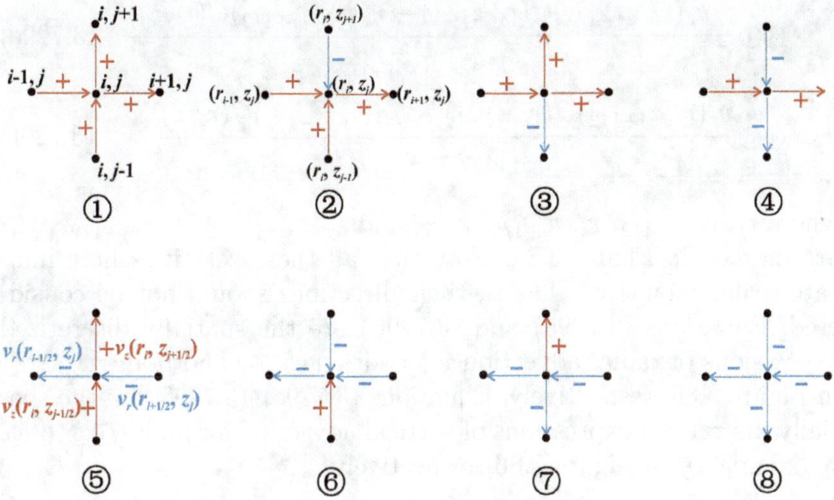

Figure 3.86. The schematic diagram of the node-dependent finite difference method, ①–⑧, describes eight kinds of combinations of inflow and outflow velocities in both radial and vertical directions. The red arrow coupled with the symbol "+" indicates the positive direction. Meanwhile, the blue arrow coupled with the symbol "−" indicate negative direction.

inflow and outflow directions. The spatially discretized expressions of radial advection for different kinds of nodes in Figure 3.86 can be written as follows:

$$①②③④\ \frac{r_{i-1/2}v_r(r_{i-1/2},z_j)C(r_{i-1},z_j)-r_{i+1/2}v_r(r_{i+1/2},z_j)C(r_i,z_j)}{r_i(r_{i+1/2}-r_{i-1/2})} \tag{3.729a}$$

$$⑤⑥⑦⑧\ \frac{r_{i+1/2}v_r(r_{i+1/2},z_j)C(r_{i+1},z_j)-r_{i-1/2}v_r(r_{i-1/2},z_j)C(r_i,z_j)}{r_i(r_{i+1/2}-r_{i-1/2})} \tag{3.729b}$$

and the spatially discretized expressions of vertical advection for different kinds of nodes in Figure 3.86 are

$$①⑤\ \frac{v_z(r_i,z_{j-1/2})C(r_i,z_{j-1})-v_z(r_i,z_{j+1/2})C(r_i,z_j)}{z_{j+1/2}-z_{j-1/2}} \tag{3.729c}$$

$$②⑥\ \frac{v_z(r_i,z_{j-1/2})C(r_i,z_{j-1})+v_z(r_i,z_{j+1/2})C(r_i,z_j)}{z_{j+1/2}-z_{j-1/2}} \tag{3.729d}$$

$$③⑦\ \frac{-v_z(r_i, z_{j-1/2})C(r_i, z_{j-1}) - v_z(r_i, z_{j+1/2})C(r_i, z_j)}{z_{j+1/2} - z_{j-1/2}} \tag{3.729e}$$

$$④⑧\ \frac{v_z(r_i, z_{j+1/2})C(r_i, z_{j-1}) - v_z(r_i, z_{j-1/2})C(r_i, z_j)}{z_{j+1/2} - z_{j-1/2}} \tag{3.729f}$$

where r_{i-1}, $r_{i-1/2}$, r_i, $r_{i+1/2}$, r_{i+1}, and z_{j-1}, $z_{j-1/2}$, z_j, $z_{j+1/2}$, z_{j+1} are marked in Figure 3.84. Note that all these velocities here indicate scalar quantity. That is, their directions should not be considered. Equations (3.729a) and (3.729b) are the spatially discretized expressions of radial advection for nodes ①②③④ and nodes ⑤⑥⑦⑧ in Figure 3.86, respectively. Equations (3.729c)–(3.729f) are the spatially discretized expressions of vertical advection for nodes ①⑤, ②⑥, ③⑦, and ④⑧ in Figure 3.86, respectively.

3.5.5 *Results and discussions*

3.5.5.1 *Particle tracking of VCW-induced flow*

The trajectories of all particles released from the injection chamber are revealed in Figure 3.87. For the sake of visualization, we choose

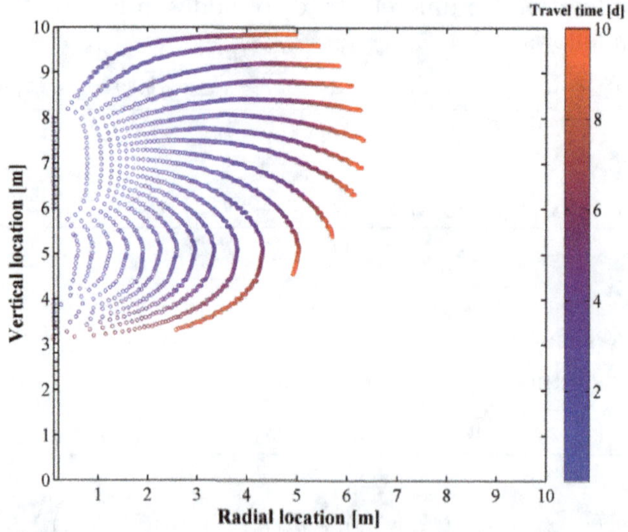

Figure 3.87. The trajectories of all particles released from the injection chamber, the colors from blue to red, indicate different time points; that is, the particles with the same color have equal travel time.

21 particles out of the total 80 particles in Figure 3.87. Except for the particle number, the other parameters are set at their default baseline values, as shown in Table 3.21. In Figure 3.87, the colors from blue to red indicate different time points. That is, the particles with the same color have equal travel times. These travel time markers are sparsely scattered near the well and closely clouded far from the well. This is because the travel time difference between the two markers is constant (0.04 day), but the particle velocity near the well is higher than that far from the well. Figure 3.87 indicates that after 10 days, there is still a sizable proportion of particles remaining in the aquifer. The proportion of particles recycled, which can be termed as recovery ratio, has its corresponding travel time.

We calculate the recovery ratio versus travel time for four cases as presented in Figure 3.88. Based on the baseline values of parameters, Case 1 increases the value of K_r from 4 times to 9 times of K_z, Case 2

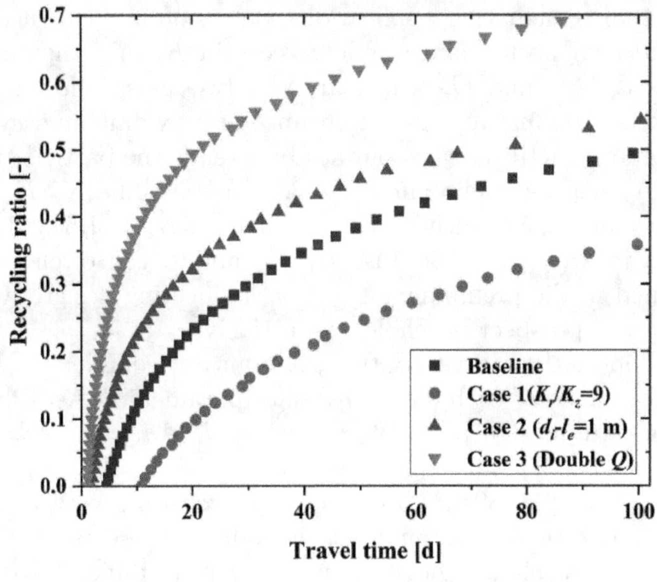

Figure 3.88. Recovery ratio versus travel time under different cases, the parameters of the baseline case are listed in Table 3.21, based on which Case 1 increases the value of K_r from 4 times to 9 times of K_z, Case 2 moves down the injection screened interval from $6\,\text{m} \leq z \leq 8\,\text{m}$ to $5.5\,\text{m} \leq z \leq 7.5\,\text{m}$ and moves up the extraction screened interval from $2\,\text{m} \leq z \leq 4\,\text{m}$ to $2.5\,\text{m} \leq z \leq 4.5\,\text{m}$, that is, $d_i - l_e$ decreases from 2 to 1; Case 3 increases the injection/extraction rate Q from 0.01 to 0.02 m^3/min, that is, doubling the Q value.

moves down the injection screened interval from $6\,\text{m} \leq z \leq 8\,\text{m}$ to $5.5\,\text{m} \leq z \leq 7.5\,\text{m}$ and moves up the extraction screened interval from $2\,\text{m} \leq z \leq 4\,\text{m}$ to $2.5\,\text{m} \leq z \leq 4.5\,\text{m}$, that is, $d_i - l_e$ decreases from 2 to 1; Case 3 increases the injection/extraction rate Q from 0.01 to $0.02\,\text{m}^3/\text{min}$, that is, doubling the Q value. Note that Case 1 involves hydrogeological conditions and Cases 2 and 3 are about design parameters in actual engineering practices. As Figure 3.88 indicates, an increasing anisotropy ratio K_r/K_z yields a lower recovery ratio, which means that the aquifer with a higher hydraulic con ductivity anisotropy ratio exerts a greater negative impact on the circula tion. For an aquifer with specified hydraulic conductivities, decreasing $d_i - l_e$ can promote the circulation; meanwhile, using a higher Q means that one is enhancing the circulating flow field more quickly. These qualitative results are accordant with those of Chen *et al.* (2010, 2011a) but with different analytical approaches: We apply the travel time-based recovery ratio and they used the plume of solute.

In actual remediation applications, one is often concerned about the circulation performance under a certain K_r/K_z, thus needs to optimize $d_i - l_e$ and Q, which are the two most relevant design parameters. To illustrate the significance of providing an easy-to-use guideline for practical usage, our study extends the particle-tracking results to such a guideline in Table 3.23. We assume K_r/K_z and B to be specific values, such as $K_r = 6\,\text{m/day}$, $K_z = 1.5\,\text{m/day}$, and $B = 8\,\text{m}$ in Table 3.23, because K_r/K_z and B can often be easily determined in the preliminary stage of engineering design by dipole flow test and prospect borehole, respectively.

We propose the recovery ratio as a quantitative index to estimate the performance of each suite of design parameters. As Table 3.23 represents, the design parameters, such as $l_i - d_i = l_e - d_e = 2\,\text{m}$, $d_i - l_e = 3\,\text{m}$, $Q = 10\,\text{m}^3/\text{day}$ and $l_i - d_i = l_e - d_e = 2\,\text{m}$, $d_i - l_e = 4\,\text{m}$, $Q = 30\,\text{m}^3/\text{day}$, generate a zero recovery ratio after 3 days. These design parameters, as well as those producing low recovery ratios, should be avoided since it is difficult to recycle the injected reagent. Thus, there is a potential to produce undesirable secondary pollution. In actual engineering practices, $l_i - d_i$ or $l_e - d_e$ may be adjusted considering specific aquifer thickness and contaminated location, and Q may be adjusted due to limited aquifer hydraulic conductivity, pump performance, or power consumption.

Table 3.23. The guideline for the optimization of design parameters, in which the injection and extraction screened intervals are assumed to be symmetrically distributed on both sides of $z = B/2$.

Hydrogeological conditions			**Design parameters**			**Recovery ratio [%] after 3 days**
K_r [m/d]	K_z [m/d]	B [m]	$l_i - d_i = l_e - d_e$ [m]	$d_i - l_e$ [m]	Q [m^3/d]	
6	1.5	8	2	1	10	19
					20	27
					30	34
					40	38
					50	41
				2	10	2
					20	11
					30	18
					40	22
					50	26
				3	10	**0**
					20	**5**
					30	9
					40	13
					50	17
				4	10	**0**
					20	**0**
					30	**0**
					40	3
					50	5
			2.5	1	10	15
					20	22
					30	29
					40	33
					50	36
				2	10	**0**
					20	**6**
					30	12
					40	17
					50	21
				3	10	**0**
					20	1
					30	2
					40	6
					50	10

(*Continued*)

Table 3.23. (*Continued*)

Hydrogeological conditions			**Design parameters**			**Recovery ratio [%] after 3 days**
K_r [m/d]	K_z [m/d]	B [m]	$l_i - d_i = l_e - d_e$ [m]	$d_i - l_e$ [m]	Q [m^3/d]	
			3	1	10	12
					20	18
					30	23
					40	27
					50	30
				2	10	0
					20	7
					30	13
					40	16
					50	19

An easy-to-use guide with respect to a target aquifer, like Table 3.23, becomes a very valuable reference example, $l_e - d_e$ or $l_e - d_e = 2\,\text{m}$, $d_i - l_e = 3\,\text{m}$, $Q = 50\,\text{m}^3/\text{day}$, which yields a recovery ratio of 17%, can be substituted by a smaller $Q = 40\,\text{m}^3/\text{day}$ coupled with $l_e - d_e$ or $l_e - d_e = 2.5\,\text{m}$, $d_i - l_e = 2\,\text{m}$, because the latter also achieves a recovery ratio of 17%. With Our MATLAB program of particle tracking, such an easy-to-use guideline like Table 3.23 can be expanded rather straightforwardly by including a wide range of design parameters, which can help engineers quickly identify the best design parameters as long as the values of K_r, K_z, and B are known.

3.5.5.2 *Validation of the NDFD method*

The NDFD model is uniformly discretized in both radial and vertical directions and the grid sizes are $r = 0.1\,\text{m}$ and $\Delta z = 0.1\,\text{m}$, respectively. The time-step Δt is $5s$. Numerical errors of the developed NDFD method should be estimated before its application. Grid Peclet number defined as $P_e = v\Delta x/D$ and Courant number defined as $C_r = v\Delta t/\Delta x$ are commonly used to analyze the effects of oscillations and numerical dispersion on numerical models, in which Δx is the spatial step [m]. The criteria of $P_e \leq 2$ and $C_r \leq 1$ should be met in all directions of the model to suppress the numerical errors (including the numerical dispersion and artificial oscillation) to a

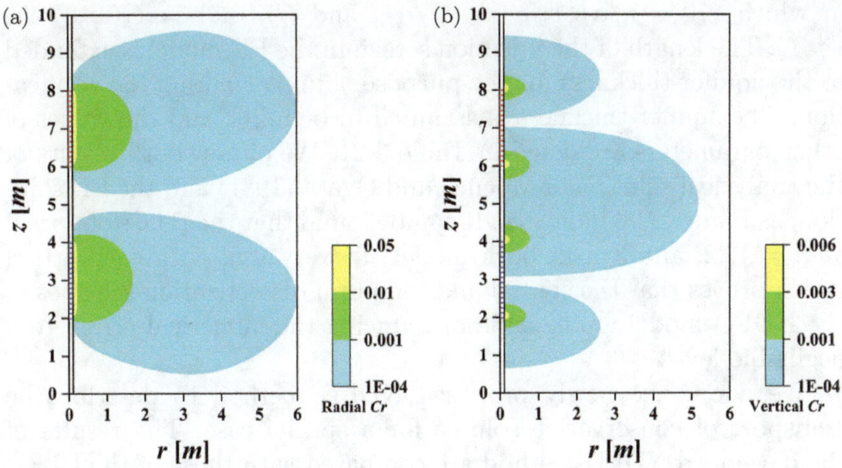

Figure 3.89. Distribution of Courant numbers in radial (a) and vertical (b) directions.

negligible level, With the given values of relevant parameters listed in Table 3.21, the radial and vertical grid Peclet numbers are calculated to be $P_{e,r} \approx \Delta r/\alpha_L = 0.5$ and $P_{e,v} \approx \Delta z/\alpha_T = 2$, respectively; as Figure 3.89 indicates, the largest values of radial Courant number $C_{r,\text{vmax}}$ and vertical one $C_{r,\text{vmax}}$ are not greater than 0.05 and 0.006, respectively Obviously, the values of P_e and C_r, in both radial and vertical directions meet the criteria, suggesting that the spatiotemporal discretization of the proposed NDFD model is capable of suppressing the numerical errors (including the numerical dispersion and artificial oscillation) to a negligible level.

In addition to the abovementioned theoretical criteria, (semi-) analytic solutions are used as benchmark solutions to provide further validation of the NDFD model. On the one hand, a Laplace domain analytical solution of Moench and Ogata (1981) for simulating divergent solute transport with a continuous constant concentration in a completely penetrating well is employed and the analytical expression of Moench and Ogata (1981) for such a problem is.

$$C_D(r,z) = C_{\text{inj}}\exp\left(\frac{r_D - r_{wD}}{2}\right)\frac{A_i\left[p^{\frac{1}{3}}\left(r_D + \frac{1}{4p}\right)\right]}{A_i\left[p^{\frac{1}{3}}\left(r_{wD} + \frac{1}{4p}\right)p\right]} \tag{3.730}$$

in which $r_D = r/\alpha_L, r_{wD} = r_w/\alpha_L$, and $C_D(r,z) = C(r,z)2\pi b\theta \alpha_L^2/Q$. The length of the injection screen in the FD model is extended to the aquifer thickness for the purpose of implementing the verification. The aquifer thickness is assumed to be unity, and the values of other parameters are listed in Table 3.21. We obtain both results of the analytical solution of Moench and Ogata (1981) and the FD solution, as Figure 3.90 indicates. It can be found that the BTCs observed at r = 1, 2, and 3 m in both models fit very well with each other, which proves that the spatial and temporal discretization schemes of the NDFD model are capable of reducing the numerical errors to a negligible level.

We adopt the newly proposed NDFD method to describe the transport of conservative solutes for a special case. The results of the developed NDFD method are compared with those of the LTFD method of Chen *et al.* (2010, 2011a). In this comparison, 10 kg conservative solute is continuously injected with an injection duration of 100 min and the other transport parameters used for both NDFD

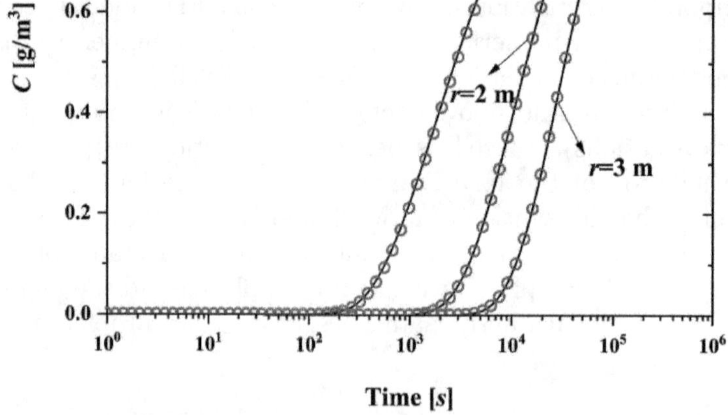

Figure 3.90. Verification of the NDFD solution with the analytical solution of Moench and Ogata (1981).

and LTFD methods are listed in Table 3.21. The BTCs of conservative solute at the top of the extraction screened interval are shown in Figure 3.91, which indicates that the solution of our NDFD method agrees well with that of the LTFD method of Chen *et al.* (2010, 2011a). Such a good agreement highlights the reliability of defining the spatially discretized advection-dispersion expressions for each node. In the baseline case, the plumes of solute at $t = 200$, 500, 1,000, and 2,000 min are represented in Figure 3.92. After cessation of solute injection at $t = 100$ min. the solute near the injection screened interval has no follow-up supplement and is flushed out to a relatively large scale. Besides, a large plume means locally diluted concentrations due to the total constant mass of solute in the system.

3.5.5.3 *Reactive transport in a VCW-induced flow field*

(a) Baseline simulation

To present the degradation performance of VCW, we conducted reactive transport modeling exercises based on Eqs. (3.717)–(3.727b).

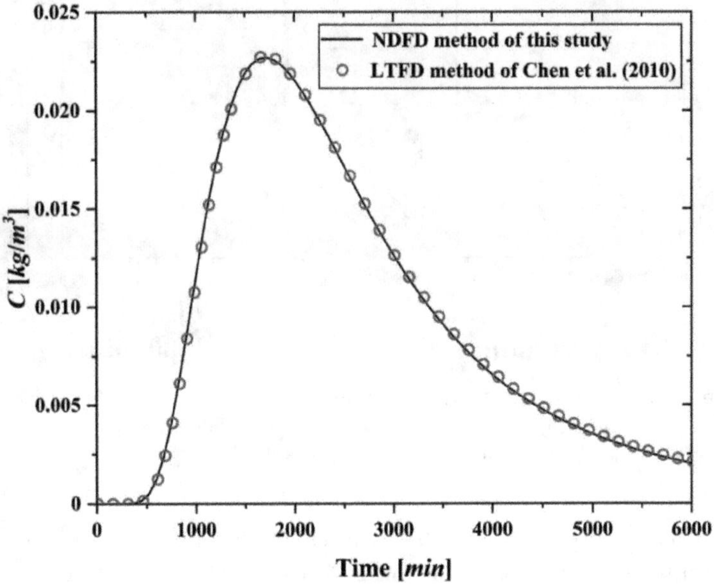

Figure 3.91. Verification of the NDFD method with the LTFD method employed by Chen *et al.* (2010) in the conservative solute transport in VCW-induced flow field.

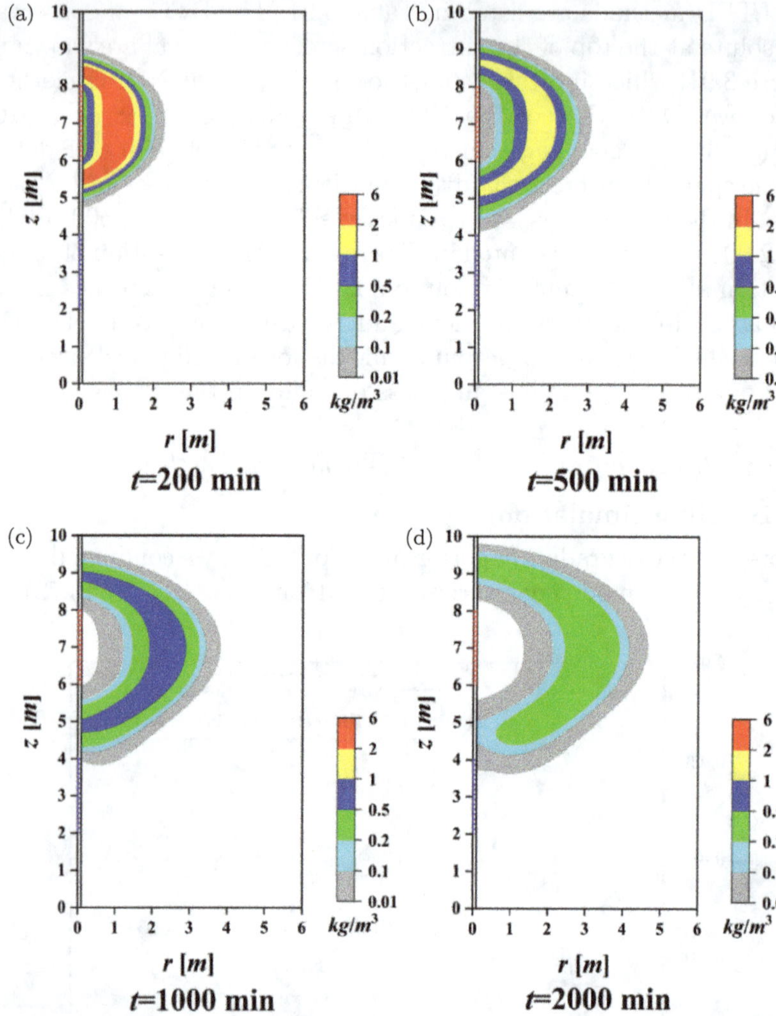

Figure 3.92. The concentration contours of conservative solute developed by the NDFD method in the baseline case at (a) t = 200 min; (b) t = 500 min; (c) t = 1000 min; (d) t = 2000 min.

Tables 3.21 and 3.22 list the values of parameters in the baseline simulation, where Table 3.21 contains the parameters related to model configuration, flow field, and solute transport properties, and Table 3.22 includes the parameters involving the source and sink terms such as aerobic redox reaction and mass transfer between

NAPL and dissolved phase. Figure 3.93 shows the removal ratio of contaminants (ED) in the baseline simulation, in which 8 kg hydrogen peroxide is injected during $t_0 = 1\,\text{h}$ at the beginning, and then the remediation performance at different times was characterized. The removal ratio indicates the mass proportion of NAPL contaminant

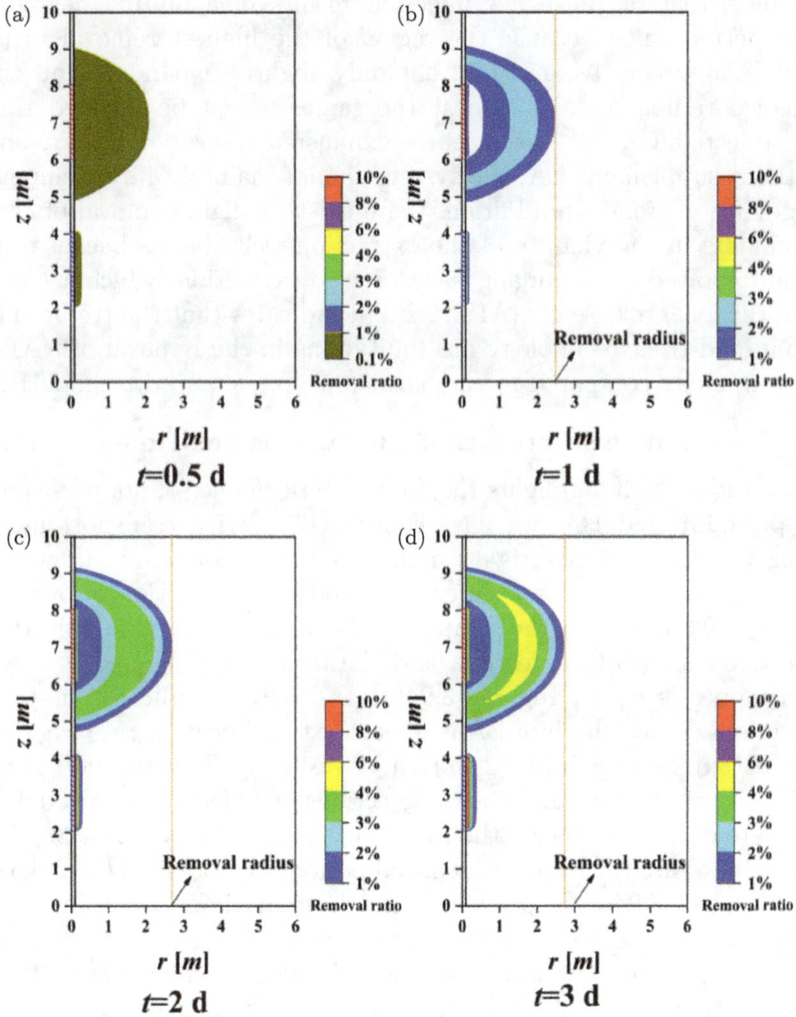

Figure 3.93. The spatial distribution of removal ratio calculated by reactive transport model in the baseline case at (a) $t = 0.5$ days; (b) $t = 1$ day; (c) $t = 2$ days; and (d) $t = 3$ days.

degraded. It can be found that the injected EA travels in the VCW-induced flow field and reacts biochemically with the dissolved ED resident in the aquifer. Over time, the removal ratio increases. The removal radius, which measures the scale of the region with a removal ratio higher than 1%, as defined in Figure 3.93, increases as well. However, comparing Figures 3.93(c) and 3.93(d) indicates that a nonpersistent or pulse EA injection results in a limited degradation performance because the region of the highest removal ratio, that is, between 4% and 6%, has only slightly expanded, and the removal radius remains almost the same. It can be inferred that the injected EA has been nearly exhausted after $t = 2$ day, and without supplement EA, the system cannot maintain its continuous degradation trend. In addition, there exists a slight removal of contaminants in the vicinity of the extraction well. Thesis because the local dissolved contaminant leaves due to extraction, which accelerates the local release of NAPL. Also, it indicates that the traditional pump-and-treat technology has limitations in the removal of NAPL contaminants compared to the biochemically enhanced technology.

(b) Injection with interval of time versus one-time injection

Since Figure 3.93 highlights the limited performance of nonpersistent or pulse-injected EA, we have conducted reactive transport modeling exercises to investigate a modified injection mode in which there exists an equal time interval between two injection durations. Figure 3.94 represents the comparison between time interval injection and one-time (pulse)injection, and the total mass injected in these two modes is equal. Figures 3.94(a) and 3.94(c) indicate one-time injections, in which all of EA is injected at an injection rate M_{EA}/t_0 of 8 kg/h during the initial 2 and 3 h, respectively. Figures 3.94(b) and 3.94(d) present the time interval injections in which EA is injected at the rate of 8 kg/h during the first hour of each day. Figures 3.94(a) and 3.94(b) are the results of removal performance at time $t = 2$ days, and Figures 3.94(c) and 3.94(d) are those at time $t = 3$ days. As Figure 3.94 indicates, using the same amount of EA, the time interval injection mode results in a generally higher removal ratio than the one-time injection mode, and the advantage of the time interval injection is increasingly notable as time goes on. Nevertheless, due to the constant mass of EA, the removal radius of the time interval injection is smaller than that of the one-time injection. That is,

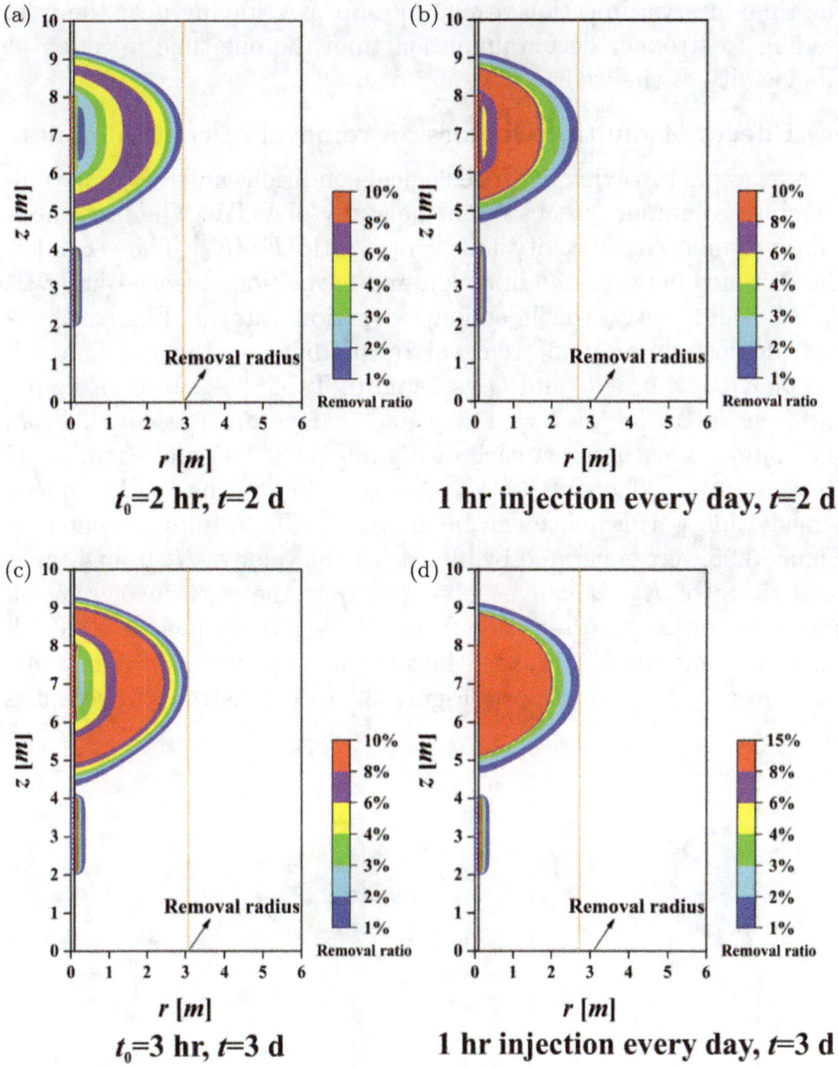

Figure 3.94. The comparison of removal performance between two injection modes, (a) and (c) indicate one-time injection mode, in which 16 and 24 kg reagent are injected within the first 2 and 3 h, respectively; (b) and (d) indicate time interval injection mode, in which 8 kg reagent is injected in the initial 1 h every day. Panels (a) and (b) are the removal ratio results at $t = 2$ days; (c) and (d) are the removal ratio results at $t = 3$ days.

the time interval injection results in more EA staying near the well, leading to stronger decontamination than the one-time injection at the vicinity of the well.

(c) Effects of multiple factors on removal of contaminants

Three factors involving hydrogeological conditions and design parameters are commonly discussed in the study of VCW. The first one is the hydraulic conductivity anisotropy ratio K_r/K_z. The second is the distance between the injection and extraction screened intervals $d_i - l_e$, and lastly, the injection/extraction rate Q. Figures 3.88–3.90 depict contaminant removal by modulating the baseline values of K_r/K_z, $d_i - l_e$ and Q as listed in Table 3.21. In each figure, the removal ratios at $t = 1$ day and 2 days are presented. From the comparisons between each of Figures 3.88–3.90 and Figure 3.93 (baseline), the effects of K_r/K_z, $d_i - l_e$ and Q on the VCW-induced remediation performance can be analyzed. The results presented in Figure 3.95 were generated by increasing the value of K_r from 4 times to 9 times of K_z. It can be observed that the pattern of removal region becomes "shorter" and "fatter" as K_r/K_z increases, which indicates that the EA injected into the aquifer prefers radial transport over vertical transport. Figure 3.95 demonstrates the results

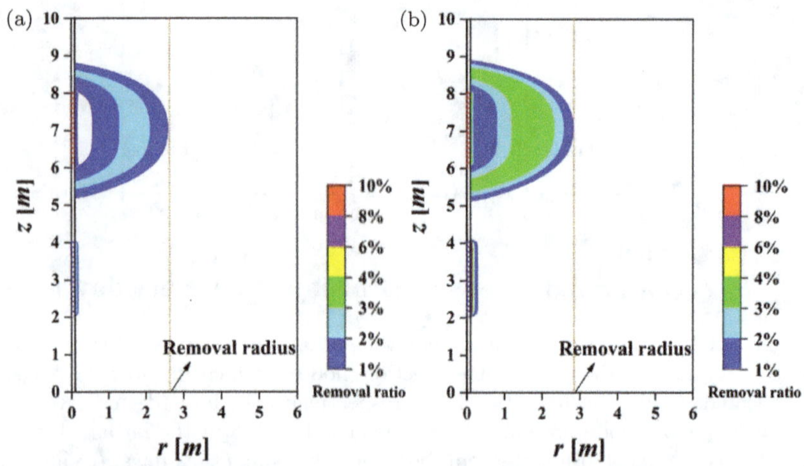

Figure 3.95. The spatial distribution of removal ratio at (a) $t = 1$ day and (b) $t = 2$ days in the case of increasing K_r/K_z to 9.

of the removal ratio by moving down the injection screened interval from 6 m $\leq z \leq$ 8 m to 5.5 m $\leq z \leq$ 7.5 m while moving up the extraction screened interval from 2 m $\leq z \leq$ 4 m to 2.5 m $\leq z \leq$ 4.5 m, thus, the value of $d_i - l_e$ decreases from 2 m in the baseline case to 1 m. The comparison between Figures 3.93 and 3.96 shows that a closer distance between injection and extraction screened intervals results in a very slight difference in removal radius, but as the red circle in Figure 3.96 indicates, a certain region closer to the extraction screened interval shows lower removal ratio compared with that of Figure 3.93. This is attributed to the closer distance between injection and extraction screened intervals results in the more unreacted reagent extracted. As for the effects of the pumping/extraction rate Q on contaminant removal, Figure 3.97 depicts the removal results by doubling the value of Q. It can be found that increasing Q results in a generally larger removal radius; however, the comparison between Figures 3.93(c) and 3.95(b) indicates that the local degradation performance with a smaller Q is better than that, with a greater Q.

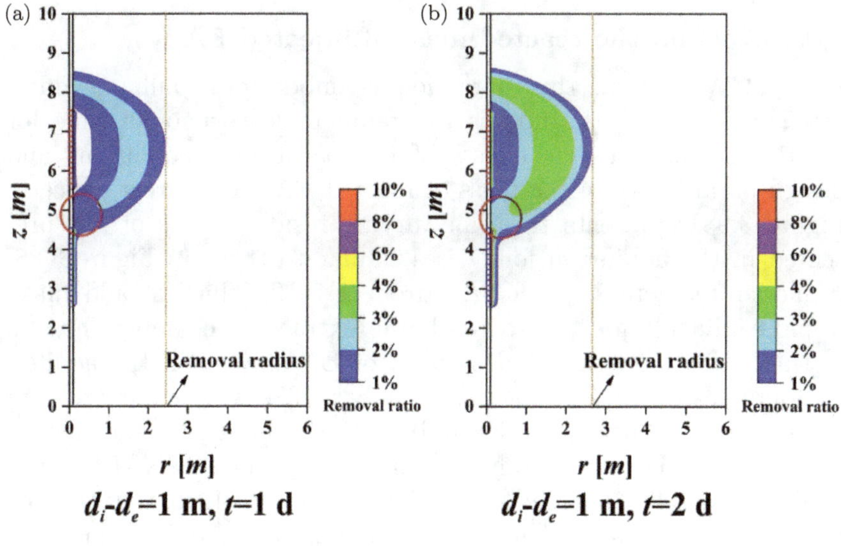

Figure 3.96. The spatial distribution of removal ratio at (a) $t = 1$ day and (b) $t = 2$ days in the case of moving down the injection screened interval from 6 m $\leq z \leq$ 8 m to 5.5 m $\leq z \leq$ 7.5 m and moving up the extraction screened interval from 2 m $\leq z \leq$ 4 m to 2.5 m $\leq z \leq 4.5i$, that is, $d_i - l_e$ decreases from 2 to 1 m.

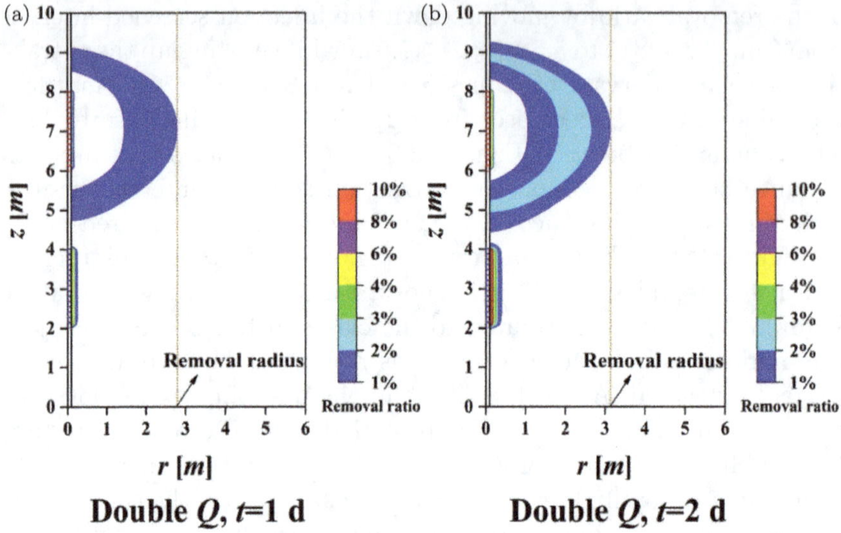

Figure 3.97. The spatial distribution of removal ratio at (a) $t = 1$ day and (b) $t = 2$ days in the case of double Q.

(d) Effects on the reacted mass of injected EA

In a VCW system, the extraction chamber may pull out unreacted EA because the flow is too rapid to leave enough time for reaction. Therefore, the effects of hydrogeological conditions and design parameters on the mass of reacted EA are of great concern. Figure 3.98 represents the time-varying residual mass of EA presented in the aquifer in four cases similar to those in Figure 3.88. A moderate increase of K_r/K_z, from 4 to 9, which should make it more difficult for EA to reach the extraction chamber, has no distinct effects on the total amount of EA staying in the aquifer. This implies that the extraction chamber collects approximately an equal amount of unreacted EA in both the baseline case and Case 1. Figure 3.99 indicates the time-varying removal mass of NAPL in the aquifer and tells that the removal mass of NAPL in the baseline case almost equals to that in Case 1. It is worth mentioning that Figure 3.98 represents the exhaustion of EA after the first day, but Figure 3.93 indicates the continued removal of NAPL until the second day. This is because the low-concentration mobile contaminant (corresponding to NAPL of high removal ratio) upstream migrates

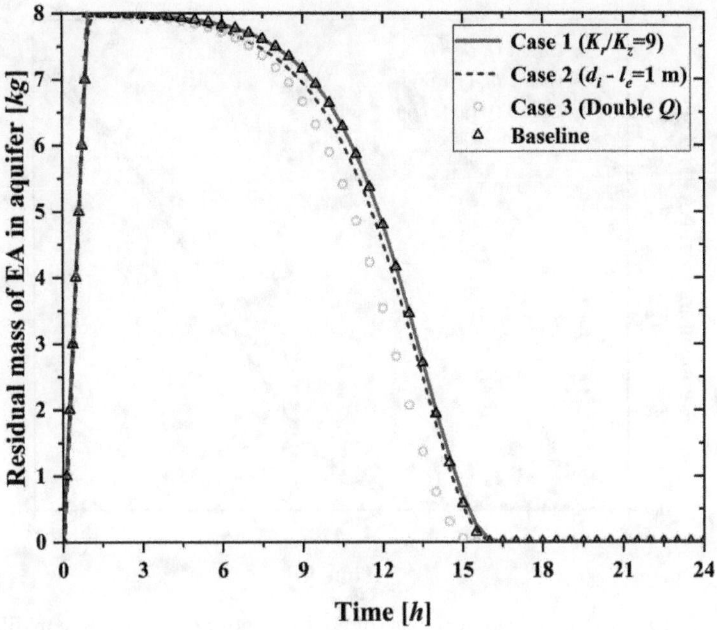

Figure 3.98. Figure of residual mass of EA in aquifer versus time in different cases.

downstream through advection-dispersion, which physically expands the removal area of NAPL with limited biochemical assistance. As for Case 2, with a closer distance between injection and extraction screened intervals, the residual mass of EA decreases faster than that in the baseline case, and the removal mass of NAP increases less than that of the baseline case. This demonstrates that a smaller $d_i - d_l$ leads to more extraction of unreacted EA, causing less removal mass of NAPL through the redox reaction. As for Case 3, though its residual mass of EA decreases most rapidly, its removal mass of NAPL also increases most rapidly as well.

It is critical to figure out the proportion of the reacted mass of injected EA with respect to such a microbially enhanced VCW remediation system. Figure 3.100 indicates the time-varying reacted mass of injected EA in the abovementioned three cases; in our NDFD method, the reacted mass can be calculated by the integral of reacted mass in each FV. Doubling Q results in the most reacted mass of EA, which is accordant to the broader removal region in Figure 3.96.

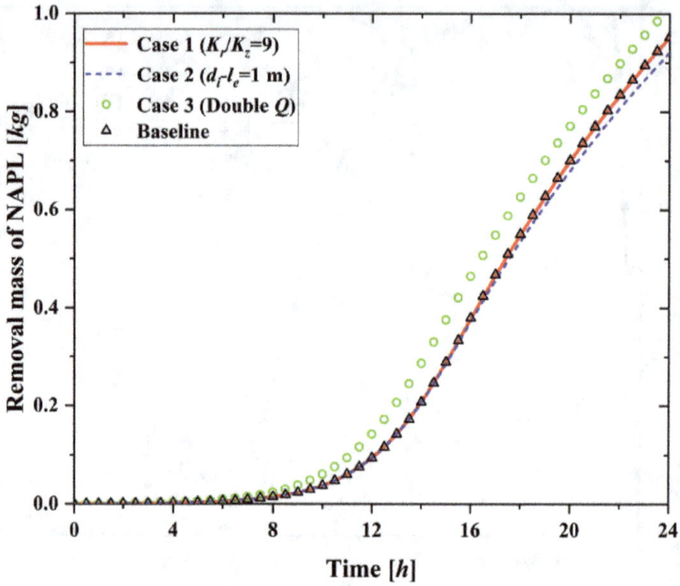

Figure 3.99. Figure of removal mass of NAPL in aquifer versus time in different cases.

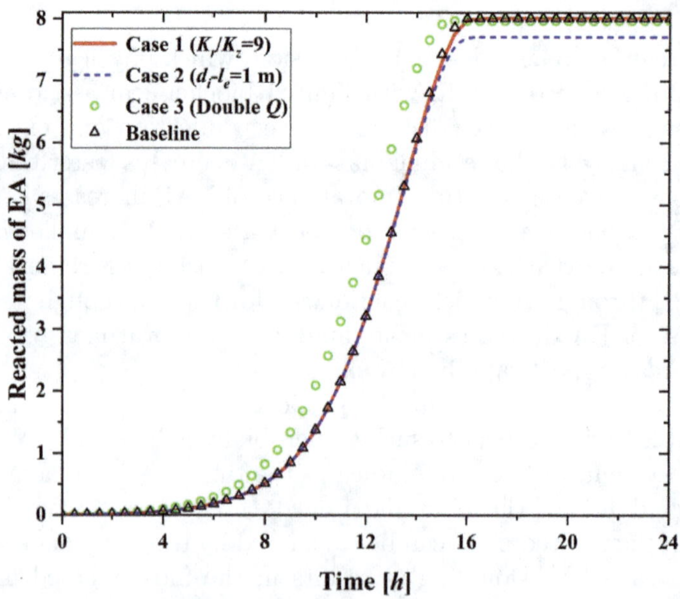

Figure 3.100. Figure of reacted mass of EA in aquifer versus time in different cases.

Note that a greater Q may extract more unreacted EA, but obviously, that does not happen because more EA has been consumed before it reaches the extraction chamber. Nevertheless, a smaller $d_i - d_l$ actually decreases the reacted mass of EA; this is because a shorter path makes more EA reach the extraction chamber without reaction. As for Case 1, the change of K_r/K_z, from 4 to 9 has an approximately equal reacted mass of EA, which means that except for the slight difference in the shape of the removal region, such a change does not affect much on the removal performance.

(4) Injection mode

In a practical VCW-induced remediation project, the total mass of EA is often limited, which means that the number of electrons transferred from ED to EA due to microbial redox reaction (which contributes to the degradation of contaminant) is limited as well. The one-time injection mode gives EA more travel time in the system than the time interval injection mode, because each EA group departing from the injection chamber is always one time interval later than the last group. Therefore, the one-time injection mode carries more EA to the outer region and leaves less EA in the vicinity of the well, which means that fewer electrons of ED near the well can be transferred to EA. The difference is that the time interval injection mode greatly increases the proportion of EA consumed in the vicinity and results in less electron transfer than the one-time injection mode in places far from the well. In a radially axisymmetric system, the volume of the contaminated region is proportional to the radius squared, which means that increasing increment r of removal radius leads to much more EA in places far from the well than in the vicinity of the well. Although the one-time injection mode can generally cover a larger region and the time interval injection mode has better removal performance at a local site, the larger region may not seem that significant, but the better removal performance is certainly more desirable. This implies that multiple wells with the time interval injection mode may be optimal because each well is capable of contaminant removal in its surrounding region.

3.5.5.4 *Hydrogeological conditions and design parameter*

The extraction chamber in VCW plays two significant roles: one is to pull out substances that are not environmentally friendly, and the other is to provide resistance to impede the upturned and especially the radial spread of EA plume due to injection. That is, the injection chamber is a source, and the extraction chamber is a sink in the system. The strength of the extraction sink to the EA plume weakens considerably if the hydraulic conductivity anisotropy ratio K_r/K_z is relatively large, thus reducing the contaminant removal efficiency of VCW. If this is true, then the single VCW system probably should be substituted by a well-to-well circulation system. In addition, design parameters should be optimized with respect to specific hydrogeological conditions. For example, if K_r/K_z is disadvantageous for the vertical transport of reagents, so adjusting the placement of screened intervals may help improve the recovery efficiency. If both K_r and K_z are relatively low, the corresponding Q should be low as well to avoid drying up the extraction chamber.

In the case of relatively high K_r and K_z, such as the baseline case, a greater Q results in the faster spread of reagent, thus ending in weaker local degradation. Suppose, the functional microbes have strong degradation capability. In that case, a moderate increase of cap Q can promote the degradation performance. Still, an excessive increase of Q may make the microbial reaction unable to follow the solute transport, resulting in a large amount of unreacted EA and overall poor performance of the VCW system. Decreasing the distance between the injection and extraction screened intervals can actually help adjust the degradation location, but a closer distance leads to a greater unreacted mass of EA. It is suggested that decreasing the distance between the injection and extraction screened intervals is coupled with a moderately smaller Q, because a shorter path associated with a smaller velocity will leave enough time for reaction.

3.5.5.5 *Particle-tracking method and the NDFD method*

The NDFD method yields a more accurate prediction of removal performance than the particle-tracking method because it describes multispecies reactive transport, which incorporates flow field, solute transport and microbially redox reaction simultaneously. Additionally, as the reactive transport results indicate, the extremely high recovery ratio obtained from a low $d_i - d_e$ and high Q may not be equivalent to a high removal ratio. Nevertheless, the usefulness of the particle-tracking method should not be overlooked in the preliminary screening of VCW operation. After all, the particle-tracking method only needs the basic information of *in situ* hydrogeological conditions and does not need the biochemical parameters, longitudinal and transverse dispersivities, as well as in situ contaminated situations. The time spent for a specific VCW design to implement one circulation of water is often of great concern. If the recovery ratio approaches 100%, it means that almost all the water departing from the injection chamber has been collected by the extraction chamber, then the total time spent can be properly calculated with the particle-tracking method. After a given time, when a number of circulations have been completed, if the recovery ratio drops to zero or too low, it can be inferred that the extraction chamber is weakening in recovering the injected EA. Consequently, a considerable portion of the injected EA may escape the capture zone of the VCW system (Philip and Walter, 1992) and become secondary pollution, which may be carried away from the source zone by the ambient groundwater flow. For those VCW facilities installed in aquifers without considerable permeability, such as the aquifer composed of fine-grained medium, a primary concern is whether the flow field can be successfully circulated or not. From Table 3.23, a proper suite of design parameters to keep VCW circulating flow in the preliminary stage can be determined. In addition, quantitative expressions like Table 3.23 are helpful and intuitive in the optimization of choosing design parameters, especially when one needs to find a substitution of design parameters due to various practical limitations.

On the other hand, it should be acknowledged that the multi-species reactive transport model based on the FD method is more helpful as long as detailed biochemical parameters, dispersivities and contaminated situations are known in advance. This is because it shows the spatiotemporal removal performance of contamination and covers all parameters involved in the remediation design. Additionally, the development of the NDFD method is of great necessity. Commonly, reagent injection and contaminant extraction take place simultaneously during remediation work in the field, in which the flow field is influenced by two hydrologic stresses with opposite driving forces and the flow directions can be bent nearly 180 degrees in some regions. This phenomenon also happens frequently near the stagnation point induced by an injection or pumping flow field superimposed by an ambient groundwater flow field. In this kind of flow field, great numerical errors can be generated if all nodes share the same upstream direction in the conventional FD method. Instead, the proposed NDFD is able to account for such a drastically bent flow field for the purpose of characterizing solute distribution because the flow direction between the neighboring nodes is strictly distinguished, ensuring an appropriate solute transfer from the upstream to the downstream node. Therefore, the NDFD method is practical and worth promoting in biologically enhanced *in situ* remediation technologies.

3.5.5.6 *Differences between the NDFD and LTFD method*

The LTFD method of Chen *et al.* (2010, 2011a) eliminated the temporal derivative in the governing equation through the Laplace transform, which yields Laplace space concentration at each discretized node. Unlike the LTFD method, the conventional FD method, which associates spatial derivative with temporal one, generates time-varying data of spatial concentration. It is known that the advection-dispersion governing equation derives from the concentration continuity equation for a REV (Bear and Cheng, 2010; Rumynin, 2011); therefore, the concentration variation of each node is attributed to advection-dispersion transfer between neighboring nodes within a time step. As for the conventional FD scheme, distinguishing "upstream node" and "downstream node" is crucial for accurate calculation of the advection fluxes between nodes because the advection direction must be accordant to the flow direction, that is, from upstream to downstream. However, the VCW-induced flow field involves eight different combinations of inflow and outflow directions in both radial and vertical directions with respect to a node (Figure 3.86). If describing VCW-induced solute transport using the commonly used FD technique in Wang *et al.* (2017) and Zhu *et al.* (2019), in which the flow is purely radial, and all nodes have the same inflow and outflow directions, a great number of numerical errors could happen. This is why we developed the NDFD method, which specifies the advection-dispersion flux into and out from each node. Though the LTFD method does not need to fit time step to spatial step, which is advantageous in computational efficiency compared to the NDFD method, the LTFD technique produced by the product of matrix and vector has a strictly closed mathematical structure and has limited chances to involve multispecies reaction. Unlike the LTFD method, the NDFD method still belongs to the conventional FD method. Thus, it is feasible to incorporate the multispecies reaction terms and still be able to depict the microbially enhanced remediation through VCW.

3.5.6 *Limitations*

The model of Zhu *et al.* (2020) elucidates the effects of hydrogeological conditions and design parameters on VCW performance for *in situ* aquifer remediation. The modeling results in this paper involve only several days. For a long-running VCW system, the biomass probably leads to bioclogging, which can remarkably reduce the porosity and even affect the hydraulic conductivities and dispersivities of the medium. In the model of Zhu *et al.* (2020). They assume that the clogging effects are not distinct during the short time of concern. In future work, we would like to involve the biologging effects in a long-term reactive transport investigation for VCW.

In addition, the model of Zhu *et al.* (2020) is based on a simplified conceptual model including homogeneous and geometrically regular aquifers as well as uniformly distributed contaminated situations. This simplification makes it difficult for the models to predict the VCW performance under more complex conditions. However, the simple model is still indispensable because of the following reasons. First, a complex transport problem involving heterogeneous media, irregular geometry, a drastically bent flow field and a multi-species reactive transport process has too many controlling factors for the system of concern. Therefore, it is advisable to simplify the problem by prioritizing those controlling factors to put more focus on understanding important issues of concern. In the model of Zhu *et al.* (2020), such an important issue of concern is the multi-species reactive transport process involving a drastically bent flow field generated by VCW, which has received very little attention before. Therefore, more focus was put on fundamentals rather than more variable factors, which are very relative from location to location, such as heterogeneity and complex geometry, especially since they have received a great deal of attention over the past decades. Second, the results of Zhu *et al.* (2020) can serve as a stepstone (or a pilot investigation for quick screening purposes instead of developing a detailed geological model) to further include other controlling factors, such as heterogeneity and complex geometry. Third, the "simple" model is directly applicable for interpreting the results of laboratory experiments with well-controlled conditions such as a homogeneous medium and a simple geometry. The "simple" model of this section can also provide useful guidelines for field applications when detailed information about aquifer heterogeneity and geometry is not available, which is not uncommon in dealing with real-world problems. Fourth, the model of Zhu *et al.* (2020) based on the homogeneity assumption can be used as an ensemble average approximation of transport in stationary heterogeneous media, and it can also be utilized to test results obtained from stochastic theories for stationary heterogeneous systems which usually has to be dealt with using Monte Carlo simulations or other stochastic methods.

The model of Zhu *et al.* (2020) assumes hydrogen peroxide to migrate in a porous medium and decompose into oxygen simultaneously, i.e., all the oxygen generated by H_2O_2 can be totally dissolved. This assumption is rational to some extent because with the spread of H_2O_2 concentration over the aquifer, while the local porous H_2O_2 concentration will not be extremely high, making the oxygen produced completely transformed into DO, which has a solubility of around 8 mg/L under normal circumstances. Besides, a time-limited injection of H_2O_2 could promote the oxygen concentration in an experimental aquifer to 23 mg/L as reported in Trötschler *et al.* (2005), which proves that in a confined aquifer environment, the oxygen concentration may be several times higher than the basic solubility of DO. However, during the injection of reagent, H_2O_2 accumulates at the vicinity of the well and the local H_2O_2 concentration is relatively high, which has a risk of generating an amount of oxygen that may not be completely dissolved. Fortunately, the area of H_2O_2 accumulation is very limited because the volume of porous medium is proportional to radius squared and the circulating water will flush the accumulated H_2O_2 downstream soon after injection. Therefore, it is reasonable to assume that this concern overall has limited effects on our results. After all, despite the limitation of oxygen solubility, the newly proposed NDFD method by Zhu *et al.* (2020) proves to be a universal method to describe multi-species reactive transport in the VCW system, which covers much more than simple aerobic biodegradation.

References

Abgaze, T.A., Sharma, P.K., 2015. Solute transport through porous media with scale-dependent dispersion and variable mass transfer coefficient. *ISH Journal of Hydraulic Engineering*, 21(3): 298–311.

Abramowitz, M., Stegun, I.A., 1968. Handbook of mathematical functions with formulas, graphs, and mathematical tables, 55. US Government Printing Office.

Adams, E.E., Gelhar, L.W., 1992. Field study of dispersion in a heterogeneous aquifer: 2. Spatial moments analysis. *Water Resources Research*, 28(12): 3293–3307.

Ahlstrom, S., Foote, H., Arnett, R., Cole, C., Serne, R., 1977. Multicomponent mass transport model: theory and numerical implementation (discrete-parcel-random-walk version), Battelle Pacific Northwest Labs., Richland, WA (United States).

Allmon, W. *et al.*, 1999. Groundwater circulating well technology assessment, Naval Research Laboratory, NRL/PU/6115-99-384, Washington, DC.

Anderson, M.P., Woessner, W.W., Hunt, R.J., 2015. *Applied Groundwater Modeling: Simulation of Flow and Advective Transport.* Academic Press.

Ashino, R., Nagase, M., Vaillancourt, R., 2000. Behind and beyond the MATLAB ODE suite. *Computers & Mathematics with Applications*, 40(4–5): 491–512.

Assayag, N., Matter, J., Ader, M., Goldberg, D., Agrinier, P., 2009. Water-rock interactions during a CO2 injection field-test: Implications on host rock dissolution and alteration effects. *Chemical Geology*, 265(1–2): 227–235.

Aubeneau, A., Hanrahan, B., Bolster, D., Tank, J.L., 2014. Substrate size and heterogeneity control anomalous transport in small streams. *Geophysical Research Letters*, 41(23): 8335–8341.

Aziz, K., 1979. Petroleum reservoir simulation. *Applied Science Publishers*, 476.

Babey, T., Dreuzy, J.-R.d., Casenave, C., 2015. Multi-rate mass transfer (MRMT) models for general diffusive porosity structures. *Advances in Water Resources*, 76(2015): 146–156.

Barbour, S.L., Hendry, M.J., Wassenaar, L.I., 2012. In situ experiment to determine advective-diffusive controls on solute transport in a clay-rich aquitard. *Journal of Contaminant Hydrology*, 131(1–4): 79–88.

Bassett, R. *et al.*, 1981. *Preliminary Data From a Series of Artificial Recharge Experiments at Stanton.* US Geological Survey, Texas, pp. 81–149.

Batu, V., 1998. *Aquifer Hydraulics: A Comprehensive Guide to Hydrogeologic Data Analysis.* John Wiley & Sons.

Bear, J., 1961. Some experiments in dispersion. *Journal of Geophysical Research*, 66(8): 2455–2467.

Bear, J., 2012. *Hydraulics of Groundwater.* Courier Corporation.

Bear, J., 2013. *Dynamics of Fluids in Porous Media.* Courier Corporation.

Bear, J., Cheng, A.H.-D., 2010. *Modeling groundwater Flow and Contaminant Transport.* Springer, p. 23.

Bekins, B.A., Warren, E., Godsy, E.M., 1998. A comparison of zero-order, first-order, and monod biotransformation models. *Groundwater*, 36(2): 261–268.

Benson, D.A., Tadjeran, C., Meerschaert, M.M., Farnham, I., Pohll, G., 2004. Radial fractional-order dispersion through fractured rock. *Water Resources Research*, 40(12).

Bentley, H., Walter, G., 1983. Two-well recirculating tracer tests at H-2: Waste Isolation Pilot Plant (WIPP), southwest New Mexico, draft paper. Hydro Geochem., Inc., Tucson, Ariz.

Berkowitz, B., Scher, H., Silliman, S.E., 2000. Anomalous transport in laboratory-scale, heterogeneous porous media. *Water Resources Research*, 36(1): 149–158.

Bharati, V.K., Singh, V.P., Sanskrityayn, A., Kumar, N., 2019. Analytical solution for solute transport from a pulse point source along a medium having concave/convex spatial dispersivity within fractal and Euclidean framework. *Journal of Earth System Science*, 128.

Bierschenk, W.H., 1959. Aquifer characteristics and ground-water movement at Hanford, General Electric Co. Hanford Atomic Products Operation, Richland, Wash.

Blanford, W.J. *et al.*, 2001. Cyclodextrin-enhanced vertical flushing of a trichloroethene contaminated aquifer. *Ground Water Monitoring and Remediation*, 21(1): 58–66.

Bodin, J. *et al.*, 2012. Predictive modelling of hydraulic head responses to dipole flow experiments in a fractured/karstified limestone aquifer: Insights from a comparison of five modelling approaches to real-field experiments. *Journal of Hydrology*, 454: 82–100.

Bredehoeft, J.D., Pinder, G.F., 1973. Mass transport in flowing groundwater. *Water Resources Research*, 9(1): 194–210.

Brusseau, M.L., 1992. Nonequilibrium transport of organic chemicals: The impact of pore-water velocity. *Journal of Contaminant Hydrology*, 9(4): 353–368.

Burnett, R., Frind, E., 1987. An alternating direction Galerkin technique for simulation of groundwater contaminant transport in three dimensions, 2, dimensionality effects. *Water Resources Research*, 23(4): 695–705.

Carrera, J. *et al.*, 1998. On matrix diffusion: Formulations, solution methods and qualitative effects. *Hydrogeology Journal*, 6: 178–190.

Cassiani, G., Kabala, Z.J., 1998. Hydraulics of a partially penetrating well: Solution to a mixed-type boundary value problem via dual integral equations. *Journal of Hydrology*, 211(1–4): 100–111.

Cassiani, G., Burbery, L.F., Giustiniani, M., 2005. A note on *in situ* estimates of sorption using push-pull tests. *Water Resources Research*, 41(3).

Chang, C.C., Chen, C.S., 2002. An integral transform approach for a mixed boundary problem involving a flowing partially penetrating well with infinitesimal well skin. *Water Resources Research*, 38(6).

Chang, C.-C., Chen, C.-S., 2003. A flowing partially penetrating well in a finite-thickness aquifer: A mixed-type initial boundary value problem. *Journal of Hydrology*, 271(1): 101–118.

Chen, C.S., 1985. Analytical and approximate solutions to radial dispersion from an injection well to a geological unit with simultaneous diffusion into adjacent strata. *Water Resources Research*, 21(8): 1069–1076.

Chen, C.S., 1986. Solutions for radionuclide transport from an injection well into a single fracture in a porous formation. *Water Resources Research*, 22(4): 508–518.

Chen, C.S., 1991. Semianalytical solutions for radial dispersion in a three-layer leaky aquifer system. *Groundwater*, 29(5): 663–670.

Chen, C.-S., 1987. Analytical solutions for radial dispersion with cauchy boundary at injection well. *Water Resources Research*, 23(7): 1217–1224.

Chen, C.-S., Woodside, G.D., 1988. Analytical solution for aquifer decontamination by pumping. *Water Resources Research*, 24(8): 1329–1338.

Chen, J.S., 2010. Analytical model for fully three-dimensional radial dispersion in a finite-thickness aquifer. *Hydrology Processing*, 24(7): 934–945.

Chen, J.-S., Chen, C.-S., Chen, C.Y., 2007. Analysis of solute transport in a divergent flow tracer test with scale-dependent dispersion. *Hydrological Processes*, 21(18): 2526–2536.

Chen, J.S., Jang, C.S., Cheng, C.T., Liu, C.W., 2010. Conservative solute approximation to the transport of a remedial reagent in a vertical circulation flow field. *Journal of Hydrology*, 390(3–4): 155–168.

Chen, J.S., Liu, C.W., Chan, Y.C., Ni, C.F., Kao, K.K., 2011a. Effect of transverse dispersion on solute transport in a vertical dipole flow test with a tracer. *Journal of Hydrology*, 402(3–4): 206–216.

Chen, J.-S., Liu, Y.-H., Liang, C.-P., Liu, C.-W., Lin, C.-W., 2011b. Exact analytical solutions for two-dimensional advection–dispersion equation in cylindrical coordinates subject to third-type inlet boundary condition. *Advances in Water Resources*, 34(3): 365–374.

Chen, K., Hongbin, Z., Yang, Q., 2017. Fractional models simulating non-fickian behavior in four-stage single-well push-pull tests. *Water Resources Research*, 53: 9528–9545.

Chen, K., Zhan, H., Zhou, R., 2016. Subsurface solute transport with one-, two-, and three-dimensional arbitrary shape sources. *Journal of Contaminant Hydrology*, 190(2016): 44–57.

Chen, Y.J., Yeh, H.D., Chang, K.J., 2012. A mathematical solution and analysis of contaminant transport in a radial two-zone confined aquifer. *Journal of Contaminant Hydrology*, 138–139: 75–82.

Cihan, A., Tyner, J.S., 2011. 2-D radial analytical solutions for solute transport in a dual-porosity medium. *Water Resources Research*, 47(4).

Cirpka, O.A., Kitanidis, P.K., 2000. Impact of biomass-decay terms on the simulation of pulsed bioremediation. *Ground Water*, 38(2): 254–263.

Cirpka, O.A., Kitanidis, P.K., 2001. Travel-time based model of bioremediation using circulation wells. *Groundwater*, 39(3): 422–432.

Claassen, H.C., Cordes, E.H., 1975. Two-well recirculating tracer test in fractured carbonate rock, *Nevada. Hydrol. Science Bulletin (United Kingdom)*, 20(3).

Cleary, R.W., Adrian, D.D., 1973. Analytical solution of the convective-dispersive equation for cation adsorption in soils. *Soil Science Society of America Journal*, 37(2): 197–199.

Cole, J.A., 1972. Some interpretations of dispersion measurements in aquifers, *Groundwater Pollution in Europe*, 86–95, Water Research Association, Reading, England.

Crane, M.J., Blunt, M.J., 1999. Streamline-based simulation of solute transport. *Water Resources Research*, 35(10): 3061–3078.

Cunningham, J.A., Hoelen, T.P., Hopkins, G.D., Lebrón, C.A., Reinhard, M., 2004. Hydraulics of recirculating well pairs for ground water remediation. *Ground Water*, 42(6): 880–889.

Dagan, G., 1988. Time-dependent macrodispersion for solute transport in anisotropic heterogeneous aquifers. *Water Resources Research*, 24(9): 1491–1500.

Danckwerts, P.V., 1953. Continuous flow systems. *Chemical Engineering Science*, 2(1): 1–13.

Danckwerts, P.V., 1995. Continuous flow systems. Distribution of residence times. *Chemical Engineering Science*, 50(24): 3857–3866.

Daniels, W., 1981. Laboratory and Field Studies Related to the Radionuclide Migration Project, October 1, 1979—September 30, 1980. Los Alamos Nat onal Laboratory Report LA-8670-PR (February 1981).

Davis, B.M., Istok, J.D., Semprini, L., 2005. Numerical simulations of radon as an *in situ* partitioning tracer for quantifying NAPL contamination using push-pull tests. *Journal of Contaminant Hydrology*, 78(1–2): 87–103.

Davis, M.E., Davis, R.J., 2002. *Fundamentals of Chemical Reaction Engineering.* Courier Corporation.

Di Dato, M., Fiori, A., de Barros, F.P.J., Bellin, A., 2017. Radial solute transport in highly heterogeneous aquifers: Modeling and experimental comparison. *Water Resources Research*, 53(7): 5725–5741.

Diersch, H.G., 2014. *FEFLOW — Finite Element Modeling of Flow, Mass and Heat Transport in Porous and Fractured Media.* Springer, Berlin Heidelberg.

Dieulin, A., 1981. *Lixiviation in situ d'un gisement d'uranium en mileu granitiqu. Draft Rep. LHM/RD/81/63. Ecole Natl. Super. des Mines de Paris.*

Doherty, J., 2002. *PEST — Model-Independent Parameter Estimation Manual* (4th Ed.).

Domenico, P.A., Schwartz, F.W., 1990. *Physical and Chemical Hydrogeology.* John Wiley & Sons, New York.

Dronfield, D.G., Silliman, S.E., 1993. Velocity dependence of dispersion for transport through a single fracture of variable roughness. *Water Resources Research*, 29(10): 3477–3483.

Egboka, B., Cherry, J., Farvolden, R., Frind, E., 1983. Migration of contaminants in groundwater at a landfill: A case study: 3. Tritium as an indicator of dispersion and recharge. *Journal of Hydrology*, 63(1–2): 51–80.

El-Kadi, A.L., 1988. Applying the USGS mass-transport model (MOC) to remedial actions by recovery wells. *Groundwater*, 26(3): 281–288.

Elder, C.R., Benson, C.H., Eykholt, G.R., 2002. Effects of heterogeneity on influent and effluent concentrations from horizontal permeable reactive barriers. *Water Resources Research*, 38(8).

Elenius, M.T., Abriola, L.M., 2019. Regressed models for multirate mass transfer in heterogeneous media. *Water Resources Research*, 20.

Falade, G., Brigham, W., 1989. Analysis of radial transport of reactive tracer in porous media. *SPE Reservoir Engineering*, 4(1): 85–90.

Feng, Q.G., Zhan, H.B., 2016. Integrated aquitard-aquifer flow with a mixed-type well-face boundary and skin effect. *Advances in Water Resources*, 89: 42–52.

Fenske, P.R., 1973. Hydrology and radionuclide transport monitoring well HT-2m, Tatum Dome, Mississippi, Nevada University, Reno (USA). Desert Research Inst.

Fetter, C.W., 2018. *Applied Hydrogeology.* Waveland Press.

Freyberg, D.L., 1986. A natural gradient experiment on solute transport in a sand aquifer: 2. Spatial moments and the advection and dispersion of nonreactive tracers. *Water Resources Research*, 22(13): 2031–2046.

Fried, J., Ungemach, P., 1971. Détermination *in situ* du coefficient de dispersion longitudinale d'un milieu poreux naturel. *Compte-rendus Acad. Sci. A*, 172: 1327.

Fried, J.J., 1975. *Groundwater Pollution.* Elsevier.

Gan, Y., Wang, Q., 2024. A model of chemical transport in a wellbore–aquifer system. *Hydrology and Earth System Sciences*, 28(6): 1317–1323.

Gao, G. *et al.*, 2010. A new mobile-immobile model for reactive solute transport with scale-dependent dispersion. *Water Resources Research*, 46(8): 5725–5741.

Gao, G.Y., Feng, S.Y., Huo, Z.L., Zhan, H.B., Huang, G.H., 2009a. Semi-analytical solution for solute radial transport dynamic model with scale-dependent dispersion. *Chinese Journal of Hydrodynamics*, 24(2): 156–163.

Gao, G.Y., Feng, S.Y., Zhan, H.B., Huang, G.H., Mao, X.M., 2009b. Evaluation of anomalous solute transport in a large heterogeneous soil column with mobile-immobile model. *Journal of Hydrologic Engineering*, 14(9): 966–974.

Garabedian, S.P., 1987. Large-scale dispersive transport in aquifers: Field experiments and reactive transport theory, Massachusetts Institute of Technology.

Gelhar, L.W., Collins, M.A., 1971. General analysis of longitudinal dispersion in nonuniform flow. *Water Resources Research*, 7(6): 1511–1521.

Gelhar, L.W., Leonhart, L., 1982. Analysis of two-well tracer tests with a pulse input, Rockwell International Corp.

Gelhar, L.W., Welty, C., Rehfeldt, K.R., 1992. A critical review of data on field-scale dispersion in aquifers. *Water Resources Research*, 28(7): 1955–1974.

Giese, S.W., Powers, S.E., 2002. Using polymer solutions to enhance recovery of mobile coal tar and creosote DNAPLs. *Journal of Contaminant Hydrology*, 58: 147–167.

Ginn, T.R., Simmons, C.S., Wood, B.D., 1995. Stochastic-convective transport with nonlinear reaction — Biodegradation with microbial-growth. *Water Resources Research*, 31(11): 2689–2700.

Goblet, P., 1982. Interpretation d'Experiences de Tracage en Milieu Granitique (Site B), Report LHM/RD/82/11, Centre d'Information Geologique, Ecole Nationale.

Gouze, P. *et al.*, 2008. Non-Fickian dispersion in porous media: 1. Multiscale measurements using single-well injection withdrawal tracer tests. *Water Resources Research*, 44(6).

Griffioen, J.W., Barry, D.A., Parlange, J.Y., 1998. Interpretation of two-region model parameters. *Water Resources Research*, 34(3): 373–384.

Grove, D., Beetem, W., 1971. Porosity and dispersion constant calculations for a fractured carbonate aquifer using the two well tracer method. *Water Resources Research*, 7(1): 128–134.

Grove, D.B., 1977. The use of Galerkin finite-element methods to solve mass-transport equations, US Dept. of the Interior, Geological Survey.

Guo, Z., Fogg, G.E., Henri, C.V., 2019. Upscaling of regional scale transport under transient conditions: Evaluation of the multirate mass transfer model. *Water Resources Research*, 55(7): 5301–5320.

Gupta, S., Tanji, K., 1976. A three-dimensional Galerkin finite element solution of flow through multiaquifers in Sutter Basin, California. *Water Resources Research*, 12(2): 155–162.

Güven, O., Molz, F.J., Melville, J.G., 1984. An analysis of dispersion in a stratified aquifer. *Water Resources Research*, 20(10): 1337–1354.

Haggerty, R., McKenna, S.A., Meigs, L.C., 2000. On the late-time behavior of tracer test breakthrough curves. *Water Resources Research*, 36(12): 3467–3479.

Haggerty, R., Schroth, M.H., Istok, J.D., 1998. Simplified method of "push-pull" test data analysis for determining *in situ* reaction rate coefficients. *Ground Water*, 36(2): 314–324.

Halevy, E., Nir, A., 1962. The determination of aquifer parameters with the aid of radioactive tracers. *Journal of Geophysical Research*, 67(6): 2403–2409.

Hall, S.H., Luttrell, S.P., Cronin, W.E., 1991. A method for estimating effective porosity and ground water velocity. *Ground Water*, 29(2): 171–174.

Hansen, S.K., Berkowitz, B., Vesselinov, V.V., O'Malley, D., Karra, S., 2016. Push-pull tracer tests: Their information content and use for characterizing non-Fickian, mobile-immobile behavior. *Water Resources Research*, 52(12): 9565–9585.

Harbaugh, A.W., Banta, E.R., Hill, M.C., McDonald, M.G., 2000. MODFLOW-2000, The US geological survey modular ground-water model-user guide to modularization concepts and the ground-water flow process. open-file report. *US Geological Survey*, (92): 134.

Harmon, T.C., Semprini, L., Roberts, P.V., 1992. Simulating solute transport using laboratory-based sorption parameters. *Journal of Environmental Engineering*, 118(5): 666–688.

Harpaz, Y., 1965. Field experiments in recharge and mixing through wells.

Hebig, K.H. *et al.*, 2015. Study of the effects of the chaser in push-pull tracer tests by using temporal moment analysis. *Geothermics*, 54: 43–53.

Hellerich, L.A. *et al.*, 2003. Bromide transport before, during, and after colloid mobilization in push-pull tests and the implications for changes in aquifer properties. *Water Resources Research*, 39(10).

Helweg, O.J., Labadie, J.W., 1977. Linked models for managing river basin salt balance. *Water Resources Research*, 13(2): 329–336.

Hendry, M.J., Kelln, C.J., Wassenaar, L.I., Shaw, J., 2004. Characterizing the hydrogeology of a complex clay-rich aquitard system using detailed vertical profiles of the stable isotopes of water. *Journal of Hydrology*, 293(1–4): 47–56.

Herrling, B., Stamm, J., 1993. Hydraulic circulation system for insitu remediation of strippable contaminants and insitu bioreclamation (gzb uvb method). In *Hydrochemistry 1993 Symposium on Hydrological, Chemical and Biological Processes of Transformation and Transport of Contaminants in Aquatic Environments*. Iahs Publications, Rostov Na Donu, Russia, pp. 347–356.

Herrling, B., Stamm, J., Buermann, W., 1991. Hydraulic circulation system for insitu bioreclamation and or insitu remediation of strippable contamination. In *International Symp on in Situ and on-Site Bioreclamation*, San Diego, Ca, pp. 173–195.

Hildebrand, F. B., 1976. Advanced Calculus for Applications, 2nd ed., Prentice Hall, Englewood Cliffs, N.J.

Hoehn, E., 1983. Advection-dispersion interpretation of tracer observations in a river/groundwater infiltration system. In *International Conference on Modern Approach to Groundwater Resources Management*, pp. 361–371.

Hoehn, E., Santschi, P.H., 1987. Interpretation of tracer displacement during infiltration of river water to groundwater. *Water Resources Research*, 23(4): 633–640.

Hoopes, J.A., Harleman, D.R., 1967b. Dispersion in radial flow from a recharge well. *Journal of Geophysical Research*, 72(14): 3595–3607.

Hsieh, P.A., 1986. A new formula for the analytical solution of the radial dispersion problem. *Water Resources Research*, 22(11): 1597–1605.

Hsieh, P.-F., Yeh, H.-D., 2014. Semi-analytical and approximate solutions for contaminant transport from an injection well in a two-zone confined aquifer system. *Journal of Hydrology*, 519: 1171–1176.

Huang, J., Goltz, M.N., 2006. Analytical solutions for solute transport in a spherically symmetric divergent flow field. *Transport in Porous Media*, 63(2): 305–321.

Huang, J.Q., Christ, J.A., Goltz, M.N., 2010. Analytical solutions for efficient interpretation of single-well push-pull tracer tests. *Water Resources Research*, 46.

Huang, C.S., Tong, C., Hu, W., Yeh, H.D., Yang, T., 2019. Analysis of radially convergent tracer test in a two-zone confined aquifer with vertical dispersion effect: Asymmetrical and symmetrical transports. *Journal of Hazardous Materials*, 377(5): 8–16.

Huang, C.-S., Yang, S.-Y., Yeh, H.D., 2014. Groundwater flow to a pumping well in a sloping fault zone unconfined aquifer. *Water Resources Research*, 50: 4079–4094.

Huyakorn, P.S., Andersen, P.F., Molz, F.J., Güven, O., Melville, J.G., 1986. Simulations of two-well tracer tests in stratified aquifers at the Chalk River and the Mobile sites. *Water Resources Research*, 22(7): 1016–1030.

Hvilshoj, S., Jensen, K.H., Madsen, B., 2000. Single-well dipole flow tests: Parameter estimation and field testing. *Ground Water*, 38(1): 53–62.

Iris, P., Adnot, J., Bonfils, R., Marchio, D., 1985. The Heliogeothermal Doublet. *International Journal of Solar Energy*, 3(2): 73–88.

Istok, J., 2012. *Push-Pull Tests For Site Characterization.* Springer Science & Business Media, p. 144.

Istok, J., Field, J., Schroth, M., 2001. In situ determination of subsurface microbial enzyme kinetics. *Groundwater*, 39(3): 348–355.

Istok, J.D., Field, J.A., Schroth, M.H., Sawyer, T.E., Humphrey, M.D., 1999. Laboratory and field investigation of surfactant sorption using single-well, "push-pull" tests. *Ground Water*, 37(4): 589–598.

Istok, J.D., Humphrey, M.D., Schroth, M.H., Hyman, M.R., O'Reilly, K.T., 1997. Single-well, "push-pull" test for *in situ* determination of microbial activities. *Ground Water*, 35(4): 619-613.

Ivanovich, M., Smith, D., 1978. Determination of aquifer parameters by a two-well pulsed method using radioactive tracers. *Journal of Hydrology*, 36(1–2): 35–45.

Jiao, J.J. *et al.*, 2010. Abnormally high ammonium of natural origin in a coastal aquifer-aquitard system in the pearl river delta, China. *Environmental Science & Technology*, 44(19): 7470–7475.

Jin, A., Wang, Q., Zhan, H., 2024. A novel four phase slug single-well push–pull test with regional flux: Forward modeling and parameter estimation. *Journal of Hydrology*, 630: 130705.

Johnsen, S.G., Whitson, C.H., 2011. On the loss of injectant and apparent dispersivity in single-well push-pull tests influenced by natural drift. *Transport in Porous Media*, 86: 475–478.

Johnson, R.L., Simon, M.A., 2007. Evaluation of groundwater flow patterns around a dual-screened groundwater circulation well. *Journal of Contaminant Hydrology*, 93(1–4): 188–202.

Jung, Y., Pruess, K., 2012. A closed-form analytical solution for thermal single-well injection-withdrawal tests. *Water Resources Research*, 48(3).

Kabala, Z.J., 1993. The dipole flow test — A new single-borehole test for aquifer characterization. *Water Resources Research*, 29(1): 99–107.

Kabala, Z.J., 2001. Sensitivity analysis of a pumping test on a well with wellbore storage and skin. *Advances in Water Resources*, 24(5): 483–504.

Karadimitriou, N.K., Joekar-Niasar, V., Babaei, M., Shore, C.A., 2016. Critical role of the immobile zone in non-Fickian two-phase transport: A new paradigm. *Environmental Science & Technology*, 50(8): 4384–4392.

Kies, B., 1982. Solute transport in unsaturated field soil and in groundwater. New Mexico State University.

Kinzelbach, W., Schäfer, W., Herzer, J., 1991. Numerical modeling of natural and enhanced denitrification processes in aquifers. *Water Resources Research*, 27(6): 1123–1135.

Klotz, D., Seiler, K.-P., Moser, H., Neumaier, F., 1980. Dispersivity and velocity relationship from laboratory and field experiments. *Journal of Hydrology*, 45(3–4): 169–184.

Knox, R.C. *et al.*, 1997. Surfactant remediation field demonstration using a vertical circulation well. *Ground Water*, 35(6): 948–953.

Konokow, L.F. 1976. Modelling solute transport in groundwater. In: *International conference on environmental sensing and assessment proceedings*, 20–30, Las Vegas, NV.

Konikow, L.F., Bredehoeft, J.D., 1974. Modeling flow and chemical quality changes in an irrigated stream-aquifer system. *Water Resources Research*, 10(3): 546–562.

Konikow, L.F., Grove, D.B., 1977. Derivation of Equations Describing Solute Transport in Ground Water. US Geological Survey, *Water Resources Division*: Reston, VA, USA.

Konikow, L.F., Hornberger, G.Z., 2006. Use of the multi-node well (MNW) package when simulating solute transport with the MODFLOW ground-water transport process. US Geological Survey, 2328–7055.

Konikow, L.F., Hornberger, G.Z., Halford, K.J., Hanson, R.T., Harbaugh, A.W., 2009. Revised multi-node well (MNW2) package for MODFLOW ground-water flow model, US Geological Survey Techniques.

Kreft, A., Lenda, A., Turek, B., Zuber, A., Czauderna, K., 1974. Determination of effective porosities by the two-well pulse method. In *Proceedings of a Symposium: Isotope Techniques in Groundwater Hydrology*, Vol. II.

Lau, L.-K.S., Kaufman, W.J., Todd, D.K., 1957. Studies of dispersion in a radial flow system. Sanitary Engineering Research Laboratory, Department of Engineering.

Le Borgne, T., Gouze, P., 2008. Non-Fickian dispersion in porous media: 2. Model validation from measurements at different scales. *Water Resources Research*, 44(6).

Le Vot, F., Abad, E., Yuste, S.B., 2017. Continuous-time random-walk model for anomalous diffusion in expanding media. *Physical review. E*, 96(3-1): 032117.

Leap, D.I., Kaplan, P.G., 1988. A single-well tracing method for estimating regional advective velocity in a confined aquifer: Theory and preliminary laboratory verification. *Water Resources Research*, 24(24): 993–998.

Lee, D.R., Cherry, J.A., Pickens, J.F., 1980. Groundwater transport of a salt tracer through a sandy lakebed 1. *Limnology and Oceanography*, 25(1): 45–61.

Leij, F.J., Priesack, E., Schaap, M.G., 2000. Solute transport modeled with Green's functions with application to persistent solute sources. *Journal of Contaminant Hydrology*, 41(1–2): 155–173.

Leland, D., Hillel, D., 1981. Scale effects on measurement of dispersivity in a shallow unconfined aquifer. In *Chapman Conference on Spatial Variability in Hydrologic Modeling.* AGU. Fort Collins. CO., July, p. 23.

Lenda, A., Zuber, A., 1970. Tracer dispersion in groundwater experiments. I*sotope Hydrology*, 1970: 619–641.

Li, L., Barry, D.A., Morris, J., Stagnitti, F., 1999. CXTANNEAL: an improved program for estimating solute transport parameters. *Environmental Modelling Software*, 14(1999): 607–611.

Li, X.T., Wen, Z., Zhu, Q., Jakada, H., 2020. A mobile-immobile model for reactive solute transport in a radial two-zone confined aquifer. *Journal of Hydrology*, 580: 124347.

Li, X.T., Wen, Z., Zhan, H., Zhan, H., Zhu, Q., 2019. Skin effect on single-well push-pull tests with the presence of regional groundwater flow. *Journal of Hydrology*, 577: 123931.

Liang, X., Zhan, H., Liu, J., Dong, G., Zhang, Y.-K., 2018. A simple method of transport parameter estimation for slug injecting tracer tests in porous media. *Science of The Total Environment*, 644: 1536–1546.

Liao, H.S., Sampath, P.V., Curtis, Z.K., Li, S.G., 2015. Hierarchical modeling of a groundwater remediation capture system. *Journal of Hydrology*, 527: 196–211.

Lin, Y., Hu, T., Yeh, H., 2019. Analytical model for heat transfer accounting for both conduction and dispersion in aquifers with a robin-type boundary condition at the injection well. *Water Resources Research*, 55(8): 7379–7399.

Lin, Z., Puls, R.W., 2000. Adsorption, desorption and oxidation of arsenic affected by clay minerals and aging process. *Environmental Geoology*, 39(7): 753–759.

Liu, C.X., Ball, W.P., 2002. Back diffusion of chlorinated solvent contaminants from a natural aquitard to a remediated aquifer under well-controlled field conditions: Predictions and measurements. *Ground Water*, 40(2): 175–184.

Liu, G., Zheng, C., Gorelick, S.M., 2007. Evaluation of the applicability of the dual-domain mass transfer model in porous media containing connected high-conductivity channels. *Water Resources Research*, 43(12).

Ma, R., Zheng, C., Tonkin, M., Zachara, J.M., 2011. Importance of considering intraborehole flow in solute transport modeling under highly dynamic flow conditions. *Journal of Contaminant Hydrology*, 123(1–2): 11–19.

Maraqa, M.A., 2001. Prediction of mass-transfer coefficient for solute transport in porous media. *Journal of Contaminant Hydrology*, 50(1–2): 1–19.

Mathias, S.A., 2010. Transient divergent flow and transport in an infinite anisotropic porous formation. *Ground Water*, 48(3): 438–441.

Mathias, S.A., Wen, Z., 2015. Numerical simulation of Forchheimer flow to a partially penetrating well with a mixed-type boundary condition. *Journal of Hydrology*, 524: 53–61.

Mathias, S.A., Butler, A.P., Zhan, H.B., 2008. Approximate solutions for Forchheimer flow to a well. *Journal of Hydraulic Engineering-ASCE*, 134(9): 1318–1325.

Matsumoto, S., Machida, I., Hebig, K.H., Zeilfelder, S., Ito, N., 2020. Estimation of very slow groundwater movement using a single-well push-pull test. *Journal of Hydrology*, 591.

Mazurek, M. *et al.*, 2011. Natural tracer profiles across argillaceous formations. *Applied Geochemistry*, 26(7): 1035–1064.

McCuen, R.H., 1985. *Statistical Methods for Engineers.* Prentice Hall, Englewood Cliffs, New Jersey.

McGuire, J.T., Long, D.T., Klug, M.J., Haack, S.K., Hyndman, D.W., 2002. Evaluating behavior of oxygen, nitrate, and sulfate during recharge and quantifying reduction rates in a contaminated aquifer. *Environmental Science & Technology*, 36(12): 2693–2700.

McMillan, L.A., Rivett, M.O., Tellam, J.H., Dumble, P., Sharp, H., 2014. Influence of vertical flows in wells on groundwater sampling. *Journal of Contaminant Hydrology*, 169: 50–61.

Meerschaert, M.M., Zhang, Y., Baeumer, B., 2008. Tempered anomalous diffusion in heterogeneous systems. *Geophysical Research Letters*, 35(17).

Mercado, A. 1966. Recharge and mixing tests at Yavne 20 well field. *Underground Water Storage Study Tech. Rep. 12*, Publ. 611. Tahal Water Plann. for Israel, Tel Aviv.

Meyer, B., Bain, C., De Jesus, A., Stephenson, D., 1981. Radiotracer evaluation of groundwater dispersion in a multi-layered aquifer. *Journal of Hydrology*, 50: 259–271.

Mobile, M.A., Widdowson, M.A., Gallagher, D.L.J.E.s., Technology, 2012. Multicomponent NAPL source dissolution. *Evaluation of Mass-Transfer Coefficients*, 46(18): 10047–10054.

Moench, A.F., 1985. Transient flow to a large-diameter well in an aquifer with storative semiconfining layers. *Water Resources Research*, 21(8): 1121–1131.

Moench, A.F., 1989. Convergent radial dispersion' a Laplace transform solution for aquifer tracer testing. *Water Resources Research*, 25(3): 439–447.

Moench, A., Ogata, A., 1981. A numerical inversion of the Laplace transform solution to radial dispersion in a porous medium. *Water Resources Research*, 17(1): 250–252.

Molinari, J., Peaudecerf, P., 1977. *Essais conjoints en laboratoire et sur le terrain en vue d'une approche simplifiée de la prévision des propagations de substances miscibles dans les aquifères réels.*

Molinari, A., Pedretti, D., Fallico, C., 2015. Analysis of convergent flow tracer tests in a heterogeneous sandy box with connected gravel channels. *Water Resources Research*, 51(7): 5640–5657.

Moltyaner, G., Killey, R., 1988. Twin Lake tracer tests: Transverse dispersion. *Water Resources Research*, 24(10): 1628–1637.

Morris, M.D., 1991. Factorial Sampling plans for preliminary computational experiments. *Technometrics*, 33(2): 161–174.

Mulligan, A.E., Ahlfeld, D.P., 1999. Advective control of groundwater contaminant plumes: Model development and comparison to hydraulic control. *Water Resources Research*, 35(8): 2285–2294.

Naymik, T.G., Barcelona, M.J., 1981. Characterization of a contaminant plume in Ground Water, Meredosia, Illinois. *Groundwater*, 19(5): 517–526.

Neuman, S.P., Mishra, P.K., 2012. Comments on "A revisit of drawdown behavior during pumping in unconfined aquifers" by D. Mao, L. Wan, T.-C. J. Yeh, C.-H. Lee, K.-C. Hsu, J.-C. Wen, and W. Lu. *Water Resources Research*, 48(2).

Neuman, S.P., Tartakovsky, D.M., 2009. Perspective on theories of non-Fickian transport in heterogeneous media. *Advances in Water Resources*, 32(5): 670–680.

Neuzil, C.E., 1986. Groundwater flow in low-permeability environments. *Water Resources Research*, 22(8): 1163–1195.

Nichols, W., Aimo, N., Oostrom, M., White, M., 1997. STOMP subsurface transport over multiple phases: Application guide, Pacific Northwest Lab., Richland, WA (United States).

Novakowski, K.S., 1992a. The analysis of tracer experiments conducted in divergent radial flow fields. *Water Resources Research*, 28(12): 3215–3225.

Novakowski, K.S., 1992b. An evaluation of boundary conditions for one-dimensional solute transport: 1. Mathematical development. *Water Resources Research*, 28(9): 2399–2410.

Novakowski, K.S., 1992c. An evaluation of boundary conditions for one-dimensional solute transport: 2. Column experiments. *Water Resources Research*, 28(9): 2411–2423.

Oakes, D., Edworthy, D., 1977. Field measurements of dispersion coefficients in the United Kingdom. *Ground Water Quality, Measurement, Prediction, and Protection*, pp. 327–340.

Office, U.S.E.P.A.T.I., Support, E.M., 1998. *Field Applications of In Situ Remediation Technologies: Ground-Water Circulation Wells.* DIANE Publishing.

Ortiz-Zamora, D., Ortega-Guerrero, A., 2010. Evolution of long-term land subsidence near Mexico City: Review, field investigations, and predictive simulations. *Water Resources Research*, 46(1): W01513.

Oya, S., Valocchi, A.J., 1998. Transport and biodegradation of solutes in stratified aquifers under enhanced *in situ* bioremediation conditions. *Water Resources Research*, 34(12): 3323–3334.

Papadopulos, I.S., Cooper Jr, H.H., 1967. Drawdown in a well of large diameter. *Water Resources Research*, 3(1): 241–244.

Papadopulos, S., Larson, S., 1978. Aquifer storage of heated water: Part II—Numerical simulation of field results. *Groundwater*, 16(4): 242–248.

Paradis, C.J., McKay, L.D., Perfect, E., Istok, J.D., Hazen, T.C., 2017. Push-pull tests for estimating effective porosity: Expanded analytical solution and *in situ* application. *Hydrogeology Journal*, 26(2).

Paradis, C.J., McKay, L.D., Perfect, E., Istok, J.D., Hazen, T.C., 2018. Correction: Push-pull tests for estimating effective porosity: Expanded analytical solution and *in situ* application. *Hydrogeology Journal*, 27(1).

Parker, J., Valocchi, A., 1986. Constraints on the validity of equilibrium and first-order kinetic transport models in structured soils. *Water Resources Research*, 22(3): 399–407.

Parker, J.R.C., Genuchten, M.T.v., 1984. Determining transport parameters from laboratory and field tracer experiments. *Virginia Agricultural Experiment Station Bulletin*, 1–96.

Paster, A., Bolster, D., Benson, D.A., 2014. Connecting the dots: Semi-analytical and random walk numerical solutions of the diffusion–reaction equation with stochastic initial conditions. *Journal of Computational Physics*, 263: 91–112.

Pebesma, E.J., 2004. Multivariable geostatistics in S: The gstat package. *Computers & Geosciences*, 30(7): 683–691.

Pedretti, D., Fernàndez-Garcia, D., Bolster, D., Sanchez-Vila, X., 2013. On the formation of breakthrough curves tailing during convergent flow tracer tests in three-dimensional heterogeneous aquifers. *Water Resources Research*, 49(7): 4157–4173.

Pedretti, D., Fernàndez-Garcia, D., Sanchez-Vila, X., Bolster, D., Benson, D., 2014. Apparent directional mass-transfer capacity coefficients in three-dimensional anisotropic heterogeneous aquifers under radial convergent transport. *Water Resources Research*, 50(2): 1205–1224.

Phanikumar, M.S., McGuire, J.T., 2010. A multi-species reactive transport model to estimate biogeochemical rates based on single-well push-pull test data. *Computers & Geosciences*, 36(8): 997–1004.

Philip, J., 1994. Some exact solutions of convection-diffusion and diffusion equations. *Water Resources Research*, 30(12): 3545–3551.

Philip, R.D., Walter, G.R., 1992. Prediction of flow and hydraulic-head fields for vertical circulation wells. *Ground Water*, 30(5): 765–773.

Pickens, J.F., Grisak, G.E., 1981a. Modeling of scale-dependent dispersion in hydrogeologic systems. *Water Resources Research*, 17(6): 1701–1711.

Pickens, J.F., Grisak, G.E., 1981b. Scale-dependent dispersion in a stratified granular aquifer. *Water Resources Research*, 17(4): 1191–1211.

Pinder, G.F., 1973. A Galerkin-finite element simulation of groundwater contamination on Long Island, New York. *Water Resources Research*, 9(6): 1657–1669.

Poeter, E.P., Hill, M.C., 1999. UCODE, a computer code for universal inverse modeling. *Computers & Geosciences*, 25(4): 457–462.

Pollock, D.W., 1988. Semianalytical computation of path lines for finite-difference models. *Groundwater*, 26(6): 743–750.

Pruess, K., Oldenburg, C.M., Moridis, G., 1999. TOUGH2 user's guide version 2, Lawrence Berkeley National Lab. (LBNL), Berkeley, CA (United States).

Rabinowitz, D.D., Gross, G.W., 1972. *Environmental Tritium as a hydrometeorologic Tool in the Roswell Basin.* New Mexico.

Rajaram, H., Gelhar, L.W., 1991. Three-dimensional spatial moments analysis of the Borden tracer test. *Water Resources Research*, 27(6): 1239–1251.

Reinhard, M. *et al.*, 1997. In situ BTEX Biotransformation under enhanced nitrate- and sulfate-reducing conditions. *Environmental Science & Technology*, 31(1): 28–36.

Rivera, J.A., Blum, P., Bayer, P., 2016. Influence of spatially variable ground heat flux on closed-loop geothermal systems: Line source model with nonhomogeneous Cauchy-type top boundary conditions. *Applied Energy*, 180: 572–585.

Roberts, P.V., 1985. Advection-dispersion sorption models for simulating the transport of organic pollutants. *Ground Water Quality*, 425–445.

Robertson, J., Barraclough, J., 1973. Radioactive and chemical waste transport in groundwater of National Reactor Testing Station: 20-year case history and digital model. In *2nd International Symposium of Underground Waste Management, Artif. Recharge Prepration. Paper*, pp. 291–322.

Robertson, J.B., 1974. Digital modeling of radioactive and chemical waste transport in the aquifer underlying the Snake River Plain at the National Reactor Testing Station, Idaho. US Geological Survey, Open-File Report, pp. 74–1089.

Robson, S.G., 1974. Feasibility of digital water-quality modeling illustrated by application at Barstow, California. US Geological Survey, WRD, California District, p. 46.

Robson, S.G., 1978. Application of digital profile modeling techniques to ground-water solute transport at Barstow, California, 2050. US Government Printing Office.

Rockhold, M.L., Yarwood, R., Niemet, M., Bottomley, P.J., Selker, J.S., 2005. Experimental observations and numerical modeling of coupled microbial and transport processes in variably saturated sand. *Vadose Zone Journal*, 4(2): 407–417.

Rousselot, D., Sauty, J.P., Gaillard, B., 1977. Etude hydrogéologique de la zone industrielle de Blyes-Saint-Vulbas, rapport préliminaire no. 5: Caracteristiques hydrodynamiques du système aquifère. *Rep. Jal 77/33.* Bur. de Rech. Geol. et Min., Orleans, Franc.

Rumynin, V.G., 2011. Advection and dispersion of dissolved species in aquifers. Advection and dispersion of dissolved species in aquifers. In *Subsurface Solute Transport Models and Case Histories. Theory and Applications of Transport in Porous Media.* pp. 3–76.

Sauty, J.-P., 1977. *Contribution à l'identification des paramètres de dispersion dans les aquifères par interprétation des expériences de traçage*, Éditeur inconnu.

Sauty, J.-P., 1980. An analysis of hydrodispersive transfer in aquifers. *Water Resources Research*, 16(1): 145–158.

Sauty, J., Gringarten, A., Landel, P., 1978. The effects of thermal dispersion on injection of hot water in aquifers. In *Proceedings of Second Invitational Well Testing Symposium.* Lawrence Berkeley Laboratory, Berkeley, California, pp. 122–131.

Schaad, D.E., 1998. Dipole flow test with a tracer: A new methodology for characterizing subsurface parameters. Duke University.

Schroth, M.H., Istok, J.D., 2005. Approximate solution for solute transport during spherical-flow push-pull tests. *Ground Water*, 43(2): 280–284.

Schroth, M.H., Istok, J.D., 2006. Models to determine first-order rate coefficients from single-well push-pull tests. *Ground Water*, 44(2): 275–283.

Schroth, M.H., Istok, J.D., Haggerty, R., 2000. In situ evaluation of solute retardation using single-well push-pull tests. *Advances in Water Resources*, 24(1): 105–117.

Schroth, M.H., Istok, J.D., Haggerty, R., 2001. In situ evaluation of solute retardation using single-well push–pull tests. *Advances in Water Resources*, 24: 105–117.

Schumer, R., Benson, D.A., Meerschaert, M.M., Baeumer, B., 2003. Fractal mobile/immobile solute transport. *Water Resources Research*, 39(10).

Schwartz, R., McInnes, K., Juo, A., Wilding, L., Reddell, D., 1999. Boundary effects on solute transport in finite soil columns. *Water Resources Research*, 35(3): 671–681.

Segol, G., Pinder, G.F., 1976. Transient simulation of saltwater intrusion in southeastern Florida. *Water Resources Research*, 12(1): 65–70.

Shampine, L., Watts, H., Davenport, S., 1975. Solving nonstiff ordinary differential equations-the state of the art. *SIAM Rev.*, 18: 376–411.

Shampine, L.F., Reichelt, M.W., 1997. The MATLAB ODE suite. *SIAM Journal on Scientific Computing*, 18(1): 1–22.

Shampine, L.F., Reichelt, M.W., Kierzenka, J.A., 1999. Solving index-I DAEs in MATLAB and simulink. *SIAM Review*, 41(3): 538–552.

Shi, W., Wang, Q., Zhan, H., 2020. New simplified models of single-well push-pull tests with mixing effect. *Water Resources Research*, 56.

Shi, W., Wang, Q., Zhan, H., Zhou, R., Yan, H., 2023. A general model of radial dispersion with wellbore mixing and skin effects. *Hydrology and Earth System Sciences*, 27(9): 1891–1908.

Silliman, S.E., Simpson, E.S., 1987. Laboratory evidence of the scale effect in dispersion of solutes in porous media. *Water Resources Research*, 23(8): 1667–1673.

Simon, M., Hough, B., Li, T., Kabala, Z., 2000. Field evaluation of dipole method to measure aquifer anisotropy. In *Second International Conference on Remediation of Chlorinated and Racalcitrant Compounds*, pp. 43–49.

Snodgrass, M.F., Kitanidis, P.K., 1998. A method to infer *in situ* reaction rates from push-pull experiments. *Groundwater*, 36(4): 645–650.

Starr, R.C., Gillham, R.W., Sudicky, E.A., 1985. Experimental investigation of solute transport in stratified porous media: 2. The reactive case. *Water Resources Research*, 21(7): 1043–1050.

Stehfest, H., 1970a. Algorithm 368: Numerical inversion of Laplace transforms. *Communication ACM*, 13(1): 47–49.

Stehfest, H., 1970b. Remark on algorithm 368: Numerical inversion of Laplace transforms. *Communication ACM*, 13(10): 624.

Sudicky, E., Cherry, J., Frind, E., 1983. Migration of contaminants in groundwater at a landfill: A case study: 4. A natural-gradient dispersion test. *Journal of Hydrology*, 63(1–2): 81–108.

Sudicky, E.A., 1986. A natural gradient experiment on solute transport in a sand aquifer: Spatial variability of hydraulic conductivity and its role in the dispersion process. *Water Resources Research*, 22(13): 2069–2082.

Sudicky, E.A., 1988. A natural gradient experiment on solute transport in a sand aquifer: Spatial variability of hydraulic conductivity and its role in the dispersion process. *Water Resources Research*, 24(7): 1211–1216.

Sudicky, E.A., Gillham, R.W., Frind, E.O., 1985. Experimental investigation of solute transport in stratified porous media: 1. the nonreactive case. *Water Resources Research*, 21(7): 1035–1041.

Suk, H., Han, W.S., Chen, J.-S., Yang, M., 2023. Semi-analytical solution of single-well push–pull test under transient flow conditions. *Journal of Hydrology*, 620: 129542.

Sutton, D., Kabala, Z., Schaad, D., Ruud, N., 2000. The dipole-flow test with a tracer: A new single-borehole tracer test for aquifer characterization. *Journal of Contaminant Hydrology*, 44(1): 71–101.

Sykes, J., Pahawa, S., Ward, D., Lantz, D., 1983. The *Validation of SWENT, a Geosphere Transport Model; W: Scientific Computing, red. R. Stapleman i in; IMAES*. North Holland, Amsterdam, The Netherlands.

Sykes, J., Pahwa, S., Lantz, R., Ward, D., 1982. Numerical simulation of flow and contaminant migration at an extensively monitored landfill. *Water Resources Research*, 18(6): 1687–1704.

Tang, D., Babu, D., 1979. Analytical solution of a velocity dependent dispersion problem. *Water Resources Research*, 15(6): 1471–1478.

Tang, D.H., Frind, E.O., Sudicky, E.A., 1981. Contaminant transport in fractured porous media: Analytical solution for a single fracture. *Water Resources Research*, 17(3): 555–564.

Tang, D.H.E., Peaceman, D.W., 1987. New analytical and numerical solutions for the radial convection-dispersion problem. *SPE (Society of Petroleum Engineers) Reserv. Engineering*, 2(3): 343–359.

Tang, Y., Aral, M.M., 1992. Contaminant transport in layered porous media: 1. general solution. *Water Resources Reserarch*, 28(5): 1389–1397.

Thullner, M., Schroth, M.H., Zeyer, J., Kinzelbach, W., 2004. Modeling of a microbial growth experiment with bioclogging in a two-dimensional saturated porous media flow field. *Journal of Contaminant Hydrology*, 70(1–2): 37–62.

Tindall, J.A., Weeks, E.P., Friedel, M., 2005. Part 2: A field study of enhanced remediation of toluene in the vadose zone using a nutrient solution. *Water, Air, Soil Pollution*, 168: 359–389.

Toride, N., 1995. The CXTFIT code for estimating transport parameters from laboratory or field tracer experiments. Version 2.0. U.s.salinity Laboratory Res.rep.

Trefry, M.G., Muffels, C., 2007. FEFLOW: A finite-element Ground Water flow and transport modeling tool. *Groundwater*, 45(5): 525–528.

Trigueros, D.E., Módenes, A.N., Kroumov, A.D., Espinoza-Quiñones, F.R., 2010. Modeling of biodegradation process of BTEX compounds: Kinetic parameters estimation by using Particle swarm global optimizer. *Process Biochemistry*, 45(8): 1355–1361.

Trötschler, O., Haslwimmer, T., Sagner, A., Tiehm, A., Koschitzky, H.-P., 2005. ENA of heterocyclic hydrocarbons by adding hydrogen peroxide in groundwater circulation wells-A field based study on a large physical model scale. In *Proceedings (CD) of the 9th International FZK/TNO Conference on Soil-Water Systems.* Bordeaux, pp. 1452–1462.

Vaccaro, J., Bolke, E., 1983. Evaluation of water-quality characteristics of part of the Spokane aquifer, Washington and Idaho, using a solute-transport digital model. *US Geological Survey*, Open-File Report, 82–769.

Valocchi, A.J., Roberts, P.V., Parks, G.A., Street, R.L., 1981. Simulation of the transport of ion-exchanging solutes using laboratory-determined chemical parameter values. *Groundwater*, 19(6): 600–607.

van Genuchten, M.T., Wierenga, P.J., 1976. Mass transfer studies in sorbing porous media. I: Analytical solutions. *Soil Science Society of America Journal*, 40(4): 473–480.

Veling, E., 2001. *Analytical Solution and Numerical Evaluation of the Radial Symmetric Convection-Diffusion Equation with Arbitrary Initial and Boundary Data.* IAHS Publication, pp. 271–276.

Voss, C.I., 1984. SUTRA (Saturated-Unsaturated Transport). A Finite-Element Simulation Model for Saturated-Unsaturated, Fluid-Density-Dependent Ground-Water Flow with Energy Transport or Chemically-Reactive Single-Species Solute Transport, Geological Survey Reston Va Water Resources Division.

Walter, G., 1983. Convergent flow tracer test at H-6: Waste isolation pilot plant (WIPP), southeast New Mexico (draft). Hydro Geochem, Inc., Tucson, Ariz.

Wang, Q., Gu, H., Zhan, H., Shi, W., Zhou, R., 2019. Mixing effect on reactive transport in a column with scale dependent dispersion. *Journal of Hydrology*, 124494.

Wang, Q. *et al.*, 2021. Revisiting simplified model of a single-well push–pull test for estimating regional flow velocity. *Journal of Hydrology*, 601: 126711.

Wang, Q., Shi, W., Zhan, H., Gu, H., Chen, K., 2018. Models of single-well push-pull test with mixing effect in the wellbore. *Water Resources Research*, 54(12): 10155–10171.

Wang, Q., Wang, J., Zhan, H., Shi, W., 2020. New model of reactive transport in a single-well push–pull test with aquitard effect and wellbore storage. *Hydrology and Earth System Sciences*, 24(8): 3983–4000.

Wang, Q., Zhan, H., 2013. Radial reactive solute transport in an aquifer–aquitard system. *Advances in Water Resources*, 61: 51–61.

Wang, Q., Zhan, H., 2019. Reactive transport with wellbore storages in a single-well push–pull test. *Hydrology and Earth System Sciences*, 23(4): 2207–2223.

Wang, Q., Zhan, H., Tang, Z., 2013. A new parameter estimation method for solute transport in a column. *Groundwater*, 51(5): 714–722.

Wang, Q., Zhan, H., Tang, Z., 2014. Forchheimer flow to a well-considering time-dependent critical radius. *Hydrology and Earth System Sciences*, 18(6): 2437–2448.

Wang, Q.R., Zhan, H.B., 2015. On different numerical inverse Laplace methods for solute transport problems. *Advances in Water Resources*, 75: 80–92.

Wang, Q.R., Zhan, H.B., Wang, Y.X., 2017. Single-well push-pull test in transient Forchheimer flow field. *Journal of Hydrology*, 549: 125–132.

Warner, J.W., Tamayo-Lara, C., Khazaei, E., Manghi, F., 2006. Stochastic management modeling of a pump and treat system at the Rocky Mountain Arsenal near Denver, Colorado. *Journal of Hydrology*, 328(3–4): 523–537.

Webster, D.S., Procter, J.F., Marine, J.W., 1970. Two-well tracer test in fractured crystalline rock. U.S. Geological Survey, Water Supply Paper, 1544–1574.

Welch, G., Bishop, G., 1995. *An Introduction to Kalman Filter*. SIGGRAPH 2001.

Wen, Z., Huang, G.H., Zhan, H.B., 2009. A numerical solution for non-Darcian flow to a well in a confined aquifer using the power law function. *Journal of Hydrology*, 364(1–2): 99–106.

Werner, A., 1983. Nutzung von Grundwasser fur Warmepumpen, Versickerrungstest Aefligen, Versuch 2, 1982/83, *Water and Energy Management*. Agency of the State of Bern, Switzerland.

White, M.D., Oostrom, M., 1996. STOMP: Subsurface transport over multiple phases, Richland, WA.

Wiebenga, W., Ellis, W., Seatonberry, B., Andrew, J., 1967. Radioisotopes as groundwater tracers. *Journal of Geophysical Research*, 72(16): 4081–4091.

Wilson, L., 1971. Investigations on the subsurface disposal of waste effluents at inland sites. US Department of the Interior.

Wood, W.W., 1981. A geochemical method of determining dispersivity in regional groundwater systems. *Journal of Hydrology*, 54(1–3): 209–224.

Wood, W.W., Ehrlich, G.G., 1978. Use of baker's yeast to trace microbial movement in Ground Water. *Groundwater*, 16(6): 398–403.

Xia, Q., Zhang, Q., Xu, M., Tang, Y.G., Teng, H.W., 2019. Visualizing hydraulic zones of a vertical circulation well in presence of ambient flow. *Desalination and Water Treatment*, 159: 151–160.

Xiang, J., Kabala, Z.J., 1997. Performance of the steady-state dipole flow test in layered aquifers. *Hydrological Processes*, 11(12): 1595–1605.

Yan, X.-P., Kerrich, R., Hendry, M.J., 2001. Distribution of the rare earth elements in porewaters from a clay-rich aquitard sequence, Saskatchewan, Canada. *Chemical Geology*, 176(1–4): 151–172.

Yang, Q. *et al.*, 2014. Groundwater hydrogeochemistry in injection experiments simulating CO2 leakage from geological storage reservoir. *International Journal of Greenhouse Gas Control*, 26: 193–203.

Yang, S.-Y., Yeh, H.-D., 2009. Radial groundwater flow to a finite diameter well in a leaky confined aquifer with a finite-thickness skin. *Hydrological Processes*, 23(23): 3382–3390.

Yates, S., 1988. Three-dimensional radial dispersion in a variable velocity flow field. *Water Resources Research*, 24(7): 1083–1090.

Yeh, H.D., Chang, Y.C., 2013. Recent advances in modeling of well hydraulics. *Advances in Water Resources*, 51: 27–51.

Yeh, T.-C., Khaleel, R., Carroll, K.C., 2015. *Flow Through Heterogeneous Geologic Media*. Cambridge University Press.

Yin, M. *et al.*, 2020. A distributed-order time fractional derivative model for simulating bimodal sub-diffusion in heterogeneous media. *Journal of Hydrology*, 591(2020): 125504.

You, K., Zhan, H., 2013. New solutions for solute transport in a finite column with distance-dependent dispersivities and time-dependent solute sources. *Journal of Hydrology*, 487(2013): 87–97.

Zhan, H., Wen, Z., Gao, G., 2009a. An analytical solution of two-dimensional reactive solute transport in an aquifer-aquitard system. *Water Resources Research*, 45(10): W10501.

Zhan, H., Wen, Z., Huang, G., Sun, D., 2009b. Analytical solution of two-dimensional solute transport in an aquifer-aquitard system. *Journal of Contaminant Hydrology*, 107(3–4): 162–174.

Zhang, Y., Meerschaert, M.M., Baeumer, B., LaBolle, E.M., 2015. Modeling mixed retention and early arrivals in multidimensional heterogeneous media using an explicit Lagrangian scheme. *Water Resources Research*, 51(8): 6311–6337.

Zheng, C., Wang, P.P., 1999. MT3DMS: A modular three-dimensional multispecies transport model for simulation of advection, dispersion, and chemical reactions of contaminants in groundwater systems; documentation and user's guide, Alabama University.

Zheng, L., Wang, L., James, S.C., 2019. When can the local advection–dispersion equation simulate non-Fickian transport through rough fractures? *Stochastic Environmental Research and Risk Assessment*, 33(3): 931–938.

Zhou, R., Zhan, H., Chen, K., 2017. Reactive solute transport in a filled single fracture-matrix system under unilateral and radial flows. *Advances in Water Resources*, 104: 183–194.

Zhu, Q., Wen, Z., Liu, H., 2019. Microbial effects on hydraulic conductivity estimation by single-well injection tests in a petroleum-contaminated aquifer. *Journal of Hydrology*, 573: 352–364.

Zhu, Q., Wen, Z., Zhan, H., Yuan, S., 2020. Optimization strategies for *in situ* groundwater remediation by a vertical circulation well based on particle-tracking and node-dependent finite difference methods. *Water Resources Research*, 56(11): e2020WR027396.

Zlotnik, V., Ledder, G., 1996. Theory of dipole flow in uniform anisotropic aquifers. *Water Resources Research*, 32(4): 1119–1128.

Zlotnik, V.A., Zurbuchen, B.R., 1998. Dipole probe: Design and field applications of a single-borehole device for measurements of vertical variations of hydraulic conductivity. *Ground Water*, 36(6): 884–893.

Zlotnik, V.A., Eisenhauer, D.E., Schlautman, D.J., Zurbuchen, B.R., Van Peursem, D., 2007. Entrapped air effects on dipole flow test in sand tank experiments: Hydraulic conductivity and head distribution. *Journal of Hydrology*, 339(3–4): 193–205.

Chapter 4

Heat Transport

4.1 Introduction

Heat transport in the subsurface plays a crucial role in numerous hydrogeological investigations and practical applications and has received much attention in recent decades. Notably, it is employed in various areas like geothermal resource evaluation, development, thermal energy storage, as well as simulating and managing the utilization of geothermal groundwater (Klepikova *et al.*, 2016a; Lin *et al.*, 2019; Park *et al.*, 2015, 2018; Saar, 2011; Shi *et al.*, 2023a), quantification of subsurface heat flow (Bredehoeft and Papaopulos, 1965; Constantz *et al.*, 2006; Hatch *et al.*, 2006; Lin *et al.*, 2022; Reeves and Hatch, 2016), estimation of hydraulic and thermal properties (McCallum *et al.*, 2012; Park *et al.*, 2015), characterization of vertical streambed fluxes (Caissie *et al.*, 2014; Rau *et al.*, 2015; Shi and Wang, 2023), and determination of lateral groundwater inflow to streams (Chen *et al.*, 2022; Shi *et al.*, 2023b). One major advantage of using heat as a natural tracer, compared to hydraulic approaches, is that subsurface thermal characteristics change over a narrower range than hydraulic properties, potentially reducing uncertainty in heat transport parameter estimations such as groundwater flow rate and thermal conductivity (Simon *et al.*, 2021; Stonestrom and Constantz, 2003). Also, the temperature can be measured much more easily than chemical tracers due to the wide availability of robust and cost-effective temperature sensors (Selker *et al.*, 2006; Simon *et al.*, 2021, 2023; Vogt *et al.*, 2010). In the critical zones, quantifying the exchange of groundwater with surface water is a crucial endeavor encountered by researchers and practitioners in multiple environmental science fields, including hydrology, hydrogeology, ecology, climatology, and oceanography (Bardini *et al.*, 2013; Boano *et al.*, 2010; Constantz *et al.*, 1994, 2006; Constantz and Thomas, 1996, 1997; Kurylyk *et al.*, 2019; Tripathi *et al.*, 2021). The application of heat as a natural tracer offers a robust approach for determining groundwater fluxes and presents several advantages over alternative methods due to its ubiquitous presence, ability to quantify multidirectional flow, continuous monitoring capability, and simplicity in measurement, as shown in the reviews of Anderson (2005), Constantz (2008), Rau *et al.* (2014), Halloran *et al.* (2016), Irvine *et al.* (2017), Ren *et al.* (2018), Tripathi *et al.* (2021), and so on. The principles underlying these techniques are rooted in Fourier's and Darcy's laws, which regulate the conduction of heat and the flow of groundwater in saturated media, respectively.

The purpose of this chapter is to present some advanced heat transport models in fractured and porous aquifers. More specifically, in Section 4.2, we introduced the general heat transport model, followed by discussions on SWPP analytical models for fracture–matrix systems and porous media aquifers in Sections 4.3 and 4.4, respectively. These models can be applied as rapid-screening tools to estimate thermal parameters by analyzing on-site temperature measurements.

4.2 Mathematical Models of Heat Transport

4.2.1 *General model of heat transport*

4.2.1.1 *Governing equation*

Heat transport in the subsurface can be described using the differential heat transport equation (HTE), which considers both convective and conductive transport:

$$\frac{\partial T}{\partial t} = \nabla \cdot (D \nabla T) - V \cdot \nabla T + \mathcal{T}_s \tag{4.1}$$

where T is temperature [°C]; t is time [s]; D is effective thermal diffusivity [m^2/s] (also referred to as thermal dispersion tensor); V is thermal front velocity [m/s]; $\mathcal{T}_s$ is heat sources and sinks [W/m^2].

Equation (4.1) is the mathematical expression to simulate heat transport on the macroscopic scale (Bear, 1972). Therefore, the HTE equation assumes a representative elementary volume (REV) large enough that the heat transport can be described with volume-averaged parameters. The first term on the right-hand side of Eq. (4.1) represents the combination of the thermal diffusivity and thermal dispersion processes, and one has:

$$D = D_c + D_{l,t} \tag{4.2}$$

where D_c is thermal diffusivity [m^2/s] and $D_{l,t}$ is thermal dispersion [m^2/s]. The second term on the right-hand side of Eq. (4.1) represents the convective heat transport, which is described in more detail below.

4.2.1.2 *Conductive heat transport*

The definitions of the thermal diffusivity (D_c) in HTE (Eq. (4.2)) can be described as

$$D_c = \frac{k}{\rho c} \tag{4.3}$$

$$k = k_w\theta + k_s(1-\theta) \tag{4.4}$$

$$\rho c = \theta\rho_w c_w + (1-\theta)\rho_s c_s \tag{4.5}$$

where k is the thermal conductivity [W/(m·K)] of saturated medium; k_s and k_w are the thermal conductivities [W/(m·K)] of solid and fluid phases, respectively; $\rho_w c_w$ and ρc are the volumetric heat capacities [J/(m$^3\cdot$K)] of water and saturated medium, respectively; $\rho_s c_s$ is the specific volumetric heat capacity of the solid phase [J/(m$^3\cdot$K)]. ρ, ρ_w and ρ_s are the densities [kg/m^3] of the saturated medium, water and solid, respectively; c, c_w and c_s are the specific heat capacities [J/(kg·K)] of saturated medium, water and solid, respectively; θ is the total porosity [Dimensionless].

The thermal conductivity k (Eq. (4.4)) is of particular importance for heat transport calculations, and consequently empirical models for the calculation of porous media thermal conductivity have been developed. For instance,

$$k = k_w^{\theta} \cdot k_w^{1-\theta} \tag{4.6}$$

where k is the geometric mean of both phases weighted by the volume fraction θ. Equation (4.6) is a simple model for a two-phase

Table 4.1. Representative values of thermal conductivity of porous materials.

Porous materials	**Thermal conductivity W/(m · K)**
Dry sediments (e.g., sand, loam, clay)	0.18–0.26
Saturated sediments (e.g., sand, loam, clay)	1.4–2.2
Shale	1–2
Sandstone	2–4.5
Limestone	2–4
Granite	3–4
Basalt	1.5–2.5

Source: From Anderson (2005).

medium with a random distribution presented by Woodside and Messmer (1961). Table 4.1 lists the representative values of k of porous materials.

4.2.1.3 *Convective heat transport*

The thermal front velocity can be determined by

$$V = \frac{\rho_w c_w}{\rho c} q = \frac{\rho_w c_w}{\rho c} \theta_e v \tag{4.7}$$

where q is Darcy velocity (or specific discharge) [m/s]; θ_e is effective porosity [–]. It is significant to point out that the effective porosity (θ_e) is used instead of the total porosity (θ). This is because it calculates the void ratio impacted by flowing water while excluding dead-end pores from the total porosity.

In addition, note that the thermal front velocity (V) is retarded compared to the pore water velocity (v) because $\frac{\rho_w c_w}{\rho c}\theta_e$ is generally lower than 1.

4.2.1.4 *Thermal dispersion*

In natural systems, the mechanical thermal dispersion process has traditionally been considered analogously to solute transport, and an analogous mathematical description is used (e.g., Bear, 1972; De Marsily, 1986; Anderson, 2005). The effective thermal diffusivity (D) with thermal dispersion in HTE (Eq. (4.1) is given as (e.g.,

Table 4.2. The values of β_l used in literature.

References	Model	Value	Unit
Lautz and Ribaudo (2012); Park *et al.* (2018)		0.001	
Park and Lee (2021)		0.00132	
Hopmans *et al.* (2002)		0.0022	
Keery *et al.* (2007); Ronan *et al.* (1998)	$D_{l,t} = \beta_{l,t}\left\|\frac{\rho_w c_w}{\rho c}q\right\|$	0.01	m
Rau *et al.* (2010)		0.06	
Anibas *et al.* (2011); Rau *et al.* (2010)		0.1	
Cuthbert and Mackay (2013)		0.1–0.2	

Bear, 1972; De Marsily, 1986))

$$D_{l,t} = \beta_{l,t}\left|\frac{\rho_w c_w}{\rho c}q\right| \tag{4.8}$$

where subscripts l and t denote longitudinal and transverse directions, respectively. The right-hand side of Eq. (4.8) is thermal dispersion due to convection. $\beta_{l,t}$ represents the longitudinal and transverse thermal dispersivity [m]. Some representative values of β_l are listed in Table 4.2. As for the values of the transversal dispersivity (β_t), which is set to one-10th of the longitudinal one.

The importance of thermal dispersivity and its contribution to the overall process of heat transport has long been discussed in the literature (e.g., De Marsily, 1986; Anderson, 2005; Hatch *et al.*, 2006; Keery *et al.*, 2007; Rau *et al.*, 2010; Park *et al.*, 2018; Shi and Wang, 2023; Shi *et al.*, 2024).

4.2.2 *The relationship between solute and heat transport models*

There are some differences and relationships between solute transport and heat transport models. In the governing equations, the effective thermal diffusivity is composed of thermal conduction (the first term on the right-hand side of Eq. (4.2)) and mechanical thermal dispersion (the second term on the right-hand side of Eq. (4.2)), and hydrodynamic dispersion is composed of mechanical dispersion and molecular diffusion. Molecular diffusion is often overlooked because it is so small in comparison to mechanical dispersion (Chen *et al.*, 2017; Wang *et al.*, 2018, 2020b), while thermal conduction is usually much larger and should not be neglected in heat transport (Rau *et al.*, 2014; Schneidewind *et al.*, 2016; van Kampen *et al.*, 2022; Vandersteen *et al.*, 2015). When it comes to aquifer heterogeneity, a solute tracer is more effective in quantifying the seepage velocity, whereas a heat tracer is more effective at estimating the Darcy fluxes because transverse diffusive heat flux in the preferential flow pathway of a heterogeneous aquifer at low hydraulic gradient results in the smoothing out of the thermal front relative to the solute front (Irvine *et al.*, 2015). In other words, using heat as a tracer may have more significant advantages than using a solute tracer when estimating Darcy fluxes.

According to Eq. (4.7), we use the term convection instead of advection to distinguish the heat transport process from solute transport. As pointed out by Anderson (2005), both terms are used in the literature to describe heat transport. Engineering literature is more stringent and refers to heat transport induced by water flow due to external pressure gradients as forced convection. In contrast, heat transport driven by density differences is termed free convection. The latter is excluded from this book.

4.3 The SWPP Analytical Models in a Fracture–Matrix System

4.3.1 *Mathematical models*

Geothermal energy is a clean, sustainable, and renewable resource. A common approach to exploit this type of energy is by injecting cold water into hot, dry rock (HDR) first and then pumping the heated water out for usage (Eidesgaard *et al.*, 2019; Kocabas, 2005). As the water storage capacity of intact HDR is sometimes very low (Taron and Elsworth, 2009; Zimmermann *et al.*, 2011), the cold water has to be injected under high pressure to generate and expand rock fractures, and to improve the recovery efficiency of the geothermal systems. The exploitation of geothermal energy could be completed by either inter-well tests or single-well push-pull (SWPP) tests. Inter-well tests need two wells at least, where the cold water is injected into the fracture–matrix system by one well and extracted by the other well(s). The SWPP test is also named the single-well injection-withdrawal test (Kocabas, 2005), and the injection, pumping and observation are completed in the same well. As the SWPP test needs only one well, it is more efficient than the inter-well test in terms of reducing operational time and cost. Therefore, this book only concerns the SWPP test.

Two important assumptions are found in previous mathematical models of the SWPP thermal test: the thermal mixing effect in the wellbore and the fracture thermal conductance are negligible (Jung and Pruess, 2012; Li *et al.*, 2010). Where wellbore mixing effect refers to the influence of mixing processes between the injected water and original water in the well on heat transport in the aquifer. More specifically, the water temperature in the wellbore is initially equal to the fracture temperature. Due to the mixing processes between the injected water and the original water in the recharge well, the temperature varies with time. When the water with constant temperature is injected into the wellbore continuously, the temperature of the water in the wellbore will reach the same as the temperature of the injected water. Some previous studies in environmental science have demonstrated when the volume of the original water in the wellbore is great, or the injection rate is small, the mixing effect becomes obvious and cannot be ignored (Chen *et al.*, 2003; Shi *et al.*, 2020; Wang *et al.*, 2018, 2020b). Note that, as for the SWPP thermal test, the mixing process also happens during the extraction phase.

These two assumptions mentioned above are not correct in many applications and must be revoked. Meanwhile, the computational cost could be forbiddingly expensive in previous three-stage solutions of the mathematical model, which needs improvement for practical purposes. The three stages refer to injection, resting and extraction. To advance the SWPP thermal test investigation with more realistic conditions, we established a new three-stage mathematical model by specifically including the thermal wellbore mixing effect and the fracture thermal conductance.

A realistic fractured aquifer could be extremely complex as multiple sets of fractures can intercept each other, and each of those fractures could involve rough surfaces with spatiotemporally variable apertures. With often limited information about those fractures, one usually must adopt some simplification to make the problem amendable. One of those simplifications involving a single rock fracture is to use a constant hydraulic aperture to deal with a realistic fracture with rough surfaces and variable aperture distribution.

The basic idea of the hydraulic aperture is that, despite the spatial variability of aperture and flow, one can still employ the so-called cubic law (a form of Darcy's law for flow in a single fracture with a constant aperture) to describe the volumetric flux in the fracture (Lucas *et al.*, 2019; Zheng *et al.*, 2008). In this section, the conceptual model of the aquifer system used in Jung and Pruess (2012) will be employed, where the well fully penetrates a single fracture with a uniform thickness, as shown in Figure 4.1. The x-axis of the coordinate system is parallel to the fracture face, and the z-axis is perpendicular to fracture. The origin of the coordinate system is

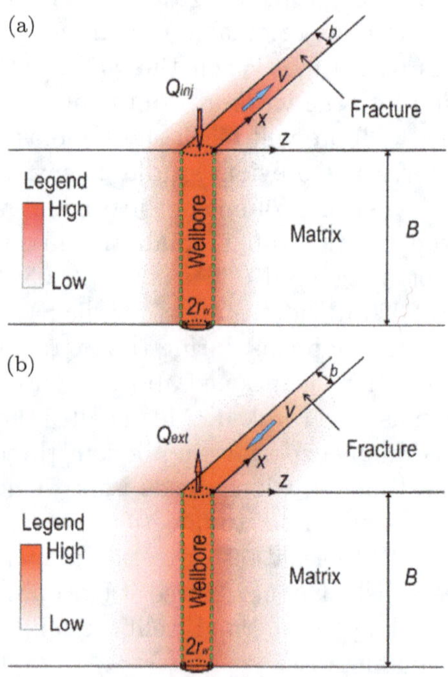

Figure 4.1. Schematic diagram of the thermal SWPP test in a fracture–matrix system with wellbore storage. (a) In the injection phase and (b) in the extraction phase.

located at the well screen/fracture interface and the middle elevation of the fracture–matrix system. The fracture can be either naturally occurring or generated by overpressure of injected fluid, such as those in hydraulic fracturing operations. The fracture aperture could be smaller and equal to the diameter of the wellbore. The buoyancy flow in the vertical fracture is neglected, and flow is horizontal during the entire operation of the SWPP test.

To date, this conceptual model has been widely used to describe groundwater flow (Jenkins and Prentice, 1982; Wen *et al.*, 2011), solute transport (Lucas *et al.*, 2019; Zhou and Zhan, 2018) and heat transport (Jung and Pruess, 2012; Kocabas, 2005) in a fracture–matrix system. Due to the difference in the physical properties of the wellbore, the fracture and the matrix (e.g., conductivity, porosity), the mathematical models have three parts, e.g., in the fracture, the matrix, and the wellbore.

4.3.1.1 *Governing equations*

The governing equations of a mathematical model of the SWPP thermal test in a fracture–matrix system incorporate two distinct HTEs, with one focused on characterizing heat transport processes within the fracture and the other describing heat transport processes within the matrix. The governing equations are

$$\rho_f c_f \frac{\partial T_f(x)}{\partial t} = k_f \frac{\partial^2 T_f(x)}{\partial x^2} - \rho_w c_w v \theta_f \frac{\partial T_f(x)}{\partial x} + \frac{k_m}{b} \left. \frac{\partial T_m}{\partial z} \right|_{z=0}, \quad x \geq r_w \tag{4.9a}$$

$$\rho_m c_m \frac{\partial T_m(x,z)}{\partial t} = k_m \frac{\partial^2 T_m(x,z)}{\partial z^2}, \ z > 0 \tag{4.9b}$$

where T_f and T_m are the temperatures [°C] of fracture and rock matrix, respectively; $\rho_f c_f = (1-\theta_f)\rho_m c_m + \theta_f \rho_w c_w$; c_f is the average specific heat [J/(m^3 · K)] of rock and fluid in fracture; ρ_f is the average density [kg/m^3] of rock and fluid in fracture; θ_f is the porosity [dimensionless] of fracture; ρ_m is the density [kg/m^3] of rock matrix; c_m is the specific heat [J/(m^3 · K)] of rock matrix; b is the half fracture aperture width [L]; k_f is the average thermal conductivity [W/(m · K)] of rock and fluid in fracture; k_m is the thermal conductivity [W/(m · K)] of rock matrix; x is the distance [m] along the flow direction, which is parallel to the fracture; z is the distance [m] normal to flow direction; v is the steady-state flow velocity [m/s] in the fracture, and the value of v could be calculated by

$$v = \frac{Q}{4bB\theta_f} \tag{4.10}$$

where Q is the flow rate [m^3/s] of the well, and it is positive for injection and negative for extraction; B is the height [m] of the fracture.

Equation (4.9a) is a thermal energy conservation statement in the fracture. In this equation, the term on the left is the heat accumulation rate. The first and second terms on the right represent the heat conductance and convection, respectively, and the last term on the right is the heat exchange between the fracture and matrix. Equation (4.9b) is the HTE in the matrix.

In the resting stage of the SWPP thermal test, the flow velocity is zero, and the governing equation of heat transport in the fracture becomes

$$\rho_f c_f \frac{\partial T_f(x)}{\partial t} = k_f \frac{\partial^2 T_f(x)}{\partial x^2} + \frac{k_m}{b} \left. \frac{\partial T_m}{\partial z} \right|_{z=0}, \quad x \geq r_w \tag{4.11}$$

Contrary to the injection stage, the direction of advection flux is reversed in the extraction stage, and the governing equation (Eq. (4.9a)) is modified as

$$\rho_f c_f \frac{\partial T_f(x)}{\partial t} = k_f \frac{\partial^2 T_f(x)}{\partial x^2} + \rho_w c_w v \theta_f \frac{\partial T_f(x)}{\partial x} + \frac{k_m}{b} \left. \frac{\partial T_m}{\partial z} \right|_{z=0}, \quad x \geq r_w \tag{4.12}$$

Unlike in Eqs. (4.9a) and (4.11)–(4.12), the model of Jung and Pruess (2012) ignored the heat conductance term in the fracture. Therefore, the governing equations of the mathematical model of the SWPP thermal test of Jung and Pruess (2012) are as follows.

In the injection phase, the governing equations are

$$\rho_f c_f \frac{\partial T_f(x)}{\partial t} = -\rho_w c_w v \theta_f \frac{\partial T_f(x)}{\partial x} + \frac{k_m}{b} \left. \frac{\partial T_m}{\partial z} \right|_{z=0}, \quad x \geq r_w \tag{4.13a}$$

$$\rho_m c_m \frac{\partial T_m(x,z)}{\partial t} = k_m \frac{\partial^2 T_m(x,z)}{\partial z^2}, \quad z > 0 \tag{4.13b}$$

In the resting phase, the governing equation in the fracture is

$$\rho_f c_f \frac{\partial T_f(x)}{\partial t} = \frac{k_m}{b} \left.\frac{\partial T_m}{\partial z}\right|_{z=0}, \quad x \geq r_w \tag{4.14}$$

In the extraction phase, the governing equation in the fracture is

$$\rho_f c_f \frac{\partial T_f(x)}{\partial t} = \rho_w c_w v \theta_f \frac{\partial T_f(x)}{\partial x} + \frac{k_m}{b} \left.\frac{\partial T_m}{\partial z}\right|_{z=0}, \quad x \geq r_w \tag{4.15}$$

Note that the governing equations in the matrix at the stages of resting and extraction are the same as Eq. (4.13b).

4.3.1.2 *Initial condition*

We assumed that the temperature was initially uniform throughout the system. For a three-stage SWPP thermal test, the initial condition is

$$T_f(x, t = 0) = T_m(x, z, t = 0) = T_0 \tag{4.16}$$

where T_0 is the initial temperature [°C] before the SWPP thermal test.

The solution at the end of the injection stage is the initial condition in the resting stage. Similarly, the temperature distribution at the end of the resting stage serves as the initial condition for the extraction stage.

4.3.1.3 *Boundary conditions without the wellbore mixing effect*

Many previous investigations assumed that the temperature of the injected fluid is constant during the injection phase. The laterally infinite boundary condition is always the same as the initial temperature. Then, the boundary conditions can be expressed as

$$T_f(x \to \infty, t) = T_0 \tag{4.17}$$

$$T_m(z \to \infty, t) = T_0 \tag{4.18}$$

$$T_f(x = 0, t) = T_{\text{inj}} \tag{4.19}$$

where T_{inj} is temperature [°C] of injected fluid.

The temperature is assumed to be continuous at the fracture–matrix interface, and one has:

$$T_f(x,t) = T_m(x, z = 0, t) \tag{4.20}$$

Equations (4.13)–(4.20) compose the mathematical model of the SWPP thermal test without wellbore mixing effect in a fracture–matrix system, and the relevant analytical solution is shown in Section 4.3.2.

4.3.1.4 *Boundary conditions with wellbore mixing effect*

As for the boundary condition in the wellbore, the model of temperature variation in the wellbore will be established based on the energy conservation law considering the wellbore mixing effect. Assuming the injected tracer instantaneously and completely mixes with the original water in the wellbore, namely wellbore mixing effect, the temperature variation in the wellbore is given by the following equation:

$$T_f(x = 0, t) = T_w(t) \tag{4.21}$$

where $T_w(t)$ is the temperature [°C] variation in the wellbore.

In the injection stage, the model describing temperature variation ($T_w(t)$) in the wellbore can be written as

$$c_w \rho_w V_{w,\mathrm{inj}} \frac{\partial T_w}{\partial t} = Q_{\mathrm{inj}}[c_w \rho_w T_{\mathrm{inj}} - c_w \rho_w T_w(t)],\ 0 < t \le t_{\mathrm{inj}} \tag{4.22a}$$

$$T_w(t = 0) = T_0 \tag{4.22b}$$

where t_{inj} is the ending time [s] of injection; $V_{w,\mathrm{inj}}$ represents the volume [m^3] of water from the wellbore in the injection stage, and $V_{w,\mathrm{inj}} = \pi r_w^2 h_{w,\mathrm{inj}}$, in which $h_{w,\mathrm{inj}}$ is the water level [m] in the wellbore in the injection stage; Q_{inj} represents the injection rate [m^3/s]. After injecting cold water into a wellbore holding hot water, T_w declines with time until it is equal to T_{inj}.

In the resting stage, the pumping rate is zero, and the convective heat transport in the fracture is neglected. Therefore, the temperature variation in the wellbore could be expressed as

$$\frac{\partial T_w}{\partial t} = 0, \; t_{\text{inj}} < t \leq t_{\text{res}} \tag{4.23a}$$

$$T_w(t = t_{\text{inj}}) = T_f|_{x=0,t=t_{\text{inj}}} \tag{4.23b}$$

where t_{res} is the ending time [s] of the resting stage.

In the extraction stage, the temperature variation could be described as

$$c_w \rho_w V_{w,\text{ext}} \frac{\partial T_w}{\partial t} = 2bBk_f \frac{\partial T_w}{\partial x}, \; t > t_{\text{res}} \tag{4.24a}$$

$$T_w|_{t_{\text{ext}}} = T_f|_{x=0,t=t_{\text{res}}} \tag{4.24b}$$

where $V_{w,\text{ext}}$ is the volume [m^3] of water in the wellbore in the extraction stage, and $V_{w,\text{ext}} = \pi r_w^2 h_{w,\text{ext}}$, in which $h_{w,\text{ext}}$ is the water level in the wellbore in the extraction stage; t_{ext} is the ending time [s] of the extraction stage.

Equations (4.9)–(4.12) and (4.16)–(4.24), except Eq. (4.19), compose the mathematical model of the SWPP thermal test with wellbore mixing effect in a fracture–matrix system, and the relevant analytical solution is displayed in Section 4.3.3.

4.3.2 *Analytical models without wellbore mixing effect*

As for the mathematical model of a three-stage SWPP thermal test without wellbore mixing effect in a fracture–matrix system (Eqs. (4.13)–(4.20)), the relevant analytical solutions using the Laplace and Fourier transform methods.

4.3.2.1 *Dimensionless transform*

To simplify the analysis, the dimensionless parameters are defined as follows:

$$T_{fD} = \frac{T_0 - T_f}{T_0 - T_{\text{inj}}}, \; T_{mD} = \frac{T_0 - T_m}{T_0 - T_{\text{inj}}}, \; x_D = \frac{k_m x}{c_w \rho_w v \theta_f b^2}, \; z_D = \frac{z}{b},$$

$$\Theta = \frac{c_m \rho_m}{c_f \rho_f}, \; t_{\text{inj}D} = \frac{k_m t}{c_f \rho_f b^2}, \; t_{\text{res}D} = \frac{k_m (t - t_{\text{inj}})}{c_f \rho_f b^2}, \; \text{and}$$

$$t_{\text{ext}D} = \frac{k_m (t - t_{\text{inj}} - t_{\text{res}})}{c_f \rho_f b^2} \tag{4.25}$$

where the subscript D represents the dimensionless terms hereinafter; $t_{\text{inj}D}$, $t_{\text{res}D}$ and $t_{\text{ext}D}$ represents the dimensionless form of the total injection time, total resting time, and total extraction time, respectively.

Using the dimensionless variables, Eqs. (4.11) and (4.9b) can be rewritten:

$$\frac{\partial T_{fD}}{\partial t_D} + \frac{\partial T_{fD}}{\partial x_D} - \left.\frac{\partial T_{mD}}{\partial z_D}\right|_{z_D=0} = 0 \tag{4.26a}$$

$$\Theta\frac{\partial T_{mD}}{\partial t_D} - \frac{\partial^2 T_{mD}}{\partial z_D^2} = 0 \tag{4.26b}$$

In the resting phase, Eq. (4.14) can be rewritten:

$$\frac{\partial T_{fD}}{\partial t_D} - \left.\frac{\partial T_{mD}}{\partial z_D}\right|_{z_D=0} = 0 \tag{4.27}$$

In the extraction phase, Eq. (4.15) can be rewritten:

$$\frac{\partial T_{fD}}{\partial t_D} - \frac{\partial T_{fD}}{\partial x_D} - \left.\frac{\partial T_{mD}}{\partial z_D}\right|_{z_D=0} = 0 \tag{4.28}$$

It needs to be noted that the dimensionless form of Eq. (4.9b) is the same in the resting and extraction phases, as is Eq. (4.26b).

The initial and boundary conditions in the dimensionless form are

$$T_{fD}(x_D, t_D = 0) = T_{mD}(x_D, z_D, t_D = 0) = 0 \tag{4.29}$$

$$T_{fD}(x_D \to \infty, t_D) = 0 \tag{4.30}$$

$$T_{mD}(z_D \to \infty, t_D) = 0 \tag{4.31}$$

$$T_{fD}(x_D = 0, t_D) = 1 \tag{4.32}$$

$$T_{mD}(x_D, z_D = 0, t_D) = T_{fD}(x_D, t_D) \tag{4.33}$$

Note that the dimensionless temperatures T_{fD} and T_{mD}, as defined in (4.25), decrease from 1 to 0 as the original temperatures increase from the cold injection temperature T_{inj} to the hot initial rock temperature T_0. Adopting this unusual convention makes the initial conditions homogeneous, as shown in (4.26a), which simplifies solving the governing equations by Laplace transforms.

4.3.2.2 *Solutions in the injection phase*

The solution for this nonisothermal injection condition is available (Jung and Pruess, 2012; Lauwerier, 1955). In the injection phase, the analytical solutions of the SWPP thermal test are:

$$T_{fD}(x_D, t_{\text{inj}D}) = \text{erfc}\left(\frac{\sqrt{\Theta}x_D}{2\sqrt{t_{\text{inj}D} - x_D}}\right) H(t_{\text{inj}D} - x_D) \tag{4.34a}$$

$$T_{mD}(x_D, z_D, t_{\text{inj}D}) = \text{erfc}\left(\frac{\sqrt{\Theta}(x_D + z_D)}{2\sqrt{t_{\text{inj}D} - x_D}}\right) H(t_{\text{inj}D} - x_D) \tag{4.34b}$$

where $H(\cdot)$ is the Heaviside step function.

4.3.2.3 *Solutions in the resting phase*

By superposition, the solution to Eq. (4.26b) satisfies the following:

$$T_{mD} = T_{mD,1} + T_{mD,2} \tag{4.35}$$

where $T_{mD,1}$ and $T_{mD,2}$ are solutions to the two problems:

$$\Theta\frac{\partial T_{mD,1}}{\partial t_D} - \frac{\partial^2 T_{mD,1}}{\partial z_D^2} = 0 \tag{4.36a}$$

$$T_{mD,1}(x_D, z_D = 0, t_{\text{res}D}) = 0, \quad 0 < t_{\text{res}D} < \infty \tag{4.36b}$$

$$\begin{aligned} T_{mD,1}(x_D, z_D, t_{\text{res}D} = 0) &= \text{erfc}\left[\frac{\sqrt{\Theta}(x_D + z_D)}{2\sqrt{t_{\text{inj}D} - x_D}}\right] H(t_{\text{inj}D} - x_D) \\ &= f(x_D, z_D), \quad 0 \le z_D < \infty \end{aligned} \tag{4.36c}$$

and

$$\Theta\frac{\partial T_{mD,2}}{\partial t_D} - \frac{\partial^2 T_{mD,2}}{\partial z_D^2} = 0 \tag{4.37a}$$

$$\begin{aligned} T_{mD,2}(x_D, 0, t_{\text{res}D}) = T_{fD}(x_D, t_{\text{res}D}) = g(x_D, t_{\text{res}D}),& \\ 0 < t_{\text{res}D} < \infty& \end{aligned} \tag{4.37b}$$

$$T_{mD,2}(x_D, z_D, t_{\text{res}D} = 0) = 0, \quad 0 \le z_D < \infty \tag{4.37c}$$

First, the solution to Eq. (4.36) is obtained using the method of images; the problem domain is extended to $z_D = -\infty$. Calling the extended function f_{ext}, we choose f_{ext} to be the odd extension of $f(x_D, z_D)$, which automatically satisfies the boundary condition Eq. (4.36b). Then, applying the Fourier transform to Eq. (4.36a) with respect to z_D, one has:

$$\frac{1}{\Theta}(i\omega)^2 \hat{T}_{mD,2} = \frac{\partial \hat{T}_{mD,2}}{\partial t_D} \tag{4.38}$$

where $\hat{T}$ is the temperature in the Fourier space. The general solution for Eq. (4.38) is

$$\hat{T}_{mD,2} = A \exp\left(-\frac{1}{\Theta}\omega^2 t_{\text{res}D}\right) \tag{4.39}$$

where A is constant. Applying the Fourier transformed boundary condition of Eq. (4.36c), we obtain:

$$A = \hat{f}_{\text{ext}}(x_D, \omega) \tag{4.40}$$

Substituting Eq. (4.40) to Eq. (4.39) and then using the Fourier convolution principle, the solution for Eq. (4.36) is

$$\begin{aligned} T_{mD,1}(x_D, z_D, t_{\text{res}D}) &= f(x_D, z_D) * \frac{\sqrt{\Theta}}{2\sqrt{\pi t_{\text{res}D}}} \exp\left[-\frac{\Theta z_D^2}{4t_{\text{res}D}}\right] \\ &= \frac{\sqrt{\Theta}}{2\sqrt{\pi t_{\text{res}D}}} \int_{-\infty}^{\infty} f(x_D, \eta) \exp\left[-\frac{\Theta(z_D - \eta)^2}{4t_{\text{res}D}}\right] d\eta \\ &= \frac{\sqrt{\Theta}}{2\sqrt{\pi t_{\text{res}D}}} \int_{0}^{\infty} f(x_D, \eta) \left\{\exp\left[-\frac{\Theta(z_D - \eta)^2}{4t_{\text{res}D}}\right] \right. \\ &\quad \left. - \exp\left[-\frac{\Theta(z_D + \eta)^2}{4t_{\text{res}D}}\right]\right\} d\eta \end{aligned} \tag{4.41}$$

where η is an integral variable of x_D.

As shown in Eq. (4.41), the solution is expressed only over the actual domain, $0 \le z_D < \infty$, and the fictitious extension is eliminated.

Next, applying the Laplace transform with respect to $t_{\mathrm{res}D}$, Eq. (4.37) becomes:

$$\frac{1}{\Theta}\frac{\partial^2 \overline{T}_{mD,2}}{\partial z_D^2} = s\overline{T}_{mD,2} \tag{4.42}$$

where over bar denotes the term in the Laplace domain hereinafter; $\overline{T}$ is the temperature in the Laplace space; s denotes the Laplace transform parameter with respect to the dimensionless time t_D.

The general solution for Eq. (4.42) is

$$\overline{T}_{mD,2} = B\exp(-\sqrt{\Theta s} z_D) \tag{4.43}$$

where B is constant. Applying the boundary condition Eq. (4.47), we obtain:

$$B = \bar{g}(x_D, s) \tag{4.44}$$

Substituting Eq. (4.44) for Eq. (4.43) and using the Laplace convolution theorem, the solution for Eq. (4.37) is

$$\begin{aligned} T_{mD,2} &= g(x_D, t_{\mathrm{res}D}) * \frac{\sqrt{\Theta} z_D}{2\sqrt{\pi (t_{\mathrm{res}D})^3}} \exp\left[-\frac{\Theta z_D^2}{4t_{\mathrm{res}D}}\right] \\ &= \int_0^{t_{\mathrm{res}D}} g(x_D, \tau)\frac{\sqrt{\Theta} z_D}{2\sqrt{\pi (t_{\mathrm{res}D}-\tau)^3}} \exp\left[-\frac{\Theta z_D^2}{4(t_{\mathrm{res}D}-\tau)}\right] \mathrm{d}\tau \end{aligned} \tag{4.45}$$

where τ is an integral variable of time.

Therefore, the solution of Eq. (4.35) is

$$\begin{aligned} T_{mD}(x_D, z_D, t_{\mathrm{res}D}) &= \frac{\sqrt{\Theta}}{2\sqrt{\pi t_{\mathrm{res}D}}} \int_0^{\infty} f(x_D, \eta)\left\{\exp\left[-\frac{\Theta (z_D-\eta)^2}{4t_{\mathrm{res}D}}\right]\right. \\ &\left. - \exp\left[-\frac{\Theta (z_D+\eta)^2}{4t_{\mathrm{res}D}}\right]\right\}\mathrm{d}\eta + \int_0^{t_{\mathrm{res}D}} g(x_D, \tau)\frac{\sqrt{\Theta} z_D}{2\sqrt{\pi (t_{\mathrm{res}D}-\tau)^3}} \\ &\times \exp\left[-\frac{\Theta z_D^2}{4(t_{\mathrm{res}D}-\tau)}\right]\mathrm{d}\tau \end{aligned} \tag{4.46}$$

Applying the Laplace transform to Eq. (4.27) with respect to $t_{\text{res}D}$, we obtain:

$$s\bar{T}_{fD} - T_{fD}(x_D, t_{\text{res}D} = 0) - \left.\frac{\partial \bar{T}_{mD}}{\partial z_D}\right|_{z_D=0} = 0 \tag{4.47}$$

From Eqs. (4.35) and (4.45), one has:

$$\left.\frac{\partial \bar{T}_{mD}}{\partial z_D}\right|_{z_D=0} = \left.\frac{\partial \bar{T}_{mD,1}}{\partial z_D}\right|_{z_D=0} - \sqrt{s\Theta}\bar{g}(x_D, s) \tag{4.48}$$

Substituting (4.48) for (4.47) and then rewriting (4.47), one has:

$$(s + \sqrt{s\Theta})\bar{T}_{fD} - T_{fD}(x_D, t_{\text{inj}D}) - \left.\frac{\partial \bar{T}_{mD,1}}{\partial z_D}\right|_{z_D=0} = 0 \tag{4.49}$$

Then, yields:

$$\bar{T}_{fD} = \frac{1}{s + \sqrt{s\Theta}}\left[T_{fD}(x_D, t_{\text{inj}D}) + \left.\frac{\partial \bar{T}_{mD,1}}{\partial z_D}\right|_{z_D=0}\right] \tag{4.50}$$

By applying the inverse Laplace transform and using the Laplace convolution principle, one has:

$$\begin{aligned} T_{fD}(x_D, t_{\text{res}D}) &= \exp(\Theta t_{\text{res}D})\text{erf}c(\sqrt{\Theta t_{\text{res}D}}) \times T_{fD}(x_D, t_{\text{inj}D}) \\ &\quad + \int_0^{t_{\text{res}D}} \exp[\Theta(t_{\text{res}D} - \tau)]\text{erf}c(\sqrt{\Theta[t_{\text{res}D} - \tau]}) \\ &\quad \times \left.\frac{\partial \bar{T}_{mD,1}}{\partial z_D}\right|_{z_D=0} \mathrm{d}\tau \end{aligned} \tag{4.51}$$

From Eq. (4.41), one has:

$$\left.\frac{\partial \bar{T}_{mD,1}}{\partial z_D}\right|_{z_D=0} = \frac{\sqrt{\Theta}}{2\sqrt{\pi t_{\text{res}D}}}\int_0^{\infty} f(x_D, \eta) \times \frac{\Theta\eta}{t_{\text{res}D}} \exp\left(-\frac{\Theta\eta^2}{4t_{\text{res}D}}\right) \mathrm{d}\eta \tag{4.52}$$

Therefore, substituting Eq. (4.52) and the initial condition $T_{fD}(x_D, t_{\text{res}D} = 0)$ into Eq. (4.51) yields the solution for the fracture temperature in the real domain during the resting phase:

$$
\begin{aligned}
T_{fD}(x_D, t_{\text{res}D}) = {} & \exp(\Theta t_{\text{res}D})\text{erf}c(\sqrt{\Theta t_{\text{res}D}}) \times T_{fD}(x_D, t_{\text{res}D} = 0) \\
& + \int_0^{t_{\text{res}D}} \exp[\Theta(t_{\text{res}D} - \tau)]\text{erf}c(\sqrt{\Theta[t_{\text{res}D} - \tau]}) \\
& \times \frac{\sqrt{\Theta}}{2\sqrt{\pi\tau}} \int_0^\infty T_{mD}(x_D, \eta, t_{\text{res}D} = 0) \\
& \times \frac{\Theta\eta}{\tau} \exp\left(-\frac{\Theta\eta^2}{4\tau}\right) \text{d}\eta \text{d}\tau
\end{aligned} \tag{4.53}
$$

The solution for the matrix temperature during the resting phase is obtained from Eqs. (4.36c), (4.37b), and (4.46):

$$
\begin{aligned}
T_{mD}(x_D, z_D, t_{\text{res}D}) = {} & \frac{\sqrt{\Theta}}{2\sqrt{\pi t_{\text{res}D}}} \int_0^\infty T_{mD}(x_D, \eta, t_{\text{res}D} = 0) \\
& \times \left\{ \exp\left[-\frac{\Theta(z_D - \eta)^2}{4t_{\text{res}D}}\right] - \exp\left[-\frac{\Theta(z_D + \eta)^2}{4t_{\text{res}D}}\right] \right\} \text{d}\eta \\
& + \int_0^{t_{\text{res}D}} T_{fD}(x_D, \tau) \frac{\sqrt{\Theta} z_D}{2\sqrt{\pi(t_{\text{res}D} - \tau)^3}} \exp\left[-\frac{\Theta z_D^2}{4(t_{\text{res}D} - \tau)}\right] \text{d}\tau
\end{aligned} \tag{4.54}
$$

4.3.2.4 *Solutions in the extraction phase*

The derivation procedure of T_{mD} in the extraction phase is identical to that of T_{mD} in the resting phase, and therefore, the solution of Eq. (4.26b) for the extraction phase is

$$
\begin{aligned}
T_{mD}(x_D, z_D, t_{\text{ext}D}) = {} & \frac{\sqrt{\Theta}}{2\sqrt{\pi t_{\text{ext}D}}} \int_0^\infty T_{mD}(x_D, \eta, t_{\text{ext}D} = 0) \\
& \times \left\{ \exp\left[-\frac{\Theta(z_D - \eta)^2}{4t_{\text{ext}D}}\right] - \exp\left[-\frac{\Theta(z_D + \eta)^2}{4t_{\text{ext}D}}\right] \right\} \text{d}\eta \\
& + \int_0^{t_{\text{ext}D}} T_{fD}(x_D, \tau) \frac{\sqrt{\Theta} z_D}{2\sqrt{\pi(t_{\text{ext}D} - \tau)^3}} \exp\left[-\frac{\Theta z_D^2}{4(t_{\text{ext}D} - \tau)}\right] \text{d}\tau
\end{aligned} \tag{4.55}
$$

Then, applying the Laplace transform to Eq. (4.26a) with respect to $t_{\text{ext}D}$, yields:

$$s\bar{T}_{fD} - T_{fD}(x_D, t_{\text{ext}D} = 0) - \frac{\partial \bar{T}_{fD}}{\partial x_D} - \left.\frac{\partial \bar{T}_{mD}}{\partial z_D}\right|_{z_D=0} = 0 \quad (4.56)$$

Similar to Eq. (4.46), one has:

$$(s + \sqrt{s\Theta})\bar{T}_{fD} - T_{fD}(x_D, t_{\text{res}D}) - \left.\frac{\partial \bar{T}_{mD,1}}{\partial z_D}\right|_{z_D=0} = 0 \quad (4.57)$$

Here the method of images is applied to extend the problem domain to $x_D = -\infty$. The initial condition $T_{fD}(x_D, t_{\text{res}D})$ is assumed to be the even function. Applying the Fourier transform to Eq. (4.57) with respect to x_D:

$$(s + \sqrt{s\Theta})\hat{\bar{T}}_{fD} - \hat{T}_{fD}(x_D, t_{\text{res}D}) - i\omega\hat{\bar{T}}_{fD} - \left.\frac{\partial \hat{\bar{T}}_{mD,1}}{\partial z_D}\right|_{z_D=0} = 0 \quad (4.58)$$

Then, one has:

$$\hat{\bar{T}}_{fD} = \frac{1}{s + \sqrt{s\Theta} - i\omega}\left[\hat{T}_{fD}(x_D, t_{\text{res}D}) + \left.\frac{\partial \hat{\bar{T}}_{mD,1}}{\partial z_D}\right|_{z_D=0}\right] = 0 \quad (4.59)$$

By applying the inverse Fourier transform and using the Fourier convolution principle, the solution is obtained and re-expressed over the actual domain, $0 < x_D < \infty$, one has:

$$\begin{aligned}
\bar{T}_{fD} &= H(-x_D)\exp[(s + \sqrt{s\Theta})x_D] * \left[T_{fD}(x_D, t_{\text{res}D}) + \left.\frac{\partial \bar{T}_{mD,1}}{\partial z_D}\right|_{z_D=0}\right] \\
&= \int_{-\infty}^{\infty} H(-\xi)\exp[(s + \sqrt{s\Theta})\xi] \\
&\quad \times \left[T_{fD}(x_D - \xi, t_{\text{res}D}) + \left.\frac{\partial \bar{T}_{mD,1}(x_D - \xi, z_D, s)}{\partial z_D}\right|_{z_D=0}\right] \mathrm{d}\xi \\
&= \int_{-\infty}^{\infty} \exp[-(s + \sqrt{s\Theta})\xi] \\
&\quad \times \left[T_{fD}(x_D + \xi, t_{\text{res}D}) + \left.\frac{\partial \bar{T}_{mD,1}(x_D + \xi, z_D, s)}{\partial z_D}\right|_{z_D=0}\right] \mathrm{d}\xi
\end{aligned} \quad (4.60)$$

where ξ is integral variable of z_D.

The inverse Laplace transform of Eq. (4.60) with respect to s is

$$T_{fD} = \int_0^{t_{\text{ext}D}} \frac{\sqrt{\Theta}\xi}{2\sqrt{\pi(t_{\text{ext}D}-\xi)^3}} \exp\left[-\frac{\Theta\xi^2}{4(t_{\text{ext}D}-\xi)}\right] \times T_{fD}(x_D+\xi, t_{\text{ext}D}=0)\mathrm{d}\xi + \int_0^{t_{\text{ext}D}} \int_0^{\tau} \frac{\sqrt{\Theta}\xi}{2\sqrt{\pi(t_{\text{ext}D}-\xi)^3}} \exp\left[-\frac{\Theta\xi^2}{4(\tau-\xi)}\right] \times \left.\frac{\partial T_{mD,1}}{\partial z_D}\right|_{z_D=0} \mathrm{d}\xi\mathrm{d}\tau \tag{4.61}$$

Similar to Eq. (4.52), one has:

$$\left.\frac{\partial \bar{T}_{mD,1}}{\partial z_D}\right|_{z_D=0} = \frac{\sqrt{\Theta}}{2\sqrt{\pi t_{\text{ext}D}}} \int_0^{\infty} T_{mD}(x_D+\xi, \eta, t_{\text{ext}D}=0) \times \frac{\Theta\eta}{t_{\text{ext}D}} \exp\left(-\frac{\Theta\eta^2}{4t_{\text{ext}D}}\right) \mathrm{d}\eta \tag{4.62}$$

Therefore, substituting Eq. (4.52) and the initial condition $T_{fD}(x_D, t_{\text{ext}D}=0)$ into Eq. (4.61) yields the solution for the fracture temperature in the real domain during the extraction phase:

$$T_{fD}(x_D, t_{\text{ext}D}) = \int_0^{\min(t_{\text{ext}D}, t_{\text{res}D}-x_D)} \frac{\sqrt{\Theta}\xi/2}{\sqrt{\pi(t_{\text{ext}D}-\xi)^3}} \exp\left[-\frac{\Theta\xi^2}{4(t_{\text{ext}D}-\xi)}\right] \times T_{fD}(x_D+\xi, 0)\mathrm{d}\xi + \int_0^{t_{\text{ext}D}} \int_0^{\min(\tau, t_{\text{res}D}-x_D)} \frac{\sqrt{\Theta}\xi/2}{\sqrt{\pi(\tau-\xi)^3}} \exp\left[-\frac{\Theta\xi^2}{4(\tau-\xi)}\right] \times\Theta \int_0^{\infty} T_{mD}(x_D+\xi, \eta, 0) \times \frac{\sqrt{\Theta}\eta/2}{\sqrt{\pi(t_{\text{ext}D}-\tau)^3}} \exp\left[-\frac{\Theta z_D^2}{4(t_{\text{ext}D}-\tau)}\right] \mathrm{d}\eta\mathrm{d}\xi\mathrm{d}\tau \tag{4.63}$$

Similar to Eq. (4.54), one has:

$$T_{mD}(x_D, z_D, t_{\text{ext}D}) = \frac{\sqrt{\Theta}}{2\sqrt{\pi t_{\text{ext}D}}} \int_0^\infty T_{mD}(x_D, \eta, t_{\text{ext}D} = 0)$$
$$\times \left\{ \exp\left[-\frac{\Theta(z_D - \eta)^2}{4t_{\text{ext}D}}\right] - \exp\left[-\frac{\Theta(z_D + \eta)^2}{4t_{\text{ext}D}}\right] \right\} d\eta$$
$$+ \int_0^{t_{\text{ext}D}} T_{fD}(x_D, \tau) \frac{\sqrt{\Theta} z_D}{2\sqrt{\pi(t_{\text{ext}D} - \tau)^3}} \exp\left[-\frac{\Theta z_D^2}{4(t_{\text{ext}D} - \tau)}\right] d\tau \tag{4.64}$$

Equations (4.34a)–(4.34b), (4.53)–(4.54), and (4.63)–(4.64) are solutions of the SWPP thermal test without wellbore mixing effect in the injection, resting, and extraction stages, respectively.

4.3.3 *Analytical models with wellbore mixing effect*

As for the mathematical model of the three-stage SWPP thermal test with wellbore mixing effect in a fracture–matrix system (Eqs. (4.9)–(4.12) and (4.16)–(4.24), except Eq. (4.19)), the relevant analytical solutions can be derived using the Laplace and Green's function methods.

4.3.3.1 *Dimensionless transform*

By substituting these dimensionless parameters defined in Eq. (4.25) into Eqs. (4.9a)–(4.9b), one has:

$$\frac{\partial T_{fD}}{\partial t_D} + \frac{\partial T_{fD}}{\partial x_D} - \frac{1}{\phi}\frac{\partial^2 T_{fD}}{\partial x_D^2} - \left.\frac{\partial T_{mD}}{\partial z_D}\right|_{z_D=0} = 0 \tag{4.65a}$$

$$\Theta\frac{\partial T_{mD}}{\partial t_D} - \frac{\partial^2 T_{mD}}{\partial z_D^2} = 0 \tag{4.65b}$$

where $\phi = \frac{(c_w \rho_w v \theta_f b)^2}{k_f k_m}$.

In the resting phase of the SWPP thermal test, the dimensionless governing equation of heat transport in the fracture becomes:

$$\frac{\partial T_{fD}}{\partial t_D} - \frac{1}{\phi}\frac{\partial^2 T_{fD}}{\partial x_D^2} - \left.\frac{\partial T_{mD}}{\partial z_D}\right|_{z_D=0} = 0 \tag{4.66}$$

Similarly, the dimensionless governing equation in the fracture is

$$\frac{\partial T_{fD}}{\partial t_D} - \frac{\partial T_{fD}}{\partial x_D} - \frac{1}{\phi}\frac{\partial^2 T_{fD}}{\partial x_D^2} - \left.\frac{\partial T_{mD}}{\partial z_D}\right|_{z_D=0} = 0 \tag{4.67}$$

The initial and boundary conditions in the dimensionless form are the same as Eqs. (4.29)–(4.33), except for Eq. (4.32). Equation (4.32) should be rewritten as

$$T_{fD}(x_D = 0, t_D) = T_{wD}(t_D) \tag{4.68}$$

In the injection stage, the dimensionless form of the inner boundary conditions (Eqs. (4.22a)–(4.22b)) becomes:

$$\alpha_{\text{inj}}\frac{\partial T_{wD}}{\partial t_D} = 1 - T_{wD} + \frac{\theta_f}{\phi}\frac{\partial T_{wD}}{\partial x_D},\ 0 < t_D \le t_{\text{inj}D} \tag{4.68a}$$

$$T_{wD}(t_D = 0) = 0 \tag{4.68b}$$

where $\alpha_{\text{inj}} = \frac{V_{w,\text{inj}}k_m}{Q_{\text{inj}}c_w\rho_w b^2}$.

In the resting stage, the dimensionless form of the inner boundary condition (Eq. (4.23)) becomes:

$$\frac{\partial T_{wD}}{\partial x_D} = 0,\ t_{\text{inj}D} < t_D \le t_{\text{res}D} \tag{4.69a}$$

$$T_{wD}(t_D = 0) = T_{wD}(t_D = t_{\text{inj}D}) \tag{4.69b}$$

Similarly, in the extraction stage, the dimensionless form of inner boundary condition (Eq. (4.24)) becomes:

$$\alpha_{\text{ext}}\frac{\partial T_{wD}}{\partial t_D} = \left.\frac{\partial T_{wD}}{\partial x_D}\right|_{x_D=0},\ t_D > t_{\text{res}D} \tag{4.70a}$$

$$T_{wD}(t_D = 0) = T_{wD}(t_D = t_{\text{res}D}) \tag{4.70b}$$

where $\alpha_{\text{ext}} = \frac{V_{w,\text{ext}}(c_w\rho_w)^2 v}{4bBc_w\rho_w k_f}$.

4.3.3.2 *Solutions in the injection phase*

Conducting Laplace transform to Eqs. (4.65a)–(4.65b), one has:

$$s\overline{T}_{fD} + \frac{\partial \overline{T}_{fD}}{\partial x_D} - \frac{1}{\phi}\frac{\partial^2 \overline{T}_{fD}}{\partial x_D^2} - \left.\frac{\partial \overline{T}_{mD}}{\partial z_D}\right|_{z_D=0} = 0 \tag{4.71a}$$

$$\Theta s\overline{T}_{mD} - \frac{\partial^2 \overline{T}_{mD}}{\partial z_D^2} = 0 \tag{4.71b}$$

Conducting Laplace transform to Eqs. (4.29)–(4.31), and (4.33), one has:

$$\overline{T}_{fD}(x_D, s=0) = \overline{T}_{mD}(x_D, z_D, s=0) = 0 \tag{4.72}$$

$$\overline{T}_{fD}(x_D \to \infty, s) = 0 \tag{4.73}$$

$$\overline{T}_{mD}(x_D, z_D \to \infty, s) = 0 \tag{4.74}$$

$$\overline{T}_{mD}(x_D, z_D = 0, s) = \overline{T}_{fD}(x_D, s) \tag{4.75}$$

In the injection stage, conducting Laplace transform to Eqs. (4.68a)–(4.68b), one has:

$$(s\alpha_{\text{inj}} + 1)\overline{T}_{wD} - \frac{\theta_f}{\phi}\frac{\partial T_{wD}}{\partial x_D} = \frac{1}{s} \tag{4.76}$$

Thus, Eqs. (4.71a)–(4.76) compose a model of the second-order ordinary differential equation (ODE) with boundary conditions. The general solution of Eq. (4.71b) is

$$\overline{T}_{mD} = A_1 \exp(-\sqrt{\Theta s} z_D) + A_2 \exp(\sqrt{\Theta s} z_D) \tag{4.77}$$

In terms of the boundary condition given by Eq. (4.74), we can obtain that A_2 equals to zero. Then, Eq. (4.77) becomes:

$$\overline{T}_{mD} = A_1 \exp(-\sqrt{\Theta s} z_D) \tag{4.78}$$

Substituting Eq. (4.78) into Eq. (4.75), one has:

$$\overline{T}_{mD} = \overline{T}_{fD}(x_D, s) \exp(-\sqrt{\Theta s} z_D) \tag{4.79}$$

Substituting Eq. (4.79) into Eq. (4.71a), one has:

$$\frac{\partial^2 \overline{T}_{fD}}{\partial x_D^2} - \phi\frac{\partial \overline{T}_{fD}}{\partial x_D} - \phi(s + \sqrt{\Theta s})\overline{T}_{fD} = 0 \tag{4.80}$$

Hence, the general solution of Eq. (4.80) is

$$\overline{T}_{fD} = C_1 \exp\left[\frac{1}{2}(\phi + \delta)x_D\right] + C_2 \exp\left[\frac{1}{2}(\phi - \delta)x_D\right] \tag{4.81}$$

where $\delta = \sqrt{\phi^2 + 4(s + \sqrt{\Theta s})\phi}$, C_1 and C_1 are constants that the boundary conditions could determine.

Substituting Eq. (4.81) into Eqs. (4.73) and (4.76), one has:

$$C_1 = 0 \tag{4.82a}$$

$$C_2 = \frac{1}{s(s\alpha_{\text{inj}} + 1)} + \frac{\theta_f(\phi - \delta)}{2\phi(s\alpha_{\text{inj}} + 1)} \tag{4.82b}$$

Substituting Eqs. (4.82a)–(4.82b) into Eq. (4.81), one has:

$$\overline{T}_{fD} = \left[\frac{1}{s(s\alpha_{\text{inj}} + 1)} + \frac{\theta_f(\phi - \delta)}{2\phi(s\alpha_{\text{inj}} + 1)}\right] \exp\left[\frac{1}{2}(\phi + \delta)x_D\right] \tag{4.83}$$

The solution of the SWPP thermal test in the matrix is

$$\overline{T}_{mD} = \left[\frac{1}{s(s\alpha_{\text{inj}} + 1)} + \frac{\theta_f(\phi - \delta)}{2\phi(s\alpha_{\text{inj}} + 1)}\right] \exp\left[\frac{1}{2}(\phi + \delta)x_D\right] \times \exp(-\sqrt{\Theta s} z_D) \tag{4.84}$$

4.3.3.3 *Solutions in the resting phase*

Conducting Laplace transform to Eqs. (4.65b) and (4.66), one has:

$$s\overline{T}_{fD} - T_{fD}(t_{\text{inj}D}) = \frac{1}{\phi}\frac{\partial^2 \overline{T}_{fD}}{\partial x_D^2} + \left.\frac{\partial \overline{T}_{mD}}{\partial z_D}\right|_{z_D=0} \tag{4.85a}$$

$$\frac{\partial^2 \overline{T}_{mD}}{\partial z_D^2} - \Theta s \overline{T}_{mD} = -\Theta T_{mD}(t_{\text{inj}D}) \tag{4.85b}$$

Conducting Laplace transform to Eq. (4.69a), one has:

$$\frac{\partial \overline{T}_{wD}}{\partial x_D} = 0 \tag{4.86}$$

Equations (4.74)–(4.75) and (4.85b) compose a model of the second-order ODE with boundary conditions; however, the governing equation and the boundary condition are inhomogeneous differential equations. In this section, we use the Green's function method to derive the analytical solution of Eq. (4.85b). Let $\overline{T}_{mD} = \Psi(z_D) + \delta_1 + \delta_2 z_D$, and substituting them into Eqs. (4.74)–(4.75) yields:

$$\Psi(z_D = 0) = 0 \tag{4.87a}$$

$$\Psi(z_D)|_{z_D \to \infty} = 0 \tag{4.87b}$$

where $\delta_1 = \overline{T}_{fD}(x_D, s)$ and $\delta_2 = -\frac{\overline{T}_{fD}(x_D,s)}{z_{eD}}$, z_{eD} represents $z_D \to \infty$.

Defining the spatial operator: $L = -[\frac{d^2}{dz_D^2} - \Theta s]$, one has:

$$L\overline{T}_{mD} = L[\Psi(z_D) + \delta_1 + \delta_2 z_D] = \Theta T_{mD}(t_{\text{inj}D}) \tag{4.88}$$

Let $L\Psi(z_D) = \Theta T_{mD}(t_{\text{inj}D}) - L[\delta_1 + \delta_2 z_D]$, one has:

$$\frac{\partial^2 \Psi(z_D)}{\partial z_D^2} - \Theta s \Psi(z_D) = -f(z_D) \tag{4.89}$$

where $f(z_D) = \Theta T_{mD}(t_{\text{inj}D}) - \Theta s(\delta_1 + \delta_2 z_D)$.

Hence, the model with inhomogeneous boundary conditions becomes a homogeneous boundary. The general solution of Eqs. (4.87a)–(4.87b) and (4.89) is:

$$\Psi(z_D, \Theta; s; \eta) = \int_0^\infty g(z_D, \Theta; s; \eta) f(\eta) d\eta \tag{4.90}$$

where η is a positive value varying between 0 and ∞; $f(\eta) = \Theta T_{mD}(t_{\text{inj}D}) - \Theta s(\delta_1 + \delta_2 \eta)$; $g(z_D, \Theta; s; \eta)$ is the Green's function, and could be expressed as

$$g(z_D, \Theta; s; \eta) = \begin{cases} B_1 \exp(-\sqrt{\Theta s} z_D) + B_2 \exp(\sqrt{\Theta s} z_D) \; 0 \le z_D \le \eta \\ B_3 \exp(-\sqrt{\Theta s} z_D) + B_4 \exp(\sqrt{\Theta s} z_D) \; \eta < z_D \le \infty \end{cases} \tag{4.91}$$

where B_1, B_2, B_3 and B_4 are unknown coefficients, which the following conditions could obtain (Chen and Woodside, 1988).

(a) $g(z_D, \Theta; s; \eta)$ is satisfied with the boundary conditions Eqs. (4.87b)–(4.87b),
(b) $g_1(z_D, \Theta; s; \eta) = g_2(z_D, \Theta; s; \eta)$;
(c) $\left.\frac{dg_2}{dz_D}\right|_{z_D=\eta^+} - \left.\frac{dg_1}{dz_D}\right|_{z_D=\eta^-} = -1.$

According to the boundary condition is given by Eq. (4.87b), the value of B_4 is zero. Based on the condition (a), one has:

$$B_1 + B_2 = 0 \tag{4.92}$$

According to the condition of (b), one has:

$$B_1 \exp(-\sqrt{\Theta s}\eta) + B_2 \exp(\sqrt{\Theta s}\eta) = B_3 \exp(-\sqrt{\Theta s}\eta) \tag{4.93}$$

According to the condition of (c), one has:

$$\begin{aligned} &-\sqrt{\Theta s}B_3 \exp(-\sqrt{\Theta s}\eta) + \sqrt{\Theta s}B_1 \exp(-\sqrt{\Theta s}\eta) \\ &-\sqrt{\Theta s}B_2 \exp(\sqrt{\Theta s}\eta) = -1 \end{aligned} \tag{4.94}$$

In the resting stage, the values of B_1, B_2, B_3, and B_4 could be determined by Eqs. (4.92)–(4.94), namely:

$$\begin{aligned} B_1 &= -\frac{1}{2\sqrt{\Theta s}\exp(\sqrt{\Theta s}\eta)} \\ B_1 &= -B_2 \\ B_3 &= B_1 - B_1 \exp(2\sqrt{\Theta s}\eta) \text{ and } B_4 = 0. \end{aligned}$$

Therefore, the solutions of the SWPP test in the matrix are

$$\overline{T}_{mD} = \int_0^\infty g(z_D, \Theta; s; \eta) f(\eta) d\eta + \delta_1 + \delta_2 z_D \tag{4.95}$$

Substituting Eq. (4.95) into Eq. (4.85a), one has:

$$\frac{\partial^2 \overline{T}_{fD}}{\partial x_D^2} - \phi\left(s + \frac{1}{z_{eD}}\right)\overline{T}_{fD} = \phi T_{fD}(t_{\text{inj}D}) \tag{4.96}$$

Similarly, we use the Green's function to drive the solution of Eq. (4.96). The general solution of Eq. (4.96) could be expressed as

$$\overline{T}_{fD} = \int_0^{\infty} G_1(x_D, s; \chi) f_0(\chi) d\chi \tag{4.97}$$

$$G_1(x_D, s; \chi) = \begin{cases} C_1 \exp\left[-\sqrt{\phi(s + \frac{1}{z_{eD}})} x_D\right] \\ \quad + C_2 \exp\left[\sqrt{\phi(s + \frac{1}{z_{eD}})} x_D\right] \; 0 \le x_D \le \chi \\ C_3 \exp\left[-\sqrt{\phi(s + \frac{1}{z_{eD}})} x_D\right] \\ \quad + C_4 \exp\left[\sqrt{\phi(s + \frac{1}{z_{eD}})} x_D\right] \; \chi < x_D \le \infty \end{cases} \tag{4.98}$$

where χ is a positive value varying between 0 and ∞; $f_0(\chi) = \phi T_{fD}(t_{\text{inj}D})$; $G_1(x_D, s; \chi)$ is the Green's function, C_1, C_2, C_3 and C_4 are the coefficients. According to the boundary condition is given by Eq. (4.73), C_4 must be zero. Similarly, based on the conditions, the unknown coefficients C_1, C_2 and C_3 could be obtained. The values of C_1, C_2, C_3, and C_4 are

$$C_1 = C_2 = \frac{1}{2\sqrt{\phi\left(s + \frac{1}{z_{eD}}\right)} \exp\left(\sqrt{\phi\left(s + \frac{1}{z_{eD}}\right)}\chi\right)},$$

$$C_3 = C_1 + C_1 \exp\left(2\sqrt{\phi\left(s + \frac{1}{z_{eD}}\right)}\chi\right) \quad \text{and} \quad C_4 = 0.$$

4.3.3.4 *Solutions in the extraction phase*

Contrary to the injection stage, the direction of advection flux is reversed in the extraction stage. The governing equation in Laplace domain is

$$s\overline{T}_{fD} - \frac{\partial \overline{T}_{fD}}{\partial x_D} - \frac{1}{\phi}\frac{\partial^2 \overline{T}_{fD}}{\partial x_D^2} - \left.\frac{\partial \overline{T}_{mD}}{\partial z_D}\right|_{z_D=0} = T_{fD}(x_D, t_D = t_{\text{res}D}) \tag{4.99a}$$

$$\Theta s\overline{T}_{mD} - \frac{\partial^2 \overline{T}_{mD}}{\partial z_D^2} = \Theta T_{mD}(x_D, t_D = t_{\text{res}D}) \tag{4.99b}$$

In the extraction stage, the inner boundary condition in the Laplace domain could be described as follows:

$$\left[\frac{\partial \overline{T}_{wD}}{\partial x_D} - \alpha_{\text{ext}} s\overline{T}_{wD}\right]\Bigg|_{x_D=0} = -\alpha_{\text{ext}} T_{fD}(x_D = 0, t_D = t_{\text{res}D}) \tag{4.100}$$

Similarly, the governing equation and the boundary condition are inhomogeneous differential equations; we use the Green's function method to derive the analytical solution of Eq. (4.99b).

Let $\overline{T}_{mD} = M(z_D) + a + bz_D$, and one has:

$$M(z_D)|_{z_D=0} = 0 \tag{4.101a}$$

$$M(z_D)|_{z_D\to\infty} = 0 \tag{4.101b}$$

where $a = \overline{T}_{f2D}(x_D, s)$, and $b = -\frac{\overline{T}_{f2D}(x_D,s)}{z_{eD}}$.

$$L\overline{T}_{mD} = L[M(z_D) + a + bz_D] = \Theta T_{mD}(x_D, t_D = t_{\text{res}D}) \tag{4.102}$$

Let $f_1(z_D) = \Theta T_{mD}(x_D, t_D = t_{\text{res}D}) - L[a + bz_D]$, one has:

$$\frac{d^2 M}{dz_D^2} - \Theta s M = -f_1(z_D) \tag{4.103}$$

where $f_1(z_D) = \Theta T_{mD}(x_D, t_D = t_{\text{res}D}) - \Theta s(a + bz_D)$.

Right now, the model with inhomogeneous boundary conditions becomes a regular Sturm–Louisville problem. Then, the general solution of Eq. (4.103) is

$$M(z_D, \Theta; s; \varsigma) = \int_0^\infty g(z_D, \Theta; s; \varsigma) f_1(\varsigma) d\varsigma \tag{4.104}$$

where ς is a positive value varying between 0 and ∞, $g(z_D, \Theta; s; \varsigma)$ is the Green's function, which could be expressed as

$$g(z_D, \Theta; s; \varsigma) = \begin{cases} g_1(z_D, \Theta; s; \varsigma) = D_1 \exp(-\sqrt{\Theta s} z_D) \\ \qquad + D_2 \exp(\sqrt{\Theta s} z_D) \;\; 0 \le z_D < \varsigma \\ g_2(z_D, \Theta; s; \varsigma) = D_3 \exp(-\sqrt{\Theta s} z_D) \\ \qquad + D_4 \exp(\sqrt{\Theta s} z_D) \;\; \varsigma < z_D \le \infty \end{cases} \tag{4.105}$$

where $f_1(\varsigma) = \Theta T_{mD}(\varsigma, t_D = t_{\text{res}D}) - \Theta s(a + b\varsigma)$, D_1, D_2, D_3 and D_4 are the coefficients which could be determined by the boundary conditions and character of the Green's function.

According to the boundary condition given by Eq. (4.74), the value of D_4 is zero. Based on the conditions (a), (b) and (c), one has:

$$[D_1 \exp(-\sqrt{\Theta s} z_D) + D_2 \exp(\sqrt{\Theta s} z_D)]|_{z_D=0} = 0 \tag{4.106}$$

$$\begin{aligned} &[D_1 \exp(-\sqrt{\Theta s} z_D) + D_2 \exp(\sqrt{\Theta s} z_D)]|_{z_D=\varsigma^+} \\ &\quad = [D_3 \exp(-\sqrt{\Theta s} z_D)]|_{z_D=\varsigma^-} \end{aligned} \tag{4.107}$$

$$\begin{aligned} &[-\sqrt{\Theta s} D_3 \exp(-\sqrt{\Theta s} z_D)]|_{z_D=\varsigma^-} - [-\sqrt{\Theta s} D_1 \exp(-\sqrt{\Theta s} z_D) \\ &\quad + \sqrt{\Theta s} D_2 \exp(\sqrt{\Theta s} z_D)]|_{z_D=\varsigma^+} = -1 \end{aligned} \tag{4.108}$$

The solution in the resting stage, the values of D_1, D_2, D_3 and D_4 could be determined by Eqs. (4.106)–(4.108), and one has:

$$D_1 = -D_2, \quad D_2 = \frac{1}{2\sqrt{\Theta s} \exp(\sqrt{\Theta s} \varsigma)},$$

$$D_3 = \frac{\exp(\sqrt{\Theta s} \phi) - \exp(-\sqrt{\Theta s} \varsigma)}{2\sqrt{\Theta s}}, \quad \text{and} \quad D_4 = 0.$$

Then, Eq. (4.99a) becomes:

$$\frac{\partial^2 \overline{T}_{fD}}{\partial x_D^2} + \phi \frac{\partial \overline{T}_{fD}}{\partial x_D} - \left(s + \frac{1}{z_{eD}}\right) \phi \overline{T}_{fD} = -\phi T_{fD}(x_D, t_D = t_{\text{res}D}) \tag{4.109}$$

Similarly, we use the Green's function to drive the solution of Eq. (4.109). Let $\overline{T}_{fD} = N(x_D) + m + n x_D$, and one has:

$$\left[\frac{\partial N}{\partial x_D} - \alpha_{\text{ext}} s N\right]\Bigg|_{x_D=0} = 0 \tag{4.110a}$$

$$N(x_D)|_{x_D \to \infty} = 0 \tag{4.110b}$$

where $n = -\frac{\alpha_{\text{ext}} T_{fD}(x_D=0, t_D=t_{\text{res}D})}{1+\alpha_{\text{ext}} s x_{eD}}$, $m = -n x_{eD}$ and x_{eD} represents $x_D \to \infty$.

Define the spatial operator: $L_1 = -\left[\frac{d^2}{dx_D^2} + \phi\frac{d}{dx_D} - (s + \frac{1}{z_{eD}})\phi\right]$, one has:

$$L_1\overline{T}_{fD} = L_1[N(x_D) + m + nx_D] = \phi T_{fD}(x_D, t_D = t_{\text{res}D}) \quad (4.111)$$

Let $f_2(x_D) = \phi T_{fD}(x_D, t_D = t_{\text{res}D}) - L_1[m + nx_D]$, then:

$$\frac{d^2N}{dx_D^2} + \phi\frac{dN}{dx_D} - \left(s + \frac{1}{z_{eD}}\right)\phi N = -f_2(x_D) \quad (4.112)$$

where $f_2(x_D) = \phi T_{fD}(x_D, t_D = t_{\text{res}D}) + \phi n - \phi(s + \frac{1}{z_{eD}})(m + nx_D)$. Thus, the general solution of Eq. (4.112) could be expressed as:

$$\overline{T}_{fD} = \int_0^\infty G(x_D, s; \varphi) f_2(\varphi) d\varphi \quad (4.113)$$

where φ is a value varying between 0 and ∞, $G(x_D, s; \varphi)$ is the Green's function:

$$G(x_D, s; \varphi) = \begin{cases} E_1 \exp\left[\frac{1}{2}(\varepsilon - \phi)x_D\right] + E_2 \exp\left[-\frac{1}{2}(\varepsilon + \phi)x_D\right] & 0 \le x_D < \varphi \\ E_3 \exp\left[\frac{1}{2}(\varepsilon - \phi)x_D\right] + E_4 \exp\left[-\frac{1}{2}(\varepsilon + \phi)x_D\right] & \varphi < x_D \le \infty \end{cases} \quad (4.114)$$

where $\varepsilon = \sqrt{\phi^2 + 4(s + \frac{1}{z_{eD}})\phi}$ and $f_2(\varphi) = \phi T_{fD}(\varphi, t_D = t_{\text{res}D}) + \phi n - \phi(s + \frac{1}{z_{eD}})(m + n\varphi)$. Similarly, based on the conditions, the unknown coefficients E_1, E_2, E_3, and E_4 could be obtained. The values of E_1, E_2, E_3, and E_4 are

$$E_1 = WE_2, \quad E_2 = \frac{1}{W\varepsilon \exp\left[\frac{1}{2}(\varepsilon - \phi)\varphi\right]},$$

$$E_3 = 0, \quad E_4 = \frac{W \exp\left[\frac{1}{2}(\varepsilon - \phi)\varphi\right] + \exp\left[-\frac{1}{2}(\varepsilon + \phi)\varphi\right]}{\exp\left[-\frac{1}{2}(\varepsilon + \phi)\varphi\right]} E_2 \quad \text{and}$$

$$W = \frac{\frac{1}{2}(\varepsilon + \phi) + \alpha_{\text{ext}} s}{\frac{1}{2}(\varepsilon - \phi) - \alpha_{\text{ext}} s}.$$

Based on the determined coefficients, the analytical solutions of the SWPP thermal test in the extraction phase are

$$\overline{T}_{fD} = \int_0^\infty G(x_D, s; \varphi) f_2(\varphi) d\varphi + m + n x_D \tag{4.115}$$

$$\overline{T}_{mD} = \int_0^\infty g(z_D, \Theta; s; \varsigma) f_1(\varsigma) d\varsigma + a + b z_D \tag{4.116}$$

Equations (4.83)–(4.84), (4.95), (4.97), and (4.115)–(4.116) are solutions of the SWPP thermal test in the Laplace domain in the injection, resting, and extraction stages, respectively. It is not easy to analytically transform them into real-time domain solutions. Fortunately, many numerical inverse Laplace transform methods are available, and the De Hoog method and the Stehfest method have been found to be accurate and efficient in solving problems related to solute transport (Wang and Zhan, 2015). In this book, the Stehfest method will be employed for the inverse Laplace transform on Eqs. (4.83)–(4.84), (4.95), (4.97), and (4.115)–(4.116).

4.3.4 *Comparisons of analytical models*

The new solutions described in Section 4.3.3 are the extension of many previous models. For instance, when the wellbore mixing effect and the fracture conductance effect are ignored (e.g., $r_w = 0$ and $k_f = 0$) the new solution reduces to the solution of Jung and Pruess (2012). If the resting time is set as zero, the three-stage solution becomes the two-stage solution, as studied by Kocabas (2005). As for the computational efficiency, the computational time of the three-stage solution of this section is much less than that involving six-dimension integrals (about 1 week), as mentioned by Maier and Kocabas (2013). For the cases discussed in the following sections, the computational time is typically a few seconds on a personal computer. Therefore, the new solutions are more efficient and easier to implement.

Comparing with the model of Jung and Pruess (2012), the new solutions described in Section 4.3.3 include two factors: the mixing effect in the wellbore and the fracture thermal conductance. To test the influences of these two factors on the results, the comparisons of models with the following constraints have been conducted:

(a) The model without the mixing effect and the fracture thermal conductance ($r_w = 0$ and $k_f = 0$);
(b) The model with the mixing effect but without the fracture thermal conductance ($r_w = 0.5$ m and $k_f = 0$);
(c) The model without the mixing effect but with the fracture thermal conductance ($r_w = 0$ and $k_f = 2.1$ W/(m · K));
(d) The model with the mixing effect and the fracture thermal conductance ($r_w = 0.5$ m and $k_f = 2.1$ W/m/°C).

Such comparisons are shown in Figure 4.2. The base values of parameters for the sensitivity analysis are the same as those in Table 1 of Jung and Pruess (2012): $k_f = 2.1$ W/(m · K), $k_m = 2.1$ W/(m · K), $\rho_m = 2650$ kg/m^3, $c_m = 1000$ Jm^{-1}kg^{-2}K, $\rho_w = 1000$ kg/m^3, $v = 4 \times 10^{-4}$ m/s, $c_w = 4200$ Jm^{-1}kg^{-2}K, $\theta_f = 0.5$, $b = 0.01$ m, $B = 50$ m, $t_{\text{inj}} = 5$ h, $t_{\text{res}} = 5$ h. According to Eq. (4.9), one has $Q_{\text{inj}} = -Q_{\text{ext}} = 4bB\phi_f v = 1.44$ m^3/h. The

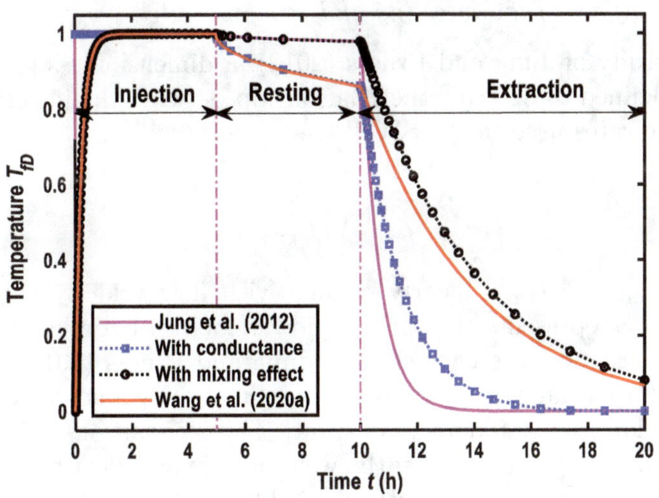

Figure 4.2. Comparison of BTCs in the wellbore among. "Jung and Pruess (2012)" ignores both mixing effect and the fracture conductance, "with mixing effect" considers the mixing effect but ignores the fracture conductance, "with conductance" ignores the mixing effect but considers the fracture conductance, and "Wang *et al.* (2020a)" which considers both mixing effect and fracture conductance.

other parameters are: $r_w = 0.5$ m, $h_{w,\text{inj}} = 65$ m, $h_{w,\text{res}} = 62.5$ m and $h_{w,\text{ext}} = 60$ m.

Several observations could be made from Figure 4.2. Firstly, the differences in breakthrough curves (BTCs) among the four scenarios mentioned above are obvious in the extraction stage. Therefore, both the wellbore mixing and fracture thermal conductance factors could not be ignored in general. Secondly, the solution of "$r_w = 0.5$ m and $k_f = 0$" is above others in the injection stage, while the solution of Jung and Pruess (2012) is the smallest after 1 h in the extraction stage. It implies that the solution of "$r_w = 0.5$ m and $k_f = 0$" overestimates BTCs, and the solution of Jung and Pruess (2012) underestimates BTCs. Thirdly, the solution of "$r_w = 0.5$ m and $k_f = 0$" is close to the solution of "$r_w = 0.5$ m and $k_f = 2.1$ W/(m · K)" in the injection stage, while the solution of "$r_w = 0$ and $k_f = 2.1$ W/(m · K)" is almost the same as the solution of "$r_w = 0$ and $k_f = 0$".

4.3.5 *Discussions*

4.3.5.1 *Effect of the injecting and pumping rates*

In the study of Jung and Pruess (2012), a dimensionless parameter of λ is defined to describe the relationship between the injection and pumping rates used in the SWPP test:

$$\lambda = \frac{Q_{\text{inj}}}{Q_{\text{ext}}} \tag{4.117}$$

As Q_{inj} and Q_{ext} linearly change with flow velocity (v) in the fracture, λ represents the ratio of flow velocities between the injection and extraction stages as well. Jung and Pruess (2012) pointed out that thermal BTCs are independent of the values of Q_{inj} and Q_{ext} as long as λ remains constant. This conclusion is based on two important presumptions: the wellbore mixing effect and fracture thermal conductance are both negligible. In this section, the relationship between BTCs and λ will be revisited based on the solution of this section.

Figure 4.3 shows thermal BTCs in the wellbore for different v but the same λ. Obviously, the difference between thermal BTCs could not be negligible. A smaller Q_{inj} results in a smaller T_{fD} in the

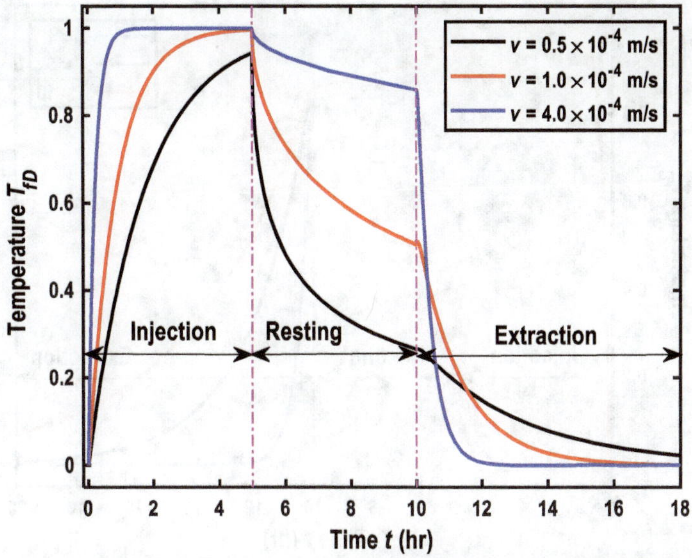

Figure 4.3. Comparison of BTCs at the wellbore for different v when $\lambda = 1$.

injection and resting stages. This observation could be explained by Eq. (4.22a). As T_{inj} is specific, the value $T_w(t)$ is dependent on the variation of $\frac{V_{w,\text{inj}}}{Q_{\text{inj}}}\frac{\partial T_w}{\partial t}$. When the term of $\frac{V_{w,\text{inj}}}{Q_{\text{inj}}}\frac{\partial T_w}{\partial t}$ is zero, it means the mixing effect is ignored, and $T_w(t)$ is constant and equal to T_{inj}, or $T_{fD}(x = 0, t) = 1$. Mathematically, it needs to take a longer time for $T_{fD}(x = 0, t)$ to be equal to 1 when Q_{inj} is smaller. Physically, a smaller Q_{inj} means slower exchange of the heat flux between the wellbore and the fracture–matrix system, and therefore T_{fD} in the injection and resting stages is smaller either. As for the thermal BTCs in the extraction stage, the values are small for a smaller Q_{inj} because of the heat distributions at the end of the resting phase, but the rate of decline of thermal BTCs is small.

Figure 4.4 shows thermal BTCs in the wellbore for different λ but the same Q_{inj}, implying that Q_{ext} is different. As the injection rate remains constant, thermal BTCs are the same for different λ in both injection and resting stages. In the extraction stage, a greater λ value results in larger thermal BTCs. This is because a greater λ value means a smaller Q_{ext} value, resulting in greater $T_{fD}(x = 0, t)$.

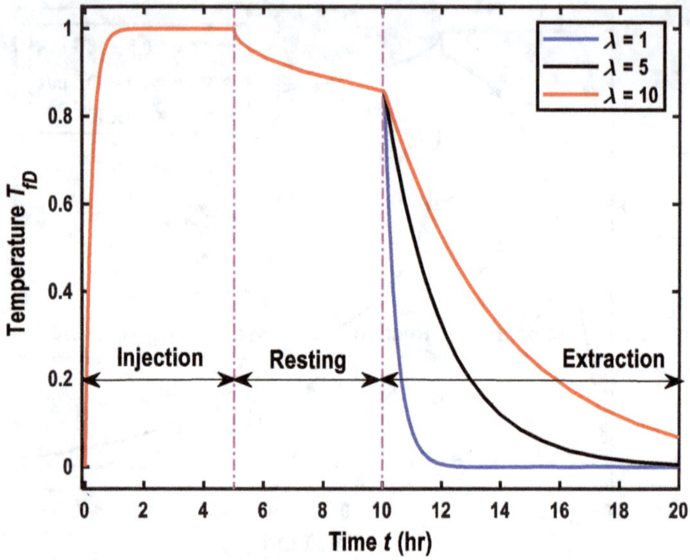

Figure 4.4. Comparison of thermal BTCs at the wellbore for different λ when $Q_{\text{inj}} = 1.44$ m^3/h.

4.3.5.2 *Effect of fracture parameters*

Figures 4.5 and 4.6 show the influence of θ_f and ϕ on BTCs. The parameters used in this section are the same as ones used in Figure 4.2. Figure 4.5 shows BTCs for different θ_f, which controls the flow velocity in the fracture, as shown in Eq. (4.10). A greater θ_f means a smaller flow velocity, and the heat transport becomes slower. Therefore, the values of thermal BTCs are smaller for smaller θ_f values in the injection and resting stages. In the extraction stage, the values of thermal BTCs are the smallest at the beginning, due to the effect of the initial condition of the extraction stage. As the slopes of thermal BTCs are smaller for smaller θ_f, the value of thermal BTC with the smallest θ_f is above others with time increases.

Figure 4.6 shows the thermal BTCs for different ϕ. The influence of ϕ on BTCs in the wellbore is quite similar to the influence of θ_f. For instance, the value of BTC for the smallest ϕ is below others in the injection and resting stages, and the value of BTC is the smallest at the beginning of the extraction stage, then reaches a maximum one hour later. This is because the parameter of ϕ also controls the flow velocity in the fracture.

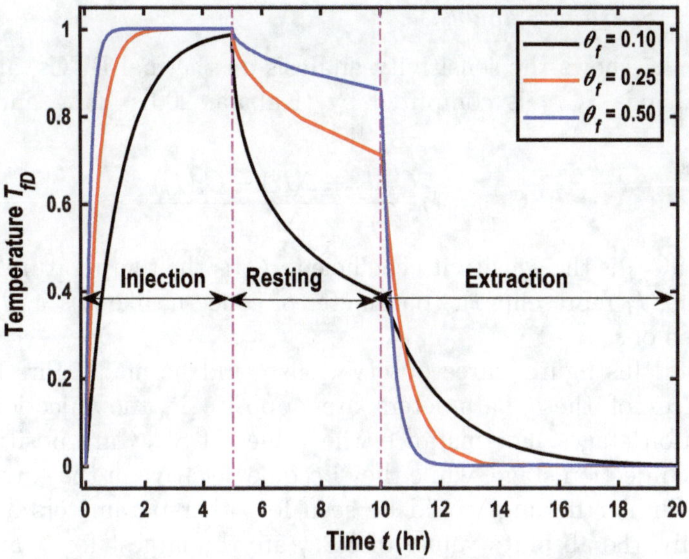

Figure 4.5. BTCs at the wellbore for different θ_f.

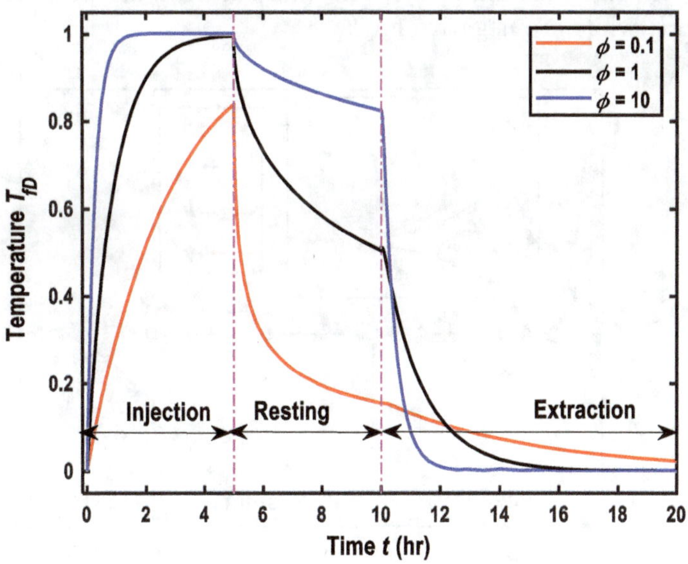

Figure 4.6. BTCs at the wellbore for different ϕ.

4.3.5.3 *Sensitivity analysis*

Figure 4.7 shows the sensitivity analysis on thermal BTCs, and the vertical axis $\mathrm{SC}_{i,j}$ is computed by (Kabala, 2001; Yang and Yeh, 2009):

$$\mathrm{SC}_{i,j} = I_j \frac{C_i(I_j + \Delta I_j) - C_i(I_j)}{\Delta I_j} \tag{4.118}$$

where $\mathrm{SC}_{i,j}$ is the sensitivity coefficient; C_i is the temperature at the ith time; I_j represents the parameter of concern, like h_w, v, θ_f, and b and so on.

From this figure, three observations could be made. Firstly, the influences of these parameters are opposite in the injection and extraction stages. For instance, the values of $\mathrm{SC}_{i,j}$ are positive for b in the injection stage, while they become negative in the extraction stage. Similar findings could be seen for other parameters as well. Secondly, the absolute values of $\mathrm{SC}_{i,j}$ are the largest for b, and the smallest for v. It implies that the most sensitive parameter is fracture aperture, and the least sensitive parameter is flow velocity in the fracture. Meanwhile, h_w is the second most sensitive parameter. Thirdly, in the injection phase, the values of $\mathrm{SC}_{i,j}$ are positive for v, θ_f, and b, while are negative for h_w.

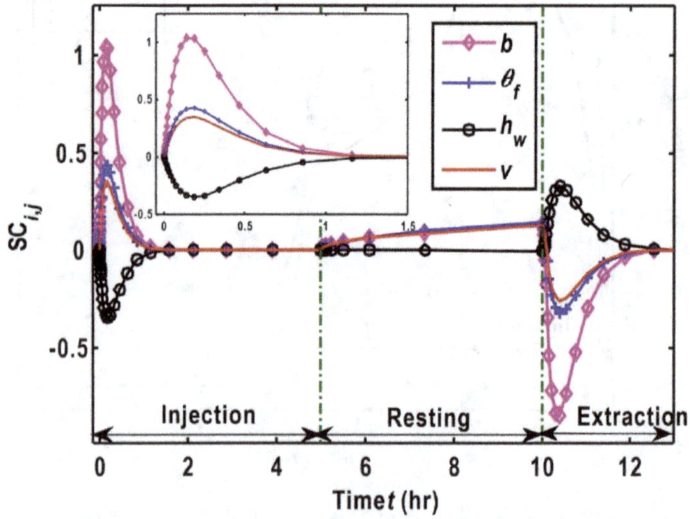

Figure 4.7. $\mathrm{SC}_{i,j}$ of the parameters b, ϕ_f, h_w, and v at the wellbore.

4.3.6 *Limitations and future works*

Although the new model of this section substantially extends previous models by removing the two presumptions, some other assumptions remain and should be elaborated on, which will be explained below.

Firstly, the mixing effect is assumed to be completed instantaneously. Generally, the radius of the injection pipe for the cold water is much smaller than the radius of the wellbore; thus, the injection process may cause a non-uniform distribution of heat inside the wellbore. In other words, the instantaneous and complete mixing of injected cold water with the pre-injection wellbore hot water may be an oversimplification. Indeed, to achieve the desirable instantaneous and complete mixing effect, additional apparatus such as an auto-mixer is recommended in the wellbore.

Secondly, this section only concerns a single fracture bounded by a homogeneous matrix, and the well fully penetrates the vertical fracture. When the spatial distribution of a complex fracture is specifically known, numerical simulations are recommended.

Thirdly, the roughness of the fracture has not been considered. Actually, a real fracture usually has irregular surfaces with different roughness and aperture; thus, flow in a fracture is three-dimensional with commonly seen channeling flow (Tsang and Tsang, 1987; Zhang *et al.*, 2019). In other words, flow in a realistic fracture is often conducted through discrete channels rather than uniform over the entire thickness. Besides, flow inside the fracture may deviate from the linear Darcy's law if a sufficiently large Reynold number has been reached, and the Navier-Stokes equations have commonly described it (Dou *et al.*, 2018; Zhang *et al.*, 2019). This section does not include the complexity of fracture flow, which must be examined in great detail in the future, probably using a high-resolution three-dimensional fracture flow simulator.

4.4 The SWPP Analytical Models in a Porous Aquifer System

With the rise in energy consumption and mounting concerns about greenhouse gas emissions, the aquifer thermal energy storage (ATES) system has received increasing attention in recent decades (Klepikova *et al.*, 2016a; Lin *et al.*, 2019). The application of ATES can reduce dependency on conventional heating and cooling resources, thereby lowering costs and CO_2 emissions. Compared to other types of thermal energy storage (TES) systems like tank TES, pit TES, and borehole TES, ATES is more suited for large-scale applications such as utility buildings, shopping malls, and residential areas (Kocabas, 2005; Lin *et al.*, 2019). The ATES system operates in an open-loop manner, combining water flow and heat transport processes, temporarily storing heat and cold in subsurface aquifers through injection or extraction wells, making it an indispensable component in sustainable energy management frameworks (Daniilidis *et al.*, 2022; Tang and Rijnaarts, 2023). In the ATES system, the cold water could be injected into the aquifers via wells, facilitating the absorption of heat from high-temperature aquifers, and then the heated water could be extracted for utilization. The thermal SWPP tests are commonly applied to simulate and evaluate geothermal energy storage (Jung and Pruess, 2012; Klepikova *et al.*, 2016a; Wang *et al.*, 2020a).

Similar to the heat transport in the fracture–matrix system, the wellbore mixing effect in the previous mathematical model of the SWPP thermal test in porous aquifers is also ignored. In addition, the following assumptions are still included in the previous mathematical model of the SWPP thermal test in porous aquifers:

(a) assuming a constant initial condition for the aquifer (Klepikova *et al.*, 2016b; Li *et al.*, 2010);
(b) a constant well flow rate during the operation (Lin *et al.*, 2019);
(c) a constant temperature in the injected water (Jung and Pruess, 2012; Klepikova *et al.*, 2016b);
(d) ignoring the thermal dispersion for heat transport in the aquifer (Jung and Pruess, 2012; Li *et al.*, 2010).

Each of these assumptions may not be valid for field problems. Some of them, such as the wellbore mixing effect, have been solved explicitly in the solute transport models. To design an efficient ATES system, accurate modeling of the thermal storage characteristics of storage media is critical. In this section, we proposed a new mathematical model of heat transport considering the wellbore mixing effect under the multi-step well flow rate. Figure 4.8 shows the conceptual model that describes heat transport around a well fully

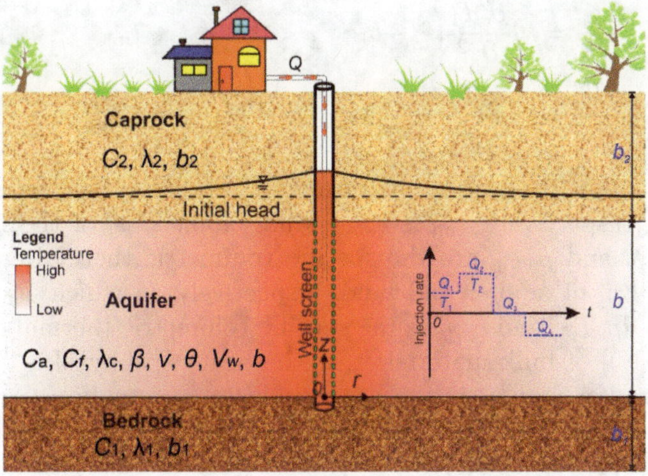

Figure 4.8. Schematic diagram of heat transport around the well with multi-step injection rates in an aquifer-rock matrix system. Q_i ($i = 1, 2, 3,$ and 4) and T_i ($i = 1$ and 2) represent the injection rates and temperature of the injected water, respectively.

penetrating a confined aquifer bounded by an underlying bedrock and an overlying caprock.

A cylindrical coordinate system is adopted, with the origin located at the center of the wellbore and the bottom of the aquifer. The r-axis is horizontal and toward infinity from the center of the well, and the z-axis is vertically upward.

4.4.1 *Mathematical models*

4.4.1.1 *Governing equations*

The following heat transport processes are considered: thermal conduction, convection, and thermal dispersion in the aquifer; thermal conduction in overlying and underlying rock matrix. The governing equations describing the temperature distribution at the stage of injection in the aquifer-rock system are as follows (Lin *et al.*, 2019):

$$C_a \frac{\partial T_a}{\partial t} = \frac{1}{r}\frac{\partial}{\partial r}\left\{ r\left[(k_c + k_d)\frac{\partial T_a}{\partial r} - \rho_w c_w \theta v T_a \right] \right\} - \frac{k_1}{b}\frac{\partial T_1}{\partial z}\bigg|_{z=0} + \frac{k_2}{b}\frac{\partial T_2}{\partial z}\bigg|_{z=b}, \quad r \geq r_w,\ 0 < t \leq t_{\text{inj}} \tag{4.119a}$$

$$C_1\frac{\partial T_1}{\partial t} = k_1\frac{\partial^2 T_1}{\partial z^2}, \quad -b_1 \le z < 0, 0 < t \le t_{\text{inj}} \tag{4.119b}$$

$$C_2\frac{\partial T_2}{\partial t} = k_2\frac{\partial^2 T_2}{\partial z^2}, \quad b < z \le b + b_2, 0 < t \le t_{\text{inj}} \tag{4.119c}$$

where T_a, T_1 and T_2 are temperatures [°C] of the aquifer, bedrock and caprock, respectively; t_{inj} is the ending moment [s] of the injection phase; C_a and $\rho_w c_w$ are the specific volumetric thermal capacities $[\text{J}/(\text{m}^3 \cdot \text{K})]$ of aquifer and fluid, respectively, in which $C_a = (1 - \theta)c_a\rho_a + \theta\rho_w c_w$, and $c_a\rho_a$ is the specific volumetric thermal capacity $[\text{J}/(\text{m}^3 \cdot \text{K})]$ of bulk material, c_a is the heat capacity $[\text{Jm}^{-1}\text{Kg}^{-2}\text{K}]$ of the aquifer and ρ_a is the density $[\text{kg}/\text{m}^3]$ of aquifer; C_1 and C_2 are the specific volumetric thermal capacities $[\text{J}/(\text{m}^3 \cdot \text{K})]$ of bedrock and caprock, respectively; k_c, k_1 and k_2 are thermal conductivities $[\text{W}/(\text{m} \cdot \text{K})]$ of the aquifer, bedrock and caprock, respectively; $k_d = \beta\rho_w c_w v\theta$; β is thermal dispersivity [m]; r is the radial distance [L] from the origin and z is the vertical distance [L] from the bottom of the aquifer; t is time [s]; t_{inj} is the ending moment [s] of the injection phase; b is aquifer thickness [m]; b_1 and b_2 are the thicknesses [m] of bedrock and caprock, respectively; v is flow velocity [m/s] and it is equal to $\frac{Q_w}{2\pi\theta br}$; Q_w is well flow rate $[\text{m}^3/s]$.

Equation (4.119a) is a thermal energy conservation statement in the aquifer. The first and second terms on the right represent heat conductance and convection, respectively, in which $\frac{k_c}{C_a} + \frac{k_d}{C_a}$ composes the effective thermal diffusivity $[\text{m}^2/\text{s}]$ of the aquifer, where $\frac{k_c}{C_a}$ is thermal conduction $[\text{m}^2/\text{s}]$, while k_d/C_a described thermal dispersion $[\text{m}^2/\text{s}]$ due to thermal convection; one can see that $\frac{k_d}{C_a} = \beta\frac{C_f q}{C_a} = \beta\frac{\rho_w c_w v\theta}{C_a}$, and q is Darcian velocity [m/s]. The third and fourth terms on the right represent heat exchange between the aquifer, bedrock, and caprock, respectively. Equations (4.119b) and (4.119c) represent heat transport in the bedrock and caprock, respectively.

The SWPP thermal test is a classic example of thermal energy exploration. It involves injection, chasing, resting and extraction phases. Equation (4.119a) is the governing equation for the multi-stage SWPP thermal test model, while the governing equations for the stages of chasing, resting and extraction differ from Eq. (4.119a).

The chasing stage of the SWPP thermal test is governed by the identical equations as Eq. (4.119a), except that the chasing temperature and injection phase are different. In the resting stage of the SWPP thermal test, the flow velocity is zero, and the governing equation of heat transport in the aquifer becomes:

$$C_a \frac{\partial T_a}{\partial t} = \frac{1}{r}\frac{\partial}{\partial r}\left\{r\left[(k_c + k_d)\frac{\partial T_a}{\partial r}\right]\right\} - \frac{k_1}{b}\frac{\partial T_1}{\partial z}\bigg|_{z=0} + \frac{k_2}{b}\frac{\partial T_2}{\partial z}\bigg|_{z=b},$$

$$r \geq r_w,\ t_{\text{cha}} < t \leq t_{\text{res}} \tag{4.120a}$$

$$C_1 \frac{\partial T_1}{\partial t} = k_1 \frac{\partial^2 T_1}{\partial z^2},\ -b_1 \leq z < 0,\ t_{\text{cha}} < t \leq t_{\text{res}} \tag{4.120b}$$

$$C_2 \frac{\partial T_2}{\partial t} = k_2 \frac{\partial^2 T_2}{\partial z^2},\ b < z \leq b + b_2, t_{\text{cha}} < t \leq t_{\text{res}} \tag{4.120c}$$

where t_{cha} is the ending moment [s] of the chasing phase, and t_{res} is the ending moment [s] of the resting phase.

Contrary to the injection stage, the direction of advection flux is reversed in the extraction stage, and the governing equation (Eq. (4.119a)) is modified as

$$C_a \frac{\partial T_a}{\partial t} = \frac{1}{r}\frac{\partial}{\partial r}\left\{r\left[(k_c + k_d)\frac{\partial T_a}{\partial r} + \rho_w c_w \theta v T_a\right]\right\} - \frac{k_1}{b}\frac{\partial T_1}{\partial z}\bigg|_{z=0}$$

$$+ \frac{k_2}{b}\frac{\partial T_2}{\partial z}\bigg|_{z=b},\quad r \geq r_w,\quad t > t_{\text{res}} \tag{4.121a}$$

$$C_1 \frac{\partial T_1}{\partial t} = k_1 \frac{\partial^2 T_1}{\partial z^2},\quad -b_1 \leq z < 0, t > t_{\text{res}} \tag{4.121b}$$

$$C_2 \frac{\partial T_2}{\partial t} = k_2 \frac{\partial^2 T_2}{\partial z^2},\quad b < z \leq b + b_2, t > t_{\text{res}} \tag{4.121c}$$

Lin *et al.* (2019) presented a new analytical solution considering both thermal conduction and thermal dispersion, but their model only considered the injection process (Eq. (4.119)). An innovative heat transport model considering the wellbore mixing effect under a multi-step flow rate in an aquifer system is proposed by Shi *et al.* (2023a), where thermal conduction, thermal dispersion, and convection occur in the aquifer, and the wellbore mixing effect is included in the wellbore.

4.4.1.2 *Initial condition*

The temperature initially distributed in the aquifer system is T_0. The initial conditions of heat transport in an aquifer system are

$$T_a(r,t)|_{t=0} = T_1(r,z,t)|_{t=0} = T_2(r,z,t)|_{t=0} = T_0 \quad (4.122)$$

As for the multi-stage SWPP thermal test models, the initial condition in the chasing stage is the solution at the end of the injection stage, the initial condition in the resting phase is the solution at the end of the chasing phase, and the temperature distribution at the end of the resting stage serves as the initial condition for the extraction phase.

4.4.1.3 *Boundary condition without wellbore mixing effect*

Similar to the treatments of Section 4.3.1, the laterally infinite boundary condition is always the same as the initial temperature. In addition, consider that the lower boundary of bedrock and the upper boundary of caprock are both under a constant temperature condition. The associated boundary conditions are given by

$$T_a(r,t)|_{r\to\infty} = T_1(r,z,t)|_{z=-b_1} = T_2(r,z,t)|_{z=(b+b_2)} = T_0, \ t \geq 0 \quad (4.123)$$

The temperature is assumed to be continuous at the interfaces between the aquifer and the overlying and underlying rock matrix. One has:

$$T_a(r,t) = T_1(r,z,t)|_{z=0} = T_2(r,z,t)|_{z=b}, \quad t \geq 0 \quad (4.124)$$

In addition, the boundary condition in wellbore without wellbore mixing effect can be described as

$$\left[v r_w \rho_w c_w T_a - k_c \frac{\partial T_a}{\partial r} \right] \bigg|_{r=r_w} = T_{in} \quad (4.125)$$

where T_{in} is temperature [°C] of injected fluid.

Equation (4.125) represents the Robin-type boundary condition for the flux between the wellbore and aquifer, and it is a constant (T_{in} Eqs. (4.119a)–(4.119c) and (4.123)–(4.125) compose the mathematical model of a single-well injection thermal test without a wellbore mixing effect in a porous aquifer-rock system, and the relevant analytical solution is displayed in Section 4.4.2.

4.4.1.4 *Boundary condition with wellbore mixing effect*

When considering the wellbore mixing effect, Eq. (4.125) is changed to

$$\left[v r_w \rho_w c_w T_a - k_c \frac{\partial T_a}{\partial r}\right]\Bigg|_{r=r_w} = f_w(Q_w, T_w, t) \tag{4.126}$$

where $f_w(Q_w, T_w, t)$ is water temperature [°C] in the wellbore, and it is a function of Q_w, and the injected water temperature T_w.

The well flow rate commonly varies with time in field practices associated with thermal energy exploitation or conservation. For instance, hot water is injected into the aquifer first, and then cool water is pumped. It involves at least two processes (i.e., injection and extraction). The step function shown below is employed to describe the variable flow rate:

$$Q_w = \begin{cases} Q_1, & 0 < t \le t_1 \\ Q_2, & t_1 < t \le t_2 \\ & \vdots \\ Q_n, & t_{n-1} < t \le t_n \end{cases} \tag{4.127}$$

where $Q_i (i = 1, 2, \ldots, n)$ is well flow rate [m^3/s] in the period of $(t_{i-1}, t_i]$; n is total time intervals [-]. The positive value of Q_w. The positive one represents injection, while the negative one is the extraction. As shown in Figure 4.8, the temperature of the injected water could vary over different periods.

For simplicity, the model of the four-step flow rates is considered in this section, and Q_w and T_w are constant within each step. The case of the four-step flow rate represents a complete operation of thermal energy exploitation: an injection phase with temperature T_{in}, a chasing phase with temperature T_{ch}, a resting phase ($Q_w = 0$), and an extraction phase. T_{in} and T_{ch} represent the temperatures [°C] of the injecting and chasing water, respectively. We denote Q_{inj}, Q_{cha}, and Q_{ext} as the well flow rates [m^3/s] in the injecting, chasing, and extraction phases, respectively. In the injection phase: $Q_w = Q_{\mathrm{inj}}$ and $T_w = T_{in}$; in the chasing phase: $Q_w = Q_{\mathrm{cha}}$ and $T_w = T_{ch}$; in the resting phase: $Q_w = 0$; in the extraction phase: $Q_w = Q_{\text{ext}}$. It is similar to a complete SWPP test of solute transport in environmental science (Chen *et al.*, 2017; Wang *et al.*, 2017, 2020b; Wang and Zhan, 2019).

In the injection phase, based on the energy conservation law and mass balance, $f_w(Q_w, T_w, t)$ could be written as

$$f_w(Q_w, T_w, t) = \frac{1}{2\pi b\theta r_w}\left\{Q_{\text{inj}}\rho_w c_w T_{in} - Q_{\text{inj}}\rho_w c_w T_{w,\text{inj}} - V_{w,\text{inj}}\rho_w c_w \left.\frac{\partial T_{w,\text{inj}}}{\partial t}\right|_{r=r_w}\right\}, \quad 0 < t \le t_{\text{inj}} \tag{4.128a}$$

$$T_{w,\text{inj}}(r,t)|_{t=0} = T_0, \; r = r_w \tag{4.128b}$$

where $T_{w,\text{inj}}(t)$ is wellbore water temperature [°C] in the injection phase; $V_{w,\text{inj}}$ is the volume [m^3] of water in the wellbore, and $V_{w,\text{inj}} = \pi r_w^2 h_{w,\text{inj}}$; $h_{w,\text{inj}}$ is water level [m] in the wellbore which could be measured during the experiment.

Similarly, $f_w(Q_w, T_w, t)$ in the chasing phase could be written as

$$f_w(Q_w, T_w, t) = \frac{1}{2\pi b\theta r_w}\left\{Q_{\text{cha}}\rho_w c_w T_{ch} - Q_{\text{cha}}\rho_w c_w T_{w,\text{cha}} - V_{w,\text{cha}}\rho_w c_w \left.\frac{\partial T_{w,\text{cha}}}{\partial t}\right|_{r=r_w}\right\}, \quad t_{\text{inj}} < t \le t_{\text{cha}} \tag{4.129a}$$

$$T_{w,\text{cha}}(r,t)|_{t=0} = T_{w,\text{inj}}(r,t)|_{t=t_{\text{inj}}}, \quad r = r_w \tag{4.129b}$$

where $T_{w,\text{cha}}(t)$ is wellbore water temperature [°C] in the chasing phase; $V_{w,\text{cha}}$ is the volume [m^3] of water in the wellbore in the chasing phase, and $V_{w,\text{cha}} = \pi r_w^2 h_{w,\text{cha}}$; $h_{w,\text{cha}}$ is the wellbore water level [m] in the chasing phase.

In the resting phase, $f_w(Q_w, T_w, t)$ could be written as:

$$f_w(Q_w, T_w, t) = \left.\left[\rho_w c_w v T_a - \frac{1}{2\pi r_w b\theta} V_w \frac{\partial T_{w,\text{res}}}{\partial t}\right]\right|_{r=r_w}, \quad t_{\text{cha}} < t \le t_{\text{res}} \tag{4.130a}$$

$$T_{w,\text{res}}(r,t)|_{t=0} = T_{w,\text{cha}}(r,t)|_{t=t_{\text{cha}}} \tag{4.130b}$$

where $T_{w,\text{res}}(t)$ is the wellbore water temperature [°C] in the resting phase; V_w is the volume [m^3] of water in the wellbore in the resting phase, and $V_w = \pi r_w^2 h_0$; h_0 refers to the initial water level [m] in the wellbore.

In the extraction phase, $f_w(Q_w, T_w, t)$ is

$$f_w(Q_w, T_w, t) = \frac{1}{2\pi b\theta r_w}\left\{\rho_w c_w V_{w,\text{ext}}\frac{\partial T_{w,\text{ext}}}{\partial t} + |Q_{\text{ext}}|\,\rho_w c_w T_{w,\text{ext}}\right\},$$
$$t_{\text{res}} < t \leq t_{\text{ext}} \tag{4.131a}$$

$$T_{w,\text{ext}}(r,t)\big|_{t=0} = T_{w,\text{res}}(r,t)\big|_{t=t_{\text{res}}}, \quad r = r_w \tag{4.131b}$$

where $T_{w,\text{ext}}(r,t)$ is wellbore water temperature [°C] in the extraction phase; t_{ext} is the end moment [s] of the extraction phase; $V_{w,\text{ext}}$ is the volume [m^3] of water in the wellbore in the extraction phase, and $V_{w,\text{ext}} = \pi r_w^2 h_{w,\text{ext}}$; $h_{w,\text{ext}}$ is wellbore water level [m] in the extraction phase.

Equations (4.119)–(4.131), except Eq. (4.125), compose the mathematical model of heat transport around a well fully penetrating the aquifer system, and the relevant analytical solution is displayed in Section 4.3.2.

4.4.2 *Analytical models without wellbore mixing effect*

Equations (4.119a)–(4.119c) and (4.123)–(4.125) compose the mathematical model of single-well injection thermal test without wellbore mixing effect in a porous aquifer-rock system, the relevant analytical solutions using the Laplace transform method. The detailed derivation is as below.

Letting $T_{aD} = T_a - T_0$, $T_{1D} = T_1 - T_0$, $T_{2D} = T_2 - T_0$, $T_{inD} = T_{in} - T_0$, and applying the Laplace transform to Eq. (4.119a) results in

$$\frac{C_a}{k_c}s\bar{T}_{aD} = \frac{(r+\beta Pe)}{r}\frac{\partial^2 \bar{T}_{aD}}{\partial r^2} + \frac{(1-Pe)}{r}\frac{\partial \bar{T}_{aD}}{\partial r} - \frac{k_1}{k_c b}\frac{\partial \bar{T}_{1D}}{\partial z}\bigg|_{z=0} + \frac{k_2}{k_c b}\frac{\partial \bar{T}_{2D}}{\partial z}\bigg|_{z=b} \tag{4.132}$$

where $Pe = \frac{\rho_w c_w Q_w}{2\pi b k_c}$. Similar to the previous section, the overbar represents the variables in the Laplace domain from now on; s is the Laplace transform parameter with respect to time t.

Similarly, Eqs. (4.119b)–(4.119c), (4.125), and (4.123) are respectively transformed as

$$C_1 s\bar{T}_{1D} = k_1 \frac{\partial^2 \bar{T}_{1D}}{\partial z^2} \tag{4.133}$$

$$C_2 s\bar{T}_{2D} = k_2 \frac{\partial^2 \bar{T}_{2D}}{\partial z^2} \tag{4.134}$$

$$r_w \frac{\partial \bar{T}_{aD}}{\partial r} = Pe\left(\bar{T}_{aD} - \frac{T_{in}}{p}\right), \quad r = r_w \tag{4.135}$$

$$\frac{\partial \bar{T}_{aD}}{\partial r} = 0, \quad r \to \infty \tag{4.136}$$

$$\bar{T}_{1D} = 0, \quad z = -b_1 \tag{4.137}$$

$$\bar{T}_{2D} = 0, \quad z = b + b_2 \tag{4.138}$$

Finally, Eq. (4.124), the continuity requirement at the interface between the aquifer and the rock, are respectively transformed as

$$\bar{T}_{1D} = \bar{T}_{aD}, \quad z = 0 \tag{4.139a}$$

$$\bar{T}_{2D} = \bar{T}_{aD}, \quad z = b \tag{4.139b}$$

The solutions for $\bar{T}_{1D}$ in Eq. (4.133) and $\bar{T}_{2D}$ in Eq. (4.134) with the conditions Eqs. (4.137)–(4.139) can be respectively obtained as

$$\bar{T}_{1D} = \frac{1}{2}\exp[-Az][\exp(2A(z + b_1)) - 1] \times [\coth(b_1 A) - 1]\bar{T}_{aD} \tag{4.140}$$

$$\bar{T}_{2D} = \frac{1}{2}\exp[-B(b + z)][\exp(2B(b + b_2)) - \exp(2Bz)] \times [\coth(b_2 B) - 1]\bar{T}_{aD} \tag{4.141}$$

where $A = \sqrt{C_1 s/k_1}$ and $B = \sqrt{C_2 s/k_2}$.

Substituting Eqs. (4.140)–(4.141) into Eq. (4.132) yields:

$$\frac{C_a}{k_c}s\bar{T}_{aD} = \frac{(r+\beta Pe)}{r}\frac{\partial^2\bar{T}_{aD}}{\partial r^2} + \frac{(1-Pe)}{r}\frac{\partial\bar{T}_{aD}}{\partial r}$$
$$-\frac{\mathrm{A}k_1[1+\exp(2Ab_1)][\cot h(Ab_1)-1]}{2bk_c}\bar{T}_{aD}$$
$$-\frac{\mathrm{B}k_2[1+\exp(2Bb_2)][\cot h(Bb_2)-1]}{2bk_c}\bar{T}_{aD} \quad (4.142)$$

For the sake of convenience, Eq. (4.142) is arranged as

$$\Lambda\bar{T}_{aD} = \frac{(r+\beta Pe)}{r}\frac{\partial^2\bar{T}_{aD}}{\partial r^2} + \frac{(1-Pe)}{r}\frac{\partial\bar{T}_{aD}}{\partial r} \quad (4.143)$$

where $\Lambda = \frac{Ak_1[1+\exp(2\mathrm{A}b_1)][\cot h(\mathrm{A}b_1)-1]}{2bk_c} - \frac{Bk_2[1+\exp(2Bb_2)][\cot h(Bb_2)-1]}{2bk_c} + \frac{C_a}{k_c}s$.

The general solution of equation Eq. (4.143) is

$$\bar{T}_{aD} = (r+\beta Pe)^{Pe}\exp(-(r+\beta Pe)\sqrt{\Lambda})[A_1 U(\gamma,\eta,\xi(r))$$
$$+ A_2 L(\gamma,\eta,\xi(r))] \quad (4.144)$$

where A_1 and A_2 are undetermined coefficients; $U(\cdot)$ is the confluent hypergeometric function; $L(\cdot)$ is generalized Laguerre polynomial. Imposing the boundary conditions, Eqs. (4.135)–(4.136), the coefficients in equation Eq. (4.144) can then be obtained as

$$A_1 = \frac{T_{in}}{s}\frac{\exp((r_w+\beta Pe)\sqrt{\Lambda})Pe(r_w+\beta Pe)^{1-Pe}}{(Pe^2\beta + r_w^2\sqrt{\Lambda} + r_w\beta Pe\sqrt{\Lambda})U(\gamma,\eta,\xi(r_w)) + r_w\gamma\xi(r_w)U(\gamma+1,\eta+1,\xi(r_w))} \quad (4.145)$$

$$A_2 = 0 \quad (4.146)$$

where $\gamma = (1+Pe-\beta Pe\sqrt{\Lambda})/2$; $\eta = 1+Pe$; and $\xi(r) = 2\sqrt{\Lambda}(r+\beta Pe)$.

Substituting these results into Eq. (4.144), the temperature solutions for the aquifer, bedrock, and caprock can be respectively obtained. The solutions of Eqs. (4.140)–(4.141) and (4.144) are in the Laplace domain, and they need to be transformed into the real-time domain. Similar to the treatments in the previous studies (Lin *et al.*, 2019; Wang *et al.*, 2018; Wang and Zhan, 2015), a numerical inverse Laplace transform approach called the Gaver-Stehfest method is employed in for the computations (Jacquot *et al.*, 1983; Stehfest, 1970).

4.4.3 *Analytical models with wellbore mixing effect*

The analytical solution of Eqs. (4.119)–(4.131), except Eq. (4.125), can be derived using the Laplace transform method, and the detailed information can be seen in the following sections.

4.4.3.1 *Dimensionless transform*

To facilitate the solution, some dimensionless parameters need to be defined as follows: $T_{aD} = T_a - T_0$, $T_{1D} = T_1 - T_0$, $T_{2D} = T_2 - T_0$, $T_{inD} = T_{in} - T_0$, $T_{chD} = T_{ch} - T_0$, $T_{wD} = T_w - T_0$, $T_{wD,\text{inj}} = T_{w,\text{inj}} - T_0$, $T_{wD,\text{cha}} = T_{w,\text{cha}} - T_0$, $T_{wD,\text{res}} = T_{w,\text{res}} - T_0$, $T_{1D,\text{cha}} = T_{1,\text{cha}} - T_0$, $T_{2D,\text{cha}} = T_{2,\text{cha}} - T_0$, $T_{1D,\text{res}} = T_{1,\text{res}} - T_0$, $T_{2D,\text{res}} = T_{2,\text{res}} - T_0$. By substituting these parameters into Eqs. (4.119a)–(4.119c), the dimensionless governing equations in the injection and chasing phases are

$$\frac{C_a}{k_c}\frac{\partial T_{aD}}{\partial t} = \frac{(r+\beta P_e)}{r}\frac{\partial^2 T_{aD}}{\partial r^2} + \frac{(1-P_e)}{r}\frac{\partial T_{aD}}{\partial r} - \frac{k_1}{k_c b}\frac{\partial T_{1D}}{\partial z}\bigg|_{z=0} + \frac{k_2}{k_c b}\frac{\partial T_{2D}}{\partial z}\bigg|_{z=b}, \quad r \geq r_w,\ 0 < t \leq t_{\text{cha}} \tag{4.147a}$$

$$C_1\frac{\partial T_{1D}}{\partial t} = k_1\frac{\partial^2 T_{1D}}{\partial z^2}, \quad -b_1 \leq z \leq 0,\ 0 < t \leq t_{\text{cha}} \tag{4.147b}$$

$$C_2\frac{\partial T_{2D}}{\partial t} = k_2\frac{\partial^2 T_{2D}}{\partial z^2}, \quad b \leq z \leq b+b_2,\ 0 < t \leq t_{\text{cha}} \tag{4.147c}$$

The dimensionless governing equations in the resting phase are

$$\frac{C_a}{k_c}\frac{\partial T_{aD}}{\partial t} = \frac{\partial^2 T_{aD}}{\partial r^2} + \frac{1}{r}\frac{\partial T_{aD}}{\partial r} - \frac{k_1}{bk_c}\frac{\partial T_{1D}}{\partial z}\bigg|_{z=0} + \frac{k_2}{bk_c}\frac{\partial T_{2D}}{\partial z}\bigg|_{z=b}, \quad r \geq r_w,\ t_{\text{cha}} < t \leq t_{\text{res}} \tag{4.148a}$$

$$\frac{C_1}{k_1}\frac{\partial T_{1D}}{\partial t} = \frac{\partial^2 T_{1D}}{\partial z^2}, -b_1 \leq z \leq 0, \quad t_{\text{cha}} < t \leq t_{\text{res}} \tag{4.148b}$$

$$\frac{C_2}{k_2}\frac{\partial T_{2D}}{\partial t} = \frac{\partial^2 T_{2D}}{\partial z^2}, b \leq z \leq b+b_2, \quad t_{\text{cha}} < t \leq t_{\text{res}} \tag{4.148c}$$

The dimensionless governing equations in the extraction are

$$\frac{C_a}{k_c}\frac{\partial T_{aD}}{\partial t} = \frac{(r-\beta P_e)}{r}\frac{\partial^2 T_{aD}}{\partial r^2} + \frac{(1+P_e)}{r}\frac{\partial T_{aD}}{\partial r} - \frac{k_1}{k_c b}\frac{\partial T_{1D}}{\partial z}\bigg|_{z=0} + \frac{k_2}{k_c b}\frac{\partial T_{2D}}{\partial z}\bigg|_{z=b}, \quad r \geq r_w, \quad t_{\text{res}} < t \leq t_{\text{ext}} \tag{4.149a}$$

$$\frac{C_1}{k_1}\frac{\partial T_{1D}}{\partial t} = \frac{\partial^2 T_{1D}}{\partial z^2}, -b_1 \leq z \leq 0, \quad t_{\text{res}} < t \leq t_{\text{ext}} \tag{4.149b}$$

$$\frac{C_2}{k_2}\frac{\partial T_{2D}}{\partial t} = \frac{\partial^2 T_{2D}}{\partial z^2}, \quad b \leq z \leq b + b_2, \quad t_{\text{res}} < t \leq t_{\text{ext}} \tag{4.149c}$$

The initial conditions become:

$$T_{aD}(r, t=0) = T_{1D}(r, z, t=0) = T_{2D}(r, z, t=0) = 0 \tag{4.150}$$

The outer boundary condition is rewritten as

$$T_{aD}(r,t)|_{r\to\infty} = T_{1D}(r,z,t)|_{z=-b_1} = T_{2D}(r,z,t)|_{z=b+b_2} = 0 \tag{4.151}$$

The interfaces between aquifer and rock boundary conditions are rewritten as:

$$T_{aD}(r,t) = T_{1D}(r, z=0, t) = T_{2D}(r, z=b, t) \tag{4.152}$$

The inner boundary conditions could be rewritten as

$$\left[v(r_w)C_f T_{aD} - k_c\frac{\partial T_{aD}}{\partial r}\right]\bigg|_{r=r_w} = f_w(Q_w, T_{wD}, t), \quad 0 < t \leq t_{\text{cha}} \tag{4.153}$$

where $f_w(Q_w, T_{wD}, t)$ is the water temperature in the wellbore, and it represents $T_{wD,\text{inj}}$ for the injection phase, $T_{wD,\text{cha}}$ for the chasing phase, respectively.

In the injection phase, the $T_{wD,\text{inj}}$ is determined by

$$\beta_{\text{inj}} \frac{\partial T_{wD,\text{inj}}}{\partial t}\bigg|_{r=r_w} = \frac{Q_{\text{inj}} C_f}{2\pi b\theta r_w} [T_{inD} - T_{wD,\text{inj}}], \quad 0 < t \le t_{\text{inj}} \tag{4.154a}$$

$$T_{wD,\text{inj}}(r,t)|_{t=0} = 0, \quad r = r_w \tag{4.154b}$$

Similarly, in the chasing phase, the $T_{wD,\text{cha}}$ is determined by

$$\beta_{\text{cha}} \frac{\partial T_{wD,\text{cha}}}{\partial t}\bigg|_{r=r_w} = \frac{Q_{\text{cha}} C_f}{2\pi b\theta r_w} [T_{ch} - T_{wD,\text{cha}}], \; t_{\text{inj}} < t \le t_{\text{cha}} \tag{4.155a}$$

$$T_{wD,\text{cha}}(r,t)|_{t=0} = T_{wD,\text{inj}}(r,t)|_{t=t_{\text{inj}}}, \quad r = r_w \tag{4.155b}$$

where $\beta_{\text{inj}} = \frac{V_{w,\text{inj}}}{Q_{\text{inj}}}$ and $\beta_{\text{cha}} = \frac{V_{w,\text{cha}}}{Q_{\text{cha}}}$.

According to Eqs. (4.130a)–(4.130b), the inner boundary conditions in the resting phase are

$$\beta_{\text{res}} \frac{\partial T_{wD,\text{res}}}{\partial t}\bigg|_{r=r_w} = \frac{\partial T_{aD}}{\partial r}\bigg|_{r=r_w}, \quad t_{\text{cha}} < t \le t_{\text{res}} \tag{4.156a}$$

$$T_{wD,\text{cha}}(r,t)|_{t=0} = T_{wD,\text{inj}}(r,t)|_{t=t_{\text{res}}}, \quad r = r_w \tag{4.156b}$$

where $\beta_{\text{res}} = \frac{V_{w,0}}{2\pi r_w b k_c}$.

According to Eqs. (4.131a)–(4.131b), the inner boundary conditions in the extraction phase are

$$\beta_{\text{ext}} \frac{\partial T_{wD}}{\partial t} = \frac{\partial T_{aD}}{\partial r}\bigg|_{r=r_w} \tag{4.157a}$$

$$T_{wD,\text{ext}}(r,t)|_{t=0} = T_{wD,\text{res}}(r,t)|_{t=t_{\text{res}}}, \quad r = r_w \tag{4.157b}$$

where $\beta_{\text{ext}} = \frac{C_f h_{w,\text{ext}}}{2bk_c}$.

4.4.3.2 *Solutions in the injection and chasing phases*

Conducting Laplace transform to Eqs. (4.147a)–(4.147c), one has:

$$\frac{\partial^2 \bar{T}_{aD}}{\partial r^2} + \frac{(1-P_e)}{r}\frac{\partial \bar{T}_{aD}}{\partial r} - \frac{k_1}{k_c b}\frac{\partial \bar{T}_{1D}}{\partial z}\bigg|_{z=0} + \frac{k_2}{k_c b}\frac{\partial \bar{T}_{2D}}{\partial z}\bigg|_{z=b}$$

$$= \frac{C_a}{k_c} s\bar{T}_{aD}, r \geq r_w \tag{4.158a}$$

$$\frac{C_1}{k_1} s\bar{T}_{1D} = \frac{\partial^2 \bar{T}_{1D}}{\partial z^2}, \quad -b_1 \leq z \leq 0 \tag{4.158b}$$

$$\frac{C_2}{k_2} s\bar{T}_{2D} = \frac{\partial^2 \bar{T}_{2D}}{\partial z^2}, \quad b \leq z \leq b + b_2 \tag{4.158c}$$

Conducting Laplace transform to Eqs. (4.153)–(4.155b), one has:

$$\left[v(r_w)C_f\bar{T}_{aD} - k_c\frac{\partial \bar{T}_{aD}}{\partial r}\right]\bigg|_{r=r_w} = \bar{f}_w(Q_w, T_{wD}, s),$$

$$0 < t \leq t_{\text{cha}} \tag{4.159}$$

$$\beta_{\text{inj}} s\bar{T}_{wD,\text{inj}} = \frac{Q_{\text{inj}}C_f}{2\pi b\theta r_w}\left[\frac{T_{inD}}{s} - \bar{T}_{wD,\text{inj}}\right] \tag{4.160}$$

$$\beta_{\text{cha}} s\bar{T}_{wD,\text{cha}} - \beta_{\text{cha}} T_{wD,\text{inj}} = \frac{Q_{\text{cha}}C_f}{2\pi b\theta r_w}\left[\frac{T_{ch}}{s} - \bar{T}_{wD,\text{cha}}\right] \tag{4.161}$$

Equations (4.159)–(4.161) are rewritten as

$$\left[P_e\bar{T}_{aD} - r_w\frac{\partial \bar{T}_{aD}}{\partial r}\right]\bigg|_{r=r_w} = \frac{P_e T_{inD}}{s\left(\beta_{\text{inj}} s + \frac{Q_{\text{inj}}C_f}{2\pi b\theta r_w}\right)} \tag{4.162a}$$

$$\left[P_e\bar{T}_{aD} - r_w\frac{\partial \bar{T}_{aD}}{\partial r}\right]\bigg|_{r=r_w} = \frac{P_e T_{chD}}{s\left(\beta_{\text{cha}} s + \frac{Q_{\text{cha}}C_f}{2\pi b\theta r_w}\right)} + \frac{\beta_{\text{cha}} P_e T_{wD,\text{inj}}}{\beta_{\text{cha}} s + \frac{Q_{\text{cha}}C_f}{2\pi b\theta r_w}} \tag{4.162b}$$

In the injection and chasing phases, the inner boundary conditions could be combined as the following equation:

$$\left[P_e T_{aD} - r_w\frac{\partial T_{aD}}{\partial r}\right]\bigg|_{r=r_w} = T_{wD,\text{inj}}\left[H(t) - H(t-t_{\text{inj}})\right]$$

$$+ T_{wD,\text{cha}} H(t - t_{\text{inj}}) \tag{4.163}$$

Conducting Laplace transform to Eqs. (4.151)–(4.152), one has:

$$\bar{T}_{aD}(r,t)\big|_{r\to\infty} = \bar{T}_{1D}(r,z,t)\big|_{z=-b_1} = \bar{T}_{2D}(r,z,t)\big|_{z=b+b_2} = 0 \tag{4.164}$$

$$\bar{T}_{aD}(r,s) = \bar{T}_{1D}(r,z=0,s) = \bar{T}_{2D}(r,z=b,s) \tag{4.165}$$

Equations (4.147b) and (4.159)–(4.165) compose a model of the second-order ODE with boundary conditions, the solution of Eqs. (4.147b)–(4.147c) with the conditions of Eqs. (4.164)–(4.165) can be respectively obtained, as shown in Eqs. (4.140)–(4.141).

Substituting Eqs. (4.140)–(4.141) into Eq. (4.158a), one has:

$$\frac{(r+\beta P_e)}{r}\frac{\partial^2 \bar{T}_{aD}}{\partial r^2} + \frac{(1-P_e)}{r}\frac{\partial \bar{T}_{aD}}{\partial r} - E_1\bar{T}_{aD} = 0, \quad r \geq r_w \tag{4.166}$$

where

$E_1 = \frac{sC_a}{k_c} + \frac{[1+\exp(2b_1A)]k_1A[coth(b_1A)-1]}{2bk_c} + \frac{[1+\exp(2b_2B)]k_2B[coth(b_2B)-1]}{2bk_c}$.

Equations (4.162a)–(4.162b), (4.164), and (4.166) compose a model of the second-order ODE with boundary conditions, the general solution of Eq. (4.166) is

$$\bar{T}_{aD} = \exp\left[-(r+\beta P_e)\sqrt{E_1}\right](r+\beta P_e)^{P_e}[N_1U(\gamma,\eta,\xi(r)) + N_2L(\gamma,\eta,\xi(r))] \tag{4.167}$$

where N_1 and N_2 are undetermined coefficients, $\gamma = (1+P_e-\beta P_e\sqrt{E_1})/2$, $\eta = 1+P_e$ and $\xi(r) = 2\sqrt{E_1}(r+\beta P_e)$. $L(\cdot)$ is generalized Laguerre polynomial, $U(\cdot)$ is the confluent hypergeometric function.

Substituting Eq. (4.167) into Eq. (4.164), we can obtain that $N_2 = 0$. Then, substituting $N_2 = 0$ into Eq. (4.167), one has:

$$\bar{T}_{aD} = N_1\exp\left[-(r+\beta P_e)\sqrt{E_1}\right](r+\beta P_e)^{P_e}U(\gamma,\eta,\xi(r)) \tag{4.168}$$

Substituting Eq. (4.168) into Eq. (4.162a), one has:

$$N_1\frac{\begin{array}{c}(P_e^2\beta + r_w^2\sqrt{E_1} + r_w\beta P_e\sqrt{E_1})U(\gamma,\eta,\xi(r_w)) + r_w\gamma\xi(r_w)\\ U(\gamma+1,\eta+1,\xi(r_w))\end{array}}{\exp[(r_w+\beta P_e)\sqrt{E_1}](r_w+\beta P_e)^{1-P_e}}$$

$$= \frac{P_eT_{inD}}{s(\beta_{\text{inj}}s + \frac{Q_{\text{inj}}C_f}{2\pi b\theta r_w})} \tag{4.169a}$$

Substituting Eq. (4.168) into Eq. (4.162b), one has

$$N_1 \frac{(P_e^2\beta + r_w^2\sqrt{E_1} + r_w\beta P_e\sqrt{E_1})U(\gamma,\eta,\xi(r_w)) + r_w\gamma\xi(r_w) U(\gamma+1,\eta+1,\xi(r_w))}{\exp[(r_w+\beta P_e)\sqrt{E_1}](r_w+\beta P_e)^{1-P_e}} = \frac{P_e T_{chD}}{s\left(\beta_{\text{cha}}s + \frac{Q_{\text{cha}}C_f}{2\pi b\theta r_w}\right)} + \frac{\beta_{\text{cha}}P_e T_{wD,\text{inj}}}{\beta_{\text{cha}}s + \frac{Q_{\text{cha}}C_f}{2\pi b\theta r_w}} \tag{4.169b}$$

Conducting Laplace transform on Eq. (4.163), one has:

$$\left[P_e\bar{T}_{aD} - r_w\frac{\partial\bar{T}_{aD}}{\partial r}\right]\Bigg|_{r=r_w} = T_{wD,\text{inj}}\frac{1-\exp(-t_{\text{inj}}s)}{s} + T_{wD,\text{cha}}\frac{\exp(-t_{\text{inj}}s)}{s} \tag{4.170}$$

Substituting Eq. (4.168) into Eq. (4.170), one has:

$$N_1 \frac{(P_e^2\beta + r_w^2\sqrt{E_1} + r_w\beta P_e\sqrt{E_1})U(\gamma,\eta,\xi(r_w)) + r_w\gamma\xi(r_w) U(\gamma+1,\eta+1,\xi(r_w))}{\exp[(r_w+\beta P_e)\sqrt{E_1}](r_w+\beta P_e)^{1-P_e}} = H \tag{4.171}$$

where $H = T_{wD,\text{inj}}\frac{1-\exp(-t_{\text{inj}}s)}{s} + T_{wD,\text{cha}}\frac{\exp(-t_{\text{inj}}s)}{s}$.

In the injection phase, substituting Eq. (4.169a) into Eq. (4.168), one has:

$$\bar{T}_{aD} = \frac{P_e T_{inD}}{s(\beta_{\text{inj}}s+1)} \frac{\exp\left[(r_w+\beta P_e)\sqrt{E_1}\right](r_w+\beta P_e)^{1-P_e} \exp\left[-(r+\beta P_e)\sqrt{E_1}\right](r+\beta P_e)^{P_e}U(\gamma,\eta,\xi(r))}{(P_e^2\beta + r_w^2\sqrt{E_1} + r_w\beta P_e\sqrt{E_1})U(\gamma,\eta,\xi(r)) + r_w\gamma\xi(r_w)U(\gamma+1,\eta+1,\xi(r_w))} \tag{4.172a}$$

In the chasing phase, substituting Eq. (4.169b) into Eq. (4.168), one has:

$$\bar{T}_{aD} = \left[\frac{P_e T_{chD}}{s(\beta_{\text{cha}} s + 1)} + \frac{\beta_{\text{cha}} P_e T_{wD,\text{inj}}}{\beta_{\text{cha}} s + 1}\right] \frac{\begin{array}{l}\exp\left[(r_w + \beta P_e)\sqrt{E_1}\right](r_w + \beta P_e)^{1-P_e} \exp\left[-(r + \beta P_e)\sqrt{E_1}\right] \\ (r + \beta P_e)^{P_e} U(\gamma, \eta, \xi(r))\end{array}}{\begin{array}{l}(P_e^2 \beta + r_w^2 \sqrt{E_1} + r_w \beta P_e \sqrt{E_1}) U(\gamma, \eta, \xi(r)) + r_w \gamma \xi(r_w) \\ U(\gamma + 1, \eta + 1, \xi(r_w))\end{array}} \tag{4.172b}$$

Substituting Eq. (4.171) into Eq. (4.168), the analytical solution combining both the injection and repulsion phases is

$$\bar{T}_{aD} = H \frac{\begin{array}{l}\exp\left[(r_w + \beta P_e)\sqrt{E_1}\right](r_w + \beta P_e)^{1-P_e} \\ \exp\left[-(r + \beta P_e)\sqrt{E_1}\right](r + \beta P_e)^{P_e} U(\gamma, \eta, \xi(r))\end{array}}{\begin{array}{l}(P_e^2 \beta + r_w^2 \sqrt{E_1} + r_w \beta P_e \sqrt{E_1}) U(\gamma, \eta, \xi(r)) \\ + r_w \gamma \xi(r_w) U(\gamma + 1, \eta + 1, \xi(r_w))\end{array}} \tag{4.173}$$

4.4.3.3 *Solutions in the resting phase*

In the resting phase, the analytical solution will be derived using the Laplace transform method and the Green's functions method, and the detailed information can be seen in the following sections.

Conducting Laplace transform on Eqs. (4.147a)–(4.147c), one has

$$\frac{\partial^2 \bar{T}_{aD}}{\partial r^2} + \frac{1}{r}\frac{\partial \bar{T}_{aD}}{\partial r} - \frac{k_1}{bk_c}\frac{\partial \bar{T}_{1D}}{\partial z}\bigg|_{z=0} + \frac{k_2}{bk_c}\frac{\partial \bar{T}_{2D}}{\partial z}\bigg|_{z=b} = \frac{C_a}{k_c}[s\bar{T}_{aD} - T_{aD}(r, t_{\text{cha}})] \tag{4.174a}$$

$$\frac{\partial^2 \bar{T}_{1D}}{\partial z^2} - \frac{C_1 s}{k_1}\bar{T}_{1D} = -\frac{C_1}{k_1} T_{1D}(r, t_{\text{cha}}) \tag{4.174b}$$

$$\frac{\partial^2 \bar{T}_{2D}}{\partial z^2} - \frac{C_2 s}{k_2}\bar{T}_{2D} = -\frac{C_2}{k_2} T_{2D}(r, t_{\text{cha}}) \tag{4.174c}$$

Notice that Eq. (4.174b) is inhomogeneous; thus, we need to homogeneous it first. Letting $\bar{T}_{1D} = \varphi(r, z, t) + \delta_1 + \delta_2 z$, and substituting them into Eqs. (4.164)–(4.165) yields:

$$\varphi(r, z = -b_1, t) = 0 \tag{4.175a}$$

$$\varphi(r, z = 0, t) = 0 \tag{4.175b}$$

where $\delta_1 = \bar{T}_{aD}(r, t)$ and $\delta_2 = \bar{T}_{aD}(r, t)/b_1$.

Defining the spatial operator: $L_1 = -[\frac{d^2}{dz^2} - \frac{C_1}{\lambda_1} s]$, one has:

$$\frac{\partial^2 \varphi}{\partial z^2} - \frac{C_1}{\lambda_1} s\varphi = -\frac{C_1}{\lambda_1} T_{1D,\text{cha}} + \frac{C_1}{\lambda_1} s(\delta_1 + \delta_2 z) \tag{4.176}$$

The general of Eq. (4.176) is

$$\varphi = \int_{-b_1}^{0} g_1(z, s; \varepsilon) f_1(\varepsilon) d\varepsilon \tag{4.177a}$$

$$f_1(\varepsilon) = \frac{C_1}{\lambda_1} T_{1,\text{cha}} - \frac{C_1}{\lambda_1} s(\delta_1 + \delta_2 \varepsilon) \tag{4.177b}$$

$$g_1(z, s; \varepsilon) = \begin{cases} g_{11}(z, s; \varepsilon) = A_1 \exp(-Az) + A_2 \exp(Az) & 0 \leq z < \varepsilon \\ g_{12}(z, s; \varepsilon) = A_3 \exp(-Az) + A_4 \exp(Az) & \varepsilon < z \leq \infty \end{cases} \tag{4.177c}$$

where ε is a positive value varying between 1 and ∞; $A = \sqrt{C_1 s / k_1}$; $g_1(z, s; \varepsilon)$ is the Green's function, A_1, A_2, A_3, and A_4 are unknown coefficients, which could be obtained based on Eqs. (4.175a)–(4.175b). The coefficients in Eq. (4.177c) can then be obtained as

$$A_1 = -A_2 \tag{4.178a}$$

$$A_2 = PA_3 \tag{4.178b}$$

$$A_3 = \frac{-\exp(\sqrt{A}\varepsilon)}{\sqrt{A}[G+P+\exp(2\sqrt{A}\varepsilon)+P\exp(2\sqrt{A}\varepsilon)]} \tag{4.178c}$$

$$A_4 = -A_3 G \tag{4.178d}$$

where $G = \exp(2\sqrt{A} b_1)$ and $P = \frac{\exp(2\sqrt{A}\varepsilon)-G}{1-\exp(2\sqrt{A}\varepsilon)}$.

Thus, the solution of Eq. (4.174b) is

$$\bar{T}_{1D} = \int_{-b_1}^{0} g_1(z, s; \varepsilon) f_1(\varepsilon) d\varepsilon + \delta_1 + \delta_2 z \tag{4.179}$$

As for the analytical solution of caprock, one could use a similar approach to the analytical solution of bedrock to get it. One could derive the solution of Eq. (4.174c), and the general solution of Eq. (4.174c) could be described as

$$\bar{T}_{2D} = \int_{b}^{b+b_2} g_2(z, s; \varepsilon) f_2(\varepsilon) d\varepsilon + a_1 + a_2 z \tag{4.180a}$$

$$g_2(z, s; \varepsilon) = \begin{cases} g_{21}(z, s; \varepsilon) = B_1 \exp(-Bz) \\ \qquad + B_2 \exp(Bz)\ b \le z < \varepsilon \\ g_{22}(z, s; \varepsilon) = B_3 \exp(-Bz) \\ \qquad + B_4 \exp(Bz)\ \varepsilon < z \le b + b_2 \end{cases} \tag{4.180b}$$

$$f_2(\varepsilon) = \frac{C_2}{\lambda_2} T_{2D,\text{cha}} - \frac{C_2}{\lambda_2} s(a_1 + a_2 \varepsilon) \tag{4.180c}$$

where $B_1 = -\exp(2Bb)B_2$, $B_2 = \frac{\exp[B(b+b_2)] - \exp(2B\varepsilon)}{2B\exp(B\varepsilon)\{\exp[B(b+b_2)] - \exp(2Bb)\}}$, $B_3 = -\exp(B(b+b_2))B_4$, $B_4 = B_2 - \frac{\exp(B\varepsilon)}{2B}$, $a_1 = (1 + \frac{b}{b_2})\bar{T}_{aD}$ and $a_2 = -\bar{T}_{aD}/b_2$.

As for the aquifer, substituting Eqs. (4.179)–(4.180a) into Eq. (4.174a), one has:

$$\frac{\partial^2 \bar{T}_{aD}}{\partial r^2} + \frac{1}{r}\frac{\partial \bar{T}_{aD}}{\partial r} - E\bar{T}_{aD} = -\frac{C_a}{k_c} T_{aD}(r, t_{\text{cha}}) \tag{4.181}$$

where $E = \frac{C_a}{k_c} s + \frac{k_1}{k_c b b_1} + \frac{k_2}{k_c b b_2}$.

Conducting Laplace transform on Eq. (4.156), one has:

$$\beta_{\text{res}} s \bar{T}_{wD,\text{res}} - \beta_{\text{res}} T_{wD,\text{cha}} = \left.\frac{\partial \bar{T}_{wD}}{\partial r}\right|_{r=r_w} \tag{4.182}$$

Similar to the treatment of Eqs. (4.175a)–(4.175b), letting $\bar{T}_{aD} = \mathcal{H}(r,t) + \sigma_1 + \sigma_2 r$, and then substituting them into Eqs. (4.182) and

(4.164) yields:

$$\frac{\partial \mathcal{H}(r_w,t)}{\partial r} - \beta_{\text{cha}} s \mathcal{H}(r_w,t) = 0 \tag{4.183a}$$

$$\mathcal{H}(r \to \infty, t) = 0 \tag{4.183b}$$

where $\sigma_1 = -\sigma_2 r_e$ and $\sigma_2 = \frac{\beta_{\text{res}} T_{wD,\text{cha}}}{\beta_{\text{res}} s r_w - \beta_{\text{res}} s r_e - 1}$.

Then defining the spatial operator: $L = -[\frac{\partial^2}{\partial r^2} + \frac{1}{r}\frac{\partial}{\partial r} - E]$, and one has:

$$\frac{\partial^2 \mathcal{H}}{\partial r^2} + \frac{1}{r}\frac{\partial \mathcal{H}}{\partial r} - E\mathcal{H} = -f(r) \tag{4.184a}$$

$$f(r) = \frac{C_a}{\lambda_c} T_{aD}(r, t_{\text{cha}}) - \left[\frac{\sigma_2}{r} - E(\sigma_1 + \sigma_2 r)\right] \tag{4.184b}$$

The general solution of Eq. (4.184a) is

$$\mathcal{H} = \textstyle\int_{r_w}^{\infty} g(r, E; \chi) f(\chi) d\chi \tag{4.185a}$$

$$g(r, E; \chi) = \begin{cases} g_1(r, E; \chi) = \mathcal{T}_1 I_0(\sqrt{E} r) + \mathcal{T}_2 K_0(\sqrt{E} r) & r_w \le r \le \chi \\ g_2(r, E; \chi) = \mathcal{T}_3 I_0(\sqrt{E} r) + \mathcal{T}_4 K_0(\sqrt{E} r) & \chi \le r \le \infty \end{cases} \tag{4.185b}$$

where χ is a positive value varying between 1 and ∞, $g(r, E; \chi)$ is the Green's function, $f(\chi) = \frac{C_a}{\lambda_c} T_{aD}(\chi, t_{\text{cha}}) - [\frac{\sigma_2}{\chi} - E(\sigma_1 + \sigma_2 \chi)]$. $I_0(\cdot)$ and $K_0(\cdot)$ are the modified Bessel functions of the first and second kinds of zero order, respectively; $\mathcal{T}_1$, $\mathcal{T}_2$, $\mathcal{T}_3$ and $\mathcal{T}_4$ are unknown coefficients. Substituting Eq. (4.185b) into Eq. (4.183b), one has:

$$\mathcal{T}_3 = 0 \tag{4.186}$$

Substituting Eq. (4.185b) into Eq. (4.183a), one has:

$$\mathcal{T}_1 = \frac{\sqrt{E} K_1(\sqrt{E} r_w) + \beta_{\text{res}} s K_0(\sqrt{E} r_w)}{\sqrt{E} I_1(\sqrt{E} r_w) - \beta_{\text{res}} s I_0(\sqrt{E} r_w)} \mathcal{T}_2 \tag{4.187}$$

Similar to the treatments of Eq. (4.93), one has:

$$\mathcal{T}_1 I_0(\sqrt{E}\chi) + \mathcal{T}_2 K_0(\sqrt{E}\chi) = \mathcal{T}_4 K_0(\sqrt{E}\chi) \tag{4.188}$$

Similar to the treatments of Eq. (4.94), one has:

$$-\mathcal{T}_4\sqrt{E}K_1(\sqrt{E}\chi)-\mathcal{T}_1\sqrt{E}I_1(\sqrt{E}\chi)+\mathcal{T}_2\sqrt{E}K_1(\sqrt{E}\chi)=-1 \quad (4.189)$$

Imposing the Eqs. (4.186)–(4.189), one has:

$$\mathcal{T}_4=\frac{1}{\sqrt{E}K_1(\sqrt{E}\chi)+FW\sqrt{E}I_1(\sqrt{E}\chi)-F\sqrt{E}K_1(\sqrt{E}\chi)},$$

$$\mathcal{T}_2=F\mathcal{T}_4,\quad \mathcal{T}_1=W\mathcal{T}_2,\quad F=\frac{K_0(\sqrt{E}\chi)}{WI_0(\sqrt{E}\chi)+K_0(\sqrt{E}\chi)},\ \text{and}$$

$$W=\frac{\sqrt{E}K_1(\sqrt{E}r_w)+\beta_{\text{res}}sK_0(\sqrt{E}r_w)}{\sqrt{E}I_1(\sqrt{E}r_w)-\beta_{\text{res}}sI_0(\sqrt{E}r_w)}.$$

Thus, the solution of the SWPP thermal test in the resting phase can be determined by:

$$\bar{T}_{aD}=\int_{r_w}^{\infty}g(r,E;\chi)f(\chi)d\chi+\sigma_1+\sigma_2 r \quad (4.190)$$

4.4.3.4 *Solutions in the extraction phase*

Conducting Laplace transform to Eqs. (4.149b)–(4.149c), one has:

$$\frac{(r-\beta P_e)}{r}\frac{\partial^2\bar{T}_{aD}}{\partial r^2}+\frac{(1+P_e)}{r}\frac{\partial\bar{T}_{aD}}{\partial r}-\frac{k_1}{k_cb}\frac{\partial\bar{T}_{1D}}{\partial z}\bigg|_{z=0}+\frac{k_2}{k_cb}\frac{\partial\bar{T}_{2D}}{\partial z}\bigg|_{z=b}$$
$$=\tfrac{C_a}{k_c}s\bar{T}_{aD}-\tfrac{C_a}{k_c}T_{aD,\text{res}} \quad (4.191\text{a})$$

$$\frac{C_1}{k_1}s\bar{T}_{1D}-\frac{C_1}{k_1}T_{1D,\text{res}}=\frac{\partial^2\bar{T}_{1D}}{\partial z^2},\quad -b_1\le z\le 0 \quad (4.191\text{b})$$

$$\frac{C_2}{k_2}s\bar{T}_{2D}-\frac{C_2}{k_2}T_{2D,\text{res}}=\frac{\partial^2\bar{T}_{2D}}{\partial z^2},\quad b\le z\le b+b_2 \quad (4.191\text{c})$$

Similar to the solutions of Eqs. (4.179)–(4.180), one could derive the solution of Eqs. (4.191b)–(4.191c), and the general solution of

Eqs. (4.191b)–(4.191c) could be described as

$$\bar{T}_{1D} = \int_{-b_1}^{0} g_1(z,s;\ell) f_3(\ell) d\ell + n_1 + n_2 z, \; -b_1 \le z \le 0 \tag{4.192a}$$

$$g_1(z,s;\ell) = \begin{cases} g_{11}(z,s;\ell) = H_1 \exp(-Az) + H_2 \exp(Az) \; 0 \le z < \ell \\ g_{12}(z,s;\ell) = H_3 \exp(-Az) + H_4 \exp(Az) \; \ell < z \le \infty \end{cases} \tag{4.192b}$$

$$f_3(\ell) = \tfrac{C_1}{k_1} T_{1D,\text{res}} - \tfrac{C_1}{k_1} s(n_1 + n_2 \ell) \tag{4.192c}$$

where $H_1 = -H_2$, $H_2 = XH_3$, $H_3 = \frac{-\exp(\sqrt{A}\ell)}{\sqrt{A}[G+X+\exp(2\sqrt{A}\ell)+P\exp(2\sqrt{A}\ell)]}$, $H_4 = -H_3 G$, $G = \exp(2\sqrt{A}b_1)$, $X = \frac{\exp(2\sqrt{A}\ell)-G}{1-\exp(2\sqrt{A}\ell)}$, $n_1 = \bar{T}_{aD}(r,t)$, $n_2 = \bar{T}_{aD}(r,t)/b_1$, ℓ is a positive value varying between 1 and ∞, $g_1(z,s;\ell)$ is Green's function.

$$\bar{T}_{2D} = \int_{b}^{b+b_2} g_2(z,s;\ell) f_4(\ell) d\ell + m_1 + m_2 z \tag{4.193a}$$

$$g_2(z,s;\ell) = \begin{cases} g_{21}(z,s;\ell) = N_1 \exp(-Bz) + N_2 \exp(Bz) \\ \qquad b \le z < \ell \\ g_{22}(z,s;\ell) = N_3 \exp(-Bz) + N_4 \exp(Bz) \\ \qquad \ell < z \le b + b_2 \end{cases} \tag{4.193b}$$

$$f_4(\ell) = \frac{C_2}{k_2} T_{2D,\text{res}} - \frac{C_2}{k_2} s(m_1 + m_2 \ell) \tag{4.193c}$$

where $N_1 = -\exp(2Bb)N_2$, $N_2 = \frac{\exp[B(b+b_2)]-\exp(2B\ell)}{2B\exp(B\ell)\{\exp[B(b+b_2)]-\exp(2Bb)\}}$, $N_3 = -\exp(B(b+b_2))N_4$, $N_4 = N_2 - \frac{\exp(B\ell)}{2B}$, $m_1 = (1+\frac{b}{b_2})\bar{T}_{aD}$, $m_2 = -\bar{T}_{aD}/b_2$, $g_2(z,s;\ell)$ is Green's function.

Substituting Eqs. (4.192a)–(4.193c) into Eq. (4.191a), one has:

$$\frac{(r-\beta P_e)}{r}\frac{\partial^2 \bar{T}_{aD}}{\partial r^2} + \frac{(1+P_e)}{r}\frac{\partial \bar{T}_{aD}}{\partial r} - E\bar{T}_{aD} = -\frac{C_a}{k_c} T_{aD,\text{res}} \tag{4.194}$$

With the Laplace transform of Eqs. (4.157a)–(4.157b), the boundary conditions become:

$$\beta_{\text{ext}} s\bar{T}_{wD} - \beta_{\text{ext}} T_{wD,\text{res}} = \left.\frac{\partial \bar{T}_{wD}}{\partial r}\right|_{r=r_w}, \quad t > t_{\text{res}} \tag{4.195}$$

As for the analytical solution of the aquifer in the extraction phase, one could use a similar approach to the analytical solutions of rocks to get it. One could derive the solution of Eq. (4.195), and the general solution of Eq. (4.195) could be described as

$$\bar{T}_{aD} = \int_{r_w}^{\infty} g(r, E; \hbar) f_e(\hbar) d\hbar + p_1 + p_2 r, \quad r \geq r_w \tag{4.196a}$$

$$g(r, E; \hbar) = \begin{cases} g_1(r, E; \hbar) = \mathcal{F}_1 \exp(-\sqrt{E} r) U(\gamma, \eta, \zeta) \\ \quad + \mathcal{F}_2 \exp(-\sqrt{E} r) L(-\gamma, Pe, \zeta) \quad r_w \leq r \leq \hbar \\ g_2(r, E; \hbar) = \mathcal{F}_3 \exp(-\sqrt{E} r) U(\gamma, \eta, \zeta) \\ \quad + \mathcal{F}_4 \exp(-\sqrt{E} r) L(-\gamma, Pe, \zeta) \quad \hbar \leq r \leq \infty \end{cases} \tag{4.196b}$$

$$f_e(\hbar) = \frac{C_a}{k_c} T_{aD,\text{res}} + \frac{(1 + P_e)}{\hbar} p_2 - E(p_1 + p_2 \hbar) \tag{4.196c}$$

where $\hbar$ is a positive value varying between r_w and ∞ (e.g., $r_w \leq \hbar \leq \infty$); $g(r, E; \hbar)$ is the Green's function. $p_1 = -p_2 r_e$, $p_2 = \frac{\beta_{\text{ext}} T_{wD,\text{res}}}{\beta_{\text{ext}} s r_w - \beta_{\text{ext}} s r_e - 1}$, $\zeta_w = 2\sqrt{\text{E}}(r_w - \beta P_e)$, $\zeta = 2\sqrt{\text{E}}(r - \beta P_e)$, $\zeta_\hbar = 2\sqrt{\text{E}}(\hbar - \beta P_e)$, $\mathcal{R} = \frac{(\beta_{\text{ext}} s + \sqrt{E}) L(-\gamma, Pe, \zeta_w) + 2\sqrt{E} L(-1-\gamma, 1+Pe, \zeta_w)}{(\beta_{\text{ext}} s + \sqrt{E}) U(\gamma, \eta, \zeta_w) + 2\gamma\sqrt{E} U(1+\gamma, 1+\eta, \zeta_w)}$,

$$\mathcal{F}_1 = -\frac{(\beta_{\text{ext}} s + \sqrt{E}) L(-\gamma, Pe, \zeta_w) + 2\sqrt{E} L(-1 - \gamma, 1 + Pe, \zeta_w)}{(\beta_{\text{ext}} s + \sqrt{E}) U(\gamma, \eta, \zeta_w) + 2\gamma\sqrt{E} U(1 + \gamma, 1 + \eta, \zeta_w)} \mathcal{F}_2,$$

$$\mathcal{F}_2 = \frac{U(\gamma, \eta, \zeta_\hbar)}{L(-\gamma, Pe, \zeta_\hbar) - \mathcal{R} U(\gamma, \eta, \zeta_\hbar)} \mathcal{F}_3,$$

$$\mathcal{F}_3 = \frac{\exp(\sqrt{E}\zeta_\hbar)\hbar}{\begin{array}{l}\sqrt{E}(\chi - \beta P_e)[(\mathcal{R}\omega + 1) U(\gamma, \eta, \zeta_\hbar) + (2\gamma + 2\mathcal{R}\omega\gamma) \\ U(\gamma + 1, \eta + 1, \zeta_\hbar) - \omega L(-\gamma, Pe, \zeta_\hbar) \\ -2\omega L(-1 - \gamma, 1 + Pe, \zeta_\hbar)]\end{array}},$$

$$\mathcal{F}_4 = 0 \quad \text{and} \quad \omega = \frac{U(\gamma, \eta, \zeta_\hbar)}{L(-\gamma, Pe, \zeta_\hbar) - \mathcal{R} U(\gamma, \eta, \zeta_\hbar)}.$$

In summary, the solutions in the Laplace domain in the injection and chasing phases are

$$\bar{T}_{aD} = H \frac{\exp[(r_w + \beta P_e)\sqrt{E}_1](r_w + \beta P_e)^{1-P_e} \exp[-(r + \beta P_e)\sqrt{E}_1](r + \beta P_e)^{P_e} U(\gamma, \eta, \xi(r))}{(P_e^2 \beta + r_w^2 \sqrt{E}_1 + r_w \beta P_e \sqrt{E}_1) U(\gamma, \eta, \xi(r)) + r_w \gamma \xi(r_w) U(\gamma + 1, \eta + 1, \xi(r_w))},$$
$$r \geq r_w \tag{4.197a}$$

$$\bar{T}_{1D} = \frac{1}{2} \exp[-Az] [\exp(2A(z + b_1)) - 1] [\cot h(b_1 A) - 1] \bar{T}_{aD},$$
$$-b_1 \leq z \leq 0 \tag{4.197b}$$

$$\bar{T}_{2D} = \frac{1}{2} \exp[-B(b + z)] [\exp(2B(b + b_2)) - \exp(2Bz)]$$
$$[\coth(b_2 B) - 1] \bar{T}_{aD}, \quad b \leq z \leq b + b_2 \tag{4.197c}$$

The solutions in the Laplace domain in the resting phase are

$$\bar{T}_{aD} = \int_{r_w}^{\infty} g(r, E; \chi) f(\chi) d\chi + \sigma_1 + \sigma_2 r, \quad r \geq r_w \tag{4.198a}$$

$$\bar{T}_{1D} = \int_{-b_1}^{0} g_1(z, s; \varepsilon) f_1(\varepsilon) d\varepsilon + \delta_1 + \delta_2 z, \quad -b_1 \leq z \leq 0 \tag{4.198b}$$

$$\bar{T}_{2D} = \int_{b}^{b+b_2} g_2(z, s; \varepsilon) f_2(\varepsilon) d\varepsilon + a_1 + a_2 z, \quad b \leq z \leq b + b_2 \tag{4.198c}$$

The solutions in the Laplace domain in the extraction phase are

$$\bar{T}_{aD} = \int_{r_w}^{\infty} g(r, E; \hbar) f_e(\hbar) d\hbar + p_1 + p_2 r, \quad r \geq r_w \tag{4.199a}$$

$$\bar{T}_{1D} = \int_{-b_1}^{0} g_1(z, s; \ell) f_3(\ell) d\ell + n_1 + n_2 z, \; -b_1 \leq z \leq 0 \tag{4.199b}$$

$$\bar{T}_{2D} = \int_{b}^{b+b_2} g_2(z, s; \ell) f_4(\ell) d\ell + m_1 + m_2 z, \; b \leq z \leq b + b_2 \tag{4.199c}$$

The solutions of Eqs. (4.197)–(4.199) are in the Laplace domain, and they need to be transformed into the real-time domain. In this book, the Stehfest method will be employed for the inverse Laplace transform on the solutions of Eqs. (4.197)–(4.199).

4.4.4 Comparisons of analytical models

The present solution of Eqs. (4.197)–(4.199) is an extension of many previous studies (Chen and Reddell, 1983; Li *et al.*, 2010; Lin *et al.*, 2019; Yang *et al.*, 2010), by considering the wellbore mixing effect, the thermal conduction and the thermal dispersion of aquifer, and the multi-step flow rate of the well. When the wellbore mixing effect is ignored (e.g., $r_w = 0$) and the flow rate is a positive constant (e.g., Q_{inj} = a positive constant, and $t_{\text{cha}} = t_{\text{res}} = t_{\text{ext}} = 0$), the present solution reduces into the model of Lin *et al.* (2019), who extended the models of Chen and Reddell (1983), Li *et al.* (2010) and Yang *et al.* (2010). If the value of thermal conduction is zero ($k_c/C_a = 0$, implying that the molecular diffusion in the solute transport model is negligible), $t_{\text{cha}} = t_{\text{res}} = 0$, $k_d/C_a \neq 0$ and the underlying and overlying layers are ignored ($k_1 = k_2 = 0$), the mathematical expression of Eqs. (4.197)–(4.199) are the same as Eqs. (19) and (20) in Wang *et al.* (2018), which was used to describe the solute transport in the aquifer.

By setting "$\frac{k_d}{C_a} = 0$" or "$\frac{k_d}{C_a} \ll \frac{k_c}{C_a}$" in Eqs. (4.197)–(4.199), the effective thermal diffusivity can be approximated as thermal diffusivity (k_c/C_a), implying that the thermal dispersion effect is ignored, the present solution is simplified to a conduction-dominant solution. Similarly, if "$\frac{k_c}{C_a} \ll \frac{k_d}{C_a}$", the thermal conduction effect is ignored, and the present solution is simplified to a dispersion-dominant solution.

This section develops a semi-analytical solution of heat transport around the well by considering the wellbore mixing effect under a multi-step flow rate. For simplicity, only four phases are included: injection, chasing, resting and extraction. Consider that the initial temperature is uniformly distributed in the aquifer system. The solution could be further extended to more general cases and was then explained as follows.

The solution of this section is composed of three sub-models: heat transport with uniform initial condition (injection and chasing phases), heat transport with non-uniform initial condition (resting phase) and zero flow rate, and heat transport with non-uniform initial condition and non-zero flow rate (extraction phase). Three sub-models are integrated by the continuity of temperature between two phases; for instance, the initial condition of the model in the resting phase begins from the result at the ending time of the chasing phase. Because the uniform temperature over space is a special case for the initial condition, and the zero-flow rate is also a special case for well discharge, the analytical solution in the extraction phase is a general form of the solutions in the other phases. For example, when substituting the initial condition of $T_a(r,t)|_{t=0}$, $T_1(r,z,t)|_{t=0}$, and $T_2(r,z,t)|_{t=0}$ and flow rate of Q_{inj} into Eqs. (4.199a)–(4.199c) instead of $T_a(r,t)|_{t=t_{\text{res}}}$, $T_1(r,z,t)|_{t=t_{\text{res}}}$, and $T_2(r,z,t)|_{t=t_{\text{res}}}$ and Q_{ext}, the extraction-phase solution reduces into the injection-phase solution. $T_{aD,\text{res}}(r)$, $T_{1D,\text{res}}(z)$, and $T_{2D,\text{res}}(z)$ in the functions of f_e, $f_3(\ell)$, $f_4(\ell)$ in Eqs. (4.199a)–(4.199c) represent the initial conditions of the aquifer, overlying and underlying layers, respectively. Therefore, the present solution could be easily extended to cases with more or less than four phases.

As for the cases where the well flow rate is a continuous function of time, a step function could be used to approximate it. Here, we present an example that has only one injection phase involved in the exploitation of the thermal energy and non-uniform initial conditions. Namely, $T_a(r,t)|_{t=0}$, $T_1(r,z,t)|_{t=0}$, and $T_2(r,z,t)|_{t=0}$ vary with spatial location. Meanwhile, both injection rate ($Q_w(t)$) and the temperature of the injected water ($T_w(t)$) continuously vary with time. Using the step function to approximate them, one has:

$$Q_w(t) \approx \begin{cases} Q_1, & 0 < t \leq t_1 \\ Q_2, & t_1 < t \leq t_2 \\ & \vdots \\ Q_n, & t_{n-1} < t \leq t_n \end{cases} \tag{4.200a}$$

$$T_w(t) \approx \begin{cases} T_1, & 0 < t \leq t_1 \\ T_2, & t_1 < t \leq t_2 \\ & \vdots \\ T_n, & t_{n-1} < t \leq t_n \end{cases} \tag{4.200b}$$

where t_n represents the ending time of the injection phase; n is the total number of the segments (or subdomains) used to approximate the continuous functions by the step functions.

The solutions are

$$\bar{T}_{aD} = \begin{cases} \bar{T}_{aD_1}, & 0 < t \le t_1 \\ \bar{T}_{aD_2}, & t_1 < t \le t_2 \\ & \vdots \\ \bar{T}_{aD_n}, & t_{n-1} < t \le t_n \end{cases} \tag{4.201a}$$

$$\bar{T}_{1D} = \begin{cases} \bar{T}_{1D_1}, & 0 < t \le t_1 \\ \bar{T}_{1D_2}, & t_1 < t \le t_2 \\ & \vdots \\ \bar{T}_{1D_n}, & t_{n-1} < t \le t_n \end{cases} \tag{4.201b}$$

$$\bar{T}_{2D} = \begin{cases} \bar{T}_{2D_1}, & 0 < t \le t_1 \\ \bar{T}_{2D_2}, & t_1 < t \le t_2 \\ & \vdots \\ \bar{T}_{2D_n}, & t_{n-1} < t \le t_n \end{cases} \tag{4.201c}$$

where the mathematical expression of $\bar{T}_{aD_i}$, $\bar{T}_{1D_i}$, and $\bar{T}_{2D_i}$ are the same as Eqs. (4.199a)–(4.199c), while the initial condition $t_{i-1} \le t < t_i$ is different for various sub-domains. For first time sub-domain (e.g., $(0, t_1]$), the initial condition is given and one has $T_{aD,\text{res}}(r) = T_a(r,t)|_{t=0}$, $T_{1D,\text{res}}(z) = T_1(r,z,t)|_{t=0}$, and $T_{2D,\text{res}}(z) = T_2(r,z,t)|_{t=0}$. For the other time subdomains, the initial condition of $(t_i, t_{i+1}]$ is computed at the ending time of $(t_{i-1}, t_i]$.

4.4.5 *Field application*

A field experiment is conducted in an alluvial-plain site in Zhengzhou city, China, to test the present solution. The confined aquifer is made up of fine sand soil, as shown in Figure 4.9. The lithologies of the overlying and underlying layers are clay and silty clay, respectively. The well radius is 0.15 m, and it penetrates the overlying clay layer (10.0 m thick), the horizontally confined aquifer (4.1 m thick) and the underlying silty clay layer (6.0 m thick). The well screen is 4.1 m in length and crosses the entire horizontal confined aquifer.

The initial water level is 15 m from the bottom of the wellbore. Three observation wells (0.05 m in radius), OBS_W1, OBS_W2, and

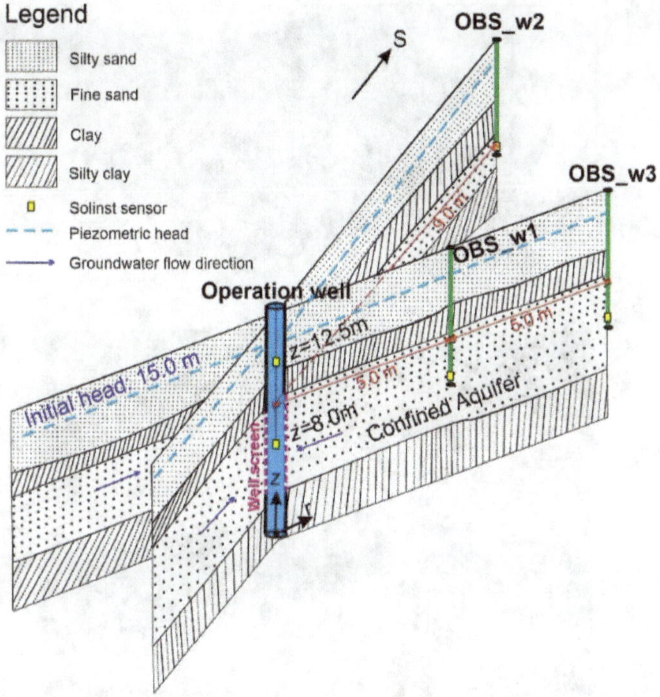

Figure 4.9. Fence diagram of the subsurface hydrologic system where the operation well is located at the intersection of two hydrogeological profiles. This is only a schematic diagram, and the thickness of each layer does not follow the actual thickness.

OBS_W3, are 5 m, 9 m, and 10 m away from the operation well, respectively. These three observation wells are not used in this section because the three observations are so far from the operational well that the heat was not transported to these locations. The initial temperature of groundwater is 17.71°C in the entire aquifer system. The temperature and water level are measured by a Solinst Levelogger (LTC 5M, Solinst Canada Ltd). The injection point of the pump is located at the bottom of the wellbore. To test whether the assumption of mixing effect in the mathematical models could be satisfied (the temperature uniformly distributed along the wellbore), two Solinst sensors are installed in the operation wellbore. They are located at the $z = 8.00$ m and $z = 12.50$ m, respectively. The origin of the coordinate system is at the center of the wellbore and the bottom of the aquifer.

Figure 4.10. The location of field site (a) and the bird view of the experiment site (b).

In this field experiment, four stages include injection, chasing, resting and extraction phases. The detailed procedure is as follows. Firstly, two water pools are prepared to fill water with different temperatures, e.g., 23.89°C and 29.39°C, as shown in Figure 4.10. Secondly, the 23.89°C water is injected into the confined aquifer with a constant flow rate of 1.46 m^3/h. After 62 min (t_{inj} = 62 min), the 29.39° water is injected using the same flow rate for 211 min (t_{cha} = 273 min).

In order to release the energy of the injected water into the aquifer, the resting phase is needed where the flow rate is zero. In this section, the resting time is 59 min (t_{res} = 332 min). Finally, the cool water is pumped from the same well with constant flow rate of −10.06 m^3/h after the resting phase.

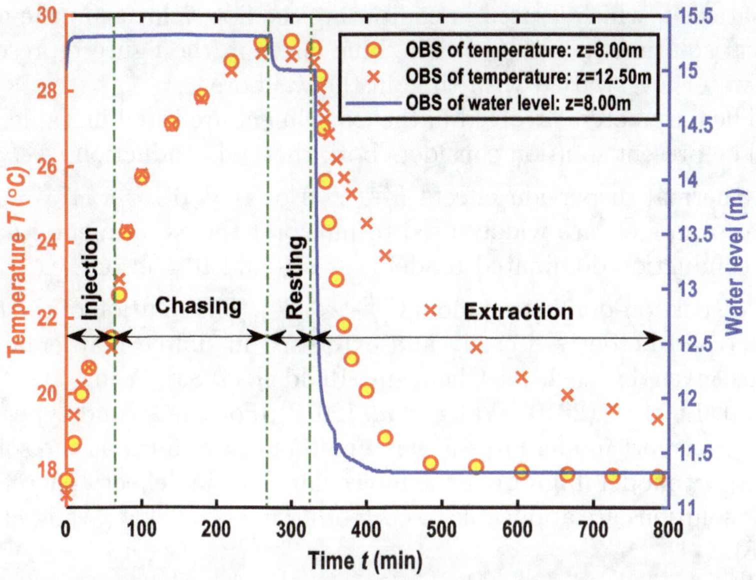

Figure 4.11. The observed temperature BTCs (scattered points) and the observed water level (blue curve) in the wellbore, where the observed BTCs at $z = 12.50$, in the extraction phase cannot represent the true water temperature variation because the water level is below the sensor location. It is only to show the complete mixing in earlier phases (injection, chasing and resting phases).

Figure 4.11 shows the observed BTCs at the wellbore. The temperatures measured at $z = 8.00$ m and 12.50 m are denoted as dot-curves. Several interesting observations could be found. Firstly, the temperature in the wellbore increases with time in the injection and chasing phases due to the wellbore mixing effect, indicating that such mixing effect could not be ignored. Secondly, one may find that two dot-curves are almost identical in the injection, chasing, and resting phases, demonstrating that the assumption involved in the mixing effect models could be satisfied in these phases. However, the difference between the observed temperatures at two locations increases with time in the extraction phase. This is because the water level is below the sensor location (namely $h_{w,\text{ext}} < z = 12.50$), and the measured data there could not represent the real water temperature in the wellbore. Thirdly, the time that the transient state approaches the quasi-steady state for flow is very short, so the assumption could be satisfied, which is used in computing the flow velocity in the analytical solution ($v = \frac{Q}{2\pi\theta br}$). The blue curve in the figure represents the water level varied with time in the wellbore.

The parameters involved in the experiment are listed in Table 4.3.

The present solution considers both thermal conduction ($\frac{k_c}{C_a} \neq 0$) and thermal dispersion effects ($\frac{k_d}{C_a} \neq 0$ or $\beta \neq 0$). Yet, two other types of models are widely used to interpret the experimental data: the conduction-dominated model ($\frac{k_c}{C_a} \neq 0$ and $\beta = 0$; or $\frac{k_c}{C_a} \gg \frac{k_d}{C_a}$) and dispersion-dominated model ($\frac{k_c}{C_a} \ll \frac{k_d}{C_a}$). The solution of the first type could be derived easily and expressed in simple mathematical forms given by such as Chen and Reddell (1983), Yang and Yeh (2008), Li *et al.* (2010), Yang *et al.* (2010). For the second type, the heat transport model has the same mathematical form as the solute transport model if ignoring the molecular diffusion effect. Therefore, their solutions are applicable to each other in both fields (Wang *et al.*, 2018).

The predictions from these three types of models are compared with the experimental data to examine the effects of thermal conduction and thermal dispersion on the output temperature. The results of parameter estimation are $\beta = 0.45$ m and $k_c = 2.30$ W/(m · °C) for the solution of the conduction-dispersion model, $\beta = 0.45$ m and $k_c = 0.23$ W/(m · °C) for the solution of the dispersion-dominant model, and $\beta = 0.00$ m and $k_c = 2.30$ W/(m · °C) for the solution of the conduction-dominant model. The values of other parameters used to plot this figure are listed in Table 4.3. The estimated λ_c is physically reasonable based on the literature. For instance, Buscheck *et al.* (1983) pointed out k_c is 2.29 W/(m·°C) for the fine sand aquifer and 2.56 W/(m · °C) for the clay aquitards. Goto (2005) found that the range of k_c is $1 - 6$ W/(m·°C) for the ordinary marine sediment and hydrothermal deposit. Sebok and Müller (2019) investigated the conduction of the sandy aquifer and found $k_c = 0.55 - 2.96$ W/(m · °C). Stonestrom and Constantz (2003) reported that the values of k_c for quartz, saturated sand and saturated clay loam were 8.4 W/(m·°C), 2.6 W/(m · °C) and 3.2 W/(m · °C), respectively.

Figure 4.12 shows that both conduction-dispersion and dispersion-dominant solutions perform well in interpreting the experimental data. The *RMSE* is also included in this figure, and both conduction-dispersion and dispersion-dominant solutions have

Table 4.3. The parameters used in the field experimental of thermal SWPP test.

Experimental parameters	Symbol	Value	Unit
Injection rate	Q_{inj}	1.46	m^3/h
Chasing rate	Q_{cha}	1.46	m^3/h
Extraction rate	Q_{ext}	10.06	m^3/h
The end moment of the injection phase	t_{inj}	62	min
The end moment of the chasing phase	t_{cha}	273	min
The end moment of the resting phase	t_{res}	332	min
Initial temperature	T_0	17.71	°C
Injection temperature	T_{in}	23.89	°C
Chasing temperature	T_{ch}	29.39	°C
Initial hydraulic head	h_0	15.0	m
Wellbore water lever in the injection phase	$h_{w,\text{inj}}$	15.3	m
Wellbore water lever in the chasing phase	$h_{w,\text{cha}}$	15.3	m
Wellbore water lever in the resting phase	$h_{w,\text{res}}$	15.0	m
Wellbore water lever in the extraction phase	$h_{w,\text{ext}}$	11.23	m
The thickness of the overlying clay layer	b_2	10	m
The thickness of the confined aquifer	b	4.1	m
The thickness of the underlying clay silty layer	b_1	6.0	m
The radius of the well	r_w	0.15	m
Confined aquifer porosity	θ	0.32	—
Thermal conductivity of overlying clay layer	k_2	1.5	W/(m · °C)
Thermal conductivity of the confined aquifer	k_c	2.3	W/(m · °C)
Thermal conductivity of underlying clay layer	k_1	1.5	W/(m · °C)
The specific volumetric thermal capacity of water	C_w	4.2	W/(m · °C)
The specific volumetric thermal capacity of the overlying clay layer	C_2	2.5	$\text{MJ}/(\text{m}^3 \cdot °\text{C})$
The specific volumetric thermal capacity of confined aquifer	C_a	2.38	$\text{MJ}/(\text{m}^3 \cdot °\text{C})$
The specific volumetric thermal capacity of the underlying clay layer	C_1	2.65	$\text{MJ}/(\text{m}^3 \cdot °\text{C})$
Thermal dispersivity of a confined aquifer	β	0.45	m

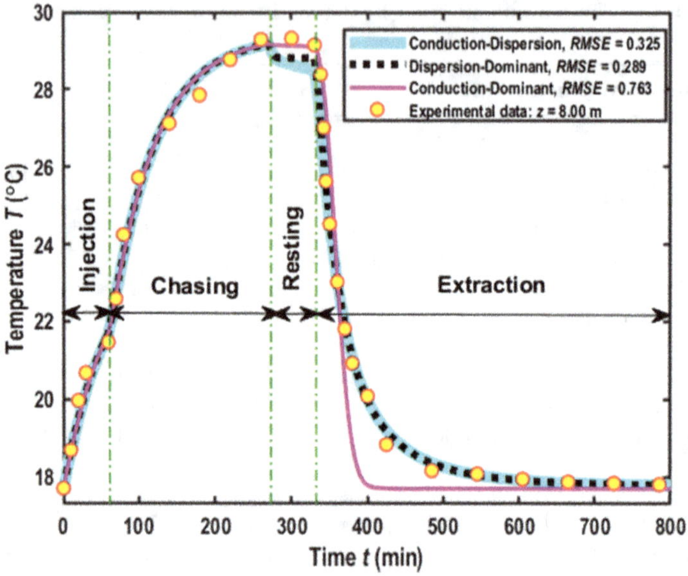

Figure 4.12. Fitness of the observed BTC by different models.

a lower $RMSE$ than the conduction-dominant solution. In addition, the figure indicates that the parameter λ_c is insensitive to the results. The key reason could be explained based on the HTE. By checking the coefficient of $\frac{\partial T_a}{\partial r}$ in Eq. (4.119a), one has $\frac{k_c+k_d}{C_a} = \frac{k_c}{C_a} + \beta\frac{C_f v\theta}{C_a} = \frac{2.3}{2.38\times10^6} + \frac{0.45\times4.20\times10^6\times0.32}{2.38\times10^6} \times v$, where $v = 1.189$ m/d at the well screen in injection and chasing phases. Therefore, the term of $\frac{k_c}{C_a}$ is much smaller than the term of $\beta\frac{C_f v\theta}{C_a}$. In this experiment, the resting phase is very short (59.00 min). Therefore, it is not surprising that the solutions of the dispersion-dominant model and conduction-dispersion model are almost identical.

When using a conduction-dominant solution to fit the observed BTCs, we find that the thermal BTCs computed by the conduction-dominant solution do not match well with the observed data in the later stage of the extraction phase. This results from the fact that a larger conduction quickly releases the energy of the injected water into aquifers and thus quickly decreases temperature in the extraction phase.

4.4.6 *Discussions*

4.4.6.1 *Effect of the parameters on BTCs*

Some previous studies have used numerical/analytical and field experiments to investigate the effect of thermal conductivity/thermal conduction on heat transport (McCallum *et al.*, 2012; Park *et al.*, 2018; Park and Lee, 2021; Rau *et al.*, 2014; Schneidewind *et al.*, 2016; van Kampen *et al.*, 2022; Vandersteen *et al.*, 2015). Rau *et al.* (2012) found that the effect of thermal dispersion was negligible in heat transport when flow Darcian velocity is less than 10 m/d, which has been ignored in quantifying hyporheic exchange in streambeds using heat tracer (McCallum *et al.*, 2012; Park *et al.*, 2018; Park and Lee, 2021; Rau *et al.*, 2014; Schneidewind *et al.*, 2016; van Kampen *et al.*, 2022; Vandersteen *et al.*, 2015). However, in the ATES system, particularly near injection/extraction well, Darcian velocity is usually greater than 10 m/d. For instance, as shown in Section 4.4.5, the Darcian velocity in the extraction phase is greater than 10 m/d near the well screen and can reach 62.5 m/d at the well screen. To test the influences of the thermal dispersion on heat transport in the ATES system, three sets of β (i.e., 0.0, 0.1 and 0.5 m) are employed, the other parameters are the same as those listed in Table 4.3.

Figure 4.13 shows the thermal BTCs in the injection well for different β. The difference between them is obvious in both the resting and extraction phases.

A larger β results in a lower temperature in the resting phase and at the beginning of the extraction phase but a higher temperature in the extraction phase. This is because a larger β increases thermal spreading, resulting in less energy near the injection well and more energy away from the injection well. In the extraction phase, a smaller β. This means that the more energy stored in the wellbore, the higher the temperature at the beginning; consequently, the temperature attenuates quickly over time. Figure 4.13 demonstrates that the thermal dispersion processes have a significant impact on thermal transfer in the ATES system.

4.4.6.2 *Global sensitivity analysis*

From the semi-analytical model of Eqs. (4.197)–(4.199), thermal BTCs are affected by the following parameters: V_w, β, b, Q_{ext}, k_c,

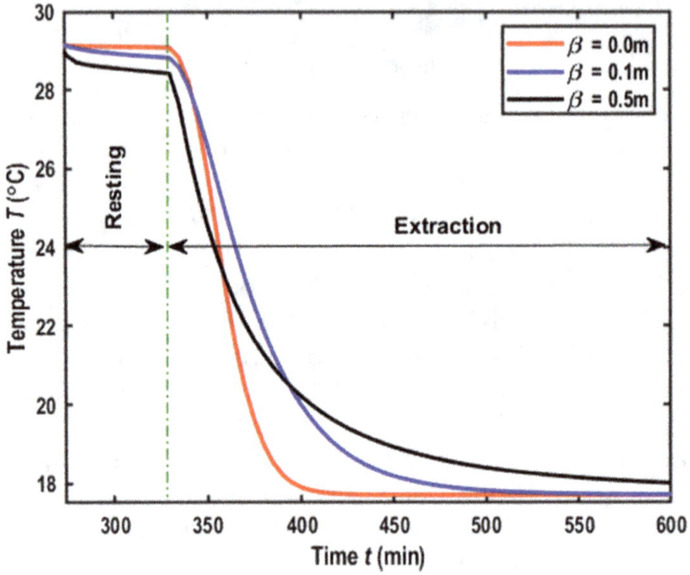

Figure 4.13. The thermal BTCs in the injection well for different β during the resting and extraction phases.

$k_{1,2}$, C_a, $C_{1,2}$, C_f, and θ. The global sensitivity analysis by Morris (1991) is employed to analyze the response of aquifer temperature to the change in each of the parameters (V_w, β, b, Q_{ext}, k_c, $k_{1,2}$, C_a, $C_{1,2}$, C_f, and θ), and the mathematical model can be described by

$$\mu_k^* = \sum_{l=1}^{m}(EE_k^l/m), \quad k = 1, 2, \ldots, n \tag{4.202a}$$

$$\mu_k = \sum_{l=1}^{m}(|EE_k^l|/m), \quad k = 1, 2, \ldots, n \tag{4.202b}$$

$$\sigma_k = \sqrt{\frac{1}{m}\sum_{l=1}^{m}(EE_k^l - \mu_k^*)^2}, \quad k = 1, 2, \ldots, n \tag{4.202c}$$

where μ_k^*, μ_k and σ_k represent the mean, the absolute mean and the standard deviation, respectively; μ_k and σ_k are used to represent the significance of the input parameters to the output results, in which a larger μ_k implies a greater sensitive influence of the kth

Table 4.4. Parameter range used in sensitivity analysis.

Parameters	Unit	Range	Remarks
V_w	m^3	0.10–1.90	[a]
β	m	0.01–0.50	[a], [b]
b	m	4.00–10.00	[a]
θ	—	0.1–0.5	[a], [b]
Q_{ext}	m^3/h	0.50–10.00	[a]
k_c	$W/(m \cdot °C)$	0.5–8.4	[b], [c], [d]
$k_{1,2}$	$W/(m \cdot °C)$	1.00–5.00	[b], [c], [d]
C_a	10^6 $J/(m^3 \cdot °C)$	1.8–3.00	[d], [e], [f], [g]
C_f	10^6 $J/(m^3 \cdot °C)$	1.00–2.00	[d], [e], [f], [g]
$C_{1,2}$	10^6 $J/(m^3 \cdot °C)$	1.50–2.50	[d], [e], [f], [g]

Note: [a]. Shi *et al.* (2023a), [b]. Lin *et al.* (2019), [c]. Stonestrom and Constantz (2003), [d]. Goto (2005), [e]. Buscheck *et al.* (1983), [f]. Li *et al.* (2010), [g]. Sebok and Müller (2019).

parameter on the output, while a larger σ_k implies stronger interaction effects and nonlinear effects than others. m is the number of the total sampling points (chosen as 50 in the present study); n refers to the number of the total parameter of interest $n = 12$; k is the kth parameter; $EE_k^l = \frac{T_a(P_1,P_2,\ldots,P_k+l\Delta,\ldots,P_n)-T_a(P_1,P_2,\ldots,P_k,\ldots,P_n)}{l\Delta}$; P_i represents the random value of the ith parameter in the range of $(P_{i,0}, P_{i,lim})$; $P_{i,0}$ and $P_{i,lim}$ are, respectively, the smallest and largest values of P_i, as shown in Table 4.4; Δ is a small increment defined as $1/(m-1)$.

Figures 4.14(a) and 4.14(b) show the variations of μ_k and σ_k with time in the injection well in the extraction phase, respectively. Figure 4.14(a) shows that the values of μ_k from large to small are V_w, Q_{ext}, β, k_c, b, $C_{1,2}$, C_a, $k_{1,2}$ and θ. It demonstrates the output temperature is more sensitive to V_w, Q_{ext}, and β than the other parameters in the extraction phase. Specifically, the most sensitive parameter is V_w in the early stage of the extraction phase, with time going on, the influence of V_w decreases with time, and the most sensitive parameter becomes Q_{ext}.

Figures 4.15 and 4.16 show the effect of V_w and Q_{ext} on thermal BTCs in the injection well, respectively. The other parameters used in these two figures are listed in Table 4.3. In Figure 4.15, one may find that the thermal BTCs decrease with increasing V_w in the injection, chasing, and resting phases. In the extraction phase, a higher

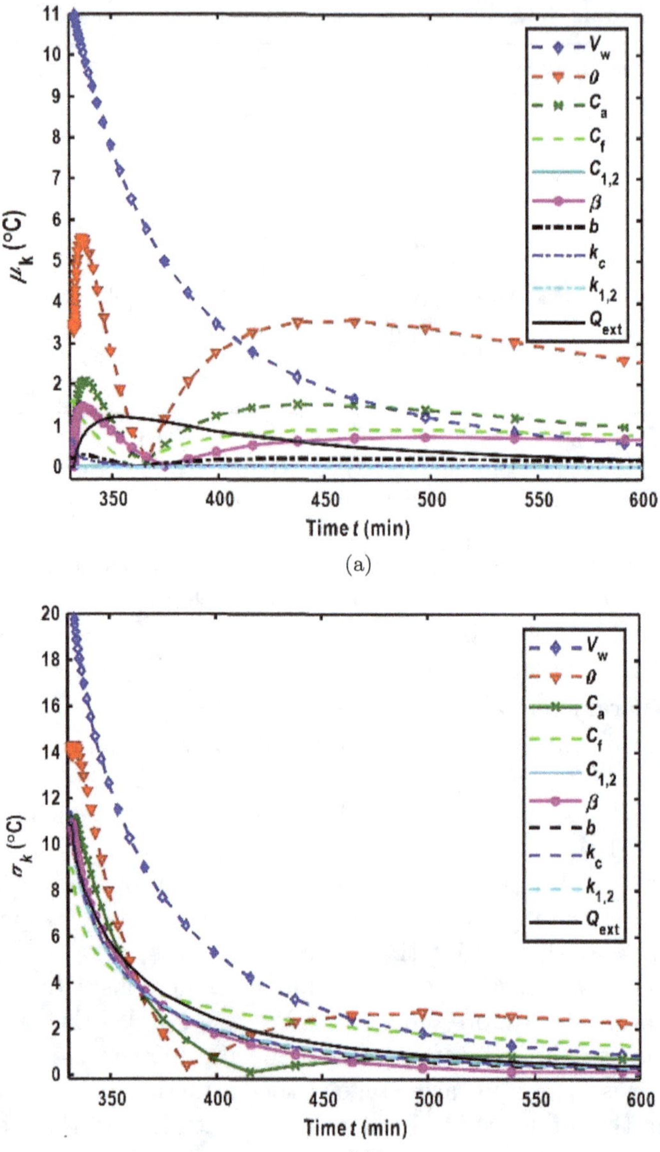

Figure 4.14. Sensitivity analysis of the input parameters on the output parameters in the extraction phase. (a) Variation in μ_k with time in the injection well. (b) Variation in σ_k with time in the injection well.

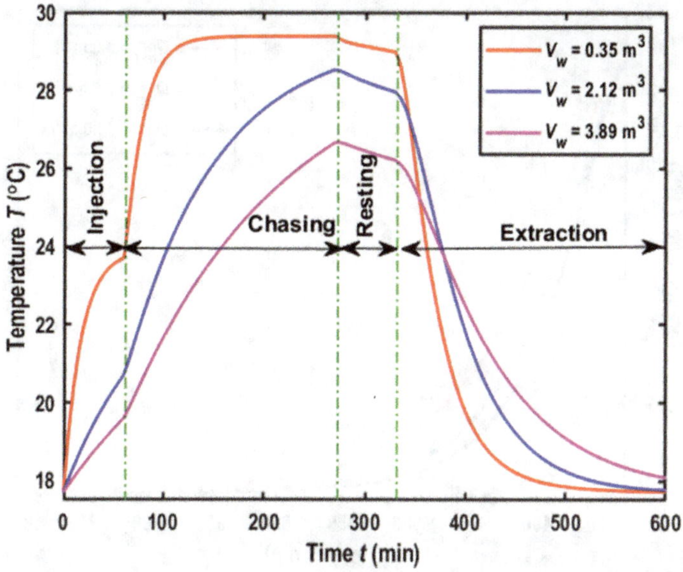

Figure 4.15. The thermal BTCs in the injection well for different V_w, the groundwater volume in the wellbore during the injection, chasing and extraction phases are $V_{w,\mathrm{inj}} = V_{w,\mathrm{cha}} = V_w + 0.07$ m^3 and $V_{w,\mathrm{ext}} = V_w - 0.07$ m^3, respectively.

V_w results in a slower fall of thermal BTC. This is because a higher V_w means a greater mixing effect in the wellbore. Figure 4.16 indicates that the thermal BTCs decrease with increasing Q_{ext} in the extraction. This is because Q_{ext} represents the thermal advection, and a greater Q_{ext} means a faster thermal front velocity, resulting in faster thermal recovery. Figure 4.14(b) shows that the parameters of β, Q_{ext} and k_c have a higher interaction effect and nonlinear effects than the others in the extraction phase.

4.4.7 *Limitations and future works*

Although the present solution is an extension of previous solutions, it still contains several assumptions:

First, the aquifer is homogeneous and with uniform thickness.

Second, the wellbore mixing effect is completed instantaneously.

Third, the regional flow is assumed to be negligible. Fourth, the density variation and the phase changes of water are ignored.

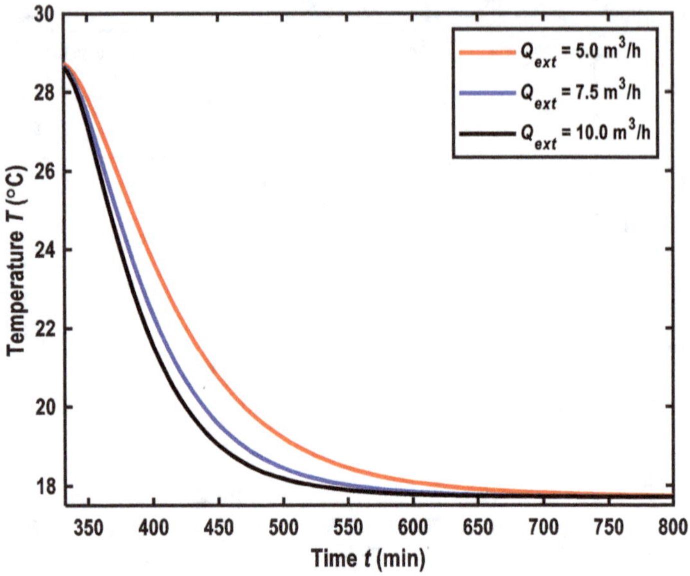

Figure 4.16. The thermal BTCs in the injection well during the extraction phase for different Q_{ext}.

Fifth, the flow field is in a quasi-steady state during the entire operation (Wang *et al.*, 2017).

Sixth, the porous media and fluid are in thermal equilibrium. These assumptions also exist in the previous analytical solutions (Bodvarsson and Tsang, 1982; Chen and Reddell, 1983; Jung and Pruess, 2012; Kocabas, 2005; Lin *et al.*, 2019; Wang *et al.*, 2020a).

References

Anderson, M.P., 2005. Heat as a ground water tracer. *Groundwater*, 43(6): 951–968.

Anibas, C., Buis, K., Verhoeven, R., Meire, P., Batelaan, O., 2011. A simple thermal mapping method for seasonal spatial patterns of groundwate' surface water interaction. *Journal of Hydrology*, 397: 93–104.

Bardini, L. *et al.*, 2013. Small-scale permeability heterogeneity has negligible effects on nutrient cycling in streambeds. *Geophysical Research Letters*, 40(6): 1118–1122.

Bear, J., 1972. *Dynamics of Fluids in Porous Media.* Elsevier, New York, 1972.

Boano, F., Demaria, A., Revelli, R., Ridolfi, L., 2010. Biogeochemical zonation due to intrameander hyporheic flow. *Water Resources Research*, 46(2): W02511.

Bodvarsson, G.S., Tsang, C.F., 1982. Injection and thermal breakthrough in fractured geothermal reservoirs. *Journal of Geophysical Research*, 87(NB2): 1031–1048.

Bredehoeft, J.D., Papaopulos, I.S., 1965. Rates of vertical groundwater movement estimated from the Earth's thermal profile. *Water Resources Research*, 1(2): 325–328.

Buscheck, T.A., Doughty, C., Tsang, C.F., 1983. Prediction and analysis of a field experiment on a multilayered aquifer thermal energy storage system with strong buoyancy flow. *Water Resources Research*, 19(5): 8.

Caissie, D., Kurylyk, B.L., St- Hilaire, A., El-Jabi, N., MacQuarrie, K.T.B., 2014. Streambed temperature dynamics and corresponding heat fluxes in small streams experiencing seasonal ice cover. *Journal of Hydrology*, 519: 1441–1452.

Chen, C.S., Reddell, D.L., 1983. Temperature distribution around a well during thermal injection and a graphical technique for evaluating aquifer thermal properties. *Water Resources Research*, 19(2): 351–363.

Chen, C.-S., Woodside, G.D., 1988. Analytical solution for aquifer decontamination by pumping. *Water Resources Research*, 24(8): 1329–1338.

Chen, J.-S., Liu, C.W., Chen, C.-S., Liao, C.-M., 2003. Effect of well bore mixing volume on non-axisymmetrical transport in a convergent radial tracer test. *Journal of Hydrology*, 277(1–2): 61–73.

Chen, K., Hongbin, Z., Yang, Q., 2017. Fractional models simulating non-Fickian behavior in four-stage single-well push-pull tests. *Water Resources Research*, 53: 9528–9545.

Chen, K. *et al.*, 2022. Estimating lateral groundwater inflow to rivers using heat as a tracer. *Journal of Hydrology*, 617(2023): 128965.

Constantz, J.E., 2008. Heat as a tracer to determine streambed water exchanges. *Water Resources Research*, 44: W00D10.

Constantz, J.E., Su, G.W., Hatch, C.E., 2006. Heat as a ground-water tracer at the russian river RBF facility, Sonoma county, California.

Constantz, J.E., Thomas, C.L., 1996. The use of streambed temperature profiles to estimate the depth, duration, and rate of percolation beneath arroyos. *Water Resources Research*, 32: 3597–3602.

Constantz, J.E., Thomas, C.L., 1997. Stream bed temperature profiles as indicators of percolation characteristics beneath arroyos in the Middle Rio Grande Basin, USA. *Hydrological Processes*, 11(12): 1621–1634.

Constantz, J.E., Thomas, C.L., Zellweger, G.W., 1994. Influence of diurnal variations in stream temperature on streamflow loss and groundwater recharge. *Water Resources Research*, 30: 3253–3264.

Cuthbert, M.O., Mackay, R., 2013. Impacts of nonuniform flow on estimates of vertical streambed flux. *Water Resources Research*, 49: 19–28.

Daniilidis, A., Mindel, J.E., De Oliveira Filho, F., Guglielmetti, L., 2022. Techno-economic assessment and operational CO_2 emissions of high-temperature aquifer thermal energy storage (HT-ATES) using demand-driven and subsurface-constrained dimensioning. *Energy*, 249: 123682.

de Marsily, G., 1986. Quantitative Hydrogeology: Groundwater Hydrology for Engineers. Academic Press, Inc., Orlando, FL.

Dou, Z., Zhou, Z.F., Wang, J.G., Huang, Y., 2018. Roughness scale dependence of the relationship between tracer longitudinal dispersion and Peclet number in variable-aperture fractures. *Hydrological Processes*, 32(10): 1461–1475.

Eidesgaard, O.R., Schovsbo, N.H., Boldreel, L.O., Olavsdottir, J., 2019. Shallow geothermal energy system in fractured basalt: A case study from Kollafjorour, Faroe Islands, NE-Atlantic Ocean. *Geothermics*, 82: 296–314.

Goto, S., 2005. Thermal response of sediment with vertical fluid flow to periodic temperature variation at the surface. *Journal of Geophysical Research*, 110(B1).

Halloran, L.J.S., Rau, G.C., Andersen, M.S., 2016. Heat as a tracer to quantify processes and properties in the vadose zone: A review. *Earth-Science Reviews*, 159(2016): 358–373.

Hatch, C.E., Fisher, A.T., Revenaugh, J.S., Constantz, J.E., Ruehl, C.R., 2006. Quantifying surface wate' groundwater interactions using time series analysis of streambed thermal records: Method development. *Water Resources Research*, 42(10).

Hopmans, J.W., ¡inonek, J., Bristow, K.L., 2002. Indirect estimation of soil thermal properties and water flux using heat pulse probe measurements: Geometry and dispersion effects. *Water Resources Research*, 38: 7-1–7-14.

Irvine, D.J., Lautz, L.K., Briggs, M.A., Gordon, R.P., McKenzie, J.M., 2015. Experimental evaluation of the applicability of phase, amplitude, and combined methods to determine water flux and thermal diffusivity from temperature time series using VFLUX 2. *Journal of Hydrology*, 531(3): 728–737.

Irvine, D.J. *et al.*, 2017. Using diurnal temperature signals to infer vertical groundwater-surface water exchange. *Ground Water*, 55(1): 10–26.

Jacquot, R.G., Steadman, J.W., Rhodine, C.N., 1983. The Gaver-Stehfest algorithm for approximate inversion of Laplace transforms. *IEEE Circuits & Systems Magazine*, 5: 4.8.

Jenkins, D.N., Prentice, J.K., 1982. Theory for aquifer test analysis in fractured rocks under linear (nonradial) flow conditions. *Groundwater*, 20(1): 12–21.

Jung, Y., Pruess, K., 2012. A closed-form analytical solution for thermal single-well injection-withdrawal tests. *Water Resources Research*, 48(3): W03504.

Kabala, Z.J., 2001. Sensitivity analysis of a pumping test on a well with wellbore storage and skin. *Advances in Water Resources*, 24(5): 483–504.

Keery, J., Binley, A., Crook, N., Smith, J.W.N., 2007. Temporal and spatial variability of groundwater surface water fluxes: Development and application of an analytical method using temperature time series. *Journal of Hydrology*, 336(1–2): 1–16.

Klepikova, M.V. *et al.*, 2016a. Heat as a tracer for understanding transport processes in fractured media: Theory and field assessment from multiscale thermal push-pull tracer tests. *Water Resources Research*, 52(7): 5442–5457.

Klepikova, M.V. *et al.*, 2016b. Heat tracer test in an alluvial aquifer: Field experiment and inverse modelling. *Journal of Hydrology*, 540(2016): 812–823.

Kocabas, I., 2005. Geothermal reservoir characterization via thermal injection backflow and interwell tracer testing. *Geothermics*, 34(1): 27–46.

Kurylyk, B.L., Irvine, D.J., Bense, V.F., 2019. Theory, tools, and multidisciplinary applications for tracing groundwater fluxes from temperature profiles. *Wires Water*, 6(1): e1329.

Lautz, L.K., Ribaudo, R.E., 2012. Scaling up point-in-space heat tracing of seepage flux using bed temperatures as a quantitative proxy. *Hydrogeology Journal*, 20: 1223–1238.

Lauwerier, H.A., 1955. The transport of heat in an oil layer caused by the injection of hot fluid. *Applied Science Research*, 5: 145–150.

Li, K.Y., Yang, S.Y., Yeh, H.D., 2010. An analytical solution for describing the transient temperature distribution in an aquifer thermal energy storage system. *Hydrological Processes*, 24(25): 3676–3688.

Lin, Y., Chang, C., Tsai, J., 2022. Analytical solution for estimating transient vertical groundwater flux from temperature-depth profiles. *Journal of Hydrology*, 610(2022): 127920.

Lin, Y.C., Hu, T.F., Yeh, H.D., 2019. Analytical model for heat transfer accounting for both conduction and dispersion in aquifers with a Robin-type boundary condition at the injection well. *Water Resources Research*, 55(8): 7379–7399.

Lucas, M., Cantareira, G.D., Wendland, E., 2019. Solute transport performance analysis of equivalent apertures in a single undisturbed basaltic fracture. *Hydrogeology Journal*, 27(6): 1999–2010.

Maier, F., Kocabas, I., 2013. Comment on "A closed-form analytical solution for thermal single-well injection-withdrawal tests" by Jung and Pruess. *Water Resources Research*, 49(1): 640–643.

McCallum, A.M., Andersen, M.S., Rau, G.C., Acworth, R., 2012. A 1-D analytical method for estimating surface water-groundwater interactions and effective thermal diffusivity using temperature time series. *Water Resources Research*, 48(11): W11532.

Morris, M.D., 1991. Factorial sampling plans for preliminary computational experiments. *Technometrics*, 33(2): 161–174.

Park, B.-H., Lee, K.K., 2021. Evaluating anisotropy ratio of thermal dispersivity affecting geometry of plumes generated by aquifer thermal use. *Journal of Hydrology*, 602(2021): 126740.

Park, B.-H., Bae, G.O., Lee, K.K., 2015. Importance of thermal dispersivity in designing groundwater heat pump (GWHP) system: Field and numerical study. *Renewable Energy*, 83: 270–279.

Park, B.-H., Lee, B.-H., Lee, K.K., 2018. Experimental investigation of the thermal dispersion coefficient under forced groundwater flow for designing an optimal groundwater heat pump (GWHP) system. *Journal of Hydrology*, 562(2018): 385–396.

Rau, G.C., Andersen, M.S., Acworth, R., 2012. Experimental investigation of the thermal dispersivity term and its significance in the heat transport equation for flow in sediments. *Water Resources Research*, 48(3): W03511.

Rau, G.C., Andersen, M.S., McCallum, A.M., Acworth, R., 2010. Analytical methods that use natural heat as a tracer to quantify surface water–groundwater exchange, evaluated using field temperature records. *Hydrogeology Journal*, 18(2010): 1093–1110.

Rau, G.C., Andersen, M.S., McCallum, A.M., Roshan, H., Acworth, R., 2014. Heat as a tracer to quantify water flow in near-surface sediments. *Earth-Science Reviews*, 129(2014): 40–58.

Rau, G.C., Cuthbert, M.O., McCallum, A.M., Halloran, L.J.S., Andersen, M.S., 2015. Assessing the accuracy of 1-D analytical heat tracing for estimating near-surface sediment thermal diffusivity and water flux under transient conditions. *Journal of Geophysical Research*, 120: 1551–1573.

Read, T. *et al.*, 2013. Characterizing groundwater flow and heat transport in fractured rock using fiber-optic distributed temperature sensing. *Geophysical Research Letters*, 40(10): 2055–2059.

Ronan, A.D., Prudic, D.E., Thodal, C.E., Constantz, J.E., 1998. Field study and simulation of diurnal temperature effects on infiltration and variably saturated flow beneath an ephemeral stream. *Water Resources Research*, 34(9): 2137–2153.

Roshan, H., Cuthbert, M.O., Cuthbert, M.O., Andersen, M.S., Acworth, R., 2014. Local thermal non-equilibrium in sediments: Implications for temperature dynamics and the use of heat as a tracer. *Advances in Water Resources*, 73(2014): 176–184.

Saar, M.O., 2011. Review: Geothermal heat as a tracer of large-scale groundwater flow and as a means to determine permeability fields. *Hydrogeology Journal*, 19: 31–52.

Schneidewind, U. *et al.*, 2016. LPMLE3: A novel 1-D approach to study water flow in streambeds using heat as a tracer. *Water Resources Research*, 52(8): 6596–6610.

Sebok, E., Müller, S., 2019. The effect of sediment thermal conductivity on vertical groundwater flux estimates. *Hydrology and Earth System Sciences*, 23(8): 3305–3317.

Selker, J.S. *et al.*, 2006. Distributed fiber-optic temperature sensing for hydrologic systems. *Water Resources Research*, 42(12): W12202.

Shi, W., Wang, Q., 2023. An analytical model of multi-layered heat transport to estimate vertical streambed fluxes and sediment thermal properties. *Journal of Hydrology*, 625(2023): 129963.

Shi, W. *et al.*, 2023a. Semi-analytical model of heat transport around a well with wellbore mixing effect under multi-step flow rate. *Journal of Hydrology*, 625(2023): 130068.

Shi, W. *et al.*, 2024. BDFLUX: A new bounded domain model for quantifying vertical fluxes in streambeds using heat as a tracer. *Journal of Hydrology*, 633(2024), 130981.

Shi, W., Wang, Q., Zhan, H., 2020. New simplified models of single-well push-pull tests with mixing effect. *Water Resources Research*, 56(8): e2019WR026802.

Shi, W., Zhan, H., Wang, Q., Xie, X., 2023b. A two-dimensional closed-form analytical solution for heat transport with nonvertical flow in riparian zones. *Water Resources Research*, 59(8): e2022WR034059.

Simon, N. *et al.*, 2023. Monitoring groundwater fluxes variations through active-DTS measurements. *Journal of Hydrology*, 622(2023): 129755.

Simon, N. *et al.*, 2021. Numerical and experimental validation of the applicability of active-DTS experiments to estimate thermal conductivity and groundwater flux in porous media. *Water Resources Research*, 57(1): e2020WR028078.

Stehfest, H., 1970. Algorithm 368: Numerical inversion of Laplace transforms. Commun. *ACM*, 13(1): 47–49.

Stonestrom, D.A., Constantz, J.E., 2003. Heat as a tool for studying the movement of ground water near streams. Circular. US Department of the Interior, US Geological Survey.

Tang, D.W., Rijnaarts, H.H., 2023. Dimensionless thermal efficiency analysis for aquifer thermal energy storage. *Water Resources Research*, 59(11): e2023WR035797.

Taron, J., Elsworth, D., 2009. Thermal–hydrologic–mechanical–chemical processes in the evolution of engineered geothermal reservoirs.

International Journal of Rock Mechanics and Mining Sciences, 46(5): 855–864.

Tripathi, M., Yadav, P.K., Chahar, B.R., Dietrich, P., 2021. A review on groundwater surface water interaction highlighting the significance of streambed and aquifer properties on the exchanging flux. *Environmental Earth Sciences*, 80: 1–16.

Tsang, Y.W., Tsang, C.F., 1987. Channel model of flow through fractured media. *Water Resources Research*, 23(3): 467–479.

van Kampen, R. *et al.*, 2022. LPMLE n- A frequency domain method to estimate vertical streambed fluxes and sediment thermal properties in semi-infinite and bounded domains. *Water Resources Research*, 58(3): e2021WR030886.

Vandersteen, G. *et al.*, 2015. Determining groundwater surface water exchange from temperature-time series: Combining a local polynomial method with a maximum likelihood estimator. *Water Resources Research*, 51(2): 922–939.

Vogt, T., Schneider, P., Hahn-Woernle, L., Cirpka, O.A., 2010. Estimation of seepage rates in a losing stream by means of fiber-optic high-resolution vertical temperature profiling. *Journal of Hydrology*, 380(1–2): 154–164.

Wang, Q., Zhan, H., 2019. Reactive transport with wellbore storages in a single-well push–pull test. *Hydrology and Earth System Sciences*, 23(4): 2207–2223.

Wang, Q., Zhan, H., Tang, Z., 2014. Forchheimer flow to a well-considering time-dependent critical radius. *Hydrology and Earth System Sciences*, 18(6), 2437–2448.

Wang, Q., Shi, W., Zhan, H., Xiao, X., 2020a. New model of single-well push-pull thermal test in a fracture–matrix system. *Journal of Hydrology*, 585(2020): 124807.

Wang, Q., Shi, W., Zhan, H., Gu, H., Chen, K., 2018. Models of single-well push-pull test with mixing effect in the wellbore. *Water Resources Research*, 54(12): 10155–10171.

Wang, Q., Wang, J., Zhan, H., Shi, W., 2020b. New model of reactive transport in a single-well push pull test with aquitard effect and wellbore storage. *Hydrology and Earth System Sciences*, 24: 3983–4000.

Wang, Q., Zhan, H., Wang, Y., 2017. Single-well push-pull test in transient Forchheimer flow field. *Journal of Hydrology*, 549: 125–132.

Wang, Q.R., Zhan, H.B., 2015. On different numerical inverse Laplace methods for solute transport problems. *Advances in Water Resources*, 75: 80–92.

Wen, Z., Huang, G.H., Zhan, H.B., 2011. Solutions for non-Darcian flow to an extended well in fractured rock. *Ground Water*, 49(2): 280–285.

Woodside, W.H., Messmer, J.H., 1961. Thermal conductivity of porous media. I. Unconsolidated sands. *Journal of Applied Physics*, 32(9): 1688–1699.

Yang, S., Yeh, H., 2008. An analytical solution for modeling thermal energy transfer in a confined aquifer system. *Hydrogeology Journal*, 16(8): 1507–1515.

Yang, S., Yeh, H., Li, K., 2010. Modelling transient temperature distribution for injecting hot water through a well to an aquifer thermal energy storage system. *Geophysical Journal International*, 183(1): 237–251.

Yang, S.-Y., Yeh, H.-D., 2009. Radial groundwater flow to a finite diameter well in a leaky confined aquifer with a finite-thickness skin. *Hydrological Processes*, 23(23): 3382–3390.

Zhang, Q., Luo, S.H., Ma, H.C., Wang, X., Qian, J.Z., 2019. Simulation on the water flow affected by the shape and density of roughness elements in a single rough fracture. *Journal of Hydrology*, 573: 456–468.

Zheng, Q., Dickson, S.E., Guo, Y., 2008. On the appropriate "equivalent aperture" for the description of solute transport in single fractures: Laboratory-scale experiments. *Water Resources Research*, 44(4): W04502.

Zhou, R.J., Zhan, H.B., 2018. Reactive solute transport in an asymmetrical fracture-rock matrix system. *Advances in Water Resources*, 112: 224–234.

Zimmermann, G., Blöcher, G., Reinicke, A., Brandt, W., 2011. Rock specific hydraulic fracturing and matrix acidizing to enhance a geothermal system — Concepts and field results. *Tectonophysics*, 503(1): 146–154.

Appendices

In this section, we provide three fundamental MATLAB computational programs, which are discussed in Chapters 2, 3, and 4 of Wang *et al.* (2014, 2018, 2020), respectively.

Appendix A: The computational program of Wang *et al.* (2014)

The following code is used to calculate the model of Wang *et al.* (2014): Wang, Q., Zhan, H., Tang, Z., 2014. Forchheimer flow to a well-considering time-dependent critical radius. *Hydrology and Earth System Sciences*, 18(6): 2437–2448. This model is discussed in Section 2.5 of Chapter 2, this code can be copied directly into MATLAB for calculations.

```
%======================Main function=============================================
function Region_Forchheimer
%This program was developed based on the Finite Difference Method
% for Darcian Flow Solution (eig1=1)
% for New Solution of Two Region model with moving interface(eig1=2)
% for Full Non-Darcain flow solution Solution (eig1=3)
% eig1 value can be changed in two FUNCTIONS:
%      function [dhdt]=odefun(...)
%      function [h,vnew1,hnew1]=FDModelFun(...)
%================Input parameters================================================
    rw=0.001;
    rc=0.001;
    re=1e0;
    RcD_old=0;
    RcD_new=0;
    v0=10;
    N=2000;
```

```
N1=N+1;
h0=zeros(N1,1);
vnew=zeros(N1,1);
hnew=zeros(N1,1);
t0=0;
tn=101;
i=[0:N]';
rBND=10.^(log10(rw)+i*(log10(re)-log10(rw))/N);
r=(rBND(1:N)+rBND(2:N+1))/2;
Time=[logspace(-6,1,tn+1)];
for ti=1:(tn+1)/2
    if ti==1
          t=[linspace(Time(ti),Time(ti+1),20)]'; % Time(ti-1)is the initial condition
          t0=0;
    else
    t0=t(end);
    for ij=1:tn-1
          if Time(ij)==t(end)
               timNo=ij;
              t=[linspace(Time(timNo+1),Time(timNo+2),20)]';
         break
         end
    end
end
DR=1;%Intial condition
ijk=0;
while DR>0.000000001
ijk=ijk+1;
RcD_old=RcD_new;
[h,vnew1,hnew1]=FDModelFun(t,re,rw,rc,r,N,RcD_old,h0,N1,t0,vnew,hnew);
 for ii=2:length(vnew)
      if (vnew1(ii-1)>=v0 & vnew1(ii)<=v0)
      RcD_new=r(ii);
      end
 end
 DR=abs(RcD_old-RcD_new);
%-------------------------------------Convergent conditions---------------------------------
 rsDijk(ijk)=RcD_new;
 if ijk>2
       if (rsDijk(ijk)==rsDijk(ijk-2))
             RcD_new=(rsDijk(ijk)+rsDijk(ijk-1))/2;
             DR=0.0000000001;
       elseif (rsDijk(ijk-1)==rsDijk(ijk-2))
             DR=0.0000000001;
       elseif ijk>5000
             DR=0.0000000001;
       end
       end
        KKK(ijk,1)=ijk;
        KKK(ijk,2)=RcD_new;
 end
 KKK;
 h0=hnew1;
 RR(ti,1)=t(end);
 RR(ti,2)= RcD_new; % h0(100);
 end
```

```
    KKK;
    r1=[0;r];
    t(end)
    semilogx(RR(:,1),RR(:,2),'k')
%=============================================================================================
   function
   [h,vnew1,hnew1]=FDModelFun(t,re,rw,rc,rPt,N,RcD_old,h0,N1,t0,vnew,hnew)
      i=[0:N]';
      rBND=10.^(log10(rw)+i*(log10(re)-log10(rw))/N);
      r=(rBND(1:N)+rBND(2:N+1))/2;
      rIn=rBND(1:N);
      rOut=rBND(2:N+1);
      pars.N=N;
      pars.r=r;
      pars.rIn=rIn;
      pars.rOut=rOut;
      pars.re=re;
      pars.rw=rw;
      pars.rc=rc;
      pars.RcD_old=RcD_old;
      pars.h0=h0;
      pars.N1=N1;
      pars.t0=t0;
      pars.vnew=vnew;
      pars.hnew=hnew;
      N=N1;
      d=[-1 0 1];
      JPat=spdiags(ones(N,3),d,N,N);
      options=odeset('JPattern',JPat);
      t=[t0;t];
      %Calculate the Head and rsDii
      [t,h]=ode15s(@odefun,t,h0,options,pars,t(end));
      hnew=h(length(t),:);
      load rsDii
      rsDii=rsDii+2;
      i=2:N;
      ijj=0;
      r1=[0;r];
%=============================================================================================
      %Calculate the Velocity
      eig1=2; %eig=1 Darcy, eig1=2, New solution, eig1=3 non-darcian
      BB=0.01;
      for jj=2:N-1
        if eig1==1 %Darcy
           vnew(jj-1)=abs((h(length(t),jj-1)-h(length(t),jj))...
                  /(r1(jj)-r1(jj-1)));
        end
        if eig1==2 %New solution
          if jj<=rsDii
           BetaD=BB;
           vnew(jj-1)=((1+4*BetaD*abs((h(length(t),jj-1)-h(length(t),jj))...
                /(r1(jj)-r1(jj-1))))^0.5-1)/(2*BetaD);
          else
           vnew(jj-1)=abs((h(length(t),jj-1)-h(length(t),jj))...
               /(r1(jj)-r1(jj-1)));
          end
```

```
        end
        if eig1==3 %Full non-darcian
           BetaD=BB;
           vnew(jj-1)=((1+4*BetaD*abs((h(length(t),jj-1)-h(length(t),jj))...
                /(r1(jj)-r1(jj-1))))^0.5-1)/(2*BetaD);
        end
      end
      vnew1=vnew;
      hnew1=hnew;
%==========================================================================================
  function [dhdt]=odefun(t,h,pars,tMax)
      BB=0.01;
      hw=h(1,:);
      h(1,:)=[];
      N=pars.N;
      r=pars.r;
      rIn=pars.rIn;
      rOut=pars.rOut;
      re=pars.re;
      rw=pars.rw;
      rc=pars.rc;
      RcD_old=pars.RcD_old;
      h0=pars.h0;
      i=2:N;
      dhdrIn(i,:)=(h(i,:)-h(i-1,:))./(r(i)-r(i-1));
      dhdrOut(i-1,:)=dhdrIn(i,:);
      dhdrOut(N,:)=(0-h(N,:))./(re-r(N));
      dhdrIn(1,:)=(h(1,:)-hw)./(r(1)-rw);
      qIn=-dhdrIn;
      qOut=-dhdrOut;
      rsDii=1;
      eig1=2; %eig=1 Darcy, eig1=2, this paper, eig1=3 non-darcian
      if eig1==1
         for iii=1:length(qIn)-1
            qIn(iii+1)=qIn(iii+1);
            qOut(iii)=qOut(iii);
         end
      end
      if eig1==2
        for iii=1:length(qIn)-1
         if r(iii)<=RcD_old
           BetaD=BB;
           qIn(iii+1)=sign(qIn(iii+1)).*((1+4*BetaD*abs(qIn(iii+1)))^0.5-1)/(2*BetaD);
           qOut(iii)=sign(qOut(iii)).*((1+4*BetaD*abs(qOut(iii)))^0.5-1)/(2*BetaD);
         else
          qIn(iii+1)=qIn(iii+1);
          qOut(iii)=qOut(iii);
         end
         if (r(iii)<=RcD_old) & (r(iii+1)>=RcD_old)
            rsDii=iii;
         end
        end
       end
       if eig1==3
         for iii=1:length(qIn)-1
          BetaD=BB;
```

```
        qIn(iii+1)=sign(qIn(iii+1)).*((1+4*BetaD*abs(qIn(iii+1)))^0.5-1)/(2*BetaD);
        qOut(iii)=sign(qOut(iii)).*((1+4*BetaD*abs(qOut(iii)))^0.5-1)/(2*BetaD);
      end
    end
    save rsDii rsDii
    dhdt=(rIn.*qIn-rOut.*qOut)./r./(rOut-rIn);
    dhwdt=2/rc^2*(1-rw*qIn(1,:));
    dhdt=[dhwdt;dhdt];
%==============================================================================
```

Appendix B: The computational program of Wang *et al.* (2018)

The following code is used to calculate the model of Wang *et al.* (2018): Wang, Q., Shi, W., Zhan, H., Gu, H., Chen, K., 2018. Models of single-well push-pull test with mixing effect in the wellbore. *Water Resources Research*, 54(12): 10155–10171. This model is discussed in Section 3.3.2.2 of Chapter 3, this code can be copied directly into MATLAB for calculations.

```
%===================Input parameter============================================
  R_m=1.0;                      % Retardation factor                   [-]
  Alpha=1;                      % Dispersivity                         [cm]
  theta_m=0.38;                 % Porosity                             [-]
  miu_m=0.0000001;              % First-order reaction rate            [1/s]
  Q_inj=45;                     % Injection rate                       [cm3/s]
  Q_ext=45;                     % Extraction rate                      [cm3/s]
  t_inj=100;                    % Injection time                       [s]
  t_ext=400;                    % Extraction time                      [s]
  B=30;                         % Aquifer thickness                    [cm]
  hw_inj=35;                    % Water level in the injection phase  [cm]
  h0=32;                        % Initial water level                  [cm]
  hw_ext=33.5;                  % Water level in the extraction phase [cm]
  rw=2.5;                       % Well radius                          [cm]
  re=100;                       % Length of aquifer                    [cm]
%=================Dimensionless transform =====================================
  rwD=rw/Alpha;
  reD=re/Alpha;
  r=logspace(log10(rwD),log10(reD),50);
  M=length(r);
  tinj=logspace(-5,log10(t_inj),50);
  L=length(tinj);
  text=logspace(-1,log10(t_ext),100);
  LL=length(text);
  Vw=pi*rw^2*hw_inj;
  V0=pi*rw^2*h0;
  fai=(Vw-V0)/Vw;
  epsem=2*pi*rw*theta_m*B;
  A=Q_inj/2/pi/B/theta_m;
```

```
  bate_injD=Vw*rwD/Alpha/epsem/R_m;
  miu_mD=Alpha^2/A*miu_m;
  Vwe=pi*rw^2*hw_ext;
  epsem=2*pi*theta_m*B;
  Ae=Q_ext/2/pi/B/theta_m;
  bate_extD=Vwe*rwD/Alpha/epsem/R_m;
  CD_m=zeros(L,M);
  CD_ext_m=zeros(LL,M);
  C_ext_m=zeros(LL,M);
%====================Injection phase===================================================
  for ii=1:L
      for j=1:M
           rD=r(j);
           tD=A*tinj(ii)/Alpha^2/R_m;
           N=18;
              for i=1:N
                   a=fix((i+1)/2);
                   b11=min(i,N/2);
                   x=0;
             for k=a:b11
                  x=x+k^(N/2)*factorial(2*k)/(factorial(N/2-k)*factorial(k)...
                  *factorial(k-1)*factorial(i-k)*factorial(2*k-i));
             end
             V(i)=(-1)^(N/2+i)*x;
         p(i)=log(2)*i/tD;
         E=p(i)+miu_mD;
         yw=rwD+1/4/E;
         y=rD+1/4/E;
         a0=(bate_injD*p(i)*fai+1)/p(i)/(bate_injD*p(i)+1)/exp(yw/2)...
             /(airy(0,E^(1/3)*yw)/2-E^(1/3)*airy(1,E^(1/3)*yw));
         CD0=a0*exp(y/2)*airy(0,E^(1/3)*y);
         f(i)=CD0;
         yy(i)=log(2)/tD*V(i)*f(i);
         end
         CD_m(ii,j)=sum(yy);
     end
   end
      save('CD_inj_m.mat','CD_m');
      load('CD_inj_m')
      CD_m(CD_m<1e-5) = 0;
      CD_m(isnan(CD_m)) = 0;
      CD1_m=CD_m(end,:);
%====================Extraction phase==================================================
  for i = 1:LL
      for j = 1:1
            tD=Ae*text(i)/Alpha^2/R_m;
            rD=r(j);
            syms s;
            syms gamma;
            syms x;
            E=@(s) s+miu_mD;
            y=@(s) E(s)^(1/3)*(rD+1/4/E(s));
            yw=@(s) E(s)^(1/3)*(rwD+1/4/E(s));
            yx=@(s,gamma) E(s)^(1/3)*(gamma+1/4/E(s));
            Bm=@(x) f_incub_dis(r,CD1_m,x);
            Bim=@(x) f_incub_dis(r,CD1_im,x);
```

```
          W=@(s) ((bate_extD*s+0.5)*airy(2,yw(s))-E(s)^(1/3)*airy(3,yw(s)))...
                /((bate_extD*s+0.5)*airy(0,yw(s))-E(s)^(1/3)*airy(1,yw(s)));
          C2=@(s,gamma) pi*airy(0,yx(s,gamma))/E(s)^(1/3);
          C1=@(s,gamma) -C2(s,gamma)*W(s);
          C3=@(s,gamma) C2(s,gamma)*(airy(2,yx(s,gamma))...
                /airy(0,yx(s,gamma))-W(s));
          b0=@(s) (bate_extD*exp(rwD/2)*CD1_m(1,1))...
             /((bate_extD*s+0.5)*rwD-1-(bate_extD*s+0.5)*reD);
             aa=@(s) -reD*b0(s);
          f=@(s,gamma) exp(gamma/2)*gamma*Bm(gamma)...
              -(gamma*E(s)+0.25)*(aa(s)+b0(s)*gamma);
          G1=@(s,gamma) C1(s,gamma)*airy(0,y(s))...
             +C2(s,gamma)*airy(2,y(s));
          G2=@(s,gamma) C3(s,gamma)*airy(0,y(s));
             V=@(s) aa(s)+b0(s)*rD;
              CD_ext_m =@(s) exp(-rD/2)*(integral(@(gamma)
           G2(s,gamma).*f(s,gamma),...
           rwD,rD,'ArrayValued',true)+integral(@(gamma)
           G1(s,gamma).*f(s,gamma),rD,reD, 'ArrayValued',true))+exp(-
           rD/2)*V(s);
           C_ext_m(i,j) = stehfest(tD,10,CD_ext_m);
       end
    end
       save('CD_extm.mat','C_ext_m');
       BTC_m=[CD_m;C_ext_m];
       t=[tinj,text+t_inj];
%====================Drawing===============================================================
  figure(1)
      hh=plot(t,BTC_m(:,1),'r-');
      set(hh,'LineWidth',1.5);
      set(gcf,'Color',[1 1 1]);set(gca,'FontSize',12, 'fontname','Arial Narrow',...
          'LineWidth',1.5,'Fontweight','bold');
      xlabel('Time {\itt} (s)','FontSize',14,'FontName','Arial Narrow',...
          'Fontweight','bold');
      ylabel('Concentration {\itC_D}','FontSize',14,'FontName',...
          'Arial Narrow','Fontweight','bold');
      set(gca,'XLIM',[0,t_ext+t_inj]);
      set(gca,'ylim',[-0.05,1.04]);
      legend('BTC');
   hold on

%====================Stehfest==============================================================
  function f_real=stehfest(t,N,f_lap)
      sum_f=0;
  for i=1:N
      sum_vi=0;
          for j=floor((i+1)/2):min(i,N/2)
               sum_vi=sum_vi+j^(N/2)*factorial(2*j)/(factorial(N/2-j)...
                  *factorial(j)*factorial(j-1)*factorial(i-j)*factorial(2*j-i));
          end
      vi=(-1)^(N/2+i)*sum_vi;
      sum_f=sum_f+vi.*f_lap(log(2)*i./t);
   end
      f_real=sum_f*log(2)./t;
   end
%==================== Spline Interpolation Functions====================================
```

```
function y=f_incub_dis(r,c,x)
y = spline(r,c,x);
    if (y < 0)
        y = 0;
    end
end
%==========================================================================
```

Appendix C: The computational program of Wang *et al.* (2020)

The following code is used to calculate the model of Wang *et al.* (2020): Wang, Q., Shi, W., Zhan, H., Xiao, X., 2020. New model of single-well push-pull thermal test in a fracture-matrix system. *Journal of Hydrology*, 585(2020): 124807. This model is discussed in Section 4.3. In order to compute this model, four script programs are required to be created. These four scripts are used to compute the results of injection phase, resting phase and extraction phase, as well as to draw the results of the computation.

The following code needs to be saved as a separate script for calculating the results of the injection phase:

```
%==================Injection phase=========================================
%-------------------------------Input parameters---------------------------
  T0=200;
  Ti=20;
  rhom=2650;
  cm=1000;
  rhow=1000;
  cw=4200;
  km=2.1;
  kf=2.1;
  b1=0.01;
  u1=0.00031623;
  theta=0.5;
  cfrhof=(1-theta)*rhom*cm+theta*rhow*cw;
  A=cm*rhom/cfrhof;
  rw1=0.5;
  h_inj=65;
  B=50;
  Q_inj=4*b1*B*theta*u1;
  V_inj=pi*rw1^2*h_inj;
  lamda=cw^2*rhow^2*u1^2*theta^2*b1^2/kf/km
  re=1000;
  rw=1E-5;
%Grid system (Develop r-axis)
   Nr=50;
   i=[0:Nr]';
```

```
    rBND=10.^(log10(rw)+i*(log10(re)-log10(rw))/Nr);
    r=(rBND(1:Nr)+rBND(2:Nr+1))/2;
    x1=r;
    M=length(x1);
    tinj=5*3600;
    t=logspace(-1,log10(tinj),200);
    L=length(t);
    tsmall=1e-4;
%-----------------------------------------------------------------------------------------
   for i=1:L
    for j=1:M
       xD=x1(j);
       tD=t(i);
       tau_inf_small=tsmall;
       alpha_inj=V_inj*km/(Q_inj*cfrhof*b1^2);
       syms s;

       E(s)=(lamda^2+4*lamda*(s+(s*A)^0.5))^0.5;
       TwD(s)=1/s/(1+s*alpha_inj);
       TD(s)=TwD(s)*exp(0.5*(lamda-E(s))*xD);
       matlabFunction(TD,'file','TD');
       T1D(i,j) = de_hoog('TD',tD-tau_inf_small);
       delete('TD.m');
     end
    end
    save('TD_inj.mat','T1D');
    save Twinj.txt T1D -ascii
%=========================================================================================
```

The following code needs to be saved as a separate script for calculating the results of the resting phase:

```
%=======================Resting phase=====================================================
  clear all;clc;
  load('TD_inj');
  TDinj=T1D(end,:);
  lamda=10;
  re=1000;
  rw=1E-5;
  %Grid system (Develop r-axis)
  Nr=50;
  i=[0:Nr]';
  rBND=10.^(log10(rw)+i*(log10(re)-log10(rw))/Nr);
  r=(rBND(1:Nr)+rBND(2:Nr+1))/2;
  xe=re;
  x1=r;
  M=length(x1);
  ze=500;
  zeD=ze;
  Time_num=100;
  tres=5*3600;
  t=logspace(-2,log10(tres),Time_num)';
  L=length(t);
  for i =1:L
     for j =1:M
        xD=x1(j);
        xeD=xe;
        tD=t(i);
```

```
%-----------------------------------------------------------------------------------------
        syms s;
        syms gama;
        syms x;
        B1=@(x) f_incub_dis(x1,TDinj,x);
        E=@(s) lamda*s;
        f=@(s,gama) lamda*B1(gama);
        H1=@(s,gama) 1/2/(lamda*s)^0.5/exp((lamda*s)^0.5*gama);
        H2=@(s,gama) H1(s,gama);
        H3=@(s,gama) H1(s,gama)+H1(s,gama)*exp(2*(lamda*s)^0.5*gama);
        G1=@(s,gama) H1(s,gama)*exp(-(lamda*s)^0.5*xD)+H2(s,gama)*exp((lamda*s)^0.5*xD);
        G2=@(s,gama) H3(s,gama)*exp(-(lamda*s)^0.5*xD);
        T2D=@(s) integral(@(gama)
G2(s,gama).*f(s,gama),0,xD,'ArrayValued',true)...
           +integral(@(gama)
G1(s,gama).*f(s,gama),xD,xeD,'ArrayValued',true);
        TresD(i,j) = stehfest(tD,10,T2D);
        TresD(TresD<0)=0;
      end
    end
    save('TD_res.mat','TresD');
    save Twres.txt TresD -ascii
%-----------------------------------------------------------------------------------------
function f_real=stehfest(t,N,f_lap)
    sum_f=0;
    for i=1:N
       sum_vi=0;
       for j=floor((i+1)/2):min(i,N/2)
          sum_vi=sum_vi+j^(N/2)*factorial(2*j)/(factorial(N/2-
j)*factorial(j)*factorial(j-1)*factorial(i-j)*factorial(2*j-i));
       end
       vi=(-1)^(N/2+i)*sum_vi;
       sum_f=sum_f+vi.*f_lap(log(2)*i./t);
      end
      f_real=sum_f*log(2)./t;
      % f_real=f_lap(1);
end
%-----------------------------------------------------------------------------------------
function y=f_incub_dis(r,c,x)
y = spline(r,c,x);
if (y < 0)
    y = 0;
end
end
%-----------------------------------------------------------------------------------------
%=========================================================================================
The following code needs to be saved as a separate script for calculating the results
of the extraction phase:
%======================Extraction phase===================================================
  clear all;clc;
  load('TD_res');
  TDres=TresD(end,:);
  T0=200;
  Ti=20;
  rhom=2650;
  cm=1000;
```

```
  rhow=1000;
  cw=4200;
  km=2.1;
  kf=2.1;
  b1=0.01;
  u1=0.00031623;
  theta=0.5;
  lamda=cw^2*rhow^2*u1^2*theta^2*b1^2/kf/km;
  cfrhof=(1-theta)*rhom*cm+theta*rhow*cw;
  A=cm*rhom/cfrhof;
  rw1=0.5;
  h_ext=60;
  B=50;
  Q_ext=b1*B*theta*u1;
  V_ext=pi*rw1^2*h_ext;
  % alpha_ext=0;
  alpha_ext=V_ext*(cw*rhow)^2*u1*theta/b1/B/kf/cfrhof/2;
  re=1000;
  rw=1E-5;
  %Grid system (Develop r-axis)
  Nr=50;
  i=[0:Nr]';
  rBND=10.^(log10(rw)+i*(log10(re)-log10(rw))/Nr);
  r=(rBND(1:Nr)+rBND(2:Nr+1))/2;
  xe=re;
  x1=r;
  M=length(x1);
  ze=10000;
  zeD=ze;
  Time_num=100;
  text=10*3600;
  t=logspace(-2,log10(text),Time_num)';
  L=length(t);
%-------------------------------------------------------------------------------------------
  for i =1:L
     for j = 1
        xD=x1(j);
        xeD=xe;
        tD=t(i);
        syms s;
        syms gama;
        syms x;
        B1=@(x) f_incub_dis(x1,TDres,x);
        E=@(s) (lamda^2+4*lamda*(s+1/zeD))^0.5;
        W=@(s) (0.5*(E(s)+lamda)+alpha_ext*s)/(0.5*(E(s)-lamda)-alpha_ext*s);
        b0=@(s) -alpha_ext*TDres(1,1)/(1+alpha_ext*s*xeD);
        a0=@(s) -xeD*b0(s);
        V=@(s) a0(s)+b0(s)*xD;
        f=@(s,gama) lamda*B1(gama)+lamda*b0(s)-
lamda*(s+1/zeD)*(a0(s)+b0(s)*gama);

        H=@(s,gama) exp(0.5*(E(s)-lamda)*gama)/exp(-0.5*(E(s)+lamda)*gama);

        H1=@(s,gama) 1/E(s)/exp(0.5*(E(s)-lamda)*gama);
        H2=@(s,gama) H1(s,gama)/W(s);
        H4=@(s,gama) H2(s,gama)+H1(s,gama)*H(s,gama);
```

```
        G1=@(s,gama) H1(s,gama)*exp(0.5*(E(s)-lamda)*xD)+H2(s,gama)*exp(-
0.5*(E(s)+lamda)*xD);
         G2=@(s,gama) H4(s,gama)*exp(-0.5*(E(s)+lamda)*xD);
         T3D=@(s) integral(@(gama)
G2(s,gama).*f(s,gama),0,xD,'ArrayValued',true)...
           +integral(@(gama)
G1(s,gama).*f(s,gama),xD,xeD,'ArrayValued',true)+V(s);
        TD(i,j) = stehfest(tD,10,T3D);
        TD(TD<0)=0;
      end
    end
    save('TD_ext.mat','TD');
    save Tw.txt TD -ascii
%-----------------------------------------------------------------------------------------
function f_real=stehfest(t,N,f_lap)
   sum_f=0;
   for i=1:N
    sum_vi=0;
    for j=floor((i+1)/2):min(i,N/2)
        sum_vi=sum_vi+j^(N/2)*factorial(2*j)/(factorial(N/2-j)*factorial(j)*factorial
(j-1)*factorial(i-j)*factorial(2*j-i));
    end
    vi=(-1)^(N/2+i)*sum_vi;
    sum_f=sum_f+vi.*f_lap(log(2)*i./t);
   end
   f_real=sum_f*log(2)./t;
   % f_real=f_lap(1);
end
%-----------------------------------------------------------------------------------------
function y=f_incub_dis(r,c,x)
y = spline(r,c,x);
if (y < 0)
    y = 0;
end
end
%-----------------------------------------------------------------------------------------
%=========================================================================================
```

The following code needs to be saved as a separate script for drawing the results of the calculation:

```
%===================Drawing===============================================================
  clear all;clc;
  load('TD_inj');
  load('TD_res');
  load('TD_ext');
  tinj=5*3600;
  tres=5*3600;
  text=10*3600;
  t_inj=logspace(-1,log10(tinj),200);
  t_res=logspace(-2,log10(tres),100);
  t_ext=logspace(-2,log10(text),100);
  BTC_m=[T1D(:,1);TresD(:,1);TD(:,1)];
  t=[t_inj,t_res+tinj,t_ext+tinj+tres];
  re=1000;
  rw=1E-5;
  Nr=50;
```

```
  i=[0:Nr]';
  rBND=10.^(log10(rw)+i*(log10(re)-log10(rw))/Nr);
  r=(rBND(1:Nr)+rBND(2:Nr+1))/2;
  x1=r;
  M=length(x1);

  figure(1)
  hh=plot(t/3600,BTC_m,'b-');
  set(hh,'LineWidth',1.5);
  set(gcf,'Color',[1 1 1]);set(gca,'FontSize',12,'fontname','Arial
Narrow','LineWidth',1.5,'Fontweight','bold');
  xlabel('Distance {\itx} (m)','FontSize',14,'FontName', 'Arial
Narrow','Fontweight','bold');
  ylabel('Temperature {\itT_f_ _D}','FontSize',14,'FontName', 'Arial
Narrow','Fontweight','bold');
  set(gca,'XLIM',[0,20]);
  set(gca,'ylim',[-0.05,1.05]);
  hold on
%==========================================================================================
```

Index

M

N

P

R

S

T

U

V

W